4C-I-29

# A GLOSSARY OF GEOGRAPHICAL TERMS

# A GLOSSARY OF
# Geographical Terms

BASED ON A LIST
PREPARED BY
A COMMITTEE OF THE BRITISH ASSOCIATION
FOR THE ADVANCEMENT OF SCIENCE

EDITED BY
SIR DUDLEY STAMP
AND
AUDREY N. CLARK

**LONGMAN**
LONDON AND NEW YORK

LONGMAN GROUP LIMITED
London

*Associated companies, branches and representatives
throughout the world*

Distributed in the United States of America
by Longman Inc., New York

© Geographical Publications Ltd 1961, 1966, 1979

All rights reserved. No part of this publication may be reproduced, stored in a retrieval system, or transmitted in any form or by any means, electronic, mechanical, photocopying, recording, or otherwise, without the prior permission of the Copyright owner.

First published 1961
Reprinted with minor corrections 1962
Third (American) printing 1962
Fourth printing with additions 1963
Second edition 1966
New impression 1970
Third edition 1979

ISBN 0 582 35258 4

**British Library Cataloguing in Publication Data**
British Association for the Advancement of Science
A glossary of geographical terms.—3rd ed.
1. Geography—Terminology
I. Stamp, *Sir* Laurence Dudley
II. Clark, Audrey N.
910'.3     G107.9     79–40238

ISBN 0–582–35258–4

Printed and bound at
William Clowes & Sons Limited, Beccles and London

# CONTENTS

| | PAGE |
|---|---|
| THE BRITISH ASSOCIATION RESEARCH COMMITTEE | vi |
| PREFACE TO THE THIRD EDITION | vii |
| PREFACE TO THE FIRST EDITION | viii |
| TO OUR READERS | xiv |
| PREFACE TO THE SECOND EDITION | xv |
| THE SPELLING OF FOREIGN WORDS | xvi |
| LIST OF STANDARD WORKS | xvii |
| LIST OF CORRESPONDENTS AND COLLABORATORS | xxv |
| ABBREVIATIONS | xxviii, xxx |
| THE GLOSSARY | 1 |
| APPENDIX I: Greek and Latin Roots commonly used in construction of terms | 531 |
| APPENDIX II: Lists of words in Foreign Languages which have been absorbed into English literature | 547 |
|     African languages | 547 |
|     Afrikaans | 547 |
|     Arabic | 548 |
|     Australian | 549 |
|     Burmese | 550 |
|     Chinese | 550 |
|     Danish-Norwegian | 551 |
|     Dutch | 551 |
|     Eskimo | 551 |
|     French | 551 |
|     German | 552 |
|     Hawaiian | 553 |
|     Hebrew | 553 |
|     Hungarian | 554 |
|     Icelandic | 554 |
|     Indian languages | 554 |
|     Indonesian languages | 555 |
|     Iranian | 556 |
|     Irish | 556 |
|     Italian | 556 |
|     Japanese | 557 |
|     Malay | 557 |
|     Maori and other New Zealand words | 557 |
|     Polish | 558 |
|     Portuguese (and Brazilian) | 558 |
|     Russian | 558 |
|     Scottish (including Gaelic) | 559 |
|     Southern Slav languages (Serbo-Croat) | 559 |
|     Spanish | 560 |
|     Swedish | 560 |
|     Thai | 561 |
|     Welsh | 561 |
| APPENDIX III: Some Stratigraphical Terms | 563 |

# BRITISH ASSOCIATION FOR THE ADVANCEMENT OF SCIENCE

## SECTION E (GEOGRAPHY)
### RESEARCH COMMITTEE

to prepare a glossary of geographical terms with agreed definitions in English, including reference to origin and to current use and misuse.

## BRITISH ASSOCIATION RESEARCH COMMITTEE

Membership as reconstructed at the Liverpool Meeting of the Association, September 1953.

### *Chairman*
Professor Emeritus L. Dudley Stamp, C.B.E., D.LIT., D.SC., EKON.D., LL.D., F.R.S.G.S., F.R.G.S.
University of London, The London School of Economics.

### *Secretary*
G. R. Crone, M.A., F.R.G.S.
Librarian and Map Curator, Royal Geographical Society.

### *Members*
Professor H. C. Darby, O.B.E., PH.D., LITT.D., F.R.G.S.
University of London, University College.

Professor W. Gordon East, M.A., F.R.G.S.
University of London, Birkbeck College.

E. O. Giffard, M.B.E., F.R.S.A., F.R.G.S.
George Philip & Son, Ltd., The London Geographical Institute.

A. F. Martin, M.A., F.R.G.S.
University of Oxford.

Professor A. Austin Miller, D.SC., F.R.G.S.
University of Reading.

C. J. Robertson, PH.D., F.R.G.S.
University of Edinburgh.

R. A. Skelton, M.A., F.S.A., F.R.G.S.
Superintendent, Map Room, British Museum.

C. T. Smith, M.A., F.R.G.S.
University of Cambridge.

G. T. Warwick, M.B.E., PH.D., F.R.G.S.
University of Birmingham.

Professor S. W. Wooldridge, C.B.E., D.SC., F.R.S., F.R.G.S.
University of London, King's College.

### *Research Officer (June 1954–July 1955)*
W. T. W. Morgan, M.SC.(ECON.), PH.D., F.R.G.S.

### *American Adviser*
Professor Chauncy D. Harris, PH.D., M.A., D.ECON.
Professor of Geography, University of Chicago.

## PREFACE TO THE THIRD EDITION

Geographical studies have continued to spread into the territories of neighbouring academic disciplines since Sir Dudley Stamp enlarged the *Glossary* for the second edition in 1965. The ramifications of human geography in particular have further penetrated economics, political science and psychology, absorbing some of the terminology of those subjects in the process.

Thus many of the terms introduced in this much extended third edition cover the new approaches, concepts and techniques applied under the general heading 'human geography' especially in the areas of spatial analysis, behavioural analysis, environmental studies and quantitative techniques. There are also new entries relating to cartography, biogeography, climatology and the wide range of studies classified as physical geography. And some of the terms included in previous editions have been augmented.

The policy outlined in previous prefaces has been closely followed: terms strictly proper to other disciplines, which will be found elsewhere in appropriate glossaries and specialized dictionaries such as Kendall and Buckland, *A Dictionary of Statistical Terms*, Longman, and 'dictionary words used in geographical writings with their normal meaning' have not been included.

I am enormously grateful to Dr. Christopher Board of The London School of Economics for his advice in the drawing-up of the final list of additional entries; and to Mr. Harry J. Barlow, then a postgraduate student at The London School of Economics, who so ably undertook the necessary library research under Dr. Board's guidance.

Inevitably terms cherished by readers will have been omitted, and detested ones included: adverse and constructive criticism will, as ever, be warmly welcomed.

*January*, 1978         AUDREY N. CLARK
                        Geographical Publications Ltd.

# PREFACE TO THE FIRST EDITION

At the Birmingham Meeting of the British Association for the Advancement of Science, held from August 30 to September 6, 1950, a discussion on geographical terminology was initiated by Professor Eva G. R. Taylor, and other contributors included Professor R. H. Kinvig and Mr. E. O. Giffard. The discussion revealed the concern which many geographers had come to feel at the varying use of a large number of geographical terms as well as at the rapidly increasing introduction of new terms. On the recommendation of the Committee of Section E (Geography) the Council of the Association set up a special Committee to study the problem. The Committee's terms of reference were 'To prepare a glossary of geographical terms with agreed definitions in English including reference to origin and previous usage and to current use and misuse.' The original committee consisted of Professor Eva G. R. Taylor (Chairman), Professor S. W. Wooldridge (Secretary), Professor W. Gordon East, Mr. E. O. Giffard, Mr. A. F. Martin, Dr. C. J. Robertson, Mr. A. E. Smailes, Mr. J. C. Swayne and Dr. G. T. Warwick. This is recorded in the *Advancement of Science*, **7** (No. 27), 357. A later number of the *Advancement of Science*, **8** (No. 31), issued in December 1951, records on p. 345 that the Council had 'authorized a Committee consisting of Lord Rennell, Professor Dudley Stamp, Professor Eva Taylor and Professor S. W. Wooldridge to approach the appropriate officers of the Royal Geographical Society and to invite the cooperation of that Society in preparing a geographical glossary'.

The Committee in fact held its first meeting on March 2, 1951, at the House of the Royal Geographical Society. This was followed by a second meeting on June 5. Later that year the Committee was reappointed with the formal addition of Mr. F. George (at that time Assistant Editor of the *Geographical Journal*) who had previously been co-opted, and Mrs. J. A. Steers. Mr. Swayne, who had in fact been acting as Secretary, was confirmed formally in that office.

At the meeting in June 1951 it was agreed to deal first with certain general terms 'likely to enter into other definitions'. At subsequent meetings definitions of these were approved and lists considered of terms used in physical geography, economic geography and agricultural geography.

Early in its existence, however, the Committee came up against a number of difficulties. The first was the size of its task and the obvious need for at least one person working either full time or at least devoting most of his time to the preparation of the basic

## PREFACE TO THE FIRST EDITION

material. Since the members of the Committee were all busy professional people and the Committee had no funds to employ a research worker, the position seemed impossible. In the second place it immediately proved very difficult to get agreed definitions even of the commonest terms. But under the stimulating lead of Professor Taylor progress was made, and the first list of agreed definitions was published in the *Geographical Journal* for December 1951 (*Geographical Journal*, **117**, part 4, 458). The definitions were brief and according to the prefatory note were 'presented for general criticism'. Other lists followed in successive issues of the *Journal* in March 1952 (physical terms), June 1952 (agricultural terms) and September 1952 (economic terms).

At the Belfast Meeting of the Association in 1952 the Committee did not seek reappointment. The lists issued had been criticized as inadequate and the simple definitions as not fulfilling the requirements laid down in the terms of reference. The Council of the Association however disliked the idea of abandoning a project for which there was an obvious need, and at the Liverpool Meeting of 1953 a new Committee was set up under the Chairmanship of Professor L. Dudley Stamp and with the Librarian of the Royal Geographical Society, Mr. G. R. Crone, as Secretary. Its membership is recorded on page vi.

At a well-attended meeting of the new Committee held in the rooms of the Royal Geographical Society on March 31, 1954, a plan was worked out for the completion of the Committee's labours and this has formed the basis for the Glossary which is now presented.

More than half a century ago a glossary was prepared by the late Dr. Hugh Robert Mill on behalf of the Royal Geographical Society. It actually reached galley-proof stage but was never published—apparently because of criticism and disagreements. The meeting of March 1954 agreed to use Mill's unpublished work as a starting point, eliminating words proper to other disciplines such as geology and meteorology and adding thereto the many terms which have been proposed or have come into use in the intervening years. At that meeting the Chairman agreed to try to raise funds for the employment of a research officer to prepare the basic material and especially to seek the essential references with definitions for the words selected by the Committee. Fortunately, in May 1954 the services became available of Dr. W. T. W. Morgan, M.Sc.(Econ.), who, after a period of geographical research at The London School of Economics, had spent two sessions at Northwestern University, Evanston, Ill. Later The London School of Economics generously agreed to pay half his salary for the year August 1, 1954, to July 31, 1955, provided that the other half was contributed from other sources. This the Chairman undertook to find, and in due course the Council of the British Association agreed to this receipt of funds from outside sources. The Royal Geographical Society provided accommodation for Dr. Morgan, and vitally important secretarial assistance. This had the great advantage that nearly all the references required were ready to hand in the Library of the Society, and the Committee owes much to the wide knowledge of its Secretary, Mr. Crone, Librarian of the Royal Geographical Society.

No praise could be too high for the meticulous basic work of Dr. Morgan. In considerable measure this Glossary is his work. Members of the Committee have, according to their special interests, checked his entries and have suggested additions and deletions, so that the responsibility is shared.

The Glossary is limited to terms used in current geographical literature written in English. Foreign words are included only if they are in use in their original form, untranslated, in works written in English. Even so it is surprising how many foreign words are thus used. Almost without thinking, writers dealing with the Indian sub-continent or with the Far East, or Arabic-speaking countries, or South Africa, tend to use words familiar in everyday speech, forgetting that the meaning of *kharif* or *doab* or *taungya* or *kop*, to cite examples at random, may be unknown to a reader in England or North America. An effort has been made to include such words, provided they are found in standard works or recognized journals.

Where the meaning of a term is clear, as when the common usage is as defined in *The Oxford English Dictionary*, the entry will be found to be correspondingly brief. Where, however, a term has changed its meaning, or is obscure, or used in different senses, quotations from original and standard sources with full references have been given. New terms, or those defined in recent years have, wherever possible, been traced back to their originators and the original definition given. With many terms, however, this has proved impossible. Many have slipped into currency without ever being precisely defined—such as 'neomalthusianism'. In other cases common words have been given a specialized meaning: this is particularly the case with foreign words imported into English. Many old words (such as stow) have been given special new meanings —often with confusing and unfortunate results.

In one important respect the Committee (and in consequence this Glossary) departed from its precise terms of reference. Rather than attempt an agreed definition—which all the members of the Committee could accept—we have quoted original definitions and standard authors, though venturing in some cases to add comments.

No one is more fully aware than the Editor of the omissions which will be detected by the specialist in this Glossary. The attempt has been made, however, to include all words found in other geographical glossaries (except those proper to other disciplines) and to include at least those of later origin suggested by the members of the

## PREFACE TO THE FIRST EDITION

Committee. In the course of my work as Editor I have, however, approximately quadrupled the number of words which were considered in the lists prepared by Dr. Morgan and must accept final responsibility for the whole.

It is our hope that this Glossary will prove of wide usefulness, and I have kept constantly in mind the needs of workers whose native tongue is not English. As the English language becomes increasingly the normal medium for scientific writing and publication, it is inevitable that two difficulties should arise. When English is used by those to whom it is not the native tongue, recourse must be had to dictionaries which often fail to give specialized or technical meanings, with the unfortunate consequence that many mis-translations result. The position is much the same when translations are made by those who may be well versed in language but who know little of the subject concerned and its technical vocabulary. Occasionally translators may seek to get an accurate meaning by going to the original roots of the words, but at times this may be disastrous. An international group of geographers visiting a small town in an Eastern country were once greeted by a large display banner which read, 'Well came, O earth writers'.

It is true, of course, that a very large number of terms now used in English are derived from Greek or Latin roots—sometimes unfortunately from a mixture of the two. It seemed therefore useful to include in an Appendix the principal roots from those two classical languages which have been used in the development of a geographical terminology. It is hoped that these lists will not afford undue encouragement to inventors of new terms. Admittedly one reader when faced in the field by the rather formless mass of part of the Eastern Townships of Canada—a continuation of New England—decided it might be described as an aschemagenetic mass.

Reference has already been made to a second difficulty—the tendency of workers living in a foreign country to absorb local words as part of their ordinary vocabulary, especially when those words have no exact English equivalent, or describe things not familiar in the home landscape or English context. As explained above, such foreign words as have become widely current in English literature are included in the main glossary, but in a series of appendices I have also listed them under each language. In preparing these lists I have relied upon the help of the many specialists whose help is acknowledged in each case.

Some explanation is needed of the policy which has been adopted in the exclusion or inclusion of words and terms.

What may be called ordinary dictionary words, used in geographical writings with their normal meaning, have *not* been included. This applies to simple nouns such as 'thermometer' or 'barometer' or to such adjectives as 'annual', 'secular', or 'periodic'. An attempt has been made, apart from this limitation, to cover the several branches of geography—physical, human, social, economic—but not to include terms proper to surveying, mathematical geography and astronomical geography. Names of map projections have been excluded. Although the Committee individually and collectively were often far from sympathetic to the use of many terms which have been introduced and some wished to show disapproval by omitting them, the majority view was that if a term had been introduced and used, a reader would legitimately expect to find it explained in a glossary. In some cases we have indicated that a particular word or term has not been generally adopted or is better avoided or that ambiguity exists.

Turning to words shared with, or proper to, other disciplines, we have included terms shared by physical geographers and geologists, but not the names of minerals, rocks or geological periods and beds. An exception has been made for certain rock names commonly misused or misunderstood and names applied to broad groups. The more important terms applied to orogenesis and orogenetic periods have been included.

In meteorology and climatology emphasis has been on words and terms likely to be found in regional geographical descriptions (such as local names of winds) rather than on the terminology of the theoretical meteorologist. Similarly in soil science and ecology an attempt has been made to introduce as much of the specialist terminology as is likely to be found in geographical works. In such cases major entries will be found under ecology, soil nomenclature, vegetation terms etc., to serve as a guide.

The length of the entry is by no means proportional to the importance of the word or term. Where there is no doubt as to the meaning the definition has been quoted, without comment, from *The Oxford English Dictionary (O.E.D.)* or *Supplement (O.E.D. Suppl.)* or from Mill's *Dictionary, Merriam-Webster* or other standard source. Where words have extended or changed their meaning this is shown by quotations primarily selected to illustrate the change. Where appropriate a final comment has been added, sometimes by an expert who has been specially consulted or by an appropriate member of the Committee, acknowledged by initials. The Editor is responsible for any comments not otherwise indicated. The Committee is greatly indebted to the Secretary to the Delegates of the Clarendon Press for permission to make the necessary quotations from *The Oxford English Dictionary* in this way and to the publishers of the many works from which quotations have been made. Unfortunately the work of the Committee was almost complete before attention was drawn to the *Glossary of Geology and Related Sciences*, edited by J. V. Howell and published by the American Geological Institute, Washington, 1957. This comprehensive work with its invaluable

## PREFACE TO THE FIRST EDITION

references has not in fact been used in compiling the present Glossary (except for a few late entries) and in many respects is complementary.

Place names and native names of features are not included except incidentally or for special reasons; the glossary in no sense covers the ground of Knox (1904).

Finally, there must be universal awareness of divergencies of usage between North American and English English. Every effort has been made to meet this difficulty. Quotations have been made about equally from American or British sources, and the Committee is greatly indebted to Professor Chauncy D. Harris of the University of Chicago for consenting to act as American adviser. He has seen the whole work only in proof stage and so cannot be held responsible for the basic form of the entries.

L. DUDLEY STAMP

*February*, 1961

# TO OUR READERS

It is perhaps inevitable, despite the care exercised by all members of the Committee, that there should be imperfections in the first edition. We know that destructive criticism is always easy: that any reader dipping into this Glossary will find omissions—many of which may have been made deliberately by the Committee—as well as interpretations and definitions which will be called into question. The quotations selected for inclusion have frequently been chosen to illustrate divergence in usage and do not indicate that the Committee is in agreement with each. In some cases a comment has been added to guide the reader; in others the decision as to the best usage is left open.

In the first printing we made an earnest appeal to all users of the Glossary to assist in its perfection by sending comments, corrections and additions, especially exact references and quotations to authorities in their own fields. We are grateful to those who have responded to this request and we have made many minor corrections and numerous additions. Continued assistance will be greatly appreciated.

A surprisingly large number of correspondents asked for the suppression of words which they personally disliked. One even wanted removed a word introduced by himself and subsequently used by others. It must be pointed out that the Committee took the view that if a word had been used in print the reader of the Glossary had a right to find it listed and defined, though in a few cases terms considered undesirable are so noted.

We would like to record our particular thanks for lengthy comments to Professor John Leighly (California), Dr. J. K. Wright (New York), the Department of Geography, University of Auckland, N.Z., Professor S. Kiuchi (Tokyo), Professor D. H. K. Amiran (Jerusalem), Professor Alan Wood, Dr. W. J. Stiller, Mr. J. Melhuish, Dr. C. A. M. King and Mr. D. A. Preston.

## PREFACE TO THE SECOND EDITION

So many additions have been made to this new printing that it can properly be called a new edition, although the bulk of the work of course remains unchanged.

In the first place there is the addition of a series of entries in the field of mathematical geography. As explained on p. xii, these were previously excluded, but Professor E. G. R. Taylor protested that mathematical geography is basic to the whole field of study and she drew up a list of forty or fifty terms which should be included. At a later stage she commented most helpfully on the proposed entries. In the meantime the newly formed International Cartographic Association set up a Commission on the Definition, Classification and Standardisation of Technical Terms, to which the Cartography Sub-Committee of the British National Committee for Geography contributed a glossary with brief definitions. I have cross-checked with this to secure agreement in usage.

The growth of quantification in geographical studies has resulted in the introduction of some new terms and redefinitions of others. An attempt has been made to make appropriate additions. Mr. W. A. Barnard, Chief Cartographer of the Geographic Office, Ontario Department of Lands and Forests, went very carefully through the whole glossary and offered a number of most helpful suggestions, especially regarding North American usage of various terms.

Numerous correspondents have called my attention to omissions, errors and especially to new terms. I am in particular indebted to Miss Mary Marshall (Oxford), to my graphicate colleagues, Professor W. G. V. Balchin and Miss Alice Coleman, to Dr. E. C. Willatts, Professor Patrick Horsbrugh and Professor D. H. K. Amiran.

Over the past year I have edited the new *Longmans Dictionary of Geography* which will be found to include what many readers have asked for—a brief definition of the chief terms based on the evidence collected for the *Glossary*. The attempt to provide these brief definitions has, however, confirmed the considerable divergence of views even amongst the experts and has to my mind justified the policy followed in the *Glossary* of quoting authorities and giving comments rather than judgments.

Many correspondents have overlooked that it was the policy of the *Glossary* Committee to exclude commodities. These I have now included in *Longmans Dictionary of Geography*, making that work complementary in another way to the present.

*December*, 1965                                           L. D. S.

# THE SPELLING OF FOREIGN WORDS

The Editor has been asked by many correspondents that guidance should be given in the spelling of foreign words. Linguistic purists have asked that anglicized forms should be condemned and that the spelling should always be that used in the country of origin including all diacritical marks, native forms of the plural, use of capitals in the case of German nouns and so on. However desirable this may be the fact remains that diacritical marks are not provided on typewriters of British or American origin and are often only available in printers' type with special instructions. Inevitably they tend to get dropped and foreign words appear in works in the English language in forms which are, strictly speaking, incorrect. It would seem to be quite impossible to prevent this; indeed the tendency is ever towards simplification. Thus the French now usually drop accents over capital letters and print ETUDES GEOGRAPHIQUES.

It would seem that the simplest rule to follow, and one generally used by many authors, is first to decide whether or not the word concerned has been definitely absorbed into the English language. If so it would be printed in the same type as the surrounding text, *not* in italics or between inverted commas or quotes, given an accepted anglicized spelling and normal English plural. Thus one would write plateaus *not* plateaux, loess *not* Löss, fiords *not* fjordar and so on. A special difficulty arises in agreeing the correct anglicized form of words of Scandinavian and east European origin since there is no real equivalent in English of Ä, Å, Ø, Ł and others. If the words are printed without the diacritical marks, as they frequently are, both meaning and pronunciation are completely changed. Indian words, widely used by both Indian and English writers on India and the East, are commonly printed without diacritical marks though what may be regarded as the more correct forms are given in the list in Appendix II.

If a word is regarded by an author as a foreign word, not yet absorbed into the English language and used because there is no convenient English equivalent, it should either be printed in a different type (*e.g.* italics) or put between inverted commas (single or double) and given its correct form in the language of origin. It should be noted that the majority of foreign words given in this Glossary are included because they *have* been absorbed into English so that English plural forms are permissible. This would make it permissible to use inselbergs, and grabens, unless one insists that the words have not been absorbed into English when one should use *Inselberge* and *Gräben*.

L. D. S.

# LIST OF STANDARD WORKS
to which reference is made under the abbreviated titles given.

| | |
|---|---|
| *O.E.D.* | *The Oxford English Dictionary* (corrected re-issue). Oxford: Clarendon Press, 12 vols., 1933. |
| *O.E.D. Suppl.* | *The Oxford English Dictionary, Supplement and Bibliography.* Oxford: Clarendon Press, 1933. |
| *S.O.E.D.* | *The Shorter Oxford English Dictionary.* Oxford: Clarendon Press, 2nd revised edition, 2 vols., 1936. |
| *Webster* and *Merriam-Webster* | *A Compendious Dictionary of the English Language,* 1806; *An American Dictionary of the English Language,* 1828; *Webster's International Dictionary,* 1890; *Webster's New International Dictionary,* 1909; *Webster's New International Dictionary, Second Edition,* Springfield, Mass., G. & C. Merriam, 1934 with addenda to 1954 in later printings. Unless otherwise stated quotations are from the 1959 printing which should properly be referred to as *Merriam-Webster*. |
| *Dict. Am.* | *Dictionary of Americanisms, on Historical Principles.* Edited by Mitford M. Matthews, Chicago: Univ. of Chicago Press, 1951. |
| *S.N.D.* | *Scottish National Dictionary* (as far as issued: A–Ke). Edinburgh: S.N.D.A. Ltd., 1931– |
| *D.O.S.T.* | *Dictionary of the Older Scottish Tongue* (as far as issued: A–La). London: O.U.P.; Chicago: Univ. of Chicago Press, 1931– |
| Mill, *Dict.* | The unpublished Dictionary of Geographical Terms prepared by Hugh Robert Mill for the Royal Geographical Society about 1900–1910 and of which the Society's Library has a set of galley proofs. |
| Comm. 1951–2 | The Committee's definitions as published in the *Geographical Journal,* 1951–1952; reprinted in *A Handbook for Geography Teachers.* 2nd ed., edited by G. J. Cons. London: Methuen, 1955. |
| *E.B.* | *Encyclopædia Britannica.* 14th ed., 1929, and earlier or subsequent editions as quoted. |

## STANDARD WORKS

| | |
|---|---|
| *L.D.G.* | *Longmans Dictionary of Geography.* London: Longmans, 1966. |
| Abler, Adams and Gould 1971 | Abler, R., Adams, J. S. and Gould, P., *Spatial Organization: the Geographer's View of the World.* Englewood Cliffs, New Jersey: Prentice-Hall, 1971. |
| *A.G.I.* | American Geological Institute, *Dictionary of Geological Terms.* New York: Dolphin Books, 1962 (abridged and revised from *Glossary of Geology and Related Sciences,* 2nd ed., Washington, 1960). |
| *Adm. Gloss.,* 1953 | Admiralty Hydrographic Department, Professional Paper No. 11, *Glossary of Terms used on Admiralty Charts and in Associated Publications.* 2nd ed. London, 1953. |
| Abercrombie *et al.,* 1951 | Abercrombie, M., Hickman, C. J. and Jenson, M. L., *A Dictionary of Biology.* Harmondsworth: Penguin, 1951: 2nd ed. 1952. |
| Ahmad, 1958 | Ahmad, Nafis, *An Economic Geography of East Pakistan.* London: O.U.P., 1958. |
| Baulig, 1956 | Baulig, H., *Vocabulaire Franco-Anglo-Allemand de Géomorphologie.* Paris: Les Belles Lettres, 1956. |
| Bencker, 1942 | Bencker, H., Maritime Geographical Terminology Relating to the Various Subdivisions of the Globe. *The Hydrographic Review,* **19**, 1942, 60–74. |
| Berry and Horton, 1970 | Berry, B. J. L. and Horton, F. E., *Geographic Perspectives on Urban Systems, with Integrated Readings.* Englewood Cliffs, New Jersey: Prentice-Hall, 1970. |
| Bryan, 1946 | Bryan, K., Cryopedology—the study of frozen ground and intensive frost action with suggestions on nomenclature. *Am. J. Sci.,* **244**, 1946, 622–642. |
| Burke, 1952 | Burke, J., *Stroud's Judicial Dictionary.* 3rd ed., 4 vols. London: Sweet & Maxwell, 1952–1953. |
| Burrows, 1942 | Burrows, R., *Words and Phrases Judicially Defined.* London: Butterworth, 1943–5. |
| Butlin, 1961 | Butlin, R. A., Some Terms used in Agrarian History: A Glossary. *Agricultural History Review,* **9**, 1961, 98–104. |
| Campbell, 1929 | Campbell, M. R., The River System: a study in the use of technical geographic terms. *Science Monthly,* **28**, 1929, 123–128. |

## STANDARD WORKS

| | |
|---|---|
| Carpenter, 1938 | Carpenter, J. R., *An Ecological Glossary*. Norman: University of Oklahoma Press; London: Kegan Paul, 1938. |
| Carson and Kirkby, 1972 | Carson, M. A. and Kirkby, M. J., *Hillslope Form and Process*. Cambridge: C.U.P., 1972. |
| Charlesworth, 1957 | Charlesworth, J. K., *The Quaternary Era*. London: Macmillan, 1957. |
| Chorley and Haggett, 1967 | Chorley, R. J. and Haggett, P. (editors), *Models in Geography*. London: Methuen, 1967. |
| Chorley and Kennedy, 1971 | Chorley, R. J. and Kennedy, B. A., *Physical Geography: a Systems Approach*. London: Prentice-Hall International, 1971. |
| Close and Winterbotham, 1925 | Close, C. F. and Winterbotham, H. St. J. L., *Text-book of Topographical and Geographical Surveying*. 3rd ed., London: H.M.S.O., 1925. |
| Cole and King, 1968 | Cole, J. P. and King, C. A. M., *Quantitative Geography: Techniques and Theories in Geography*. New York: John Wiley, 1968. |
| Cotton, 1922 | Cotton, C. A., *Geomorphology of New Zealand, Part 1—Systematic. An introduction to the study of landforms*. New Zealand Board of Science and Art, Manual No. 3. Wellington, New Zealand: Dominion Museum, 1922. |
| Cotton, 1941 | Cotton, C. A., *Landscape as Developed by the Processes of Normal Erosion*. Cambridge: C.U.P., 1941. |
| Cotton, 1942 | Cotton, C. A., *Climatic Accidents in Landscape-making*. Christchurch, N.Z.: Whitcombe and Tombs, 1942. |
| Cotton, 1945 | Cotton, C. A., *Geomorphology. An Introduction to the Study of Landforms*. 4th ed. Christchurch, N.Z.: Whitcombe and Tombs, 1945. |
| Davis, 1909 | Davis, W. M., *Geographical Essays*. Edited by Douglas Wilson Johnson. New York: Ginn, 1909. |
| Dickinson, 1947 | Dickinson, R. E., *City, Region and Regionalism*. London: Kegan Paul, 1947. |
| Dunbar, 1955 | Dunbar, M. J., *Ice Terminology* in Kimble, G. H. T. and Good, D., *Geography of the Northlands*. New York: Amer. Geog. Soc., Spec. Publ. No. 32, 1955 (following mainly Koch, L., 1945, The East Greenland Ice, *Medd. om Grönl.*, **130**, 2). |

| | |
|---|---|
| Dury, 1959 | Dury, G. H., *The Face of the Earth*. Harmondsworth: Penguin Books, 1959. |
| Ekwall, 1940 | Ekwall, E., *The Concise Oxford Dictionary of English Place Names*. 2nd ed. Oxford: Clarendon Press, 1940. |
| FAO, 1963 | Description of the Soil Associations of Europe, FAO: European Commission on Agriculture, Sub-Commission on Land and Water Use, Soil Map of Europe. Edited René Tavernier, 1963. |
| Fay, 1920 | United States Bureau of Mines, Bulletin 95, *Glossary of the Mining and Mineral Industry*. Edited by A. H. Fay. Washington, 1920. |
| Fischer, 1950 | Fischer, E., and Elliott, F. E., *A German and English Glossary of Geographical Terms*. New York: Amer. Geog. Soc., 1950. |
| Flint, 1947 | Flint, R. F., *Glacial Geology and the Pleistocene Epoch*. New York: John Wiley and Sons, 1947. |
| Geikie, A., 1865 | Geikie, Sir Archibald, *The Scenery of Scotland viewed in connection with its Physical Geology*. London: Macmillan, 1865. |
| Geikie, J., 1874 | Geikie, J., *The Great Ice Age and its relation to the Antiquity of Man*. London: W. Isbister, 1874. |
| Geikie, J., 1898 | Geikie, J., *Earth Sculpture or The Origin of Land Forms*. London: John Murray, 1898. |
| Geikie, J., 1905 | Geikie, J., *Structural and Field Geology for Students of Pure and Applied Science*. Edinburgh: Oliver and Boyd, 1905. |
| Gilbert, 1877 | United States Department of the Interior: Geographical Survey of the Rocky Mountain Region. *Report on the Geology of the Henry Mountains*, by G. K. Gilbert. Washington: Government Printing Office, 1877. |
| Glentworth, 1954 | Glentworth, R., *The Soils of the Country round Banff, Huntly and Turriff*. Memoirs of the Soil Survey of Great Britain: Edinburgh: H.M.S.O., 1954. |
| G.S.G.S., 1943 | Geographical Section, General Staff, War Office, *Short Glossaries for Use on Foreign Maps*, 1943–44. |
| Haggett, 1965 | Haggett, P., *Locational Analysis in Human Geography*. London: Arnold, 1965. |
| Haggett, 1972 | Haggett, P., *Geography: a Modern Synthesis*. Harper International Edition. New York and London: Harper and Row, 1972. |

## STANDARD WORKS

| | |
|---|---|
| Hare, 1953 | Hare, F. K., *The Restless Atmosphere*. London: Hutchinson, 1953. |
| Harrison-Church, 1957 | Harrison-Church, R. J., *West Africa*. London: Longmans, 1957. |
| Hartshorne, 1936 | Hartshorne, R., Suggestions on the Terminology of Political Boundaries. *Mitt. Ver. Geog.*, Univ. Leipzig, **14/15**, 1936, 180–192. |
| Hartshorne, 1939 | Hartshorne, R., *The Nature of Geography*. Lancaster, Penn.: Association of American Geographers. Fifth Printing, 1956. (Originally published in the *Ann. Assoc. Amer. Geog.*, **29**, 1939, parts 3–4.) |
| Harvey, 1969 | Harvey, D., *Explanation in Geography*. London: Edward Arnold, 1969. |
| Hatch and Rastall, 1938 | Hatch, F. H. and Rastall, R. H., *The Petrology of the Sedimentary Rocks*. 3rd ed., London: Allen and Unwin, 1938. |
| Hatch and Wells, 1937 | Hatch, F. H., and Wells, A. K., *The Petrology of the Igneous Rocks*. London: Murby, 1937. |
| Hills, 1940, 1953 | Hills, E. S., *Outlines of Structural Geology*. 1st ed. London: Methuen, 1940; 3rd ed., 1953. |
| Himus, 1954 | Himus, G. W., *A Dictionary of Geology*. Harmondsworth: Penguin, 1954. |
| Hinks, 1942 | Hinks, A. R., *Maps and Survey*. Cambridge, C.U.P., 4th ed., 1942. |
| Holmes, 1928 | Holmes, A., *The Nomenclature of Petrology*. London: Murby, 1928. |
| Holmes, 1944 | Holmes, A., *Principles of Physical Geology*. London: Nelson, 1944. |
| Howell, 1957, 1962 | Howell, J. V. (Editor), *Glossary of Geology and Related Sciences*. Amer. Geol. Inst., 1957. Also issued in an abridged edition as a *Dictionary of Geological Terms*, New York: Dolphin Books, 1962. See above, *A.G.I.* |
| *Ice Gloss.*, 1956, 1958 | Armstrong, T. E., and Roberts, B. B., Illustrated Ice Glossary, *Polar Record*, **8**, No. 52, 1956, pp. 4–12, and **9**, No. 59, 1958, pp. 90–96. Cambridge: Scott Polar Research Inst. |
| Jacks, 1954 | Jacks, G. V., *Multilingual Vocabulary of Soil Science*. Food and Agriculture Organization of the United Nations. Rome, 1954. |

| | |
|---|---|
| James, 1959 | James, Preston E., *Latin America*. 3rd ed. New York: Odyssey, 1959. |
| Johnson, 1919 | Johnson, Douglas Wilson, *Shore Processes and Shoreline Development*. New York: John Wiley, 1919. |
| Jones, 1930 | Jones, C. E., *South America*. New York: Holt, 1930. |
| Kendrew, 1922, 1953 | Kendrew, W. G., *The Climates of the Continents*. Oxford: Clarendon Press, 1st ed., 1922; 4th ed., 1953. |
| King, L. J., 1969 | King, L. J., *Statistical Analysis in Geography*. Englewood Cliffs, New Jersey: Prentice-Hall, 1969. |
| Knox, 1904 | Knox, A., *Glossary of Geographical and Topographical Terms*. London: Stanford, 1904. |
| Kubiëna, 1953. | Kubiëna, W. L., *The Soils of Europe*. London: Murby, 1953. |
| Küchler, 1947 | Küchler, A. W., Localizing Vegetation Terms. *Ann. Assoc. Amer. Geog.*, 37, 1947, 197–208. |
| Kuenen, 1955 | Kuenen, P. H., *Realms of Water*. London: Cleaver-Hume, 1955. |
| Lake, 1958 | Lake, P., *Physical Geography*. 4th ed. under editorship of J. A. Steers. Cambridge: C.U.P., 1958. |
| Lobeck, 1939 | Lobeck, A. K., *Geomorphology: An Introduction to the Study of Landscapes*. New York: McGraw-Hill, 1939. |
| Maxson and Anderson, 1935 | Maxson, J.H., and Anderson, G. H., Terminology of Surface Forms of the Erosion Cycle. *J. Geol.*, 43, 1935, 88–96. |
| *Met. Gloss.*, 1944 | Air Ministry, Meteorological Office. *The Meteorological Glossary*. 3rd ed. London: H.M.S.O., reprinted with amendments, 1944. |
| Millar and Turk, 1943 | Millar, C. E., and Turk, L. M., *Fundamentals of Soil Science*. New York: Wiley, 1943. |
| Miller, 1931, 1953 | Miller, A. Austin, *Climatology*, 1st ed., 1931; 8th ed., 1953. London: Methuen. |
| Monkhouse and Wilkinson, 1952 | Monkhouse, F. J., and Wilkinson, H. R., *Maps and Diagrams*. London: Methuen, 1952. |
| Monkhouse, 1954 | Monkhouse, F. J., *The Principles of Physical Geography*. London: Univ. of London Press, 1954. |
| Moore, 1949 | Moore, W. G., *A Dictionary of Geography*. Harmondsworth: Penguin, 1949. |

| | |
|---|---|
| Muller, 1947 | Muller, S. W., *Permafrost*. Ann Arbor: Edwards, 1947. |
| Muma and Muma, 1944 | Muma, M. H., and Muma, K. E., *A Glossary of Speleology*. Bull. Nat. Spel. Soc., **6**, 1944, 1–10. |
| Page, 1865 | Page, D., *Handbook of Geological Terms*. 2nd ed. Edinburgh: Blackwood, 1865. |
| Palgrave, 1894 | Palgrave, Sir R. H. I., *Dictionary of Political Economy*. 3 vols. London: Macmillan, 1894–1899. |
| Plaisance, 1958 | Plaisance, G., et Cailleux, A., *Dictionnaire des Sols*. Paris: La Maison Rustique, 1958. |
| Powell, 1875 | Powell, J. W., *Exploration of the Colorado River of the West and its Tributaries. Explored in* 1869, 1870, 1871 *and* 1872, *under the direction of the Secretary of the Smithsonian Institution*. Washington: Government Printing Office, 1875. |
| Rice, 1941 | Rice, C. M., *Dictionary of Geological Terms (exclusive of stratigraphic formations and paleontologic genera and species)*. Ann Arbor, Mich.: Edwards, 1941. |
| Rietz, 1930 | Rietz, J. E. du, Classification and Nomenclature of Vegetation. *Proc. 5th Internat. Bot. Congr.*, Cambridge, 1930, 74. |
| Raunkiaer, 1934 | Raunkiaer, C., *The Life Forms of Plants and Statistical Plant Geography*. Oxford, 1934. |
| Robinson, 1949 | Robinson, G. W., *Soils: Their Origin, Constitution and Classification*. 3rd ed. London: Murby, 1949. |
| Russell, 1950 | Russell, E. J. (revised by Russell, E. W.), *Soil Conditions and Plant Growth*. 8th ed. London: Longmans, 1950. |
| Salisbury, 1907 | Salisbury, R. D., *Physiography*. London: Murray, 1907. |
| Shackleton, 1958 | Shackleton, M. R., *Europe: A Regional Geography*, 5th ed. London: Longmans, 1958. |
| Shepard, 1937 | Shepard, F. P., Revised Classification of Marine Shorelines. *J. Geol.*, **45**, 1937, 602–624. |
| Shepard, 1952 | Shepard, F. P., Revised Nomenclature for Depositional Coastal Features. *Bull. Amer. Assoc. Petroleum Geologists*, **36**, 1952, 1902–1912. |
| Sloan and Zurcher, 1949 | Sloan, H. S., and Zurcher, A. J., *A Dictionary of Economics*. New York: Barnes and Noble, 1949. |

## STANDARD WORKS

| | |
|---|---|
| Smailes, 1953 | Smailes, A. E., *The Geography of Towns*. London: Hutchinson, 1953. |
| *Soils and Men*, 1938 | United States Department of Agriculture, Soils and Men. *Yearbook of Agriculture*, 1938. |
| Spate, 1954 | Spate, O. H. K., *India and Pakistan*. London: Methuen, 1954. |
| Stamp, 1923 | Stamp, L. D., *An Introduction to Stratigraphy*. London: Murby, 1923. |
| Stamp, 1948 | Stamp, L. D., *The Land of Britain: Its Use and Misuse*. London: Longmans, 1948. |
| Strahler, 1946 | Strahler, A. N., Geomorphic Terminology and Classification of Land Masses. *J. Geol.*, **54**, 1946, 32–42. |
| Strahler, 1951 | Strahler, A. N., *Physical Geography*. New York: Wiley, 1951. |
| Sverdrup, 1942 | Sverdrup, H. U. and others, *The Oceans*. New York: Prentice-Hall, 1942. |
| Swayne, 1956 | Swayne, J. C., *A Concise Glossary of Geographical Terms*. London: Philip, 1956. |
| Tansley, 1939 | Tansley, A. G., *The British Islands and their Vegetation*. Cambridge: C.U.P., 1939. |
| Tator, 1952 | Tator, B. A., Pediment Characteristics and Terminology. *Ann. Assoc. Amer. Geog.*, **42**, 1952, 295–317. |
| Taylor, 1951 | Taylor, T. Griffith (Editor), *Geography in the Twentieth Century*. London: Methuen, 1951. |
| Thornbury, 1954 | Thornbury, W. D., *Principles of Geomorphology*. New York: Wiley, 1954. |
| Trent, 1959 | Trent, C., *Terms Used in Archæology*. London: Phoenix House, 1959. |
| Twenhofel, 1939 | Twenhofel, W. H., *Principles of Sedimentation*. New York: McGraw-Hill, 1939. |
| Tyndall, 1860 | Tyndall, J., *The Glaciers of the Alps*. London: John Murray, 1860. |
| Umbgrove, 1947 | Umbgrove, L. H. F., *The Pulse of the Earth*. The Hague, 1947. |
| *U.S. Ice Terms*, 1952 | United States Navy Hydrographic Office, Pub. No. 609. *A Functional Glossary of Ice Terminology*. Washington, D.C., 1952. |
| von Engeln, 1942 | von Engeln, O. D., *Geomorphology*. New York: Macmillan, 1942. |

CORRESPONDENTS AND COLLABORATORS xxv

| | |
|---|---|
| Washburne, 1943 | Washburne, C. W., Some Wrong Words. *J. Geol.*, **51**, 1943, 496–497. |
| Wellington, 1955 | Wellington, J. H., *Southern Africa*, 2 vols. Cambridge: C.U.P., 1955. |
| Wills, 1929 | Wills, L. J., *The Physiographical Evolution of Britain*. London: Arnold, 1929. |
| Wooldridge and Morgan, 1937 | Wooldridge, S. W., and Morgan, R. S., *Physical Basis of Geography*. London: Longmans, 1937. |
| Wright, 1914 | Wright, W. B., *The Quaternary Ice Age*. London: Macmillan, 1914. |

*See also*

Science Library Bibliographical Series No. 707 (1952), *Technical Glossaries and Dictionaries*. London, 1952.

U.N.E.S.C.O. (1951), *Bibliography of Interlingual Scientific and Technical Dictionaries*. Paris, 1951.

*S.Y.B.*   *Statesman's Year Book*. London: Macmillan, Annual.

**NOTE**

**In many cases there are later editions of the books listed above but as it is the purpose of this Glossary to establish early use of words in standard works, the reference is normally to the FIRST Edition, except where important definitions have been added in later editions.**

## LIST OF CORRESPONDENTS AND COLLABORATORS

Geographers in many parts of the world have been most helpful in answering queries, especially in regard to terms of local use and to foreign words which have been used in writings in English. In some cases their comments or definitions have been quoted and are acknowledged by initials.

| | |
|---|---|
| H.W.A. | Professor Hans W:son Ahlmann, President of the International Geographical Union, 1956–1960, Stockholm (Scandinavian words). |
| K.S.A. | Professor Kazi S. Ahmad, University of Lahore (words in use in Indian subcontinent). |
| N.A. | Professor Nafis Ahmad, formerly of University of Dacca (words in use in Bangladesh). |
| D.H.K.A. | Professor David H. K. Amiran, Hebrew University of Jerusalem (words in use in Israel). |
| S.J.K.B. | Professor S. J. Kenneth Baker, formerly of University College of East Africa, Makerere, Uganda (words in use in East Africa). |

## CORRESPONDENTS AND COLLABORATORS

| | |
|---|---|
| H.B. | Professor Hans Boesch, University of Zürich (German and German-Swiss and other words). |
| E.G.B. | Professor Emrys G. Bowen, University College of Wales, Aberystwyth (Welsh words). |
| H.C.B. | Dr. H. C. Brookfield, Australian National University, Canberra (Australian words). |
| C.A.C. | Professor Sir Charles A. Cotton, Victoria University of Wellington, New Zealand (geomorphological terms). |
| G.B.C. | Professor George B. Cressey, University of Syracuse (Chinese and other words). |
| E.D. | Dr. Elwyn Davies, University of Wales, Swansea (Welsh words). |
| S.G.D. | Professor S. G. Davis, formerly of University of Hong Kong (Chinese words). |
| P.D. | Professor Pierre Deffontaines, Institut Français de Barcelone (Spanish and French words). |
| M.D. | Professor Max Derruau, University of Clermont-Ferrand (French words). |
| H.P.D. | Mrs. Hem P. Devi, Patna (Bihari words). |
| E.H.G.D. | Professor E. H. G. Dobby, formerly University of Malaya (Malay words). |
| C.H.E. | Professor C. H. Edelman, Netherlands (Dutch words and soil terms). |
| T.H.E. | Professor T. H. Elkins, University of Sussex (German words). |
| E.E.E. | Professor E. Estyn Evans, Queen's University, Belfast (Irish and Welsh words). |
| P.G. | Professor Pierre Gourou, Universities of Brussels and Paris (French words and tropical terms). |
| F.K.H. | Professor F. Kenneth Hare, University of Toronto (meteorological terms). |
| R.J.H-C. | Professor R. J. Harrison-Church, formerly of The London School of Economics (words in use in West Africa). |
| E.S.H. | Professor E. S. Hills, University of Melbourne (Australian words). |
| R.H. | Professor Robert Ho, University of Malaysia, Kuala Lumpur (Malay words). |
| G.V.J. | Mr. G. V. Jacks, Commonwealth Soil Bureau, Rothamsted (soil terms). |

## CORRESPONDENTS AND COLLABORATORS

| | |
|---|---|
| E.K. | Professor Edgar Kant, University of Lund (Scandinavian and other words). |
| G.H.T.K. | Professor G. H. T. Kimble, formerly of Indiana University (miscellaneous words). |
| G.K. | Professor George Kuriyan, formerly of University of Madras (words in use in India). |
| E.D.L. | Dr. E. D. Laborde (French words). |
| J.H.G.L. | Professor John H. G. Lebon, University of Khartoum (Sudanese and Arabic words). |
| M.W.M. | Professor Marvin W. Mikesell, University of Chicago (Moroccan Arabic words). |
| R.M. | Professor Ronald Miller, University of Glasgow (Scottish words). |
| S.D.M. | Dr. S. D. Misra, University of Madras (words in use in India). |
| W.H.N. | Dr. W. H. Nieuwenhuija, Food and Agriculture Organisation, Rome (forestry terms). |
| A.C.O'D. | Professor A. C. O'Dell, University of Aberdeen (Scottish words). |
| L.P. | Dr. Leo Peeters, Institut Universitaire des Territoires d'Outre-Mer, Antwerp (tectonic terms). |
| M.B.P. | Professor Maneck B. Pithawalla, formerly of University of Karachi (words in use in Pakistan). |
| J.C.P. | Professor J. C. Pugh, King's College, London (words in use in Nigeria). |
| P.W.R. | Professor P. W. Richards, University College of North Wales, Bangor (vegetation terms). |
| I.S. | Professor Isaac Schapera, London School of Economics (Afrikaans words). |
| P.S.G. | Dr. P. Sen Gupta, National Research Council, Delhi (words in use in India). |
| P.S. | Professor P. Serton, University of Stellenbosch (Afrikaans words). |
| M.S. | Dr. Muhammad Shafi, Moslem University, Aligarh (words in use in India). |
| A.H.S. | Dr. A. H. Siddiqi, formerly of University of Dacca (words used in the Indian subcontinent). |
| R.L.S. | Professor R. L. Singh, Banaras Hindu University (words used in India). |
| U.S. | Dr. U. Singh, University of Patna (words used in India). |
| K.A.S. | Dr. Karl A. Sinnhuber, Wirtschaftsuniversität, Wien (German words). |

## ABBREVIATIONS

| | |
|---|---|
| H.O'R.S. | Professor Hilgard O'Reilly Sternberg, formerly of University of Rio de Janeiro (Portuguese words). |
| E.G.R.T. | Professor Eva G. R. Taylor, Birkbeck College, London (Mathematical Geography). |
| F.T. | Professor Fumio Tada, University of Tokyo (Japanese words). |
| G.T. | Professor Emeritus T. Griffith Taylor, Sydney, N.S.W. (various terms). |
| T.K. | Professor Daw Thin Kyi, University of Rangoon (Burmese words). |
| T.R.T. | Dr. T. R. Tregear, formerly of University of Hong Kong (Chinese words). |
| G.T.T. | Professor Glenn T. Trewartha, University of Wisconsin (Japanese words). |
| C.T. | Professor Carl Troll, University of Bonn (German words). |
| J.H.W. | Professor J. H. Wellington, University of the Witwatersrand, Johannesburg (South African words). |
| W.W-O. | Professor William William-Olsson, University of Stockholm (Scandinavian words). |
| P.J.W. | Dr. P. J. Williams (Norwegian words). |

Other initials are those of members of the Glossary Committee.

Under the auspices of the Agricultural History Society there is in preparation an International Lexicon of Agrarian Terminology. The Committee is greatly indebted to Dr. R. H. Hilton of the School of History of the University of Birmingham for placing at its disposal material collected for this Lexicon relative to terms used in rural historical geography (acknowledged as I.L.A.T.).

## ABBREVIATIONS

In addition to the abbreviations used in the Glossary, which are listed on p. xxx, the following initials will be commonly found in geographical works.

| | |
|---|---|
| A.A.G. | Association of American Geographers. |
| A.A.A.S. | American Association for the Advancement of Science. |
| A.C. | Alpine Club (London). |
| A.G.F. | Association de Géographes français. |
| A.G.S. | American Geographical Society. |
| B.A. | British Association for the Advancement of Science (also B.A.A.S.). |
| B.C.I. | Bibliographie cartographique internationale. |
| B.G.I. | Bibliographie géographique internationale. |

# ABBREVIATIONS

| | |
|---|---|
| B.M. | British Museum. |
| C.N.R.S. | Centre Nationale de la Recherche Scientifique (Paris). |
| C.O.I. | Central Office of Information (London). |
| D.M.S. | Directorate of Military Survey. |
| D.O.E. | Department of the Environment. |
| D.O.S. | Directorate of Overseas Surveys. |
| FAO | Food and Agriculture Organization of United Nations (Rome); also F.A.O. |
| F.I.D.S. | Falkland Islands Dependencies Survey. |
| G.A. | Geographical Association (also Geologists' Association). |
| G.S.G.S. | Geographical Section, General Staff (War Office). |
| H.M.S.O. | Her Majesty's Stationery Office. |
| I.B.G. | Institute of British Geographers. |
| I.C.A.O. | International Council of Aeronautical Organizations. |
| I.C.S.U. | International Council of Scientific Unions. |
| I.G.G.U. | International Geodetic and Geophysical Union (also I.U.G.G.). |
| I.G.U. | International Geographical Union. |
| I.G.Y. | International Geophysical Year. |
| J.I.B. | Joint Intelligence Bureau. |
| L.C. | Library of Congress (Washington). |
| M.A.F.F. | Ministry of Agriculture, Fisheries and Food (England and Wales). |
| M.H.L.G. | Ministry of Housing and Local Government (England and Wales). (Now Department of the Environment: D.O.E.) |
| M.O. | Meteorological Office (Air Ministry). |
| O.S. | Ordnance Survey. |
| P.R.O. | Public Records Office (London). |
| R.G.S. | Royal Geographical Society. |
| R.I.I.A. | Royal Institute of International Affairs (also known as Chatham House). |
| R.S. | Royal Society. |
| R.S.G.S. | Royal Scottish Geographical Society. |
| S.P.R.I. | Scott Polar Research Institute (Cambridge). |
| U.N. | United Nations. |
| UNESCO | United Nations Educational Scientific and Cultural Organization; also U.N.E.S.C.O. (Paris). |
| U.N.O. | United Nations Organization. |
| W.H.O. | World Health Organization. |
| W.M.O. | World Meteorological Organization. |
| W.O. | War Office (London). |

# ABBREVIATIONS

*Note:* As the Glossary quotes from numerous dictionaries and glossaries, the abbreviations may not always be uniform. The following short list should help to remove any difficulties.

| | |
|---|---|
| A.A.A.G. | *Annals of the Association of American Geographers* |
| A.-S. | Anglo-Saxon |
| adj. | adjective |
| coll., col. | colloquial |
| Da. | Danish |
| dial. | dialect |
| esp. | especially |
| f. | formed on; from |
| fam. | familiar, familiarly |
| F., Fr. | French |
| fn. | footnote |
| G., Ger. | German |
| Gr. | Greek |
| I.G.U. | International Geographical Union |
| Icel. | Icelandic |
| L. | Latin |
| lit. | literally |
| N.R., n.r. | no reference |
| N.W.T. | North West Territories |
| Nor. | Norwegian |
| O.E. | Old English |
| obs. | obsolete |
| pl. | plural |
| pron. | pronounced |
| quot. | quotation |
| sb. | substantive |
| sing. | singular |
| tr. | translation |
| U.A.R. | United Arab Republic |

# A

**Aa, a-a** (Hawaiian; *pron.* ah-ah)
Webster. *Petrog.* Rough, scoriaceous lava;— contrasted with *pahoehoe. Hawaii.*
Mill, *Dict.* Rough scoriaceous lava flows (Maunaloa, Hawaii).
Rice, 1941. A Hawaiian word specially introduced into American usage to describe jagged scoriaceous lava flows. It is contrasted with pahoehoe (Fay).
Dana, J. D., 1887, History of Changes in the Mt. Loa Craters in Hawaii, *Am. J. of Sci.*, 33, 1887, 433–451 *et seq.* 'An *aa* or *arate* lava stream consists of detached masses of lava... the masses are of very irregular shapes and confusedly piled up to nearly a common level, although often covering areas many miles long and half to a mile or more wide. The size of the masses in the coarser kind varies from a few inches across to several yards.'
See also Wilkes, C., 1845, *Narrative of the Expedition 1838-1842*, vol. iv; Washington, H. S., 1923, The Formation of Aa and Pahoehoe, *Am. J. of Sci.*, 5th Ser., **6**, 408–423; Emerson, O. H., 1926; *Ibid.*, **12**, 109–114; Cotton, C. A., 1944, *Volcanoes*, 27, 113, 132; Holmes, 1944, 448.
*Comment.* A term originating from Hawaii introduced into American literature and now becoming widespread partly through the influence of Cotton. It should be noted that petrologically *aa* and *pahoehoe* lava are often identical and may be from the same flow. (G.T.W.)

**Abi** (Indo-Pakistan: Urdu)
Land irrigated by basket or Persian wheel from tanks, pools or streams; Abi-sailābā, also flooded by river.

**Abime**
1. An early form of abyss (*q.v.*).
2. As abîme used currently in French but not internationally for a disused pothole. (G.T.W.)
3. The Commission on Karst Phenomena of the International Geographical Union in its *Report* (1956) says 'Parmi les jamas (*q.v.*) les uns aboutissent à un réseau de cavités souterraines; les autres se terminent assez rapidement. On peut appeler les premiers des *abîmes*; les seconds des *avens*.' In this sense an abime is a pothole or natural shaft in karst country which is bottomless because it opens into an underground channel. See Karst terminology.

**Ablation, Ablation moraine**
*O.E.D.* 1. The action or process of carrying away or removing; removal.
4. Geol. The wearing away or superficial waste of a glacier by surface melting, or of a rock by the action of water.
1860 Tyndall, Glaciers, 11, par. 32, 418. The ablation of the ice must be less than what is generally supposed. 1863 Ball, J. Guide to West Alps, Introd., 70. The vast amount of ablation, or loss, which a glacier annually undergoes through the melting of the surface.
Rice, 1941. (1) The formation of residual deposits by the washing away of loose or soluble minerals. (2) The wearing away of rocks, or the surface melting of glaciers (Fay). Fay quotes (1) Kemp, Handbook of rocks, 1904; (2) 20th C. Standard Dict.
Cotton, 1942. '... ice falls..., often with crowded ridges and pinnacles sharpened by ablation' (p. 138). Footnoted: '"Ablation" includes wastage by evaporation as well as melting.'
Cotton, 1945. 'The zone of ablation, where the surface ice is wasting away by melting and evaporation...' (p. 281).
'The larger glaciers of the eastern side of the Southern Alps are so full of rock debris that for the last few miles of their courses they are completely covered by sheets or coalescing heaps of such ablation moraine.' (p. 289)
Flint, 1947. '... melting and evaporation (together constituting the process of ablation), resulting in losses of snow or neve.' (p. 21)
*Comment.* It seems desirable to restrict the meaning to the surface phenomenon of the wasting of snow and ice, especially glacier ice, by melting and evaporation. French usage however is still much wider (see Baulig 1956) as was the older usage in English (Rice, *op. cit. sup.*). Some recent writers refer to ablation of sandy areas by wind.

**Abo, Abos** (Australia)
A common contraction of Australian Aborigine or Aborigines used familiarly in Australia. See Taylor, T. Griffith, 1958,

*Journeyman Taylor*, London, Robert Hale, 148.

## Aborigines
*O.E.D.* A purely Latin word, applied to those who were believed to have been the inhabitants of a country *ab origine*, *i.e.* from the beginning.... 1. The original inhabitants of a country; originally the race of the first possessors of Italy and of Greece, afterwards extended to races supposed to be the first or original occupants of other countries. 2. spec. The natives found in possession of a country by Europeans who have gone thither as colonists. 3. occas. used also of animals and plants.

*Webster* 1. The earliest known inhabitants of a country; the native race, esp. as contrasted with the invading or colonizing race. 2. The original fauna and flora of a geographical area.

Mill, *Dict.* Representatives of a human stock native in a region at the beginning of its ascertained history.

*Comment.* 'Aborigines' is now applied especially to the Australian Aborigines but the adjective 'aboriginal' is more widely used.

## Abrasion, abrade
*O.E.D.* 1. The act or process of rubbing off or away, wearing down by friction. lit. and fig. 1656 Blount *Glossogr.*, Abrasion, a shaving away. 1837 Babbage, *Bridgw. Treat.* K 250. Let us suppose, that from the abrasion of the channel, the later tide arrives... earlier than before.

Page, 1865. 'The operation of wearing away by rubbing or friction. Currents of water laden with sand, shingle, and other rock-debris are the chief abrading agents in nature. Abrasion may also result from the passage of icebergs, the descent of glaciers, etc....' (p. 56)

Geikie, J., 1898, Glossary. 'The operation of wearing away by aqueous or glacial action.'

Mill, *Dict.* (1) The operation of wearing away of a surface by river, wave, ice, or wind action. (2) The attack of the sea on the land (*Gregory, J. W.*)

Cotton, 1942, refers to abrasion of, as well as by, pebbles contained within a glacier. (p. 312)

Holmes, 1944. 'Abrasion is the scraping and scratching of rock surfaces by debris frozen into the sole of a glacier or ice sheet. [p. 215]... Armed with the sand grains thus acquired, the wind near the ground becomes a powerful scouring or abrading agent. The resulting erosion is described as wind abrasion.' (p. 254)

Flint, 1947. 'Abrasion, the scour of rock by rock...' (under heading 'Quantitative Effects of Glacial Erosion') (p. 76).

*Comment.* A general term which should not be restricted to the work of any one agent. Abrasion may be brought about by moving debris whether it be in a stream, the sea, ice, wind, or in moving regolith. One point at issue is whether or not it includes the wear of particle against particle (attrition). (G.T.W.)

## Abrasion platform, abrasion plane
Wooldridge and Morgan, 1937. 'In other cases it [the wave-cut beach-bench] is continued seaward in a flatter surface, the abrasion platform, over which a thin offshore "veneer" of finer material may be spread.' (p. 322)

Umbgrove, J. H. F., 1947, *Pulse of the Earth.* 'An abrasion-plane formed by the erosive action of the breakers.' (p. 100)

Thornbury, 1954, refers to it as the *abrasion platform* (p. 434).

*Comment.* Some writers prefer wave-cut platform as indicating the agent of abrasion.

## Abrasion tableland
Mill, *Dict.*, A tableland with outcrops of various rocks which have been reduced to approximately the same level by denuding agents.

*Comment.* No recent use has been noted.

## Abri (French)
*O.E.D.* n.r.

*Webster.* A cavity in a hillside.

Davies, G. M., 1932, *A French-English Vocabulary in Geology and Physical Geography*, London: Murby. Shelter, rock-shelter.

Hughes, T. McKenny, 1887, On Caves, *Trans. Vict. Inst.* 'Some of the most interesting caves, in respect of their contents and the light they throw on the history of primeval man, are only rock-shelters—abris—such as are seen in the Dordogne district.'

Oakley, K. P., 1949, *Man the Toolmaker*, gives in Fig. 15 a 'Diagrammatic section of ideal rock-shelter (shallow cave or abri)....'

*Comment.* A French word occasionally used in English literature; not in standard works.

### Absolute
*O.E.D.* Pure and simple, mere; in the strictest sense.

For absolute drought see Drought; for absolute humidity, see Humidity.

Absolute maximum, absolute minimum: 'the highest and lowest temperatures during the whole period of observations are called the absolute extremes.' (*Met. Gloss.*)

### Absolute age, absolute chronology
*A.G.I.* Involves dating of geologic events in years.

*Comment.* The table on p. 531 shows the end of the Cambrian to be 500 million years ago. The oldest known rocks of the earth's crust are believed to be 6,500 million years old.

### Absondering
Rice, 1941. Von Leonard's term for the jointing in igneous rocks dividing them into more or less regular posts. This jointing is caused by cooling, contraction etc. (p. 1)

### Abstraction
Gilbert, 1877. 'A stream which for any reason is able to corrode its bottom more rapidly than do its neighbours, expands its valley at their expense, and eventually "abstracts" them.' (p. 141)

Cotton, 1922. 'Interfluves may be cut through by planation, and capture (abstraction) of smaller streams by their larger neighbours may take place as a result.' Similar usage in 4th ed. 1945.

Fay, 1920. 'In geology, the withdrawal of a stream from a lower portion of its course by an adjoining stream having more rapid corrosive action.' (Source: *Standard Dictionary*, 1910.)

Thornbury, 1954. 'The term *abstraction* is often applied to the simplest type of piracy, which results from competition between adjacent consequent gullies and ravines.' (p. 152)

*Comment.* No geographical usage is given in *O.E.D.* and there is no reference in Davis, 1909; Geikie, J., 1898; Wooldridge and Morgan, 1937; Holmes, 1944; Lobeck, 1939. Salisbury, 1907, refers to piracy by rivers. It seems better to avoid a specialized meaning for a common word, especially as it is frequently used in a general sense. For example J. K. Charlesworth, 1957, *The Quaternary Era*, p. 1354, says 'the growth of the *mers-de-glace* and glaciers was necessarily attended by an abstraction of water from the sea and a uniform glacio-eustatic fall of the ocean level'.

### Abysmal
In scientific writings this word has given place to abyssal for most purposes. Mill, *Dict.*, notes 'Abysmal Region of the Ocean—the limited regions where the ocean is very deep (more than 300 fathoms).' J. Johnstone, 1923, *An Introduction to Oceanography*, refers to abysmal life in the deepest parts of the oceans. (p. 89)

### Abyss, abysm, abime
*O.E.D.* from Latin abyssus and Greek αβυσσος, bottomless, the deep. The older forms in English were abime, abysme from the French. The Latin abyssus was adopted as a more learned word and thus the word has had five variants, abime, abysm, abysmus, abyssus, abyss. Used from 1398 as the great deep, the primal chaos, from 1639 as a bottomless gulf, any apparently unfathomable cavity.

Mill, *Dict.* A deep gulf or chasm in the ocean. A portion of a lake deeper than 100 fathoms.

*Comment.* Obsolete or obsolescent in this sense; replaced by Deep (*q.v.*); survives in the adjectival forms abyssal and abysmal.

### Abyssal
*O.E.D.* Of unsearchable depth, unfathomable; belonging to the lowest depths of ocean. Abyssal zone, the bottom strata of the sea, the belt of water below 300 fathoms, 1830 Lyell *Princ. Geol.* (1875), II, III, xlix, 589. The coral fauna of the deep and abyssal sea. 1872 Nicholson *Palaeont.* 23 The abyssal mud of the Atlantic is to a very large extent composed of the microscopic shells of Formanifera.

Holmes, 1944. No definition, but used in sense of deep ocean floor as distinct from continental shelf. No depth given.

Murray, J., 1913, *The Ocean*, London: Williams and Norgate. 'A great submerged plain—the floor of the ocean-basins (below about 1750 fathoms) is called the abyssal area.' (p. 230)

Sverdrup, 1942. (In relation to animal life) 'finally (below 500 fathoms) there is the abyssal zone' (p. 60). 'The benthic division includes all of the bottom terrain from flood-tide level to the greatest depths. The benthic division may be subdivided into the littoral and the deep-sea systems... (this) is divided into an upper (archibenthic) and a lower (abyssalbenthic) zone.' (pp. 275-276)

*Comment.* The current trend is to use abyssal as restricted to the depths of the ocean. With geologists abyssal has been dropped in favour of plutonic as applied to igneous rocks which have consolidated at great depths.

**Abyssal deposits**

Moore, 1949. The solid matter which covers the floor of the Abyssal region of the ocean.

**Abyssal rocks** (Brögger)

A term now largely discarded in favour of plutonic rocks (*q.v.*) thus avoiding confusion with abyssal deposits. The term hypabyssal rocks (*q.v.*) is, however, still generally used. Hatch and Wells, 1937, refer to 'the injection of an abyssal wedge of magma into the lower crust.' (p. 8)

**Access, accessibility**

Stamp, 1948. 'We may use the words access and accessibility as virtually synonymous, though a distinction is sometimes drawn between the local access, *i.e.* the existence of a road fit for motor traffic direct to a farm—an access road—indicating actual physical means of approach to a given point, in contrast to accessibility, which is used to cover the broader aspects of the relationship between one region in one part of the country and another' (p. 201). The question is discussed at length, pp. 201-215. See also Jonasson, O., 1926, Agricultural Regions of Europe, *Econ. Geog.*, **2**, 19-48.

*Comment.* The economic element in accessibility is explicitly introduced in the FAO definition of accessible forests as 'forests which are now within reach of economic management or exploitation as sources of forest products'. FAO, 1955, *World Forest Resources* (p. 112) (C.J.R.) Access road often carries a connotation of a private road especially one giving access to fields.

**Accident** (W. M. Davis, 1894)

Salisbury, 1923, *Physiography for Schools.* 'Accidents to streams—streams are subject to many accidents. If the land through which they flow sinks, ... they flow less rapidly, or may even cease to flow altogether' (pp. 119 *et seq*.). Goes on and includes Drowning, Rejuvenation, Ponding, Piracy under this heading.

Davis, 1909 (1899). 'There are two other classes of departure from the normal or ideal cycle ... these are changes of climate and volcanic eruptions, both of which occur so arbitrarily as to place and time that they may be called accidents' (p. 274). (Similar use in Essay published 1894 (*J. of Geol.*)—see Essays, p. 180.)

See also Cotton, 1942 *Climatic Accidents in Landscape-making.*

**Accidented relief**

*O.E.D.* Accident. 5. An irregular feature in a landscape; an undulation. 1870.

*Comment.* The phrase, 'accidented relief' or 'highly accidented relief', commonly and loosely used to describe rugged and irregular relief, is probably a literal translation of the common French phrase 'relief accidenté'. Baulig, 1956, translates terrain accidenté as broken country. (A.F.M.). There is also a possible connection with such 'accidents' as those listed above.

**Acclimate, acclimatement, acclimation**

*Webster.* To habituate or cause habituation, to a climate not native; acclimatize.

*Comment.* American usage; English is invariably acclimatize.

**Acclimatize, acclimatization**

Mill, *Dict.* To habituate or inure to a new climate, or to one not natural.

*Webster.* To cause to become acclimated; to adapt to a new environmental condition, as of soil.

*Comment.* Usually restricted to climate; reference to soil unusual.

**Accommodation land**

Stamp, 1948. 'The so-called "accommodation land" for the final fattening of cattle before slaughter often means particularly high rentals near large towns' (p. 201).

Refers to small enclosed fields especially on town fringes where animals can conveniently be 'accommodated' after being purchased for slaughter.

**Accommodation unit**

A convenient planners' term for a unit of housing occupied by one household (whether an individual, or individuals sharing, as a family or otherwise, irrespective of whether that unit is a separate house, maisonette, flat or apartment and whether permanent or temporary). It should not be confused with a *structure* considered as a unit, whether occupied by an individual (*e.g.* a small shack), or a group of families (*e.g.* a large farm unit or block of flats) which is the sense of the French unité d'habitation or German Wohneinheit. (C.J.R.)

**Accordance of Summit Levels**

*O.E.D.* Accordance: 1. The action or state of agreeing; agreement; harmony: conformity.

Davis, 1909. 'Repeated views of the uplands from various hilltops impress me much more with the relative accordance of their altitudes than with their diversity.' (p. 353, from paper pub. 1899)

Daly, R. A., 1905, The Accordance of Summit Levels among the Alpine Mountains: The Fact and its Significance, *J. of Geol.*, 13 (2), 105–125.

'The word "accordance" is used advisedly. "Equality" of height is not meant ... for limited areas "subequality" of the summits is a fact, but over wider stretches, and especially over the whole of a single range, even subequality fails, and the accordance takes the form of sympathy among the peaks whose tops in companies or in battalions rise or fall together in imaginary surfaces often far removed from the spheroidal curve of the earth.' (p. 106)

Wooldridge and Morgan, 1937. 'Accordance of summit-level or bevelling of hill-tops of itself does not prove a former peneplain' (p. 213). See also Summit-planes.

Cotton, 1942. '... accordance of summit levels may indicate that a peneplain has been destroyed by erosion, a number of peaks still reaching to about a common level, though the tops of each peak may be some little distance below the position of the former surface.' (p. 138)

See also Summit-plane, Gipfelflur.

**Accordant drainage**

*L.D.G.* Accordant drainage describes the conditions when the surface drainage is directly related to the dip of the underlying strata.

**Accordant junctions**

*O.E.D.* See Accordance of summit levels above.

Playfair, 1802. *Illustrations of the Huttonian Theory of the Earth.* 'Every river appears to consist of a main trunk, fed from a variety of branches, each running in a valley proportioned to its size, and all of them together forming a system of vallies, communicating with one another, and having such a nice adjustment of their declivities that none of them join the principal valley either on too high or too low a level; a circumstance which would be infinitely improbable if each of these vallies were not the work of the stream that flows in it.' (p. 102) (Reprinted in facsimile, 1956)

Cotton, 1945. '... Playfair's law of accordant junctions.' (p. 38)

Wooldridge and Morgan, 1937. '... that rivers excavate the valleys in which they flow, was perceived by Playfair in 1802, in presentation of the prescient, but then little known, work of Hutton. In his so-called Law of Accordant Junctions he emphasized the fact that tributaries (with a few readily explicable exceptions) join the main stream at exactly the appropriate level which, in modern terminology, implies their power to grade their courses to the level of their outfall.' (p. 152)

**Accumulated temperature**

*Met. Gloss.*, 1944. 'The integrated excess or deficiency of temperature measured with reference to a fixed datum over an extended period of time.'

'The adoption of 42°F. (6°C.) as the basic temperature follows the practice of Angot and others who have considered that value to be the critical temperature above which the growth of vegetation in a European climate is initiated and maintained. For the study of heating problems a basic temperature of 60°F. has been adopted by American and British engineers.'

Miller, A. A., 1931, *Climatology*, London: Methuen. '... the duration of temperatures above a certain minimum amount, *e.g.* the minimum for growth; ... Experiments at Rothamsted showed that 42° may be considered the basal temperature for wheat; each degree in excess of the mean daily temperature over this critical value may be called a "day degree", and if they are added together for the period of growth they give a measure of the accumulated temperature.' (pp. 10–11)

**Accumulation, mountains of, accumulation-mountains**

Geikie, 1898. Of original or tectonic mountains: 'Some of these have been piled or heaped up at the surface—they have grown into heights by gradual accumulation, and may therefore be termed accumulation-mountains. This group naturally includes all volcanic cones and hills, geyser mounds, mud volcanoes, etc.' (p. 272). The similarity of some eminences formed by epigene agents, *e.g.* moraines and sand-dunes, is considered but they are usually too small to be termed 'mountain'.

Comment. In a simple classification, mountains of accumulation offer a contrast to mountains of circumdenudation and, according to some authors, to tectonic mountains. No reference in Lyell, 1830; Salisbury, 1907; Cotton, 1922; Mill, *Dict.*; Wooldridge and Morgan, 1937; Holmes, 1944; Fay, 1920; Rice, 1941.

## Acid rock

Hatch, F. H., and Wells, A. K., 1937, *The Petrology of the Igneous Rocks*. 'Arranged in order of their silica content, the igneous rocks fall into three groups: 1 Acid, with silica content above 66% ... These limits are ... entirely arbitrary, but ... they permit a convenient separation in accordance with existing records of rock types' (p. 139).

Igneous rocks, whether plutonic, hypabyssal or volcanic, which have such a high proportion of silica that there is free silica (quartz) in addition to silica-rich felspars and other minerals. As orthoclase felspar has 66 per cent of silica, the total in an acid magma is in excess of this amount, hence the alternative terms oversaturated or persilicic rocks which, though more accurate, are not as commonly used as acid. The typical acid plutonic rock is granite; the hypabyssal is quartz porphyry, the volcanic is rhyolite. Acid lavas are very sticky even at high temperatures and tend to form plugs and solid cones rather than lava flows. Many acid lavas solidify in a glassy rather than crystalline form and are then known as obsidian, but some glassy lavas are less acid in composition (L.D.S.). See also Intermediate, Basic.

## Acid soil

Uvarov, E. B., and Chapman, D. B., 1952, *Dict. of Science*, 2nd ed. Acid: Substance which forms hydrogen ions in solution; substance which contains hydrogen which may be replaced by a metal to form a salt. (p. 9)

Millar, C. E., and Turk, L. M., 1943, *Fundamentals of Soil Sc*. 'Types of Soil Acidity— for a soil to give an acid reaction there must be an excess of H ions over OH ions in the solution surrounding the colloidal complex. The concentration of these H ions ... constitutes the *active* acidity of the soil.... The H ions on the acidoid (colloidal complex) constitute a reserve supply which is known as the *reserve* or *potential* soil acidity' (p. 96). 'Acid Soil.

A soil giving an acid reaction (below pH 7·0).' (p. 431)

Comment. A soil is said to be acid when its pH value (Hydrogen ion concentration) is below 7·2 which is neutrality. On raw peat with free organic acids values as low as 3·0 may occur; British soils usually tend to be on the acid side between 6·4 and 6·9. See Russell, E. J., *The World of the Soil*, London: Collins, 1957.

## Aclinic structure, Acline

*O.E.D.* Aclinic, from Greek, 'unbending'. Without inclination. Applied to the magnetic equator, or line surrounding the earth and cutting the terrestrial equator, on which the magnetic needle has no dip but lies horizontal.

Rice, 1941. 'Acline. Beds without dip; unfolded; in a horizontal position (Cox-Dake-Muilenburg).'

Comment. Rarely used. Not recorded in Lyell, 1830; Geikie, 1898; Cotton, 1922; Wooldridge and Morgan, 1937; Holmes, 1944; Lobeck, 1939. From the simple meaning 'without inclination' it is obvious that the adjective could be applied in several ways. In practice its use has been limited to magnetism and geology.

## Acre

The statutory acre is the legal measure of land in Britain and is an area of 4 roods each of 40 rods, poles or perches and equivalent to 4840 square yards. Originally the acre ( 'customary acre' ) was theoretically the amount of land a plough team could plough in a day. See Butlin, 1961

See Yardland

## Actic Region

Mill, *Dict.* 1. The zone of comparatively steep slope from the edge of the continental shelf down to the depths of the ocean. 2. zone of great tectonic change.

Comment. Obsolete; *O.E.D.*, n.r.

## Action space

Wolpert, J., 1965, Behavioural Aspects of the Decision to Migrate, *Papers & Proc. Reg. Sci. Assoc.*, **15**, 159–159. '... action space is the set of place utilities which the individual perceives and to which he responds.' (p. 163)

Horton, F. E. and Reynolds, D. R., 1971, Effects of Urban and Spatial Structure on Individual Behaviour, *Econ. Geog.*, **47**, 36-88. 'Action space is the collection of all urban locations about which the individual has information and the subjective utility or preference he associates with these locations.' (p. 37)

**Active**
*O.E.D.* 4. Opposed to quiescent or extinct; existing in action, working, effective.
This common word is frequently used with special meanings in geographical literature especially in relation to karstic phenomena, frozen ground and volcanoes.
*Active cave, active swallow,* etc.
Balch, H. C., 1948. 'If we desire to describe a swallet with running water entering it, we prefer the word "active".' Also used to refer to caves and other karstic phenomena in which running water is still at work in developing them. (G.T.W.)
*Active layer, active permafrost*
Muller, S. M., 1947, *Permafrost.* 'Above the permafrost is a layer of ground that thaws in the summer and freezes again in winter. This layer represents the seasonally frozen ground and is called the *active layer*' (p. 6). 'Active permafrost (active permanently frozen ground)—permafrost which, after having been thawed due to natural or artificial causes, is able to return to permafrost under the present climate. Also called annually thawed layer' (p. 213).
*Active volcano*
Mill, *Dict.* A volcano from which periodical eruptions still occur. Contrast with dormant and extinct.

**Adiabatic**
*Webster.* Occurring without loss or gain of heat. Adiabatic gradient. The rate at which the temperature of an ascending or descending body of air is changed by expansion or compression—being about 1·6°F. for each 300 feet of change of height; also a curve representing this.
Mill, *Dict.* Relating to the changes which occur in the pressure, volume, and temperature of a mass of gas (*e.g.* the atmosphere) which is neither parting with nor receiving heat during the process.

**Adit**
*O.E.D.* 1. An approach; spec. a horizontal opening by which a mine is entered or drained.
Mill, *Dict.* A more or less horizontal working for the extraction of a mineral vein from a hillside.
Rice, 1943. A nearly horizontal passage from the surface by which a mine is entered and unwatered.
*Comment.* Hence adit-mining, adit-mine. In the lead-mining districts of the Pennines, adit would refer to a mine or a drainage level. The latter is there also called a 'sough'. (G.T.W.)

**Admiralty chart**—see Chart

**Adobe**
*O.E.D.* From Spanish, adopted in U.S. from Mexico, and popularly made into dobie. An unburnt brick dried in the sun, 1834.
*Webster.* 1. An unburnt brick dried in the sun;—a name common where the Spanish have settled, as in Texas and New Mexico. 2. Earth from which unburnt bricks are made, esp. in the arid regions of the western United States; hence, any alluvial or playa clay in desert or arid regions. 3. A house or structure made of such bricks or clay.
*Dict. Am.* 1. Sun-dried mud or clay, or a crude cement made of this. Also a brick or brick-like piece of such material used in building (citations back to 1759).
Mill, *Dict.* Sundried bricks (building materials, Mexico), etc.
Holmes, 1928 (quoting Russell, 1889). A loess-like deposit occurring in the plains and basins of the Western States, and in the arid parts of Spanish America.
Fay. 1920. 1. (Sp.) A sun-dried brick; often shortened to adob and even 'dobe. 2. The mixed earth or clay of which such bricks are made. 3. In mining a brick of pulverized ore mixed with clay, as in quicksilver metallurgy (*Standard Dict.*, 1910). 4. The Mexican silver dollar. (Also a name for the mud cap placed over an explosive charge in blasting—W.M.)
Rice, 1941. A sandy, often calcareous, clay used in the west and south-west for making sun-dried brick (Ries and Watson). Adobe flat. A generally narrow plain, having a slope of 5 to 20 feet to the mile, built of fine sandy clay or adobe brought down by an ephemeral stream, having a smooth surface that is usually unmarked by stream channels, but where so marked the channels are insignificant (Bryan).
Lobeck, 1939. 'The adobe soils of the southwestern United States and of Mexico, used for making bricks, have some of the characteristics of loess. They are largely water deposited, that around Mexico City being old lake clays' (p. 397).
Holmes, 1944. 'In the semi-arid regions of the western States and in the Mississippi valley there are thick deposits of *adobe*

which corresponds in all essentials to the loess of Europe and Asia. Here again material has been sifted by the wind from the rock waste of deserts and glaciated areas, and blown far outside the regions of supply, to find lodgement on surfaces, like the prairies, protected by vegetation. Similar deposits also occur in the pampas of South America' (p. 269).

*Comment.* Originally referred to sun-dried bricks which is still the current and commonest use. Later extended to the loess-like material from which the bricks were made and to soils derived from that material.

**Adolescent**

*Lit.* becoming adult, growing from childhood to maturity; used especially by Griffith Taylor in his classification of towns. See Urban hierarchy.

**Adret**

*O.E.D.* n.r.

Mill, *Dict.* Adret or Adra. The mountain slopes facing more or less southward (Alps). Fr. endroit; Ger. sonnenseite.

Miller, A. A., 1931, *Climatology*, London: Methuen. 'Exposure with respect to the midday sun makes a very considerable difference to surface temperature and has a profound effect on the habitability of a slope outside the tropics. Owing to the transparency of the air the shaded slope lies in heavy shadow while the sunny slope is bathed in brilliant light and warmth. The significance of the contrast is witnessed by the distinguishing names which these receive: in German sonnenseite and schattenseite, in French l'adret and l'ubac, in Italian adretto and opaco.' (p. 254)

Peattie, R., 1936, *Mountain Geography*, Cambridge: Harvard University Press. 'The matter of sunlight is so important that each mountain speech or dialect has its special terms for the sunny and shady slopes. Some of these are herewith given:

|  | Sun | Shade |
|---|---|---|
| German: | Sonnenseite | Schattenseite |
|  | Sonnenberg | Schattenberg |
|  | Sommerseite | Winterseite |
| French: | Adret | Ubac |
|  | Endroit | Envers |
|  | Soulane | Ombrée |
| Catalan: | Sola | Baga |
|  | Solana | Ubach |
|  | Soula | Ubago |
| Italian: | Indretto | Inverso |
|  | Adritto | Opáco' |

Also pp. 90 and 188, 'Secondary adret' as sunny slope on 'ubac' side of valley.

Spate, O. H. K., 1954, *India and Pakistan*, London: Methuen. Tailo = adret and Saylo = ubac (p. 403).

de Martonne, E., *Shorter Physical Geography*. 'Ubac, or the wrong side, and adret, or the right side.' (p. 76)

Garnett, Alice, Insolation and Relief, *Trans. Inst. Brit. Geog.* 5 (esp. note climatological inaccuracy of terms).

*Comment.* Adret and ubac are now well established in literature in English; they were originally French dialect words. See also Aspect, Exposure.

**Adritto** (Italian)—see Adret

**Advection**

Webster. *Meteorol.* The horizontal shifting of a mass of air, considered in its causal relation to changes of temperature or other changes therein—distinguished from *convection*.

*Met. Gloss.*, 1944. 'The process of transfer by horizontal motion. The term is more particularly applied to the transfer of heat by horizontal motion of the air. The transfer of heat from low to high latitudes is the most obvious example of advection.'

*Comment.* Also applicable to all fluids and used not only of heat. Sverdrup uses it of 'changes that take place in currents' (oceanographic). Contrast convection—vertical movement.

**Advection fog**

Kendrew, *Climatology*, 1944. 'Another type [of fog] is brought by moving air and may be termed "advection-fog". The essential feature is the advance of an air-mass over a cooler surface, so that its lower layers are chilled below dew-point.' (p. 275)

Swayne notes the sea-fogs of the Grand Bank of Newfoundland.

**Adventitious Rural Population** (L. D. Stamp, 1949)

Stamp, L. D., 1949, The Planning of Land Use, Pres. Address, Section E, British Association, *The Advancement of Science*, 6 (No. 23), 224–233.

Stamp, L. D., 1955, *Man and the Land*, Collins New Naturalist Series. 'To simplify a study of the problem [of rural depopulation] I suggested a division of rural

populations into three groups: (a) primary rural—farmers and farm workers (b) secondary rural—existing to service the first (c) adventitious—the population living in the country by choice' (p. 233).
See also Vince, S. W. E., 1952, *Trans. Inst. Brit. Geog.*, **18**, 53.

### Adventive cone, adventive crater

*O.E.D.* Adventive = Adventitious. Appearing casually or in unusual places.

Cotton, 1945. 'Basalt cones may perhaps be regarded as embryonic domes... Hundreds of such "adventive" cones are dotted over the great dome of Etna, ...' (pp. 372–373).

Cotton, 1944. 'Parasitic.' Small cones on the backs of larger volcanic domes (pp. 146–151).

Daly, R. A., 1914, *Igneous Rocks and their origin*, New York: McGraw-Hill. *Adventive* (parasitic or lateral) *craters* are those opened on the flanks of great cones. (p. 144)

Similarly Rice, 1941, quoting Fay, 1920; Daly, 1914.

### Aeolian, æolian, eolian

*O.E.D.* 2. Of Aeolus, the mythic god of the winds; hence of, produced by, or borne on the wind, or by currents of air; aerial.

1879 Rutley, Study of Rocks, 14, 275. Rounded by attrition, the result of their transport by water, or in the case of aeolian rocks, of their transport by wind.

*Webster.* Borne, deposited, produced, or eroded by the wind; as *aeolian* sand; *aeolian* rock sculpture.

Mill, *Dict.* Formed by the action of the wind.

Geikie, J., 1898. Chapter XII is entitled 'Land-forms modified by aeolian action.'

Holmes, 1928. A term applied to deposits whose constituents have been carried by, and laid down from, the wind. (p. 121)

Moore, 1949. Relating to, or caused by, the wind. Aeolian deposits are materials which have been transported and laid down on the earth's surface by the wind, and include *Loess*, and the sand of deserts and dunes.

*Comment.* Rice, 1941, uses the spelling eolian though even in American literature aeolian is still common.

### Aeration, zone of (O. E. Meinzer)

Meinzer, O. E., 1942, *Hydrology*, New York. 'The permeable rocks (including the soil) that lie above the zone of saturation may be said to be in the zone of aeration (a relatively new term that was proposed by the writer).' (p. 343)

'In a paper by the author (*Trans. Amer. Geophys. Soc.*, 1939) the following terms of Greek origin were proposed for international use: "kremastic water" for all the water in the zone of aeration; "rhizic water", for the water in the belt of soil water, or root zone; "argic water" for the water in the intermediate belt and "anastasic water" for the water in the capillary fringe.' (p. 394)

Tolman, C. F., 1937, *Ground Water.* 'Aeration, Zone of, or Zone of Suspended Water. The zone above the water table in which the interstices are partly filled with ground air except in the saturated portion of the capillary fringe.' (p. 557)

*Comment.* The normal meaning of 'aeration' is 'exposure to the chemical action of the air' (*O.E.D.*, from 1836) and the special use in Zone of Aeration given above is mainly American. No reference in Robinson, 1932; Russell, 8th ed., 1950; Fay, 1920; also known as the Vadose Zone.

### Aerobe, aerobic

*S.O.E.D.* A microbe living on free oxygen derived from air (1879).

*Comment.* The opposite is anaerobic and the two terms aerobic and anaerobic are now commonly used of two groups of soil organisms, the one living where air penetrates into the interstices of the soil and the other where it does not. See E. J. Russell, 1957, *The World of the Soil*, London: Collins, p. 58. See also Robinson, 1949, p. 201.

### Aerography

*O.E.D.* Description of the atmosphere. Rare.

### Aerology

*O.E.D.* That department of science which treats of the atmosphere.

*Webster.* a. The branch of physics treating of the atmosphere. b. *Meteorol.* Specif. the description and discussion of the phenomena of the free air as revealed by kites, balloons, airplanes, and clouds.

*Met. Gloss.*, 1944. A word denoting the study of the atmosphere and including the upper air as well as the more general studies understood by the word meteorology. It is frequently used as limiting the study to the upper air.

### Aerosphere

*Webster.* The atmosphere.

Swayne, 1956. The entire atmosphere surrounding the earth.

Comment. This term appears to have been introduced because of the restriction of 'atmosphere' to the lower layers and the use of stratosphere and other terms for other layers but it has not been generally adopted. Not in O.E.D.

**Aesthenosphere**, error for **asthenosphere** (Barrell, 1914)

Barrell, J., 1914, The Strength of the Earth's Crust, *Jour. Geol.*, **22**. 'A thick and somewhat plastic zone beneath the more rigid lithosphere.' (p. 680)

Davies, A. Morley, 1918, The Problem of the Himalaya and the Gangetic Trough, *Geog. Jour.*, **51**. 'The modern geological conception [of a liquid sub-crust] is perhaps best expressed by Professor Barrell's "Asthenosphere"—a thick earth-shell marked by a capacity to yield readily to long-enduring strains of limited magnitude though transmitting earthquake waves like a rigid body' (pp. 175-183). See The Strength of the Earth's Crust, *Journ. Geol.* (Chicago), **22**, 23, 1914-1915 (p. 177 *fn.*).

*Asthenospheric shell*

Daly, R. A., 1940, *Strength and Structure of the Earth*. 'Asthenosphere, an abbreviation of the more formally correct asthenospheric shell, is a useful name for the underlying, weak shell, though the word literally implies complete absence of strength [underlying the lithosphere (*q.v.*)]' (p. 13).

Bucher, 1933. 'Barrell calls this zone of weakness the "asthenosphere". It underlies the lithosphere from which it differs by "its inability to resist stress-differences above a certain small limit," ' (p. 38). (J. Barrell, *J. Geol.*, **22**, 1914, 729-741; **23**, 1915, 27-44.)

**Affluent**

O.E.D. A stream flowing into a larger stream or lake; a tributary stream; a feeder. 1833. Penny. Cycl. I, 433. The great Missouri with its affluent the Mississippi (*sic.*).

Mill, *Dict.* A stream which flows into a large stream or a lake.

Jackson, Colonel, 1834. Hints on the Subject of Geographical Arrangement and Nomenclature, *J. of R.G.S.*, IV. 'I would make a still further distinction, calling generally by the name of tributary any stream which directly or indirectly contributes its waters to the main trunk; reserving the term affluent for such only as flow immediately into another river mentioned.' (p. 78)

*Note:* a confluent would be a stream formed of two others, neither of whose names it bears.

Comment. Obsolescent; the word tributary would normally be used.

**Afforestation, afforestable**

O.E.D. The action or result of converting into forest or hunting-ground. 1615 Manwood, Lawes of Forest. The disafforestation of the new afforestations aforesaid. Other examples 1649, 1751, 1862.

O.E.D. Suppl. Afforestable: capable of being afforested. 1928 Britain's Industr. Future (Liberal Ind. Inquiry) Index 489 Estimate of afforestable land.

Fleure, H. J., 1918, *Human Geography in Western Europe*, London: Williams and Norgate. '... our new poverty may force us to afforest our immense waste lands ...' (p. 49).

Brown, R. N. R., 1939, *The Principles of Economic Geography*, 4th ed., London: Pitman. 'Most timber-growing countries are now trying to replant devastated areas, and protect young trees. Afforestation, however, is a slow task ...' (p. 21).

Tansley, 1939. 'It has often been pointed out that a "forest" in this sense was not necessarily a "forest" in the sense that the ground was covered with trees, and that the terms "afforestation" and "disafforestation" do not mean planting with trees and clearing forest, but placing a tract of land under, or freeing it from, forest law.' (p. 176)

Stamp, 1948 (referring to Deer Forests after 1850). 'Estate after estate was "afforested" which meant essentially clearing of sheep and cattle ... the use of land as deer forests is incompatible with commercial afforestation.' (p. 165)

Moore, 1949. The process of transforming an area into a forest, usually when trees have not previously grown there. See Reforestation.

Comment. Although the older meaning is that explained by Tansley and in the first sense given by Stamp above, the usual present-day meaning is planting with trees. Accordingly, the word will be found in geographical literature used in at least three senses (a) the old legal sense, (b) the legal sense with special reference to deer forests and (c) the modern botanical-commercial sense. See also Forest, reforestation.

**Africander**—see Afrikaner

**Afrikaner, Africander, Afrika(a)nder**
Afrikaner cattle or sheep are varieties derived from the animals originally found in the possession of the Hottentots. They are, however, not the simple continuation of the native forms, but are the result of long selective breeding by Europeans. Both are specializations. Afrikaner cattle are not the most profitable as milk- or meat-producers, but unsurpassed as draught-oxen. Afrikaner sheep have hair instead of wool, but they are valuable for their meat and for the fat contained in their remarkably thick tail. The majority of cattle and sheep in South Africa belong to other, imported, breeds. (P.S.)

*Comment.* Wellington, J. H., 1955, *Southern Africa*, uses the anglicized form Africander, also written Afrikander. There is much confusion in the use of the several words (see *O.E.D. Supplement*) since Africander has been applied to (1) white South Africans especially those of Dutch descent, (2) South Africans of mixed blood. Some writers use Afrikander or Africander for animals, but Afrika(a)ner for human beings, *i.e.* those speaking Afrikaans. Professor Serton says, however, that whatever foreigners may write the position in Afrikaans is clear. '*Afrikaander* is given by the official dictionary as an antiquated form for *Afrikaner* which is given as the correct form in all three different meanings, *viz.* (1) a man speaking Afrikaans as his mother-tongue; (2) an animal belonging to a variety of cattle or sheep derived from the original native varieties; (3) certain flowers.' Serton also notes that the younger English-speaking South Africans are using Afrikaner in all three meanings (MS. communication).

**After-glow**
*O.E.D.* A glow or refulgence that remains after the disappearance of any light, esp. that which lights the western sky after sunset.

**Aftout** (West Africa: Mauritania)
Harrison-Church, 1957. 'The country is composed of clayey plains (aftouts) and of north-west–south-west trending stable dunes (sbar). The latter are infertile but clay depressions between them can be more intensively used as water becomes available' (p. 231).

**Agglomerate** (i) (Geology)
*O.E.D.* Geol. A mass consisting of volcanic or eruptive fragments, which have united under the action of heat; as opposed to a conglomerate, composed of waterworn fragments, united by some substance in aqueous solution.
1830 Lyell. Princ. Geol. (1875), II, 11. xxvii, 72. 'This great overlying deposit ... is a white tufaceous agglomerate.' 1881. Geikie. ...

Mill, *Dict.* Rough, irregularly grouped blocks cemented together to form rock; *e.g.* material filling old volcano funnels.

Page, 1959. 'A term employed by Sir Charles Lyell to designate those accumulations of angular fragments of rock which are thrown up by volcanic eruptions and showered to a greater or less distance around the cone of eruption; when they are carried to a distance by running water and get worn and rounded they become conglomerates.' (p. 61)

Geikie, J., 1920, *Structural and Field Geology*, 4th ed., Edinburgh: Oliver and Boyd. 'Volcanic Agglomerate is the name given to a coarse admixture of large and small blocks and stones set in a matrix of comminuted rock debris and grit, which may be either abundant or meagre.' (p. 53)

Hatch and Wells, 9th. ed., 1937. 'Pyroclastic accumulations in which there are large blocks ... usually found actually in the vent itself or not far removed from it. Rocks composed of smaller fragments are termed volcanic breccias.' (p. 256)

Rice, 1941. 'Contemporaneous pyroclastic rocks containing a predominance of rounded or subangular fragments greater than 32 mm. in diameter, lying in an ash or tuff matrix and usually localized within volcanic necks (vent agglomerates) or at a short distance therefrom. The form of the fragments is in no way determined by the action of running water, as in volcanic conglomerates, but is a primary feature determined during the actual eruption' (Wentworth-Williams, 1930–32).

Holmes, 1944 [Of fragments blown from volcano during eruption] 'The coarser fragments ... fall back near the crater rim and roll down the inner or outer slopes, forming deposits of agglomerate or volcanic breccia, the latter term implying that the blocks consist largely of country rocks from the foundations of the volcano.'

**Agglomerate (2) (Soil science)**
Robinson, G. W., 1949, *Soils*, London: Murby. '*Agglomeratic* (aggregates) ... grains uncoated or with loosely-attached friable coatings.... Plasmatic material present as aggregates in the pore-spaces. Aggregation undeveloped or only slightly developed.' (p. 256)

**Agglomeration**
*O.E.D.* 1. The action of collecting in a mass, or of heaping together. 1874. Helps Soc. Press., ii, 18. The agglomeration of too many people on one spot of ground.
2. A mass formed by mere mechanical union or approximation; an unmethodical assemblage; or clustering or cluster. 1859. Jephson Brittany, xiii, 215. It was an agglomeration of forbidding-looking granite houses.
Geddes, P., 1915, *Cities in Evolution*, London: Williams and Norgate. (Captions to illustrations) 'Fig. 13.—Lancashire towns agglomerating as "Lancaston".' (p. 31)
Aurousseau, M., 1920, *Geog. Rev.*, **10**. 'The following forms of rural agglomerations are recognizable: small towns, villages, true village communities, and hamlets.' (p. 224)
Fawcett, C. B., 1932, *Geog. Jour.*, **79**. '.... the French census distinguishes, in some areas, between "population agglomérée" and "population dispersée".' (p. 100 *fn.*)
Dickinson, R. E., 1951, *The West European City*, London: Routledge. 'The first type includes a central city with a fringe of satellite urban areas ... Paris, the greatest urban agglomeration in France, belongs to this type. Included also are many smaller centres which consist of two or more administrative units. ... The second main type of agglomeration consists of several separate urban centres which, by expansion, have coalesced to form a single continuous area of brick and mortar and other urban land uses, or a group of towns which are within a few miles of each other and have become closely interrelated in function.' (p. 451)
'The word "city" could hardly be applied to haphazard agglomerations created at the command of a chieftain ...' (p. 471).

Zimmermann, E. W., 1951, *World Resources and Industries*, 2nd ed., New York: Harper. 'An industrial agglomeration is a group of associated and dependent industries.' (p. 618)
*Comment.* This term, following French usage, is applied to the grouping of both people and buildings, especially with reference to settlement types. Its adjectival use might well be abandoned in favour of 'compact' or 'nucleated' where these more fittingly apply. (W.G.E.)
Although loosely used of people or buildings, there is usually, but not always, an underlying suggestion of formlessness or lack of planning. Industrial complex would be a more usual description for the phenomenon mentioned by Zimmermann.

**Aggrade, aggradation, aggradational (1)**
*O.E.D. Suppl.* Aggrade. Geol. To fill up (a bay, valley, etc.) with detritus (the opposite of degrade). Hence aggraded, aggrading; also aggradation, whence aggradational.
1905 Chamberlin and Salisbury. Geol. I. 2. The deposition of material, whether on the land or in the sea, is aggradation.
*Ibid*, 318. The degradational and aggradational work of the sea are greatest near its shores.
Davis, 1909. 'If an initial consequent stream is for any reason incompetent to carry away the load that is washed into it, it cannot degrade its channel, but must aggrade instead (to use an excellent term suggested by Salisbury).' (p. 259; also see p. 392)
Wooldridge and Morgan, 1937. 'In the recent past active aggradation by coarse debris was in progress, while in a still more recent period, extending into historical times, conditions of flood alluviation have prevailed.' (p. 171 *fn.*)
Rice, 1941. 1. In geology, the natural filling up of a bed of a water course by deposition of sediment. 2. Specifically, the building up by streams in arid regions of fan-like graded plains, by reason of the shifting streams and the loss of the water in the dry soil. Contrasted with degradation (Fay).
Cotton, 1945. 'When ... a stream deposits in and so builds up its channel to establish or maintain grade, it is said to aggrade; and the process is termed aggradation.' (p. 59)

'Aggraded valley plains (on deposits of alluvium).' (p. 193)
'Plains of aggradation in semi-arid conditions.' (p. 252)
Lewis, W. V., 1936, *Geog. Jour.*, **88**. 'The bank (offshore bar) is, for the most part, a simple one showing no aggradation ridges. . . . Only at the entrance to Hornafjördur were small aggradation ridges observed.' (p. 438)
Spate, 1954. 'Monotonous aggradational plains' of Northern India. (p. 1)
Committee, List 2. 'The process of raising a surface by the deposition of rock waste.'
See also Grade, Degrade, Degradation.

### Aggrade, aggradation (2)
Muller, S. W., 1947, *Permafrost*, Ann Arbor. 'In some areas, instead of disappearing, permafrost may actually be spreading or appearing for the first time. This spread or appearance of permanently frozen ground is called *aggradation* of permafrost.' (p. 22)
'Aggradation—growth of permafrost under the present climate due to natural or artificial causes. Opposite to degradation.' (p. 213)

### Aggregate (soil)
Millar and Turk, 1943. A single mass or cluster of soil particles, such as a clod, crumb, or granule (p. 431) (American usage).
Robinson, 1949. Does not give a formal definition, but gives seven types. See also Russell, 1950, p. 71.

### Agistment
Butlin, 1961. 'The pasturing of one person's animals on another's land on payment of a due.'

### Agonic line
*O.E.D.* The irregular line passing through the two magnetic poles of the earth along which the magnetic needle points directly north or south; the line of no magnetic variation (quot.: Atkinson, tr. Ganot's *Physics*, 1870, para. 674).
Salisbury, R. D., *Physiography*, 1923. 'A line connecting places of no declination.' (p. 303)

### Agrarian
*O.E.D.* 1. Relating to the land. 2. Connected with landed property, or with cultivated land, or its cultivation. 1792.
*Comment.* In geographical literature agrarian is found much less frequently than agricultural but it is commonly used in economic, sociological and political literature. There would appear to be a distinction in modern usage between 'agricultural' and 'agrarian'. The latter appears to connote matters of land tenure or considerations arising from land ownership (*e.g.* Agrarian Revolution). French and German usage of the comparable words are, however, confused. (C.J.R.)

### Agricultural area
Committee, List 3. The area of land utilized for farming, including arable, improved or unimproved grassland, and other pasture.
F.A.O. Yearbook. 11, Part 1, 1957 'Arable land and land under tree crops plus permanent meadows and pastures = agricultural area.' (p. 3 also p. 370)
*Comment.* A difficulty arises with regard to unimproved grassland. Where this corresponds to 'range land' or the wide open natural grasslands of the tropics it would not normally be included.

### Agricultural Region
Baker, O. E., 1926. The Agricultural Regions of North America, *Econ. Geog.* **2**. 'The term agricultural region, as used in this paper, refers to a large (sub-continental) area of land characterized by homogeneity of agricultural conditions, especially of grown crops, and sufficient dissimilarity from conditions in adjacent territory as to be clearly recognizable' (p. 468). Compare Jonasson, O., 1926, Agricultural Regions of Europe, *Econ. Geog.* **2**, 19–48.
Whittlesey, D., 1927, The Major Agricultural Regions of the Earth, *A.A.A.G.*, **26**, 199–240.
*Comment.* Baker's usage corresponds to 'section' in United States economic and political history. (C.J.R.)
See also Region, Natural region.

### Agriculture
*O.E.D.* The science and art of cultivating the soil; including the allied pursuits of gathering in the crops and rearing livestock; tillage, husbandry, farming (in the widest sense). 1603 Holland *Plutarch's Mor.* 9. Such tooles as pertaine to Agriculture and husbandrie.
b. restricted to Tillage, rare.
1862 Stanley *Jew. Ch.* (1877) I, xii, 228. The lands . . . were not fields for agriculture, but pastures for cattle.
Mackinder, H. J., 1902, *Britain and the British Seas*, London: Heinemann. 'This is expressed agriculturally in the prevalence of pasture. . . .' (p. 315)

Herbertson, A. J., 1899, *Commercial Geography of the British Isles*, London: Chambers. Used as Farming as distinct from Ranching, Lumbering, Mining, etc.

Brown, R. N. Rudmose, 1936, *The Principles of Economic Geography*, 3rd ed., London: Pitman. 'Agriculture leads to a denser population than pasturage or hunting.' (p. 89)

Watson, S. J., *Chambers's Encyclopedia*, 1950. 'The original meaning of the term agriculture is the culture of the soil but this is much too narrow an interpretation. The growing of crops and the rearing of livestock are as much a part of agriculture as the original cultivation of the soil.... One simple definition of agriculture is the science and practice of farming, using the word in its broadest sense.'

Cambridge Econ. Hist. of Europe, 1941, Vol. 1 (Aut., R. S. Smith, North Carolina). 'An important chapter in agrarian history is the relation between agriculture and grazing.' (p. 351)

Zimmermann, E. W., 1951, *World Resources and Industry*, New York: Harper. 'Agriculture covers those productive efforts by which man, settled on the land, seeks to make use of and, if possible, accelerate and improve upon the natural genetic or growth processes of plant and animal life, to the end that these processes will yield the vegetable and animal products needed or wanted by man' (p. 148). 'It includes animal husbandry, tree culture, forestry, irrigation, fish hatcheries, fur farming, perhaps even hydroponics....' (p. 147)

Childe, V. Gordon, Neolithic Culture of the British Isles. *Listener*, June 24, 1954. '...that neolithic Britons were pure pastoralists and practised no sort of agriculture.' (p. 1090)

Cobbett, W., [1832], *Rural Rides* (Everyman Edition), London: Dent. 'They are not agricultural counties; they are not counties for the producing of bread, but they are counties made for the express purpose of producing meat...' (p. 279) (30 Sept., 1832). Also 'This then is not a country of farmers, but a country of graziers;...'

Committee, List 3. The science and art of cultivating the soil, including the rearing of livestock.

Comment. The word is most commonly used as previously defined by the Committee, and by Watson (above) but still by some writers in the restricted sense, rarely in the very wide sense suggested by Zimmermann. For the latter see agro-forestal and agro-silvo-pastoral. Note the Ministry of Agriculture, Fisheries and Food—covering all aspects of farming—and F.A.O. (Food and Agriculture Organization of United Nations) which includes forestry and fisheries.

**Agro-forestal, agroforestal**

Pertaining to both agriculture and forestry. Since agriculture is not normally taken to include the practice of forestry or the management and cultivation of trees the word agro-forestal has been coined to cover both the agricultural and forestal uses of land.

Comment. This word appears to have been introduced into English by the International Institute of Agriculture in 1939 (League of Nations, European Conference on Rural Life, 1939, Document No. 1, *Population and Agriculture*, p. 12, English Edition) though it had previously been used by the Italian *Catasto Agrario* of 1929 (as *superficie agroforestale*). It is a useful category when the total of agricultural and forestal area or production is differentiated from unproductive or built-over areas. It is also a convenient and now widely used category of land-use for statistical purposes when the distinction between agricultural and forest land is not well defined—as when woodland is used for grazing (C.J.R.). Certain members of the Committee objected to this as an 'ugly and unnecessary term' but it is firmly established. Some writers prefer to use the more accurate 'agro-silvo-pastoral'. (L.D.S.)

**Agronomy, agronomic**

S.O.E.D. The management of land; rural economy.

Webster. That branch of agriculture dealing with the theory and practice of field-crop production and soil management.

Swayne, 1946. Agricultural economy, including the theory and practice of crop production, soil management, and every form of husbandry.

**Agro-silvo-pastoral**

Lit. related to agriculture, forestry and grazing. See Comment under agro-forestal.

**Agrostology**

S.O.E.D. That part of botany that treats of grasses. The study of grasses.

**A-horizon** (Soil science)—see Horizon, soil

**Ahqāf** (Arabia: Arabic)
Stamp, L. D., 1959, *Asia*, London: Methuen, 10th ed. 'Very soft dune country... cannot be crossed except in narrow belts ...rare in Arabia' (p. 145).

**Aiguille** (French)
*O.E.D.* A slender, sharply-pointed peak (of rock); especially the numerous peaks of the Alps so named.
Charlesworth, 1957, 'The recession of inosculating cirques creates zig-zag knife edges, with *dents* and *aiguilles* so well described by J. Ruskin (*Modern Painters*, 4, 1856, ch. 15). They are the culmination of the cuspate septum between two cirques. With their sharply projecting towers, the 'atahoma of Russell and *gendarmes* of Alpine literature they serrate the skyline in the Hohe Tatra, Lofotens and Snowdonia. Sharp peaks project from the main watershed; such are the Finsteraarhorn, the *Aiguilles* of Mont Blanc. ... the Cobbler of Argyll and other peaks in Arran, Mull and Skye.' (p. 306)
*Comment.* Charlesworth does not appear to regard *aiguilles* as being synonymous with *gendarmes*, the latter appearing to be less pointed and more tower-like.

**Air**
*O.E.D.* 1. The transparent, invisible, inodorous, and tasteless gaseous substance which envelopes the earth, and is breathed by all land animals....
*Met. Gloss.*, 1944. The mixture of gases which form the atmosphere. ... The dust is regarded as an impurity and the water, whether in the form of vapour or cloud, as an addition to the air.

**Air drainage**
*Webster. Meteorol.* The downward flow of surface air, as down a steep slope, caused by its relatively high density produced by contact cooling, a phenomenon prevalent on still, clear nights.
Swayne, 1956. The movement of cold air from higher to lower altitudes owing to its greater density.
*Comment.* In the absence of other factors, cold air behaves much like cold water in the way in which it drains down valleys, is apt to be held up by obstructions and drains into hollows to form frost pockets etc. See Stamp, 1948, 263-265.

**Air gap**
*O.E.D.* A gap or hole through which air passes. No specific geographical use given.
Mill. *Dict.* A wind-gap (*q.v.*).
Cotton,' 1945. 'The former gorge, or water gap, through this stratum is now no longer traversed by a stream, and becomes an "air gap".' (p. 73)
*Comment.* Less commonly used than wind-gap (*q.v.*).

**Airglow**
Beynon, W. J. G., 1964, International Years of the Quiet Sun, 1964-65, *Nature*, **204**, 623. 'The phenomena of the aurora are a striking and impressive consequence of the interaction of solar particle radiation with the Earth's atmosphere in polar latitudes. But there is also a much weaker and less variable radiation of light from the high atmosphere which goes on all the time and at all places. This is the airglow or "light of the night sky" as it used to be called.'

**Air hole**
*Webster.* A spot not frozen over in the ice, esp. one caused by a spring or current in a river or pond.
Kuenen, 1955. Open water in the middle of a sheet of frozen water resulting from wind action. 'The wind is constantly pushing the water over the air hole and under the ice crust again, on the leeshore. To compensate for this loss, new water from underneath the ice is drawn up on the wind-side. ... It may even be that, in spite of severe frost, an air-hole grows larger towards the windward side.' (p. 245)

**Air mass**
*Met. Gloss.*, 1944. Broadly this term is used in synoptic meteorology to denote the mass of air which is bounded by frontal surfaces (often smoothed out into transitional zones) and often extending in well marked examples from the ground to the stratosphere.
Swayne, 1956. A mass of air bounded by frontal surfaces ... the air is almost homogeneous over a large area ... if the air is followed in three dimensions the potential temperature and the water content per unit mass are independent of adiabatic processes (since there is no condensation or evaporation) and the wet bulb potential temperature remains constant. Regions where homogeneous air masses are usually produced are known as 'source areas'.

Comment. Contrasted main types of air mass are distinguished and denoted by letter symbols. Examples are:
mT maritime tropical: warm and moist
cT continental tropical; hot and dry
mP maritime polar: cold and moist
cP continental polar: cold and dry
mA maritime arctic or antarctic: very cold and moist
cA continental arctic or antarctic: very cold and dry.

### Air port, airport
An aerodrome or airfield designed for use by civil, as opposed to Service, aircraft, and equipped with facilities for the handling of passengers and goods. There is an official international classification of airports, revised frequently, according to facilities available.

### Air saddle (Australian)
Fay, 1920. 'A surface saddle or depression produced by erosion at the top of an anticline.'
Comment. An Australian miner's term, now rarely used and best avoided.

### Ait, eyot
O.E.D. An islet or small isle; especially one in a river, as the aits or eyots of the Thames.
Comment. Some aits are still so named on O.S. maps and the word has been revived by some geomorphologists. (E.O.G., G.T.W.)

### Albedo
Webster. Astron. The ratio which the light reflected from an unpolished surface bears to the total light falling upon it.
Zeuner, F. E., 1945, *The Pleistocene Period*, London. '... the loss of heat by reflection called *albedo*....'
Comment. The albedo of the earth is about 0·4 which means that four tenths of the sun's radiation is reflected into space. (A.A.M.)

### Alcove, Alcove lands
Powell, 1875. A landscape deeply cut by narrow valleys into gloomy alcoves (p. 151).
Cotton, 1944, Volcanoes. 'Some of the largest springs that drain from the lava field north of the Snake River now discharge into the heads of short, rather deep, amphitheatre-headed and steep-sided canyons, known as the Snake River "alcoves" as a result of ... headward erosion ... the springs have sapped back into the basalt terrain' (p. 344). 'Though spring sapping has rarely been recognized elsewhere, except on some very weak terrains, this "alcove" type of headward erosion, albeit slow in action, may have operated very generally as the predominant process developing insequent valleys in the protracted youthful stage of the dissection of lava plateaus' (p. 346).
Thornbury, 1954. 'Spring alcoves: boxhead canyons along the Snake River Canyon, U.S.A., formed by solutional spring sapping of the basalt' (p. 507).
Comment. Cotton has extended the use of this local term from the Snake River area. It is not clear whether or not it should be confined to lava plateaus. (G.T.W.)

### Alidade, Alidad
O.E.D. The index of an astrolabe, quadrant, or other graduated instrument, carrying the sights or telescope, and showing the degrees cut off on the arc of the instrument.
Webster. a. That part of any optical, surveying or measuring instrument which comprises the indicator, verniers, microscopes etc. for taking on accurate reading or observation.
b. A rule equipped with sights, used in surveying for the determination of direction.
c. the frame which supports the vernier of an angle-measurement instrument.
Comment. The O.E.D. definition, would be improved by the insertion of 'or arm or rule' after index; the usual meaning is b. of Webster. (E.G.R.T.)

### Alio (French)
Mill, Dict. The ferruginous crust formed at a more or less constant depth (3 to 6 feet) by the deposition of salts of iron carried by subterranean water. The crust is impervious, and finally prevents cultivation (Landes of Gascony). See Hard pan, Iron pan.

### Alkali, alkaline
S.O.E.D. Alkali. A series of bases analogous to, and including soda, potash, and ammonia, highly soluble in water, producing solutions which neutralize acids.
O.E.D. Suppl. Alkaline 3. of soils or areas: charged or permeated with alkali. U.S. (examples 1869, 1870).
Webster. Geol. and Agric. A soluble salt, or a mixture of soluble salts (esp. the sulphates and chlorides of sodium, potassium, and magnesium), present in some

soils of arid or semiarid regions in quantity detrimental to ordinary agriculture.

*Soils and Men*, 1938. 'Alkali soil—a soil containing alkali salts, usually sodium carbonate (with a pH value of 8·5 and higher). The term is frequently used loosely to include both alkali soil and saline soil as here defined. Where applied to saline soil the expression "white alkali" is used in some localities and the expression "black alkali" is used for alkali soil as here defined, with or without the presence of neutral salts.

Alkaline soil. Any soil that is alkaline in reaction (precisely, above pH 7·0; practically, above pH 7·3).'

Holmes, 1928. Alkali rocks. Igneous rocks in which the abundance of alkalis in relation to other constituents has impressed a distinctive mineralogical character; generally indicated by the presence of soda pyroxenes, soda amphiboles, and/or felspathoids. *Cf.* calcalkali rocks.

Millar and Turk, 1943. (Similarly Russell, 1950):

*Alkali soil.*—A soil containing alkali salts, usually sodium carbonate or in which the colloidal complex contains a large amount of sodium, with a pH value of 8·5 and higher.

*Alkaline soil.*—Any soil that is alkaline in reaction; *i.e.* above pH 7·0.

Swayne, 1956. Soils found in arid and semiarid regions, where owing to lack of rainfall, the soluble salts have not been washed away (Alkali soils).

*Comment.* Alkali and alkaline have many meanings (see *O.E.D.*). Holmes is quoted to illustrate the specialized petrographical usage. In geographical writings the meanings are usually as given above from *Soils and Men*. Alkaline soils are the Solonetz of the Russians (*q.v.*). The American distinction between alkali and alkaline soils should be noted. See also Acid soils, where neutrality is given as 7·2.

**Alkali flat**

Thornbury, 1954. Referring to dried-up *playas* in arid regions 'such salt-covered surfaces are called *alkali flats*.' (p. 234)

Moore, 1949. An alkaline, marshy area in an arid region into which one or more desert streams lead. In the dry season, when all the water has evaporated, it becomes a barren area of hard mud covered with alkali; after heavy rainfall, it becomes a shallow, muddy lake.

**Alkali soil**

FAO, 1963. '*Alkali soils* are soils in which more than 15% of the exchangeable bases consist of sodium ions or which contain sodium carbonate in the upper horizon; the pH value is 8·5 or higher.'

see also above (Millar and Turk).

**Allocthon, allocthone, allocthonous**

The antonym of *autochton* (*q.v.*). Not in *O.E.D.*

*Webster. Geol.* Formed elsewhere than in situ, not autochthonous—said of coal deposits. Hence pert. to or designating the theory that coal deposits are formed of carbonaceous material transported from a distance.

Twenhofel, W. H., 1939, *Principles of Sedimentation*. 'Before fossil plants may be used to reconstruct a physical environment it must be first determined whether the plant material is *autochthonous*, that is whether it grew where it is found; or *allocthonous*, that is where it was transported to where it is found' (pp. 161–162).

Steers, J. A., 1932, *The Unstable Earth*. 'Nappes characteristic of the true geosyncline which have often been moved far from their place of origin, and to which the term *allocthonous* is applied' (p. 114).

*Comment.* A general definition might be 'transported, not in place, applied to transported fossil plants, to organic deposits formed from them, and to nappes which have travelled a considerable distance from their place of origin, the opposite of autochthonous'. (G.T.W.)

**Allogenic, allogene, allothigenous, allothigene**

*Webster. Geol.* Having a diverse or different and usually distant origin, as certain sedimentary rocks; not authigenic.

Geikie, Sir A., 1923, *Text Book of Geology*, 4th ed., London: Macmillan. 'Some eruptive rocks abound in corroded or somewhat rounded or broken crystals which obviously have belonged to some previous state of consolidation. Such crystals, which are obviously more ancient than those forming the general mass of the rock, have been called allogenic, while those which belong to the time of formation of the rock, or to some subsequent change within the rock, are known as authigenic' (p. 90).

Wooldridge and Morgan, 1937. [Of the English Chalk country] 'Many, but not all, of these streams are "allogenic", deriving much of their water from terrains beyond the Chalk outcrops' (p. 277). Also used without quotation marks (p. 290).

Rice, 1941. Originating elsewhere; whether applied to the components of a clastic rock or to xenoliths. The contrasted form is authigenic (Kemp).

Holmes, 1920. Allothigenous, or Allogenic, Kalkovsky, 1880. Terms, meaning generated elsewhere, applied to those constituents that came into existence outside of, and previously to, the rock of which they now constitute a part; *e.g.* the pebbles of a conglomerate. Contrast Authigenous.

*Comment.* Not to be confused with allogeneous, of different nature, diverse in kind (*O.E.D.*).

### Allometric growth

Used by Newling, B. E., 1966, Urban Growth and Spatial Structure: Mathematical Models and Empirical Evidence, *Geog. Rev.*, **56**.2, 213–223, to relate population growth to increasing distance from the city centre. He felt that population growth should increase with distance from the centre. Tests of the relationship in Chicago have shown intra-urban growth to be allometric up to 13 to 15 miles from the centre. After this the rate of population growth declines with increasing distance from the city centre. See Berry and Horton, 1970, p. 282.

*Comment.* The adjective 'allometric' refers to the systematic differential growth of parts within a complex growth structure. See Reeve, E. C. R., Some Problems in the Study of Allometric Growth in Le Gros Clark, W. E. and Medawar, P. B., 1945, *Essays on Growth and Form presented to D'Arcy Wentworth Thompson*, Oxford, pp. 121–156, and Gould, S. J., 1966, Allometry and Size in Ontogeny and Phylogeny, *Biological Reviews*, **41**, 587–640.

### Allotment

*O.E.D.* 4. esp. A portion of land assigned to a special person, or appropriated to a particular purpose. spec. A small portion of land let out for cottage cultivation (examples from 1674).

*Comment.* Under the enclosure awards in England and Wales certain portions of land were set aside for the use of the villagers who would otherwise have become landless. They were given the right of pasturage for certain animals and certain other rights on the Common land (*q.v.*) but in addition 'allotments' were usually set aside for specific purposes such as peat lands as 'fuel allotments'; sometimes simply allotments for cultivation for the poor. From the latter has come the principal modern use of the word—small pieces of land, privately or publicly owned, let annually for cultivation. Hence allotments may be regarded as rented gardens for cultivation at a distance from the house (Stamp, 1948, p. 23). See Report of the Royal Commission on Common Lands, 1955–58)

### Alluvial

*O.E.D.* Of, pertaining to, or consisting of alluvium; deposited from flowing water or pertaining to such a deposit.

Hence alluvial bench, alluvial clay, alluvial cone, alluvial deposit, alluvial fan, alluvial flat, alluvial mud, alluvial plain, alluvial terrace and others.

### Alluvial bench

Hobbs, W. H., 1931, *Earth Features and their Meaning*, 2nd edn. (1st 1912). After describing alluvial cones or dry deltas in desert regions he goes on to say: 'As they enlarge their boundaries, the neighbouring deltas eventually coalesce and so form an *alluvial bench* or "gravel piedmont" at the foot of the range. Only the larger streams are able to entirely cross this bench of parched deposits. . . .' (p. 214)

### Alluvial cone

*Webster.* The alluvial deposit of a stream where it issues from a gorge upon an open plain.

(Note: same as alluvial fan.)

Mill, *Dict.* 'Or Cone of Dejection' . . . A fan-shaped deposit of sediment with the apex pointing up the watercourse.

Russell, I. C., 1898, *Rivers of North America*. 'In America such piles are termed *alluvial cones*. In India and in Europe they are generally known as *alluvial fans*, in reference to their fan-like forms when seen from above. As suggested by Gilbert, there is an advantage to be gained from retaining both of these terms, employing cone when the angle of slope is high and fan when it is low' (p. 101).

Holmes, 1944, distinguishes between Fans and Cones, *e.g.* 'If the circumstances are such . . . practically the whole of the load is dropped and the structure rapidly gains height and becomes an alluvial cone' (p. 201). See under Fan.

Wooldridge and Morgan, 1937. Example of Alpine torrents from Surell, 1841, '... while their lower course is across the surface of a *cône de déjection* (alluvial cone or fan).' (pp. 169-170)

Cotton, 1945. 'Very steep fans are called alluvial cones..., and there is a transition through these from alluvial fans to talus slopes.' (p. 198)

Hobbs, W. H., 1951, *Earth Features and their Meaning*, 2nd edn. 'Dependent upon its steepness of slope, this delta is variously referred to as an *alluvial fan*, or *apron* or as an *alluvial cone*' (p. 213).

*Comment.* Some writers use alluvial cone and alluvial fan as synonymous (Rice, 1943; Moore, 1949) but there is a general tendency to follow the usage of Holmes and Cotton. See also Gilbert, 1877.

### Alluvial fan

*Webster.* The alluvial deposit of a stream where it issues from a gorge upon an open plain.

Mill, *Dict.* A similar formation to that of an alluvial cone, but lower and less steep. Also used for both cone and fan.

Davis, W. M., and Snyder, W. H., 1898, *Physical Geography*, Boston: Ginn. Of torrents: 'The coarser part of the waste then accumulates in a cone-like form, known as an alluvial fan, spreading with even slope from the ravine mouth into the main valley.' (pp. 275-276)

Cotton, 1945. 'As rivers emerge fully loaded from eroded valleys, in which they may have been degrading, into wide depressions where the slope is so gentle that the streams are compelled to aggrade in order to prolong their graded slopes... they deposit part of their load in such a manner as to build alluvial fans.' (p. 197)

Holmes, 1944. 'The obstructed stream—as in delta formation—divides into branching distributaries, and the heap of debris spreads out as an *alluvial fan*.' (p. 201)

Cotton, 1941. 'Alluvial cone has been used in same sense as alluvial fan (G. K. Gilbert, Henry Mts., 1877). "Fan" used by J. Haarst (Form. of Canterbury Plains, 1864), restricting its use to subaerial parts of deltas, and using term "half-cone" for the majority of what are now called "fans", because of their greater steepness.' (p. 174 *fn.*)

See also Alluvial cone, Fan.

### Alluvial flat

Mill, *Dict.* Alluvial Flats or Plains—The flat tracts bordering the side of a river which receive deposit during floods.

Fairey, J., 1811, *General View of the Agriculture of Derby*, referring to the Trent, Dove and Derwent valleys, 'alluvial flat of loam or sandy loam has accumulated upon the gravel, by the sediments of the floods which occasionally overflow them, from one to several feet in thickness generally without the admixture of stones or other heavy bodies' (p. 133). Also refers to 'alluvial clays' from these flats being used for brick making. (p. 452) (G.T.W.)

### Alluvial terrace

Mill, *Dict.* The remains of an alluvial flat which was formed prior to the rejuvenescence of the river (see Rejuvenated River). Their materials are often described as river gravels.

Geikie, J., 1898. 'In a broad river-basin alluvial terraces and plains usually occur at various heights, marking successive levels...' (p. 40).

Davis, 1909 (of New England). 'The terraces that occupy so many of our valleys are known as river terraces, drift terraces, or alluvial terraces.' (Follows distinction from those of seashores, lake shore, structural rock benches, rock terraces of Rhine.)

Wooldridge and Morgan, 1937. 'In such valleys, moreover, rejuvenation will commonly lead to the incision of the river-course below aggraded flood plains. In this fashion "paired" alluvial terraces are formed.' (p. 222)

Holmes, 1944. Similar = river terrace.

'A typical terrace is a platform of bedrock thickly veneered with a sheet of river-gravel and sand passing upward into finer alluvium.' (p. 196)

*Comment.* If the term alluvium is restricted to the finer grained deposits, a large number of river terraces would not be alluvial terraces because they consist essentially of gravel; Mill's statement is also incorrect.

### Alluviation

The process of filling up with alluvium, aggradation with alluvium. See above, Aggradation, Wooldridge and Morgan.

### Alluvion

*O.E.D.* 1. The wash or flow of the sea against the shore, or of a river on its banks.

2. An inundation or overflow; a flood, especially when the water is charged with much matter in suspension.

3. The matter deposited by a flood or inundation.

4. The matter gradually deposited by a river = alluvium.

5. *Law.* The formation of new land by the slow and imperceptible action of flowing water.

Comment. A word with an interesting evolution as the O.E.D. shows but now almost obsolete and best avoided. See also Diluvion.

### Alluvium, alluvia

*O.E.D.* Plural, alluvia, alluviums. From L. alluvius, washed against. A deposit of earth, sand, and other transported matter left by water flowing over land not permanently submerged; chiefly applied to the deposits formed in river valleys and deltas (examples from 1665-6 include both sea and river action.)

Mill, *Dict.* Material brought down by rivers in suspension and deposited on their flood plain or in deltas. Term used by Lyell to include volcanic cinder-mud.

Lyell, 1830. 'We restrict the term alluvium to such transported matter as has been thrown down, whether by rivers, floods, or other causes, upon land not permanently submerged beneath the waters of lakes or seas.' (The results of landslides are included but the inclusion of blown sand and volcanic ejecta is not so certain.)

Page, 1865. 'Matter washed or brought together by the ordinary operations of water is said to be alluvial, and the soil or land so formed is spoken of as alluvium.... All mud-deposits, as silt, warp, and the like, when converted into dry land, constitute alluvium.'

Geikie, J., 1898. Glossary. 'A deposit resulting from the action of rivers or of tidal currents.'

Fay, 1920. 1. Lyell's name for the deposit of loose gravel, sand and mud that usually intervenes in every district between the superficial covering of vegetal mould and the subjacent rock. The name is derived from the Latin word for inundation. It was employed by Naumann as a general term for sediments in water as contrasted with eolian rocks. It is generally used today for the earthy deposit made by running streams, especially during times of flood (Kemp). Also equivalent to Alluvion, 3, or 'a consolidated volcanic cinder-mud'. (W.M.)

Holmes, 1928. 'A general term for all ... deposits resulting from the operations of modern rivers, thus including the sediments laid down in river-beds, flood plains, lakes, fans at the foot of mountain slopes, and estuaries.'

Comment. Though little justified by historical usage there is a marked tendency to restrict alluvium to fine grained deposits (in texture, silt or silty clay) and not to include gravel and sand. Indeed this distinction is made on the maps of the British Geological Survey. For examples of early use and meaning see J. Fairey, 1811, *General View of the Agriculture of Derby*; p. 109 with William Smith's views. As late as 1936 H. H. Read in 23rd Edition of Rutley's *Mineralogy* seems to include marine deposits. Charlesworth 1957, p. 616, credits W. Smith with first distinguishing alluvium from solid rock and glacial drift and claims Buckland and Mantell distinguish diluvium from alluvium. (G.T.W.) See also Diluvium.

### Almacantar (*pl.*)

*O.E.D.* Small circles of the sphere parallel to the horizon, cutting the meridian at equal distances; parallels of altitude. (The horizon itself was reckoned as the first almacantar.)

### Almanac

*O.E.D.* An annual table or (more usually) book of tables, containing a calendar of months and days, with astronomical data and calculations, ecclesiastical and other anniversaries, besides other useful information, and, in former days, astrological and astrometeorological forecasts ....

*Webster.* A book or table containing a calendar of days, weeks and months, to which astronomical data and various statistics are often added, such as the times of the rising and setting of sun and moon, changes of the moon, eclipses, hours of full tide etc.

The *Ephemeris and Nautical Almanac*, issued by the chief governments two or three years in advance, contains elaborate tables of accurately predicted positions of the heavenly bodies, with times of celestial phenomena, and other data, for use by astronomers and navigators.

Comment. Also popularly and commonly used for a simple pictorial calendar, also for an annual work of reference (notably *Whitaker's Almanack*, using the old spelling.)

## Alp, Alps

*O.E.D.* (in *pl. Alps*, a. Fr. *Alpes*: L. *Alpes*) name of a mountain system in Switzerland and adjacent countries; said by Servius to be of Celtic origin, and variously explained as meaning 'high' (*cf.* Gaelic *alp*, a high mountain, Irish *ailp*) and 'white' (*cf.* L. *albus*)

1. *pl.* Proper name of the mountain range which separates France and Italy, etc. *sing.* A single peak (applied in Switzerland to the green pastureland on the mountain side).
1860 Tyndall. *Glac.* I, 3, 27. After a rough ascent over the Alp we came to the dead crag.
2. Any high, especially snow-capped, mountains.
1598 Hakluyt (plural) 1667 Milton (singl.) 1856 Ruskin (singl.)

Mill, *Dict.* (repeats *O.E.D.* then:) Applied in Switzerland to the high terraced pastures.

Peattie, R., 1936, *Mountain Geography*, Cambridge: Harvard U.P. 'The alp is, of course, the upland pasture.... There are the following equivalents, or partial equivalents:

| alpe | berg | galen |
| alpo | berge | pla |
| arpe | olbe | planina |
| alpage | monti | plaroure |
| albe | montaigne | jasse |
| alm | montagna | |

Some American physiographers have selected the term alb for use.' (p. 129)

## Alpides, alpids

Mill, *Dict.* Folds belonging to the great system which stretches from the Black Sea to Gibraltar (Suess). The Alpides belong to the posthumous Altaides (*q.v.*).

*Webster.* One of a group of mountain ranges in which folding continued into Tertiary time, as the Alps. The Alpids have the same general trend as the Altaids.

*Comment.* See -ides. A general term applied to the Alpine fold ranges.

## Alpine

*O.E.D.* (see alps) of or pertaining to the Alps, hence, of any lofty mountains.
*O.E.D. Suppl.* B. An alpine plant, or one that grows on high ground.
Mill, *Dict.* With the characteristics of the higher part of a mountain system.
Daly, R. A., 1905, The Accordance of Summit Levels, *J. of Geol*, **13**, 2. 'In the present paper the term "alpine" is used to signify a range possessing not only the rugged, peak-and-sierra form of the Swiss Alps, but, as well, the internal structures incidental to intense crumpling, metamorphism and igneous intrusion as exemplified in the Swiss Alps.' (pp. 106–107)
Davis, 1909. '... the Alps and other strongly glaciated ranges have been taken as types of mountain form from a time when there was no question at all of glacial erosion; and as a consequence the exceptional quality of many Alpine forms has passed relatively unnoticed.'
Cotton, 1942 (after describing 'inosculation' of cirques). 'The landscape now becomes "Alpine" ... This is the stage of glacial dissection referred to in Chapter XIII as a "fretted upland".' (p. 190)
Flint, 1947. Of cirque erosion, horns, arêtes, etc. 'Because the sequence of landforms just described is well developed in the Alps, where the study of glacial erosion began, these forms are often referred to as Alpine. In Alpine sculpture the details of the valleys are the work of glacial erosion, but equally important details of the cirques and higher spurs and ridges are largely the work of frost wedging and other processes of mass-wasting.' (p. 99 *et seq.*)

## Alpine glacier

*Webster.* A glacier formed among summits and descending a mountain valley.
Rice, 1941. A type of glacier occurring about the peaks and in the valleys and gorges of mountains, originating above in various amphitheatres, terminating below by melting, or by spreading out into piedmont glaciers; an ice river.

## Alpine orogeny, alpine revolution

See Orogeny. The great mountain-building movements which took place in the Tertiary period, culminating in the mid-Tertiary and resulting in the creation of the Alpides or Alpine system of the Alps, Carpathians, Balkan Mountains, Pyrenees, Atlas and other great chains.

## Altaides, Altaids, Altaid orogeny (Suess)

Mill, *Dict.* The great systems of fold mountains which have arisen in long waves, one behind the other, from the ancient Baikalian vortex of the Eurasian mass. They include most of the mountains of Europe and Asia.

*Webster. Geol.* One of a group of mountain ranges, of world-wide extent, in which folding, with compression, generally along nearly north-south lines, began in Carboniferous time; the Hercynian chain of mountain folds. The Altai range of Asia is the type.

Wills, L. J., 1929, *Physiographical Evolution of Britain*, London: Arnold. 'The earth movements gave rise to the system of mountain-chains which Suess named the *Altaides*. They stretch through Central Europe and Asia. In Europe two important chains can be distinguished. The Armorican, trending from South Ireland through South England, North-West France and the Central Massif; the Variscan or Hercynian trending from Belgium eastwards through Central Europe. The term Hercynian is, however, also applied sometimes to the whole of this mountain-system of Europe. At present the stumps of the mountains are largely buried below newer rock, but appear in the European horsts...' (p. 71)

*Comment.* Webster seems in error in suggesting the trend is generally north-south. East-west would be nearer the truth.

### Altimeter
Any instrument for measuring altitude.

### Altimetric frequency curve
*O.E.D.* Altimetric is not given, but altimetrical: (rare). Pertaining to the measurement of heights or altitudes. Blount, Gloss. 1681.

Wooldridge and Morgan, 1937. 'If the map is divided into small uniform squares, we may estimate the elevation of the highest point in each square, or use simply the spot heights shown on the map, providing that these, in general, mark elevated points. A curve may then be constructed showing altitudinal distribution of such points over the region.' (p. 266)

See also Baulig, H., *Le Plateau Central*, pp. 563–574 and pp. 539–543; The Changing Sea Level. *I.B.G. Pub. No. 3*, 1935, p. 41.

*Comment.* There are several variations in the method of constructing the curve. See Monkhouse and Wilkinson, (p. 34) and Miller, A. A., *The Skin of the Earth*, London: Methuen.

### Altiplanation (H. M. Eakin)
Eakin, H. M., The Yukon-Koyukuk Region, Alaska, *U.S. Geol. Survey* Bull. 631. 'The author has suggested the term "altiplanation" to designate a special phase of solifluction that, under certain conditions, expresses itself in terrace-like forms and flattened summits and passes that are essentially accumulations of loose rock materials.' (p. 78). Quoted by Thornbury, 1954, p. 412.

*Comment.* The same features are also known as equiplanation terraces. (G.T.W.)

### Altiplano (Spanish: South America)
*Webster* (Amer. Sp.). An elevated plateau; specif. [*cap.*], the high plateau of western Bolivia.

James, 1959. 'The highlands of Bolivia are made up of three distinct parts: the Western Cordillera; a string of high intermont basins known collectively as the *Altiplano*; and the Eastern Cordillera' (pp. 208–209).

### Altitude
Swayne, 1956. (a) The height of a hill or mountain measured perpendicularly from the base to the summit; (b) height above mean sea level; (c) the angular height in degrees of a heavenly body above the astronomical horizon.

*Comment.* Vertical distance above or below a datum. The datum commonly used is Mean Sea-Level (*q.v.*). Measurement is commonly in feet or metres.

### Alvar (Swedish)
1. Bare (or covered by a thin earthen layer) treeless limestone surface or area (in Estonian lood).
2. Equivalent to alvarmark, rendzina of the Baltic Sea island of Gotland, Öland, Oesel (Saaremaa) Moon (Muhu), and North-western Estonia. (E.K.)

### Āmān (Indo-Pakistan: Bengali)
Winter rice or rice-crop.

Johnson, B. L. C., 1958, Crop Association Regions in East Pakistan, *Geography*, 43, 86–103. 'The hemantic crop, *āmān* paddy, is called a "cold weather" crop as it is harvested at the beginning of the cool dry season.' (p. 87). See also Ahmad, Nafis, *An Economic Geography of East Pakistan*, O.U.P., 1958, 122. Aman season = hemantic season: June–November.

### Amerind
*Dict. Am.* A word composed of the first syllables of 'American Indian' used to designate the race of man inhabiting the

New World before its occupancy by Europeans (1907. Hodge, *Amer. Indians* I. 49) (Also used by Powell in *Rep. Bureau Amer. Ethnol.* I. xlviii, 1897–8).

*Webster.* An individual or one of the native races of America; an American Indian or Eskimo.

**Amin** (Indo-Pakistan: Urdu)
A surveyor employed for making village maps. Also called patwari or lekhpal.

**Anabatic**
*Webster.* Ascending; spec. b. *Meteorol.* Upward-moving; as, an *anabatic* wind.
*Met Gloss*, 1944. 'Referring to the upward motion of air due to convection. A local wind is called anabatic if it is caused by the convection of heated air: as for example, the breeze that is supposed to blow up valleys when the sun warms the ground.'
No reference *O.E.D.* (except medical); Mill, *Dict.*

**Ana-branch, anabranch**
*O.E.D. Suppl.* Anabranch. *Australia* (f. Ana +Branch sb.; suggested by Anastomosis) A branch stream which turns out of a river and re-enters it lower down, forming a branch island.
1834 *J. R. Geog. Soc.* IV, 79. 'Thus, such branches of a river as after separation re-unite, I would term anastomosing branches; or, if a word might be coined, ana-branches.'
*Webster.* A diverging branch of a river which re-enters the main stream; also, a branch which loses itself in sandy soil.
Mill, *Dict.* Ana-branch = Anastomosing branch.—A river branch which leaves the main stream and joins it lower down.
Given in Rice, 1941, quoting Fay, 1920, who gives it as Australian. Not used in modern texts.

**Anaclinal** (J. W. Powell, 1875)
*Webster. Geol.* Opposed to the dip—applied to a stream or valley whose course is against the dip of the underlying rocks.
Mill, *Dict.* Applied to streams which run contrary to the main direction of dip in a country of faulted structure (Powell). See Cataclinal and Diaclinal.
Powell, 1875. 'Anaclinal, valleys that run against the dip of the beds.' (p. 160)
Davis, 1909. '...a distinct though gentle slope to the north-west is apparent in the unconsumed surface of the past plain; but this strong river runs southeastward against the slope; it is an anaclinal stream.' (*i.e.* this could mean that an anaclinal stream is one which flows in a direction opposite to the tilt of a raised past plain = peneplain?). (p. 511)
Lobeck, 1939. 'The term anaclinal is applied to an antecedent stream flowing on a surface which has been slowly tilted in a direction opposite to the flow of the stream. If sufficiently vigorous, such a stream maintains its course.' (quotes Davis). (p. 173)
Cotton, 1941. Anti-dip or anaclinal streams are those flowing in the reverse direction to the dip. (p. 25)
*Ibid.* 'A third class, "obsequent", perhaps better termed "anaclinal", is recognised.' (p. 62)
Fay, 1920. Descending in a direction opposite to the dip of the strata, as an anaclinal river. Opposed to Cataclinal.
No reference in Geikie, 1898; Cotton, 1922; Cotton, 1945; Wooldridge and Morgan, 1937; Holmes, 1944; Salisbury, 1907; Worcester, 1939.
*Comment.* As with the comparable terms, cataclinal and diaclinal also introduced by Powell, now rarely used. For full discussion see Baulig, H., *Jour. Geomorphology*, **1**, 224–9.

**Anaerobic**—see Aerobe, aerobic

**Anaglyph**
*Webster.* 1. Any sculptured, chased or embossed ornament worked in low relief, as a cameo.
2. A picture combining two images of the same object ... one image in one colour being superposed upon the second image in a contrasting colour. Viewed through an anaglyphoscope, it produces a stereoscopic effect. Helava, U.V., 1958, *Canadian Geog.*, **11**, 'using the anaglyphic method three-dimensional maps, models and other illustrations can be included in books and other publications' (p. 44).
*Comment.* An anaglyphic map is one printed in two complementary colours, commonly red and green, which do not register but in such a way that when studied with a twin eyepiece of the same colours a realistic impression of relief is given stereoscopically.

**Analemma**
*O.E.D.* L.: analemma, the pedestal of a sundial, hence the sundial itself. 1. orig. A sort of sundial. Obs. (and perhaps never in Eng.)

2. An orthographical projection of the sphere made on the plane of the meridian, the eye being supposed to be at an infinite distance and in the east or west point of the horizon; used in dialling, etc.
3. A gnomon or astrolabe, having the projection of the sphere on a plate of wood or brass, with a horizon or cursor fitted to it, formerly used in solving certain astronomical problems.
4. A scale of the sun's daily declination drawn from tropic to tropic on artificial terrestrial globes.

Raisz, E., 1948, *General Cartography*. 2nd ed., New York: McGraw-Hill. 'This is a line which connects the points on the earth's surface where the sun is directly overhead throughout the year when it is twelve o'clock local time on the meridian upon which the analemma is centred.' (p. 269)

## Analogue model

Chorley, R. J., in Chorley and Haggett, 1967. 'Analogue models involve radical changes in the media of which the model is constructed. They have much more limited aims than scale models in that they are intended to reproduce only some aspects of the structure or web of relationships recognized in the simplified model or idealized system of the real world segment. [Such transformations are obviously rather difficult and great potential sources of 'noise' (i.e. extraneous confusion), in that great and often questionable assumptions must be made regarding the appropriateness of the changes of the media involved].' (p. 68)

See Lewis, W. V. and Miller, M. M., 1955, Kaolin Model Glaciers, *Jour. Glaciology*, **2**.18, 533–538, for an example of an analogue model. See Chorley, R. J., 1964, Geography and Analogue Theory, *Ann. Assoc. Am. Geographers*, **54**.1.

## Anastomosis, Anastomosing streams

*O.E.D.* Anastomosis. mod. L., from Greek, to furnish with a mouth or outlet. Intercommunication between two vessels, channels, or distinct branches of any kind, by a connecting cross branch. Applied originally to the cross communications between the arteries and veins, or other canals in the animal body; whence to similar cross connections in the sap-vessels of plants, and between rivers or their branches; and now to cross connections between the separate lines of any branching system, as the branches of trees, the veins of leaves, or the wings of insects. Anastomosing. Communicating by anastomosis; inosculating.

1842. Blackw. Mag. LII, 170. A Flemish landscape, irrigated by anastomosing ditches.

Cotton, 1941. 'In streams with these latter characteristics, typically flowing in anastomosing, or braided, courses, as contrasted with the well-defined meandering courses. ...' (The characteristics are the deposition of such quantities of coarse waste as to cause the stream to flood frequently and spill over into new channels across the flood plains.) (p. 131)

Rice, 1941. Intercommunicating by branching, usually producing a netlike appearance.

Cullingford, C. H. D. (Ed.), 1953, *British Caving*. 'Anastomosis; an intricate network of random tubular channels of various sizes in a plane (either the bedding plane, or more rarely a joint plane).' This usage was introduced by J. H. Bretz. *Jour. Geol.*, **50**, 1942, pp. 675–811.

No reference in Lyell, 1830; Geikie, J., 1898, Davis, 1901; Cotton, 1922; Salisbury, 1907; Wooldridge and Morgan, 1937; Holmes, 1944; Lobeck, 1937.

**Anchor ice**—see Ice, sundry terms

## Andesite

Himus, 1954. A volcanic rock, belonging to the intermediate group (containing 55 to 66 per cent silica) composed essentially of a plagioclase felspar, with a pyroxene, hornblende or biotite, and more or less of a glassy base.

## Andesite line

The geographical boundary between the circum-Pacific petrographical province characterized by andesites (Pacific suite) and the mid-Pacific province characterized by basalts (*e.g.* Hawaii).
See also Atlantic suite.

## Andosol, ando-soil (Soil Science)

An Americanization of the Japanese word 'ando' used to signify a dark-coloured volcanic soil occurring in Japan and possibly similar to what is known in Chile as 'trumao'. (G.V.J.) See J. Thorp and G. D. Smith, *Soil Sc.*, **67**, 1949, 117.

## Angerdorf (German; *pl.* Angerdörfer: *lit.* green village)

A village type largely found in eastern Germany dating back to the period of eastern colonization of the Germans and characterized by a central village green (usually of oblong shape) which on two sides is continued by the village street; generally associated with open fields (K.A.S.). 'Village or hamlet in which the homesteads cluster around a rectangular or spindle-shaped green. In the simplest form the green extends along a single street, but where the green is broader several roads may enter and cross the central open space which often has one or more ponds. This form of settlement is common in eastern England, Germany east of the Elbe and in Poland, and was widespread in Denmark and Skåne before enclosure. Although many *Angerdörfer* in Germany east of the Elbe sprang up as colonial settlements during vigorous German expansion and in the twelfth and thirteenth centuries, evidence from Poland and eastern England strongly suggests that similar forms were in existence much earlier.' (H. Thorpe in MS.)

**Angulate pattern of drainage**
Thornbury, 1954. 'A variant of rectangular drainage is the angulate pattern. It develops where faults or joints join each other at acute or obtuse angles rather than at right angles.' (p. 123)

**Anicut** (India: Anglo-Indian corruption from Tamil). Also annicut, anaicut.
A weir, or dam across a river over the crest of which surplus water is discharged. G.K. (Southern India).
Spate, 1954. 'anicuts or weirs' (p. 714).
*O.E.D.* 'In the Madras Presidency, the dam constructed across a river to fill, and regulate the supply of, the channels drawn off from it. (Col. Yule).'

**Annular drainage**
Lobeck, 1939. 'Annular Drainage Pattern. Some of the streams which drain a maturely dissected dome follow circular paths around the dome in conformance with the outcrops of the weaker belts. These are subsequent streams.... An annular drainage pattern is simply a variation or special form of the trellis pattern.' (p. 175)

**Anœkumene, anœcumene, anecumene**
Mill, *Dict*. A community or district which is essentially dependent on other regions as regards population, supplies, or influence. Most parts of America are anœkumenic, especially as regards population. The English channel, with its population of passengers and crews of boats, is a type of anœkumene.
*Comment.* Whilst oecumene (*q.v.*) in the sense of the habitable parts of the earth's surface is now commonly used, anoekumene or anoecumene is rare. There is no reference in *O.E.D.* and Mill's definition is quoted in full. His understanding of the term is not, however, very clear and it is doubtful whether many authorities would agree with him. The term is best avoided.

**Anomaly**
*O.E.D.* from Greek, irregular, uneven
2b. Nat. Sc. Deviation from the natural order.
1860 Maury, Phy. Geog. XV, 669. A low barometer... was considered an anomaly peculiar to the regions of Cape Horn.
*Met. Gloss*, 1944. 'The departure of a meteorological element from its normal value. In meteorology the word is used chiefly in connexion with temperature, to indicate the departure from the mean value for the latitude. Places that are relatively warm for their latitude... have a positive temperature anomaly; places that are relatively cold... have a negative temperature anomaly.'
Mill, *Dict*. Thermal Anomaly. The difference between the mean temperature of any place and the mean temperature of its latitude.
Jefferies, H., 1950, *Earthquakes and Mountains*, 2nd ed. 'In computing gravity anomaly we compare observed gravity with a standard formula corresponding to a spheroidal level surface with an average value of the ellipticity. If we correct this formula for the height above sea-level, simply allowing for the fact that the place of observation is further from the earth's centre, we obtain the "free-air anomaly"; if we allow for the attraction of the visible excess or defect of matter we get the Bouguer anomaly.... If we also allow in our theoretical estimate, for the disturbance due to any form of compensation, we are left with an "isostatic anomaly". It is unfortunate that in published work the last is the only one given. To determine the earth's external field we need the free-air anomaly; while all the information gravity can give about the distribution of density is summed up in the Bouguer anomaly. The isostatic anomaly is really of no interest except to test a particular hypothesis....' (p. 78) See also Holmes, 1944, 404.
*Comment.* Applied also in geographical

literature to various deviations from the normal.

**Antarctic**
*O.E.D.* Adj. Opposite to the Arctic, pertaining to the south polar regions; southern. As noun, The south pole or the regions adjacent.
*Comment.* See comment on arctic. The same remarks apply to antarctic.

**Antarctic circle**
*O.E.D.* The parallel of 66° 32' South, which separates the South Temperate and South Frigid Zones. 1556 Recorde Cast. Knowl. 27. The Antartike circle is equall and equidistant to the Arctike circle.
*Comment.* The Antarctic and Arctic Circles are commonly drawn respectively at 66°30' North and 66°30' South. See remarks under Arctic circle.

**Antarctic convergence**
Gould, L. M., 1958, *The Polar Regions in their relation to Human Affairs*, New York: Amer. Geog. Soc., 1958. 'an unbroken and rather sharply defined boundary around the continent whose position shifts slightly from season to season ... caused by the fact that the cold waters of the northward-moving Antarctic intermediate current are denser than the sub-Antarctic waters and sink sharply below them with but little mixing. Sea and air temperatures, water analyses, the character of the plankton and the sea birds, all reveal this boundary ... regarded as the most notable natural boundary relating to the totality of South Polar conditions ... roughly parallels the isotherm of 50°F. for the warmest month.' (p. 11)

**Antarctica**
The name now generally given to the continental land area which surrounds the South Pole and has an estimated area of about 4,410,000 square miles. Owing to the great thickness of the ice-cap in certain areas its base may be below sea-level and it is possible that the actual land may be in two parts, East and West Antarctica. But, in the prevailing conditions, ice is a rock and to be considered as part of the land area.

**Antecedent**
*S.O.E.D.* Preceding in time or order.
Hence the specialized uses in geographical writings especially antecedent drainage, boundaries.
*Webster. Phys. Geog.* Established before the deformation of a surface and persisting after the deformation has taken place and in spite of it—said of drainage, a stream, or a valley, and contrasted with *consequent*.

**Antecedent boundary**
Jones, S. B., 1937, The Cordilleran Section of the Canada-United States Borderland, *Geog. Jour.*, **89**. 'Nevertheless, there are many dissimilarities, most of which meet without transition at the totally antecedent boundary.' (p. 439)
*Comment.* The word antecedent has been borrowed by political geographers: Antecedent political boundaries, established before the development of cultural differentiation. Totally antecedent or pioneer boundaries are found where the line was drawn before any settlement took place. Classified by Hartshorne, see Boggs, S. W., 1940, *International Boundaries*, New York: Columbia U.P., pp. 28–29. See also *A.A.A.G.*, 26, 1936.

**Antecedent drainage, antecedent river** (Powell, 1875)
Powell, 1875. '... the drainage was established antecedent to the corrugation or displacement of the beds by faulting and folding ... I propose to call such valleys, including the orders and varieties before mentioned, antecedent valleys.' (p. 163)
Mill, *Dict.* Antecedent River. A river which flows across some barrier through which it has been able to cut its way as rapidly as that barrier was formed by the differential movement of the Earth's crust (W. M. Davis).
Davis, 1909. '... rivers have held their courses through mountain ridges that slowly rose across their path; ... This idea first came into prominence through Powell's report on the Colorado River of the West (1875), in which he gave the name "antecedent" to rivers of this class.' (p. 81 *et seq.*)
Wooldridge and Morgan, 1937. 'The essence of the antecedent relationship is a successful contest waged by rivers against localized uplift.' (p. 209)
Committee, List 2. Antecedent drainage. A drainage system maintaining its original direction across a line of localized uplift.
See also Rice, 1941; contrast epigenetic, superimposed.

**Anteconsequent rivers** (C. A. Cotton, 1917)
Cotton, C. A., *Amer. J. of Sc.*, 4th Series, **44** (whole number 194) 1917, '... Streams

that were consequent upon the form of the surface assumed as the result of early movements, but are antecedent to later movements of the same series. Such streams might perhaps be appropriately termed antecedent consequents or antec-consequents' (p. 253).

*Comment.* The spelling in the definition is clearly a misprint and anteconsequent is used later in the paper. Wooldridge and Morgan, 1937, p. 209, quote Cotton but the term is rarely used, doubtless due to the practical difficulties of differentiating the effects of stages in tectonic movements in most areas.

**Anthracite**

*Webster.* A hard, compact variety of natural coal, of high luster, differing from bituminous coal in containing only a small amount of volatile matter, in consequence of which it burns with a nearly non-luminous flame. Pennsylvania anthracite contains 85-93 per cent of carbon; those of South Wales, 88-95.

Himus, 1954. A lustrous variety of coal, which does not soil the fingers. Contains 92 per cent and over of carbon and yields not more than 8 per cent. of volatile matter when heated out of contact with air. Burns smokelessly and slowly, without flame.

**Anthropogeography**

*O.E.D. Suppl.* (F. Anthropo + Geography, after German anthropogeographie (F. Ratzel, 1882). That department of geography which treats of the relations of the earth to mankind as its inhabitants.
1652 Hermeticall Banquet, 120. The new Anthropogeographicall Map.
1899 *Geog. Jour.* Feb., 171. Anthropogeography is a convenient term under which to include all those aspects of geography that deal with the relations of humanity... to the earth.... 'Applied Geography' might be taken as an alternative term.

Page, *Handbook of Geol. Terms,* 1865. Anthropography—that branch of physical geography which treats of the distribution of the human species, as distinguished by physical features, language, institutions and customs.

Mackinder, H. J., 1895, Presidential address to Section E (Geography) of British Association, Sept. 12, 1895. Reprinted *Geog. Jour.,* 1895, pp. 367-379. (of Biogeography), 'It has three sections—phytogeography, or the geography of plants; zoogeography, or the geography of animals; and anthropogeography, or the geography of men.' (p. 375)

Mill, *Dict.* 'The Science of the distribution of human affairs on the Earth.

Hartshorne, 1939. 'The term anthropogeography... is misleading. It suggests the geography of man in terms of individuals and races, anthropological geography; whereas the major objects of Ratzel's concern were the works of man, particularly... the products of man's social life in relation to the earth.' (p. 90)

Wooldridge, S. W. and East, W. G. *The Spirit and Purpose of Geography,* London: Hutchinson, 1951, suggest 'Anthropogeography' as geographical studies in physical and social anthropology, noting the wider usage of the term derived from Ratzel.

*Comment.* The general tendency is to drop this term, especially in view of the differing opinions as to whether it does or does not coincide with human geography. In Italian, for example, *antropogeografia* is human geography. (C.J.R.) See also Racial geography, Human geography.

**Anthropology**

*L.D.G.* Anthropology is literally the study of man but there was formerly a tendency to restrict the term to physical anthropology or the physical measurements of man, such as skull shapes and sizes, and the evolution of the species from anthropoid (ape-man) ancestors. It is now generally agreed that anthropology should embrace also ethnology—with the development of technical skills in the making of implements and the utilitarian arts, clothing etc., social organization, religion and other aspects of culture, linguistics, folklore and what has been called ethnography (the geographical distribution of races and peoples, or anthropogeography). Archaeology (*q.v.*) which deals generally with the material remains of early man is also included.

**Anticlinal bend**

Hills, E. S., 1953, *Structural Geology,* 3rd Ed. 'An upwardly convex flexure in which one limb dips gently towards the apex and the other limb more steeply away from it.' (p. 76)

**Anticlinal ridge**

A ridge or uplifted tract of country which

corresponds with an anticlinal fold in the underlying strata.

Rice, 1943. Mountains that occupy the sites of anticlines.

**Anticlinal valley**

Powell, 1875. Valleys. 'which follow anticlinal axes.' (p. 160)

**Anticline, anticlinal**

*O.E.D.* Geol. An anticlinal fold. Anticlinal: Forming a ridge, in which strata lean against each other, and whence they slope down, or dip, in opposite directions. The opposite of *synclinal*.

1876 Page. *Advd. Textbk. Geol.* iv, 83. When strata dip ... like the roof of a house ... the strata are spoken of as forming an anticline or saddleback.

B. sb. Geol. (by ellipsis) An anticlinal fold, axis, crest or line; a line whence strata dip in opposite directions, 1869. Phillips *Vesuv.* IX, 255. Anticlinals and synclinals, in the earth's crust.

*Webster.* Anticlinal: inclining in opposite directions; specif. *Geol.* Having or pertaining to a fold in which the sides, or limbs dip from a common line or crest, called the anticlinal axis.

Mill, *Dict.* Upfold where the strata dip in opposite directions away from a central line.

Geikie, J., 1920, *Structural and Field Geology,* 4th ed. Edinburgh: Oliver and Boyd. 'When the strata dip away from the axial plane on either side at approximately the same angle, the structure is known as an Anticline or Saddleback.' (p. 137) (The First Edition, 1902, correctly refers to this as a Symmetrical Anticline. p. 134.)

Rice, 1943. An upfold or arch in the rocks. As a rule the beds dip outwards in two or more directions from the crest.

Holmes, 1944. 'When beds are upfolded into an arch-like form (with the lower beds within the upper) the structure is called an anticline. ...'

Hills, E. S., 1953, *Structural Geology,* 3rd ed. 'An upwardly convex flexure in which a given bed intersects the same horizontal plane in both limbs.' (p. 76)

*Comment.* The *O.E.D.* definition is misleading, the strata do not 'lean against each other' and whereas an anticline when first uplifted may have formed a ridge, denudation so frequently attacks the uplifted portion that the centre of an anticline becomes a valley in which the older rocks in the core of the anticline are exposed (see anticlinal valley). It should be noted that the older form for the noun was anticlinal and was so used by Lapworth, 1913 (*British Assoc. Handbook, Birmingham*) and Kinvig, 1928 (in *Great Britain: Regional Essays*). Anticline is later but is now used exclusively as the substantive, reserving anticlinal as the adjective. Mill's simple definition does not cover the overfolded anticline and this is a difficulty with many definitions including Hill's, but is allowed for by Rice. L. J. Wills, 1956, *Concealed Coalfields,* uses anticlinal and synclinal for indefinite anticlines and synclines. (p. 7)

**Anticlinorium**

*O.E.D. Suppl.* A mountain range or region in which the folds of the strata are chiefly anticlinal.

1893. 13th Ann. Rep. U.S. Geol. Surv. II, 220. It often happens that the result of the combination of many anticlines and synclines is to form a complex structure, which, regarded as a whole, is either synclinal or anticlinal. The former is called a synclinorium, the latter an anticlinorium.

*Webster. Geol.* A series of anticlines and synclines, so grouped that, taken together, they have the general outline of an arch—opposed to *synclinorium*.

Mill, *Dict.* Anticlines with many secondary foldings and frequently intense deformation. Also called geo-anticlines (Dana).

Geikie, J., 1920, *Structural and Field Geology,* 4th ed., Edinburgh: Oliver and Boyd. '... one great arch composed of numerous subordinate wrinkles or minor folds and flexures. A complex arch of this kind is termed an Anticlinorium. If the arch be simple—a broad anticlinal fold with no conspicuous wrinklings or flexures —it is known as a Geanticline.

Holmes, 1944. '... the puckers and smaller folds are superimposed on broad anticlinal and synclinal folds of a much larger order. An anticlinal complex of folds of different orders is called an anticlinorium....' (p. 74)

*Comment.* The *O.E.D.* definition is erroneous. Anticlinorium is a complex structure and has nothing to do with a mountain range or region and in an anticlinorium, correctly defined by Geikie and Holmes, there are likely to be just as many synclines as anticlines. See also synclinorium. The plural is anticlinoria but anticlinoriums may sometimes be seen. Webster is correct.

**Anticyclone** (Francis Galton, 1861)

*O.E.D.* Meteor. The rotatory outward flow of air from an atmospheric area of high pressure; also, the whole system of high pressure and outward flow.

Mill, *Dict.* Anticyclone, or High-Pressure System, or High. A system of atmospheric pressure in which the isobars indicate a relatively high pressure in the interior of the system and decreasingly lower pressures towards the exterior. It gives rise to winds that blow spirally outwards with a clockwise movement in the northern hemisphere and the opposite movement in the southern (Galton). Quotes Hann's distinction between thermal anticyclones and dynamical anticyclones.

*Met. Gloss.*, 1944. An anticyclone is a region in which the barometric pressure is high relative to its surroundings. First introduced by Sir Francis Galton in 1861.

*Comment.* Modern meteorologists regard the latter part of Mill's definition as untrue; the winds blow parallel to the isobars.

**Anti-dip stream**

A stream flowing in approximately the opposite direction to the regional dip. Frequently but not necessarily an obsequent stream. See also Anaclinal.

**Antidune** (G. K. Gilbert, 1914)

Gilbert, G. K., 1914, Transportation of Debris by Running Water, *U.S. Geol. Surv. Prof. Paper* 86. 'Finally a third stage [following the 'dune' and 'smooth' stages with increasing velocity of the water] was reached in which the bed was characterized by waves of another type. These are called *antidunes*, because they are contrasted with dunes in their direction of movement; they travel against the current instead of with it. Their downstream slopes are eroded and their upstream slopes receive deposit. They travel much faster than the dunes, and their profiles are more symmetrical' (p. 31).

Twenhofel, W. H., 1939, *Principles of Sedimentation.* 'The *antidune phase* which develops in highly-loaded streams from the smooth phase on increase in velocity. In the *antidune phase*, ridgelike structures develop that are eroded on the downcurrent side. These have also been termed sand waves. The sand wave or antidunes move up-current as the individual sands move down-current. Antidunes seem to have more symmetrical slopes than dunes and they may also be much higher. In the San Juan River of Arizona they have been observed to be 5 to 6 meters from crest to crest and 0·9 meter from crest to trough. Some have been seen with a height of 1·8 meters. Enormous quantities of sediments are moved in this form of traction' (p. 190).

**Antimeridian**—see Meridian

**Antipodes, antipodal**

*O.E.D.* From L. antipodes, from Greek, 'having the feet opposite'.

3. Places on the surfaces of the earth directly opposite to each other, or the place which is directly opposite to another; esp. the region directly opposite to our own.

Mill, *Dict.* The further extremity of a common diameter of the sphere.

*Comment.* Since Australia and New Zealand lie roughly on the opposite side of the earth to the British Isles, British writers have often referred to them as 'the Antipodes', a term accepted in good part by the people of those lands. Antipodes Island (approximately 49°30′S. and 178°30′E.) is the nearest piece of land to the exact antipodes of London—actually of the Channel Islands.

**Antipodes day**—see Meridian day

**Anti-trade wind, Anti-trades**

*O.E.D.* A wind that blows steadily in the opposite direction to the trade-wind, that is, in the northern hemisphere from S.W., and in southern hemisphere from N.W. 1853. Sir J. Herschel, *Pop. Lect.* iv, 19, 1873, 157. The great and permanent system of winds known as the 'trades' and 'anti-trades.'

1867. E. Denison, *Astron. without Math.* 40. This secondary or anti-trade wind prevails from about 30° to 60° latitude at sea.

1875. Croll, *Climate and Time*, ii, 28. The south-west wind to which we owe so much of our warmth in this country, is the continuation of the anti-trade.

*Webster. Meteorol.* a. The prevailing westerly winds of middle latitudes. b. The westerly winds above the trade winds.

Mill, *Dict.* 1. Poleward winds in the upper regions of the atmosphere of the tropics created by the expansion and rise of superheated air above the equator. 2. The westerly storm winds of higher latitudes (Herschel). Much confusion results from the double use of the term.

*Met. Gloss.*, 1944. 'Anti-trades. At a height of 2,000 metres or more above the surface in trade-wind regions the wind direction is sometimes reversed.... These winds are believed to be the return currents carrying the air of the trade winds back to higher latitudes, hence they are termed "anti-trades" or counter-trades, but they are not regularly developed.'

Trewartha, G. T., 1937, *Introduction to Weather and Climate*, New York: McGraw-Hill. Uses it in sense of Mill's (1) (upper air).

Shanahan, E. W., 1933, *A Modern World Geography*, London: Methuen. '... two bands of regions, ... which are always under the influence of the Prevailing Westerlies (or Anti-Trades) with their succession of cyclonic storms.' (p. 37)

Kendrew, W. G., 1949, *Climatology*, Oxford University Press. 'The upper, westerly, winds above the trades are commonly called the anti-trades or counter-trades.' (p. 124)

*Comment.* In nearly all the older textbooks the winds now called the 'Westerlies' in middle latitudes were called anti-trades as in *O.E.D.* The modern usage is quite different, as in Mill (1), and *Met. Gloss.*, and Trewartha.

**Apartheid** (Afrikaans, *lit.* separation, separateness)
The doctrine of racial segregation translated into policy by the Nationalist government of the Union of South Africa which came into power in 1948. 'Used primarily with reference to the segregation policy of the South African Government to imply that so far as possible there shall be residential, economic and social separation of Whites and Blacks (and other non-White races) in South Africa.' (Professor Isaac Schapera in MS. communication) Used in both Afrikaans and English literature.

**Aphinite, aphanite**
*Webster.* Aphanite. *Petrog.* Any dark rock of such close texture that its separate grains are invisible to the naked eye.

Himus, 1954. A term applied to any fine-grained igneous rock or groundmass the constituents of which cannot be distinguished by the naked eye.

**Aphelion**
*O.E.D.* Pl. aphelia. Græcized form of mod. L. aphelium. Aphelium was also the earlier form in English. That point of a planet's or comet's orbit at which it is farthest from the sun.

**Aphotic**—see Photic

**Apogee**
*O.E.D.* 1. The point in the orbit of the moon, or of any ... planet, at which it is at its greatest distance from the earth; also the greatest distance of the sun from the earth when the latter is in aphelion. (A term in the Ptolemaic astronomy, which viewed the earth as the centre of the universe; in modern astronomy strictly used in reference to the moon and popularly said of the sun in reference to its apparent motion.)
2. The greatest altitude reached by the sun in his apparent course; his meridional altitude on the longest day. Obs.

**Apophysis** (*pl.* apophyses)
*Webster.* Literally, an offshoot; specif. *Geol.* An offshoot from an intrusive body of igneous rock.

Himus, 1954. Veins, dykes and tongues which are offshoots from larger igneous intrusions, the connexion with which can be directly traced.

**Appalachian revolution, orogeny**
Himus, 1954. An era of intense mountain-building movements which took place during the interval between the Permian and Trias, in which deposits in the Appalachian and Cordilleran geosynclines were folded to form the Appalachian and Palaeocordilleran mountains. The revolution was equivalent to but later than the Armorican and Hercynian movements in Europe.

**Apparent time**
*Webster.* The time of day at any given place, reckoned by the diurnal motion of the true sun, so that *apparent noon* is the instant of transit of the sun's center over the meridian.

Mill, *Dict.* Time regulated by the apparent motion of the sun; time as shown by a properly adjusted sundial: solar time.

U.S. War Dept. Tech. Manual TM5-235, 1940, *Surveying*. 'Solar time is further sub-divided into apparent time, mean time and standard time' (p. 361). Refers to real or apparent motion being non-uniform from day to day—hence necessity for "mean sun"—"angular difference between the position (of the "real" and "mean" sun) ... at any time is termed the equation of time' (p. 365).

Close and Winterbotham, 1925, *Textbook of Topographical and Geographical Surveying*,

3rd ed. 'Apparent or solar day is the interval between two successive upper transits of the sun's centre across the same meridian, and apparent time (A.T.) at any instant is the hour angle of the sun measured westward from the meridian in hours (0 to 24)' (p. 171).

*Field Astronomy*, 1932 refers to 'The local apparent time (L.A.T.) at any instant is the hour angle of the sun at that instant measured westwards from its transit across the meridian of the place in question from 0 to 24 hours.' (p. 23)

**Applied Geography**

L.D.G. Applied geography involves the application of geographical methods of survey and analysis, including especially the use of maps and cartograms, towards the solution of both world and local problems, notably in underdeveloped areas and in such fields as town and country planning.

**Apposed glacier**

Swayne, 1956. A single glacier formed by the junction of two glaciers.

**Apron**

*Webster. Geol.* An outspread alluvial deposit, with generally low outward slope, deposited by a stream or streams in front of a glacier; an alluvial fan.

An outwash plain (*q.v.*).

**Aquiclude** (C. F. Tolman)

Tolman, C. F., 1937, *Ground Water*. A formation which, although porous and capable of absorbing water slowly, will not transmit it fast enough to furnish an appreciable supply for a well or spring. An aquiclude is porous but not pervious to water moved by gravity. It contains *fixed ground water*. (pp. 36–37)

**Aquifer, aquafer**

*O.E.D.* no reference; Aquiferous. Conveying or yielding water.

*Webster.* One who or that which bears water; specif. *Geol.*, a water-bearing bed or stratum of earth, gravel, or porous stone.

Mill, *Dict.* Water-bearing layer (Aquiferous Stratum). A rock stratum which absorbs water and is prevented from passing it on downwards by the presence of an impervious layer below it.

Wooldridge and Morgan, 1937. 'Rocks which will hold and permit the passage of water are called aquifers... most argillaceous rocks, while distinctly porous, are essentially impervious. The pore spaces are too small to allow free passage of water, which is firmly held by surface tension.' (p. 268)

Moore, 1949. Aquafer or Aquifer. A porous, tilted layer of rock, lying between impermeable layers, so that surface water may percolate through it and travel long distances.

Fay, 1920. A porous rock stratum that carries water (Lowe).

Rice, 1941. Rocks or formations that are water-bearing (Reiss and Watson).

Kuenen, 1955. 'A reservoir full of ground water.' (p. 196)

Tolman, C. F., 1937. *Ground Water*, Aquifer. A geologic formation or structure which transmits water in sufficient quantity to supply pumping wells or springs. (p. 557)

*Comment.* Should not be used in the sense of Moore or Kuenen; neither dip nor adjacent layers are essential. Spelling should be aquifer.

**Aquifuge**

Tolman, C. F., 1937, *Ground Water*. Aquifuge [1] is a rock which is impervious to water. Examples are granite, quartzite and completely cemented sedimentary rocks. An aquifuge is neither porous nor pervious; the enclosed aquifers are shear zones, fissure zones, slately cleavage, fissures and faults, and porous zones in lavas.

[1]. Footnote. The term was originally suggested by Dr. W. M. Davis. As defined by him it included both aquifuge and aquiclude. The term is used in this book in the more restricted sense given above and this modification was approved by him.

*Comment.* An American term which has not come into general use.

**Arable land**

*O.E.D.* Arable. a. Capable of being ploughed, fit for tillage, opposed to pasture or woodland.

1577...

1628. Coke On Litt. 536. If the tenant convert arable land into wood....

1866. Rogers Agric. and Prices, I, ii, 15. Half the arable estate, as a rule, lay in fallow.

b. Arable land

1576, Lambarde Peramb. Kent (1826), 3. Consisting indifferently of arable, pasture, meadow and woodland.

Committee, List 3. Arable land. Agricultural land which is tilled for crops, not necessarily each year, *e.g.* plough-land, gardens, vineyards, temporary fallow and rotation grass.

Stamp, 1948. 'It rightly suggests land fit for ploughing, or land able to be ploughed, but is commonly used to indicate land which is actually ploughed and cropped. There are pitfalls in the use of the word since in certain farm leases the indication that certain fields are arable refers not to the actual use made of them but to permitted use. It is not possible to construct a land utilisation map from such farm leases.' (p. 83)

In the instructions to the surveyors of the Land Utilisation Survey it was laid down that Arable or tilled land should include rotation grass and fallow together with market gardens, but not orchards or gardens.

Millar and Turk, 1943. 'A broad and variable interpretation of the term arable is possible, depending upon experience and viewpoint. Strictly speaking ... land which is suitable for plowing or cultivation. No mention is made ... of whether such land may be cultivated profitably at any set price level for agricultural products or whether the land will deteriorate rapidly under cultivation unless special tillage practices and cropping systems are used ... tendency to apply ... from a purely physical viewpoint and to consider any cleared land arable which is not too stony, of too rough topography, too badly in need of drainage to grow crops or in an area of very limited rainfall during the growing season.' (p. 422)

*Comment.* British agricultural statistics for summary purposes distinguish only (a) arable (b) permanent grass (c) rough grazing (d) woodland, but for other purposes orchards and gardens may be excluded. In certain cases 'tillage' (land actually under crops) may be distinguished from arable. Millar and Turk give an American view.

### Archaean, archean

*O.E.D.* Geol. Of or belonging to the earliest geological period.

Holmes, 1944. 'The term archaean refers to the oldest Pre-Cambrian crystalline rocks of a given region.' (p. 104)

*Comment.* A convenient omnibus term for all Pre-Cambrian crystalline rocks though various restrictions in meaning have been proposed and used, and usage tends to be different in Britain and America where some geologists restrict it to the oldest stratified rocks. Archean is the American spelling. The *O.E.D.* definition is incorrect.

### Archaeology

L.D.G. Archaeology may be used for ancient history generally, a systematic study of antiquities, but is commonly applied to the study of the remains left behind by prehistoric man, notably stone tools. Earliest are the primitive stone tools with some chipping believed, though not by all archaeologists, to be the result of man's work. These are eoliths, hence eolithic, meaning the dawn of the Stone age. Where first studied seriously in England and France flint is the commonest material for stone implements, especially in the Palaeolithic or Old Stone Age where the implements are chipped but again show an evolution of type from primitive Strépyan, though bold Chellean and Acheulian to side scrapers (Mousterian), made by Neanderthal man, of whom we have a few remains; he was a man with projecting jaws and beetling brows. The finest of all chipped implements are the Aurignacian, Solutrian and Magdelanian. Then comes the Neolithic or New Stone Age of polished stone implements, followed closely by the Bronze and Iron Ages. In Britain at least the greatest extent of the Ice Age was after the Acheulian stage (River Drift Man) when Mousterian man was driven to seek shelter in caves.

### Archaeomagnetism

Cook, R. M. and Belshé, J. C., 1958. Archaeomagnetism: A Preliminary Report on Britain, *Antiquity*, 32, 167–178. 'The existence of permanent magnetism in igneous rocks and fired clays has been known since the middle of the 19th century.' (p. 167)

See also Palaeomagnetism.

### Archibenthic

See Abyssal (quotation from Sverdrup).

### Archipelago (*pl.* archipelagos or archipelagoes) (1502)

*O.E.D.* Also: archpelago, archipelagus,—pelage.

From Italian: arcipelago; arci, chief, principal; pelago, deep, abyss, gulf, pool. It is a true Italian compound and not found in ancient or medieval Greek.

1. The Ægean Sea, between Greece and

Asia Minor (examples: 1502–1847). Hence (as this is studded with many isles):
2. Any sea, or sheet of water, in which there are numerous islands; and transf. a group of islands (examples include: 1600 Hakluyt, Voyages, 1810, 111. 'There broken lands and Islands being very many in number, do seeme to make there an Archipelagus.')
Mill, *Dict.* Originally an island-studded sea; now a group of islands.
*Comment.* The original meaning is obsolete.

### Arctic
*O.E.D.* From Greek, ἄρκτος, the Bear. 1. adj. Of or pertaining to the north pole or north polar regions: northern. 2. sb. The north pole or north polar regions; the arctic circle.
*Comment.* Arctic is really the adjective. Its use as a substantive (The Arctic) is colloquial though accepted. As an adjective it is applied to climate, vegetation, fauna and lands etc., as well as specifically to the Arctic Circle (*q.v.*) (A.A.M.).
See also Antarctic. The Arctic may be defined as that portion of the earth's surface, whether land or water, lying within the Arctic Circle, but is commonly used loosely to cover areas where climatic conditions resemble those of the Arctic. Various definitions have been given of an 'Arctic Climate'.

### Arctic circle
*O.E.D.* (see Antarctic). Arctic Circle of the heavens (obs.) Arctic Circle of the Earth: the fixed parallel of 66°32′ North.
*Comment.* The Arctic Circle is commonly accepted as 66°30′ North. Owing to the inclination of the earth's axis there is one day of the year (June 21) on which the sun does not sink below the horizon and one day of the year (December 22) on which it fails to rise above the horizon. The number of such days increases towards the North Pole until at the pole itself there is six months of daylight and six months of night. At any given date the conditions are the reverse of those in the Antarctic.

### Arctic Prairies see Tundra

### Arcuate delta
Swayne, 1956. A delta with a rounded outer margin.

### Areal differentiation
Haggett, 1965, quotes Hartshorne, 'Geography is concerned to provide accurate, orderly and rational description and interpretation of the variable character of the earth's surface' (Hartshorne, 1959, p. 21). In order to perform this considerable task, Hartshorne argues that geographers are primarily concerned with region construction, with what he terms 'the areal differentiation of the earth's surface.' (p. 10)
*Comment.* Problem as to which areas the geographer should study: the difficulty in defining regions is closely connected with this term. See Hartshorne, R., 1959, *Perspective on the Nature of Geography*, Chicago: Rand McNally, for a full discussion on the use of the term areal differentiation (pp. 12–21). The phrase 'areal differentiation' was introduced by Sauer in 1925 in paraphrasing Hethner's statement of his concept of geography. See Sauer, C. O., 1927, Recent Developments in Cultural Geography in Hayes, E. G. (ed.) *Recent Developments in the Social Sciences*, Philadelphia.

### Areg (Arabic, *pl.* of erg, *q.v.*)

### Areic (E. de Martonne, as aréique, 1928)
Wellington, 1955. 'The terms (i) areic, (ii) endoreic, and (iii) exoreic (employed by E. de Martonne and derived from the Greek verb *rhein*, to flow) will be used ... with the meanings: (i) without flow, (ii) with interior flow, and (iii) with exterior flow (*i.e.* to the sea).' (p. 339)
Applied to drainage basins: 'the Molopo-Nossob drainage, formerly a part of the Orange system, is now endoreic.' (p. 339)
Taylor, T. Griffith, 1940, *Australia*. 'Areic ... means a region of such small rainfall that no real stream valleys have developed and the rain mostly sinks into the ground or evaporates.' (p. 238)
*Comment.* For de Martonne's original use see Union géographique internationale, Publ. No. 3, 1928, p. 4 '... les régions aréiques qui correspond à peu près aux déserts.' The form aretic is also used, *e.g.* Fisher, W. B., *The Middle East*.

### Arena (E. J. Wayland, 1934)
A term introduced by Wayland, working in East Africa, to describe a shallow basin surrounded or almost surrounded by a rim marking the outcrop of resistant rocks round the central area where less resistant rocks have been eroded away.
McMaster, D. N., 1956, in *Natural Resources, Food and Population in Inter-Tropical*

*Africa*, I.G.U. 'Grant Bay is a shallow basin breached and invaded by Lake Victoria. The differential erosion of a central granite mass and of the folded quartzite rim largely account for the topography. Similar granite-floored basins are a recurrent feature of the landscape of Ankole where they have been termed "arenas".' (p. 79)

### Arenaceous
*S.O.E.D.* f. L. arenaceus, f. arena, sand.
Having the appearance of sand; sandy largely composed of sand or quartz grains.
Rice, 1943. A term applied to rocks containing sand and having a sandy texture. A general term applicable to several grades or types.
Himus, 1954. Arenaceous or psammitic rocks. Sedimentary rocks consisting of grains of sand which may be loose or cemented.
Contrast argillaceous.

### Arête (French: Classed by *O.E.D.* as an alien, but commonly naturalized)
*O.E.D.* A sharp ascending ridge or 'edge' of a mountain. The local name in French Switzerland, whence it has become a technical term with mountain climbers.
1862. Land Rev., 23 Aug., 164. The Weisshorn ... is formed of three great ridges, like the edges of a bayonet, culminating in a beautiful pyramidal point. Two of the arêtes are probably impracticable.
*Webster. Geog.* An acute and rugged crest of a mountain range or of a subsidiary ridge between two mountain gorges.
Mill, *Dict.* Arête—a sharp ridge.
Knox, 1904. A sharp rocky crest.
Wooldridge and Morgan, 1937. 'The arête is a sharp-edged ridge made of bare rock, with its summit either roughly horizontal or inclined.' (p. 375)
Cotton, 1945. 'The steep walls of adjacent cirques may meet in a ragged comb ridge or arête.' (p. 310)
Strahler, A. N. 1951, *Physical Geography*, New York: Wiley. 'Where two cirque walls intersect from opposite sides, a jagged, knife-like ridge, called an arête, results.' (p. 180)
Hobbs, W. H., 1910, *Geog. Jour.*, **35**. For the term arête or grat proposes to use the word 'comb-ridge'. (p. 161)

### Argillaceous
*S.O.E.D.* f. L. argillaceus, f. argilla, clay. Of the nature of clay; largely composed of clay; clayey.
Rice, 1943. An adjective meaning clayey and applied to rocks containing clay. A general term applicable to several grades or types.
Himus, 1954. Argillaceous or pelitic rocks. Sedimentary rocks characterized by an abundance of clay minerals, and a predominance of the mud grades.
North, F. J., 1930. *Limestones.* 'Argillaceous limestone ... contains an appreciable yet still relatively small proportion of clay in addition to calcite (according to Twenhofel, the clay content should be under 50%).' (p. 26)
Contrast arenaceous.

### Argol, argal, argul (China: Mongolia)
*Webster.* (Mongolian). Dry dung, as of camels, cattle, sheep, etc. often used as fuel, esp. in the steppe regions of Central Asia.
Cressey, G. B., 1955, *Land of the 500 Million*. New York: McGraw-Hill, 'cattle dung, known as argol, provides the nomad's only fuel' (p. 34).

### Arid, aridity
*O.E.D.* 1. Dry, without moisture, parched, withered: c. of the ground or climate. Hence, barren, bare.
1656. Blount, Glossogr., Arid, dry barren, withered, unfruitful.
*Webster.* Without moisture; parched; dry: barren; specif., having insufficient rainfall to support agriculture without irrigation.
Mill, *Dict.* . . . with deficient precipitation.
*Met. Gloss*, 1944. A climate in which the rainfall is insufficient to support vegetation is termed arid. (Quotes Köppen's formula ½R and explains construction.)
*Comment.* For various purposes it has been necessary to attempt to define 'arid' precisely—as for example when United Nations Educational, Scientific and Cultural Organization (UNESCO) set up an Advisory Committee on Arid Zone Research and initiated its Major Project on Scientific Research on Arid Lands. For this purpose UNESCO accepted the maps drawn by Peveril Meigs, 1952 (*Distribution of arid homoclimates*, Maps Nos. 392 and 393, United Nations) showing extremely arid, arid and semi-arid lands and which are reproduced as small scale maps as the end-papers of *The Future of Arid Lands*, edited by Gilbert F. White, Amer. Assoc. Adv. Sci., Washington 1956. Meigs used the criteria laid down by C. W.

Thornthwaite, An Approach toward a Rational Classification of Climate, *Geog. Rev.* 38, 1948, 55–94. However, the position is still as stated by H. L. Shantz in Gilbert White. *op. cit. sup.* p. 3. 'The arid zone has not been precisely defined.'

**Aridity, index of**
de Martonne, E; 1926, Aréisme et indice d'aridité, *C.R. Acad. Sci.*, **182**, 23, 1395–6; $I = P/(T+10)$ where $P$ = precipitation in mm, $T$ = Temperature in degrees Centigrade. This gives a sliding scale whereas Köppen's formula has rigid break points. See *Die Klimate der Erde* (1923).
Thornthwaite, C. W., 1948, *Geog. Rev.*, **38**, 55–94. 'De Martonne's index of aridity, $I = P/(T+10)$, is a slight refinement of Lang's (the Lang Rain Factor, $I = P/T$). Köppen's three formulae for delimiting the dry climates, $I = 8P/(5T+120)$, $I = 2P/(T+33)$, and $I = P/(T+7)$, presented in 1918, 1923, and 1928 respectively, are similar to those of Lang and De Martonne. All use annual values of precipitation and temperature given in the metric system.' (p. 74) Thornthwaite then goes on to give his own Moisture Index and Index of Aridity as $I_h = 100s/n$ and $I_a = 100d/n$ where $s$ is water surplus, $d$ is water deficiency and $n$ is water need. See discussion in Miller, 1953, pp. 84–5.

**Aries, first point of**—see Zodiac

**Arkose** (1839)
*Webster. Petrog.* A sandstone derived from the disintegration of granite or gneiss, and characterized by feldspar fragments.
*S.O.E.D.* A sandstone containing felspar and quartz.
Himus, 1954. A coarse-grained, highly felspathic sandstone or grit, formed by the rapid disintegration of granite or gneiss. It differs from felspathic grit and sandstone in that the high percentage of felspar has suffered little or no alteration by weathering.

**Armorican, armoricanoid**
From Armorica, that part of Gaul or France now called Bretagne or Brittany. The adjective (not given in *O.E.D.* in this sense) is applied in several distinct senses:
1. The Armorican revolution or orogeny or earth-building movements which took place at the end of Carboniferous and in early Permian times. It resulted in the formation of great mountain chains stretching from western Ireland through southern Britain (especially the South-western Peninsula) and from Brittany and Central France through Belgium (Ardennes) into Germany (especially the Harz Mountains, whence the alternative name Hercynian).
2. Armorican times—the period when the main folding took place: can be applied to volcanic and other igneous activity or products of the time, *e.g.* Armorican granites of Devon and Cornwall.
3. Armorican massifs or mountain blocks—those formed by the movements and now forming conspicuous elements in the structure of Europe, *e.g.* Brittany.
4. Armorican trend (lines) or strike. In Southern Britain the general trend of folding is west to east; in France it swings round to north-west to south-east, and then north-eastwards from Central France, and thence east-north-eastwards through the Southern Rhineland. (Himus, 1954); hence parallel to these directions.
See also Malvernian, Pennine, Lancastrian, Charnian.
Suess, E., *Das Antlitz der Erde* (*The Face of the Earth*; trans H.B.C. Sollas, Vol. II, 1906), who introduced the term used it in a restricted sense—'Armorican' and 'Variscan' were two main arcs of the 'Hercynian' orogeny, uniting in syntaxis in eastern France. Referring to the folded mountains of southern Ireland he says 'They are only part of a greater arcuate range... continued into Wales, England and Belgium. This great arc, striking from east to west ... we will designate the Armorican Arc.' (p. 83; p. 101 of German original of 1886)
Wills 1929, p. 76, refers to Prof. Chas. Lapworth's term 'Armoricanoid' for E-W folds... 'The termination *-oid*, shows that direction only is implied, whereas the suffix *-ian* should properly indicate that the fold belonged to the particular orogeny' (Lapworth *Rept. Br. Ass.* for 1913. (1914). p. 598).
Compare use of Charnoid (NW-SE), Malvernoid (N-S), Caledonoid (NE-SW).
*Comment.* Because of the evolution in the meaning of Armorican and the vast literature involved, an attempt has been made above to summarize present usage.

**Arris, arridge**
A local (Lake District) term for an arête (*q.v.*)

**Arroyo** (Spanish)
*O.E.D.* A rivulet or stream; hence the bed

of a stream, a gulley (in U.S.).
*Dict. Am.* A brook or creek, also a channel, gully, dry wash, stream bed or valley. (Western U.S.) (citations back to 1806).

Mill, *Dict.* A streamway, ordinarily dry, in which water occurs only immediately after torrential rains (Spanish America).

*Comment.* In contrast to *rio*, a stream with permanent water, an *arroyo* is periodically dry. *Cf.* nala, wadi, etc.

### Artesian, basin, well (Lyell, 1830)

*O.E.D.* Artesian: Of or pertaining to Artois, or resembling the wells made there last century, in which a perpendicular boring into a synclinal fold or basin of the strata produces a constant supply of water rising spontaneously to the surface of the ground 1830. Lyell, Princ. Geol.
1860. Tristam, Gt. Sahara, XVII, 287. . . .

Page, 1865. 'Wells sunk by boring perpendicularly through the solid strata, and in which the subterranean waters rise to the surface or nearly so. . . .'

Mill, *Dict.* A channel of outlet for confined subterranean waters; it is generally constructed by boring through impervious rocks in order to reach down into pervious layers, into which water has filtered from the sides.

Tolman, C. F., *Ground Water*, 1937. 'A well tapping a confined or artesian aquifer in which the static water level stands above the water table. The term is sometimes used to include all wells tapping confined water, in which case those wells with water level above the water table are said to have *positive artesian head* (pressure) and those with water level below the water table *negative artesian pressure*' (p. 557).

Kuenen, 1955. 'The ideal case is the aquifer between two watertight beds, the whole curved in the shape of a basin . . . sometimes artesian water derives pressure from tapering of the porous layer or from later cessation of the porosity.' (p. 198)

*Comment.* Artesian should strictly be used in the sense given in *O.E.D.* from Lyell, 1830, but is frequently applied to any deep well in which water rises under pressure, not necessarily to the surface (*cf.* Page, 1865). Even Lyell used the word later in this loose sense when he refers in the 1875 edition of his *Principles of Geology* to 'artesian borings at Calcutta'. Some writers accordingly distinguish 'sub-artesian' from artesian or flowing wells. Mill's definition is inadequate. Meinzer has introduced the term 'piestic water' for water occurring under artesian conditions with division into hyperpiestic, mesopiestic and hypopiestic.

### Artifact, artefact

Webster. Archaeol. A product of human workmanship; esp. one of the simpler products of primitive art as distinguished from a natural object.

### Ås, (Swedish *pl.* åsar)

The Swedish term ås is equivalent to esker.

Geikie, J., 1872, *The Great Ice Age*, 'Long winding ridges, which are known in Sweden as åsar (Ås singular; similar ridges in Norway are termed Raer). They generally rise abruptly to a height that may vary from 50 to 100 feet.' (p. 385)

*Comment.* The Swedish word has been superseded by esker in most English works.

### Ascension

*O.E.D. Astr.* The rising of a celestial body. *Right ascension* of the sun or a star; the degree of the equinoctial or celestial equator reckoned from the first point in Aries which rises with it in a right sphere, or which comes with it to the meridian.

### Ash cone, cinder cone

Webster. Geol. A conical height formed by the accumulation of volcanic ash around a vent.

Moore, 1949. The conical hill or mountain built up with the ejected material from a volcano.

### Aspect

*O.E.D.* II 2. A looking, facing, or fronting, in a given direction; exposure 1667; the side or surface which fronts in any direction 1849.

Carpenter, 1938. 'Aspect. 1. The seasonal impress on a community, *e.g.* the spring aspect. 2. The direction toward which a slope faces, as designated by the points of the compass' (p. 27).

Tansley, A. G., 1946, *Introduction to Plant Ecology*. 'Steepness of slope . . . increases the effect of aspect on exposure' (p. 168).

Miller, A. A., 1953, *Climatology*. 'South aspect' is the main natural asset of many popular resorts such as Nice and Torquay' (p. 50).

*Comment.* See also under Exposure, Adret, Ubac, etc. Some writers (*cf.* Peattie, Tansley and Miller) regarded aspect and exposure as synonymous. Latest usage (*cf.* Brooks) seems to give aspect the meaning 'direction of slope', and exposure 'degree of openness to wind, sun and weather'. Compare forestry, in which exposure is taken to mean principally exposure to wind, especially drying winds (A.F.M.).

**Assart**

Webster. O. Eng. Law. a. Act of grubbing up trees or bushes, as in converting forest land into arable land. b. A piece of land so cleared.

Swayne, 1956. A clearing made by the removal of trees and bushes so as to provide cultivable land (used of mediaeval England).

**Association (ecological)**

Webster. Ecology. Specif. an assemblage of plants having ecologically similar growth requirements and including one or more dominant species, from which it derives a definite aspect; as, a deciduous forest *association*, heath *association*, etc.—often explicitly called *plant association* and, formerly, *plant formation*.

Mill, *Dict*. Association (plant or animal). An assemblage of plants or animals typically found together and living in close inter-dependence and in correlation with certain physical conditions.

Tansley, 1939. 'A formation is composed of associations, each with different dominants and at least some different subordinate species. The difference of formation-types is a difference of dominant species—a "floristic" as distinct from a vegetational difference.' (See also p. 232, for confusions.) Also *Intro. to Plant Ecology*, e.g. European Deciduous Summer Forest formation; oak-beech wood association; oak consociation. (p. 34)

Flahault and Schröter 'report to the Brussels Congress, 1910' quoted in Tansley, *J. of Ecol.*, **8**, 1921. 'An association is a plant community of definite floristic composition, presenting a uniform physiognomy, and growing in uniform habitat-conditions. The association is the fundamental unit of synecology.' (p. 127)

Pearse, A. S., 1939, *Animal Ecology*, 2nd ed., New York: McGraw-Hill. ch. VII. In interest of 'readily understandable' names but based on terminology of Clements, 1936, has drawn up a Synopsis of Communities:

> Superrealm *e.g.* Aquatic
> Realm—Marine

Province—Littoral
Formation—Hard Beaches
Association—Rockwalls
Fasciation—Zone above high tide

Ginsberg, M., 1930, 'Association' in *Encyclopedia of Social Sciences*. '.... a group of individuals united for a specific purpose or purposes and held together by recognized or sanctioned modes of procedure and behaviour... to be distinguished from the institution....'

Carpenter, 1938. Provides three pages of definitions from various writers, under the heads:
1. Sense of climax. 2. Sense including climax and developmental stages. 3. Sense of a community of no stated developmental rank (a) floristic criteria (b) habitat criteria. 4. Phytosociological sense, somewhat similar to 1. 5. Aggregational sense, 6. Sense of structure. Also: Adams, 1915. Bull. Ill St. Lab. Nat. Hist. 11:15. The animals found living together in a given combination of environmental conditions form an animal association or a social community, and the study of the responses of such a community is the province of associated ecology.

*Comment.* Association is an example of a very common word with many different meanings which has been given highly specialized meanings by plant and animal ecologists, whence its main use in geographical literature. In human geography it will also be found with the meaning defined above by Ginsberg.

**Asthenosphere**—see Aesthenosphere

**Atavism, atavistic**

From the Latin *atavus* an ancestor; resemblance to remote ancestors or grandparents rather than to parents; tendency to reproduce ancestral features in animals and plants; used also of other types of reversion as in an atavistic reversion to a primitive peasant community.

**Atlantic climate**

L.D.G. The term is used in a special sense for a sudden development of a mild moist phase following the dry cold boreal phase during the retreat of the Great Ice Age.

**Atlantic Suite**

Some geologists have distinguished among world regions of abundant igneous rocks 'petrographical provinces'. Petrologists have used the term Atlantic suite to denote a series of igneous rocks rich in alkalis—

potassium and sodium contrasted with the Pacific suite of igneous rocks rich in calcium. The Atlantic suite is associated with the Atlantic type of coastline and with faulting of rock masses, the Pacific suite with Pacific coastlines and folding. A third suite, the Spilitic, is sometimes distinguished associated with submarine volcanic action and slow submergence. See Stamp, 1923, 9–10, summarizing A. Harker, Pres. Add. to Section C British Assoc. 1911 and later works.

**Atlantic type coastline**

Mill, *Dict*. A diversified ocean boundary, due to the cutting across of ancient mountain lines, the foundering of plateaux, and the fracturing of table-lands.

Comment. Applied to coastlines the contrast is between Atlantic and Pacific. A map of the Pacific Ocean shows the surrounding rim of mountain chains in general parallel to the main coastlines. On the other hand mountain chains such as the Atlas appear to be truncated by the ocean shore and this is repeated in many areas around the Atlantic, hence Atlantic type coastlines.

**Atlas, atlases**

*O.E.D.* From the Latin name of the Greek god.
3. A collection of maps in a volume (this use of the word is said to be derived from a representation of Atlas supporting the heavens placed as a frontispiece to early works of this kind, and to have been first used by Mercator in the 16th c.) *e.g.* 1636 (title) Atlas; or a Geographic Description of the World, by Gerard Mercator and John Hondt.
4. A similar volume containing illustrative plates, large engravings, etc., or the conspectus of any subject arranged in tabular form; *e.g.* 'an atlas of anatomical plates', 'an ethnographical atlas'.
5. A large square folio resembling a volume of maps; also called atlas-folio.

Comment. Just as the god Atlas was supposed to support the heavens so were the mountains of North Africa, hence the name Atlas for those mountains and the adjective Atlantic now applied to the neighbouring ocean. The term 'atlas' first appeared on the general title-page of Mercator's 'Atlas', 1595 (G.R.C.).

A *national atlas* depicts different aspects of a single country; a *regional atlas* those of a well-defined part of a country or continent.

**Atmosphere**

*O.E.D.* The mass of aeriform fluid surrounding the earth; the whole body of terrestrial air. (Originated as the exhalation of planets and the moon, then applied to earth.)
1638. Wilkins New World.
6. A pressure of 15 lbs. on the square inch, which is that exerted by the atmosphere on the earth's surface.
1830. Lyell, *Princ. Geol.* I, 396. Congealed under the pressure of many hundred, or many thousand atmospheres.

Mill, *Dict*. The mass of gases, water-vapour, and dust which forms the outer envelope of the globe.

*Met. Gloss*, 1944. A name given to the air which surrounds us, and is carried along with the earth. The pressure of the atmosphere at sea level is equal to a pressure of about $14\frac{1}{2}$ lbs. per sq. in.

Comment. The atmosphere is now commonly divided into concentric layers: the lower atmosphere is the *troposphere*, separated by the tropopause from the *stratosphere* (*q.v.*), above which is the *mesosphere* (*q.v.*) then the *ionosphere* (*q.v.*).

**Atmospheric Weathering**—see Weathering

**Atoll, atollon**

*O.E.D.* (adaption of the native name Atollon, Atoll, applied to the Maldive Islands, which are typical examples of this structure; prob. Malayalam adal, 'closing, uniting').
A coral island consisting of a ring-shaped reef enclosing a lagoon. Darwin's theory now generally accepted is that the lagoon occupies the place of a submerged island.
1625. Purchas Pilgrims . . . Atollon
1832. Lyell, Princ. of Geol. . . . atoll
1859. Darwin, Orig. Spec.

Mill, *Dict*. A surface reef or series of such reefs built up of coral limestone with or without islands, surrounding a lagoon.

Knox, 1904. '(Maldive anglicised), one or any greater number of coral islands of little height above the sea, situated on a strip or ring of coral surrounding a central lagoon.'

Atollon (Fr.), an atoll
Atollon (Eng.), a small atoll on the margin of a larger one.

Kuenen, 1955. 'A circular reef round a deep

lagoon.' (p. 310) He gives a section of an atoll of the Maratua type on p. 311, stressing the wearing down of the rim and the filling of the lagoon with sediment which he connects with fluctuations of sea level during the Ice Age.

Comment. Charles Darwin's *Coral Reefs*, 1842, is the classic work but many other theories have since been advanced. See also J. Johnstone, 1923, *Introduction to Oceanography*, p. 81.

**Attrition**

*O.E.D.* 1. The act or process of rubbing one thing against another.

Rice, 1943. Act of rubbing together; friction; act of wearing, or state of being worn; abrasion.

Holmes, 1944. 'Attrition is the wear and tear suffered by the transported materials themselves, whereby they are broken down, smoothed and rounded. The smaller fragments and the finer particles liberated as by-products are then more easily carried away.' (p. 151)

Thornbury, 1954. 'Attrition is the wear and tear that rock particles in transit undergo through mutual, rubbing, grinding.' (p. 48)

Comment. As defined by Holmes and Thornbury, attrition is not a synonym for abrasion.

**Aufeis** (German)

Muller, S. W., 1947, *Permafrost*. 'As the winter freezing of a river sets in, the live channel beneath the ice gradually becomes constricted and eventually becomes too small for the entire volume of flowing water.... The impeded water creates hydrostatic pressure, which forces the water to permeate the porous alluvium both sides of the stream. The gradually increasing hydrostatic pressure ultimately causes the water to force its way to the surface either through the river ice or through the ground.... Water issuing from such a break spreads out and freezes in successive sheets of ice, which may cover an area of several square kilometers and commonly attain a total thickness of 1 to 4 meters. These ice incrustations or "icefields" are known as *icings* or *aufeis*, in Russian "*naledee*" (singular "*naled*"). The icing that forms over the river ice is generally referred to as *river icing*... icing formed on the surface of the ground is known as *ground icing*'. (p. 24)

Comment. Term in use in periglacial studies of German origin. 'Icing' as proposed by Muller is preferable. (G.T.W.).

**Aureole**—see Metamorphic aureole

**Aurora**

*O.E.D.* 5. A luminous atmospheric phenomenon, now considered to be of electrical character occurring in the vicinity of, or radiating from, the earth's northern or southern magnetic pole, and visible from time to time by night over more or less of the adjoining hemisphere, or even of the earth's surface generally: popularly called the Northern (or Southern) Lights.

Aurora borealis (or septentrionalis) (Northern); aurora australis (Southern).

**Aūs** (Indo-Pakistan: Bengali)

Autumn rice or rice-crop.

Johnson, B. L. C., 1958, Crop-Association Regions in East Pakistan, *Geography*, 43, 86–103. 'Aūs paddy is grown on higher ground [than āmān] where soils are generally coarser... for at least half its growing period it depends on pre-monsoon rainfall (March–May).' (p. 90) See also Ahmad, Nafis, *An Economic Geography of East Pakistan*, O.U.P., 1958, 123.

**Ausland** (German)

Stilgenbauer, F. A., Detroit's Expansion into the Umland..., *Papers of the Michigan Academy of Sciences*, 33, 1947. 'That part of the Umland (*q.v.*) of a city which lies in a foreign country. Part of the Umland of Detroit is in Canada.' (p. 215)

Comment. An example of a simple and common foreign word being given a specialized meaning when used in English.

**Auster**

*O.E.D.* The south wind; *hence* the south.

**Austral**

*O.E.D.* Belonging to the south, southern.

**Australite**—see Tektite

**Australoid** (T. Griffith Taylor)

*Webster*. Of the ethnic type of the Australian blacks.

'Australian is used both for Europeans and for aborigines, so I coined Australoid for the latter.' (G.T. in MS.)

**Autan, vent d'**

Mill, *Dict*. 'A strong hot dry wind blowing out from South France into the centres of low pressure which come from the ocean into the Bay of Biscay.'

**Autarchy**

*O.E.D.* Autarchy (1) 1691. 1. Absolute

sovereignty, despotism. 2. Self-government.
*Webster.* a. Absolute sovereignty; absolute or autocratic rule. b. Economic self-sufficiency — by confusion of Greek *autarkeia* with *autarchia.*
See Autarky.

### Autarky
*O.E.D.* Autarky (2) (better -arky, or -arcie), 1643. Self-sufficiency.
*Webster* (in addenda). National economic self-sufficiency.
Maull, Otto, 1925, *Politische Geographie*, Berlin: Borntraeger. See section entitled Die Bedeutung der Autarkie für die Staaten. (p. 533)
Dickinson, R. E., 1943, *The German Lebensraum*, Harmondsworth: Penguin. 'The second five-year plan aimed at Autarkie or national self-sufficiency, that is, the greatest possible measure of self-sufficiency, so as to enable the country to withstand a wartime blockade.' (p. 141)
Committee, List 4. 'Economic self-sufficiency as the object of a national policy; to be distinguished from autarchy ( = self-government).'
*Comment.* Although the distinction between autarky and autarchy is justified by the quite different concepts in the original αρκετν and αρχειν, in practice as the *O.E.D.* points out they are used commonly as alternative spellings of the same word.

### Autecology
Carpenter, 1938. The relation of individual plants to their habitats. The ecology of the individual organism. In contrast to synecology. See Ecology.

### Authigenous, authigenic
*Webster. Petrog.* Formed in place—said of mineral particles of rocks formed by crystallization in the place they now occupy. Opposed to *allothogenic, allothigenous.*
Himus, 1954. Formed *in situ*; applied to the constituents of a rock that came into existence with or after the formation of the rock of which they form a part. For example, the primary and secondary constituents of an igneous rock, or the cement of a sedimentary rock.

### Autobahn (German; *pl.* Autobahnen)
*Webster.* In Germany, a road with double traffic lanes in each direction separated by a park strip and with no restriction upon speed [and generally without surface cross roads].
Motor road or motorway, *i.e.* a road specially constructed for and restricted to motor traffic. Until the passing of legislation giving power to restrict new roads to be used only by certain types of traffic (there can be no such restriction on the 'King's Highway') this German word, usually printed with a small a, was widely used in England. Since the opening of the first Motorway, M1, officially so-called, between London and the Midlands in 1959 the German word is being dropped. An important feature is the principle of controlled access. Experience has led in Britain to the imposition of a speed limit.
See also Highway.

### Autochthon, autochthone, autochthonous (1805)
*O.E.D.* = Autochthonic (fr. Greek 'sprung from the land itself') Native to the soil, aboriginal, indigenous.
1805. Taylor, W. in *Ann. Rev.* III, 309 'If the English have this great prediliction for autochthonous bread and butter.'
Mill, *Dict.* Autochthones—Aborigines.
*Encyclopaedia Britannica*, 14th ed., 1929. 'Autochthones—sons of the soil (1646), the original inhabitants of a country as opposed to settlers.... The idea that they were autochthones was a source of great pride to the Athenians,...'
Churchward, A., 1921, *Origin and Evolution of the Human Race*, London: Allen and Unwin. '... the theory "of the Autochthonous", which implies that man has been generated at several points simultaneously....' (p. 300; also p. 372)
Holmes, 1928. 'Gümbel, 1888. A term applied to rocks such as rocksalt and stalactite, denoting that they and their constituents have been formed *in situ.* cf. Allochthonous.'
Jackson, B. D., 1928, *A Glossary of Botanic Terms*, 4th ed. London: Duckworth, Autochthon: an aboriginal form; a native plant, not an introduction; Autochthonous theory, the theory that each species originated where now found (L. H. Bailey). (p. 40)
Wills, L. J., 1929. *The Physiographic Evolution of Britain*, London: Arnold. Applied to rocks either as nappes or in their natural positions that originally formed part of

the belt of country in which they are now found. Others are exotic or parautochthonous. (p. 185)

Russell and Russell, 1950. 'S. Winogradsky ... suggested the name autochthonous ... for this common microflora of short and coccoid rods ... feeding on the soil humus.' (p. 151)

*Comment.* It is interesting to note how a word denoting originally a native of the soil has been applied successively to a special theory of the origin of mankind, to plants by botanists and to rocks by geologists (in several different senses) and even to soil organisms by pedologists.

**Autoclastic**
Himus, 1954. Term applied to rocks such as Crush-breccias (fault-breccias) that have been brecciated mechanically in place.

**Autogenetic drainage and topography**
*O.E.D. Suppl.* 2. Phys. Geogr. of or pertaining to a system of drainage developed by erosion of the stream itself.

*Webster. Geol.* Drainage by streams whose courses have been determined by the conditions of the land surface over which they flow independently of conditions upon any older, higher land surface. Cf. *epigenetic drainage.*

Cotton, 1941. Falls are called autogenetic when the result of headward erosion or 'plunge-pool back-scour'. (pp. 31–32)

Fay, 1920. Autogenetic drainage. Drainage due to erosion caused by the waters of the constituent streams (Standard). Autogenetic topography. Conformation of land due to the physical action of rain and streams (Standard).

No reference in Lyell, 1830; Powell, 1875; Gilbert, 1877; Geikie, J., 1898; Davis, 1909; Salisbury, 1907; Wooldridge and Morgan, 1937; Holmes, 1944; Lobeck, 1939; Moore, 1949.

**Autogenic change**
Self-induced change produced by natural processes.

**Auto-geosyncline** (Marshall Kay, 1944)
See comment and reference under Geosyncline; also Umbgrove, 1947.

**Autonomy, autonomous**
The right of self-government; self-government; self-governing.

**Autopiracy, auto-piracy**
Simplification of a stream by capture of an upper section by a lower. A cut-off meander is strictly speaking auto-piracy. (G.T.W.)

**Autostrada** (Italian)
Autobahn; motorway.

**Autumn**
*O.E.D.* 1. The third season of the year, or that between summer and winter, reckoned astronomically from the descending equinox to the winter solstice; *i.e.* in the northern hemisphere from September 21 to December 21. Popularly, it comprises, in Great Britain, August, September and October (Johnson); in North America, September, October and November (Webster); in France 'from the end of August to the first fortnight of November' (Littré); in the southern hemisphere it corresponds in time to the northern spring.

*Comment.* British farmers and meteorologists use the months of September, October and November. See also Spring, summer, winter.

**Available relief**
Glock, W. S., 1932, *J. of Geol.*, **40**, 74–83. 'Available relief is defined as the vertical distance from an original, fairly flat upland down to the initial grade of the streams.' (p. 74) (Also defines Low, Medium and High Available Relief, Critical Relief, Drainage Relief.)

Johnson, D., 1933, *J. of Geol.*, **41**, 293–305. 'Available relief is the vertical distance from the former position of an upland surface down to the position of adjacent graded streams.' (p. 295)

Cotton, 1941. 'Without accepting, or attempting to frame, such a definition as will make possible a precise quantitative statement of the available relief of any district, one may take it to be at any particular place and time the height of the surface undergoing dissection above the local base-level controlling dissecting streams.' (p. 60)

See also Monkhouse and Wilkinson, pp. 77–78; G-H. Smith, 1935, *Geog. Rev.*, **25**. The Relative Relief of Ohio; A. A. Miller, *Inst. Brit. Geog.* Publ. 14, 1949, p. 3 (with suggested modifications). The term 'local relief' is another synonym.

**Avalanche** (French)
*O.E.D.* 1. A large mass of snow, mixed with earth and ice, loosened from a mountain side, and descending swiftly into the valley below (examples, 1765–1870).

Mill, *Dict.* A mass of snow, nevé or ice gliding down or falling from a mountain side. When the snow glides over a rock or earth surface it is sometimes distinguished as a ground avalanche, while the term dust or wind avalanche is applied to one that glides or is driven by the wind over a crust of old snow more or less dispersed into a powdery shower.

*U.S. Ice Terms*, 1952. 'Masses of snow detached from great heights in the mountains and acquiring enormous bulk by fresh accumulations as they descend. When they fall into the valleys below, they often cause great destruction.'

Fay, 1920. 2. Falling masses of rock and earth, sometimes called avalanches, are better designated landslides (Standard).

Kuenen, 1955, describes a number of types of avalanche and discusses causation.

Bagnold, 1941. '... sand avalanches slipping down an advancing slip face.' (p. 240)

*Ice Gloss.*, 1958. Mass of *snow* which becomes detached and slides down a mountain side, often acquiring great bulk by fresh accumulations as it descends.

Comment. W. M. Davis, 1898, *Physical Geography* derives this French dialect word from à val—to the valley. *Cf.* also *avaler* = to let down. Various classifications have been proposed for different types of avalanche; see, *inter alia*, J. E. Church in Meinzer, 1942, 134–5; also Charlesworth, 19–22 (G.T.W.) The word originally referred to snow but can be used, especially if qualified, to refer to other moving materials such as rocks and sand.

**Avalanche cone**
Mill, *Dict.* The mass deposited where an avalanche falls, consisting not merely of snow, nevé, or ice, but including everything torn away and carried along by the avalanche (Penck).

**Aven (French)**
Mill, *Dict.* A swallow hole (*q.v.*) or dolina. In the Causses region, the well pierced at the bottom of a sotch or dolina (*q.v.*) (France).

Wooldridge and Morgan, 1937. 'Vertical or highly inclined shafts, leading from swallow-holes, or direct from the surface to underground caves, are... "avens" in France, while many other terms are used.' (p. 289)

de Martonne, E., 1927, *A Shorter Physical Geography*, Trans. E. D. Laborde. London: Christophers. 'At the bottom of a sotch... sometimes, instead of a basin, there is a gaping hole like a well, usually surrounded by trees and bushes.' (p. 168)

Ormsby, H., 1950, *France*, 2nd ed., London: Methuen. Of the Causses region. 'The smaller streams, however, entrench themselves in deep narrow trenches, known as avens.' (p. 64) (distinct from sotches).

Harrap's *Fr.-Eng. Dict.* 'Geol: Swallow (-hole); swallet.'

Davies, G. M., 1932, *A French-English Vocabulary in Geology and Physical Geography*, London: Murby. 'Sink, Swallet.'

Martel, E. A., 18, *Geog. Jour.*, **10**, 500–511. '... a real pit (aven), that is to say a narrow natural well.' (p. 504)

Joanne, P., Dictionnaire Géographique et Administrative de la France. Paris-Hachette, 1890. 1. Equivalent to swallow-holes, etc. 2. A Breton name for river.

Warwick, G. T., 1953, in British Caving. 'Seepage is facilitated by the development of a large cave system, and sometimes this is reflected in the enlargement of roof joints which act as feeders to the main cave. These enlargements taper upwards and are known as *avens* and are presumed to result from the accelerated flow of water near its exit causing a greater amount of solution to occur than higher up the joint. ... These features may... lead up into surface depressions, but not all of the latter lead down into avens.... Sometimes the term "aven" is given to roof features which may later prove to be vertical tributaries of the cave system, *e.g.* Bar Pot in Gaping Gill.' (p. 45)

Comment. A local term in the south of France for a deep pothole in limestone regions, which has been extended to cover similar features elsewhere and to associated cave systems below. In England the term refers to enlarged vertical joints in the roofs of cave passages which normally narrow off upwards. This English usage was probably fostered by the visit of E. A. Martel, who named such features in Peak Cavern 'avens' (see E. A. Martel, *Ireland et les Cavernes Anglaises*, 1897 and *The Caverns of Derbyshire*, 1914—the Derbyshire chapters of the above book, translated by F. A. Winder and S. C. Phillips). (G.T.W.)

**Avulsion**
Webster. *Law.* The sudden removal of land from the estate of one man to that of another by an inundation or a current, or

by a sudden change in the course of a river by which a part of the estate of one man is cut off and joined to the estate of another. The property in the part thus separated continues in the original owner. 3. *Law.* Hence, by analogy, in New York, a sudden change in the shore of the sea whereby boundaries are affected.

Swayne, 1956. (*a*) The action of forming a meander-lake or the lake itself; (*b*) The separation of part of an estate by the action of flood, river, or sea-water (Legal; United States).

**Aÿacut** (India: Urdu)

Irrigable area (southern India)

Spate, 1954. 'Paddy is practically confined to ayacuts (areas supplied from anicuts or weirs) in the major valley-bottoms.' (p. 714)

**'Azbeh** (Arabic: Israel)

A summer residence for those in charge of the herds and flocks sent down to graze on the plain (D.H.K.A.).

**Azimuth**

*O.E.D.* 1. An arc of the heavens extending from the zenith to the horizon which it cuts at right angles; the quadrant of a great circle of the sphere passing through the zenith and nadir, called an *azimuth circle.* 2. The angular distance of any such circle from a given limit, *e.g.* a meridian.

Admiralty *Navigation Manual.* 'The azimuth of a heavenly body is the angle at the zenith between the observer's meridian and the vertical circle through the heavenly body... All great circles passing the observer's zenith are necessarily perpendicular to the celestial horizon and are known as vertical circles'.

*Comment.* Also used as the equivalent of bearing (*q.v.*). An azimuthal map-projection is one in which all bearings are laid off correctly from the central point of the map. (E.G.R.T.)

**Azoic**

Literally without life; hence applied to pre-Cambrian strata with no recognizable organic remains.

**Azonal soil**

Robinson, 1949. '... including skeletal and alluvial soils.' (p. 440)

Millar and Turk, 1943. 'Any group of soils without well-developed profile characteristics, owing to their youth or conditions of parent material or relief that prevent the development of normal soil-profile characteristics.' (p. 430)

Russell and Russell, 1950. '... have not been under the influence of soil-forming processes long enough for their parent material to be appreciably affected, and alluvial soils in the temperate regions are typical examples of this group.' (p. 512)

# B

**Bach** (New Zealand)
Beach or lakeside cottage used at weekends and on holiday occasions. Similarly 'baching,' or roughing it and living like a bachelor. Bach is a North Island term and is replaced by crib in the South Island.

**Backfolding**
Jefferies, H., 1950, *Earthquakes and Mountains*, 2nd Edn. Cambridge: C.U.P. 'Back-folding is shown by folds with their fronts in the opposite direction to the main nappes.... The amount of back-folding in comparison with the main nappes would depend on the degree of asymmetry in the primary elevation.' (p. 156)

**Backing**
*Met. Gloss.* 1944. A wind is said to 'back' when it is changing in the opposite direction to the hands of a watch. Thus the wind backs at a place north of the centre of a depression travelling eastwards in the northern hemisphere. The same definition of backing applies to the southern hemisphere but as wind direction in relation to systems of closed isobars is there reversed, a backing wind in the southern hemisphere is equivalent to a veering wind in the northern hemisphere, and conversely.

**Backland**
Mill, *Dict.* A region behind mountain ranges whether occupied by sea or by land, *e.g.* The Atlantic ocean is described as a backland of the great fold mountains. *cf.* Foreland.

Du Toit, A. L., 1937, *Our Wandering Continents*, Edinburgh: Oliver and Boyd. [Of the four structural elements in mountain folding] '...the more variable and unstable backland marking the side from which the pressures have come.'

*Comment.* In the sense used by Mill the word has become obsolete and should not be confused with back country or hinterland. Du Toit's use of the word is not general. There is no reference in Lyell, 1830; Fay, 1920; Knox, 1904; Rice, 1941; Moore, 1949. Baulig, 1956, gives as synonymous with back-swamp.

**Backshore**
King, C. A. M., 1959, *Beaches and Coasts*, London: Arnold. 'The term "beach" will be used to include the backshore [above high water], foreshore [between the tide marks] and offshore zones [from the uppermost point always covered by water to a depth at which substantial movement of beach material ceases].' (p. 48)

Wooldridge and Morgan, 1937 (following D. W. Johnson). 'We may thus distinguish the "foreshore", the region extending between the ordinary tidal limits, from the "backshore", lying immediately at the cliff-foot.' (p. 322)

*Back-shore terraces. Ibid.*, 'terraces on the higher part of the beach.' (p. 322)

**Back slope**
Cotton, 1941. 'A tilted block is limited by a fault scarp on the uplifted side only, and from the crest-line of this scarp an inclined *back-slope* descends' (p. 241).

Swayne, 1956. The gentle inclination at right angles to an escarpment; the side of a ridge with the gentler slope.

*Comment.* The majority of writers use the term dip-slope, but as the back slope rarely coincides exactly with the dip of the rocks the term back slope is strictly more accurate. As defined by Cotton, back slope may be unrelated to the dip of the rocks.

**Backswamp deposits, back-swamp**
Thornbury, 1954. 'Backswamp deposits are those that were laid down in the flood basins back of natural levees. They consist of extensive layers of silt and clay.' (p. 170)

Fisk, H. N., 1947. *Fine Grained Alluvial Deposits and their effects on Mississippi River Activity*. This work includes a very full treatment of backswamp deposits which 'constitute the long-continued accumulation of floodwater sediments in flood basins marginal to meander belt ridges (p. 43)... deposited as thin layers... because of their coarse, crumbly structure when dry locally referred to as "buckshot" clays (p. 49)... often oxidized near the surface and have a typically reddish-brown and gray color.' (p. 57)

**Backwash, back-wash**
*O.E.D.* The motion of a receding wave; a backward current.

Kuenen, 1955. After a wave has broken on a shore 'the water particles have expended their energy. Some now trickle through the beach sand, thus eventually returning to the sea, while the remainder run down the slope back to the sea, not as a wave movement, but as "back-wash".' (p. 58) *Comment. Cf.* swash (forward motion) and plunge (downward)—see Johnson, 1919. (A.A.M.)

**Backwasting**

Thornbury, 1954. Recession of an ice front. Ablation results in both 'backwasting' and 'downwasting'. (p. 363)

**Backwater**

*S.O.E.D.* 2. Water dammed back in its course, or that has overflowed in time of flood, 1629.

3. Water dammed back for any purpose. 1792.

4. A piece of water without current, parallel to a river, and fed from it at the lower end by a back flow. 1803.

5. A creek or arm of the sea parallel to the coast, separated by a narrow strip of land from the sea, and communicating with it by barred outlets. 1867.

6. A backward current of water. 1830.

*Comment.* Used generally in the sense of 4 above, but loosely for any tranquil piece of water connected with but little affected by a main stream: hence figuratively.

**Backwoods**

*Dict. Am.* The woods lying back from the more settled areas; the uncleared forest to the west of the earlier settlements (citations back to 1709).

*Webster.* The forests or partly cleared grounds on the frontiers or removed from the centers of population, esp. in the United States and Canada.

*S.O.E.D.* Wild, uncleared forest land; *e.g.* that of North America.

Swayne, 1956. A region of wild and partially cleared forest, particularly in the Appalachian Mountains; any sparsely settled country of this type far from centres of population.

**Badlands, bad lands**

*Dict. Am.* 1. pl. (cap.) A region in South Dakota and Nebraska where erosion has resulted in varied and fantastic land masses. 1882 *Cent. Mag.* xxiv, 510. The term Bad Lands does not apply to the quality of the soil. The Indian name was accurately rendered by the early French voyageurs as *Mauvaises Terres pour traverser*—bad lands to cross.

*Webster.* Regions characterized by the intricate and sharply erosional sculpture of generally soft rocks. These are usually nearly horizontal sedimentary beds, but badland topography may develop in decomposed granite, loess, or other soft material. Vegetation is scanty or absent. The hills are steep, furrowed, and often of fantastic form. The drainage is intricately labyrinthine and the watercourses or arroyos are normally dry.

Powell, 1875. 'These are dreary wastes—naked hills, with rounded or conical forms, composed of sand, sandy clays, or fine fragments of shaly rocks, with steep slopes, and, yielding to the pressure of the foot, they are climbed only by the greatest toil....' (pp. 149–151) (Note: Distinguishes 'alcove lands')

Gilbert, 1877. 'Where a homogeneous, soft rock is subjected to rapid degradation in an arid climate, its surface becomes absolutely bare of vegetation and is carved into forms of great regularity and beauty.' (p. 120) (Long description, emphasizing regularity.)

Mill, *Dict.* Bad Lands. A tangle of changing crests and ravines (Dakota, U.S.A.).

Davis, W. M. and Snyder, W. H., 1898, *Physical Geography*, Boston: Ginn. 'Bad Lands. When fine-textured, unconsolidated deposits, such as lake silts, suffer dissection in an arid climate, they acquire an extremely irregular surface; hence their name, bad lands, originally applied on account of the difficulty of travelling over them.' (p. 303) '... minute imitation of a maturely dissected plateau.' (Reference to S. Dakota and Wyoming.)

Fay, 1920. 'A region nearly devoid of vegetation where erosion instead of carving hills and valleys of the ordinary type, has cut the land into an intricate maze of narrow ravines and sharp crests and pinnacles. Travel across such a region is almost impossible, hence the name (*U.S. Geol. Surv. Bull.*, **613**, p. 182). Specifically, the Badlands of the Dakotas.'

Cotton, 1945. 'Bare ground is carved by rain-wash into innumerable closely-spaced, steep-sided ridges and valleys of miniature dimensions, producing an almost impassable land surface generally referred to as "badlands" or badland erosion.' (p. 31)

*Comment.* Written as one word 'badlands' has come to mean a distinctive type of erosional landscape. Because of the attractive colouring of the sediments and the bizarre shapes, the original 'badlands' of South Dakota have been constituted a National Park and are visited by many thousands of tourists annually. There Cotton's mention of 'miniature dimensions' scarcely applies but otherwise his and Fay's definitions may be accepted. The contrast is often between the level plateau and its fantastic dissection. (L.D.S.)

**Bādōb** (Sudan: Arabic)
The stoneless clay soils east of the White Nile in the Sudan which are seamed by gaping cracks during the dry season. (J.H.G.L.)

**Baer's law** (K. E. von Baer)
Mill, *Dict*. The theory according to which the rotation of the Earth causes asymmetrical lateral erosion of river beds. Not in *O.E.D.* or current standard works.

**Baguio**
*Met. Gloss.*, 1944. A local name by which the tropical cyclones experienced in the Philippine Islands are known. A number of the cyclones or typhoons of the Western Pacific cross the Philippines, and in addition there is a class of cyclone which is especially associated with these islands, occurring from July to November.

**Bahada, Bajada** (Spanish)
*Dict. Am.* A descent, slope or downgrade. Also an alluvial fan (citations back to 1866).
*Webster.* A long outwash detrital slope at the base of a mountain range. Southwestern U.S.
Mill, *Dict*. Bajadas (Spanish-American). Gradually descending slopes (Spanish America), sometimes limited to slopes of degradational or aggradational origin. (*R. T. Hill*)
Tolman, C. F., 1909, Erosion and Deposition in S. Arizona Bolson Region, *J. Geol.*, 17, 136–163, introduced the word into the literature to prevent confusion between detrital slopes and the rock slopes of the mountains. '... The flanking detrital slopes, built up by terrestrial deposition, the aggradational equivalent of the active erosion above.' (pp. 141–142)
Wooldridge and Morgan, 1937. '"Bajada," plainly due to aggradation by floods debouching from the mountain valleys.' (p. 311)
Cotton, 1941. 'Where a number of streams emerge from mountains undergoing dissection and build fans along their front, the fans if large become laterally confluent, and thus form a continuous apron built of waste bordering the mountains, the surface of which is a piedmont alluvial plain or bahada' (thus pronounced, and now generally written, but originally the Spanish form was bajada).
*Comment.* Contrast pediment, cut in rock.

**Bahru** (Malay)
Town. (baharu = new)

**Baid, Baids, Baid lands** (Indo-Pakistan: Bengali)
Ahmad, 1958. 'The whole forest area (Dacca-Mymensingh) consists of higher ground or knolls called *chalas*, intersected by numerous depressions or valleys known as *baids*—cultivated with paddy.' (p. 200)

**Bailiwick**
*O.E.D.* A district or place under the jurisdiction of a bailie or bailiff. Used in Eng. Hist. as a general term including sheriffdom, and applied to foreign towns or districts under a vogt or bailli.

**Baite**
Mill, *Dict*. A shelter in the upper limit of the high Alpine pastures, and used for a short time for sheltering hay, for converting wood into charcoal, etc. Often merely a natural hollow covered or completed by branches, etc. Geographers speak of the Zone of Baite, that of thin pasture, up to 2,000–2,400 metres.

**Bai-u rains** (Japanese)
Late spring or early summer rainfall in southern Japan.
Trewartha, G. T., 1945, *Japan: A Physical, Cultural and Regional Geography*, Univ. of Wisconsin. 'From mid-June to about mid-July there is a high frequency of relatively weak, slow-moving cyclones and fronts which travel slowly north-eastward, or even stagnate over subtropical Japan.... Much cloudiness, abundant rain, high humidity, and high sensible temperatures make the so-called bai-u or plum rains a very uncomfortable and gloomy season.' (p. 42)
*Comment.* So called because they come when plums are ripening. (G.T.T.) Better known in English as 'plum-rains'. (A.A.M.)

**Bajada**—see Bahada

**Bajir**
Mill, *Dict*. A lake among sandhills in Central Asia (*bayir* in Lop basin).

**Balagh** (Sudan; Arabic)
Land in the Gash delta in the Sudan which is

## Balance of Payments

Kindleberger, C. P., 1953, *International Economics*. 'A systematic record of all economic transactions between the residents of the reporting country and residents of foreign countries' (p. 16). As the balance is theoretically built on double-entry book-keeping the total debits must always equal the total credits. It includes all goods and services, both visible, such as merchandise, bullion and specie, and invisible, such as payments for transportation, banking and insurance services, travel abroad, interest payments, as well as the flow of capital.

The balance of payments is broken down into the following categories:

The current account, including merchandise exports and imports, and services or invisible items, such as payments for shipping and freight, banking and insurance, travel abroad, income on investments, rents, royalties, donations.

The capital account, including all charges or claims on or of a country owned or owed by other countries.

The gold account (bullion and specie) deals with monetary gold and is separate from the merchandise account because of its monetary function in the banking system of most countries and in the balance of payments.

Errors and omissions are a balancing item. (C.J.R.)

## Balance of Trade

The relation between the total values of a country's merchandise or visible exports and imports.

Kindleberger, C. P., 1953, *International Economics*. 'In many countries its significance is limited and the use of the term may be misleading.' (p. 21)

## Balk, Baulk

Royal Commission on Common Land, 1955–1958. *Report*. 'A piece of unploughed land in an open field used for grazing and giving access to the ploughed portions. In many open fields each man's strip was separated from his neighbour's by such a balk, though in many cases also a double furrow was used instead as a boundary. The larger balks were primarily grass-covered "occupation roads" and were called "the town balk" or "the common balk" to distinguish them from the minor balks between the strips.' (p. 272) See also C. S. and C. S. Orwin, *The Open Fields*, Oxford; Clarendon; also Butlin, 1961.

## Balleh (Somaliland)

Greenwood, J. E. G. W., 1957, The Development of Vegetation Patterns in Somaliland Protectorate, *Geog. Jour.*, **123**, 465–473. 'The surface deposits contain a comparatively high percentage of clay, either throughout the soil profile or in a relatively impermeable illuvial "B" horizon, a fact demonstrated by the occurrence of *ballehs* or temporary ponds.' (p. 465)

## Ballon (French)

A dome-shaped hill formed by erosion and on the edge of an escarpment. Origin: Alsace Vosges. (E.D.L.) See also Baulig, 1956.

## Ballstone

Rice, 1941. A Shropshire term for an irregular lenticular mass of unstratified limestone occurring in the Wenlock and other Palaeozoic limestones. Examples vary in dimensions up to 60 feet or more and are found to consist of colonies of corals and stromatoporoids (in the position of growth) enveloped in a matrix of calcareous mud.

## Balma (Spanish)

A concave cliff which forms a shelter under the overhanging rock (P.D.).

## Balneology

*Webster*. The science of bathing, or esp. of the therapeutic use, external and internal, of natural mineral waters.

Swayne, 1956. The study of mineral springs and baths and their healing effects.

## Baltic shield

Mill, *Dict*. The Archaean platform of Finland and E. Scandinavia. See also Shield.

## Bamboo Curtain

By analogy with Iron Curtain (*q.v.*) the limit of Chinese Communist influence and so including within it the Chinese People's Republic, North Korea, North Vietnam.

## Bañada, bañado

Mill, *Dict*. Bañada: Marshes in the porous loess (Pampas).

James, 1959. Bañado: 'Chaco—the Southern Border. Croplands, irrigated by the annual floods, are known in Argentina as *bañados*.' (p. 315)

## Banana hole

Vaughan, 1914 . . . 'abundant sub-aerial pot-

holes, known as banana holes, formed by the solution of the limestone'—referring to the Barrier Reef of Andros—'when drowned known as "blue holes".'
Drew, 1914. 'Andros Island.... In the numerous "pot-holes" which occur all over the island a small deposit of black leaf mould can be found and these "pot-holes" are the favourite places for the cultivation of sugar cane and bananas.'
Comment. A local term current in the Bahamas. (G.T.W.)

**Bāngar, bhāngar** (Indo-Pakistan: Urdu-Hindi; Panjabi)
Old alluvium of northern India, usually with Kankar. See Ahmad, 1958, p. 60. Contrast with Khādar.
High area within a river valley built up of old alluvium and beyond the reach of river floods. Also called manjha in Panjabi.

**Banjar** (Indo-Pakistan: Urdu)
Assessed waste land (Government land) in India. (G.K.) Land which has not been cultivated for the last four harvests is known as banjar jadid (new fallow); if it is not cultivated for 8 harvests it becomes banjar kadim (old fallow) but the latter term also includes all 'cultivable' waste whether it has ever been under the plough or not. See also Ahmad, 1958, p. 74; see also Chachar.

**Banjir** (Java)
Robequain, C., *Le Monde Malais* (translated Laborde, E. D., 1954). 'sudden, violent spates or *banjir*, in which masses of water heavy with boulders and mud sweep down into the valley' (p. 49).

**Bank** (Marine)
*O.E.D.* A shelving elevation in the sea or the bed of a river, rising to or near the surface, composed of sand, mud, gravel, etc. Also a bed of oysters, mussels or the like.
*Adm. Gloss.*, 1953. A detached elevation of the sea bottom over which the depths do not normally constitute a danger to surface navigation. A shallow area connected with the shore. Such a bank may be of any material, but not usually of rock or coral.
*Comment.* The definition given in *Adm. Gloss.*, 1953, is exemplified by the Dogger Bank (North Sea) and the Grand Bank (often wrongly written Grand Banks) off Newfoundland.

**Bank (2)**
*O.E.D.* The sloping margin of a river or stream; the ground bordering upon a river. A raised edge of a pond, lake, etc.

*Comment.* Also used in the north of England in the sense of hill or hillside. (A.F.M.)

**Banke** (Afrikaans)
*Pl.* of Bank, but in this sense seldom used in sing. Especially used in the compound term Bankeveld: *antiq*: banken.
Moderately high escarpments formed by resistant rock layers, often occurring in pairs or larger numbers with broad longitudinal valleys in between.
The term cuesta, used in America, indicates a somewhat similar feature, but is mostly limited to flatter ridges of less than 200 ft. high. The South African banke are usually higher (up to 1500 or 1600 ft. above the corresponding valleys) thus forming high hills or even low mountains. (P.S.)

**Banket** (South Africa)
*O.E.D.* Dutch *banket* banquet, also a confection resembling almond hardbake. A gold-bearing conglomerate found in the Witwatersrand.
Wellington, J. H., 1955, *Southern Africa*, II. 'The Dutch word meaning confectionery. The weathered conglomerate [the gold-bearing quartz-pebble conglomerates of the Witwatersrand] resembles somewhat an almond confection.' (p. 124)

**Banner cloud**
*Webster. Meteorol.* A cloud touching and extending out from the lee side of a mountain peak, like a banner.
Swayne, 1956. A long, flag-like cloud which forms on the lee side of a hill or mountain, *e.g.* at Crossfell Range (the Helm Cloud) and Table Mountain (the Table Cloth).

**Banto faro** (West Africa: Gambia)
Harrison-Church, 1957. 'These are Manding words meaning "beyond swamps". They are grasslands which are submerged in the flood season but are above water in the dry season, when their coarse grasses wither.' (p. 222)

**Bār** (Indo-Pakistan: Panjabi, Hindi)
The higher and more arid parts of the doabs (*q.v.*) beyond the bāngar (*q.v.*) of northern India—usually unirrigated and waste (see Spate, 1944, p. 463). Anglicized as 'bār-lands,' country of rolling sand-dunes with patches of grass and deep-rooted trees and of plains glistening with salt. The Thal between the Indus and Jhelum was such an area until reclaimed.

**Bar** (Meteorology)
Mill, *Dict.* The accepted unit of atmospheric pressure equivalent to 29·53 inches or

750·1 mm of mercury at 0°C. in latitude 45°.

The unit commonly used is a millibar—a thousandth part of a bar.

**Bar (1)**

Shepard F P., 1952, Revised Nomenclature for Depositional Coastal Features, *Bull. Ass. Am. Petr. Geol.*, 36, pp. 1902-1912. 'Submerged deposits which extend along the coast' (p. 1902). 'One of the most confusing features of present terminology is the indiscriminate use of "bar". Originally this term as applied to the sea referred to slightly submerged sand ridges such as those existing at the entrances to harbors. Tennyson's immortal poem "Crossing the Bar" no doubt helped to establish this meaning for the term. However on the plea that submerged bars may become elevated into beaches or islands, "offshore bar" has been used by Johnson (1919) and others to refer to sandy islands which extend for as much as a hundred miles along the coast... we should use different names for such widely different features. The word "barrier" as in "barrier beach" or "barrier island" offers a substitute.... Here following Price, 1951, the term bar is confined to sand shoals covered at least at high tide and "barrier" refers to the sand masses which extend along the coast and rise above high-tide level.' (pp. 1902 and 1904)

*Comment.* An initial confusion results from the use of the word bar for (a) coastal features and (b) river features, hence the separation here into Bar (1) and Bar (2). Thornbury indexes Bar (1) as 'bars, marine', whilst Bar (2) is described under meander bar and delta bar.

**Bar (2)**

Thornbury, 1954. 'The floodplain.... Along the stream channel are likely to be found numerous deposits of sand and gravel designated by such names as channel bars, meander bars, and delta bars. *Channel bars* are located in the stream course... *meander bars* form on the inside of meanders and extend into meander curves ... *delta bars* are formed by tributary streams building deltas into the main stream channel.' (pp. 165-166)

*Comment.* This use of bar is American and from it have been derived many terms such as bar plain (*q.v.*).

**Bar, cuspate**

Shepard, F. P., 1952, Revised Nomenclature for Depositional Coastal Features, *Bull. Ass. Am. Petr. Geol.*, 36, pp. 1902-1912. 'Cuspate bars and sandkeys: Where the cusp form is submerged, but may be seen either from the air through clear water or is known from soundings or surf patterns, the feature may be called a "cuspate bar." These bars may be built up above the surface to form islands which should then be called "cuspate sandkeys".' (p. 1911)

**Bar, longshore**

Shepard, F. P., 1952, Revised Nomenclature for Depositional Coastal Features, *Bull. Ass. Am. Petr. Geol.*, 36, pp. 1902-1912. 'Longshore Bar: This name refers to the sand ridges which extend generally parallel with the shoreline and are submerged at least by high tides. Longshore bars have been referred to in the past as "ball" in "ball and low" (Johnson, 1919, the elongate character of these bars is not well indicated by "ball" which suggests a round object (p. 486)).'

**Bar, lunate**

Shepard, F. P., 1952, Revised Nomenclature for Depositional Coastal Features, *Bull. Ass. Am. Petr. Geol.*, 36, pp. 1902-1912. 'Lunate bars and sandkeys: The lunate or crescent shape of bars and small islands is perhaps more common than the cuspate form. Such "lunate bars" and "lunate sandkeys" are found commonly off passes between barrier islands or off the entrances to harbors. Where the waves are large the lunate bar may have a deep crest, for example "Five Fathom Shoal" outside the Golden Gate at San Francisco. Where the waves are small the bar is likely to be very shallow or even stand above the surface. At both ends of the bar or sandkey there is ordinarily a channel maintained by the tidal currents. Elsewhere the channel may cut through the center of the bar or island. Another type of lunate bar is found outside some beaches of seas with a small tidal range such as the Mediterranean and the Gulf of Mexico. Most of these bars occur off straight beaches so that they are not related to beach cusps. The lunate bars appear to be modifications of longshore bars since they may be traced laterally into ordinary longshore bars and since they are not directly connected to the beach. So far as is known, this type of lunate bar does not build above the surface.' (pp. 1911-1912)

**Bar, offshore**
*O.E.D.* Bar. II. 15. 'A bank of sand, silt, etc., across the mouth of a river or harbour, which obstructs navigation.' Examples 1586–1868.
Mill, *Dict.* Bar. (1) An accumulation of sand or gravel rising to near the surface in a river or along the shore. Inner bar, the bar formed at the upper end of the flood channel; outer bar, that formed at the mouth of the ebb channel of a river estuary (Krümmel).
Johnson, 1919. Contains a long discussion and a history of the debate. 'As soon as the submarine bar lying offshore has been raised above the surface of the water, we can distinguish an outer and an inner shoreline; the first bordering the seaward side of the offshore bar, or barrier beach as it is often called. ... Between the mainland and the bar lies a lagoon. ... A narrow bar lying parallel to, but some distance from, a gently sloping sandy shore.'
The examples quoted are above the surface of the water. (p. 350)
Lewis, W. V., 1936, *Geog. Jour.*, **88**, 431–447. Describes the offshore bar of S. Iceland, which is above the surface, seaward of the low-water mark and includes a beach.
Wooldridge and Morgan, 1937. No reference to offshore bar, but a *bay-bar* lies above the surface, *e.g.* Loe Pool (p. 335). An 'offshore beach terrace' contains no lagoon and equals Johnson's 'shoreface terrace'. (p. 322)
Bencker, 1942. 'Bar = Submarine obstacle formed at the mouth of a river or at the entrance to a port, generally sandy and mobile, and forming a rise which renders the passage of ships difficult and even dangerous.' (p. 65)
(The Admiralty Glossary contains a similar, submarine, definition.)
Shepard, F. P., 1952. Revised nomenclature for depositional coastal features. *Bull. Amer. Ass. Petrol. Geol.*, **36**. States that bar has been used by Johnson (1919) to refer to sandy islands above high tide level. It is proposed to call such a 'barrier', thus:
Barrier beach (called an 'offshore bar' by Johnson (1919), p. 259), 'a single elongate sand ridge, rising slightly above the high-tide level and extending generally parallel with the coast, but separated from it by a lagoon.' Also defines longshore bars as '... the sand ridges which extend generally parallel with the shoreline and are submerged at least by high tides.' (This is equivalent to the 'ball' of 'ball and low' of Johnson, p. 486.) (pp. 1902–1912)
See also Barrier and Bar (1).

**Bar plain** (F. A. Melton, 1936)
Melton, F. A., 1936. An Empirical Classification of Flood-plain streams, *Geog. Rev.* **26**. 'Streams with neither a low water channel nor an alluvial cover ... have a relatively smooth flood plain, the topographical detail of which is a network of "bars" elongate in shape and irregular in size, built from the tractional and suspended load in the declining stages of the last flood ... the plains will be called bar plains.' (pp. 595–596)

**Bar, point or meander**
Thornbury, 1952. 'The term *point bar* was applied to the bar that develops on the inside of a meander bend and grows by the slow addition of individual accretions accompanying migration of the meander. It is roughly equivalent to what has been called a meander bar, meander scroll (Davis, 1913) or scroll meander (Melton, 1936).' (p. 168)
See also Fisk, 1947 (quoted under Backswamp deposits).

**Bar, reticulated**
Shepard, F. P., 1952, Revised Nomenclature for Depositional Coastal Features, *Bull. Ass. Am. Petr. Geol.*, **36**, pp. 1902–1912. 'Reticulated bars: A criss-cross pattern of bars, with both sets diagonal to the shoreline. ... As observed by the writer, these reticulated bars are found only in bays and in lagoons on the inside of barrier islands ... they may occur in other environments.' (p. 1909)

**Bar, transverse**
Shepard, F. P., 1952, Revised Nomenclature for Depositional Coastal Features, *Bull. Ass. Am. Petr. Geol.*, **36**, pp. 1902–1912. '*Transverse bars* ... bars which extend at right angles to shorelines. ... They have been described elsewhere as "sand waves" (Gibson, 1951) and as "giant ripples", (Van Straaten, 1950), but these terms fail to indicate their important relationship to the shoreline ...' (p. 1909).
See also Gibson, W. M., 1951, Sand Waves

in San Francisco Bay, *Coast and Geodetic Survey Jour.* No. 4, pp. 54–55; and Van Straaten, L. M. J. U., 1950, Giant Ripples in Tidal Channels, *Tidjkschrift. Kon. Nederl. Aardr. Gen.*, **67**, pp. 76–81.

**Bārāni** (Indo-Pakistan: Urdu-Hindi; Panjabi)
Cultivated or crop land entirely dependent upon rainfall.
*Comment.* In the land-use survey of the Indus Basin carried out in 1954–55 Cropland was separated into four categories—perennial irrigation, seasonal irrigation from inundation canals, sailaba (land watered by surface run off trapped in field bunds) and barani (land dependent for water supply on normal rainfall in all seasons) (L.D.S.).

**Barbed drainage patterns**
Thornbury, 1954. Where 'the tributaries join the main stream in "boat hook bends" which point upstream.' (p. 123)

**Barchan, Barkhan, Barkan, Barchane**
Barkhan, originally a Turki word meaning a sandhill in the Kirgiz steppe, Barkhan is the spelling given by Murzaev and Vasmer; balqan by Ushakov (see Appendix, Russian words).
*O.E.D. Suppl.* also Barkham, Barkan (native name)
A crescent-shaped dune of shifting sand such as occur in the deserts of Turkestan.
Mill, *Dict.* Barkhan (called in the Sahara sif (plur. siuf) = Medan of South America). A sand-dune of crescentic outline with convex windward side and steeper concave leeside.
Knox, 1904. Barkhan (Cent. Asia), a sandhill.
Bagnold, R. A., 1941, *The Physics of Blown Sand and Desert Dunes*, London: Methuen. 'Dunes assume two fundamental shapes (i) the *barchan* or crescentic dune, and (ii) the longitudinal dune, whose Arabic name *seif* or "sword" will be used in default of any universally recognised name. (p. 189) The *barchan dune* occurs when the wind is nearly *uni-directional*.' (p. 195)
Moore, 1949. Barkhan. 'An isolated, crescent-shaped sand dune, with the horns of the crescent projecting down wind, caused by the sand being blown round the edges as well as out the top of the heap; it is common in the desert areas of Turkestan....'
*Comment.* If Bagnold's suggested terminology is accepted, Mill's reference to sif requires correction. The spelling barchane is used by many American writers (Lobeck, von Engeln).

**Barrage**
*O.E.D.* The action of barring; the formation of an artificial bar in a river or watercourse to increase the depth of water; the artificial bar thus formed, especially those in the Nile.
Swayne, 1956. A natural or artificial obstacle to the flow of a stream.
*Comment.* With the growth of large scale irrigation in many parts of the world the term 'barrage' is commonly applied to large scale masonry structures across the major rivers designed to conserve a large body of water. The word 'dam' seems to be preferred if water-storage is combined with the generation of electric power, 'barrage' if it is not, but there is no hard and fast rule.

**Barran** (West Yorkshire)
Mill, *Dict.* Stones collected off a field and heaped up on bosses of bare rock.

**Barranca** (Spanish; used in United States), **barranco**
*O.E.D.* A deep ravine with precipitous sides.
Lyell, 1872, *Principles of Geology*, 11th Edn. Vol. i. In describing the outer or northern edge of M. Somma on Vesuvius. 'From the crest of the great escarpment of the "Atrio," or what the Spaniards would call the "Caldera," deep ravines or "barrancos" very near each other radiate outwards in all directions... very shallow near the summit, but becoming rapidly deeper and having precipitous sides towards their terminations... at the upper end of the ravine-like portion of several of these valleys is frequently seen a precipice over which a cascade falls in the rainy season when the channel of the torrent above is full of water....' (p. 635)
*Dict. Am.* A deep break or hole made by a heavy rain, a ravine (citations back to 1836).
*Webster.* A ravine with steep sides; also, a steep bank or a bluff.
von Engeln, 1942. 'Originally numerous, subequal, consequent-stream furrows on the sides of a (volcanic) cone are in time replaced by a comparatively few, deep and wide valleys, *barrancas*.' (p. 608)
*Comment.* Mill gives two meanings, the first from Spain, the second from the Canaries. German writers use barranco. (Cotton, 1944, 362)

**Barren**

**O.E.D.** A tract of barren land; *spec.* applied in N. America to: *a.* elevated plains on which grow small trees and shrubs, but no timber, classed as *oak-barrens, pine-barrens*, etc., according to the trees growing on them; *b.* in Kentucky, to certain fertile tracts; *c.* An open marshy space in the forest in Nova Scotia or New Brunswick.

*Dict. Am.* A tract of relatively open land, having little or no vegetation and usu. regarded by the earliest pioneers as lacking in fertility. Often pl. (Citations back to 1651.)

*Webster*. 1. A tract of barren land. 2. Usually in *pl*. Level tracts of land, poorly forested and commonly having light sandy soil; as *pine barrens*.

*Comment*. The Barrens is a name given to the northernmost tundra plains of Canada. Being a derogatory term, Stefansson introduced 'arctic prairies'.

**Barrier**

Shepard, F. P., 1952, Revised Nomenclature for Depositional Coastal Features, *Bull. Ass. Am. Petrol. Geol.*, **36**, pp. 1902–1912. 'Distinct separations are made between submerged deposits which extend along the coast, here referred to as "bars," and emerged deposits referred to as "barriers"' (p. 1902). ... "barrier" refers to the sand masses which extend along the coast and rise above high-tide level' (p. 1914). He then goes on to define various combinations of barrier, given below:

*Barrier beach* ... a single elongate sand ridge rising slightly above the high-tide level and extending generally parallel with the coast, but separated from it by a lagoon. The term should apply to islands and spits so far as they agree with the rest of the definition. The name has been used previously, but was not restricted to single ridges (pp. 1904 and 1906). Called an 'offshore bar' by Johnson, 1919, p. 259; footnote p. 1904; and Twenhofel, 1939, p. 28.

*Barrier island* ... similar to a barrier beach but consists of multiple instead of single ridges and commonly has dunes, vegetated zones and swampy terranes extending lagoonwards from the beach. The inner ridges as Price has indicated may be eliminated by the encroachment of a series of dunes. The D. W. Johnson terminology includes this in the catch-all 'offshore-bar' (p. 1906). Name suggested by Price, W. A., 1951, Barrier Island not 'Offshore-Bar,' *Science*, **113**, pp. 487–488.

*Barrier-spit*. Since many of the barriers are connected at one end to the mainland, 'barrier spit' seems appropriate to add to the terminology although the difference between 'barrier island' and 'barrier spit' lies only in a connection to the mainland which may be shortlived. It should be noted that the term 'barrier spit' is not necessary where there is a single ridge since the term 'barrier beach' is suggested for islands and spits alike (pp. 1906 and 1908).

*Barrier chain*. A series of barrier islands, barrier spits and barrier beaches which extend along a considerable length of coast. The term is similar to 'barrier' since the latter refers to all types previously defined, but 'barrier chain' indicates continuity, thus providing the same contrast as between 'mountain' and 'mountain range' (p. 1908).

Note also used of ice cliff *e.g.* Ross Ice Barrier.

**Barrier reef**

*O.E.D.* (Quotations only, inc. Flinders, 1805; De La Beche, 1853) 1877. Green, *Phys. Geol.*, iv. 3, 136. 'A mighty wall of coral rock, separated from the land by a deep and broad channel, and bounded on the seaward side by a face almost vertical and of enormous height. Such a reef is called a Barrier reef.'

Mill, *Dict*. A coral reef fronting the shore, at some distance from it, and separated from it by a lagoon or a navigable channel of moderate depth.

Knox, 1904. 'Barrier reef (English), a reef fronting a coast line or encircling an island or group of islands, leaving a deep channel between it and the shore. *cf.* Fringing reef.'

Johnstone, J., 1923, *Introduction to Oceanography*, refers to smaller reefs as fringing reefs having 'much the same characters represented on a much smaller scale than the great Barrier Reefs'. (p. 80)

Salisbury, 1923. 'Coral reefs. Those which are far enough from the land to leave a somewhat wide and deep belt of water (lagoon) inside are *barrier reefs*: those close to land are *fringing reefs*.' (p. 517)

*Comment*. Green's definition is fanciful. Used as defined by Mill and applied especially to the Great Barrier Reef of eastern Australia (exposed at low tides). The extension of the term 'barrier reef' from the Great Barrier Reef of Australia to other examples appears to be due to Charles Darwin, Coral Reefs, 1842.

**Barton**
O.E.D. 2. A farm-yard. 3. A demesne farm; the demesne lands of a manor, not let out to tenants, but retained for the lord's own use.

Comment. In the latter sense the word survives in the names of country properties.

**Barysphere** (cf. Bathysphere)
O.E.D. Suppl. From Greek, 'heavy sphere'. The internal substance of the earth enclosed by the lithosphere.
1901. Science, 15th Nov., 747.
1926. Enc. Brit. New Suppl. II. 'The bulk of the earth consists of a nickel-iron mass, the barysphere, which is enclosed by a rocky crust, the lithosphere.' (p. 172)
Mill, Dict. The dense, hot mass occupying the centre of the earth and bounded by the lithosphere.
Fay, 1920. The central or deep interior portions of the earth, presumably composed of heavy metals or minerals. It is contrasted with Lithosphere, the stony shell (Kemp). Also called Pyrosphere.
No reference Lyell, 1830; Page, 1865, Geikie, 1898.

**Basal conglomerate**
Rice, 1941. A conglomerate or coarse sandstone forming the lowest member of a series of related strata which lie unconformably on older rocks. It records the progressive encroachment of the sea-beach on the former dry-land.

**Basal platform**—see Basal surface

**Basal surface** (B. P. Ruxton and L. Berry, 1959)
Mabbutt, J. A., 1961, Proc. Geol. Assoc., 72, 357, 'Ruxton and Berry (Proc. Geol. Assoc., 70, 285–290) have suggested "basal surface" for the boundary between weathered and unweathered rock, in place of "basal platform" as used by Linton (Linton, D. L., 1955, The Problem of Tors, Geog. Jour., 121, 470–486).... the writer considers this term inapt and proposes the term "weathering front" to define the interface of fresh and weathered rock. It is noted that Büdel has used "Tiefenfront" with this meaning.'

**Basalt**
Webster. Petrog. A dark-gray to black dense to fine-grained igneous rock, the extrusive equivalent of gabbro, consisting of basic plagioclase (usually labradorite), augite, and usually magnetite. Olivine or basalt glass or both may be present. The rock is often vesicular, and the cavities may be filled with secondary minerals.

A fine-grained volcanic rock belonging to the basic group with between 45 and 55 per cent of silica. When poured out as a molten magma from a volcanic vent or a fissure basalt is very fluid and tends to form even sheets of lava covering large areas. The great lava plains and plateaus of the world are mainly of basalts. It may solidify into perfect hexagonal columns (good examples are Fingal's Cave, Staffa and Giants' Causeway). In the past the word has been used very loosely and care should be taken not to use old definitions.

**Base, base exchange**
Uvarov and Chapman, 1952. Base (chemistry). Substance which reacts with an acid to form a salt and water only; substance which has a tendency to accept protons; substance which yields hydroxyl ions if dissolved in water. Basic (chemistry) of the nature of a base; opposite to acidic; reacting chemically with acids to form salts. Base exchange, Base exchange capacity, Base status, Base saturation. See Robinson, 1949.

Miller and Turk, 1943. Base exchange. Denotes an exchange of bases. Refers to the reaction which certain insoluble elements of soils undergo when they are in contact with a salt solution. The cations of the salt replace the bases from the soil (p. 431).

Russell and Russell, 1953. 'Unless one chooses to define the base-exchange capacity of a clay as the permanent negative charge carried by its particles, it can only have an arbitrary meaning'. (p. 98).

**Base flow**
Todd, D. K., 1959, Ground Water Hydrology, New York: Wiley. 'That portion of streamflow coming from ground water discharge is base flow'. (p. 152).

**Base level**
Webster. The level below which a land surface cannot be reduced by running water. Sea level is the plane to which all erosion by running water tends; but the level of a lake or a stream determines the base level of streams flowing into it.

Powell, 1875. A level 'below which the rocks on either side of the river, though exceedingly friable, cannot be degraded ... an imaginary surface, inclining slightly in all its parts towards the lower end of the principal stream draining the area through which the level is supposed to extend, or having the inclination of its parts varied in direction as determined by tributary streams'. (p. 203) It is recognized that there may be local base levels due to lakes and 'natural dams' and that the sea is the 'grand base level'.

Mill, *Dict.* Base level of Erosion. The ideal surface of the valleys of a river system which has attained to a stable condition, *i.e.* when the thalweg of each constituent forms a profile of equilibrium (Powell). French equivalent for this is *surface de base*. Also used sometimes as equivalent of the French term *niveau de base*, the level on attaining which the independent erosive action of a river ceases, *i.e.* sea-level for a main stream, or the junction of a tributary with a larger stream in the case of the tributary.

Committee, List 2. The level below which erosion is unable to proceed; the general and ultimate base-level for the land surface is sea-level, but other local or temporary base-levels may exist.

See also Salisbury, 1923, p. 92 (for temporary base-level); W. M. Davis, Base-level, Grade and Peneplain, *J. of Geol.*, **10**, 1902, 77–111; Twenhofel, 1939, p. 8.

**Base Level of Deposition**

Twenhofel, 1939. '*The Base Level of Deposition* is the highest level to which a deposit can be built ... many local base levels of deposition, owing to local conditions. The base level of deposition due to marine agencies coincides with the base level of erosion.' (p. 8)

**Base (-line)**

O.E.D. Base. 17. *Surv.* A line on the earth's surface or in space, of which the exact length and position are accurately determined, and which is used as a base for trigonometrical observations and computations.

See Close and Winterbottom, 1925, p. 5. Also refers to 'check bases'.

**Base-map**

*Webster.* A map, having only essential outlines, used for indicating specialized data of various kinds.

Monkhouse and Wilkinson, 1952. 'Base map is an outline map used for plotting information.' (p. 10)

*Comment.* A base-map is not necessarily an outline map. Better defined as 'a map of any kind on which additional information is plotted for a particular purpose.' The term 'basic map' has occasionally been used. (E.O.G.) A common base-map may be prepared for a series of distributional maps, *e.g.* for the Ordnance Survey Archaeological Series (G.R.C.)

**Basement, basement complex, basal complex**

*O.E.D.*, 1. The lowest or fundamental portion of a structure ... 1843, J. Portlock, *Geology*, 97. The augitic rock which forms the basement of the promontory.

*Webster.* basement complex: *Geol.* a. The assemblage of metamorphic and igneous rocks underlying the stratified rocks in any particular region. b. The Archean rocks. See fundamental complex.

Scott, W. B., 1907, *Introduction to Geology*, 2nd ed. 'The Archaean includes the most ancient rocks, often spoken of as the basement, or basal complex' (p. 534).

Daly, R. A., 1940, *Strength and Structure of the Earth*. 'At the surface the average density of the Basement (Archean) Complex of each continent is close to 2·70 ...' (p. 17).

**Basic industry**—see Industry, basic

**Basic rock**

Hatch and Wells, 1937. 'Arranged in order of their silical content, the igneous rocks fall into three groups: 1. Acid ... 2. Intermediate ... 3. Basic, with silica content below 55% ...' (p. 139).

*Comment.* See Acid rock. The typical basic plutonic rock is gabbro; the hypabyssal is dolerite; the volcanic is basalt. It is usual to distinguish also a fourth class, the ultrabasic, with a still lower percentage of silica.

**Basin**

O.E.D. 9. A hollow receptacle, natural or artificial, containing water.

10. A dock ... b. part of a river or canal widened and furnished with wharfs.

11. A land-locked harbour; a bay.

12. *Phys. Geog.* The tract of country drained by a river and its tributaries, or which drains into a particular lake or sea. 1792. Young. 1830. Lyell. 1860. Maury.

13. *gen.* A circular or oval valley or hollow. c. 1854. Stanley. 1860. Tyndall.

14. *Geol.* A circumscribed formation in which the strata dip inward from all sides to the centre; the stratified deposit, especially of coal, lying in such a depression. 1821. 1850. Lyell 'Deposits lying in a

hollow or trough, formed of older rocks.' 1877. Green. 'If the beds dip everywhere towards a centre....'

Webster. 8. Geol. a. An area in which the strata dip usually from all sides towards a center; as the Richmond coal *basin*. b. An area that does not drain to the ocean (as the Great Basin). 9. *Phys. Geog.* a. A large or small depression in the surface of the land, the lowest part of which may be occupied by a lake or pond; as the *basin* of Lake Michigan; also, a similar depression in the ocean floor; as the Tuscarora deep. b. An area largely enclosed by higher lands, even though it has an outlet and is drained; as the Big Horn *basin*. c. The entire tract of country drained by a river and its tributaries—called specif. *river basin*. d. A great depression in the surface of the lithosphere, occupied by an ocean—called specif. *ocean basin*.

Mill, *Dict*. In general, any hollow in the Earth's surface. In particular, river-basin = area drained by a river; lake-basin = basin filled by water of lake. In geology, a region in which the strata near the surface all dip towards the centre; may also be defined as a short syncline.

Knox, 1942. 'Basin (Eng.), used in suboceanic relief, for a depression of approximately round form, in which the horizontal diameters are about equal.'

Bencker, 1942. 'Basin—large submarine cavity of more or less round or oval form.' (p. 62) A small feature is a cauldron.

Geikie, 1898 (also in Structural and Field Geol., 1905, p. 414). '... to classify hollows according to the mode of their formation.' '1. Tectonic basins. 2. Volcanic basins. 3. Dissolution basins. 4. Alluvial basins. 5. Æolian basins. 6. Rockfall basins. 7. Glacial basins.' (p. 223)

*Comment*. The two principal uses in geographical works are as in *O.E.D.* 12 (a river basin) and 14 (a tectonic basin *e.g.* coal-basin). The concept of a basin is often loosely interpreted. In addition to 10a and b of *O.E.D.* note use of tidal basin either a dock which is subject to tidal movement, or a basin filled with water at high tide which is then retained to be released at low tide, when the force of the outflowing water scours the neighbouring harbour.

See also S. N. Dicken, 1935, *J. of Geol.*, **43**, 716 (used for a large doline in Karst country); E. S. Hills, 1953, *Structural Geology*, p. 76 (basins exhibit *periclinal structure*). Note also American use of watershed for drainage basin.

**Basin of deposition**

An original basin or area of deposition of a group of sedimentary rocks. Because of confusion with other uses of basin, the word *cuvette* (*q.v.*) has been used by L. J. Wills and others.

**Basining**

Rice, 1941. In geology, a settlement of the ground in the form of basins, in many cases at least due to the solution and transportation of underground deposits of salt and gypsum. Such basining produces numerous depressions from those of a few square yards to those of 50 sq. miles in area, in the high plains region east of the Rocky mountains.

**Basin cultivation**

Stamp, L. D., 1953, *Africa, A study in Tropical Development*, New York: Wiley. 'In regions of heavy rainfall the land may be ridged by hoes into a succession of little basins.... This is a most effective means of preventing soil erosion.' (pp. 198–199; of Nigeria). See also *Geog. Rev.*, **28**, 1938, 42–45.

**Basin irrigation**

The type of irrigation associated especially with the Nile in Egypt and the Sudan whereby flood waters are led off into specially prepared 'basins' which vary in size from a few dozen acres to many square miles. The basins are separated from one another by earth banks. The system permits some control of the flood waters and is the traditional method of irrigation in Egypt but has been progressively replaced by permanent canal irrigation.

**Basin-lister**

A type of plough developed in the United States which by 'jumping' at intervals produces a series of 'basins' instead of one long continuous furrow. The object is to minimize the danger of soil erosion when a sudden rainstorm may turn a continuous furrow into a watercourse. The basins produced resemble those made by hand for the same purpose in Nigeria and elsewhere (see Burges, A. E., 1936 *Soil Erosion Control*, Atlanta, Ga., Smith Co., pp. 127–129).

**Basin range**
*Webster.* A mountain range that owes its present elevation essentially to faulting and tilting; a tilted fault block from the Great Basin region of the western United States, where examples occur.

**Basket-of-eggs topography**
Holmes, 1944. 'Swarms of whale-backed mounds called *drumlins*. Being distributed more or less en *échelon* the mounds give rise to what has been aptly described as "basket of eggs" topography' (p. 230) For an explanation of possible origin of drumlins and this type of relief see Wooldridge and Morgan, 1937, p. 385.

**Bass** (Russian)
Russian abbreviation of basseyn (basin), referring usually to a coal basin and used as a suffix e.g. Donbass.

**Bastion, rock**
von Engeln, O. D., 1937, *Geog. Rev.*, 27. '"Rock Bastions"... where tributary glaciers joined main glaciers the outlet of the hanging valleys are fronted by rock masses that project far out into the main valley.' (pp. 478–482) (Review of Otto Flückiger, Glaziale Felsformen, *Pet. Mitt.*, No. 218, 1934—who uses the term 'Felsvorbau', literally = rock screen.)
Charlesworth, 1957. 'Pronounced salients of solid rock which project from a few trough walls below or at the lips of hanging valleys are termed bastions (Ger. *Felsvorbauten*; *Stufenvorbauten*): a fine example occurs at the mouth of the Trient valley. Their presence indicates a weakening of the erosive power of the trunk glacier as a result of thrust exerted at right angles to it from the tributaries, of a reduction in the rate of flow due to accumulation of debris at the confluence of the glaciers, or at a discordant junction.' (p. 314)
Note: Also Cotton, 1942, footnote—also Felsbastion, but this is used by Richter in a different sense.

**Batāi** (Indo-Pakistan: Urdu-Hindi)
The system common in Pakistan of farming the land on a metayage basis whereby the tenant (hari) keeps half the produce but provides seeds, fertilizers, implements and cattle. Also in east Punjab and western U.P.

**Bat furan** (Arabic), **bat hiddan** (Arabic)
The season respectively of 'sea open' and 'sea closed' corresponding to the winter or period of the North-East Monsoon and summer or period of the South-West Monsoon. In the one case the Arabian Sea with comparatively light winds is open to native shipping, in the other the storms are too dangerous. See Miller, 1953, 153–154.

**Batholith, Bathylith, Batholite** (Suess, defined in German, 1895)
*O.E.D. Suppl.* Geology. Also (now disused) bathylith (from Gr. *bathos* depth). Each of a series of masses of granite or anorthosite in a mountain ranged in the direction of its axis, having been brought up from great depths.
Hatch and Wells, 1937. 'This term was introduced by Suess to connote the major deep-seated intrusive masses of very large size occurring typically in great mountain ranges. Their roofs are dome-shaped, their walls steeply inclined... when exposed by denudation they either maintain their diameter or grow broader with increasing depth. Suess defined a batholith as a stock-shaped or shield-shaped mass that has originated by the melting, and assimilation by the intruding magma, of a portion of the invaded formation. The universal acceptance of Suess' term, however, by no means implies general agreement as to the mechanism responsible for the origin of batholiths.' (p. 7)
Wooldridge and Morgan, 1937. 'Bathyliths... are large dome-shaped masses whose sides plunge steeply to unknown depths; their base is never seen... very commonly, though not invariably, formed of granite. (pp. 106–108)
Holmes, 1944. 'Batholiths are gigantic masses of essentially igneous rocks with highly irregular dome-like roofs, and walls that plunge downwards so that the intrusions enlarge in depth and appear to be without visible foundations.' (p. 89)
*Comment.* The *O.E.D.* misses the essentials though it is true that many of the great batholiths of the world are elongated (*e.g.*, the Coast Batholith of British Columbia). Bathylith and Batholith are both current spellings. See also laccolith and dome.

**Bathyal**
*Webster.* Of or pertaining to the deeper parts of the ocean; deep-sea. *Bathyal district* or *zone. Oceanography.* The slope from the continental shelf at one hundred fathoms to the abyssal district at one thousand fathoms. The upper half is the *typical*,

the lower half the *deeper, bathyal district* or *zone.*
Holmes, 1944. 'The muds, etc., of the continental slope, and of similar depths around oceanic islands belong to the bathyal zone; while the oozes of the deep ocean floor belong to the abyssal zone.' (pp. 313-314) On Fig. 171 Holmes shows the Bathyal zone as corresponding to the continental slope. Note also bathyal deposits.
Twenhofel, 1939. 'The bathyal environment is that part of the sea bottom between the depths of 100 and 1,000 fathoms.' (p. 122)
### Bathygraph
Mill, *Dict.* Curves showing the areas of the ocean or any portion thereof at different depths below sea-level (Penck).
Comment. Not in *O.E.D.* nor supplement, nor in standard works.
### Bathymetry
Atlas of Canada, 1959, 'Plate II—this plate shows the depths, or bathymetry, of the waters in and around Western Canada.'
### Bathy-orographical
An adjective applied to maps which show both the relief of the land and submarine depths.
Webster. Of or pertaining to ocean depths and mountain heights; as, a *bathyorographical* map.
*O.E.D. Suppl.* gives 'bathyorographically, *adv.* so as to depict the variations in level in the depths of the sea.' This definition needs modification.
### Bathysphere
Lake, 1958. '... many writers limit the term lithosphere to the outer part of the earth, where the rocks are more or less similar to those which are visible at the surface, and separate the inner portion under the name bathysphere, centrosphere or barysphere.' (p. 3) See Barysphere.
No reference in Holmes, 1920; Fay, 1920; Rice, 1941.
### Baulk—see Balk
### Bauxite
Webster (F. fr. *Baux* or *Beaux*, near Arles). Mineral. A ferruginous aluminum hydroxide, essentially $Al_2O_3 2H_2O$, but consisting of several minerals occurring in oölitic masses and in earthy form.
Himus, 1954. Aluminium hydroxide, $Al_2O_3 2H_2O$, found as an amorphous earthy, granular or pisolitic mineral of dirty white, greyish, brown, yellow or reddish colour, and containing oxides of iron and phosphate as impurities. It is the chief source of aluminium. Probably results from the weathering of igneous rocks under tropical conditions; may be residual or transported.
### Bay
*O.E.D.* 1. An indentation of the sea into the land with a wide opening.
2. An indentation or rounded projection of the land into the sea. Obsolete (perhaps a distinct word from Bey, to bend).
3. An indentation, recess in a range of hills, etc.
4. (In *U.S.*): a. An arm of a prairie extending into, and partly surrounded by, woods. b. A piece of low, marshy ground producing large numbers of Bay-trees.
Mill, *Dict.* A wide indentation of the sea into the land.
*Adm. Glos.*, 1953. A comparatively gradual indentation in the coast-line, the seaward opening of which is usually wider than the penetration into the land; as distinct from a Gulf, Loch or Firth.
Moore, 1949. A wide indentation into the land formed by the sea or by a lake.
Thornbury, 1954. 'The Carolina "Bays." The term bay has long been applied to elliptical-shaped shallow depressions which are particularly numerous in the coastal plain of the Carolinas but extend from Florida to New Jersey.' (p. 518) They are described at length, with a map, and it is noted that their origin is still uncertain—they may be the scars of a swarm of meteorites, sink holes, salt domes etc.
### Bay bar, bay-bar
Wooldridge and Morgan, 1937. 'Continued lengthwise growth may convert a spit into a *bay-bar* (Fig. 209). A similar effect might clearly arise from the opposed or convergent growth of two spits.' (p. 335) Fig. 209 is of Looe Pool, a bay-bar in south Cornwall.
Johnson, 1919. ... adds that a bay-bar may also be formed by an offshore bar developing across a bay or being driven inwards to block a bay. (pp. 300-301)
See Bay barrier, barrier, bar.
### Bay barrier, bay-barrier
Shepard, 1952. 'Where a spit has grown entirely across the mouth of a bay so that the bay is no longer connected to the main body of water the term "bay barrier" is proposed. This replaces "bay bar" a confusing term which fails to indicate whether or not the sand ridge is submerged or stands above the water level.' (p. 1908)
See Bar, barrier.
### Bay-head delta
Cotton, 1941. 'Deltas are common at the

heads of, or quite filling, estuaries, into which rivers have discharged as a result of a late partial submergence of the land margin. Such bay-head deltas . . .' (p. 180)

**Bay ice**—see Ice, sundry terms

**Bayou**

*Dict. Am.* (Amer. F. from Choctaw *bayuk*, river, creek). A sluggish stream or body of water, often connecting larger waters or emptying into adjacent streams (citations from American French back to 1699 and in English back to 1766). 1914, R. S. Tarr *Physiography*, 150 'Some abandoned [river] courses along the lower Mississippi, are called *bayous*.'

*Webster.* In general, a creek, secondary watercourse, or minor river, tributary to another river or other body of water—the regular term in the lower Mississippi basin and in the Gulf-coast region of the United States. Along with the general sense many specific senses are found as: a. A large stream or creek, or a small river, characterized by a slow or imperceptible current; a sluggish stream, esp. along a tortuous course, through alluvial lowlands, swamps, plantations, etc.—the typical watercourse of the Pleistocene or Quaternary area of the Mississippi Embayment, where *bayou* is the term in almost exclusive use. b. A clear brook or rivulet arising in the hills—a sense surviving in a few counties in northern Arkansas and southern Missouri. c. An effluent branch of a main stream, esp. one sluggish or stagnant, as a natural canal connecting any two bodies of water, a by-channel of a river, enclosing a low island, or a branch discharging through a delta. d. An intermittent, partly closed, or disused watercourse, sluggish or stagnant, as a partly closed channel of a river delta, a lake or pool in an abandoned channel of a stream (called also *bayou lake*), a swampy or miry offshoot of a lake or river subject to overflow, an outlet for a coastal lake or swamp, a slough in a salt marsh, or a shallow or stagnant inlet opening into a bay, lake, or river. e. An estuarial creek or inlet on the Gulf coast; a small bay, open cove, or harbor; also, a lagoon, lake, or bay, as in a sea marsh or among salt-marsh islands. f. A passage connecting two bodies of open water, as bays, or a navigable channel through sand bars or mud flats.

*O.E.D.* (Prob. a corruption of Fr. boyau, gut). The name given (chiefly in the southern States of N. America) to the marshy off-shoots and overflowings of lakes and rivers.

Mill, *Dict.* Bayou, Girt. The outlet of a delta lake or one of the several distributaries of a river through its delta; a sluggish watercourse.

Knox, 1904. (U.S.A.). A lake or intermittent stream formed in an abandoned channel of a river; one of the half-closed channels of a river delta.

Salisbury, 1923. 'Such lakes are called *oxbow lakes* or *bayous*.' (p. 132)

**Beach**

*O.E.D.* (Origin unknown: apparently at first a dialect word, meaning, as it still does in Sussex, Kent and the adjacent counties, the shingle or pebbles worn by the waves.)
1. (Usually collect., formerly occas. with pl.) The loose water-worn pebbles of the sea-shore; shingle (quots. 1553-1884)
2. A ridge or bank of stones or shingle. Obs.
3. The shore of the sea, on which the waves break, the strand; spec. the part of the shore lying between high- and low-water mark; also applied to the shore of a lake or large river. In Geol. an ancient sea-margin.
1880. Geikie, Phy. Geog. iii, xvii, 154. The strip of sand, gravel or mud, which is alternately covered and laid bare by the rise and fall of the tidal undulation is called the beach.

Mill, *Dict.* 1. The part of the shore lying between high and low water marks. 2. = batch. A steep-sided valley with cwm at the head (Shropshire).

Gilbert, G. K., Lake Bonneville, *U.S. Geol. Surv. Mon.* I, 39, 1890. '. . . the zone occupied by the shore drift in transit.'

Johnson, 1919. '. . . the deposit of materia which is in more or less active transit, along shore or on-and-off shore.' (p. 162)

Wooldridge and Morgan, 1937. 'The term *beach* is applied to debris temporarily accumulated in the shore zone. It rests on a wave-cut bench. It may be built seaward into the off-shore zone as an "off-shore beach terrace".' (p. 322)

*Adm. Gloss.*, 1953. Any part of the shore over which shore debris, such as mud, sand, shingle, pebbles or smooth stones, is accumulated in a more or less continuous sheet. The term Beach is not used to describe areas of jagged reef, rocks or coral.

King, C. A. M., 1959, *Beaches and Coasts*, London: Arnold. 'A beach is an accumulation of loose material around the limit of wave action ... may be taken to extend from the extreme upper limit of wave action to the zone where the waves, approaching from deep water, first cause appreciable movement of the bottom material.' (p. 1)

*Comment.* Wooldridge and Morgan, and King, go back to the original meaning of the word.

## Beach cusp

*Webster.* Sand and gravel deposits formed by wave action into points that project seaward along the coast.

Shepard, 1952. 'Along many straight open beaches ... varying from 30 to 200 feet between points, the distance increasing in general with wave height. Most of these as indicated by Kuenen (1948) have depressions seaward from the cusps and elevations seaward from the troughs. Following common practice these relatively small features are being referred to as "beach cusps".' (p. 1909)

See also Kuenen, P. H., 1948, The Formation of Beach Cusps, *J. of Geol.*, **56**, 34–40.

## Beach material

The loose materials which constitute a beach (*q.v.*) range from mud to stones and boulders. See classification by size according to the British Standards Institution, Wentworth and Inman, discussed by C. A. M. King, *Beaches and Coasts*, pp. 3–5.

**Beach, raised**—see Raised beach

## Beacon

*O.E.D.* 5. A conspicuous hill commanding a good view of the surrounding country, on which beacons were (or might be) lighted. Still applied to such hills in various parts of England, *e.g.* Brecon Beacons. ...
b. A division of a wapentake; probably a district throughout which a beacon could be seen, or which was bound to furnish one. Obsolete.
6. A lighthouse or other conspicuous object placed upon the coast or at sea, to warn vessels of danger or direct their course.

Notes on the Making of Plans and Maps. School of Military Engineering, Chatham, 1937. 'A beacon is any object which is erected over the selected station and which may be intersected by theodolite from other stations.' (p. 3)

## Beaded esker

Wooldridge and Morgan, 1937. 'If the ice were retreating rapidly the products of a single sub-glacial stream would be distributed over the country as a ridge or trail ... each pause, or ice-stand, will be marked by a swelling of the ridge, and thus a series of fans or deltas may be strung along the "feeding esker", like beads on a string. To such a form the term "beaded esker" is applied.' (p. 388)

See Esker.

## Beaded lakes

Swayne, 1956. Strings of long, narrow lakes between sand dunes.

## Bearing

*O.E.D.* 13. The direction in which any point lies from a point of reference, esp. as measured in degrees from one of the quarters of the compass. In pl. the relative positions of surrounding objects.
14. The direction of any line on the earth's surface in relation to a meridian.

Middleton, R. E. and Chadwick, O., 1941, *A Treatise on Surveying*, 5th ed., London: Spon. Vol. 2. 'The "bearing" of a line is the angle which the plane of a "great circle" containing it makes with some standard "great circle", usually a meridian passing through the middle of the area to be surveyed.' Is contrasted with azimuth. (p. 231)

See also Close and Winterbottom, 1925, pp. 37–8 and standard works on surveying.

*Comment.* Perhaps better defined as the horizontal angle measured clockwise from a specific reference line which for a *true bearing* is the meridian so that the bearing is the angle from true North. Hence also *magnetic bearing* measured clockwise from magnetic north; *compass bearing* from the north as indicated by the compass; *grid bearing* from the north–south grid lines on a grid-map. A *reverse* or *reciprocal bearing* is the reverse or reciprocal of a given bearing.

**Beastgate**—see Cattlegate

## Beaufort Scale

A scale devised by Captain (later Admiral) Beaufort in the early part of the nineteenth century (modified in 1926) to indicate wind strengths from 0, calm, to 12, a hurricane, based on estimated velocity at 10 metres above the ground.

| | Beaufort Wind Number | | kph | mph |
|---|---|---|---|---|
| Light winds | 0 | Calm | 0 | 0 |
| | 1 | Light air | 1·5–5 | 1–3 |
| | 2 | Light breeze | 6–12 | 4–7 |
| | 3 | Gentle breeze | 12·5–20 | 8–12 |
| Moderate winds | 4 | Moderate breeze | 21–30 | 13–18 |
| | 5 | Fresh breeze | 31–39 | 19–24 |
| | 6 | Strong breeze | 40–50 | 25–31 |
| | 7 | Moderate gale | 51·5–61 | 32–38 |
| Gales | 8 | Fresh gales | 62·5–74 | 39–46 |
| | 9 | Strong gale | 75–87 | 47–54 |
| | 10 | Whole gale | 88–102 | 55–63 |
| | 11 | Storm | 103–121 | 64–75 |
| | 12 | Hurricane | 121 | over 75 |

Note that 7 (moderate gale) is classed for statistical purposes with 'moderate winds'.

**Beaver-dam, beaver-meadow**
A dam across a stream resulting from the activities of beavers which gnaw through trees and cause them to fall making an obstruction. They build their homes nearby. An area of soft moist ground (beaver meadow) may also result (Canada).

**Beck**
*O.E.D.* A brook or stream: the ordinary name in those parts of England from Lincolnshire to Cumberland which were occupied by the Danes and the Norwegians; hence, often used specifically in literature to connote a brook with stony bed, or rugged course, such as are those of the North country.
*Comment.* Sometimes used to emphasize a *small* stream. (G.T.W.)

**Bed**
*O.E.D.* 9. The bottom of a lake or sea, or of the channel of a river or a stream. a. 1586–1830. Lyell.
13. A layer, a stratum; a horizontal course. b. Geol. A layer or stratum of some thickness. 1684 ... 1878, Huxley.
Mill, *Dict.* 1. The land covered by the waters of a river, lake or ocean. 2. In geology, a layer or stratum of rock.
Page, 1865. '... a stratum of considerable thickness, and of uniform homogeneous texture.... Originally and strictly, however, the term bed referred to the surface-junction of two different strata, and seam to the line of separation between them.'

Geikie, J., 1905 (In section on sedimentary rocks). 'Bed or stratum is the term applied to any sheet-like mass which has a more or less definite petrographical character, and is separated by well-marked parallel division-planes from overlying and underlying rocks.' (p. 108)
Fay, 1920. includes: '1. The smallest division of a stratified series, and marked by a more or less well-defined divisional plane from its neighbours above and below (Kemp). 2. A seam or deposit of mineral, later in origin than the rock below, and older than the rock above; that is to say, a regular member of the series of formations, and not an intrusion (Raymond). A deposit, as of ore (or coal) parallel to the stratification (Standard Dict.),' etc.
Manual of Instruction for the Survey of Dominion Lands, Ottawa, 1918. 'The *bed* of a body of water has been defined as the land covered so long by water as to exclude from it vegetation, or as to mark a distinct character upon the vegetation and upon the soil itself ... the limit of the bank is the line where vegetation ceases or where the character of the vegetation and soil changes.' (p. 34)

**Bedding**
Swayne, 1956. The arrangement of rock strata in bands of various thickness and character; stratification.
For different types of bedding—cross, current, false, graded, etc., see under these headings; also Hills, E. S., *Structural Geology*, 3rd ed., 1953.

**Bedding cave**
Gemmell, A. and Myers, J., 1949, *Adventures Underground*. 'A horizontal cave passage formed along a bedding joint or shale bed in the rock; usually wide and low, with a flat roof.'
Warwick, G. T., 1953, *British Caving*. 'Caves developed along bedding-planes tend to be wide with low roofs.' (p. 8)
*Comment.* A term used to describe caves or portions of caves developed from a bedding plane, and in Yorkshire often associated with washed-out shale bands. (G.T.W.)

**Bedding plane**
*Webster.* The surface that separates each successive layer of a stratified rock from its preceding layer; depositional plane; plane of stratification.
*O.E.D. Suppl.* Bedding-plane = bed-plane (examples 1908, 1920) bed-plane. Geol., the junction between two layers or strata.

Mill, *Dict*. Bedding-plane. The surface on which the material building up a stratum was originally deposited.
Fay, 1920. The planes or surfaces separating the individual laminae or beds of a sedimentary rock. (La Forge). See also Stratification planes.
*Comment*. Used as in Fay not as in Mill. Bed-plane is now rare.

### Bed rock, bed-rock
*O.E.D.* (*Geol.*) The solid rock underlying alluvial and other superficial formations.
Geikie, J., Structural and Field Geology. Edinburgh: Oliver and Boyd, 3rd ed., 1912. 'Bed-Rock Soils: Bed-rock. Just as the soil passes down gradually into the subsoil, so it is often hard to say where subsoil ends and "living rock" begins.' (pp. 383–384)

### Bedouin (many spellings)
*O.E.D.* A dweller in the desert; an Arab of the desert.
Forde, C. D., 1934. *Habitat Economy and Society*. London: Methuen. '... the most famous of all pastoral peoples, the Badawin Arabs..., the occupants of *el Badia* (the desert), are set apart from other '*arab* by a real difference in economy.' (p. 308–309)

### Behavioural environments
Kirk, W., 1963, Problems of Geography, *Geography*, **48**, 357-371. 'It is a concept developed by the school of Gestalt psychology in work on perception.... The Behavioural Environment is thus a psycho-physical field in which phenomenal facts are arranged into patterns or structures (*gestalten*) and acquires values in cultural contexts. It is the environment in which rational human behaviour begins and decisions are taken which may or may not be translated into overt action in the Phenomenal Environment.' (p. 365)
See also Phenomenal environment and Gestalt psychology.

### Behavioural geography
Colledge, R. G., Brown, L. A. and Williamson, F., 1972, Behavioural Approaches in Geography: An Overview, *The Australian Geographer*, **12**, 1. 'The behavioural approach in geography represents a point of view. The researcher is interested in the processes leading to observed spatial patterns. These may include two or three dimensional patterns of the distribution of the phenomenon being studied or the spatial characteristics of behaviour itself. In either case, emphasis is placed on processes relating to human actors of a system rather than on the system elements. ... the behavioural point of view involves (a) the researcher viewing the real world from a perspective of those individuals whose decisions affect locational or distributional patterns and (b) trying to derive sets of empirically and theoretically sound statements about individual, small group, or mass behaviours.' (p. 59)

### Behead, Beheading; beheaded rivers
*O.E.D.* Behead. 2. of things: to deprive of the top or foremost part. Rare.
Mill, *Dict*. Beheading—the action of a river in cutting back its valley so as to intercept and divert the water of a second river.
Davis, 1909. 'Terminology of Rivers changed by Adjustment' from *The Rivers and Valleys of Pennsylvania*. N.G.M., 1889, I, pp. 183-253.
Kuenen, 1955. '*Piracy or Beheading*. A river may abruptly capture a considerable portion of its neighbour's territory. This happens when headward erosion, as it cuts back, succeeds in completely consuming the water-shed in the middle of the other's basin.... This process is sometimes called *beheading*, though the subject is often the head of a tributary to the neighbour's system.' (pp. 289–290)

### Bel, bhel (Indo-Pakistan: Panjabi)
Sandy islands in the beds of rivers especially R. Jhelum.

### Bell pit
North, F. J., 1930, *Limestones*. Early type of mining whereby a seam of coal, or chalk, was mined by removal from the base of a shallow shaft and abandoned when the roof became unsafe. Afterwards subsidence around these shafts leads to shallow depressions ... (p. 338).
*Comment*. A term used in Buckinghamshire for chalk mining and in the Black Country and North Staffordshire and Peak District for coal and other mines. A type of mining long abandoned, though the depressions remain.

### Belt
*O.E.D.* 2. A broadish strip of any kind, or a continuous series of objects encircling or girdling something. 5. A broad band

characteristically distinguished from the surface it crosses; a tract or district long in proportion to its breadth.
*Comment.* In the sense 2 above it is used in the term 'green belt' (*q.v.*); in the sense 5 for such geographical regions as the Corn Belt, Cotton Belt of the United States though there, as used by O. E. Baker, it becomes almost synonymous with region independent of shape. Also in ice terminology a long area of pack ice from a few kilometres to more than 100 km in width (*Ice Gloss.*).

**Belted coastal plain, belted outcrop plain**
Lobeck, 1939. With parallel cuestas and lowlands. (p. 452)
Wooldridge and Morgan, 1937 'On the final peneplain, the straightness of the original outcrops will be re-established.... Such a surface is termed a "belted outcrop plain".' (p. 203)

**Belukar** (Malay)
Defined on official one-inch maps of Malaya (Hind, 1035 Series) as 'secondary jungle' —the scrub which grows up in clearings as opposed to the grass lalang which may also occupy old clearings.

**Ben** (Scottish; from Gaelic *beann*)
A mountain peak; used with names of Scottish and Irish mountains, *e.g.* Ben Nevis. See Pin.

**Bench**
*O.E.D.* 7. Any conformation of earth, stone, etc., which has a raised and flat surface, *e.g.* the coping of a wall (? *obs.*); a level ledge or set-back in the slope of masonry or earthwork; in *U.S.* a level tract between a river and neighbouring hills; a horizontal division or layer of a coal-seam, cut by itself.
*O.E.D. Suppl.* b. *Geol.* A natural terrace marking the outcrop of a harder seam or stratum.
Also 'Bench-land,' U.S. Land situated in, or forming, a 'bench'.
*Webster.* A level surface of ground, rock, etc. raised and narrow, with a steep slope at the back; specif. ... c. A terrace or shelf, esp. a former shore line of a river or lake; also, a shelf formed in working an open excavation on more than one level.
*Dict. Am.* 1. A level, somewhat narrow plain bordering upon an upland region or mountain. 2. A level tract between a river or lake and neighboring mountains. (Citations back to 1803.)

Davis, W. M. and Snyder, W. H., 1898, *Physical Geography*, Boston: Ginn. '... on the projecting headlands and the outlying islands, here the sea may beat furiously, cutting a rocky bench beneath bold cliffs. ...' (p. 360) Also p. 363.
Cotton, 1941. 'In the course of a single cycle on horizontally bedded formations, ... Lower plateaus begin to emerge as fringing step-like structural benches ... long before the highest plateau is destroyed.'
Fay, 1920. 'Benches. A name applied to ledges of all kinds of rock that are shaped like steps or terraces. They may be developed either naturally in the ordinary processes of land degradation, faulting, and the like; or by artificial excavation in mines and quarries (Kemp).'
Johnson, 1919. '... wave-cut *bench*, a sloping erosion plane, inclined seaward ... it may gradually decrease in slope until it merges imperceptibly into the more extensive, nearly horizontal, plane produced by long-continued wave erosion, which is commonly called the *abrasion platform*.'
Wooldridge and Morgan, 1937. 'The *beach* ... rests on a wave-cut bench.' (p. 322)

**Bench mark, bench-mark**
*O.E.D.* A surveyor's mark cut in some durable material, as a rock, wall, gate-pillar, face of a building, etc., to indicate the starting, closing or any suitable intermediate point in a line of levels for the determination of altitudes over the face of a country. It consists of a series of wedge-shaped incisures, in the form of the 'broad-arrow' with a horizontal bar through its apex, thus ⌰. When the spot is below sea-level, as in mining surveys, the mark is inverted. [The horizontal bar is the essential part, the broad-arrow being added (originally by the Ordnance Survey) as an identification. In taking a reading, an angle iron > is held with its upper extremity inserted in the horizontal bar, so as to form a temporary bracket or *bench* for the support of the levelling-staff, which can thus be placed on absolutely the same base on any subsequent occasion. Hence the name.] 1864. *Webster* cites Francis.
Raisz, E., 1948, *General Cartography*, 2nd ed. New York: McGraw-Hill. 'Bench marks of levelling are small bronze tablets with a bar engraved in the center.'

American Society of Civil Engineers: *Definitions of Surveying Terms*, 1954. 'Bench mark. A relatively permanent material object, natural or artificial, bearing a marked point whose elevation above or below an adopted datum is known.' (p. 23)

*Comment.* On British Ordnance maps the height above O.D. (Ordnance Datum or Mean Sea Level) is instrumentally determined with great accuracy. On the map it is recorded by the letters B.M. and the height in feet to one place of decimals. Contrast Spot height.

**Benelux**
A name suggested in 1948 and subsequently widely used for *Be*lgium, the *Ne*therlands and *Lux*embourg, which formed a customs union.

**Benthic zone**
See Abyssal (quotation from Sverdrup).

**Benthos (Haeckel), Benthic**
*O.E.D. Suppl.* Haeckel's name for the flora and fauna at or near the bottom of the sea. See Plankton, Nekton.
Also archibenthic and abyssobenthic—see under Pelagic.

**Berg** (Afrikaans: *Pl.* Berge, *lit.* Mountain(s)).
In South African English often left untranslated when forming part of a proper name; *e.g.* Drakensberge (*i.e.* Dragon Mountains); Swartberge (*i.e.* Black Mountains, etc.).
English speaking people without a sufficient knowledge of Afrikaans sometimes use pleonasms like Drakensberg Mountains. This is considered a rather serious mistake. (P.S.)

**Berg** (German)
*Lit.* mountain; used in various combinations, *e.g.* Bergwind, Bergschrund etc. (*q.v.*)

**Berg**
*O.E.D.* (From Iceberg, a. Ger. eisberg—ice mountain). Short for iceberg: A (floating) mountain or mass of ice; (only used when ice is mentioned or understood in the context).
*O.E.D. Suppl.* S. Afr. (Du = OE. beorg, etc.) A mountain or hill.
Knox, 1904. Berk, Berg (Anglo-Saxon), a barrow.
Berg (Anglo-Saxon, Dch., Ger., Da., Nor., Sw.), mountain, hill.
See also Ice terminology, iceberg.

**Bergschrund** (German)
*O.E.D. Suppl.* Phys. Geogr. (G. f. berg+schrund, cleft, crevice). A crevasse or series of crevasses often found near the head of a mountain glacier. 1897 *Outing* (U.S.), XXIX, 339. Crevasses ... varying in width from fissures that a child could easily step across to yawning 'bergschrunds' ten or twelve feet wide.
1902. *Encycl. Brit.*, XXI, 24/1. At the foot of a snow or ice slope is generally a big crevasse, called a bergschrund.
Mill, *Dict*. The wide crevasse usually found at the line where a glacier touches the solid rock of the valley wall. The French term is '*Rimaye*'.
Tyndall, J., 1860, *Glaciers of the Alps*, London: Murray. 'Some valleys are terminated by a kind of mountain-circus with steep sides, against which the snow rises to a considerable height. As the mass is urged downwards, the lower portion of the snow-slope is often torn away from its higher portion, and a chasm is formed, which usually extends round the head of the valley. To such a crevasse the specific name of Bergschrund is applied in the Bernese Alps....'
*Ice Gloss.*, 1958. The *crevasse* which occurs at the head of a *cirque* or *valley glacier* and which separates the moving glacier ice from the rock wall and the *ice apron* attached to it. When the ice apron is absent the gap is known as a Randkluft.
*Comment.* Described in detail in Johnson, W. D., *J. of Geol.*, 12, 1904, 569–578. He notes that it is open to floor of the corrie, and lies between ice and rock. See also Cotton, 1942, pp. 178–180. The E.B. reference quoted by *O.E.D.* is misleading.

**Bergwind** (Afrikaans; also German) *lit.* mountain wind.
Left untranslated in South African English, but written in two words (Berg Wind). Hot dry Föhn-like wind, blowing down from the plateau towards the coastal districts and sometimes causing there oppressive weather even during the cool season. (P.S.)
In German contrast Talwind (*q.v.*).

**Bergy bit**
*Ice Gloss.*, 1956. A massive piece of sea ice or disrupted *hummocked ice*; also medium-sized piece of floating glacier ice. Generally less than 5m. above sea-level, and not more than 10m. across.
*Comment.* The authors of *Ice Gloss*, 1956, now suggest 'massive' should be replaced by 'heavy but compact'.

**Berm**
*O.E.D.* 1. A narrow space or ledge; espec.

in Fortif. a space of ground, from 3 to 8 feet wide, sometimes left between the ditch and the base of the parapet. 2. Berm-bank, the bank of a canal opposite the towing path.

*O.E.D. Suppl.* 3. A ledge or flat of land bordering either bank of the Nile and inundated when the river overflows.

*Webster.* A narrow shelf, path, or edge as at the bottom or top of a slope, or along a bank; specif.: a. *Fort.* Such a ledge between the foot of the exterior slope and the top of the scarp. b. The level space between the waste bank and the edge of a drainage ditch. c. *Local, U.S.* The bank of a canal opposite the towpath. d. *U.S.* The shoulder of a road.

*Dict. Am.* The bank of a canal opposite the towing path. Also the side or shoulder of a road. Orig. *berm-bank.* (Citations back to 1854.)

Bascom, F., 1931, Geomorphic Nomenclature, *Science*, 74. '... those terraces which originate from the interruption of an erosion cycle with rejuvenation of a stream in the mature stage of its development. Dissection, following upon elevation of the land, will leave remnants of the earlier broad valley floor of the rejuvenated stream as a terrace, or berm, and remnants of the uplifted abrasion platform as a seaward-facing terrace, or berm.' (Quoted in Thornbury, 1954, pp, 172–173.)

Rice, 1941. Flattish area cut while drainage was at grade in a previous cycle of erosion.

von Engeln, 1942. Of rejuvenation within a middle or late mature valley. 'Such a remnantal flat, which has a surface slope downstream, may be called a strath or sometimes, together with the valley shoulder, a berm.' (p. 221)

Cotton, 1945. 'The term "berm" has been introduced for any remnant of a surface developed to full maturity in a cycle that has since been interrupted. Though it has been said to be a kind of "terrace", a berm may include more than the valley floor which becomes a true river terrace, and the term cannot be considered to be synonymous with "valley-plain terrace".' (p. 242)

*Adm. Gloss.*, 1953. Part II. 'A narrow, nearly horizontal shelf or ledge above the foreshore built of material thrown up by storm waves. The seaward margin is the crest of the berm.' Part III. 'A horizontal ledge on the side of an embankment or cutting to intercept falling earth or to add strength.'

Muller, S. W., 1947, *Permafrost*, Ann Arbor. 'A bench or a horizontal ledge partway up a slope.' (p. 214)

*Comment.* Commonly used also in military circles for the banks between a road and the ditch (G.T.W.).

**Bet** properly bét (Indo-Pakistan: Panjabi) Flood-plain, usually anglicized as bet lands.

Spate, 1944. 'The immediate riverain or bet lands are agriculturally valuable but exposed to flooding.' (p. 463)

**Betrunked river**

Davis, 1909. (Physical Geography as a University Study, *J. of Geol.*, 1894) [of interruptions of cycle and adaption of rivers and streams] 'They are shortened and betrunked if the interruption is a depression. . . .' (p. 181)

Mill, *Dict.* The result of subsidence during which the main river is replaced by a coast indentation and the tributaries of the old river commence a separate existence.

**Bevel, bevelled surface, bevelling**

*O.E.D.* Bevel. 1. To cut away or otherwise bring to a slope; to reduce (a square edge) to a more obtuse angle; often with away, off, etc.

2. To recede in a slope from the right angle; to slant.

1862. Tyndall. Mountaineer, vii, 63. At one place, however, the precipice bevels off to a steep incline of smooth rock. Bevelled, beveled. Made or cut to a bevel; sloped off.

1860. Tyndall, Glac. II, 11, 292.

Wooldridge and Morgan, 1937. 'In the earlier stages of the second cycle the former peneplain is revealed in the bevelled summits of the escarpments.' (p. 210) 'Accordance of summit-level or bevelling of hilltops of itself does not prove a former peneplain.' (p. 213)

No reference Mill, *Dict.*; Gilbert, 1877; Powell, 1875; Davis, 1909; Fay, 1920; Cotton, 1922; 1941; 1942; Lobeck, 1939; Steers, 1946; Rice, 1941; Moore, 1949.

*Comment.* Some confusion is caused in geographical literature because bevelling may be used in the sense of *O.E.D.* 1 to denote the smoothing of the sharp edge between a horizontal surface (*e.g.* of a ledge) and a vertical surface (*e.g.* of a cliff face). This is quite different from the bevelling of hill *tops*. (G.T.W.)

**Bhabbar, bhābar** (Indo-Pakistan: Urdu-Hindi; Panjabi)
Band of gravelly deposits fringing the outer margin of the Siwaliks.
Spate, 1944. 'The bhabar (-"porous") is simply the great detrital piedmont skirting the Siwaliks.' (p. 495) See also Terai.

**Bhadoi** (Indo-Pakistan: Bengali)
The rainy season, April to August. See Hemantic.

**Bhāngar**—See bāngar

**Bhīl, bīl, bheel** (*antiq.*) (Indo-Pakistan: Bengali)
Oxbow or cut off lakes occupied by stagnant water in the Ganges delta of Bengal, sometimes saline.
Spate, 1954. 'The disjecta membra of dead rivers.' (p. 525) Also called jhīl or jhāor in East Pakistan.

**Bhit** (Pakistan: Sindhi)
A sand-hill or ridge (Pakistan).

**B-horizon** (soil science)
The soil horizon (*q.v.*) of illuviation (*q.v.*) below the A-horizon. It is in this horizon that deposition of material leached or washed down from the A-horizon takes place.

**Bhūr** (Indo-Pakistan: Urdu-Hindi, Sindhi, Panjabi)
Hills of wind-blown sand frequently seen capping the high banks of rivers. Also Bhurland, an area of bhurs. See also Terai.

**Bhūra** (Indo-Pakistan: Panjabi)
Ravines scoured out by surface drainage alone, having no torrent-bed.

**Bid price (rent) curves**
Bid price curves describe the trade-off of cheaper land prices with increased movement costs due to increased distance from the city centre. Haggett, P., 1972. 'Each line represents the rent values that exactly balance the increased transport costs by locating away from the city centre.... By superimposing the bid price curves on the actual rent curve for a given city, we can determine both the point where the best trade-off is reached and the corresponding rent level.' (pp. 259–260)

**Bifurcation ratio**
The ratio between the number of streams of one order to the number of streams of the next higher order. Also applied to central places, but not generally adopted. See Berry, B. J. L., 1967, *Geography and Retail Distribution*, Englewood Cliffs, New Jersey: Prentice-Hall, p. 38.

See also Strahler, A. N., 1957, Quantitative analysis of watershed geomorphology, *Trans. Am. Geophys. Union*, 38, 913–920.

**Bight**
*O.E.D.* 3. A bend or curve as a geographical feature, *e.g.* an indentation in a coast line, a corner or recess of a bay, a bend in a river, etc. (Note: 1876, Morris *Sigurd*, III, 326. Far off in a bight of the mountains). 4. The space between two headlands, a bay, generally a shallow or slightly-receding bay; *spec.* in the Bights of Benin and Biafra, and the Australian Bight; also trans. a bay-like segment. (Many of the examples given regard a bight as a smaller feature than, or contained within, a bay, W.M.)
Mill, *Dict.* An incurve larger than a bay, or less abrupt in its curvative, *e.g.* Australian bight.
*Adm. Gloss.*, 1953, Part II. 'A crescent shaped indentation of the coastline, usually of large extent and not more pronounced than a 90° section of a circle.'
*Ice Gloss.*, 1956. An extensive crescent-shaped indentation in the *ice-edge*, formed either by wind or current.

**Bill**
*O.E.D.* 3. A beaklike projection; a spur, tooth, spike. Applied to some narrow promontories, as Portland Bill, Selsea Bill.
Mill, *Dict.* A headland semi-detached from the mainland (South Coast, England).
Moore, 1949. A narrow headland, or small peninsula.
*Adm. Gloss.*, 1953. A narrow promontory.
No reference in Johnson, 1919; Steers, *Coastline of England and Wales*, 1948.
*Comment.* A term of local use in southern England rather than with a specific meaning.

**Billabong** (Australia: aboriginal word; *lit.* dead river)
Typically an elongated water hole, generally in the bed of a non-perennial stream; has been wrongly described as applying exclusively to a cut-off or ox-bow lake (E.S.D.). See Fenner, C., *Geography of South Australia*. Equated by some writers with anabranch.

**Biltong** (Afrikaans)
Raw meat preserved by being salted and dried; considered as a great delicacy in South Africa. From Dutch bil (buttock) and tong (tongue); to prepare good biltong

only choice meat from the hind leg is cut out in long narrow strips. (P.S.)

**Bing** (Scotland)
Waste heap from a mine.

**Biochore**
See Biosphere (Dansereau).

**Bioclastic** (A. W. Grabau, 1904)
*A.G.I.* Refers to rocks consisting of fragmental organic remains.

**Bioclimatic, bioclimatological**
Of or pertaining to bioclimatology. The adjectival forms are not in *O.E.D.* but are now very extensively used especially in bioclimatic limits, bioclimatic regions (=major climatic-vegetation regions); in a different sense Hopkins' 'bioclimatological law.' See the extensive discussion in Dansereau, 1957 (see under Bio-geography and the important discussion under Ombrothermic).

**Bioclimatology**
Moore, 1949. The study of climate in relation to life and health.

**Biocoenosis, biocenosis**
*Webster.* An assemblage or association of diverse organisms forming a natural ecological unit, in which there is more or less obvious dependence on mutualism; an assemblage of diverse organisms inhabiting a common biotope, a biotic community.
Dansereau, P., 1957, *Biogeography*, New York: Ronald Press, 'Biocenosis is the participation of various organisms in the nutritive elements of the environment, or better, in all the resources of the environment: space, food, shelter, etc.' (p. 238).

**Biocycle**
See Biosphere (Dansereau).

**Bio-geography, biogeography**
*O.E.D. Suppl.* Biogeography (Bio-). The science of the geographical distribution of living things, animal and vegetable.
1899. H. R. Mill, *Internat. Geogr.* 4. '.... the purpose of Biogeography is to trace out the reasons why particular species occupy the regions where they are now found.'
Mill, *Dict.* The science of the distribution of animals and plants.
Committee, List 1. Biogeography. 1. The geographical study of animate nature. 2. The geography of organic life.
Dansereau, Pierre, 1957, *Biogeography: An Ecological Perspective*, New York: Ronald Press. 'The scope of this book extends across the fields of plant and animal ecology and geography, with many overlaps into genetics, human geography, anthropology and the social sciences. All of these together form the domain of biogeography.'
See also Phytogeography and Zoogeography.
See also Mackinder under Anthropogeography and Gaussen under Ombrothermic.

**Bioherm**
*Webster.* Any body of rock built up by, or composed mainly of, sedentary organisms, as corals, algae, mollusks, etc. and enclosed or surrounded by rocks of different origin. An ancient coral reef, *cf.* biostrome.
Thornbury, 1954. A more appropriate name for a coral reef='organic mound'. (p. 480)
Twenhofel, 1939. 'Colonial corals have been among the chief contributors to the building up of the reefs or bioherms.' (p. 329)

**Biomass**
Tivy, S., 1971, *Biogeography: A Study of Plants in the Ecosphere*, Edinburgh: Oliver and Boyd. 'Plants—vegetable matter—both living and decaying, comprise the greatest bulk of the total world biomass (i.e. volume of organic material) both above and below the ground surface, and in water bodies.' (p. 3)

**Biome**
Abercrombie, M., Hickman, C. J. and Johnson, M. L., 1951, *A Dictionary of Biology.* Harmondsworth: Penguin. Major regional ecological community of plants and animals extending over large natural areas, e.g. tropical rain forest, coral reef.

**Biosphere**
*O.E.D. Suppl.* The totality of living things on the earth (quotes from Mill, 1899; Suess, trans. 1909).
*Webster.* The sphere of living organisms penetrating the lithosphere, hydrosphere, and atmosphere.
Mill, *Dict.* The living matter on the surface of the earth.
No reference in Bryan, 1933. Hartshorne, 1939; Moore, 1949; Wooldridge and East, 1951.
Dansereau, Pierre, 1957, *Biogeography*, New York: Ronald Press. 'The *biosphere* is that part of the earth's crust and atmosphere which is favorable to at least some form of life. It can be subdivided into *biocycles*, of which there are three: salt water, fresh water, and land.... In turn, the *biochore* is the *geographical* environment where certain dominant life-forms appear to be adapted to a particular conjunction of meteorological factors ... each biochore is characterized by a major type of vegetation ... within each biochore there develops one or more *formations* ...

within each of these formations *climax areas* can be distinguished. Not nearly all the space is taken up by climax vegetation, for the nature of the topography will allow a differentiation into many *habitats*. Within a habitat which harbors an ecosystem consisting of one or more *communities* there are other subdivisions, the *layer* (or synusia) and the *biotope*.' (pp. 125-127) A table of relationships to the province, (biochore), region (climax area), landform (habitat), layer (synusia) and niche (biotope) is also given.

*Comment.* Notice the complete change in meaning.

**Biostrome**
A modern coral reef.

**Biosystem** (K. Thienemann in German 1939)
See Ecosystem

**Biota**
Fauna and flora considered as a whole.

**Biotic factor**
*O.E.D.* Biotic: of or pertaining to life.
Carpenter, 1938. 'The results of interrelations of organisms to each other from an ecological standpoint.'
Biotic factor refers to the influence of living organisms, including man, in contrast to such factors as climatic and edaphic.

**Biotope**
See Biosphere (Dansereau).

**Bipolar distribution, bi-polar**
*Webster.* Pertaining to or associated with the polar regions; as, the *bipolar* distribution of certain marine organisms found only north and south of an equatorial or median zone.
Mill, *Dict.* The same species occurring towards both poles, but not in the intervening regions.

**Bird's foot delta**
Holmes, 1944. A delta extended into the sea on either side of distributaries outstretched like fingers, *e.g.* the Mississippi. (p. 171)

**Birth rate**
L.D.G. The birth rate is commonly measured by number of live births per 1,000 of the population. This is the crude birth rate or natural increase and currently ranges from 12 to 50, western countries with a high standard of living being between 15 and 20. The death rate is similarly measured in deaths per 1,000 of the population and varies from 6 to 25 or more. Net increase in population is the difference between birth rate and death rate. At the present day death rates continue to fall in most countries with the increase in medical skill and services; there is only a slight tendency for birth rates to fall. As a result the world population is increasing more rapidly than ever before in history—over 2 per cent per annum.

**Biscuit-board topography**
Lobeck, 1939. 'A topography characterized by a rolling upland, out of which cirques have been cut like so many big bites, is known as biscuit-board topography. It represents an early stage, or only a partial completion, of the process of glaciation.' (p. 265)

*Comment.* A purely American term; the analogy is with something unfamiliar in British kitchens.

**Bise, bize** (French)
*O.E.D.* A keen dry N. or NNE. wind, prevalent in Switzerland and the neighbouring parts of France, Germany, and Italy.
Mill, *Dict.* A cold piercing wind (Languedoc).

**Bitter lake**
A salt lake so called because of the taste of the water; especially the Great Bitter Lake and the Small Bitter Lake through which the Suez Canal passes.

**Bitumen**
*O.E.D.* 2. Generic name of native hydrocarbons more or less oxygenated including naphtha, petroleum, asphalt, etc.

**Bituminous**
Literally related to bitumen but commonly applied to coals of certain chemical compositions through an old and mistaken concept of their character.
Himus, 1954. Coal containing from 75 to 92 per cent. carbon, from 4·5 to 5·6 per cent. hydrogen and yielding from 15 to 45 per cent. volatile matter when heated out of contact with air.
See Coal.

**Black-band ironstone**
Himus, 1954. 'A variety of clay-ironstone which contains sufficient carbonaceous matter to admit of its being calcined without the addition of extraneous fuel.'
Twenhofel, 1939. 'Iron Carbonate. Siderite is the only iron carbonate.... Blackband iron carbonate derives the name from the colour. This is due to contained organic matter, and like the clay ironstone concretions, the blackbands are interbedded with clays and shales that contain more or less organic matter. Beds are not thick, and the common range is from less than 1 to 10 cm or more' (p. 394).

Hatch and Rastall, rev. Black, 1938. 'The blackband ironstones of the Staffordshire and Scottish coalfields contain from 10 to 20 per cent of coaly matter and are practically free from argillaceous material. They are bedded rocks, and those of Staffordshire replace the upper parts of coal seams, suggesting that they were precipitated in shallow lagoons. ... The high content of coaly matter allows them to be smelted very economically ... most valuable of the Coal Measure ores' (p. 149).

**Black box**
Chorley, R. J. in Chorley and Haggett, 1967. 'The most extreme application of general systems models involves the concept of the "black box". This requires little or no detailed information regarding the system components or subsystems within the black box system, interest being focused upon the nature of the outputs which result from differing inputs. Thus the black box is analogous to the "grey box" of partial system and the "white box" of the synthetic system.' (p. 85)
Harvey, 1969. 'System analysis frequently makes use of the concept of a black box, which is a system whose internal characteristics (structure and functioning) are unknown, but whose stimulus-response characteristics can be studied in detail. If we know the input-output relationship characteristic of a given black box, it may be possible to seek out some model of that system, whose structure is known, which achieves precisely the same behaviour.' (pp. 472–473)
*Comment*. The black box concept has been responsible for a large amount of important geomorphology in the last 100 years. For example, see Gilbert, G. K., 1877, Land Sculpture, ch. 5 in *The Geology of the Henry Mountains*, Washington: U.S. Department of the Interior, for a negative feedback black box model for land forms.

**Black Cotton Soil**—See Regur.
**Black Country**
*O.E.D.* Parts of Staffordshire and Warwickshire (*sic*.) grimed and blackened by the smoke and dust of the iron and coal trades.
*Comment*. The Black Country coincided approximately with the South Staffordshire Coalfield (Staffs.–Worc.). With the partial exhaustion of the coal and the migration of the iron industry, the character of the district has changed, but the name persists and is sometimes applied to heavy industrial areas in other parts of the world.

**Blackearth, Black Earth**
Mill, *Dict*. Black Earth (Russia). See Chernozem.
Jacks, 1954. General term including chernozem and dark plastic clays of tropics.

**Black turf** (South Africa), **Black turf soil**
A very dark soil derived from basic igneous rocks (norite) common in certain parts of the Transvaal. Notwithstanding the dark colour, which often deceives superficial observers, the percentage of organic material is very low, while clay forms from 40–60% of the soil. After good rains it is reasonably fertile. The depth of this black soil may be from 3–5 feet.
In the Netherlands the word turf means dried peat, used as fuel, but the term may sometimes be applied to the dark peat-soil itself. This is of course a typical organic soil. It seems that the memory of its dark colour has led the old pioneers to give its name to a completely different soil in South Africa. (P.S.)
Jacks, 1954. Black turf soil. Dark clay soil usually with calcareous subsoil.

**Blaen** (Welsh; *pl.* blaenau)
Head, end, source of river, upland.
Is used in contrast with bro (*q.v.*) in South Wales—the upland and the lowland, the hills and the vale (E.G.B.) see Davies, Margaret, 1951, *Wales in Maps* and Bowen, E. G., (Ed.), 1957, *Wales: A Physical, Historical and Regional Geography*.

**Blanket bog**
Tansley, 1939. '... where the rainfall is high and the air so constantly moist that bog is the climatic formation, not necessarily arising in fen basins but covering the land continuously except on steep slopes and outcrops of rocks. This is the third type of bog met with in the British Isles and may be called blanket bog, because it covers the whole land surface like a blanket. ... Blanket bog is independent of localised water supplies, depending on high rainfall and very high average atmospheric humidity.' (p. 676) Tansley's other types of bog are valley bog and raised bog. See also Bog.

**Blato**
Mill, *Dict*. A summer swamp replacing a winter karst lake.

**Blijver** (Dutch)—see Trekker
**Blind creek**
*O.E.D. Suppl.* Note (Blind creek): '1886.

J. W. Anderson, Prospector's Handbook, 115. Blind Creek, a creek, dry, except during wet weather.'

Comment. Obsolete or obsolescent and better avoided because of confusion with blind valley. Also known in U.S. as a draw (A.G.I.).

### Blind lake
Mill, Dict. See Turlough.

### Blind valley
Mill, Dict. Blind Valleys (Bouts du Monde: Côte d'Or); Reculées (French Jura). The abrupt termination, caused by the near approach of their precipitous sides, of the deep valleys or canyons formed by rivers, near the source of those rivers.

Geikie, J., 1898. 'Not less characteristic features of the karst-lands are the so-called blind-valleys and dry-valleys. Through the former a river flows to disappear into a tunnel at the closed or blind end' (p. 217).

Wooldridge and Morgan, 1937. As Geikie— 'inverted hanging valleys' (p. 292).

Cotton, 1941. As Geikie (p. 283).

von Engeln, 1942. 'Developed on uvala floors. These have steep sides and a steep wall for a terminus. At the base of the steep wall the surface flow disappears underground. The channel section is inferred to develop by successive infall from the interior of horizontal solution galleries. Blind valleys differ from pocket valleys in that the latter develop where underground water emerges in greater or less volume' (p. 578).

Comment. Essentially karstic features which may be dry or still contain a stream. They may end in a cave or a swallow-hole (open, choked or partially choked). Many swallets are miniature blind valleys, sometimes cut into the floor of major valleys. Sometimes streams disappear into the sides of these valleys. (G.T.W.) Mill's usage is now geographically obsolete.

### Blizzard
O.E.D. (A modern word, prob. more or less onomatopoeic)
2. A furious blast of frost-wind and blinding snow, in which man and beast frequently perish; a 'snow-squall'.

Webster. An intensely cold high wind filled with fine snow.

Dict. Am. A violent storm of fine driving snow accompanied by intense cold. (Use in existence as early as 1836.)

Mill, Dict. The wind in a cold wave (q.v.) spread over the northern part of the U.S.A. and thence extended to mean a cold period with heavy snowfall and high wind.

Met. Gloss., 1944. A term originally applied to the intensely cold north-westerly gales accompanied by fine drifting snow which may set in with the passage of a depression across the United States in winter. The term has come to be applied to any high wind accompanied by great cold and drifting or falling snow, especially in the Antarctic, where, however, the blizzards often cause a rise in temperature by removing the surface layer of very cold air formed during calms.

### Block diagram
Lobeck, A. K., 1924, Block Diagrams, New York: Wiley. 'A block diagram presents the relationship between the surface of the ground and the underground structure by representing an imaginary block cut out from the earth's crust. The top of the block gives a bird's eye view of the region, and the sides of the block its underlying geological structure.' (pp. 1–2)

Similarly Raisz, 1948; Monkhouse and Wilkinson, 1952.

Comment. G. K. Gilbert, 1877, is often credited with the first use of a block diagram, apparently because of his 'half-stereogram' of Mount Ellsworth as frontispiece. A block diagram does not necessarily show geological structure. T. W. Birch, Maps, 1949, says they 'may be regarded as sketches of relief models' (p. 136); C. B. Brown and F. Debenham would prefer 'perspective block' and note the use in America of 'stereogram' (Structure and Surface, 1929, p. 12). The simplified block-diagrams favoured by Griffith Taylor really deserve a distinct name (see Griffograph).

### Block faulting
Hills, E. S., 1953, Structural Geology, 3rd Ed. '... The crust may be divided into a number of small blocks, constituting a block-mosaic ... regions which are divided by faults into a number of differentially elevated or depressed blocks are said to exhibit block faulting.' (p. 70)

### Block field
Thornbury, 1954. (=Felsenmeer). Concentrations of rock blocks on many high mountain summits and on subpolar islands, believed to be caused by vigorous frost riving (p. 91). Moving downslope they become 'stone rivers'.

Charlesworth, 1957. 'The flat terrain at higher elevations in middle and higher latitudes is often covered with a sea of angular blocks variously called "block fields," "boulder fields" (Ger. *Felsenmeer, Blockmeer, Blockfelder*). Composed almost entirely of local rocks, this mountain-top frost-debris may be a few metres deep.... Transition forms glide or creep down the slopes along depressions....' (p. 574) Discussion follows.

**Block lava**
Hatch and Wells, 1937. Subaerial lavas... consist either of tumbled scoriaceous masses resembling clinker or slag (*block-lava*), or ... (p. 29).

Cotton, 1944. 'As used by many authors "block lava" is synonymous with "aa." Washington ... and, more recently, Finch (Finch, R. H., 1933, Block Lava, *J. of Geol.*, **51**, 764–770) have pointed out a contrast between the typically spinous "clinkers" formed of a true aa surface and the blocks of angular form of which some lava fields, notably those consisting of the more acid lavas, are superficially composed. Finch suggests that the descriptive term "block lava" should be reserved for these latter.' (p. 153)

**Block mountain**
*Webster.* A mountain caused by faulting and uplifting or tilting. See *Basin range*.

Mill, *Dict.* Block, Block Mountain. A part of the Earth's crust relatively higher than surrounding parts from which it is separated by faults. See Horst.
(under Horst 'Usually a large area, a small mass being often described as a block mountain').

Wooldridge and Morgan, 1937. Of the basin-range area of U.S. 'The mountains with their fault-scarp faces are true "block mountains" and to such the name should be restricted'. (p. 103) (Footnote deprecates use for ancient upland massifs of Europe for which 'Upland Massif' is sufficient.)

Cotton, 1941. '... uplifted fault blocks (block mountains)...' (p. 245)

Johnson, D. W., 1929, Geomorphological aspects of Rift Valleys, *Int. Geol. Cong.* (S.A.). 'Tilted blocks are the familiar forms commonly described simply as block mountains ... or fault block mountains, although this term has also been extended to include horsts which have no obvious monoclinal tilting.... Tilt block mountains...' (appears not to apply to 'true rift block mountain or horst', or thrust block mountains).

Strahler, A. N., 1946, Terminology and classification of land masses. *J. of Geol.*, **54**. Criticizes Davis's use of mountains to signify landmasses of disordered structure regardless of stage of erosion. Suggests replacing term 'block mountains' by 'fault blocks'.

'Cotton prefers to restrict "block mountains" to actual mountain ranges carved from large uplifted earth blocks bounded on one or both sides by fault scarps.' (*Geom.*, 1942, 154–161.)

'Marland Billings uses the terms "tilted fault block" and "horst" for structures which Davis, Johnson, or Lobeck would have classified as "block mountains".' (*Structural Geol.*, 1942, p. 37) pp. 32–42.

'In the judgment of the present writer, "block mountains" can be used as a convenient designation for actual mountain ranges produced by faulting but should not be applied to the structural class as a whole or to up-faulted blocks which have since been reduced to pediment or peneplain surfaces.' (pp. 39–40)

The comparison is between a terminology based on 'Initial Land Forms' and on 'Geologic Structure'.

**Blood rain**
*Webster.* Rain colored red by dust from the air.

Holmes, 1944. 'Red dust sometimes falls in Italy, and even in Germany, carried north by great storms from the Sahara. The "blood rains" of Italy are due to this same red dust washed down by rain from the hazy atmosphere.' (p. 253)
See also Wooldridge and Morgan, 1937, 299.

**Blossom shower**
The same as mango shower; showers which occur before the bursting of the monsoon in south-east Asia.

**Blowhole, blow hole**
Mill, *Dict.* Blowhole or Souffleur. A hole in the land near the shore through which air and water are forced by the rising tide.

McKenny Hughes, 1887. (On Caves, from *Trans. Vict. Inst.*). '"Blow Holes" or "Puffing Holes"... so when the wave rushes into a narrowing, funnel-shaped cave, with a small aperture communicating with the surface, the water is forced up through the opening, and often a spout of spray is carried high into the air.' (p. 3)

### Blowing well
Kendall and Wroot, 1924, *Geology of Yorkshire*. (in) '... the Magnesian Limestone ... wells which respond to variations of the barometer by respiring air' (p. 231); Gives examples at Solberge, Langton, etc. —see Fairly in *Proc. York. Geol. Soc.*, 7, 409–421.

### Blow-out
*Dict. Am.* A hollow made by the wind in sandy or light soil. 1892 Smith and Pound. *Bot. Survey Nebraska* II. 8. 'If a spot on a dry hill becomes bare, the loose sand is blown away, a small hollow is made, the surrounding grass dries from drought.'
*Webster.* A valley or depression blown out by the wind in areas of shifting sand or of light, cultivated soil. U.S.
*O.E.D. Suppl.* 3b. A butte, the top of which has been blown out by the wind until it resembles the crater of a volcano; a hollow in an area of shifting sand, caused by the action of the wind. Local U.S.
Bagnold, R. A., 1941, *The Physics of Blown Sand and Desert Dunes*, London: Methuen. 'If a small chance lowering occurs in the crestline of a long transverse dune, the local increase of wind velocity through the gap causes a rise in the rate of sand movement there, with a consequent removal of sand from the surface of the gap. The gap gets bigger, and a *blow-out* is formed.' (p. 206)
*Comment.* The second meaning given by *O.E.D.* as local U.S. is now common in Britain especially with reference to wind erosion in coastal dunes when vegetation is disturbed. (G.T.W.)

### Blow-well
Buckland, 1837, *Geology and Mineralogy*, 2nd Ed., Vol. I. Under the heading of Artesian wells.... 'Fountains of this kind are known by the name of Blow wells, on the eastern coast of Lincolnshire, in the low district covered by clay between the Wolds of Chalk near Louth and the Seashore. These districts were without any springs, until it was discovered that by boring through this clay to the subjacent Chalk, a fountain might be obtained, which would flow incessantly to the height of several feet above the surface' (p. 563).
Kendall and Wroot, 1924, *Geology of Yorkshire*. 'On the warplands of Holderness and the Humber Estuary there bubble up with great vigour a series of springs known as "blow wells"' (p. 626).
*Comment.* There appears to be two usages here—to a natural phenomenon and to artificial wells (G.T.W.).

### Blue ground (South Africa)
Diamond bearing dark volcanic rock (Kimberlite) filling the 'pipes' of the South African diamond mines. Most of it has the favourable property of crumbling on exposure to the air, which simplifies the extraction of the precious stones. (P.S.)
Wellington, J. H., 1955, *Southern Africa* I. 'Kimberlite, an ultra-basic magnesia-rich igneous rock, is known as "hardebank" at depth, changing nearer the surface to the softer "blue ground" which at the surface oxidizes to form the well-known "yellow ground".' (p. 28)

### Bluff
*O.E.D.* 1.b. Of a shore or coast-line: Presenting a bold and almost perpendicular front, rather rounded than cliffy in outline. Smyth, Sailor's Word-bk.
A cliff or headland with a broad precipitous face. (First used in N. America, and still mostly of American landscapes.)
*Dict. Am.* A steep river bank or shore, or the top of such a bank. (Citations back to 1687).
*Webster.* A high steep bank, as by a river or the sea, or beside a ravine or plain; a cliff with a broad face.
Mill, *Dict.* Also ... (2) The steep slopes bordering a river (U.S.A.).
*Comment.* Early uses include Cuvier (trans. Jameson) 1822 *Essay on the Theory of the Earth*, referring to sea-cliffs as 'steep-sloping bluffs or cliffs'; Lyell, *Principles of Geology* (bluffs of the Mississippi). Charlesworth, 1957, quotes G. Swallow as introducing the term 'Bluff Formation' for loess; Cotton, 1941, has diagram of meanders undercutting valley sides which the caption describes as bluffs.

### Boca (Spanish; Spanish America)
*Dict. Am.* The mouth of a river, gorge, etc.
Mill, *Dict.* The point at which a stream leaves a barranca, cañon or other precipitous gorge and enters the plain.

### Bocage (French).
*O.E.D.* (mod. French, bocage, wood) Woodland: a by-form of Boscage.

Mill, *Dict.* A region of mixed woodland and pastureland (W. France).

Monkhouse, F. J., 1959. *A Regional Geography of Western Europe*, London: Longmans. 'Most of this farmland is enclosed in small fields by banks, crowned with a thick hedge, often containing pollarded oaks and ash, though in the west these banks are replaced by dry-stone walls. This is the *bocage* country so characteristic of north-western France, and contrasting markedly with the open hedgeless *plaines*, *campagnes* and *champagnes* of the Paris Basin.' (pp. 459–460)

Comment. A type of country cut up into a chessboard pattern of little fields by hedges or lines of trees: originally from Normandy and Brittany. (E.D.L.)

**Bochorno** (Spanish)
Mill, *Dict.* The warm wind blowing up the Ebro valley.

**Bodden** (German)
Mill, *Dict.* The indentations of the Baltic coast between Mecklenburg and the Oder characterized by having islands to seaward.

Shackleton, 1958. 'Two types of coast are distinguished, the *bodden* and the *förden* ... both appear to be drowned coasts, but the *bodden* are irregularly shaped inlets in contrast to the long, straight-sided deep *förden*.' (p. 250) See Förden.

**Boerde, Börde** (German; *pl.* Boerden)
The sub-Hercynian loess zone in northern Germany or individual regions within this zone, *e.g.* Magdeburg boerde, Soest boerde, etc. (K.A.S.).

**Bog**
*O.E.D.* I 1. A piece of wet spongy ground, consisting chiefly of decayed or decaying moss and other vegetable matter, too soft to bear the weight of any heavy body upon its surface; a morass or moss. c. 1505–1846. b. (without pl.) bog-land, boggy soil. c. 1687–1861.

*Webster*. a. A quagmire filled with decayed moss and other vegetable matter; wet spongy ground, where a heavy body is apt to sink; a small, soggy marsh; a morass. b. Ecology. A wet overwhelmingly vegetable substratum which lacks drainage and where humic and other acids give rise to modifications of plant structure and function. Only a restricted group of plants, mostly mycorrhizal, can tolerate bog conditions. Cf. *swamp*, *marsh*, *meadow*.

Mill, *Dict.* (as *O.E.D.*) See Moor, Ripe Bog, Unripe Bog. Ripe Bog: a bog is said to be ripe when the upper and bottom layers of peat meet. Unripe Bog: a lake becoming modified into a bog, consisting of a layer of peat at the bottom, then the water of the bog, and finally a growing layer of peat-moss at the top.

Tansley, 1939. Moss and Bog 'the natural group of wet peat-forming and peat inhabiting communities.... The disadvantage of the word "bog" is that it is sometimes loosely used in common language for any wet soil into which the foot sinks, but this again is not a very serious drawback.... In the present work, therefore, "moss" and "bog" are used synonymously for the wet acid peat vegetation;...' (p. 674). Also defines valley bog, raised bog, blanket bog.

Fay, 1920. 'Bog (Celtic for soft). A wet spongy morass, chiefly composed of decayed vegetal matter (Power).'

Carpenter, 1938. Among others quotes: *Ecol. Ass. of Am.*, 1934, List R-1. 'The vegetation complex which develops in un- or imperfectly drained localities.'

Comment. The word is used in two senses: (a) type of surface—soft, wet, spongy ground—and (b) the vegetation complex associated with that type of surface. See the quotation under Fen. Note also quaking bog. Mill's distinction between 'ripe' and 'unripe' would seem to have been dropped; not in Tansley, 1939.

**Bogaz** (Southern Slav)
Moore, 1949. In a Karst region, a long narrow chasm, enlarged by solution of the limestone, into which a surface stream empties.

Wooldridge and Morgan, 1937. 'The Serbian term for deeper fissures is "bogaz".' (p. 289)

Not in *O.E.D.*, Mill.

**Bog iron ore**
Read, H. H., 1936, *Rutley's Mineralogy*, 23rd Ed. 'A loose porous earthy form of limonite, found in swampy and low-lying ground, often impregnating and enveloping fragments of wood, leaves, mosses, etc.' (p. 463)

Twenhofel, 1939. 'Swamp deposits may contain iron oxide, which is rarely pure and usually contains a high percentage of phosphorus and considerable silica. The beds of iron oxide may attain a maximum thickness of about 1 meter, but they are usually thinner and without stratification. Many have concretionary structure. These are bog-iron ores, known as *murram* in the tropics and *moccarrero* in Cuba. In some places they form the *iron pan*. . . . Deposits of fresh-water swamps contain little or no iron sulphide. These may be considerable in marine-swamp deposits.' (p. 79)

Himus, 1954. 'An impure ferruginous deposit, formed in a bog or swamp by oxidation by algae, bacteria, or the atmosphere of solutions containing salts of iron' (p. 16).

**Bohorok** (Sumatra)

Robequain, C., *Le Monde Malais* (translated by Laborde, E. D., 1954), 'a sort of foehn which blows down from the Karo Mountains' (p. 161).

**Boli** (Sierra Leone)

Kline, H. V. B., 1956, *IGU Report of a Symposium at Makerere*, 1955, 43. 'Wide inland marshes which exist where major rivers cross a belt of old sedimentary rocks.'

White, H. P., 1958, *Geography*, **43**, 1958. 'The Boli lands of Sierra Leone ... the bottom lands or *bolis* are flooded in the rains.' (p. 270)

**Bolson**

*Dict. Am.* S.W.U.S. A low area or basin completely surrounded by higher ground or mountains. (Citations back to 1838.)

*Webster.* A flat-floored desert valley which drains to a playa.

*O.E.D. Suppl. U.S.* (sp., augmentative of bolsa, purse)

In the south-western U.S.A. and Mexico, a basin-shaped depression surrounded by mountains.

Mill, *Dict*. Plains covered by pebbles brought down by occasional torrents in the mountain-surrounded basins of Mexico; apparently also used for the basins themselves.

Lobeck, 1939. 'Bolson plains. In southern New Mexico and in Texas the broad intermontane plains known as *bolsons* (from the Spanish word for purse) are largely wind-excavated depressions. Some of them contain deep alluvial accumulations washed into them from the surrounding mountains. But in many of the depressions the rock floor is exposed' (p. 378).

Tolman, C. F., 1937, *Ground water*. 'The area within centripetal drainage of the sunken block is designated as a *bolson*. If drained by an intermittent longitudinal stream it is a *semi-bolson*.' (p. 522) (He refers to Tolman, C. F., Erosion and Deposition in the S. Arizona Bolson Region, *J. of Geol*., **18**, 1909, 136–163.) 'Definitions. A topographic basin with centripetal drainage.' (p. 558)

*Comment.* Lobeck's 'wind-excavated depressions' is far from generally accepted.

**Boma** (East Africa, especially Tanganyika; Swahili)

*Webster.* 1. A Central African defensive enclosure of a camp or village; a stockade. Cf. *kraal*. 2. In Central Africa, a police post or an office of a district commissioner.

A place protected by an earthwork, stockade or strong hedge: a cattle kraal; a fort; government offices, especially the administrative headquarters. (S.J.K.B.)

**Bomb, volcanic**

In many types of volcanic eruption, liquid lava is thrown high into the air and rounded masses of varied size fall to the ground as bombs.

**Booly, booley** (Irish) and other spellings including bouille

*O.E.D.* A temporary fold or enclosure used by the Irish who wandered about with their herds in summer; a company of people and their herds thus wandering about. Hence booling for the practice.

**Boolyunyakh, bulgun(n)yakh** (Russian): Yakut

A hydrolaccolith (*q.v.*). Occasionally used in American translations of Russian works.

**Bora, borino**

*O.E.D.* (According to Diez, Venetian, Milanese form of It. borea, north wind. But *cf.* Illyrian, bura, storm, tempest, which may have been confounded with the Ital. in the Adriatic)

A severe north wind which blows in the Upper Adriatic.

*Webster.* An occasional violent, cold, north to northeast wind that blows over the northern Adriatic from the interior highlands.

*Met. Gloss.*, 1944. A cold, often very dry, north-easterly wind which blows, sometimes in violent gusts and squalls, down from the mountains on the eastern shore of the Adriatic.

Biel, E. R., 1944, *Climatology of the Mediterranean Area*. Chicago. 'The cold downslope winds of the eastern coast of the Adriatic Sea known as bora winds (derived from the Greek *boreas*, north wind) are the best

known representatives of a whole family of winds known as "bora-type winds".'
... 'By definition, bora winds are drifts from cold continental source regions toward warm sea shores or lowlands which slope downward from mountains of moderate altitude.' 'a sub-type "anticyclonic bora"' (p. 21). 'When the center of low pressure is passing the central Adriatic, the northern regions experience bora winds.... In such cases ("cyclonic" or "black" bora)' (p. 22). '... bora winds occur also in summer... much less frequent and weaker. They are named "borino" (little bora)' (p. 22). 'In particularly exposed areas railroads have to be protected by "bora walls"' (p. 22). 'The area of Novorossiysk... is notorious for its violent bora winds' (p. 55).

Kendrew, 1953, *Climates of the Continents*, 4th ed. 'Black Sea. When, however, a deep depression lies over the east of the sea, the lower Western end of the Caucasus is unable to keep back the N.E. wind which blusters down as the dreaded bora, an exceedingly strong, cold, dry wind from the steppes, especially prominent in the neighbourhood of Novorossiysk' (p. 266). Also gives for Adriatic, (p. 376).

Hare, F. K., 1953, *Restless Atmosphere*. 'The Bora is a wind blowing through the gap between the Alps and the high ground of the central Dinaric ranges. Severe bora winds are confined to the saddle region behind Trieste...' (p. 178).

Mori, A., 1957, *L'Italia Fisica*. '*Bora chiara* for the anticyclone type with its clear skies' and refers to the cyclone as *bora scura* (*lit.* clear and dark boras). Also refers to the 'so-called borine' (*pl.* form) for summer winds.

*Comment*. There is a local usage, sometimes given a capital *B*, and a general term. Note also the sub-types.

### Börde—see Boerde

### Border

*O.E.D.* 2. The district lying along the edge of a country or territory, a frontier; pl. the marches, the border districts.
b. The boundary line which separates one country from another, the frontier line.
3. spec. a. (Eng. and Sc. Hist.) The Border, the Borders: the boundary between England and Scotland; the district adjoining this boundary on both sides; the English and Scottish borderland. (The term appears to have been first established in Scotland, where the English border, being the only one it has was emphatically the border.) c. In U.S. The line or frontier between the occupied and unoccupied parts of the country, the frontier of civilization.

### Borderland

Umbgrove, 1947. 'In the opinion of Stille, a "borderland" generally has to be considered as part of a geosyncline that was folded during a previous epoch of compression and subsequently became a geanticlinal ridge' (p. 35).

*Comment*. A highly specialized use by structural geologists of a word normally used with a very different meaning.

### Bore

*O.E.D.* 2. A tide-wave of extraordinary height, caused either by the meeting of two tides, or by the rushing of the tide up a narrowing estuary. *cf.* Eagre.

*Webster. Phys. Geog.* a. A tidal flood which regularly or occasionally rushes with a roaring noise into certain rivers of peculiar configuration or location, in one or more waves which present a very abrupt front of considerable height, dangerous to shipping, as at the mouth of the Amazon in South America, the Hooghly and Indus in India, and the Tsientang, in China. b. Less properly, a very high and rapid tidal flow, when not so abrupt, such as occurs in the Bay of Fundy and in the Severn.

Mill, *Dict*. 1. The tubular outlet of a geyser. 2. A deep vertical hole of small diameter made to ascertain the nature of the underlying strata or to obtain water. 3. A high-crested wave caused by the rushing of the tide up a narrowing gulf, estuary or river, or by the confluence of tides.

### Boreal

*O.E.D.* Of or pertaining to the north.

*Webster*. 1. Of or pertaining to Boreas; hence, northern. 2. [cap.] Biogeog. Designating or pertaining to a terrestrial division consisting of the northern and mountainous parts of the Old and the New World —equivalent to the *Holarctic region* exclusive of the Transition, Sonoran, and corresponding areas. The term is used by American authors and applied by them chiefly to the Nearctic subregion. The

Boreal region includes approximately all of North and Central America in which the mean temperature of the hottest season does not exceed 64·4°F. Its subdivisions are the Arctic zone and the *Boreal Zone*, the latter including the area between the Arctic and Transition zones and having as subdivisions the Hudsonian and Canadian zones.

*Comment.* Applied especially to the northern coniferous forests, the boreal forests or taiga.

**Bornhardt** (Bailey Willis, 1936)

Willis, B., 1936, *East African Plateaus and Rift Valleys*, Carnegie Inst., Washington Publ., 470. 'A residual hill which possesses the peculiar, typical characteristics of a true inselberg... the peculiar type of residual which is most strikingly characterized by resemblance to an island rising from a water surface.' Bornhardt was the originator of the term 'inselberg'. (p. 121)

*Comment.* Professor Carl Troll informed the Committee that the term is obsolescent in Germany; nevertheless some English writers have attempted to resuscitate it and to give it a more precise meaning than inselberg.

**Bōrō** (Indo-Pakistan: Bengali)

Summer rice or rice-crop.

Johnson, B. L. C., Crop Association Regions in East Pakistan, *Geography*, 43, 1958. 'Bōrō paddy is grown on low (and therefore generally heavy) land close to sources of irrigation water.' (p. 88) See also Ahmad, 1958, 128.

**Borough**

Originally a fortified place or a town with a municipal organization, it was applied in England and Wales to towns with a certain status and organization conferred by Charter. A *county borough* (C.B.) reserved for large towns, has the status of an administrative county; its elected head is usually a Lord Mayor. A *municipal borough* (M.B.) has a royal charter and is headed by a Mayor but is a dependent part of an administrative county. A town having the right to send one or more members to Parliament is a *parliamentary borough*. Before the Reform Bill of 1832 certain very ancient boroughs had a tiny electorate and were thus disproportionately represented in Parliament and, being easily controlled by a few influential persons, became known as *Rotten Boroughs*. The word is also loosely used for other towns. The Scottish equivalent is *burgh* (also in older English usage).

*Webster.* 4a. In some States of the United States (as Connecticut, Pennsylvania, New Jersey, and Minnesota), a form of municipal corporation corresponding in general to the incorporated town or village of other States. b. In Greater New York, one of the five constitutent political divisions of the city. 5. In New Zealand, a village, township, or town having a special governing body called a *borough council*. 6. *Australia* a. In New South Wales, a town area as incorporated by an act of Parliament of 1857, or one holding a special charter from the crown (royal charter)—formerly distinguished from a *municipality* but now superseded by it, *municipality* being applied to an urban area as distinguished from a rural area, or *shire*. b. In other States, a municipal area of a certain minimum size and population.

**Bösche**

Thornbury, 1954. 'According to Penck, there are two basic elements in any valley slope ... —the upper, relatively steep *bösche* or *steilwand* or what Meyerhoff (1940) has called the *gravity slope*; and the lower, more gentle *haldenhang* or *wash slope* of Meyerhoff.' (p. 200)

*Comment.* Not used and not in the glossary of Czech and Boswell's translation of W. Penck, *Morphological Analysis of Land Forms* (London: Macmillan, 1953). Would seem to be an error.

**Boss**

*O.E.D.* 2. A knoll or mass of rock; in Geol. applied chiefly to masses of rock·protruding through strata of another kind.

Mill, *Dict.* A large amorphous mass of crystalline igneous rocks which have cooled or consolidated at some depth from the surface and are now exposed by denudation.

Hatch and Wells, 1937. 'Boss is the term applied to stocks of circular cross-section.' (p. 10)

Holmes, 1944. 'Smaller intrusions of similar type (to batholiths) and with a real dimensions of only a few square miles or less, are called *stocks*. When a stock has a roughly circular outline it is sometimes referred to as a *boss*.' (pp. 89–90)

*Comment.* See also batholith; the modern trend is to follow Holmes.
**Bosveld**—see Bush veld
**Bote**—see Estovers
**Bottom**
*O.E.D.* 2. The ground or bed under the water of a lake, sea or river.
4a. The bed or basin of a river. b. Low-lying land, a valley, a dell; an alluvial hollow.
*Webster.* (chiefly in *pl.*). Low land formed by alluvial deposits along a river, as the Mississippi River *bottoms.*
*Dict. Am.* 1. Level land, usu. fertile and cultivable, on the margins of rivers and creeks. (Citations back to 1634.) 2. First, second, third bottoms, the first and succeeding levels or plains along the shores of some rivers. 1788 Imlay, *Western Territory* (1797), 595 'Next to these are what is called second bottoms, which are elevated plains and gentle risings of the richest uplands, and as free from stone as the low or first bottom.' 1803 Lewis in *Journal of Lewis and Ordway*, 34 'What is called the third bottom is more properly the high benches.'
Mill, *Dict.* A dry valley floor; in the United States an alluvial plain.
*Comment.* In geographical literature applied especially to the alluvial flood plains of the Mississippi, *e.g.* Yazoo bottoms. It has also been used for valleys in karst country in the United States but this is condemned by S. N. Dicken, Kentucky Karst Landscapes, *J. of Geol.*, **43**, 1933, 713 as a loose term for what should be called ponor and doline. In Britain the word bottom often appears in place names for dry valleys, especially in chalk and other limestone regions in southern England. (G.T.W.)

**Bottom ice**
Also known as anchor ice—see Ice, sundry terms.

**Bottom-set beds**
Fay, 1920. 'The layers of finer material carried out and deposited on the bottom of the sea or a lake in front of a delta. As the delta grows forward they are covered by the fore-set beds (La Forge).'
Cotton, 1945. 'In a delta some of the silt layers are covered over by advancing fore-set beds, and then become the bottom-set beds of the delta.' (p. 206)
Kuenen, 1955. 'In front of these fore-sets fine silt comes to be deposited, serving as a foundation for the fore-sets that are extending farther foreward. These are the bottomsets.' (p. 279)
See Delta structure.

**Boulder**
*O.E.D.* 2. spec. Geol. A large weather-worn mass or block of stone, frequently carried by natural forces to a greater or less distance from the parent rock, and generally lying on the surface of the ground, or in superficial deposits; an erratic block.
*Webster.* Any detached and rounded or much worn mass of rock, from a size distinctly larger than a cobblestone to one ten or more feet in diameter. Many boulders have been carried far from their original sources, as by rivers (river boulders) or glaciers (glacial boulders); others (shore boulders) have been worn, as by waves, without much transportation; others (boulders of weathering) owe their rounded form to weathering in situ.
Humble, 1800. 'Large fragments or rounded masses of any rock found lying on the surface, or sometimes imbedded in soil, and differing from the rocks where they are found ... of no determinate size. ...' (p. 59)
Page, 1859. 'Any rounded or water-worn blocks of stone, which would not from their size be regarded as pebbles or gravel, are termed boulders. The name is usually restricted to the large, waterworn and smoothed blocks (erratic blocks) found embedded in the clay and gravels of the Drift formation of the Pleistocene epoch.' (p. 95)
Rice, 1951. A more or less rounded block of rock, commonly larger than a cobblestone. Usually boulders are rounded by being carried or rolled along by ice and water, but certain boulders have been rounded by weathering in place and are known as boulders of weathering, disintegration or exfoliation. (p. 51)
Balch, H. E., 1943. *Caves of Mendip*, Vol. III. 'Boulder is used in a special sense ... but there is no other word which so well expresses the detached masses of rock varying from 2 to 5 or 6 feet irregular cubes. ...' (p. 6)
Glentworth, R., 1954. *The Soils of the Country Round Banff, Huntly and Turriff*, Edinburgh: H.M.S.O. Rocks larger than $4\frac{1}{2}$ in.

diameter (*i.e.* larger than gravel). (p. 162)
King, C. A. M., 1959. *Beaches and Coasts*, London: Arnold, gives grade sizes of beach materials: boulders more than 200 mm (8 ins.) (British Standards Inst.), more than −8·0 units (Inman), more than 256 mm. (Wentworth). The next smallest is designated cobble.

Comment. A typical example of a very common word which has been given specialized meanings in different branches of science.

**Boulder Clay**
O.E.D. A clayey deposit of the ice-age, containing boulders.
Wright, 1914. '... the ground moraine becomes much compacted, and, when finally abandoned by the ice, forms the stiff, tenacious deposit known as boulder-clay.... Boulder-clay is the commonest of the drift deposits in all glaciated regions. In its typical form it is a stiff, tenacious, unstratified clay full of subangular stones of various sizes, and lying in all positions in the clay.' (p. 29)
Holmes. 1928. 'A tenacious unstratified deposit of glacial origin consisting of a stiff clay (rock flour) packed with subangular stones of varied sizes.'
Rastall, R. H. (revis.), 1941, Lake and Rastall's Textbook of Geology, 5th ed., London: Arnold. 'True boulder-clay ... consists of a matrix of clay, often very stiff indeed, enclosing an extremely variable proportion of blocks of rock of any size, from mere chips and pebbles to masses weighing many tons, and often showing the characteristic facetting and scratching.' (p. 121)
Geikie, J., 1898 regards it as = Till but in *The Great Ice Age* distinguishes them. Till = ground moraine. Boulder clay = submarine terminal moraine and true *moraine profonde*.
Flint, 1947. Regards Boulder Clay as English form of till (Scots) which is not good because 'some till contains no boulders, some contains little or no clay, and some (though perhaps not very much) contains neither boulders nor clay, but only silt, sand, and small stones.' (p. 103)
Comment. The distinction made by Geikie is not maintained. Whilst Wright is correct, the comments of Flint are justified because 'boulder-clay' as mapped by the British Geological Survey may contain few if any boulders and may be so sandy as not to be a clay at all. It is, in other words, synonymous with ground moraine. It follows that boulder-clays yield soil of most varied characters. That derived from chalky boulder clay (with pebbles of chalk) is very good agriculturally, whereas soils derived from very stiff clays with numerous boulders may be almost useless. Because of the extreme variability of boulder clays, some of the deposits mapped as such having no boulders and very little clay, there is a tendency to drop the term in favour of till, especially in America. *Webster* simply gives 'See *till*'. See also Charlesworth, Chapter XVIII.

**Boulevard** (French)
*Webster*. 1. *Fort.* Orig. the flat top of a bulwark or rampart. 2. A broad avenue in or around a city, esp. one decoratively laid out with trees, belts of turf, etc. The Parisian *boulevards* occupy the site of demolished fortifications. 3. An improved avenue or thoroughfare, esp. in cities, for pleasure vehicles, sometimes exclusively; often one which is a through way.
A very wide main street: properly one occupying the site of a former town wall. (E.D.L.)
Comment. In recent years this word has come to be widely used in America and Britain especially for wide new roads, *e.g.* ring roads round cities such as Nottingham, also for scenic highways (Skyline Boulevard).

**Bound** (and **Free**) **Resources**
Dansereau, Pierre, 1957, *Abs. of Papers, Ninth Pacific Science Congress*, Bangkok. 'One physical condition can block off another physical element in which case the resource is said to be "bound".' (p. 256)

**Boundary**
Committee, List 1. 1. Synonymous with *frontier*. 2. The line of delimitation or demarcation between administrative units or between geographical *regions* of various types, whether *physical* or *human*.
Boggs, S. W., 1940, *International Boundaries*, New York: Columbia Univer. Press 'International boundaries are thus sharply defined lines ...' (p. 6).
Wooldridge, S. W. and East, W. G., 1958, *The Spirit and Purpose of Geography*, London: Hutchinson ... 'The political

geographer is concerned with frontiers and boundaries, which are distinct conceptions: the frontier is a zone, the boundary a line.... Not all States have boundaries....' (p. 134)

Comment. The English language admits of this convenient distinction, which some political geographers are trying to standardize. (W.G.E.)

See also Relict boundary and Superimposed boundary.

### Bourne

*O.E.D.* Bourn, bourne. Forms: burn, burna, burne, borne, bourne, boorn/e. A variant of Burn, being the form commonly used in the south of England since the 14th. c. Originally pronounced like *burn, adjourn*: but the influence of the *r* disturbed the pronunciation, as in *mourn*, whence the mod. spelling and pronunciation. A small stream, a brook; often applied (in this spelling) to the winter bournes or winter torrents of the chalk downs. Applied to northern streams it is usually spelt Burn. Examples c. 1325–1879. Note also use of bourn as a boundary between fields, a bound, a limit, the limit or terminus of a race; the ultimate point aimed at, or to which anything tends; destination, goal.

Mill, *Dict.* Bourn or Bourne. (1) The winter torrents of the chalk and limestone heights (S. England). 2. The headwaters of a river flowing from one or more large springs (Cambridge and E. Anglia).

Knox, 1904. Bourne (Anglo-Saxon), a stream, rivulet, *e.g.* Eastbourne. *cf.* Born, Burn.

Jukes-Browne, A. J., 1904, *Cretaceous Rocks*, III, Mem. Geol. Surv. '... temporary streams which are known as bournes, nailbournes, winterbournes, woe-bournes, levants and gypsies.' (p. 431) Also refers to regular bournes (winterbournes) and 'occasional bournes' and to 'bourn-holes' also to the 'breaking-out of the bourne'. (p. 431)

Comment. Found commonly as a place name on the chalklands of southern England. The intermittent character is explained by variations in the level of the water table in the chalk. In winter when it rises the water table is above the height of the valley floor and there is a surface stream; in summer it sinks below the level of the valley floor and the stream bed becomes dry: hence winterbourne. (L.D.S.)

### Boval (*plural*; *sing.* bowé)
'Lateritic crusts with peculiar vegetation in savanna regions in west and central Africa.' (P.W.R. in MS)

### Bovate (Agricultural History)—see under Yardland

### Box-canyon
In the west of the United States, where every young valley is called a canyon, in order to distinguish those with more or less vertical walls the term 'box-canyon' is used. (*fide* Baulig).

See Cañon.

### Bradyseism
*O.E.D. Suppl.* (Milne). A slow rise and fall of the earth's crust.

### Brae (Scottish, dialect; northern Ireland)
*O.E.D.* 1. The steep bank bounding a river valley.

2. A steep, a slope, a hill-side.

A broader usage refers to an upland district. Examples are Braemar, the Braes of Balquhidder; also Brae Lochaber and Brae Moray in contrast to Nether Lochaber and Laigh of Moray respectively (C.J.R.). More usual of hillside; compare use of brow in northern England (A.F.M.).

Also used in street names for a road with a steep gradient. S.N.D. gives example Windmill Brae, Aberdeen.

### Braided river courses, braiding
*O.E.D.* Braid 4. To twist in and out, interwave, plait.

Mill, *Dict.* Braided Stream. A stream broken up into parts; it may break only when the water is low.

Salisbury, 1923. 'When a stream deposits sediment in its channel, the channel becomes smaller. In time it may become too small to hold all the water, and a part then breaks out, and follows a new course in the valley flat. This process may be repeated.... The departing streams may or may not return to the main. If they do the stream becomes a network of little streams, sometimes called a *braided stream*' (p. 118). Quotes the R. Nebraska as example—again only in low water stage.

Friedkin, J. F., 1945, *A Laboratory Study of the Meandering of Alluvial Rivers*. 'Rivers are described as braided when the channel is extremely wide and shallow and the flow passes through a number of small interlaced channels separated by bars. There is little or no erosion of the main banks. The channel as a whole does not meander, although local meandering in minor channels generally occurs ... braiding results when the banks of a channel are extremely easily eroded.' (p. 16)

Cotton, 1945. Of aggrading rivers ...'... flowing in anastomosing (or braided) courses.' (p. 194)

Kuenen, 1955. 'If bar-building takes place on an ample scale and the available valley floor is wide, the river piles the bars up to a considerable height, divides and passes around them on either side. The river is said to braid.' (p. 275)

Charlesworth, 1957. '... they anastomose (streams on *sandar*) as "braided streams" which swing down the valley in an intricate lacework of interosculating channels' (p. 436).

*Comment.* Cotton also refers to braided lava streams (*Volcanoes*, 1944, p. 131).

**Brak soils (Afrikaans)**
Afrik: brak; Engl. brackish; German brackig. Slightly salty or brackish.
In South Africa brak is used both as an adjective and as a substantive. Brak soils have a salt content high enough to damage crops, but not so high as to make all vegetation impossible. Brak (*i.e.* high salt content, so the word is used as a substantive here) forms a serious problem in many irrigated areas. (P.S.)

**Brake**
*O.E.D.* A clump of bushes, brushwood, or briers; a thicket.

**Brakeph** (T. Griffith Taylor, c. 1920)
Taylor, 1951, 'broad-headed' (p. 611). Abbreviated from brachycephalic (for person or race). Contrast Dokeph.

**Brash**
*O.E.D.* A mass or heap of fragments; applied to (a) loose broken rock forming the highest stratum beneath the soil of certain districts; rubble (b) fragments of crushed ice.
*Comment.* (a) is uncommon except in term 'cornbrash' an horizon in the English Jurassic originally so called because its brashy soils were good for corn growing. From (b) comes brash-ice. Also a forestry term.

**Brash ice**—see Ice, sundry terms

**Brave West Wind(s)**
Miller, 1953. 'In the southern hemisphere the disturbance of the planetary winds [the Westerlies of mid-latitudes] is much less [than in the northern hemisphere]; "Roaring Forties" and the "Brave West Winds" blow all the year round with considerable force and a high degree of dependability.' (p. 199)

**Break of profile**
A knickpoint (*q.v.*)—used by some writers who prefer to avoid a term of mixed German-English origin.

**Break of slope**
Any more or less sudden change in a slope, *e.g.* of a hillside.

**Breaker**
*O.E.D.* A heavy ocean-wave which breaks violently into foam against a rocky coast or in passing over a reef or shallows.
Swayne, 1956. A wave breaking on the shore.

**Breccia**
*O.E.D.* (Italian, breccia, gravel or rubbish of broken walls, from Teutonic: Old Teut., brekan, to break. Used in the name *Breccia Marble*, before its separate use in Geology). *Geol.* A composite rock consisting of angular fragments of stone, etc., cemented together by some matrix, such as lime: sometimes opposed to *conglomerate*, in which the fragments are rounded and waterworn.

Scott, W., 1832. *Stourbridge and its Vicinity.* 'Breccia has been described as: 1. Consisting of irregular masses of two or more minerals, so intimately blended as to show that they were consolidated at the same time. 2. Fragments of pre-existing rocks united by a cement. 3. Rounded, waterworn stones, consisting of various components—siliceous breccia, as that of Hertfordshire, consisting of siliceous pebbles cemented by a paste of the same substance.' (p. 493)

*Comment.* This word, of Italian origin, formerly had the wider meaning it still has on the continent and included conglomerates with rounded pebbles. It is now always used in English only when the fragments of rock are angular. See also agglomerate, autoclastic, fault-breccia.

**Breck, Breckland**
*O.E.D.* 2.=Break, 1787. Marshall, Rura Econ. E. Norfolk, Breck ... a large new-made inclosure. 1863. Morton, Cycl. Agric. II, Breck (Norf., Suff.), a large field. In Northumberland, etc. a portion of a field cultivated by itself.
*O.E.D. Suppl.* (adds few quotations, three referring to Norfolk and including:) 1897.

W. Rye, Norfolk Songs, 124. Such cramped wild country, half rough breck land and half marsh.

English Place-Name Society, *Survey of English Place-Names, Vol. I, Part II. The Chief Elements used in English Place-Names*, ed, A. Mawer. Cambridge: C.U.P., 1924. Bræc, brec., O.E., dial. brack, breck, 'strip of uncultivated land', 'strip of land taken in from a forest by royal licence, for temporary cultivation.' Difficult in Scand. England to distinguish from brekka. (p. 8)

Wallace, D., and Bagnall-Oakeley, R. P., *Norfolk*, London: Hale, 1951. Oppose the interpretation 'land broken by the plough' and suggest a connection with 'brake', a thicket or with 'bracken'. (W.M.) (p. 32)

Clarke, W. G., 1937, *In Breckland Wilds*, 2nd ed., Cambridge: Heffer. In the middle of the 19th cent. 'Breck district' became current—'breck, meaning a tract of heathland broken up for cultivation from time to time and then allowed to revert to waste'. 'Breckland' in one word as a regional name was first used by W. G. Clarke in 'A Breckland Ramble' in *Naturalists' Journal*, 1894, II, 104–133, p. 1 and p. 174.

Mosby, J. E. G., 1938, *Land of Britain, Part 70, Norfolk*, London: Geographical Publications. Acknowledges adoption of term by Clarke. (pp. 222–229)

Knox, 1904. Break (Icelandic brekka, a declivity), a hollow among hills; (Scotch) a division of land in a farm.

*Comment.* Ekwall gives from old English bræc, ground broken up for cultivation; newly cleared land (*fide* Ekwall), hence temporary clearings for cultivation in the outfield (*q.v.*).

**Breeze**

*Met. Gloss.* 1944. A wind of moderate strength (see Beaufort Scale). The word is generally applied to winds due to convection which occur regularly during the day or night.... Land and Sea Breezes, Mountain Breeze, Valley Breeze.

**Breezes, Land and Sea**

Miller, 1953. 'The difference in behaviour of land and water towards diurnal and annual temperature changes begets a difference of pressure which results in periodic diurnal and seasonal winds known as land and sea breezes and monsoons. The heating of the land during the day causes an ascent of air over the land and an indraught of oceanic air; the descent of air cooled by radiation over the land at night causes an expulsion of land air out to sea. Land and sea breezes are most noticeable and regular where temperature changes are most regular and particularly in equatorial climates.' (p. 43)

**Breva** (Italian)

Mill, *Dict*. The valley breeze blowing upward in the morning and afternoon as a warm damp wind from the Italian lakes. The opposite is tivano.

**Brick-earth, brickearth**

*O.E.D.* Brick-earth. Earth or clay suitable for making bricks; in Geol. A clayey brownish earth lying below the surface soil in the London basin.

1878. Huxley, Physiogr., xvii, 280. In many places round London the sheet of gravel is overlaid by a thin deposit of brownish loam represented on the map as brick-earth since it is largely worked by brick makers.

Mill, *Dict.* Earthy deposits with angular fragments derived from weathering of neighbouring rocks, and accumulated through action of wind, rain, etc. (S. England).

Page, 1865. 'In geological classification, the term "Brick-clay"—meaning thereby those finely-laminated clays of the Pleistocene epoch which immediately overlie the true Boulder-clay, and have evidently been derived from it by the wasting and re-assorting agency of water.'

Geikie, J., 1905. 'This rain-wash is occasionally sufficiently fine-grained and plastic to serve for brick-making purposes (brick-earth).'

Rice, 1941. 'Brick earth. Clay or earth for making bricks (Fay). Used in England. In glacial sediments it is leached clayey material produced by weathering of drift in situ or alluvial accumulation of the weathered part of drift from slightly higher adjacent areas. The two types might best be known as residual brick earth and alluvial brick earth (Antevs and MacClintock).'

Oakley, K. P., 1949, *Man the Toolmaker*, repeats Rice and MacClintock.

Note: Geological Survey spell in one word and use it for deposits associated with clay-with-flints on downs, as well as in

Middlesex (Whittaker, 1889, Woodward revised, 1922).

**Comment.** Current usage confines brickearth to the fine grained deposit (not as defined by Mill) overlying the gravels on certain of the Thames terraces (as defined by Huxley). It has been likened to loess but in a humid climate resorted by water. For an early reference see Prestwich, *Proc. Roy. Soc.*, 1867, p. 272 refers to 'Brick-Earth or Loess... result of river floods.'

**Brickfielder, Brick Fielder** (Australia)
Moore, 1949. 'The hot wind experienced in south-eastern Australia especially during the summer, and often bringing clouds of dust' (p. 26).
Kendrew, W. G., *Climates of the Continents*, 4th Ed., 1953, quoting the Australia Year Book (no year specified).... 'In Victoria the hot winds are known as "Brick Fielders", a name originally applied to the "Southerly Bursters" in Sydney because of the dust they raised from the brickfields to the south of the city. When the gold fields were discovered in Victoria the miners hailing from Sydney gave the name to the dusty winds from the opposite quarter.' (p. 552)

**Bridge-head**
*O.E.D.* A fortification covering or protecting the end of a bridge nearest the enemy.
Hence extended to an advanced post across a river in figurative sense.

**Bridge-point, bridging point**
A point at which a river is, or can be, bridged; used especially in 'lowest bridging point'—the lowest point at which a river could be bridged and often an important factor in the original location of a town.

**Bridle path, bridle-path, bridle way, bridle-way, -road**
*O.E.D.* A path fit for the passage of a horse, but not of vehicles.
**Comment.** The term has certain legal implications; there may be a right of way for pedestrians and riders on horseback but not for wheeled vehicles; the rights of the public are greater than those on a footpath but less than those on a highway.

**Brigg, brig**
Mill, *Dict.* A headland or landing-place formed by a scarp of hard rock cropping out at or near tide marks (Yorkshire and Northumberland).

Phillips, J., 1835, *Illustrations of the Geology of Yorkshire*, Pt. I. 2nd. Ed., '... the oolitic rocks emerge from the sea and form the long reef called Filey brig.' (p. 49)

**Bro** (Welsh)
Region, vale, lowland. See Blaen.

**Broad, Broads, Broadland**
*O.E.D.* B.5. In East Anglia, an extensive piece of fresh water formed by the broadening out of a river.
1787 Marshall, Norfolk (E.D.S.) Broads, fresh-water lakes (that is, broad waters; in distinction to narrow waters, or rivers).
Mill, *Dict.* A lake or sheet of water bordering or forming the direct course of a sluggish river near its estuary (East Anglia).
**Comment.** The origin of the Norfolk broads has been the subject of much discussion. It now seems certain that the working and removal of peat for fuel in past times is an important factor. See J. M. Lambert, J. N. Jennings, *et al. The Making of the Broads*, London: Royal Geog. Soc., Research Series No. 3, 1960. The district on the Norfolk-Suffolk border concerned is sometimes called Broadland.

**Brockram**
Himus, 1954. A sedimentary rock, consisting of angular blocks, which occurs in the Permian strata west of the Pennines.

**Brook**
*O.E.D.* 1. A small stream, rivulet; orig. a torrent, a strong flowing stream.
*Webster.* A natural stream of water, smaller than a river or creek; esp. a small stream or rivulet which breaks directly out of the ground, as from a spring or seep; also, a stream or torrent of similar size, produced by copious rainfall, melting snow and ice, etc.; a primary stream, not formed by tributaries, though often fed below its source, as by rills or runlets; one of the smallest branches or ultimate ramifications of a drainage system.

In England and in New England, streams which elsewhere in the United States are denominated *creeks* are regarded as small rivers, and any tributary or independent stream of smaller size is usually called a *brook*. In Maine, many of the larger brooks and intermediate watercourses are styled simply *streams*, the terms *brook* and *stream* being sometimes used interchangeably. In the northerly middle States and Lake States, populated partly from New

England, *brook* as a geographical term occurs to a varying extent, along with synonymous terms. In the southern and western States generally, the word *brook*, though freely employed in literature and in conversation, has no vogue as a geographical term. It is replaced in the easterly southern States oftenest by *run* and in the remaining southern and most of the western States somewhat uniformly by *branch*.

### Brow
*O.E.D. north dial.* A slope, an ascent = Sc. *brae*.

### Brown coal
Himus, 1954. An intermediate stage in the conversion of peat into true coal; sometimes used as synonymous with *Lignite*, the term is preferably restricted to materials whose vegetable origin is less evident than that of lignite.
See Coal, lignite.

### Brown earth, Brown Forest Soil
Jacks, 1954. 'Soil with mull horizon and having no accumulation horizon of clay and sesquioxides.' (a great soil group—see Soil classification).

FAO, 1953. '*Brown Forest Soils* have A (B) C profiles with merging boundaries to the horizons; normally the A-horizon is prominent; the base saturation of the clay in the (B)-horizon is high or medium ($>30\%$). The ratio $SiO_2:R_2O_3$ in the clay fraction is reasonably uniform throughout the profile and there is little or no evidence of clay illuviation. The (B)-horizon usually has granular structure. The humus form is generally mull. Calcareous brown forest soils are included.'

*Acid brown forest soils* have A (B) C profiles the horizons of which have merging boundaries . . . very little clay illuviation. The humus form is generally mull or moder. The base saturation of the clay fraction is low ($<30\%$) and the reaction decreases down the profile. The structure of the (B)-horizon is weakly developed sub-angular blocky (or granular).'
See also Reddish-brown soil.

### Brown Mediterranean Soil
FAO, 1963. '*Brown Mediterranean soils* have an ABC horizon sequence . . . differ from the Red Mediterranean soils in the colour of the B-horizon . . . the solum of the soils is often non-calcareous.'

### Brückner Cycle
Miller, 1953. 'There are frequencies which emerge . . . from an empirical study of the periodicity of phenomena, but whose causation remains enigmatical. The best known of these is the 35-year or Brückner Cycle, found by Brückner to recur, albeit very irregularly, in a variety of phenomena including the advance and retreat of the Alpine glaciers, variations in the dates of opening and closing of Russian rivers, the level of the Caspian Sea and the rivers emptying into it, the date of the grape harvest and the price of grain. There would appear to be irregular alternations of cold, wet periods with hot, dry periods. . . .' (p. 285)
See Brunt, D., Climatic Cycles, *Geog. Jour.* **89**, 1937, 214-238; also Cycle, below.

### Brumby, brumbie (Australian)
The wild, or more correctly the feral, horse of Australia.

### Brush
*O.E.D.* The small growing trees or shrubs of a wood; a thicket of small trees or underwood. (Especially in U.S., Canada and Australia.)

Scrub or bush; vegetation of low woody plants especially sagebrush in U.S.

### B.S.T.
British Summer Time, one hour in advance of G.M.T. (Greenwich Mean Time).

### Buffer state
From the analogy with the use of 'buffer' in engineering: a mechanical apparatus to deaden the concussion between two bodies in motion or a body in motion and one at rest. A weak state situated between two or more powerful states, usually independent. In that the territory of the buffer state is interposed it may serve to lessen the danger of direct conflict between the two major powers, but buffer states have often suffered through the passage of armies from both sides. A commonly cited example is Belgium between France and Germany. See Lord Curzon, 1907, *Frontiers*.

### Bugr (Sudan; Arabic)
Land in Khartoum Province, Sudan, behind and higher than '*saqiya*' land, and cultivated only when the rains have been good or the Nile flood exceptionally high. (J.H.G.L.)

### Built-up area
The term commonly used in planning to

denote the area in a town covered with buildings and having no space for further similar development. By replacement of existing buildings such an area can be subject to *re-development* in contrast to open land which is subject to *development*.

**Bunch-grass, bunchgrass**
*O.E.D. Suppl.* One or other of various grasses of western North America, characterized by growing in clumps.
*Webster.* Any of several grasses, chiefly of the western United States, which grow in tufts forming dense turf, as *Sporobolus aeroides, Elymus condensatus, Andropogon scoparius, Oryzopsis hymenoides*, various species of *Stipa*, etc.
*Comment.* The *O.E.D.* identified Bunch-grass as *Festuca scabrella* but from the application of the name to a single species it came to mean grasses with the characteristic growth habit, later extended from North America to descriptions of vegetation in other parts of the world. An alternative is tussock grass; both bunchgrass and tussock-grass are now used in contrast to those grasses which form a continuous sward since the bunches or tussocks are usually separated by bare ground.

**Bund (1), bundis**
*O.E.D.* Bund. (Anglo-Indian: Hindustani; of Persian origin). In India: 'Any artificial embankment, a dam, dyke, or causeway.' In the Anglo-Chinese ports 'applied specially to the embanked quay along the shore'. (Col. Yule)
Knox, 1904. (Pers.), a dam.
*Comment.* The word is extensively used in literature relating to India and applies on the one hand to large dams and embankments and on the other to the small ridges between ricefields.

**Bund (2)**
*O.E.D. Suppl.* (German: related to Band). A league or confederacy; spec. the confederation of German states.

**Buran**
*O.E.D. Suppl.* In Central Asia, a snowstorm, esp. one accompanied by high winds; a blizzard.
*Met. Gloss.*, 1944. A strong N.E. wind which occurs in Russia and Central Asia. It is most frequent in winter, when it is very cold and often raises a drift of snow, but strong N.E. winds in summer are also termed buran. The winter snow-bearing wind is also called poorga.

Kendrew, 1953. 'It is only the wild buran, the purga of the tundras, that brings danger to man and beast. . . . The buran is known and dreaded in South Russia and throughout Siberia.' (p. 268)

**Burgess's concentric ring model**—see Concentric-zone growth theory.

**Burgh** (Scottish)
*O.E.D.* Originally = Borough; now restricted to denote a town in Scotland possessing a charter. There are three classes of burghs, viz. *Royal burghs*, the charter of which is derived from the king, *Burgh of regality* and *Burgh of barony* having their charters respectively from a lord of regality and from a baron. Originally only the royal burghs sent representatives to Parliament.
See Borough.

**Burn** (Scotland, Northern Ireland, north of England)
A small stream or brook; formerly also a spring or fountain. *Cf. bourne* in south.

**Burn (2)**
*O.E.D. Suppl.* An instance of burning the vegetation on land as a means of clearing it for cultivation. A place where the trees or brush have been burned; a clearing in the woods made in this way. U.S.

**Burster**—see Southerly Burster

**Bush**
*O.E.D.* 1. A shrub, particularly one with close branches arising from or near the ground; a small clump of shrubs apparently forming one plant.
9. (Recent, and probably a direct adoption of the Dutch bosch, in colonies originally Dutch.) Woodland, country more or less covered with natural wood: applied to the uncleared or untilled districts in the British Colonies which are still in a state of nature, or largely so, even though not wooded; and by extension to the country as opposed to the towns.
*Webster.* 2b. A stretch of uncleared or uncultivated country, esp. of woodland or land covered with shrubby vegetation. Also, specif., (1) *New Zealand* (a) The forests of the North Island or (b) the undergrowth of the South Island. (2) *Australia*. The vast area or areas of arid scrub-covered country distinctive of certain interior districts.
*Dict. Am.* The wilderness or uncleared forest. often with *the*. (Examples back to 1657.)
*Comment.* The meaning varies widely according to the country to which applied and the word has become common in several

**forms** of pidgin-English. To go to bush—to go into the wild uncleared country. In U.S. applied to natural vegetation of low woody plants, scrubland, etc., *e.g.* creosote bush (compare brush). See also bushveld and veld. Over much of Africa it is used of the wilder countryside as opposed to cultivated land (*cf.* jungle in India).

**Bush veld, bushveld**
Translation of Afrikaans bosveld, a savanna formation, covering a large part of tropical and sub-tropical South Africa. It consists of grass with scattered trees, sometimes forming open park lands, in other places becoming so dense as to be nearly forest. (P.S.)

**Busti, bustee** (Bengali)
Strictly a village, but commonly applied, with the spelling bustee, to the slums of Calcutta, with single-room huts.

**Butt**
Butlin, 1961. 'A section in a common arable field which has been shortened by the irregular shape of the field, and is therefore shorter than the others in the shott.'

**Butte**
*O.E.D.* U.S. Also bute. In Western U.S.: an isolated hill or peak rising abruptly, examples 1838–1880 inc.
1845. Fremont, Rocky Mount., 145 (Bartlett). It (the word butte) is applied to the detached hills and ridges which rise abruptly, and reach too high to be called hills or ridges, and not high enough to be called mountains.
1880. Century Mag., XXIV, 510. Everything in the way of hill, rock, mountain, or clay-heap is called a butte in Montana.
1881. Geikie, In Wyoming in Macm. Mag. XLIV, 236. Here and there isolated flat-topped eminences or 'buttes', as they are styled ... rise from the plain.
Note: Butt. Obs. exc. dial (from French butte). A hillock, mound, *e.g.* 1693. Evelyn. 1862. Barnes. 1877. Peacock, N.W. Lincs. Gloss (E.D.S.) Butt-hills.

*Webster.* An isolated hill or small mountain with steep or precipitous sides and a top variously flat, rounded or pointed. A butte may be a residual mass isolated by erosion as at Butte, Montana, or a volcanic cone such as Lassen Butte, California.

Mill, *Dict.* Butte. 1. An isolated mount composed of undisturbed strata, outstanding owing to circumdenudation. The uppermost layer consists of a relatively resistant rock. It is the penultimate condition in the degradation of a tableland, the series being tableland, mesa, or table, butte, plain (*cf.* Zeuge).
2. Volcanic rocks round which the softer rocks have been worn away (*e.g.* Edinburgh Castle rock).

Geikie, J., 1898, Gloss. Buttes (Fr.) and mesas (Sp.): names given, in the Territories of the U.S. America, to conspicuous and more or less isolated hills and mountains. Buttes are usually craggy, precipitous, and irregular in outline; mesas are flat-topped or tabular. (p. 300)

Knox, 1904. (Fr.), a knoll. (U.S.A.), a lone hill rising with precipitous cliffs or steep slopes; a small isolated mesa (*q.v.*).

von Engeln, 1942. 'When the mesa has been reduced to the dimensions of a narrow isolated summit it is called a *butte*. Unfortunately "butte" is also applied to similar summits resulting from the isolation, also by differential weathering and degradation, of narrow vertical intrusions of igneous rock into overlying weaker beds. These may be referred to as *volcanic buttes*. Ship Rock, New Mexico is such a butte, Both types are regarded as monadnocks by some writers.' (p. 291)

Strahler, A. N., 1951, *Physical Geography*, New York: Wiley. '... a small steep-sided hill or peak known as a butte.' (The last remnant of a mesa.) (p. 236)

**Butte témoin** (French)
Lobeck, 1939. 'Isolated hills of limestone in poljes; name from the Causses. Also called hums, haystack or pepino hills (W. Indies).' (p. 133)

Baulig, 1956. The edge of the plateau shows reentrants, salients and outliers (avant-buttes) [Note to avant-buttes] ... 'buttes témoins (Zeugenberge) only when their surface, being at the same height as the plateau, is "witness" to the erosion accomplished'.

*Comment.* Témoin = witness, *i.e.* of a former extension of an escarpment edge or plateau surface (A.A.M.).
Used for the topographic outliers of escarpments. The 'hum' usage is only a particular case of a general term (G.T.W.). In most cases a *butte* is actually a *butte témoin* and geologically an outlier, but in view of the loose use of butte in the western United States, where it has become a naturalized American word, the two terms have been separated (L.D.S.).

**Buttress**
*O.E.D.* 3. A projecting portion of a hill or mountain looking like the buttress of a building.
*Comment.* A term used by climbers but useful to describe wall-like rock projections on a hillside, the alternative of rock-bastion (see bastion) being pre-empted. (G.T.W.)

**Buys Ballot's Law**
Moore, 1949. The law, enunciated by Buys Ballot in 1857, which states that if an observer in the northern hemisphere stands with his back to the wind, the atmospheric pressure will be lower to his left hand than to his right, the reverse being true for the southern hemisphere, Expressed in a different way the law states that in the northern hemisphere the winds move anti-clockwise round centres of low pressure and clockwise round centres of high pressure, the reverse being true, again, for the southern hemisphere.

**Bysmalith** (Iddings and Weed, 1898–1899)
Iddings and Weed, 1899, *Description of the Geology of the Gallatin Mountains*, 'the vertical dimension of the intruded mass becomes still greater as compared with the lateral dimension, so that the shape is more that of a plug or core. Such an intruded plug of igneous rock may be termed a *bysmalith*.' (p. 18)

Hatch and Wells, 1937. 'Bysmalith or Plug—as defined by Iddings (*J. of Geol.*, 6, 1898, 707) a bysmalith is an injected body, having the shape of a cone or cylinder which has either penetrated to the surface or terminates in a dome of strata like that over a laccolith.... Vertical displacement with faulting is the characteristic of this method of intrusion.... Mount Holmes, in Yellowstone Park is cited by Iddings as the type example... appears to be genetically related to the laccolith, but owes its cross-cutting relationship to the invaded formations to the relative incompetency of the latter to resist the passage of the magma, and to the greater speed of injection.... such movements are referred to as "cauldron subsidences" and they are due to "piston faulting". The classical British example occurs at Glencoe....' (p. 15)

# C

**Caatinga, catingas** (Brazil)
*O.E.D. Suppl.* (Tupi, from caa, natural vegetation, forest and tinga, white). A forest consisting of thorny shrubs and stunted trees.
Mill, *Dict*. The bush forms of spiny and fleshy xerophilous plants on deep soil in the drier parts of the E. Brazil highlands.
James, 1959. 'In the interior of the Northeast where droughts occur frequently, there is a kind of thorny deciduous drought-resistant woodland which the Brazilians call *caatinga*.' (p. 395)
*Comment*. Professor P. W. Richards writes: 'It should be made clear that this is used in Brazil for two quite different things (a) an evergreen type of low forest with distinctive structure and composition which occurs especially in the Rio Negro region of Amazonas where the climate has no severe dry season. It occurs on "white sand" (lowland tropical podsols) and is similar to the Wallaba forest of British Guiana, (b) deciduous seasonal forest in northeast Brazil. See Davis and Richards, 1934, 22, 126.'
See also Cole, M. M., 1960, Cerrado, caatinga and pantanal: the distribution and origin of the savanna vegetation of Brazil. *Geog. Jour.*, 126, 168–179.

**Cacimbo**
Kendrew, 1953. Angola Littoral. The Benguela current is responsible for the poor rainfall; it is notably foggy and the W. winds bring cool damp air and much fog from over it in winter, but little measurable rain. The fog and very low stratus cloud (which sometimes gives drizzle) are similar to those in South-west Africa; cacimbo is the local name... very frequent in July and August, spreading landward over the coastal plain as far as the plateau' (pp. 102–103).
Similar to 'smokes' of the Guinea Coast.

**Cadastral**
*O.E.D.* Cadastral (a. mod. F. cadastral. relating to the cadastre)
2. Cadastral survey: a. strictly, a survey of lands for the purpose of a cadastre; b. loosely, a survey on a scale sufficiently large to show accurately the extent and measurement of every field and other plot of land. Applied to the Ordnance Survey of Great Britain on the scale of 1 : 2,500 or 25·344 inches to a mile. So cadastral map, plan etc.
Hinks, A. R., 1942, *Maps and Survey*, 4th ed., 'But it should be remarked that the English map on the scale of 1 in 2500, commonly called the 25-inch map, shows visible hedges and fences, whereas the real boundary of the property is very frequently some feet beyond the hedge. Hence the 25-inch map is not strictly a cadastral map, though it is commonly called so.' (p. 11)
*Comment*. A cadastre (French: occasionally anglicized as cadaster) is a register of property. The word cadastral came into prominence in the 1860's when proposals were made to extend large scale mapping by the British Ordnance Survey to areas previously mapped only on the two-inch or one-inch scale.

**Caingin, Kaiñgin** (Philippines)
Shifting cultivation, *ladang* of Indonesia (*q.v.*).

**Cainozoic, Kainozoic, Cenozoic**
*O.E.D.* Of or pertaining to the third of the great geological periods (also called Tertiary), or to the remains or formations characteristic of it. [period should be era].
*Comment*. The usual spelling is Cainozoic but purists prefer Kainozoic or Cenozoic.

**Cairn**
*O.E.D.* (General Celtic origin). A pyramid of rough stones, raised for a monument or mark of some kind: a. As a memorial of some event, or a sepulchral monument over the grave of some person of distinction. b. As a boundary mark, a landmark on a mountain-top or some prominent point, or an indication to arctic voyagers or travellers of the site of a cache or depot of provisions. e. A mere pile of stones.
Found in Lowland Scots early in 16th c., and thence recently in English, as a term of prehistoric archaeology, and now widely

and popularly in connection with the piles of stones used or raised by Ordnance Surveyors.
*Comment.* The final comment above is out of date so far as the Ordnance Survey is concerned (G.R.C.).

**Cake**—see Ice Cake

**Calcareous**
*O.E.D.* Of the nature of (carbonate of) lime; composed of or containing lime or limestone.
Rice, 1941. 'Adjective applied to rocks containing calcium carbonate, or the carbonate of calcium with less than ·5% of magnesium. A general term applicable to several grades and types' (p. 58).
Lyell, C., 1872, *Principles*, 11th Ed., Vol. 1. 'Springs which are highly charged with calcareous matter ... although most abundant in limestone districts, are by no means confined to them (cites cases from crystalline rocks of the Massif Central—giving rise to travertine deposits at least 20 miles from a limestone)' (p. 396).
North, F. J., 1930, *Limestones*. 'If the proportion of sand or clay is small, the rock may be designated "sandy limestone" or "argillaceous limestone" ... when those substances predominate the terms calcareous sandstone, calcareous clay ... are applied' (p. 26).
Cotton, 1942, *Climatic Accidents*. Refers to 'calcareous dunes', *i.e.* coastal dunes made of calcareous sand and often indurated (p. 102).
Hatch and Rastall, 1938. '... *calcareous* organic sediments include the modern shell sands, coral reef deposits and foraminiferal oozes, together with chalk and most limestones amongst the older rocks' (p. 38).
Millar and Turk, 1943, Calcareous soil. '... containing sufficient calcium carbonate (often with magnesium carbonate) to effloresce visibly to the naked eye when treated with hydrochloric acid. Soil alkaline in reaction owing to the presence of free calcium carbonate' (p. 431).

**Calcareous crust** (Soil science)
Jacks, 1954. 'Indurated horizon cemented with calcium carbonate.'

**Calcicolous, calcicole**
*O.E.D. Suppl.* Growing upon limestone.
Abercrombie *et al.*, 1951 (of plants) 'growing best on calcareous soils.' (p. 38)
*Comment.* Used of vegetation or plants which require or seek a lime-rich soil. Hence Calcicole, a plant which seeks such a soil.

**Calcifuge**
Abercrombie *et al.*, 1951, 'Of plants, growing best on acid soils.' (p. 38)
Literal concept is a fugitive from lime; lime-avoiding.

**Calciphobe**
Of plants, lime-hating *cf.* calcifuge.

**Calcrete, calcicrete**
*Webster.* A limestone formed by the cementation of soil, sand, gravel, shells, etc. by calcium carbonate deposited by evaporation, or by the escape of carbon dioxide from vadose water.
Rice, 1941. Term suggested for conglomerates formed by the cementation of superficial gravels by calcium carbonate. The term calcicrete is suggested by Bonney as preferable (p. 60).
Glennie, E. A., 1953, Glossary in *British Caving*. Limestone breccia loosely cemented by calcrete (a surface phenomenon) (p. 438).
*Comment.* The term should not necessarily be restricted to either conglomerates or breccias. A coarsely clastic rock (either a breccia or conglomerate) formed from superficial rock debris, cemented by calcareous material (often only loosely) (G.T.W.).

**Caldera** (Spanish), **Caldeirão** (Portuguese)
*O.E.D. Geol.* (a. Sp. caldera = cauldron, kettle, boiler). A deep cauldron-like cavity on the summit of an extinct volcano.
1865. Lyell, Elem. Geol. (ed. 6), 632.
1875. Watts, Dict. Chem., VII, 553 (In) the valley of Furnas ... the soil is now perforated by a number of geysers. The three largest and most active of these are called 'caldeiras'.
Mill, *Dict.* Caldeira, or Caldera. Employed technically to designate an extra wide crater formed by the in-sinking of the parts of the volcano surrounding the crater proper, often caused by dry explosion (Canaries).
Knox, 1904. Caldeira (Fr. Port.) see Caldron (= minor form of sub-oceanic relief).
Geikie, J., 1898. 'Caldeirãos of Bahia' = shallow depressions in dry area.
Williams, Howel, 1941. Calderas and their Origin. *Univ. Cal. Publ., Bull. Dept. Geol. Sci.*, 25, 6, 239–346. 'Large volcanic depressions more or less circular or cirque-like in form, the diameters of which are many times greater than those of the

included vent or vents, no matter what the steepness of the walls or form of the floor ... calderas are passive forms of destruction.' (p. 242)

Cotton, 1944. 'Large depressions of volcanic origin which are of irregular outline and are not obviously centred on ancient large volcanoes are excluded from the category of calderas.'

Rice, 1941. A broad, relatively shallow, volcanic basin (Spanish for Caldron) (Willis).

Putnam, W. C., 1938, The Mono Craters, California. *Geog. Rev.*, **28**. 'Daly would restrict (the term) to explosive craters, and call forms produced by subsidence ' volcanic sinks". According to Stearns, the term may be used "in its generic sense as meaning a circular or amphitheater-shaped depression on a volcanic mountain."' (footnote p. 79 with references)

Thornbury, 1954. '... typically several miles across. Caldera has been applied by some, particularly in Germany, to any enclosed or walled basin, regardless of origin. It would simplify the problem of origin if we restricted the name to large circular basins of volcanic origin ... three types (1) explosion ... (2) subsidence ... (3) combined explosion-subsidence. Stearns has recognized a fourth class, the erosional caldera ... illogical.' (pp. 501–502)

See also Holmes, 1944 (with super-calderas); Daly, 1914; Lobĕck, 1939.

## Caledonian folds, Caledonian orogeny, Caledonides, Caledonoid

Mill, *Dict.* Lines running north-north-east to south-south-west in various parts of North-Western and Northern Europe, indicated by rock folds and hills, faults, valleys, etc. Some of the earth-movements concerned date back to early Palaeozoic times, but may have continued or may have been renewed at various epochs (Caledonian mountains). The pre-Devonian mountains of Norway, Scotland, Ireland and North Wales, and mountain folds parallel to them, and of the same or earlier date, the remnants of which are found in Europe and the Sahara (Suess). Suess gives the name Caledonides to these mountains wherever found, and the name Saharides to those of the Sahara region.

Himus, 1954. Caledonian: refers to the great mountain-building episode of late Silurian and early Devonian times.

Wills, 1929. 'The chains are known as the *Caledonides* on account of their great development in Scotland (p. 272).... *Caledonian* implies a fold produced during the post-Silurian orogeny; *Caledonoid* refers to a southwest-northeast fold of any age.' (p. 76) The latter usage follows C. Lapworth, *Rep. Brit. Assoc.* 1913 (1914), p. 598.

## Calendar

*O.E.D.* 1. The system according to which the beginning and length of successive civil years, and the subdivision of the year into its parts, is fixed; as the Jewish, Roman or Arabic Calendar.

2. A table showing the division of a given year into its months and days ... often including important astronomical data ... sometimes only facts and dates belonging to a particular profession.

## Calf

Mill, *Dict.* An islet lying off a small island (Norse). Mill gives also calva.

Similarly *O.E.D.* but used only of The Calf of Man, off the Isle of Man. Also an iceberg detached from a coast glacier, a fragment of ice detached from an iceberg or floe.

## Caliche (Spanish)

*Webster.* (Amer. Sp. fr. Sp. *caliche* a pebble in a brick, a flake of lime). a. The nitrate-bearing gravel or rock of the sodium-nitrate deposits of Chile and Peru. b. A crust or succession of crusts of calcium carbonate that forms within or on top of the stony soil of arid or semiarid regions, especially in Arizona.

Twenhofel, 1939. Caliche ... is a deposit of calcium carbonate and other salts made in semi-arid regions. The salts are brought to the surface by rising waters which on evaporation deposit the salts on the surface or in the surface materials. Caliche is also termed hardpan in some localities. The deposits with the enclosed cemented materials may attain thicknesses of a metre or more. They conform to the surface... (p. 342)

Read, H. H., 1936, Rutley's *Mineralogy*, London: Murby. Soda Nitre, etc. ... Economically important deposits are found in the Atacama Desert of northern Chile ... the soda-nitre occurs, mixed with sodium chloride, sulphate and borate, and with clayey and sandy material in beds up to six feet thick. The sodium nitrate forms 14–25 per cent of this *caliche* as the material is termed, and is accompanied by

2–3 per cent of potassium nitrate, and up to 1 per cent of sodium iodate ... these ... deposits have most likely been leached from surrounding volcanic rocks. ... (p. 194)

James, P. E., 1959, *Latin America*, 3rd ed., New York: Odyssey. Under Atacama desert. 'In the old lake beds the salts contained in the lake waters were deposited. A series of layers are recognized, one of them containing the valuable *caliche* which is composed of sodium chloride, sodium nitrate, and a variety of other substances including iodine salts. The caliche layer varies in thickness from a few inches to many feet with an average of perhaps one foot.' (p. 256)

Russell, 1950. 'The effect of increasing temperature on these grassland soils is threefold. ... In the second place the horizon of calcium carbonate deposition may both thicken and harden becoming a band up to several feet thick. This is known as caliche in parts of America and *kankar* in parts of India and is also well seen in many parts of the Mediterranean.' (p. 529)

Millar, C. E., and Turk, L. M., 1943, *Fundamentals of Soil Science* (American) 'more or less cemented deposits of calcium carbonate known as *caliche*' (p. 363).

*Comment*. A term of Spanish origin, used in U.S.A. to denote accumulations of calcium carbonate in the soils of arid and semi-arid regions, and in Chile for lacustrine evaporite deposits containing sodium nitrate and other salts. In some cases re-solution of the caliche, or its presence in non-arid soils, has been taken to indicate evidence of climatic change (using it in sense 1, the commonest form). Twenhofel (p. 445) gives synonymous with alkali (*q.v.*) and claims that whilst calcium carbonate is the dominant mineral many others occur and cites the nitrates as being economically the most valuable. If this interpretation of *caliche* is accepted the discrepancies disappear. (G.T.W.)

**Calina** (la calina or la calima, Spanish; calitgia in Catalan)
A warm, lead-coloured fog, dry and dusty, which characterizes hot days in the south of Spain (P.D.).

**Calm, calms**
*Met Gloss*, 1944. Calm: absence of appreciable wind. Notation 0 in the Beaufort Scale (*q.v.*).

Calms prevail through much of the year in certain latitudes: hence the Belt of Calms or Doldrums associated with equatorial latitudes and the Calms of Cancer and Capricorn associated with the *Horse Latitudes* about the Tropics of Cancer and Capricorn.

**Calving**
*O.E.D.* Of a glacier or ice berg.
1837. MacDougall, tr. Graah's *E. Coast Greenland*, 48. An occasional report, caused by the calving of the ice-blink.
Mill, *Dict*. The formation of an iceberg by the splitting off the end of a glacier.
*Ice Gloss.*, 1956. The breaking away of a mass of ice from a *glacier, ice front* or *iceberg*.
*Comment*. The 1837 quotation given by *O.E.D.* makes nonsense in the current use of *ice-blink* (*q.v.*).

**Camanchaca**
A twice-daily layer of dense fog on the coasts of northern Chile and Peru at about 1000 meter altitude. Cf. Cacimbo.

**Cambrian**
*O.E.D.* 2. Geol. A name given by Sedgwick in 1836 to a group or 'system' of Palaeozoic rocks lying below the Silurian, in Wales and Cumberland.
As originally defined, the Silurian of Murchison and Cambrian of Sedgwick, being established in different districts, were found on further investigation to overlap each other; ...
*Comment*. The name Ordovician was suggested by Charles Lapworth for the middle or overlapping part of the succession and has been generally adopted. The Cambrian (from Cambria, Wales) is the earliest period of the Palaeozoic era.

**Campagne, campagna**—see Champaign, also Bocage

**Campo** (Portuguese; Brazil)
*O.E.D. Suppl*. A field or plain; the Portuguese name for the grass plains of Brazil, which appear in the midst of the dense forests of the country. 1863. Bates, Nat. Amazon (1864), 176. 'The country around Santarem is a campo region; a slightly elevated and undulating tract of land, wooded only in patches, or with single scattered trees.'
*Webster*. A level open tract of grassland in South America, having scattered perennial herbs and, sometimes, stunted trees.
Mill, *Dict*. Savannahs of the uplands (Brazil).
Schimper, A. F. W., *Plant Geography* ... Oxford: O.U.P., 1903. 'The campos of

Brazil, like the llanos and the savannahs of Guiana, do not consist of a uniform formation spread over a wide area, but of a richly differentiated, undulating, park-like country, in which different forms of woodland and grassland partake, although the latter predominate.' (p. 373)

Knox, 1904, distinguishes Campos abertos, agrestes; cerrados; geraes; mimosos: veros.

James, 1959. 'A very large proportion of the total area is covered with a mixture of scrubby deciduous woodland and grass. In places the scrub woodland is predominant forming... *campo cerrado* (*lit.* a "closed grassland"). In other places are vast expanses of savanna with scattered trees or patches of forest (*campo sujo, lit.* "dirty grassland")... there are open grasslands with no trees (*campo limpo, lit.* "clean grassland")' (p. 395).

**Comment.** Like the comparable savannas of Africa the campos of Brazil have been the subject of recent studies which tend to show that the vegetation is not the natural climax but the result of human interference, especially burning. See Savanna, and references under caatinga.

**Cañada** (Spanish)

*Webster.* a. A small canyon; a glen. b. An open valley. c. *Western U.S.* A brook.

*O.E.D.* In the Western States of North America; a narrow valley or glen; a ravine or small cañon.

'Transhumance route for sheep across Spain' (Pierre Deffontaines in MS.). See also Aitken, R., *Geog. Jour.*, **106**, 1945, 59.

**Canal**

*O.E.D.* From F. canal in 16th c. Canel, cannel, chanel and channel were earlier adoptions from old French.

4. *Geog.* A (comparatively) narrow piece of water connecting two larger pieces; a strait. *Obs.*; now Channel.

6. An artificial watercourse constructed to unite rivers, lakes, or seas, and serve the purposes of inland navigation. (The chief modern sense, which tends to influence all the others.)

**Comment.** In *O.E.D.* 4 survives in Lynn Canal (Alaska)—the fiord leading to the port of Skagway except that there it is a narrow piece of water with no outlet. Ship Canal—one available to ocean-going vessels, *e.g.* Manchester Ship Canal in contrast to barge canal. Canal is also used for a main distributing channel for irrigation water (irrigation canal), and for water supply. Also used for a section of a cave passage flooded or partly flooded with stagnant or slow-flowing water (G.T.W.).

**Canale** (Italian; *pl.* canali)

Wray, D. A., 1922. Karstlands of Western Jugoslavia, *Geol. Mag.*, **59**, pp. 392–408. 'Closely related to the poljes, and of similar origin, are the long fjord-like arms of the sea along the Dalmatian coast known as "canali" which, judging by their outline and analogy, appear to constitute karst poljes drowned by the sea.'

Wooldridge and Morgan, 1937. 'drowned valleys (*canali* or *valloni*) of the Dalmatian coast emphasize its longitudinal structure'. (p. 354)

*Conosci l'Italia*: I, Italia Fisica, 1957, Ch. vi, I Mari, le coste le isole by G. Morandini. '... la parte terminale delle valli scendenti dall'-altopiano calcareo istriano è stata invasa delle acque in tempi geologici recenti, e il mare penetra profondamente entro terra (costa a "rias"), formando i caratteristici *canali* (del Quieto, di Leme)' (p. 148).

**Comment.** This is a case of taking a general term in a specific sense—the term is Italian and is used as the equivalent of a ria in Istria—the illustrations chosen by Morendini (see his fig. 69) show a typical ria at right angles to the coast. Not found in Johnson. Should not be used in the special sense attributed by Wray; simply an Italian word for ria which is best so-called. (G.T.W.)

**Cancer, Tropic of**

The Northern Tropic; approximately $23\frac{1}{2}°$N., the northern limit of the tropics or parts of the earth's surface where the sun's rays strike vertically at noon on at least one day of the year.

**Cañon** (Spanish, anglicized as **Canyon**) See Canyon

**Canopy**

Carpenter, 1938. The high leafy covering in woodlands, the uppermost layer in forests.

Richards, P. W., 1952, *The Tropical Rain Forest*, Cambridge: C.U.P. 'The term canopy is sometimes used as a synonym for stratum; a canopy means a more or less continuous layer of tree crowns of approximately even height. The closed surface or roof of the forest is sometimes loosely referred to as the "canopy"; in the Rain Forest, as we shall see, this surface may be formed by the crowns of the highest tree stratum alone or (more frequently) by the highest and the second storeys together.' (p. 23)

Similar usage in: Tansley, A. G., and Chipp, T. F., 1926, *Aims and Methods in the Study of Vegetation*, London (p. 14).

Toumey, J. W., 1928, *Foundations of Silviculture upon an Ecological Basis*, New York: Wiley, (p. 52). (Particularly with reference to shade. W.M.)

### Cantonment, cantonments

Permanent encampments for troops, or permanent winter-quarters. Applied especially in India to the military stations, usually adjoining existing cities, which were constructed during the British régime.

### Canyon

*O.E.D.* A deep gorge or ravine at the bottom of which a river or stream flows between high and often vertical sides; a physical feature characteristic of the Rocky Mountains, Sierra Nevada and the western plateaus of North America.

*Webster.* (Sp. *cañon*, tube, hollow, fr. *caño* tube, *caña* reed.) A deep valley with high steep slopes. *Western U.S. and Mex.* Canyons are characteristic of regions where, owing to aridity or to great slope, the downward cutting of the streams greatly exceeds weathering.

*Dict. Am.* A deep gorge, ravine, or valley having steep sides. (Cites examples back to 1834.)

Salisbury, 1923. Valleys which are narrow and deep, but small are often called gorges. Similar valleys of larger size are called *canyons*. The sides of gorges and young canyons are sometimes nearly vertical, but the sides of the larger canyons are rarely so. The distinction between a canyon and a valley which is not a canyon is not a very sharp one, and in regions where canyons abound, the term is often applied to all valleys. (p. 105)

Rice, 1941. A steep-walled valley or gorge in a plateau or mountainous area. The high precipitous slopes impress the observer more than the flat land which may occur along the stream, and this impression depends on the distinction between canyons and other valleys. (p. 62)

Balch, H. E., 1948, *Caves of Mendip*, Vol. III. 'Canyon is a deep and steep-sided channel in a cave worn by water along some line of weakness in the limestone and may vary from 6–30 feet in depth' (p. 6).

Hills, G. F. S., 1947, *The Formation of the Continents by Convection*. 'submarine canyons ... steep-sided ... of a winding character and have all the characteristics of valleys draining a high plateau. In certain cases, *e.g.* the Hudson canyon, they are in a very marked way the continuation of existing rivers.... The maximum depth of the Hudson canyon is 3,720 feet and its width is 6 miles there.' Veatch and Smith say: 'The major canyons and the features on the slope are certainly the result of stream erosion....' (*Geol. Soc. Am. Sp. Paper* 7). '... Du Toit (*Our Wandering Continents*, 1937, p. 226) says that the strips of the sea floor bordering the Atlantic, in which these canyons occur, are the marginal parts of the continents that have been depressed through the formation of the Atlantic rift basin and that 'These gorges would in short represent nicks sliced by the rivers through the fault line scarps before and while the downdragging took place.' (p. 61) 'If, however, as seems more likely, these submarine canyons are drowned fiords...' (p. 62).

### Capacity of a stream (G. K. Gilbert, 1914)

Gilbert, G. K., 1914. The Transport of Debris by Running Water. *U.S.G.S. Prof. Paper*, 86. 'The maximum load a stream can carry is its capacity. It is measured in grams per second. As the work of the laboratory was largely to determine capacity by measuring maximum load, the two terms are to a large extent interchangeable in the discussion of laboratory data, but the distinction is nevertheless important. The symbol for Capacity is C.' (p. 35)

Capacity is a function of various conditions such as slope and discharge... When a fully-loaded stream undergoes some change of condition affecting its capacity it becomes thereby overloaded. If overloaded, it drops part of its load, making a deposit. If underloaded, it takes on more load, thereby eroding its bed.... It is a general fact that the loads of streams flowing on bed-rock are less than their capacities. (p. 35)

Mackin, J. H., 1948, Concept of the Graded Stream, *Bull. Geol. Soc. Amer.*, 59, 463–512.

'Capacity' as defined by Gilbert (1914, p. 35) refers to the 'maximum load a stream can carry'. The experimental data on which the competence principle was based demonstrate also that, in a stream of given discharge, the velocity required for transportation varies with the quantity of any one grain size, the velocity requirement increasing with increase in the quantity or total weight of the material shed by the stream.

Gilbert used 'capacity' in discussing data relating to transportation of weighed amounts of particles in the sand-gravel size range under laboratory conditions. In spite of his explicit warning that his concept of capacity does not necessarily apply to natural streams ... there has been a tendency so to apply it; the expressions 'loaded to capacity' or 'fully loaded' or 'saturated with load' are frequently used in discussion of the graded condition in streams. These expressions are of course meaningless unless accompanied by some statement of the grain sizes which constitute the load. A stream 'loaded to capacity' with coarse sand and pebbles could carry an enormously greater tonnage of material without change in velocity if the materials making up the load were crushed to silt size.

The capacity principle, like the competence principle, probably does not apply to the transportation of very small particles. Since the maintenance in suspension of ultra-fine clay particles and colloids ... does not depend upon velocity, there is no theoretical upper limit to the amount of these materials that a stream of given size and capacity can carry. A stream 'loaded to capacity' with exceedingly fine particles would be a mud flow (Hjulström 1935, pp. 344–345) (pp. 468–469).

*Comment.* See Competence, Grade.

### Cape

*O.E.D.* adaption of F. 'cap', head, cape. 1. A piece of land jutting into the sea; a projecting headland or promontory.

*Comment.* 'The Cape' is commonly used to refer to the Cape of Good Hope or to the neighbouring part of South Africa.

### Cape Doctor (South Africa)

Humorous name given at Cape Town to the disagreeable south-east wind, very common during the summer months, and often reaching hurricane strength. It has the advantage of preventing the stagnation of air in the mountain amphitheatre in which central Cape Town has been built. This explains the name. (P.S.)

### Capillary fringe (O. E. Meinzer, 1942)

Meinzer, O. E., 1942. 'Capillary fringe: the belt immediately above the water-table in which water is held above the water by capillarity' (p. 394).

Similarly Müller, S. W., 1947, *Permafrost*.

### Capital

*O.E.D.* 2. A capital town or city; the head town of a country, province or state.

Mill, *Dict*. The seat of government of a country, province, or district.

*Comment.* Often used loosely in the sense of the chief town or city as in the phrase 'commercial capital'. Note also ecclesiastical capital, *e.g.* Canterbury. In monarchical states the site of the Royal residence does not necessarily mark the capital, *e.g.* Amsterdam is the capital of the Netherlands, not The Hague.

### Capital goods

Goods, especially manufactures such as machinery, but including also raw materials, destined to be used in the manufacture of other goods. The contrast is with *consumer goods*, destined for use by the purchasers so that they will, sooner or later be consumed or used up.

### Capricorn, Tropic of

The Southern Tropic; approximately $23\frac{1}{2}°S.$, the southern limit of the tropics. See Cancer, Tropic of and Tropics.

### Cap-rock

Swayne, 1956. (*a*) A stratum of resistant rock covering another of less resistant material. (*b*) The rock cover over the top of a salt-plug. (*c*) Unproductive rock covering valuable ore.

### Capture of rivers, River capture

Mill, *Dict*. Capture (of rivers). The action of headward erosion in one river leading to the diversion of the headwaters of another, *cf.* Beheading.

Gilbert, 1877. *cf*. '... abstraction. A stream which for any reason is able to corrode its bottom more rapidly than do its neighbors, expands its valley at their expense, and eventually "abstracts" them.' (p. 141)

Cotton, 1941. 'Diversion of headwater streams to become tributaries of other rivers is generally termed capture....

River capture may be understood as including abstraction, but is sometimes limited in its application to a process of diversion by streams developing headward at a rapid rate under certain favourable conditions so as to tap and lead off the waters of others. This process has been called also "river piracy".'

Comment. Both capture and abstraction (*q.v.*) are used by Wooldridge and Morgan, 1937, pp. 192–193.

**Carapace lateritique** (French: Soil science)
Lateritic crust or hard pan found at or near the surface of many tropical soils.

**Caravan** (from Persian Kārwān)
*O.E.D.*1. A company of merchants, pilgrims or others, in the East or northern Africa, travelling together for the sake of security, esp. through the desert.

**Caravanserai**
*O.E.D.* A kind of inn in Eastern countries where caravans put up, being a large quadrangular building with a spacious court in the middle.
See also serai

**Carbo-electricity**
Not in *O.E.D.* or *Suppl.* Electricity produced by the use of coal and, presumably, other carbonaceous materials; used in contradistinction to hydro-electricity.

**Carbonaceous**
Twenhofel, 1939. 'Sediments defined as carbonaceous include original organic tissues and subsequently produced derivations of which the composition is chemically organic. The organic matter may be carbonized (coalified) or bituminized' (p. 414).

**Carbon-dating** or **radio-carbon dating**
L.D.G. This is a means developed by W. F. Libby of Chicago of determining the age of prehistoric organic remains—wood, bone etc.—up to about 20–30,000 years. It is based on the fact that radioactive carbon or carbon[14] diminishes at a known rate after death of the organism.

**Carboniferous**
*O.E.D.* 1. . . . Applied in *Geol.* to the extensive and thick series of palaeozoic strata, with which seams of coal are associated, the *Carboniferous System* or *Formation*, lying next above the Devonian or Old Red Sandstone, and including the Coal Measures, Millstone Grit, and Mountain or Carboniferous Limestone; also to the rocks, fossils, etc. of this formation, and to the age of geological time, the *Carboniferous Age*, *Era*, or *Period*, during which these strata were deposited, and the luxuriant vegetation existed that formed the coal-beds.

Comment. Carboniferous is literally carbon-bearing (*i.e.* coal-bearing). According to current geological usage the *O.E.D.* is confused. Reference should be to the Carboniferous period (not era) and system; the three divisions in England from the bottom up are the Carboniferous Limestone, Millstone Grit and Coal Measures. The whole is the Mississippian and Pennsylvanian of North America, to which some authorities add the Permian.

**Cardinal points**
*O.E.D.* 4.a. The four points of the horizon (or the heavens) which lie in the direction of the earth's two poles (cardines), and of sunrise and sunset respectively; the four intersections of the horizon with the meridian and the prime vertical; the north, south, east and west points.

**Careenage**
*O.E.D.* A careening place (*cf.* anchorage) Careen: to turn (a ship) over on one side for cleaning, caulking or repairing.

Comment. Still used, especially in West Indies, for places where small ships can be beached or berthed.

**Carfax**—see Carrefour.

**Carn** (Cornwall)
A small hill or knoll.
*O.E.D.* Var. of Cairn, c. a pile of stones.

**Carpedolith** (Soil science)
Stone-line

**Carr, carr-lands**
*O.E.D.* 1. A pond or pool; a bog or fen; now usually, wet boggy ground; a meadow recovered by drainage from the bog.
2. A fen or bog grown up with low bushes, willows, alders, etc.; a boggy or fenny copse.
Tansley, A. G., 1939, *The British Islands and their Vegetation*, Cambridge: University Press. 'Fen carr ... the alder is the main dominant of many of the characteristic fen woods or "carrs"' (p. 460). . . . 'ultimate carr, *i.e.* the "edaphic climax" woodland of the fens' (p. 461).

**Carrefour, carfour, carfax**
A place where four roads meet. An interesting case of a word formerly naturalized (as carfour and carfax) which has become obsolete and now, if used at all, is regarded as French. It survives as Carfax, the crossroads in the heart of the city of Oxford.

**Carrying capacity of land**
Stamp, L. D., 1960. *Our Developing World*, London: Faber. 'The area of land under

cultivation to support one person—in other words the "carrying capacity" of the land in terms of population.' (p. 112) Also stock-carrying capacity (p. 119); the problem of measurement is discussed at length. See also Stamp, L. D., 1958 *Geog. Rev.*, **48**, 1–20.

**Carse** (Scottish)
*O.E.D.* The stretch of low alluvial land along the banks of some Scottish rivers.
*Comment.* The word may be connected originally with *carr* (*q.v.*) and associated with the wet fen land with numerous pools. It is now, however, associated with the agriculturally rich, fertile lands forming extensive plains (far more than river alluvium) such as the Carse of Gowrie and Carse of Stirling.
*Cf.* entry for *kjarr* in A. H. Smith, 1956, *English Place-Name Elements*, Part II, p. 4 (Cambridge). (H.C.D.)

**Carso** (Italian)—see Karst

**Carst, carstification**
Some writers prefer this spelling to the more usual karst and karstification. For some reason which is not clear the *Report of the Commission on Karst Phenomena* of the International Geological Union (1956) in one summary uses karst but carstification.

**Cartogram**
*Webster.* A map using shades, curves or the like, to show geographically statistics of various kinds.
Raisz, E., 1938, *General Cartography*, New York: McGraw-Hill, notes that some authors, especially European, call every statistical map a cartogram, but American usage speaks of dot maps, etc. 'In the following discussion the term is restricted to any highly abstracted, simplified map the purpose of which is to demonstrate a single idea in a diagrammatic way.' (p. 257) See also:
Raisz, E., 1934, The Rectangular Statistical Cartogram, *Geog. Rev.*, **24**, 292—296.
Finch, V. C., Trewartha, G. T. and Shearer, M. H., 1941, *The Earth and its Resources*, New York: McGraw-Hill. Cartograms = dot maps (p. 600).
British National Committee for Geography, 1966, *Glossary of Technical Terms in Cartography*, London: The Royal Society. 'Small diagram on the face of a map displaying quantitative data. It has also been used to describe an abstracted and simplified map for which the base is not true to scale.' (p. 11)
International Cartographic Association, Commission II, 1973, *Multilingual Dictionary of Technical Terms in Cartography*, Wiesbaden: Franz Steiner Verlag G.M.B.H. 'A small diagram on the face of a map displaying quantitative data.' (p. 90)
*Comment.* The common usage is to make the simple word 'map' or the term 'diagrammatic map' cover what some authors would call cartograms.

**Cartography**
*O.E.D.* The drawing of charts or maps.
Raisz, 1938. 'The surveyor measures the land, the cartographer collects the measurements and renders them on a map, and the geographer interprets the facts thus displayed' (p. 1).
Amer. Soc. of Civil Engineers. *Definitions of surveying terms.* 'Specifically, cartography is the art of map construction and the science on which it is based. It combines the achievements of the astronomer and mathematician with those of the explorer and surveyor in presenting a picture of the physical characteristics of the earth's surface.'
United Nations, Dept. Social Affairs, 1949, *Modern Cartography*. New York. 'Cartography is considered as the science of preparing all types of maps and charts, and includes every operation from original surveys to final printing of copies.'
International Cartographic Association, Commission II, 1973, *Multilingual Dictionary of Technical Terms in Cartography*, Wiesbaden: Franz Steiner Verlag G.M.B.H. 'The art, science and technology of making maps, together with their study as scientific documents and works of art. In this context maps may be regarded as including all types of maps, plans, charts and sections, three-dimensional models and globes representing the Earth or any celestial body at any scale.' (p. 1)
See the British Cartographic Society (1964) definition in British National Committee for Geography, 1966, *Glossary of Technical Terms in Cartography*, London: The Royal Society. (p. 11)
*Comment.* Cartography is generally used in Raisz's sense in Britain, except that his definition is too narrow. A cartographer may compile his map from materials other than 'measurements'. Not all maps are

drawn from surveys. Problems of projection are the cartographer's special concern before anything further can be done. (E.O.G., R.A.S.)

**Cartouche**

*O.E.D.* 2.c. A tablet for an inscription or for ornament, representing a sheet of paper with the ends rolled up; a drawing or figure of the same, for the title of a map, or the like; a drawn framing of an engraving, etc. Often attrib.

Raisz, 1938. 'The title, scales and descriptive material are usually collected into decorative frames called "cartouches".' (p. 40)

Lynam, E., 1944, *British Maps and Map-Makers*. London: Collins. 'This map, with its firm but delicate lines, its stately ornamental panels or "cartouches" enclosing the title and descriptive texts and its clear sloping Italic lettering must have been a revelation to Englishmen' (p. 14).

*Comment.* A panel on a map, often with decoration, enclosing the title or other legends, the scale, etc. (R.A.S.).

**Carucate** (Agricultural History)—see under Yardland

**Cascade**

*O.E.D.* (From Italian, via French, cascade, fall)

1. A waterfall. a. Usually, a small waterfall; esp. one of a series of small falls, formed by water in its descent over rocks, or in the artificial works of the kind introduced in landscape gardening.

Farey, J., 1811. 'These are *Water-falls* or natural Cascades ...' (p. 488).

Salisbury, 1923. 'Steep rapids are often called falls, and both are sometimes called *cascades*' (p. 111).

Tolman, 1937. '*Ground-water Cascade.* Descent of ground water on a steep hydraulic gradient to a lower and flatter water-table slope. A cascade occurs below a ground-water barrier or dam which may develop effluent seepage above it, and at the contact of less permeable material with more permeable material downslope.' (p. 500)

**Cascading systems**

Chorley and Kennedy, 1971. 'These are composed of a chain of subsystems, often characterised by thresholds, having both spatial magnitude and geographical location which are dynamically linked by a cascade of mass or energy. In this cascade the mass or energy output from one subsystem becomes the input for the adjacent locational subsystem. ... Cascading systems in physical geography vary markedly in magnitude, from the basic solar energy cascade, to the basic hydrological cycle cascade, and to the cascading subsystem formed by a wave moving from the deep water to the swash zone subsystem.' (p. 7) See also pp. 77–125.

**Cash crop**

A crop grown primarily for sale as contrasted with a crop grown primarily for the sustenance or use of the grower and his family.

**Cataclastic, kataclastic**

Himus, 1954. This term has been applied to (*a*) the structures produced in a rock by severe mechanical stress during dynamic metamorphism, resulting in deformation and granulation of the minerals, and (*b*) to clastic rocks which have been produced by the fracture of pre-existing rocks as a result of earth-stresses, *e.g.* crush-breccias.

See Crush-breccia.

**Cataclinal** (J. W. Powell, 1875)

Mill, *Dict.* Applied to streams which run in the main direction of dip in a country of faulted structure (Powell).

Powell, 1875. '... valleys that run in the direction of the dip.' (p. 160)

*Comment:* Now rarely used, as with the comparable terms anaclinal and diaclinal.

See Baulig, H., 1938, Questions de Terminologie, *Jour. of Geomorphology*, **1**, 224–229.

**Cataract**

*O.E.D.* A waterfall; properly one of considerable size, and falling headlong over a precipice; thus distinguished from a cascade.

Mill, *Dict.* A great waterfall, one falling headlong over a precipice. Also applied to rapids (Nile, etc.).

*Comment.* Largely because of the constant use of the word to indicate the numbered rapids of the Nile, it is now rarely used in *O.E.D.* sense.

**Catastrophism**

*O.E.D.* The theory that certain geological and biological phenomena were caused by catastrophes, or sudden and violent disturbances of nature, rather than by continuous and uniform processes.

Mill, *Dict.* The theory that the Earth's crust owes its characteristics mainly to sudden and violent changes in past time; opposed to uniformitarianism (*q.v.*).

*Comment.* Sometimes referred to as convulsionism.

### Catch crop

*O.E.D.* Catch-crop, a crop got by catching or seizing an opportunity when the ground would otherwise lie fallow between two regular or main crops; hence catch-cropping; ... catch-land. 1874. catch crop. 1887 'catch cropping' 1674. Ray, S. and E. Country Wds. Coll. 61. Catch-land, land which is not certainly known to what parish it belongeth; and the Minister that first gets the tithes of it enjoys it for that year. [obsolete]

*Webster.* A crop grown between two crops in ordinary sequence, or between the rows of a main crop, or as a substitute for a staple crop that has failed.

*Enc. Brit.*, 14th ed. introduced an apparent confusion. 'The natural successor to the wasteful fallow.... Now every vegetable or market garden represents a method of catch-cropping wholly substituted for any slowly maturing crop ... substituting "bastard" fallows for the old "bare" or winter or summer fallows.'

Stamp, 1948. '... a catch crop, which is one snatched between two of the main crops of a rotation ...' (p. 97)

### Catch meadow, catch-meadow

A meadow irrigated with water from an adjoining hill-slope collected by 'catch-drains'—terms long used in works on British farming.

### Catchment area, basin

*O.E.D.* Catchment = Catching; appropriated to the catching and collection of the rainfall over a natural drainage area, in catchment basin, area.

1878. Huxley, Physiogr. 34. 'The catchment-basin is a term applied to all that part of a river-basin from which rain is collected, and from which therefore the river is fed.'

*Met. Gloss.*, 1944. Catchment Area. Defined for administrative purposes as the area within the jurisdiction of a Catchment Board under the Land Draining Act of 1930. The term is also commonly used in the same meaning as Drainage Area. (*q.v.*) Drainage Area. The area where the surface directs water towards a stream above a given point on that stream. Gathering Ground. An area from which water is obtained by way of rainfall, drainage or percolation.

*Comment.* Catchment basin may be equated with drainage basin but under the present system in Britain a Catchment Board may deal with an area covering several drainage basins and may have boundaries which do not coincide with the natural basin. See map in Stamp, L. D. and Beaver, S. H., *The British Isles*, 4th ed. London, 1954. For American terminology see under Watershed.

### Catena (Soil science, G. Milne, 1936) or Catenary complex

Jacks, 1954. 'A sequence of different soils usually from similar parent material but varying with relief and drainage.' Described originally by Milne to cover the succession of soils found repeated over and over again in East Africa from a valley bottom to the neighbouring hill top forming a catena or chain and along what is commonly a catenary curve (L.D.S.).

Milne, G., 1936, A Provisional Soil Map of East Africa with Explanatory Memoir, *Amani Memoirs.* See also *Geog. Rev.*, 26, 1936, 522–523.

### Catinga—see Caatinga

### Catstep

Lobeck, 1939. Small, backward tilting terraces due to slumping. (p. 93)

*Comment.* American; in British usage *cf.* terracette, lynchet.

### Cattlegate, cattle gait, gait

Royal Commission on Common Land, 1955–58. *Report.* '(1) The sole or exclusive right acquired by grant from the owner of the soil or by prescription to graze a particular number of beasts on a piece of land, in the soil of which the holder of the right has no interest. (2) The right to graze a number of beasts, fixed according to the holder's gate, on a piece of land which the holder of the right owns in common with the possessors of other gates. This term is most prevalent in north-east England; there and in some places elsewhere "stint" may be an alternative, as also may "beast-gate" and "pasturegate". Both forms of gate may be bought and sold.' (p. 273)

### Cauldron subsidence (Clough, Maufe and Bailey, 1909)

Clough, C. T., Maufe, H. B. and Bailey, E. B., 1909. *Quart. Jour. Geol. Soc.*, 65, described the granite of 'Glen Coe where part of a batholith's roof has sunk along peripheral faults (a "cauldron subsidence"). As the sinking progressed the magma was squeezed up, following the faults on nearly every side of the sunken area' (quoted from summary by R. A. Daly, 1914, *Igneous Rocks*, p. 122).

Holmes, 1944. 'Cauldron subsidence. Part of the cylindrical block of country rock enclosed within a vertical ring-dyke may founder into the underlying magma reservoir while volcanoes are active at the surface' (p. 90).

### Cause and effect analysis

Harvey, D., 1969, *Explanation in Geography*, London: Edward Arnold. 'Cause-and-effect... became one of the dominant forms of explanation in geography during the nineteenth century. Its unfortunate association with mechanistic and deterministic metaphysical concepts has led to some reaction against its use in the twentieth. But... there is no need to regard cause-and-effect analysis as necessarily implying causal deterministic explanation. The search for the "factors" that govern geographic distributions is a good example of the restrained use of causal analysis at the present time.' (p. 80)

See Jones, E., 1956, Cause and effect in human geography, *Ann. Assoc. Am. Geographers*, **46**, 369–377.

### Causse, Causses (French)

*Webster*. A small limestone plateau occurring in the south-eastern portion of the central massif of France.

*O.E.D. Suppl*. Causse (French). A limestone plateau. From cau, equivalent in southern France to chaux, chalk.

Mill, *Dict*. Causses. A region in Central and Southern France resembling the Karst.

Cvijič, 1925, *C.R.Ac. Sci.*, 180, refers to Causse-type of karst—a transitional type between his Holokarst and Mesokarst. It is characterized by relatively pure, thick limestones, with an impermeable basement exposed in the bottom of the deep valleys which carve the area up into large distinct plateaus. Potholes and caves of considerable length are common though surface features (chiefly lapiés and dolines) are not so well-developed as in the type karst of Slovenia. The plateau surfaces are generally of low relief and traversed by shallow dry valleys (pp. 1038–1039) (summary—G.T.W.).

von Engeln, 1942. 'The French word *causses*, referring to areas on the south of the central upland of France, is sometimes used in the generic sense as the equivalent of "karst"' (p. 565).

Clozier, R., 1940, *Les Causses du Quercy*. 'Le mot causse... designe les plateaux calcaires, compartimentés par de profondes vallées, limités de raides escarpements et minés par l'erosion souterraine... pour le paysan du Midi de la France, le causse implique uniquement une notion de sol: la terre de causse s'oppose la terre de rivière...' (p. 13).

*Comment*. The above quotations show that the word causse or causses is used in several different senses (a) locally in southern France for a limestone soil (b) for a limestone plateau of a characteristic type and as a place name for such plateau in southern France (c) a general term for karst regions, or karst regions of this type.

### Cave

*Webster*. A hollowed-out chamber in the earth, or in the side of a cliff or hill; esp. a large natural cavity in the earth with an opening to the surface.

Muma and Muma, 1944. A natural cavity, recess, chamber, or series of chambers or galleries occurring beneath the surface of the earth and usually extending to total darkness and large enough to permit human entrance. Also loosely used as a verb meaning to collapse or fall-in. Colloquially used as a verb, to enter and explore caves. (American)

Page, 1859. Caves, Caverns (Lat. cavus: hollow). Caves occur less or more along the rocky shores of all free-flowing seas and are the results of abrasion by waves laden with gravel etc. acting upon pre-existing fissures or the softer part of the exposed rocks. The most celebrated caverns however occur in limestone strata and appear to be the result of fissuring by subterranean disturbance and partly of percolation and passage of carbonated waters.... Some are celebrated for their ... sub-fossil bones and ... known as Bone Caves or Ossiferous Caverns.... (p. 109)

*Comment*. Note the varying usages (see also Cavern): 1. For any subterranean cavity appreciably larger than a joint or bedding plane, preferably one which may be entered, though this does not permit of precise definition. 2. To explore caves (a term commoner in the south of England than the north). 3. Synonymous with collapse (or cave-in). 4. Frequently restricted to caves with essentially horizontal entrances (*cf.* the French grotte), as opposed to a pothole or pot. 5. As synonymous with cavern. 6. In U.S. for undercut banks in rock gorges—*e.g.* Powell in the Colorado—rarely used in Britain in this sense. 7. Occasionally used for artificial openings, especially those associated with hermits, or stone mines *e.g.* Chislehurst Caves, Wren's Nest Caves. 8. Generally assumed to be in limestone—otherwise requiring additional qualifying adjectives. 9. Caves may also be classified in terms of their contents, *e.g.* Bone Caves, or of their hydrology, *e.g.* Cave of Debouchure (from French), or Effluent Cave (Warwick. 1953. in *British Caving*); or Cave of Engulfment (Influent Cave of Warwick, 1953). 10. Also used as Adjective—of or pertaining to caves, *e.g.* Cave Art, Cave fauna, etc. (G.T.W.) Also sea-caves.

### Cave breccia, cave deposits, cave earth

These terms mean nothing more than breccia, earth and other deposits found in caves, but owing to the special features of cave formation, the debris left on the floor of caves often exhibits special characteristics, especially in limestone caves where water charged with calcium bicarbonate on evaporation leaves stalactites hanging from the roof, stalagmites built up on the floor. Irregular blocks may be similarly cemented to form cave breccia. Many caves were inhabited by primitive man, hence the special interest of cave deposits.

### Cave-in lake

Muller, S. W., 1947, *Permafrost*, Ann Arbor: Edwards. Cave-in lake; kettle-hole lake. A lake formed in a caved-in depression produced by the thawing of ground-ice (ice lens or ice-pipe) (p. 214).

Note also use of 'caving' to describe slumping of river banks (American usage).

### Cavern

Webster. A subterranean hollow; an underground chamber; a cave—often used, as distinguished from *cave*, with implication of largeness or indefinite extent.

von Engeln, 1942. A generally horizontal passageway through rock is, somewhat ludicrously, said to be a cavern when its diameter is sufficiently large to permit a man to crawl through. (p. 580)

Muma and Muma, 1944, p. 3. A large pretentious, natural cavity or cave. A relative term contrasted with cave.

Jansson, 1947, *Natural History* refers to use as synonymous with cave—claims geologists distinguish caverns as 'formed beneath the general ground level... where subterraneous waters seeping through the rock have dissolved it away... caves are formed by the abrasive action of surface water, caverns by dissolution of rock by subsurface waters.' (p. 440)

Gemmell, A., and Myers, 1949. A cave of large size or (plural) a large cave system.

*Comment*. Used in various ways: 1. As synonym for cave (most common), in fact 'cavern' is often preferred by older writers; perpetuated in many old cave names. 2. A large cave. 3. A large chamber in a cave—most commonly used in this way in Yorkshire (and used by M. M. Sweeting in this sense, *Geog. Jour.*, **124**, 1958). 4. A limestone cave formed by solution by subterranean water and streams (American).

Farey, J., 1811 and Salisbury, 1923, use cave or cavern as synonymous (p. 59).

See also Cave.

### Cavernous

Cotton, 1942, *Climatic Accidents*. 'Cavernous weathering... the process of differential sapping results also in various manifestations of cavernous weathering, producing the small and large recesses in rock faces for which Penck has adopted the Corsican term "tafoni"' (*q.v.*) (p. 8).

### Cay, key

*O.E.D.* A low insular bank of sand, mud, rock, coral etc.; a sandbank; a range of low-lying reefs or rocks; originally applied to such islets around the coast and islands of Spanish America.

See Steers, J. A., *Geog. Jour.*, **95**, 1940, 30.

**CBD, C.B.D.**—see Central Business District.

Murphy, R. E. and Vance, J. E., 1954, Delimiting the CBD, *Econ. Geog.*, **30**, 189–222.

Scott, Peter, 1959, The Australian CBD, *Econ. Geog.*, **35**, 290–314. 'CBDs were delimited by applying the Central Business Index Method advanced by Murphy and Vance'.

### Cedar-tree laccolith

Himus, 1954. A series of laccoliths one above the other, forming parts of a single mass of igneous rock. See Laccolith.

*Comment*. So called because in section there is a resemblance to the spreading branches of a cedar tree.

**Ceja** (Peru: Spanish)
James, 1959. 'Tropical montane rain forest. A very dense growth of broadleaf evergreen species (known in Peru as the *ceja de la montaña*)' (p. 38); 'Ceja de Montaña, or "eyebrow of the forest"' (p. 173).

**Celestial**
*O.E.D.* Of or pertaining to the sky or material heavens. *Celestial globe, map*: one representing the heavens.
*Comment.* The *O.E.D.* definition with its reference to the 'material heavens' has an archaic sound, though the sense is clear. In mathematical geography use is made of celestial latitude and longitude, the celestial poles, celestial equator, celestial meridians, celestial co-ordinates etc.

**Celt, celtic, Kelt, keltic (1)**
*O.E.D.* (Hist.). Applied to the ancient peoples of Western Europe called by the Greeks κελτοι, κέλται, and by the Romans *Celtae*.
2. A general name applied in modern times to peoples speaking languages akin to those of the ancient Galli, including the Bretons in France, the Cornish, Welsh, Irish, Manx and Gaelic of the British Isles.
*Chambers's Encyc.*, 1950. 'Celt. The name is used confusedly, sometimes for warrior peoples of the early iron age north of the Alps, sometimes for later peoples who spoke or now speak Celtic languages. It is in fact a linguistic term.'
Evans, E. E., 1958, The Atlantic Ends of Europe, *The Advancement of Science*, **15**, 54–64. 'A word heavy with romantic overtones, has been applied variously to a group of languages, a culture, an art-style, a fringe, a twilight, a "race" and a pony. It is used to designate an early field system...' (p. 54).
*Comment.* Despite the difficulties, the word is still very widely used in geographical literature.

**Celt (2)**
*O.E.D.* An implement with chisel-shaped edge of bronze or stone (but sometimes of iron).
Trent, C., 1959, *Terms used in archaeology*, 'Celt, an obsolescent term referring to prehistoric offensive weapons, whether of stone, bronze, or iron... typical of the New Stone Age.' (p. 22)

**Celtic field, Celtic field system**
Curwen, E. C., 1927, Prehistoric Agriculture in Britain, *Antiquity*, 1. That type of field, often roughly square, in use before the Saxon introduction of the strip field or statute acre, and which can still be traced over parts of southern England especially the Downs. (pp. 261–289)

**Cement, cementation** (of sediments)
Loose detrital deposits such as silt, sand and gravel in course of time become converted into hard rocks usually by the deposition from circulating waters of some cementing material such as calcite, quartz, or limonite with resulting formation of siltstone, quartzite, sandstone, conglomerate, etc.

**Cenosis**—see Coenosis

**Cenote** (Spanish)
*Webster.* (Sp. fr. Maya *conot*). A natural underground water reservoir.
Mill, *Dict.* Large hollows caused by the solvent action of rain-water on calcareous surfaces. They form centres round which population gather in Yucatan.
Cotton, 1948. 'Such lakes in dolines are natural wells like the "cenotes" of Yucatan.'
Lobeck, 1939. 'Northern Yucatan presents some karst features of unique interest... the chief source of water is in deep caverns and sink holes. Most of these are of a peculiar chimneylike structure and are known as *cenotes*.' (p. 144; diagram p. 142)

**Cenozoic**—see Cainozoic

**Central Business District (CBD)**
Murphy, R. E. and Vance, J. E., 1954, Delimiting the CBD, *Econ. Geog.*, **30**, 189–222. 'The Central Business District, frequently referred to as the CBD, is the heart of the American City. Among alternative names are Central Traffic District, Central Commerical District, Downtown Business District or, more popularly, Downtown' (p. 189).
Hartman, G. W., 1950, The Central Business District. *Econ. Geog.*, **26**. '... the primary focus of internal activities and the major contact with a tributary area was found in the "business district".'
Harris, C. D. and Ullman, E. L., 1945, The Nature of Cities, *Ann. Am. Acad. Polit. Soc. Sci.*, **242**, 7–17. 'This is the focus of commercial, social and civic life and of transportation. In it is the downtown retail district with its department stores, smart shops, office buildings, clubs, banks, hotels, theatres, museums, and organization headquarters. Encircling the down-

town retail district is the wholesale business district.' (p. 12)
See also Concentric-zone theory.

*Comment.* The concept is American and so is the nomenclature though the phenomenon is almost universal. A commonly used slang American term is the 'Loop district' derived from analogy with Chicago where the heart of the central business district lies within a rectangular loop of the overhead railway system. See also CBD, above.

**Central place theory**
Christaller, W., 1966, *Central Places in Southern Germany*, Englewood Cliffs, New Jersey: Prentice-Hall (translated from Christaller, W., 1933, *Die Zentralen Orte in Süddeutschland*, Jena, by C. W. Baskin). 'The regional part of our investigation has shown with clarity to what a great degree the market, the traffic, and the separation principles determine the distribution, sizes and number of central places. We may call these principles laws of distribution of central places, or laws of settlement, which fundamentally and often determine, with astonishing exactness, the locations of central places.' (p. 190)
Berry, B. J. L., 1967, *Geography of Market Centers and Retail Distribution*, Englewood Cliffs, New Jersey: Prentice-Hall. 'Central-Place Theory is the theory of the location, size, nature and spacing of ... clusters of activity, and is therefore the theoretical base of much of urban geography and of the geography of retail and service business.' (p. 3)

**Centrocline, centroclinal**
Page, 1876, *Advanced Textbook of Geology*. 'When strata dip ... to a common centre they are said to be centroclinal.'
Himus, 1954. A basin-like structure into which the bed dips in every direction.
*Comment.* Though a convenient term, now rarely used. Not in Holmes, 1944, Wooldridge and Morgan or other recent standard works.

**Centrography**
Poulson, T. M., 1959, Centrography in Russian Geography, *A.A.A.G.*, **49**, 326–7, 'Directed towards the establishment of laws of the distribution of phenomena based on the relationships and migrations of their "centers of gravity".'
*Comment.* The determination of centres of distributions and plotting them on maps.

**Centrosphere**
*O.E.D. Suppl.* The nucleus or central portion of the earth.
*Comment.* See Barysphere which is more usual, but 'centrosphere' does not indicate the character of the central sphere, whereas barysphere does.

**Cephalic index**
*L.D.G.* Cephalic index is a measure of skull-shape used in studies in physical anthropology. Long-headed or dolichocephalic peoples have an index less than 75, broad-headed or brachycephalic over 83. Between these limits are the mesocephalic peoples. The index is obtained by taking the maximum width of the head and dividing it by the maximum length, multiplied by 100.

**Ceramics**
The art of making pottery (*O.E.D.*).

**Ceramurgy** (N. F. Astbury, 1964)
Astbury, N. F., 1964. *Nature*, **201**, No. 4920, '*The Physics and Chemistry of Ceramics* (ed. Cyrus Klingsberg, 1963), is a valuable and welcome addition to the growing volume of sophisticated literature on what, one hopes might soon be called 'ceramurgy', to distinguish once and for all between science and aesthetics in this important field of materials' (p. 648).
*Comment.* The analogy is presumably with metallurgy: the root is εργος (ergos, working, worker) and κεραμικος (keramikos, of pottery).

**Cerrado, Cerradão** (Brazil: Portuguese)
See Campo, Savanna and references under Caatinga.

**Chachār** (Indo-Pakistan: Urdu-Hindi; Bengali) (also kachār)
Ahmad, 1958. 'According to the *Ain-i-Akbari* (1582) land was classified into four main divisions. *Polaj* was land annually cultivated. *Paranti* was left out of cultivation for some time to recuperate its strength and fertility. *Chachar* was left fallow for three or four years, while *banjar* was left uncultivated for five years or more' (p. 74).

**Chaco** (South America; Spanish)
Literally hunting ground, chase; *el gran chaco*, the great hunting ground.
*Chambers's Encyc.*, el Gran Chaco: 'the name formerly given to the extensive area of plains in South America between the Paraguay river and the Andes and north of the Pampa region of Argentina.'
James, 1959. 'Vast alluvial plain .. almost

CHĀHI 102 CHALK

featureless.' (p. 285). 'A region of deciduous scrub woodland interspersed with patches of grassy savanna' (p. 314).

**Comment.** Although like the llanos and campos the chaco is one of the great areas of savanna or scrub in South America the word has not passed in the same way into international literature as denoting a type of country, but is restricted rather to the area concerned.

**Chāhi, chāi** (Indo-Pakistan: Urdu-Hindi)
Land or crop irrigated by wells.
Chahi-sailābā—also flooded by river.
Chahi-abi—also by basket or Persian wheel.

**Chain** (Surveying)
*O.E.D.* 9. A measuring line, used in land-surveying, formed of one hundred iron rods called links jointed together by eyes at their ends. At first chains of varying length were used or proposed, but that described by Gunter in 1644 is the one now adopted; it measures 66 feet or 4 poles, divided into 100 links.
b. A chain's length, as a lineal measure, equal to 22 yards, 66 feet, or 4 poles.

**Chain** (Physiography)
A mountain chain. Whereas the word 'range' implies a single line of mountains, or a single edge, 'chain' implies a complexity of several ranges, *e.g.* the Andean Chain. In the past it has been wrongly applied to a simple upland, *e.g.* the Pennine Chain, which still appears on some maps. (L.D.S.)

**Chāk** (Indo-Pakistan: Panjabi)
A small village or settlement in the canal-irrigated areas of West Pakistan (Panjab) or more correctly as described by Spate, 1954, 'the land is divided into *chaks* or blocks fed by a single outlet, of about 2 cusecs, from a main distributary, and so far as possible villages conform to chaks' (pp. 468–469).

**Chalk**
*O.E.D.* 2. An opaque white soft earthy limestone, which exists in deposits of vast extent and thickness in the south-east of England, and forms high cliffs along the sea-shore... 3. Applied to other earths resembling chalk. Fullers chalk; Brown Chalk (umber), French chalk, Red chalk.
Page, 1865. '... it should be borne in mind that the term "chalk" has also been applied to other substances which are in no sense of the word limestones....'
Mill, *Dict.* 1. A soft earthy limestone. 2. The term is also used in a chronological sense for the upper series of rocks in the Cretaceous system.

Himus, 1954. The term has two significations, (i) lithological, connoting a soft white limestone which sometimes consists largely of the remains of foraminifera. It generally consists of very pure calcium carbonate and leaves little residue when treated with hydrochloric acid. In addition to foraminifera there are remains of echinoderms, molluscs and other marine organisms. (ii) Stratigraphical; the upper members of the Cretaceous System.

North, F. J., 1930, *Limestones*. 'As applied to a rock, the term chalk connotes a soft, earthy, light-coloured and very pure limestone, but "The Chalk" in the stratigraphical sense includes some rocks that are hard and marly and others that are sandy, although on the whole the variations are not obvious to the casual observer.' '... in England divided into Upper, Middle and Lower Chalk, the first being characterized by abundant flints and the latter by containing more marl layers. (p. 250)

Hatch, F. H. and Rastall, R. H., revised by Black, G. M., 1938, *Petrology of the Sedimentary Rocks*, 3rd. Ed., '... a soft white limestone, specially characteristic of the Upper Cretaceous of western Europe and parts of N. America ... as a rule a remarkably pure limestone ... most samples contain 97–98% $CaCO_3$ ... remainder ... terrigenous impurities.... Except in a few localities the chalk of S. England is uncemented ... in Scotland, Ireland and the North of England it is moderately well cemented—a comparatively hard limestone' (p. 167).

Twenhofel, W. H., 1939, *Principles of Sedimentation.* 'Chalk has been interpreted as fossil ooze, although in most cases foraminiferal shells are not present in abundance. Most chalks are composed of very finely divided calcium carbonate, to the formation of which it seems probable that Foraminifera made only limited contribution. The environment of deposition was largely shallow water, as shown by the presence of large shells' (p. 328).

**Comment.** It may be noted that the modern view of the origin of chalk is to emphasize the organic remains less than formerly, and to suggest that chemical precipitation has played a large part. The origin of chalk, however, is still much disputed. The old idea that it was a 'fossil ooze' of the deep oceans has been abandoned: E. B. Bailey stressed shallow water origin in clear seas free from sediment, possibly because of arid conditions over surrounding lands. It is important to avoid confusion by spelling the word with a Capital C when used in a stratigraphical sense (Mill's 2 and Himus' ii).

**Chalk marl, Chalk-marl**
Himus, 1954. The lowest division of the English Chalk; the calcareous material is mixed with up to 30 per cent muddy sediment.
*Comment.* The term is reserved for this particular stratigraphical horizon; if it is necessary to refer to a marl rich in chalk it would be called 'a chalky marl'.

**Chalk rock, Chalk-rock**
Himus, 1954. A bed of hard nodular chalk, sometimes containing green-coated calcareous or phosphate nodules, which occurs at or near the base of the Upper Chalk in England.
*Comment.* The remarks under Chalk marl apply.

**Chalybeate**
*O.E.D.* Impregnated or flavoured with iron; especially as a mineral water or spring.

**Champagne**—see Champaign, champain
The French form is preferred by some writers to the old English champaign, champain. It is, however, usually restricted to the *pays* of France—Champagne humide etc. and the old province.

**Chamaephyte**
Raunkiaer, 1934, 'a plant—characterized by having the surviving buds situated close to the ground'. (p. 17).

**Champaign, champain, champian, champion**
*O.E.D.* (From Latin Campania, 'plain, level country', via central French into Middle English).
1. An expanse of level, open country, a plain; a level field; a clearing.
2. (without pl. or article) As a species of land or landscape: flat open country, without hills, woods, or other impediments.
3. The champaign (without pl.): a. the level, open country, in opposition to the mountains and woods; also b. the country, as opposed to the town.
4. The open unenclosed land, as opposed to that partitioned into fields; the moor, fell, or down, unowned or held in common possession; the common land; = Champian. Obs.
5. The level open country as the chief scene of military operations; 'the field'. Obs.
*Comment.* A word that has fallen into disuse in ordinary speech but favoured by some geographers as describing the type of open landscape characteristic of large parts of north-eastern France in contrast to the close *bocage* of the north-west. For example A. G. Ogilvie (*Europe and its Borderlands*, Edinburgh: Nelson, 1957, pp. 151, 154 etc.) uses the form champain without italics and without explanation as an English word. The insertion of a 'g' may be due to confusion with the Italian spelling *campagna* as in Campagna romana. See also the quotation from Monkhouse, who uses the French *campagne* and *champagne*, under bocage.

**Channel**
*O.E.D.* 1. The hollow bed of running waters; also the bed of the sea or other body of water.
2. A rivulet, a stream. Obs.
3. An artificial course for running water or any liquid. a. The watercourse in a street, or by a roadway, the gutter; = Canal, Kennel. Still common locally.
4. Geog. A (comparatively) narrow piece of water, wider than a mere 'strait', connecting two large pieces, usually seas. The Channel: spec. the English Channel.
b. A navigable passage between shallows in an estuary, etc. . . . .
5. An artificial waterway for boats; = canal (quots. 1612–1683).
Mill, *Dict.* 1. The river-bed. 2. A relatively narrow piece of water connecting two larger pieces, *e.g.* two seas. 3. The deep navigable part of a bay or estuary.
*Adm. Gloss.*, 1953. A comparatively or sufficiently deep waterway, natural or dredged, in a river, harbour, strait, etc., or a navigable route between shoals, which affords the best and safest passage for vessels or boats. The name given to certain wide straits or arms of the sea; *e.g.* English Channel, Bristol Channel.
*Comment.* Actually an old form of the word canal and sometimes used interchangeably *e.g.* irrigation channels or ditches.

**Chapada** (Portuguese: Brazil)
Mill, *Dict.* Sertaos or chapadaos. High woodlands (Brazil).
Knox, 1904. Chapadas (Brazil), 'high ground', applied vaguely to elevated plateaux, low ridges or serras traversing the Campos.
Jones, C. F., 1930, *South America*, New York:

Henry Holt. 'The uplands of Matto Grosso consist mainly of broad flat-topped areas ... known as chapadas, these level expanses are potential cattle lands, produced by sandstones.' (p. 488)

Comment. James, 1959, does not use the term but refers to the 'cover of sandstone strata which still remains lying over the crystallines ... a detached outlier to the flat-topped Chapada de Araripe' (p. 410). Mill was evidently misled and so was Knox, if Jones is right in distinguishing chapadas from serras (p. 471). See Sertão.

**Chaparral** (Spanish)

*O.E.D.* U.S. (Spanish, chaparral, from chaparra, evergreen oak + -al, a common ending for a grove, plantation, or collection of trees.) Properly, a thicket of low evergreen oaks; hence, Dense tangled brushwood, composed of low thorny shrubs, brambles, briars, etc. such as abounds on poor soil in Mexico and Texas. (The word came into use in U.S. during the Mexican War, c. 1846.)

*Dict. Am.* 1846 'chapporal ... is a term applied to a species of evergreen thicket, composed of the musquit [mesquite] bush matted with vines, growing about six or seven feet long,'

*Webster*. Spec., a thicket of dwarf evergreen oaks; hence, in general sense and more common usage any dense impenetrable thicket of stiff or thorny shrubs or dwarf trees.

Mill, *Dict*. A dense, largely thorny bush (Texas, term of Spanish origin).

Küchler, A. W., 1947, Localizing Vegetation Terms. *A.A.A.G.*, **37**. 'It is suggested here to use the term Chaparral for a broad-leaf evergreen shrub formation, with or without trees, in the south-western United States and northwestern Mexico.'

Comment. The American equivalent of maquis (*q.v.*) especially characteristic of the Mediterranean rainfall area of southern California.

**Char** (Indo-Pakistan: Bengali)

*Webster*. (Hindi: car) A sandbank, a bar of sand or mud.

Spate, 1954. 'chars or diaras—the floodplain islands. ...' (p. 172)

Ahmad, 1958. '*char* lands (freshly formed silt and sand deposits)' (p. 14); 'char—newly formed alluvial tract' (p. 338). See also pp. 34–35.

Chatterjee, S. P. 'a large newly-formed island in the bed of a deltaic river' (MS. communication).

**Characteristic sheet**

Hinks, A. R., 1942, *Maps and Survey*, 4th ed., Cambridge: C.U.P. 'The characteristic sheet is the key to the system of conventional signs employed on the map.' (p. 4)

**Charnian, charnoid**

Stamp, 1923. 'The Charnian folding, so called from Charnwood Forest, is mainly pre-Cambrian, and is important in the Midlands. The axes of the folds run from N.W. to S.E.' (p. 53).

Comment. Following the principle discussed under Caledonian (C. Lapworth, 1914), folds with the same trend but of later date are termed Charnoid. Even to this day movements causing slight earthquakes are associated with Charnian fault lines.

**Chart**

*O.E.D.* 1. a. A map, *obsolete* in the genera sense. b. *Specifically*, short for *sea-chart*. A map for the use of navigators; a delineation of a portion of the sea, indicating the outline of the coasts, the position of rocks, sandbanks, channels, anchorages etc. c. An outline map.
2. A graphical representation (by means of curves or the like) of the fluctuation of any variable magnitude, such as temperature, barometric pressure, prices, population etc.

Admiralty *Manual of Seamanship*, 1922, Vol. 1. Charts are maps showing the coast-line and the depths of water in different parts of the sea, the latter marked in feet or fathoms, as stated on the chart. (p. 188)

Comment. A modern variant is the *air-chart* or map for the use of navigators of aircraft. The modern use is essentially for maps designed primarily for navigation, mainly by sea or air. Admiralty charts were prepared primarily for the Royal Navy, now world-wide in use.

**Chase**

*O.E.D.* 3. A hunting-ground, a tract of unenclosed land reserved for breeding and hunting wild animals; an enclosed park-land.

Mill, *Dict*. Chase = chace. Tract of open country intermediate between forest and park suitable for hunting (Gt. Britain).

Cox, J. C., 1905, *The Royal Forests of England*. 'A chase was, like a forest, unenclosed and only defined by metes and bounds, but could be held by a subject.

Offences committed therein were, as a rule, punishable by the Common Law and not by forest jurisdiction.... The terms "chase" and "forest" were occasionally used interchangeably, owing to a chase having been secured by the Crown, or the Crown having granted a royal forest to a subject.' (p. 2)

*Comment.* Occurs in place names and a useful clue to the former use of land, otherwise obsolete. See also the references to Turner, G. J., 1901, Barcley, M. L., 1921, and Petit-Dutaillis, C., 1915, under Forest (H.C.D.).

**Chatter mark**

*O.E.D. Suppl.* b. A mark made on a surface by a fragment of rock on the under-surface of glacier ice.

1905. Chamberlin and Salisbury, *Geol.* I. 270. Glacial striae and bruises. The block to the right shows two sets of striae; that to the left shows the peculiar curved fractures known as Chatter Marks.

Lobeck, 1939. 'Chatter marks ... are crescentic cracks, the horns of the crescent pointing in the direction of ice movement and away from the ice mass. They are formed by pressure, the crack resulting from tension. *Crescentic gouges* pointing backward are also known ... explained by pressure ... opposite ...' (p. 301).

Chamberlin, T. C., 1888, The Rock Scourings of the Great Ice Invasions, *U.S. Geol. Survey Ann. Rep.* 7, pp. 147–248. (probable source of term).

Charlesworth, 1957. '"Chattered striae" or "chatter marks" (E. Collomb's *saccades*) are curved, transverse lines or fractures arranged along an axis. Found in firm but brittle rocks, *e.g.* granite, basalt or quartzite, they are so minute and densely packed that often only close inspection detects them. Appearing as a succession of bruises, they were made by a breaking out of thin rock-surfaces through compression in front and tension in the rear of a point of application. The boulders, partially and insecurely embedded in the ice, moved vibratingly and unsteadily and struck with slow rhythmic or jerky effect' (p. 248).

*Comment.* See also Flint, 1947, p. 71; Cotton, 1947, p. 246. The *O.E.D.* is not very clear.

**Chaung** (Burmese; also -yaung, -young as suffix in place names)

A watercourse whether occupied permanently by a stream or not and so comparable with wadi (Arabic) or nālā (nullah, Hindi).

Thus Yenanyaung is yè, water; nan, stinking (or yè-nan, oil); chaung, creek or river = oil creek. (L.D.S.)

**Chaur** (Indo-Pakistan: Hindi)

Spate, 1954. 'long semi-circular marshes which develop into a vast and intricate chain of temporary lakes during the rainy season' (p. 514).

**Cheesewring**

Swayne, 1956. A mushroom rock on the east of Bodmin Moor. The name is sometimes given to similar rocks elsewhere.

*Comment.* On the south-eastern margin of the granite mass of Bodmin in Cornwall there is a granite tor in which the granite boulders of the upper part overhang a narrower 'stem'. From a fancied resemblance (inverted) to the muslin bag in which curds from sour milk were once put by country people and the moisture wrung out so as to leave a white cream cheese, this tor has long been called 'The Cheesewring'. Some writers have used the name for similar erosional features, even extending it to those due to undercutting by wind driven sand in arid regions, but the term has not been generally adopted.

**Cheiragratic coast**

Mill, *Dict.* A coast of a folded and faulted region with complex submergences, giving as a result a succession of deep gulfs and finger-like promontories.

*Comment.* From a Greek root meaning hand. The term does not appear in modern works and is apparently obsolete.

**Chelogenic** (J. Sutton, 1963)

Sutton, J., 1963, *Nature*, **198**, 'Shield-forming'; applied to major cycles in the earth's geological history.' (p. 731). See also *Nature*, **200**, 1965, 1023–1027.

**Chena** (Ceylon: Sinhalese)

The particular form of shifting cultivation (*q.v.*) practised in Ceylon.

**Chenier**

Howell, 1962. Beach ridge built upon swamp deposits.

**Cheri, chāri** (Indo-Pakistan: Urdu-Hindi)

Spate, 1954. 'segregating the untouchables in outlying *cheris* or sub-villages, sometimes located several hundred yards from the main villages ...' (p. 177).

**Chernozem, Tschernosem** and other spellings (Russian; black earth)

Mill, *Dict.* Chernozem or Chernozyom (the latter form phonetic). Russian term for fertile black soil rich in humus.

Soils and Men, 1938, Glossary. Chernozem

Soils. A zonal group of soils having a deep, dark-colored to nearly black surface horizon, rich in organic matter, which grades below into lighter colored soil and finally into a layer of lime accumulation; developed under tall and mixed grasses in a temperate to cool subhumid climate. From the Russian for black earth. Sometimes spelled Tschernosem, Tschernosiom.

Jacks, 1954. 'Chernozem. Dark, well-drained grassland soil granular and rich in humus to some depth, with or without concentration of clay in the B horizon, and calcareous below.'

Robinson, G. W., 1949, *Soils*, 3rd ed. says no simple definition and gives details (pp. 334–338). See also Russell, E. J. and E. W. *Soil Conditions and Plant Growth*, 6th ed., 1950, pp. 526–530.

FAO, 1963. '*Chernozem soils* typically have AC profiles with thick A-horizons with mull or mull-like humus form. When a B-horizon is present, as in "degraded chernozem", the ratio of bivalent to monovalent exchangeable ions in the clay fraction of the horizon is low. The soils are moderately calcareous, though the surface may be non-calcareous ...'

### Chert

*O.E.D.* Also chirt. (App. a local term which has been taken into geological use). A variety of quartz, resembling flint, but more brittle, occurring in strata; also called hornstone. Also applied to various impure siliceous or calcareo-siliceous rocks, including the jaspers.

Holmes, 1928. A more or less pure siliceous rock composed in part of fibrous and radial chalcedony with or without the remains of siliceous and other organisms such as sponge spicules or radiolaria; occurring as independent formations and also as nodules and irregular concretions in formations (generally calcareous) other than the Chalk. The fracture is generally splintery rather than conchoidal. *cf.* Flint.

*Comment.* Chert is a rock not a mineral and so cannot be described as 'a variety of quartz'. The *O.E.D.* is accordingly misleading.

See also Page, 1859, p. 114; Read, H. H., *Rutley's Elements of Mineralogy*, 23rd ed., 1936, p. 312; Twenhofel, 1939, p. 367.

### Chestnut soil

Jacks, 1954. 'Dark brown over lighter coloured soil overlying a calcareous horizon.'

FAO, 1963. '*Chestnut soils* have an ABC or an AC profile with a typical dark brown friable (chernosemic) A1-horizon that normally overlies a brown, prismatic B-horizon with lime accumulation.' Includes reddish-chestnut soils.

### Chetoi (Egypt)

The natural winter harvest of Egypt. Seed is sown in the Nile mud after the flood period in November and the harvest is from February to early May. Contrast Nili.

### Chili

Moore, 1949. The hot dry southerly *Sirocco* wind of Tunis, North Africa.

See Sirocco; also Kendrew, 1943, p. 359.

### Chimney

*O.E.D.* 6. transf. a. Applied to a natural vent or opening in the earth's surface, esp. that of a volcano.

8. A name given by mountain-climbers to a cleft in a vertical cliff by which it may be scaled, usually by pressing rigidly against the opposite sides.

Mill, *Dict.* 1. A relatively narrow cleft in a vertical cliff, especially in the English Lake District; one so narrow as to enable the cliff to be scaled by pressing rigidly against the sides. *cf.* Couloir (2). In America = English stack = a pillar of rock, such as the Old Man of Hoy; also the pipe of a volcano.

Gemmell and Myers, 1944. *Adventure Underground*. 'A vertical shaft in the roof of a cave passage, smaller than an aven, up which it is possible to climb.'

*Comment.* See also Rice, 1941, for other meanings. Best reserved for rock-climber's usage (G.T.W.).

### China Clay—see Kaolin

Himus, 1954. An almost white clay resulting from the decomposition of felspars in granite which may result from the action of ascending gases and vapours chiefly of carbon dioxide and super-heated steam from a deep seated magma.

### Chine

*O.E.D.* A fissure in the surface of the earth; a crevice, chasm. Obsolete in this general sense but still used specifically. On the Isle of Wight and Hampshire coast, a deep and narrow ravine cut in soft rock strata by a stream descending steeply to the sea. See Lyell, 1830, I, 281.

### Chinook

*O.E.D.* U.S. Native name of an Indian tribe on the Columbia river. *Chinook wind*: an ocean wind, warm in winter, cool in summer, which blows on the Pacific slope of the Rocky Mountains.

*Webster*. a. A warm, moist, southwest wind

of the coastal regions of Oregon and Washington. b. A warm, dry, foehn-like wind that descends the Rocky Mountains.

Mill, *Dict*. A wind blowing down the landward slopes of the Rocky Mountains which has been dried by its passage over the mountains and is warmed by compression as it reaches lower warmer ground. *cf*. Föhn.

Burrows, Alvin T., 1901, The Chinook Winds, *Yearbook U.S. Dept of Ag.*, Washington, 1902. 'At the present time there are three different winds called Chinooks. Each of them is essentially a warm wind, whose effect is most noticeable in winter. Under their influence snow is melted with astonishing rapidity and the weather soon becomes balmy and spring-like. The name "Chinook" is that of an Indian tribe which formerly lived near the mouth of the Columbia River. It was applied to a warm southwest wind which blew from "over Chinook camp" to the trading post established by the Hudson Bay Fur Company at Astoria, Oregon. The name soon came into general use in that locality, and as the adjacent country was settled the usage extended, so that now "Chinook" is applied not only to the warm, moist southwest winds along the Oregon and Washington coast, but to the warm, dry, descending winds east of the Cascade range in Washington and the Rocky Mountains in Montana and elsewhere. In 1895 Mr. B. S. Pague, the local forecast official at Portland, Oregon, began to call the descending southeast winds that visited western Oregon and Washington during the winter Chinooks.' (pp. 555-6)

*Met. Gloss.*, 1944. 'Chinook. A warm dry wind, similar in character to the Föhn, which occurs on the eastern side of the Rocky Mountains. It blows from the west across the mountains and is warmed adiabatically; it usually occurs suddenly, a large rise in temperature takes place, and the snow melts very rapidly.'

Ward, R. D., Brooks, C. F., Connor, A. J., 1936, *The Climates of North America*, Berlin: Borntraeger. 1. First used for moist, warm S.W. wind in W. Oregon to Washington and B.C.
3. Applied by early settlers to warm dry wind along E. base of Rocky Mts. (believed to come from Pacific).
'The name is often applied at the present time to the warm, damp south-westerly winds on the coasts of Washington and Oregon, but its use among meteorologists should be limited to the warm and dry descending winds of distinct foehn type.' There is a long discussion.

*Comment*. In modern climatology the word is exclusively used as defined by Mill and *Met. Gloss.*

**Chō** (Indo-Pakistan: Panjabi)
Along the southern margin of the Siwaliks, a rainy season torrent on the plain. The word *cho* connotes a bed of loose boulders, gravel and sand, indicating rapid erosion. Spate, 1954, says 'these chos are dry except for sudden spates—the chos country is really an immense pan-fan ... each cho is a broad river of sand, with a shallow ever-shifting bed ...' (p. 484).

**Chore**—see Landscape, cultural

**C-horizon** (Soil Science)
The lowest of the three typical horizons of a soil: commonly equated with the 'subsoil'. See Horizon.

**Chorography**
*O.E.D.* 1. The art or practice of describing, or of delineating on a map or chart, particular regions, or districts; as distinguished from geography, taken as dealing with the earth in general, and (less distinctly) from topography, which deals with particular places, as towns, etc.
2. concr. A description or delineation of a particular region or district.
3. transf. The natural configuration and features of a region (which form the subject matter of its chorography in sense 2).

Mill, *Dict*. Detailed description of a region (frequently viewed as a synonym for topography).

*Comment*. An attempt has been made to revive chorography and chorographical as terms to apply to regions rather than specific localities or small areas. A chorographical map would thus be distinguished from a topographical on a larger scale. As noted by *O.E.D.* the term, derived from the Greek for district, was much in use in the 17th century.

**Chorology**
*O.E.D.* The scientific study of the geographical extent or limits of anything.

Mill, *Dict*. 1. An inquiry into the causal relations of the phenomena belonging to a particular region. 2. The causal study of the distribution (in space) of organisms at any period of the Earth's history. The general assemblage of organisms found in one locality at one period and living under

one set of external conditions is called a 'Facies'.
*Comment.* Rarely used; but see Chorography which is much more frequently used. Chorology was used by von Richthofen in 1883 to mean the explanatory description of areas.

**Choropleth, chorisopleth**
*O.E.D.* no reference.
Monkhouse and Wilkinson, 1952. 'In this book ... the category of choropleth map will be used for all quantitative areal maps, calculated on a basis of average numbers per unit area.' This book gives a full discussion of isopleths and choropleths. The authors use or refer to chorogram, choropleth, chorisogram, chorisometers, chorisopleths, chorometrograms, chorisochores; also isopleth, isarithm, isoline, isobase, isogram, isontic line, isometric line. (pp. 28–30)
Wright, J. K., 1944. The Terminology of Certain Map Symbols, *Geog. Rev.*, **34**. '"Choropleth" has already gained some currency as the designation for an area symbol bounded by the limits of political or other statistical subdivisions of the territory mapped.' (p. 653)
See also Wright, J. K., 1938, Problems in Population Mapping, in *Notes on Statistical Mapping, with Special Reference to the Mapping of Population Phenomena*, Amer. Geog. Soc. and Pop. Ass. of Amer., pp. 1–18, ref. on p. 14.
Raisz, E., 1938, *General Cartography*, New York: McGraw-Hill. 'The term "choropleth" (quantity in area) is not necessarily limited to civil divisions. If the area were divided into squares and tinted proportionately, this would also make a choropleth map. Maps showing distribution by civil divisions may be called demopleth maps as a distinct type of choropleth map.' (p. 246)
British National Committee for Geography, 1966, *Glossary of Technical Terms in Cartography*, London: The Royal Society. 'A chorogram consisting of distinctive colour or shading applied to areas other than those bounded by isograms (usually to statistical or administrative areas).' (p. 13)
Haggett, 1972. By placing a grid over a distribution map 'we can make choropleth maps (from the Greek *choros*, area, and *plethos*, fullness or quantity) by assigning different shades of colour to each of the cells. By linking cells with similar colours we can create a general picture of the distribution.' (p. 6)
International Cartographic Association, Commission II, 1973, *Multilingual Dictionary of Technical Terms in Cartography*, Wiesbaden: Franz Steiner Verlag G.M.B.H. 'Choropleth technique. A method of Cartographic Representation which employs distinctive colour or shading applied to areas other than those bounded by Isolines. These are usually statistical or administrative areas.'

**Chott**
French spelling of Shott (1), *q.v.*

**-chow (Chinese)**
As suffix to place name denotes the chief town of a district.

**Chronology**
*O.E.D.* The science of ... recording and arranging events in the order of time. See also Geochronology.

**Chronometer** (*fr.* Gk., *lit.* time-measure)
*O.E.D.* An instrument for measuring time; spec. applied to time-keepers adjusted to keep accurate time in all variations of temperature etc.... used for determining longitude at sea, and for other exact observations.

**Chronotaxis** (L. G. Henbest, 1952)
Similarity in age geologically, in contrast to homotaxis (*q.v.*).
See Henbest, L. G., *Jour. Pal.*, **26**, 299–394.

**Chute**
*O.E.D.* (Here there appears to be a mixture of the French, chute, fall (of water, descent of a canal lock, etc.) and English shoot.)
1. A fall of water; a rapid descent in a river, or steep channel by which water escapes from a higher to a lower level.
2. A sloping channel or passage for the conveyance of water, or of things floating in water, to a lower level; in North America, an opening in a river dam for the descent of logs, etc. b. A fish way (U.S.).
4. The steep slope of a spoil-bank beside a quarry or mine, down which rubbish is shot; also a steep slope for tobogganing.
5. In the Isle of Wight, a steep cutting affording a passage from the surface above a cliff to the lower undercliff ground.
Muma and Muma, 1944. 'An inclined channel or trough' (referring to caves) (p. 36).
Fisk, 1947. 'Shortening the length of the channel [in a meandering channel] is also brought about by chute cut-offs which

**Cinder cone**
*O.E.D. Suppl.* A cone formed round the mouth of a volcano by debris cast up during eruption.
Lobeck, 1939. 'This material is termed cinders or ash even though it may not be of igneous origin.' (p. 664)
Cotton, 1944. 'Scoria mounds', 'scoria cones' or 'cinder cones' . . . 'such mounds rarely grow to very large dimensions because of the temporary and shifting nature of the vents.' (p. 141)
*Comment.* Composed of fragmental material, in distinction to lava domes or composite cones (*e.g.* Holmes, 1944, p. 452). Identical with ash cone (*q.v.*).

**Cinglos** (Spanish; cingles in Catalan)
The face of a cuesta when it is formed by high abrupt cliffs (P.D.).

**Cinque port**
*O.E.D.* A group of sea-ports (originally five, whence the name) situated on the south-east coast of England, and having jurisdiction along the coast continuously from Seaford in Sussex, to Birchington near Margate, including also Faversham, which have existed as an incorporation from an early period of English history. The five 'Ports' are in order of precedence Hastings, Sandwich, Dover, Romney, Hythe, to which were added in very early times, Rye and Winchelsea. The Lord Wardenship is now chiefly an honorary dignity.

**Circle**—see Great Circle, Small Circle

**Circulation**
East, W. G. and Moodie, A. E., 1956, *The Changing World*, London, Harrap 'The term "circulation" is used here to include all those movements of goods, passengers, news, ideas, and capital which are essential to both national and international well-being.' (p. 208)
*Comment.* This is included as a special meaning, in addition to normal use, *e.g.* atmospheric circulation.

**Circumdenudation**, also **circumerosion**
*O.E.D. Geol.* denudation all around.
1882. Geikie, Text-Bk. Geol., VII, 925. 'Eminences detached by erosion from the masses of rock . . . have been termed hills of circumdenudation.'
Geikie, J., 1898. 'To such an extent have many ancient plateaux of erosion been denuded, so deeply have they been entrenched, that their surface has become resolved into a truly mountainous region, wherein all the elevations are mountains of circumdenudation, the tops of which are the only remaining relics of the original plateau surface.' (p. 106)
'Intrusive masses formerly deeply buried are eventually exposed, and, owing to the more rapid removal of the rocks through which they rise, may come to form mountains of circumdenudation . . .' (p. 155).
p. 276 appears to equate with relict mountains, p. 116 with subsequent mountains.
*Comment.* See accumulation. The contrast is between mountains of circumdenudation and mountains of accumulation, *e.g.* volcanoes. An old concept, little used now.

**Cirque** (French)
*O.E.D.* 2. A natural amphitheatre, or rounded hollow or plain encircled by heights; esp. one high up in the mountains at the head of a stream or glacier. (So in French)
1874. Dawkins, *Cave Hunting*, 1878. A. Ramsay, Phys. Geog., xxiii.
1882. Geikie, Text-Bk. Geol., vii, 924.
Mill, *Dict.* Cirque. The rounded upper end of a valley. It is normally found in areas which have been glaciated, *e.g.* the upper end of the Glen Cloy, Arran, and many glens of the Scottish highlands, and the Cirque of Gavarnie in the Pyrenees. The term is best restricted to this signification, but it has been used as a synonym for 'corrie', 'cwm' and 'combe' (*q.v.*).
Tyndall, 1860. 'Some valleys are terminated by a kind of mountain-circus with steep sides. . . .'
Geikie, J., 1898. Cirque or corrie basin is used interchangeably for valley-head amphitheatres, and for those not at the head of a valley. Generally regarded as result of glacial or frost action, but also of running water. (p. 230 *et seq.*)
Knox, 1904. 'Cirque (U.S.A.), a glacial amphitheatre or basin.'
Hobbs, W. H., 1909, *Geog. Jour.*, **35**. Does not make the Cotton distinction, but regards terms cirque, corrie, cwm, botn, kjedel, kahr and cirkus as equivalent. Uses term 'glacial amphitheatre'. (p. 148)
Cotton, 1942, contains a discussion of use of term to describe (i) valley head amphitheatres or valley head cirques; (ii) hanging or perched cirques (not continued in a

valley). (p. 169) Cotton uses both. Also discusses use of cirque, *cf.* corrie, to no conclusion.

Holmes, 1944. 'Valley glaciers characteristically originate in deep arm-chair-shaped hollows, called *corries* or *cirques*, situated at the valley heads.' (p. 209)

Flint, 1947. Describes a cirque as the theatre-shaped head of many glaciated valleys, sculptured by a combination of nivation and glacial scour. Later discussion refers to small 'nivation cirques' etc., and deepening near the headwalls. Equivalent to corrie, cwm. Kar, botn. (p. 93)

*Comment.* Used in France for any amphitheatre; in Britain usually restricted to those of glacial origin though older writers such as W. Boyd Dawkins, 1874, *Cave Hunting* (p. 56) used the term for dolines in limestone country. Charlesworth, 1957, 244 has used *pseudo-cirques* for these and similar forms. He attributes introduction of cirque as a term to J. Charpentier (in French) *Essai sur la constitution geognostique des Pyrénées*, Paris, 1923, p. 24 (G.T.W.).

**Cirque glacier**

*Ice Gloss.*, 1958. A *glacier* which occupies a separate rounded niche which it has formed on a mountain side.

**Cist**

*O.E.D.* A chamber excavated in rock or formed of stones or hollowed tree trunks; especially a stone-coffin formed of slabs placed on edge, and covered on the top by one or more horizontal slabs.

*Comment.* Apparently from Welsh *cist* but the word has been used of Irish finds.

**City**

*O.E.D.* A long historical sketch may be summarized:

'Civitas' was applied by the Romans to the independent states or tribes of Gaul and later to their seat of civil and episcopal government. In 13th cent. cité was used for ancient towns and principal boroughs. As the episcopal sees began to settle in these, the identification of city with 'cathedral town' grew up, which was strengthened by the boroughs containing the new bishoprics of Henry VII being created 'cities'. This connexion is not necessary however and cities have been created by royal authority, without a cathedral.

In Scotland the word was introduced from English after the association with episcopal seats, and was so used, irrespective of whether the place was a burgh. By 15th cent. applied to large burghs, not bishoprics (Perth, Edinburgh) and later history similar to England. Similarly in Ireland. 'In other lands now or formerly under British rule, "city" is used sometimes more loosely, but often with more exact legal definitions than in England.' 'The distinction is unknown to other Teutonic and (now) also to Romanic languages....'

2. spec. A title ranking above that of 'town'.

2.d. in U.S. 'A town or collective body of inhabitants incorporated and governed by a mayor and aldermen' (*Webster*); but applied, in the newer states, much more loosely, and often given in anticipation.'

e. In the Dominion of Canada: a municipality of the highest class.

6. As the equivalent of Gr. πολις, L. civitas, in the original sense of a self-governing city or state with its dependencies.

Mill, *Dict.* Properly in England a town which is or has been the seat of a bishop or site of a cathedral church; in America an incorporated town; often vaguely applied to any large important town.

Smailes, A. E., 1944, Urban Hierarchy... *Geography*. Hierarchy consists of cities, major towns or minor cities, towns. (pp. 41–51)

Jefferson, Mark, 1931. What is a City? *Geog. Rev.*, 21, 446.

Queen, S. A. and Carpenter, D. B., 1953, *The American City*, New York: McGraw-Hill. 'Simply as a preliminary statement, we may say that a city is a collection of people and buildings, large for its time and place, and characterized by distinctive activities.' (p. 19) 'Distinctive activities' appears to mean non-agricultural employment, and the U.S. Census Bureau lower limit of 2,500 pop. is regarded as acceptable.

*Comment.* Enc. of Soc. Sc. does not offer a definition but uses as =town, as do other American texts. The locale of the 'urban way of life' and 'urban problems.' There must remain the distinction between English and American usage: a town in England or the Commonwealth must be accorded the legal right before it can call itself a city. Thus Nairobi became the first city in E. Africa. Nevertheless the word is used in a general sense to mean a very large town, as in Smailes.

**City circulation** or **trade area**
Dickinson, 1947, *City, Region and Regionalism*. After describing the City Settlement Area: 'The City Circulation or Trade Area is the area of wider and more extensive, more occasional circulations to and from the city....' (p. 170)
*Comment.* Not generally adopted. *Cf.* umland, urban hinterland.

**City settlement area**
Dickinson, 1947, 'The City Settlement Area embraces the urban tract and the outer zone or rural-urban fringe....' (In general a limit of one hour of journey time—20 miles—and the supply area for milk and vegetables—'zone de voisinage' of Chabot; though this is now out of date.)

**Civil Lines**
The European section of the larger Indian cities during the British period with the official residences of the local bureaucracy; there was often a distinct 'railway colony' and 'cantonments'.

**Clachan** (Gaelic, Highland and Ayrshire Scots and recent Northern Irish).
From *clach, clachan*, stone(s). D.O.S.T. and S.N.D. limit clachan to a hamlet or village with a church, which would make it the equivalent of kirk-toun. A clachan or kirkton usually had one or more craftsmen. The word is used by Scott (the 'clachan of Aberfoyle'), Tannahill and Ayrshire writers. Arthur Geddes supports the D.O.S.T. in making the church essential and points out that a broader use in certain districts of Scotland should be regarded as decadent. The word was apparently introduced into Northern Ireland by John Donaldson (1818–38) and applied to a cluster of farmhouses and their outbuildings, without any formal plan or arrangement. E. Estyn Evans has adopted clachan to describe any small rural settlement consisting of a group of small farms, probably originally occupied by members of one family. In Northern Ireland the word is commonly applied to any small settlement (*cf.* hamlet in England). (C.J.R.) See also Fairhurst, H., 1960, Scottish Clachans, *S.G.M.*, 76, 67–76.

**Clastic rock**
Himus, 1954. Rocks composed of fragments of pre-existing rocks. They include the sedimentary rocks and those formed from the dispersed consolidation products of magmas, *i.e.* the tuffs, agglomerates and volcanic ashes.

**Clatter**—see also Clitter
A scree (Dartmoor)

**Clay**
*O.E.D.* 1. A stiff viscous earth found, in many varieties, in beds or other deposits near the surface of the ground and at various depths below it; it forms with water a tenacious paste capable of being moulded into any shape, which hardens when dried, and forms the material of bricks, tiles, pottery, and 'earthenware' generally. Clay consists mainly of aluminium silicate, and is derived mostly from the decomposition of felspathic rocks. The various beds are distinguished geologically as *boulder, Kimmeridge, London, Oxford Purbeck Clay*, etc. Particular kinds of clay are known as *brick, fatty, fire, plastic, porcelain* and *potter's clay; pipe-clay*, etc. Examples c. 1000 to 1882.
2. In early use the tough, sticky nature of the substance appears to have been mainly in view, and the name was applied to other substances of this nature, as to the *bitumen* of the Vulgate....
3. Used loosely for: Earth, moist earth, mire, mud; esp. the earth covering or enclosing a dead body when buried.

Mill, *Dict.* Clay. Strictly speaking, a deposit of fine particles of silicate of alumina, *e.g.* China clay. Often used loosely of deposits of fine-grained particles of diverse composition, though these deposits are more accurately termed muds.

Holmes, 1928. 'An earthy deposit of extremely fine texture which is usually plastic when wet, and becomes hard and stone-like on being heated to redness. Chemically it is characterized by containing hydrous silicates of alumina in considerable quantity, with felspars and other silicates and quartz, and variable amounts of carbonates and ferruginous and organic matter. A proportion of the constituents is generally in the colloidal state, and then acts as a lubricant to the grains and flakes of non-colloidal material.'

Glentworth, R., 1954, *The Soils of the Country Round Banff, Huntly and Turriff*. Edinburgh: H.M.S.O. 'Mineral particles less than 0·002 mm. diameter. As a texture class applied to soil containing 30 per cent. or more of clay. A soil separate.'

*Comment.* The separate uses of the word should be distinguished:
1. in pedology, the size of the particles in a soil and resulting soils;
2. in mineralogy, a complex group of minerals to the individual members of which specific names have been given;
3. in geology, rocks consisting mainly of clay minerals and fine particles and with certain physical characters;
4. in stratigraphical geology, to specific beds at different horizons.

The *O.E.D.* attempts to list in a naïve way some of the clays under (3) and (4).

### Clay-ironstone
Himus, 1954. Sheet-like deposits of concretionary masses of argillaceous siderite (ferrous carbonate), associated with Carboniferous strata, particularly with the Coal Measures.

### Clay-slate
Himus, 1954. A metamorphosed argillaceous rock possessing good cleavage. The term is used to distinguish argillaceous slates from those derived from volcanic ash.

### Clay-with-flints
Himus, 1954. A term properly used to describe a deposit of mixed chalk-flints with clay that lies directly on the Chalk in many districts; often found in potholes or pipes. It is generally considered to be the residue left from the solution of chalk, but there may be an admixture of Tertiary materials. The clay is reddish or brown, often nearly black at the base, becoming lighter and more sandy higher up. The term is also loosely used to denote nearly all the clay-flint deposits that rest on the Chalk.

### Cleat
*Webster. Coal mining.* The joints along which coal breaks when mined.

Kendall and Wroot, 1924, *Geology of Yorkshire*, p. 130.
Principal Joints—a term usually confined to coal joints, often clean, whilst joints in the opposite direction are of limited extent, and tight (*cf.* Slyne, used in other coalfields). (p. 130)

### Cleavage
*O.E.D.* 1. The action of cleaving or splitting crystals and certain rocks along their lines of natural fissure; the state of being so cleft.
2. *Min.* Arrangement in laminae which can be split asunder, and along the planes of which the substance naturally splits; fissile structure; the property of splitting along such planes.
3. *Geol. Slaty cleavage*: the fissile structure in certain rocks, especially in clay slate and similar argillaceous rocks, whereby these split into thin laminae or 'slates' used in roofing, etc. This structure is quite distinct from, and in origin posterior to, the stratification and jointing, the cleavage-lines crossing these at any and every angle, while parallel to themselves over extensive tracts of country.

*Webster. Cryst.* Quality possessed by many crystallized substances of splitting readily in one or more definite directions and yielding surfaces always parallel to actual or possible crystal faces. Also, the direction of the dividing plane.

*Geol.* The structure possessed by some rocks by virtue of which they break more readily and more persistently in one direction or in certain directions, than in others.

Mill, *Dict.* A structure produced in rocks by pressure, causing them to split in close parallel planes, usually across the stratification.

Holmes, 1928. (a) The property of minerals, due to their atomic structure, whereby they can be readily separated along planes parallel to certain possible crystal faces. (b) The property of rocks such as slates, which have been subjected to orogenic pressure, whereby they can be split into thin sheets, the plane of cleavage being at right angles or inclined to the direction in which the pressure was applied, according to the effects produced by shearing-stress during the process.

Willis, B. and R., 1934, *Geologic Structures*. '... an ambiguous term as applied to rocks unless qualified by an adjective. We may distinguish: *Slaty cleavage*... *Fracture cleavage*... *Flow cleavage*. (p. 263)

Hills, E. S., 1953. 'Rocks that possess cleavage may be split into thin sheets along parallel or sub-parallel planes that are of secondary origin and are formed as a result of metamorphism... *Flow cleavage*... *Fracture cleavage*... *Strain-slip cleavage, slip cleavage*, or *shear cleavage*....'

*Comment.* The interpretation of cleavage is a complex problem and references to literature are given in the works quoted above.

In geographical writings the reference is usually to slaty cleavage which has an importance in the association with economically important 'slates'.

**Cleugh, clough, cleuch** (Scottish)

*O.E.D.* A gorge or ravine with precipitous and usually rocky sides, generally that of a stream or torrent.

2. The precipitous side of a gorge; a steep and rugged descent.

Mill, *Dict.* Cleugh = Clough = Cleuch. A rugged and rocky narrow glen (S. Scotland and N. England). A steep-sided tributary valley (Midlothian, Derbyshire).

*Comment.* Enters frequently into place names in Scotland (*pron.* clooch or clyooch with ch as in loch) and in Derbyshire (*pron.* cluff) (C.J.R.).

**Cliff**

*O.E.D.* 1. A perpendicular or steep face of rock of considerable height. Usually implying that the strata are broken and exposed in section; an escarpment. Examples 854–1837 (Cheddar Cliffs). b. *Esp.* (in modern use) A perpendicular face of rock on the seashore, or (less usually) overhanging a lake or river. Examples a. 1000–1879.

2. (Extension of 1.b.): Land adjacent to a sea or lake; shore, coast, strand. *Obs.* examples c. 1000–1600.

3. A steep slope, a declivity, a hill; = Cleve 3. (In Lincolnshire, the sloping and cultivated escarpment of the oolite is called the Cliff). Examples a. 1200–1870.

4. The strata of rock lying above or between coal seams. Examples 1676–1721.

*Comment.* Distinctions are often made by a prefix, the meaning of which is obvious, e.g. sea-cliff, lake-cliff, etc. For relationship to other slopes, see Wood, *Proc. Geol. Assoc.*, **103**, 1942, 129.

**Climagram**

Lee, D. H. K., 1957, *Climate and Economic Development in the Tropics*, New York: Harper, uses 'climagrams' combined with 'strain lines' in discussing human health and efficiency. These appear to be Griffith Taylor's climograms.

**Climate**

*O.E.D.* From the Greek, inclination or slope. The meaning passed in Greek through the senses of 'slope of ground, *e.g.* of a mountain range', the supposed 'slope or inclination of the earth and sky from the equator to the poles', 'the zone or region of the earth occupying a particular elevation on this slope, *i.e.* lying in the same parallel of latitude', 'a clime', in which sense it was adopted in late Latin.

1. A belt of the earth's surface contained between two given parallels of latitude. *Obs.* b. more vaguely: A region of the earth, a 'clime'. *Obs.* exc. as in 2.

2. A region considered with reference to its atmospheric conditions, or to its weather. ex. 1398–1874.

3. Condition (of a region or country) in relation to prevailing atmospheric phenomena, as temperature, dryness or humidity, wind, clearness or dullness of sky, etc. esp. as these affect human, animal, or vegetable life. ex. 1611–1880.

1860. *Cornh. Mag.* II, 566. Climate is properly the long average of weather in a single place. 1880. Haughton, *Phys. Geog.* iii, 74. 'Climate' may be defined as the complex effect of external conditions of heat and moisture upon the life of plants and animals.

*Webster.* 3. The average course or condition of the weather at a particular place, over a period of many years, as exhibited in absolute extremes, means, and frequencies of given departures from these means, of temperature, wind velocity, precipitation, and other weather elements.

Mill, *Dict.* The sum-total of meteorological phenomena which characterize the average conditions of the atmosphere at any one point on the Earth's surface. (Hann)

*Met. Gloss.*, 1944. The average weather conditions of any locality.

Manley, G., 1952, *Climate and the British Scene*, London: Collins. 'Climate may be defined as an expression of our integrated experiences of "weather".' (p. 1)

Haurwitz, B. and Austin, J. M., 1944, *Climatology*, New York: McGraw-Hill. 'Even though the changes of the weather are proverbial, it is nevertheless possible at every place to arrive at a generalization and a composite of these variations. One speaks then of the *climate of the region*.' (p. 1)

Longwell, C. R., 1954, The origin of the word Climate, *Science*, 120 (No. 3113). 'Ptolemy of Alexandria of the second century A.D., divided the world into κλίματα, a succession of zones from the equatorial regions poleward, differing in the obliquity of the sun's rays to the earth's surface within the

several zones. Ptolemy inherited this concept.... Eratosthenes of the third century B.C. probably was the true inventor.' Longwell considers the association with the inclination of the earth's axis is without foundation and hence disagrees with *Webster* and successive editions of the *Enc. Brit.* (p. 355).
*Comment.* The definitions given by Mill and the *Met. Gloss.* would seem to be more appropriate to a microclimate, since climate is commonly used of regions, often of great extent.

**Climatic geomorphology**
Stoddart, D. R., 1969, Climatic geomorphology: review and re-assessment, *Progress in Geography*, **1**, London: Edward Arnold. 'Climatic geomorphology ... postulates that different climates, by affecting processes, develop unique assemblages of landforms. Systematic climatic geomorphology is the analysis of these processes and forms, and their relationship with climate, and has the aim of defining morphogenetic regions on a world basis.' (p. 163)
See Davis, W. M., 1899, The Geographical Cycle, *The Geog. Jour.* **14**, 481–504. The term climatic geomorphology probably first used by Martonne, E. de, 1913, Le Climat Facteur du Relief, *Scientia*, 339–355.
See also Climato-genetic geomorphology.

**Climato-genetic geomorphology**
Stoddart, D. R., 1969, Climatic geomorphology: review and re-assessment, *Progress in Geography*, **1**, London: Edward Arnold. Climato-genetic geomorphology 'depends on the fact that since climate has demonstrably changed during the Tertiary and Pleistocene, and continues to change, sets of climatically-controlled landform features are being continuously superimposed on each other, so that the landforms of any one area bear the imprint of climates no longer operative.' (p. 163)
See also Climatic geomorphology.

**Climato-isophyte** (Sten Sture Paterson, 1956)
Paterson, S. S., 1956, *The Forest Area of the World and its potential productivity*, Göteborg. 'Line connecting places with equal plant growth ability resulting from climate, expressed by the CVP-index' (p. 8).

**Climatology**
*O.E.D.* That branch of physical science which deals with climate, and investigates climatic conditions (sometimes used for the conditions themselves as a subject of observation).
Moore, 1949. The science which treats of the various climates of the earth, and their influence on the natural environment.
Kendrew, W. G., 1949, *Climatology*, Oxford: O.U.P. 'In the study of climatology the primary interest lies in the facts of the climates of the earth in themselves, and as elements in the natural environment of life. The investigation of the physical causes underlying the facts is a valuable as well as an interesting side of the study, but is to be regarded as subordinate' (p. 12).
See also Miller, 1953, Chapter I, The Meaning and Scope of Climatology; Haurwitz and Austin, 1944, 'the science that discusses the climates found on the earth' (p. 1).

**Climatotherapy, climatotherapeutic, climatotherapeutics**
*O.E.D. Suppl.* The treatment of disease by a favourable climate (from 1887).

**Climax** (vegetation)
*Webster.* 6. *Phytogeog.* The culminating stage of the possible development of vegetation in a region. Its nature is usually controlled by conditions of soil (edaphic climax), as in bogs or sand dunes, or climate (climatic climax), as in tropical rain forests or northern coniferous forests.
Tansley, 1939. 'The great climax communities of the world ... represent the permanent forms of vegetation and the ... completest and most complex adaptation to the conditions of life obtaining in the various climatic regions of the world' (p. 218). Many agencies are constantly at work to destroy them and thus to produce the conditions for new successions. New bare soils are colonized by pioneer communities of various kinds ... followed by different stages of various priseres' (p. 219). Fig. 48 gives the 'Scheme of a Prisere'—starting from pioneer community through second and third stage communities to a pre-climax, then to the Climatic Climax and, with the operation of an exceptionally favourable factor, to a post-climax. Through the operation of an arresting factor a pioneer community may develop only

to a sub-climax; only when the arresting factor is renewed does the Climatic Climax follow (pp. 218-223). Apart from the main climatic climax are 'Climaxes determined by different factors ... edaphic and physiographic ... biotic ... anthropogenic' (pp. 224-225) (due respectively to soil, relief, animals, man). A factor may *deflect* the development from a pioneer community to the climatic climax and a *plagioclimax* may result.

Braun-Blanquet, J., 1932, *Plant Sociology* (Trans. G. D. Fuller and H. S. Conrad), New York: McGraw-Hill. '... development of vegetation and formation of soil tend toward a definite end point determined and limited by the local climate.... This relatively permanent final condition we have called soil and vegetative climax' (p. 322).

*Comment.* The whole field of plant ecology is involved in the concept of the Climax. For some modern views see Pierre Dansereau, 1957, *Biogeography*, New York: Ronald Press; see also Biosphere, ecosystem.

**Climograph, Climogram**

*Webster.* A graphic representation of the action of climate upon man.

Taylor, T. Griffith, Australian Meteorology, 1920, describes climograph thus: 'With wet-bulb temperatures as ordinates and relative humidities as abscissae the twelve average monthly figures are plotted for the locality required.'

Monkhouse and Wilkinson, 1952. 'A climograph (or climogram) is a diagram in which the data for elements of climate at any one station are plotted against one another, and the shape and position of the resultant graph provides an index to the general climatic character of a place' (pp. 160-161). (Examples are provided from: J. Ball, *Cairo Scientific Journal*, 4, 1910. W. Köppen, *Petermann's G. Mitt.*, 64, 1918. J. B. Leighly, *U. of California Pubs. in Geog.*, 2, 1926, E. E. Foster, *Trans. of the Amer. Geoph. Union*, 1944. *Climate and Man*, 1941. R. Lang, 1920.)

*Comment.* In a letter to the Committee (21 November, 1955) Professor Griffith Taylor says, 'In the Weather Service I was working at climatology most of the time and so produced *climograph* and *hythergraph*, 1915. The former really refers to comfort and humidity, the latter to rain and temperature. I see lots of folks use *climograph* now as a sort of catholic term, which was not my intention.'

Thus both Monkhouse and Wilkinson, 1952 and S. S. Visher in *Geog. in 20th Cent.*, 1951, regard the term as applicable to the entire class of graphs.

If a word is to be used in this wider sense it should be climogram. See also Hythergraph.

**Clinographic curve**

Mill, *Dict.* A curve representing the slope or slopes of the Earth's surface obtained by joining points in a rectangular system of coordinates determined by ordinates corresponding to the intervals of base length between isohypses and abscissae corresponding to the isohypses' (Penck).

Wooldridge and Morgan, 1937. '... a clinographic curve designed to show the actual variation of average slope in an area' (p. 264).

Hanson-Lowe, J., 1935. The Clinographic Curve, *Geol. Mag.*, 72, 180. Described but not defined.

Monkhouse, F. J. and Wilkinson, H. R., 1971, *Maps and Diagrams—their compilation and construction*, 3rd ed., London: Methuen. 'The clinographic curve seeks to illustrate the average gradient between any two contours, and to express a series of these averages in a single curve. Its chief value, therefore, is that it indicates both sudden changes and breaks in the general relief of any region, and moreover it emphasizes uniform areas such as plateaus. It gives at the same time average gradients, the percentage extent of each average gradient, and exact breaks of slope. It is much more sensitive to small changes than the hypsometric curve, and in some cases it is less misleading.' (pp. 116-117)

See Clarke, J. I., 1966, Morphometry from Maps, in Dury, G. H., *Essays in Geomorphology*, London: Heinemann, pp. 248-256, for details of the clinographic curve employed by Finsterwalder, S. and Hanson-Lowe, J.

**Clint, clent**

*O.E.D.* Clint. A hard or flinty rock; a hard rock projecting on the side of a hill or river or in the bed of a stream; a part of a crag standing out between crevices or fissures.

Mill, *Dict.* Bare open plateaus formed by horizontal limestones traversed by open

fissures or grikes, sometimes = crags (Cumbrian Lakes and Yorkshire).
See also Grike; Moore, 1949, gives grike or clint, which is incorrect

**Clitter**
O.E.D. Clitter = clatter.
Clatter. A mass of loose boulders or shattered stones; so called on Dartmoor.
Mill, *Dict.* Clitter = clatter. A scree (Devonshire).
Comment, In Devonshire especially used for sub-angular and rounded boulders of granite derived from tors, streaming down a hillside, probably detached under periglacial conditions when solifluxion made movement easier. See Worth, 1930, *Trans. Devon. Assoc.* (G.T.W.)

**Cloosian dome**
Dury, 1959. 'A distinctive type crustal movement—the upwarping of an elliptical dome. The eminent geologist Hans Cloos maintains that uplifts of this sort have occurred in several parts of the world ... East African troughs ... rift valley of the Rhine.' (p. 210) Reference is to Cloos, H., Hebung, Spaltung, Vulkanismus. *Geol. Rundschau,* 30, 1939.

**Close**
O.E.D. An enclosed place; an enclosure.
Comment. From this the word has acquired, past and present, specialized and restricted meanings especially in different parts of Britain. e.g. a small field (Midlands); a farm-yard (Kent and Scotland); an entry or passage. In general use as the space round a cathedral as cathedral close. See also Half-year Close.

**Cloud**
O.E.D. 3. A visible mass of condensed watery vapour floating in the air at some considerable height above the general surface of the ground.
Webster. A visible assemblage of particles of water or ice, formed by the condensation of vapor in the air; a fog or mist or haze suspended, generally at a considerable height, in the air; also, the material of which these masses are composed.
Mill, *Dict.* Visible water-particles, usually condensed round a minute solid nucleus.
Hare, 1953. 'Processes of condensation by which vapour passes to the cloud-droplet, and coagulation, by which these tiny and almost buoyant cloud droplets are fused together to form rain. ...' (p. 28)

Comment. No attempt is made in this glossary to deal with terms proper to meteorology such as cloud formation and types of cloud.

**Cloud Belt**
Troll, Carl, 1957, *Abs. of Papers, Ninth Pacific Science Congress,* Bangkok. The altitudinal belt for example on mountains in the tropics where the constant presence of cloud or mist results in such typical life forms as abundance of epiphytes (p. 255).

**Cloudburst, cloud-burst**
O.E.D. (U.S.) A violent storm of rain, a 'waterspout'.
O.E.D. Suppl. Orig. U.S. A torrential fall of rain.
Webster. A sudden copious rainfall, as if the whole cloud had been precipitated at once.
Dict. Am. A sudden, violent, and heavy fall of rain. 1869. Muir, *First Summer in Sierra,* 48. 'Heavy thunder-showers, called "cloud-bursts".'
Mill, *Dict.* Torrential rain or hail, usually in a well defined small area.
Cotton, 1945, *Geomorphology.* 'Infrequent and local but very heavy downpours of rain' (desert cloudbursts) (p. 257). 'intermittent streams and ephemeral floods that result from "cloudbursts" at long intervals' (p. 255).
Cotton, 1947, *Climatic Accidents.* 'The extreme infrequency of the sheetfloods ..., their local nature (due to concentration of heavy cloudburst rainfall in small areas) ...' (p. 50).
Comment. O.E.D. gives a quotation from the Chicago Times of 1881 referring to waterspout or cloudburst, but the two phenomena are in fact quite different. Cloudburst is used in newspapers in Britain, but not so commonly in geographical literature.

**Cloud Forest**
Mountain broadleaf evergreen forest. Used in contrast to Rain Forest by W. R. Barbour on South America in *A World Geography of Forest Resources,* New York: Ronald Press, 1956.
See also Mist forest.

**Cluse** (French)
Mill, *Dict.* A steep-sided transverse valley through a mountain ridge (Jura).
Baulig, 1956, translates as 'gap' (p. 140) and also notes *cluse sèche* and *cluse morte* (dry-, wind-, or air-gap) and *cluse fonctionelle* or *cluse active* (water-gap) (p. 141).
Comment. Used mainly in the Jura but also of the pre-Alps in Haute Savoie. It is not

clear whether a cluse cuts through a single ridge or through a chain of ridges (G.T.W.).

## Cluster analysis

Kendall, M. G. and Buckland, W. R., 1972, *A Dictionary of Statistical Terms*, 3rd ed., Edinburgh: Oliver and Boyd. 'A general approach to multivariate problems in which the aim is to see whether the individuals fall into groups or clusters. There are several methods of procedure but they all depend on setting up a metric to define the "closeness" of individuals.' (p. 24)

See Krumbian, W. C. and Graybill, F. A., 1965, *An Introduction to Statistical Models in Geology*, New York: McGraw-Hill (pp. 406–408)

## Coal

*O.E.D.* 5. A mineral, solid, hard, opaque, black or blackish, found in seams or strata in the earth, and largely used as fuel; it consists of carbonized vegetable matter deposited in former epochs of the world's history.

*Webster.* 3a. A black, or brownish-black, solid, combustible mineral substance formed by the partial decomposition of vegetable matter without free access of air, under the influence of moisture and, in many cases, of increased pressure and temperature. A complete series can be traced from the cellulose of wood through lignite or brown coal, and soft, or bituminous, coal, to hard coal, or anthracite, or, as a final product, to graphite. The order given is one of decreasing volatility and increasing carbon content.

*Comment.* For the very complicated subject of the classification of coals see Kendall, P. F., 1929, in *Handbook of the Geology of Great Britain*, London: Murby, pp. 267–268, following Grout—a continuous series according to percentage of fixed carbon from peat and turf (below 55), lignite (30–60), cannel (35–48), bituminous (48–83), semi-anthracite (83–93); anthracite (over 93). Humic is an alternative name to bituminous. Those who have studied the constitution of coals (especially M. E. Stopes, 1919, *Proc. Roy. Soc. (B)*, 90, 470–487) introduced the terms durain, clarain, virtain and fusain for constituent minerals. This work was extended by A. Raistrick and C. E. Marshall, 1939 (*The Nature and Origin of Coal and Coal Seams*, London). British official classification distinguishes I Low-Volatile Coals (volatile matter 20 per cent or less—anthracite and steam coals); II Medium Volatile (20–30 per cent); and III High Volatile. Coals are also referred by potential use into such classes as Caking, Coking, Gas, Household, Bunker etc. See also Brown coal.

## Coal Measures

*O.E.D.* 2.b. (Geol.) The whole of the series of rocks formed by the seams of coal and the intervening strata of clay, sandstone, etc., in a coal-field, constituting the upper division of the Carboniferous formation. Also attrib. (referring evidently to the long-established practice of naming the different seams of a coal-field by their measure or thickness).

Mill, *Dict.* The uppermost series of the Carboniferous system. The term 'measure' is applied by pitmen to strata; hence the coal-containing strata.

*Comment.* Some geologists prefer to use 'Coal Measures' only for coal-bearing horizons, for others 'Coal Measures' equals the time measure 'Upper Carboniferous' in Britain, whether the strata include coal or not. Conversely the Lower Carboniferous coal-bearing series in Scotland are not included in the 'Coal Measures'. The Coal Measures of Britain are broadly synchronous with the Pennsylvanian of North America.

See *Handbook of the Geology of Great Britain*, ed. Evans and Stubblefield, 1929, London: Murby and *The Coalfields of Great Britain*, ed. A. E. Trueman, London: Arnold, 1954.

## Coast

*O.E.D.* 4. The edge or margin of the land next the sea, the sea-shore.
a. In the full phrase, *coast of the sea*, seacoast = sea-side. Formerly sometimes land's coast. b. By ellipsis, *coast* (the ordinary use). c. *The coast* is familiarly applied in different regions to specific littoral districts, in India, esp. to the Coromandel Coast. d. Rarely, the bank of a river or pond. *Obs.*
5. The border, bound, or limit, of a country; territory on or near a boundary or frontier, borderland (chiefly *pl.*). *Obs.*
6. A tract or region of the earth; a district, place, clime, country, 'part of the world'. *Obs.*
8. A point of the compass; quarter, direction.

11. (U.S. and Canada) A (snow- or ice-covered) slope down which one slides on a sled; the act of so sliding down (originally local).

Mill, *Dict*. The belt of the land meeting open water which is under the direct influence of the waves, also that part of the sea-floor which is under the direct influence of the land.

*Adm. Gloss.*, 1953. The meeting of the land and sea considered as the boundary of the land. The narrow strip immediately landward of the high water line of mean spring tides, or sometimes a much broader zone extending some distance inland.

Johnson, 1919. 'Landward from the shore is a much broader zone of indeterminate width, which will here be called the coast.' Some have included the shore, but Johnson excludes it (p. 160).

Moore, 1949. That part of the land which borders the sea or other extensive tract of water, and so comes under the direct influence of the waves.

Stroud, 1952. '"The coasts of Scotland" shall mean and include, all bays, estuaries, arms of the sea, and all tidal waters within the distance of three miles from the mainland or adjacent islands' (30 and 31 Vict., c. 52, s. 11).

'The coast is, properly, not the sea but the land which bounds the sea; it is the limit of the land jurisdiction, and of the parishes and manors (bordering on the sea) which are part of the land of the country.'

*Comment.* Steers, 1946, appears to use 'coast' as including the shore. Johnson would have given his book the title of 'Shorelines of Britain'. In contrast to the commonly accepted meaning of coast is the interpretation by the Judicial Committee of the Privy Council in the Labrador Boundary Report, quoted in *Geog. Jour.*, **70**, 1927, 40: 'With regard to the limit in depth of the country which may be described as "coast", where that term is used in the wider sense, it is argued that the natural limit is to be found (in the absence of special circumstances) in the watershed which is the source of the rivers falling into the sea at that place; and there is much to be said in favour of that view.'

For different types of coast see Concordant, discordant, longitudinal, transverse, Atlantic, Pacific etc.

### Coastal plain

Mill, *Dict*. The zone of low-lying land between the sea and the nearest hill, also used for plains formed by deposition of sheets of rock material over the sea floor and its uplift above the sea-level (Gregory).

Geikie, J., 1898. '... of coastal plains generally it may be said that they are either directly or indirectly of fluviatile origin.' (pp. 260–261)

Salisbury, 1907. 'A narrow coastal plain may have originated in either of two ways: (1) It may be a part of the former continental shelf from which the sea has withdrawn, or (2) the sediment washed down from the land may have been deposited in the shallow water of an epicontinental sea, building up (aggrading) its bottom above the surface of the water, and thus converting it into land.... Coastal plains may also be made by the degradation of coastal lands which were once high.'

Strahler, A. M., 1946, Geomorphic Terminology and Classification of Land Masses, *Jour. of Geol.*, **54**. '... it might appear that some basis exists for finding a new term to replace "coastal plain" to designate extensive, recently emerged coastal belts underlain by appreciable thicknesses of seaward-sloping marine strata. Retention of "coastal plain" seems desirable for the following reasons: (1) The term is widely understood and is used in all branches of geology and geography.... (2) The term "coastal plain" is descriptive and does not violate accepted usage of the included words.'

Thornbury, 1954. 'One of the simplest situations under which cuestas may develop is represented by a recently emerged coastal plain underlain by seaward-dipping weak and strong strata. The Gulf coastal plain through Alabama and Mississippi has developed a series of cuestas with intervening lowlands or vales to form a *belted coastal plain*.' (p. 133) Term also used by Cotton, 1945.

*Comment.* W. M. Davis, 1909, *Geographical Essays* discusses coastal plain of the Carolinas as 'the old bottom of the Atlantic'; Johnson, 1919, criticizes use of 'coastal plain' for a 'coastal plane' and suggests greater use of 'marine plain' and 'marine plane'. (p. 166)

### Coastline

Mill, *Dict*. Coast-line. The line which follows

the main outline of the land including bays, but crosses narrow inlets and rivers at their mouths (Holdich).

Johnson, 1919. The coast line is the boundary between the shore (including 'backshore') and the coast. '...it marks the seaward limit of the permanently exposed coast.' (p. 160)

Wooldridge and Morgan, 1937. '"coastline" is to be read as cliffline or its equivalent, the margin of the land.' (p. 321)

*Adm. Gloss.*, 1953. The landward limit of the beach. The extreme limit of direct wave action (in onshore gales during equinoctial spring tides). It may be some distance above the H.W. line of mean spring tides, but for practical hydrographic purposes the two are usually regarded as coincident. A general term used to describe the appearance of the shore or coast as viewed from seaward, *e.g.* a low coastline.

*Comment.* Steers, 1946, uses coastline as a very general term.

See also Shoreline.

### Cob
*O.E.D.* A composition of clay (marl, or chalk), gravel, and straw, used, esp. in the south-west of England for building walls, etc.

### Cobble, cobble-stone
*Webster.* A naturally rounded stone larger than a pebble, esp. one from six inches to a foot in diameter.

Farey, J., 1811. 'Cobbles...are what we in London should call good round Coals, being the larger lumps picked out of what they call the Slack or waste small coals...' (p. 187).

Humble, 1860. A pebble. This word is given by Ray as belonging to the northern counties. Cobble has the same significance as boulder (p. 98). Copple-Stone—boulders, cobble-stones, which see (p. 104).

Rice, 1941. Cobblestone—a smoothly rounded stone, larger than a pebble and smaller than a boulder (p. 80).

Twenhofel, 1939. Cobbles are generally better rounded than boulders, generally have wide geographical distribution in any environment and generally have about the same range of composition as boulders.

*Comment.* The form 'copple-stone' is obsolete; 'popple' is a local variation found in Devonshire. Cobble affords another example of a common old English word which has been given a specialized meaning by some writers. In the grading of Beach material the British Standards Institution ranks cobbles (60–200 mm. or 3–8 inches) between 'coarse gravel' and 'boulders'; Inman between pebbles and boulders ($-8 \cdot 0$ to $-6 \cdot 0$ $\phi$ units) and Wentworth also between pebbles and boulders (4–64 mm.). See King, C. A. M., 1959, *Beaches and Coasts*, London: Arnold, pp. 3–4.

### Cockpit
Literally a pit where game-cocks are set to fight for sport (now illegal in most countries) and consequently applied to natural pits (sink-holes) in the karst country of the West Indies. An area of Jamaica is known as the Cockpit Country because of the abundance of these pits in a limestone plateau.

### Coenosis, cenosis
Carpenter, 1938. A population system held together by ecological factors in a state of unstable equilibrium. Also biogeocoenosis, biocoenosis, phytocoenosis, zoocoenosis. See also Biocoenosis.

### Cognitive description
Harvey, D., 1969, *Explanation in Geography*, London: Edward Arnold. 'Under this heading are included the collection, ordering and classification of data. No theory may be explicitly involved in such procedures, but it is important to note that a theory of some kind is implied.... Cognitive descriptions may thus range in quality from simple primary observations through to sophisticated descriptive statements.' (p. 79)

### Cognitive map
Downs, R. M. and Stea, D., 1973, *Image and Environment*, London: Edward Arnold. 'Cognitive mapping is a process composed of a series of psychological transformations by which an individual acquires, codes, stores, recalls, and decodes information about the relative locations and attributes of phenomena in his everyday spatial environment.' (p. 9) '...the product of this process at any point in time can be considered as a cognitive map.' (p. 10)

*Comment.* Mental map sometimes regarded as an alternative but less desirable term for cognitive map. See also Mental map.

### Col (French)
*O.E.D.* A marked depression in the summit-line of a mountain chain, generally affording a pass from one slope to the other.

A word belonging to the Romanic dialects of the Alps, which Alpine climbers and geologists have used of other regions.

1853. Th. Ross, *Humboldt's Trav.* III, xxxii, 291 note. The *Cols* or passes indicate the minimum of the height to which the ridge of the mountains lowers in a particular country. Also 1873 (Geikie).

Mill, *Dict.* Col. (1) A maximum point, the highest point between two valleys. (a) The lowest point on a ridge. (3) A neck of low pressure between two anticyclones (Abercromby).

Knox, 1904. Col, (Fr., Eng.), a neck, an elevated pass. Note also Colle (It.), a hill, etc.

*Met. Gloss.*, 1944. Col. The saddle-backed region occurring between two anticyclones and two depressions, arranged alternately. Frequently on a weather map only two anticyclones appear, then the col is the region of relatively low pressure between them and may be likened to a mountain pass between two peaks. In a similar way the col may appear when only two depressions are shown on the weather map... etc. (illustration).

Abercromby, R., and Goldie, A. H. R., 1934, *Weather*, London: Kegan Paul. '...the "col", or neck of low pressure, which lies between two adjacent anticyclones.' (Gives an example, however, similar to *Met. Gloss.*; an area of no gradient.) (pp. 98–99)

Thornbury, 1954. 'An arete or serrate ridge consists essentially of alternating sags or *glacial cols* produced by intersection of opposed cirques and pointed peaks or *horns* representing unreduced portions of the original mountain range.' (p. 373) Similarly Cotton.

*Comment*. The simple French word 'col' from the Latin *collum* means neck. Baulig, 1956, pp. 36, 44, 236 illustrates some of the difficulties in translation when the word has been used in French with certain specialized meanings and in English with other specialized meanings.

**Cold desert**

Commonly used as equivalent to tundra (*q.v.*) or of all arctic and antarctic regions where plant and also animal life is inhibited by low temperatures.

**Cold front**

*Met. Gloss.*, 1944. 'The boundary line between advancing cold air and a mass of warm air under which the cold air pushes like a wedge.'

**Cold Pole**

A term embodying a popular rather than a scientific concept though much discussed in the first half of the 18th century. It is commonly applied to the town or neighbourhood of Verkhoyansk in the eastern interior of Soviet Asia where the winter temperatures are usually extremely low. The mean temperature in January is $-58°F$., the mean minimum for the same month is $-83°F$. and on one occasion the lowest reading then taken on the surface of the earth $-94°F$. was recorded (see Kendrew, 1953, pp. 263–265). More logically the 'Cold Pole' might be the spot with the lowest mean annual temperature—with its extremely low winter figures Verkhoyansk would again seem to qualify. In later years, however, lower temperatures have been recorded in Antarctica.

**Cold wave**

Mill, *Dict.* (1) Cool or cold equatorward wind blowing in the wake of a cyclonic storm. (2) A fall of temperature below 32°F. in N.W. and 40°F. in S., with a change of at least 20°F. in twenty-four hours (U.S.A.)

*Met. Gloss.*, 1944. The fall of temperature associated with the air behind the cold front of a passing depression. The term is used in a technical sense by the United States Weather Bureau, with the meaning of a fall of temperature by a specified amount in 24 hours, to a minimum below a certain temperature. The amount of fall and the limit of minimum are different for different times of the year and for different parts of the country.

*Webster*. A sudden drop in temperature; loosely, any period of unusually cold weather.

**Collectivism**

L.D.G. Collectivism is a general term applied to a system of organization and management based on central planning by the State (as in U.S.S.R.); hence **collective farming** involving a large area of State land with labourers working as directed under a manager.

**Colloid plucking** (P. Reiche, 1950)

Reiche, P., 1950, *A Survey of Weathering Processes and Products*, rev. ed. University of New Mexico Press.

Thornbury, 1954. 'A weathering process of

uncertain importance has been called colloid plucking by Reiche. It seems probable that soil colloids may have the power to loosen or pull off small bits of rock from the surfaces with which they come into contact.'

**Colloidal complex** (Soil Science)
Robinson, 1949. '... the colloidal clay and ... the colloidal organic matter' (p. 27).

**Colluvium, Colluvial soil**
*Webster.* Colluvial. 2. *Geol.* Designating or pertaining to (1) heterogeneous aggregates of rock detritus, (2) eluvial material.
Soils and Men, 1938. 'Heterogeneous deposits of rock fragments and soil material accumulated at the base of comparatively steep slopes through the influence of gravity, including creep and local wash' (p. 1165).
Swayne, 1956. A collection of rock débris, *e.g.* screes or mud flows, derived from various sources.
Mill, *Dict.* Colluvial Soil. Soil resulting from transport merely by brooks and small streams (Hilgard).

**Colmatage**
Mill, *Dict.* The natural growth and definition of banks in the lower part of a river due to the deposition of alluvium.
*Comment.* Obsolete, but still used by reclamation experts for the artificial encouragement of this process. (C.J.R.)
Not in *O.E.D.* Compare warping.

**Colonization, colonisation.** See also Colony
*O.E.D.* Colonization: the action of colonizing or fact of being colonized; establishment of a colony or colonies. ex. 1770–1875.
Bowman, I., 1931, *The Pioneer Fringe*, New York: Amer. Geog. Soc. Used for organized settlement on a large scale, *e.g.* colonization companies in W. Canada and the appointment of a man to promote state colonization by the Oregon State Chamber of Commerce.
Harrison-Church, R. J., 1951, *Modern Colonization*, London: Hutchinson. In the preface defines forms of colonization:
a. Permanent settlement—colonies de peuplement ou d'enracinement.
b. Economic colonization—colonies d'exploitation ou d'encadrement.
c. Strategic colonies—colonies de position.
d. Settler colonies in alien lands (*e.g.* Germans in Brazil). 'All these examples illustrate a vital factor in colonization, namely, the contact of two different groups of people' (p. ix). 'Demangeon, 1923, *Colonial Geography*—to study the geographical aspects of the contact of different peoples associated by colonization.' 'Hardy, G., 1933, regards colonization as concerned with the transformation of backward regions.' (p. 12)
*Comment.* A distinction may be drawn between 'internal' and 'foreign' colonization. Note also extension of meaning in colonization by plants, etc.

**Colony**
*O.E.D.* (from Latin colon-us, tiller, farmer, cultivator, planter, settler in a new country. L. colonia = farm or settlement and esp. public settlement of Roman citizens in hostile or newly conquered country. *e.g.* 9 Roman colonies in Britain included London, Bath etc. Also used for Greek settlements—a body of emigrants who settled abroad as an independent self-governed state unconnected with the mother city save by religious ties. 'Its modern application to the planting of settlement, after Roman or Greek precedents, in newly discovered lands, was made in the 16th c. by Latin and Italian writers whose works were rendered into English by Richard Eden.')
II in modern application. 4. A settlement in a new country; a body of people who settle in a new locality, forming a community subject to or connected with their parent state; the community so formed, consisting of the original settlers and their descendants and successors, as long as the connection with the parent state is kept up. ex. 1548–1883. b. The territory peopled by such a community. 5. trans. A number of people of a particular nationality residing in a foreign city or country (especially in one quarter or district); a body of people of the same occupation settled among others, or inhabiting a particular locality. b. The district or quarter inhabited by such a body of people.
7. *Geol.* Applied by Barrarde to a group of fossil forms appearing exceptionally in a formation other than that of which they are characteristic. 8. *Biol.* An aggregate of individual animals or plants, forming a physiologically connected structure, as in the case of the compound ascidians, coral-polyps, etc.
*O.E.D. Suppl.* 5.c. An establishment in which persons are engaged to work who are otherwise unemployed or unemployable,

or are trained for some occupation or trade. exs. 1888–1897.

Stroud, 1952. In all Acts of Parliament passed after the 31st December, 1889. '"Colony" shall mean any part of Her Majesty's Dominions, exclusive of the British Islands, and of British India....' (Interpretation Act, 1889 (52 and 53 Vict., c. 63), s. 18 (3).)

Statute of Westminster (22 and 23 Geo. 5, c. 4), s. 11. '... the expression "Colony" shall not, in any Act of the Parliament of he United Kingdom passed after the commencement of this Act, include a Dominion or any Province or State forming part of a Dominion.'

**Columnar structure**

Certain volcanic rocks, especially basalt, when cooling form hexagonal columns at right angles to the surfaces of cooling. These may be very regular, and well known examples of this columnar structure are Giant's Causeway in northern Ireland and Fingal's Cave, Island of Staffa. Very similar hexagonal cracks develop when mud dries.

**Combe. Coombe, coomb, coom**

O.E.D. Gives coomb first.

a. A deep hollow or valley: in O.E. charters; not known in M.E.; but occurring from the 16th c. in the general sense of valley, and more especially of a deep narrow valley, clough or cleugh. Examples 770–1872.

b. spec. In the south of England, a hollow or valley on the flank of a hill; esp. one of the characteristic hollows or small valleys closed in at the head, on the sides of and under the chalk downs; also, a steep short valley running up from the sea coast. Examples 1674–1886.

c. In the south of Scotland and in the English Lake District '(in) such hills as are scooped out on one side in form of a crescent, the bosom of the hill, or that portion which lies within the lunated verge, is always denominated the coomb.' (Hogg. *Queen's Wake*, 1813. *Notes* xxiv.)

Mill, *Dict.* Combe. (1) The amphitheatre-like steep bank of an incised meandering river, *e.g.* of the tributaries of the upper Thames (*cf.* Welsh *cwm*). (2) Synonymous with cirque. (3) Synclinal valleys of the Jura. (4) Narrow valleys in the Jura developed along the axes or sides of anticlines (monoclinal valleys). (5) A gully or valley of the third order, a mere fold on a hillside (Côte d'Or). (6) A valley drained by a stream disappearing at lower end underground (Jura).

Knox, 1904. 'Combe (Celto-Saxon; Cymrie, cwm; A.S. comb, cumb), a hollow between two hills, valley, fingle; a bowl-shaped valley, *e.g.* Wycombe.' 'Combe (Fr.), a small valley.' 'Comba (Sp.), a valley.'

Page, 1865. 'Combe or Coomb (Sax). A common term in the south of England for an upland valley, generally narrow, and without a stream of water.'

*Encyc. Brit.*, 1906. 'A term frequently used in south-western and southern England for a short, closed-in valley, either inland on the side of a down or forming a small coastal feature.'

**Combe-rock, coombe-rock**

An irregular mass of rock debris, often partly filling dry chalk valleys in south-eastern England, probably originating as the result of nivation, *i.e.* a glacial sludge (Wooldridge and Morgan, 1937, pp. 278 and 403). See also Congeliturbation, Cryoturbation.

Charlesworth, 1957. 'The coombe-rock, a term G. A. Mantell (*Fossils of the South Downs*, 1822, p. 277) first introduced geologically, is a structureless mass of unrolled and unweathered flints, embedded in a matrix of chalky paste and disintegrated chalk ... (p. 1081) ... the solifluxion facies of chalk areas ... it was formed by sheet-flowing, not only in the coombes but over the whole of the dip-slopes.' (p. 410)

Dines *et al.*, 1940, *Geol. Mag.*, 77, p. 148. Head composed of chalk debris.

Kerney, M. P., 1965, *Proc. Geol. Assoc.*, 76, 'Coarse chalk solifluxion debris' (p. 269).

**Comb-ridge**—see Arête

**Comfort-zone**

*L.D.G.* Comfort-zone is the range of temperature and humidity in a climate within which human beings feel comfortable. As used in central heating 68°–72°F. (20°–21°C.) and [55–60 per cent relative humidity are common standards. See Sensible temperature.

**Common, Common land, Common lands**

O.E.D. 5. A common land or estate; the undivided land belonging to members of a local community as a whole. Hence, often, the patch of unenclosed or 'waste' land which remains to represent that.

b. Law (Also right of common, common right). The profit which a man has in the land or waters of another; ... = Commonage, Commonty.

Royal Commission on Common Land, 1955–58, *Report*, Cmd. 462, 1958. Common Lands. A term generally used nowadays of 'manorial waste' as well as of other, lands over which rights of common exist.

*Comment.* The terms 'cow common', 'goose common', 'hay common' and 'pasture common' are self-explanatory.

**Common, Right of; Rights of; Common rights**
*O.E.D. Law.* The profit which a man has in the land or waters of another; as that of pasturing cattle (*common of pasture*), of fishing (*common of piscary*), of digging turf (*common of turbary*), and of cutting wood for fire or repairs (*common of estovers*).

Royal Commission on Common Land, 1955–58, *Report*, Cmd. 462, 1958, '*Right of Common Appendant.* The common law right of the freeholder of a manor (i.e. the occupier of land granted as freehold arable by the lord of the manor) to the use of the manorial waste to the extent necessary for the maintenance of his husbandry. *Right of Common Appurtenant.* A right created by the grant of the owner of the soil and appertaining (*i.e.* attached) to a particular holding. *Right of Common in Gross.* Held by grant but not attached to any particular land. It descends to an heir.'

**Commonable Animal**
Royal Commission on Common Land, 1955–58, *Report*, Cmd. 462, 1958. 'Those animals which a holder of a right of common of pasture may lawfully turn out on a common. Where the right is a right appendant, only horses, sheep, and cattle (and formerly oxen to pull the plough) may be pastured, and then only as many as can be wintered (see 'levant and couchant') on the commoner's freehold. Where the right is appurtenant or in gross other animals may be pastured according to the terms of an actual grant or presumed grant.' Sheep were sometimes excluded.

**Commoner**
One who enjoys Common rights (*q.v.*).

**Common Field**
In England and parts of western Europe until inclosure in the 15th to 18th centuries, the arable was worked in large open or common fields. See Field system.

**Community**
*O.E.D.* II. A body of individuals. 7. A body of people organized into a political, municipal, or social unity: a. A state of commonwealth. 1769 Robertson *Chas. V*, I.1.66 Europe was broken into many separate communities. b. A body of men living in the same locality.

Hillery Jr., G. A., 1955, Definitions of Community: Areas of Agreement, *Rural Sociology*, **20**, '. . . basic agreement that community consists of persons in social interaction within a geographic area and having one or more additional common ties.' (p. 111)

Park, R. E., Burgess, E. W. and McKenzie, R. D., 1925, *The City*, Chicago: University of Chicago Press. '. . . the human community differs from the plant community in the two dominant characteristics of mobility and purpose, that is the power to select a habitat and in the ability to control or modify the conditions of the habitat.' (pp. 64–65)

*Comment.* Ecologists take the view that community is analytically separable from social interaction, that community is to be found primarily in relationships that people have with space. See Fairchild, H. P. (ed.), 1944, *Dictionary of Sociology*, New York: Philosophical Library. (p. 145)

See also Community (Ecology).

**Community** (Ecology)
Abercrombie *et al.*, 1951. (Of plants), unit of vegetation having distinct, recognisable features which distinguish it from others (p. 57).

**Commuter**
*O.E.D.* One who commutes. In U.S. the holder of a commutation-ticket (derived from Commute 4), to change (one kind of payment) *into* or *for* another; *esp.* to substitute a single payment for a number of payments or a fixed payment for an irregular or uncertain one.

*Webster.* One who or that which commutes; specif. a. One who uses a commutation ticket; also, one who travels back and forth between a city and an outside residence.

*Dict. Am.* One who commutes. 1865. *Atlantic Mo.* XV. 82. 'Two or three may be styled commuters' roads, running chiefly for the accommodation of city businessmen with suburban residences.

*Commute.* To travel regularly by public conveyance between two points, as between one's suburban residence and the city. *Commutation ticket.* A passenger ticket good for several rides issued at a reduced rate. 1849.'

Smailes, A. E., 1953. '... the term "commuter" more especially applies to the daily traveller making his journey to and from work.' (p. 141)

*Comment.* The nearest equivalent in Britain is season-ticket holder, but there is an increasing tendency to adopt this American term in the sense of one who travels regularly a considerable distance, usually daily, to and from his work.

**Compage** (Derwent Whittlesey, 1954)

Hartshorne, R., 1959. *Perspective on the Nature of Geography,* Chicago: Rand McNally. 'Whittlesey's committee urged the use of a new concept, the "compage", which is limited to "all the features of the physical, biotic and societal environments that are functionally associated with man's occupance of the earth".' (p. 139) The quotation is from P. E. James and C. F. Jones (eds.) 1954, *American Geography: Inventory and Prospect.* Syracuse, p. 44.

*Comment.* The revival of an obsolete word which meant 'means of joining, connecting matter' (*O.E.D.*)—an attempt to give precision to an aspect of 'region'.

**Compass, Points of**
North, South, East and West.

**Compass: the Mariner's Compass**

*O.E.D.* An instrument for determining the magnetic meridian, or one's direction or position with respect to it, consisting of a magnetized needle turning freely on a pivot; notably employed in the guidance of a ship's course at sea.

**Competence, competent** (of rocks)

*Webster. Geol.* Of a bed or stratum, strong enough to transmit effectively the thrust when strata are folded by lateral compression and capable of sustaining the weight of overlying strata when arched into an anticline.

Willis, B. and R., 1934. *Structural Geology,* 3rd ed. 'When strata are simply flexed and they take no part in transmitting the stress to adjacent masses, they are *incompetent* to do so ... offer no appreciable resistance to bending. When the pressure is aligned in the direction of the bedding, the strata act as more or less rigid struts and are accordingly more or less *competent* to transmit the stress to further points or masses. Hence the classification of flexures as *incompetent structures* and of folds due to horizontal pressure as *competent structures.*' (p. 78)

Hills, 1953. '... relatively strong ("competent") beds ...' (p. 82).

**Competence of a stream** (G. K. Gilbert, 1914), **competency, competent**

*Webster. Phys. Geog.* The ability of a stream to transport detritus, as measured by the size of the largest particle, pebble, or boulder it can move forward; contr. with *capacity,* the ability of a stream to transport detritus, as measured by the quantity it can carry past a given point in a unit of time.

Gilbert, G. K., 1914. The Transport of Debris by Running Water. *U.S.G.S. Prof. Paper,* 86.

*Competence*—Under certain combinations of controlling factors capacity is zero or negative. If then some one factor is changed just enough to render capacity positive, that factor in its new condition is said to be *competent,* or else to be a measure of the stream's competence. For example a stream at its low stage cannot move the debris on its bed; with increase of discharge a velocity is acquired such that traction begins; and the discharge is then said to be competent. A stream flowing over a too gentle slope has no capacity, but coming to a steeper slope it is just able to move debris; the steeper slope is said to be competent. A current flowing over debris of various sizes transports the finer, but cannot move the coarser; the fineness of the debris it can barely move is the measure of its competence. (p. 35)

Twenhofel, W. H., 1939, *Principles of Sedimentation.* '*Competency* is defined as the ability of currents to transport in terms of dimensions of particles. Competency depends upon velocity and turbulence ... (p. 191) ... *Capacity* is defined as the ability of the current to support in terms of quantity ...' (p. 193).

*Comment.* Gilbert's theoretical concepts of 'capacity' and 'competence' based upon laboratory experiments have been discussed and criticized by a number of writers. See especially Mackin, 1948 (quoted under capacity, *q.v.*); W. H. Rubey, 1938 (*U.S.G.S. Prof. Paper* No. 189–E);

Twenhofel, 1939; A. C. Woodford, 1951 (*Bull. Geol. Soc. Am.*, **62**, 799–852—discusses 'load' and 'overload').

**Conacre, con-acre** (Ireland)
*O.E.D.* In Irish land-system: The letting by a tenant, for the season, of small portions of land ready ploughed and prepared for a crop. 'Irish form of land-tenure—letting for 11 months.' (E.E.E.)

**Concentric-zone growth theory (Concentric growth theory)**
Introduced by Burgess, 1927, and based largely on his studies of urban growth in the Chicago area.

Haggett, 1965. 'The theory suggests that a city expands radially from its centre so as to form a series of concentric zones or annules. For Chicago the five annules were, in order from the centre outwards: (a) an inner central business district; (b) transition zone surrounding the central business district with residential areas being "invaded" by business and industry from the inner core; (c) a working-class residential district; (d) a zone of better residences with single-family dwellings; and (e) an outer zone of commuting with suburban areas and satellite cities. Burgess acknowledged that this simple annular pattern would be inevitably modified by terrain, by routes, and so on, he nevertheless considered that each inner zone extended by colonization of the next outer zone, and that therefore radial expansion along a broad front was the dominant process in shaping the pattern of the city area.' (pp. 177–178)

See Burgess, E. W., 1927, The determination of gradients in the growth of the city, *American Sociological Society Publications*, **21**, 178–184.

See also Succession.

**Concordant coast**
Mill, *Dict.* A coast-line approximately parallel to the great trendlines of the land.

Wooldridge and Morgan, 1937. 'The terms "concordant" and "discordant", proposed by Supan, are good alternatives for "longitudinal" and "transverse" in the general description of coastlines.' (p. 354 *fn.*)

Longitudinal coasts, in a classification by von Richthofen, for those parallel with the structural grain. This is also roughly equivalent to the 'Pacific Type' of Suess. (pp. 352–354)

**Condominium**
*O.E.D.* Joint rule or sovereignty.
*Comment.* A territory governed jointly by two or more countries: formerly (1898–1953) the Anglo-Egyptian Sudan was governed jointly by Britain and Egypt.

**Cone of Dejection, Dejection cone**
Mill, *Dict.* Cone, Alluvial, or Cone of Dejection. A fan-shaped deposit of sediment with the apex pointing up the watercourse.

Wooldridge and Morgan, 1937. (of erosion and deposition by Alpine torrents) 'About their upper course is the half-funnel shaped *basin de réception*, while their lower course is across the surface of a *cône de déjection* (alluvial cone or fan).' (p. 170)

*Comment.* The term in English is rarely used; it is not in Cotton, 1941, Lobeck, 1939, Geikie, 1898 or Davis, 1898. It is perhaps a bad translation of the French term. It does not accord with *O.E.D.* definition under Dejection which is: *Geol.* 'Matter thrown out from a volcano,' quoting Murchison, *Siluria* 1839. Murchison's use of volcanic dejections is obsolete. There is some doubt whether or not cône de déjection should be translated as alluvial cone or fan. Many mountain torrents do not *emerge* from a narrow valley; they plunge over a valley side bench and a cone of coarse material (scarcely alluvium) is built up at the base (G.T.W.).

**Confluence**
*O.E.D.* 1. Flowing together; the junction and union of two or more streams or moving fluids.

2. The place where two or more rivers, etc., unite.

3. A body of waters produced by the union of several streams; a large body of water, or other fluid, flowing together; a combined flood.

4. The running or flocking together of persons; the act of crowding to a place; concourse.

5. A numerous concourse or assemblage (of people); a 'multitude crowded into one place'.

Mill, *Dict.* A point at which a stream flows into another or where two streams converge and meet.

*Comment.* The term is also used, by analogy, for the junction or coming together of routeways.

**Confluent**
*O.E.D.* A stream which unites and flows with another: properly applied to streams of nearly equal size; but sometimes loosely used for affluent, *i.e.* a smaller stream flowing into a larger.
Also used as adjective.

**Conformable**
*O.E.D.* 4. *Geol.* Having the same direction or plane of stratification; said of strata deposited one upon another in parallel planes.
Fay, 1920. When beds or strata lie upon one another in unbroken and parallel order, and this arrangement shows that no disturbance or denudation has taken place at the locality while their deposition was going on, they are said to be conformable. But if one set of beds rests upon the eroded or the upturned edges of another, showing a change of conditions or a break between the formations of the two sets of rocks, they are said to be unconformable. (Roy. Comm. on Min. Res. of Ontario, 1890.)

**Conformal**
*O.E.D.* 1. Conformable. 2. Corresponding so as to fit or suit etc.
Raisz, E., 1938. *General Cartography.* 'Conformal, or orthomorphic, projections are those on which any small area has the same shape as on the globe.' (p. 82)

**Congelifraction** (Kirk Bryan, 1946)
Bryan, K., 1946, Cryopedology ... *Am. Jour. Sc.*, 244. 'Frost-splitting or frost-riving.' 'Congelifract = the individual fragment produced by frost-splitting.' 'Congelifractate = a body of congelifracts, a mass of material of any grain size produced by congelifractation.' (p. 640)
*Comment.* Numerous new terms were introduced by Kirk Bryan in the paper quoted. They include, in addition to the above, congeliturbation, cryopedology and cryoplanation. As they have been adopted by other writers they cannot be ignored though some members of the Committee object strongly to this multiplication of terms. It is pointed out by others, however, that each term has been precisely defined and their use should promote scientific accuracy. (L.D.S.)

**Congeliturbation** (Kirk Bryan, 1946)
Bryan, K., 1946, Cryopedology ... *Am. Jour. Sc.*, 244, 1946. '... frost action including frost-heaving and differential and mass movements. Includes solifluction, sludging, etc.'
'Congeliturbate = a body of material disturbed by frost action = warp, trail, head Combe rock, Erdfliessen, Brödelerde, etc. These materials are characterized by surface forms: structure, soils, soil stripes, block-fields, mounds, etc. In places, structures characteristic of the surface forms are recognizable in the congeliturbate.' (Latin: congelare, to freeze; and turbare, to stir up.)

**Conglomerate**
*O.E.D. Geol.* (= *conglomerate rock*). A composite rock consisting of rounded and waterworn fragments of previously existing rocks, united into a compact mass by some kind of cement; often called *pudding-stone*.
*Comment.* A consolidated shingle or gravel bed: see notes under breccia.

**Congost** (Spanish)
Corresponds to the French cluse, either structural or erosional (P.D.).

**Coniferous**
*O.E.D.* Bearing cones; belonging to the botanical order *Coniferae* pertaining to or consisting of conifers.
*Comment.* Coniferous forests occur in many conditions but the vast continuous area is that of the Northern Coniferous Forest or Taïga, also called the boreal forests. See especially *A World Geography of Forest Resources*, ed. Stephen Haden-Guest *et al.* New York: Ronald Press, 1956.

**Conjunction**
*O.E.D.* 3. *Astron.* An apparent proximity of two planets or other heavenly bodies; the position of these when they are in the same, or nearly the same, direction as viewed from the earth.

**Connate water**—see Fossil water
*Webster. Geol.* Sea water held in the interstices of sedimentary deposits and sealed in by the deposition of overlying beds.

**Conoplain**
*Webster. Geol.* An erosional plain which surrounds and slopes radially away from a central mountain mass.
Lobeck, 1939. 'At the base of many mountains and especially in arid regions, there is a flat zone of bedrock one mile to several miles in width, only slightly veneered with alluvium and which slopes away to the adjacent basins.... Such a feature is

**Consequent drainage, river, valley; consequents** (Powell, 1875)

*O.E.D. Suppl.* 8. Geol. No definition, but gives three quot. inc.: 1905. Chamberlin and Salisbury, Geol. I, 74. 'Streams and valleys, the courses of which are determined by the original slope of the land, are said to be consequent.'

...own as a *rock pediment, rock plane* or *conoplain* and may be thought of as a local peneplane beveling the rock structure... the *zone of planation*.' (p. 245)
*Comment.* Not in common use.

Mill, *Dict.* Consequent River. A river which owes its direction to uplift, and on the whole takes a course parallel to the greatest slope of the surface produced by that uplift (Powell). On a glacial drift surface the course is determined by the slope and relief of the drift. The primitive river (de la Noë).

Powell, 1875. '... valleys are found having directions dependent on corrugation. I propose to call these consequent valleys.' (p. 163)

Gilbert, 1877. 'Streams will rise along the crest of each anticlinal, will flow from it in the direction of the steepest dip, will unite in the synclinals, and will follow them lengthwise. The axis of each synclinal will be marked by a watercourse; the axis of each anticlinal by a watershed. Such a system is said to be consequent on the structure.' (p. 143)

Davis, 1909. The Geographical Cycle, *Geog. Jour.*, 1899, **14**, 481–504). 'All the changes which directly follow the guidance of the ideal initial forms may be called consequent; thus a young form would possess consequent divides, separating consequent slopes, which descend to consequent valleys, the initial troughs being changed to consequent valleys in so far as their form is modified by the action of the consequent drainage.' (p. 257)

Committee, List 2. 'Consequent drainage. A drainage system directly related to the initial slope of the surface.'
*Comment.* By a 'consequent' is understood a consequent stream.

**Conservancy**

*O.E.D.* Official conservation; the office of conservators, a board of official conservators.

Zimmermann, E. W., 1951. 'Conservancy means reduction in the rate of exhaustion of a natural resource which is not sought for its own sake but is incidental to the exercise of economy.' (p. 807)
*Comment.* Britain had long established the *Thames Conservancy, Mersey Conservancy* to regulate the port and river and in 1949 set up the *Nature Conservancy* to establish and control nature reserves. The word has thus become identified with the commission in each case and divorced from their work of conservation.

**Conservation**

*O.E.D.* The action of conserving; preservation from destructive influences, natural decay, or waste.

*Comment.* Without being redefined this word has undergone a recent evolution in meaning. With the realization especially in the United States in the nineteen-thirties of the evil results of soil erosion, it came to mean particularly soil-conservation—the protection of soil from erosion—and has since been extended to the protection from wasteful use and destruction of natural resources, including wild life. Summarized by Pierre Dansereau in 1959 as 'maintaining a favourable population—resource balance in an environment' (*Trans. N.E Wildlife Conf.*, Montreal). See, inter alia, *The Conservation of Natural Resources*, 2nd ed. Guy-Harold Smith, New York: Wiley, 1959.

Zimmermann, E. W., 1951, *World Resources and Industries*, 2nd ed., New York: Harper. 'Conservation is any act reducing the rate of consumption or exhaustion for the avowed purpose of benefiting posterity.' (p. 807) See also Economancy.

**Consociation**

*L.D.G.* A unit of natural vegetation dominated by a single species. Compare association.

**Constant slope**—see Waxing slope, waning slope.

**Consumer goods**—see capital goods
*Webster.* Consumers' goods. *Econ.* Economic goods that directly satisfy human wants or desires, such as food, clothes, etc.—opposed to *producers' goods*. Cf. *capital goods*.

**Contact fields**

Morrill, R. L., 1970, *The Spatial Organization of Society*, Belmont, California: Wadsworth Publishing Co. 'Refers to the

spatial distribution of acquaintances of an individual or group.' (p. 241)
*Comment.* Originator of contact fields was Hägerstrand. See his Swedish work on diffusion: Hägerstrand, T., 1967, *Innovation Diffusion as a Spatial Process*, Chicago: University of Chicago Press (translated from *Innovationsförloppet ur korologisk synpunkt*, 1953, Lund: C. W. K. Gleerup).
See also Neighbourhood effect.

### Contact metamorphism
See Metamorphism, Dynamo-metamorphism.

### Continent
*O.E.D.* 3. A connected or continuous tract of land. *Obs.*
b. The land as opposed to the water, etc.; 'terra firma'; the earth. *Obs.*
c. The 'solid globe' or orb of the sun or moon. *Obs.*
4. esp. The main land, as distinguished from islands, islets or peninsulas; mainland. *Obs.* exc. as in b. or when referring to one of the recognized continents of modern Geography: see 5.
b. esp. The Continent: the mainland of Europe, as distinguished from the British Isles. (Orig. a specific use of 4; now commonly referred to 5.)
5. One of the main continuous bodies of land on the earth's surface.
b. *trans.* A continuous mass or extent of land of any kind, of ice, or the like.
6. Amer. Hist. Applied during and immediately after the War of Independence as a collective name for the revolting colonies (which ultimately became the United States).
Mill, *Dict.* A large area of land (and shallow sea) distinctly separated from other land.
*Comment.* Seven Continents are commonly recognized: North and South America, Europe, Asia, Africa, Australia, Antarctica. In this sense the continents include their neighbouring islands and by extension of this idea we have the 'continent' of Australasia (Australia and New Zealand) or even 'Oceania' including with Australasia the scattered Pacific islands.
Although 'The Continent' is used colloquially by people of the British Isles for the mainland of Europe, Scandinavians in particular do not like being referred to as 'continentals' (G.T.W.).

### Continental basin
*Webster.* A region in the interior of a continent. It may comprise one or several closed basins.

### Continental borderland
Thornbury, 1954. 'Shepard and Emery (1941) proposed the term *continental borderland* to describe the type of shelf . . . off the coast of California, where there is a series of submarine basins and ridges similar in origin to the faulted structures landward from them.' (pp. 462–463) See *Geol. Soc. Amer. Spec. Paper* 31.

### Continental climate
*Cf.* Continentality.
*O.E.D.* (quotation). 1880. Geikie, *Phys. Geog.*, v, 351. 'A continental climate, *i.e.* one where the summer is hot, the winter cold, and where the rainfall is comparatively slight.'
*Webster.* A climate of large daily and annual ranges of temperature, as in the interior of a continent.
Mill, *Dict.* A climate of extremes, with maxima and minima occurring soon after the summer and winter solstices respectively; the climatic characteristic of the interior of continents in the temperate zone, especially in the northern hemisphere, and of coasts and islands in the same zone exposed in winter to prevailing winds from the interior.
*Comment.* The climate of continental interiors or other places protected from or unaffected by maritime influences. The effect of continentality is most marked in temperate latitudes but it is also effective in high and low latitudes. There are two fundamental types of tropical climate, continental and marine, the one with, the other without, a pronounced dry season. (A.A.M.)

### Continental conditions, deposition, deposits, formations
Himus, 1954. Conditions characterized by low rainfall, scarcity of rivers, salt lakes, which obtain in the interior of large land masses remote from the sea (continental conditions). The laying down of rocks by sub-aerial agency and in shallow temporary lakes under continental conditions is referred to as *Continental Deposition*.
Mill, *Dict.* Continental formations. Non-marine deposits accumulated on a continental area. *Cf.* terrestrial deposit.

### Continental divide
*Webster.* A divide separating streams which

flow to opposite sides of a continent.

**Dict. Am.** The divide which separates the watersheds of the Pacific from those of the Atlantic.

**Continental drift, continental displacement**
The hypothesis that the present distribution of the continental masses is the result of the fragmentation of one or two greater land-masses followed by the drifting apart of the fragments in the course of geological time.

**Holmes, 1944.** 'The continents are essentially thin slabs of sial, distributed to form a northern pair, Laurasia, and a more scattered southern group, Gondwanaland. To what extent these primary units and their arrangement have been stable during geological time is one of the fundamental problems of geology.... The suggestion that there might have been lateral displacements on a gigantic scale is generally ascribed to F. B. Taylor in America (1908) and to Alfred Wegener in Germany (1910). Actually the same idea occurred to Antonio Snider (Paris, 1858). (pp. 487–509)

*Comment.* There is a huge and still growing literature on the subject. New contributions are being made by the study of palaeomagnetism.

**Continental glacier**
**Webster. Geol.** An ice sheet or icecap that covers a considerable part of a continent.

**Continental island**
**Webster.** An island near, and geologically related to, a continent—contrasted with *oceanic island*.

**Mill, Dict.** Islands formed of the same rocks as those which build up the neighbouring continents to which they were once joined (*Gregory*).

Contrast oceanic island.

**Continental period**
In geological time a period when the area under consideration formed part of a continent or was above sea-level.

**Continental platform**
**Moore, 1949.** The approximately level part of the earth's crust which is raised above the depressions in which the oceans lie, and includes both the lower land areas of the continents themselves and also the *Continental Shelf* which borders the continents.

**Swayne, 1956.** Part of the continental shelf lying next to the land and reaching to the nearest edge of the mud-line.

*Comment.* As the continents plus their surrounding shelves, in contrast to the ocean basins, the term is used by Salisbury, 1907; Holmes, 1944 (p. 10). Geikie, J., 1898 calls this the continental plateau. The definition quoted from Swayne makes it clear that some confusion exists as to the meaning of the term. See also continental shelf, slope; Mill uses continental block or platform.

**Continental river**
**Mill, Dict.** Rivers with no outlet to the sea; the water of these rivers is lost by percolation or evaporation.

*Comment.* Now rare. See Endoreic and inland drainage.

**Continental sea**
**Mill, Dict.** Seas separated from the ocean save for a narrow and fairly shallow strait, *e.g.* Mediterranean sea, Baltic sea, Red sea. Characterized by regular' coasts, disappearance of estuaries and formation of deltas, abnormalities of currents, tides, salinity, temperature, etc.

*Comment.* Note also epicontinental or shelf seas with a somewhat different meaning. (Holmes, 1944, 12.) Also 'enclosed seas' and 'partially enclosed seas' in text-books of oceanography. See also Sea.

**Continental shelf**
**O.E.D.** Shelf. 4.b. Continental shelf, the relatively shallow belt of sea-bottom bordering a continental mass, the outer edge of which sinks rapidly to the deep ocean-floor.
1892. H. R. Mill, *Realm of Nature*, 201.

**Mill, Dict.** The gently sloping submarine margin of a continent from the coast to the brow of the Actic Region (*q.v.*).

**Webster. Phys. Geog.** A submarine plain of variable width, forming a border to nearly every continent. The water above it is comparatively shallow (usually less than 100 fathoms). The rapid descent from it to the ocean depths is known as the continental slope.

**Bencker, 1942.** 'Immersed zone bordering the continent running from the line of low water to a place where there is a marked declivity towards the ocean depths, in general at a depth of 200 metres.' (p. 64)

**Sverdrup, H. U.** *et al.*, 1942, *The Oceans*, New York: Prentice-Hall. 'The continental shelf is generally considered to extend to depths of 100 fathoms, or 200 m.; but Shepard found that the limit should be somewhat less than this; namely, between

60 and 80 fathoms (110 and 146 m.). (p. 20)

Holmes, 1944. 'The continental platform includes a submerged outer border, known as the *continental shelf*, which extends beyond the shore zone to an average depth of about 100 fathoms.' (p. 11)

*Adm. Gloss.*, 1953. The comparatively shallow platform around the continents extending from the low water line to the depths at which there is a marked increase of slope to greater depths. Conventionally, its edge is taken at 100 fathoms (or 200 metres), but it may be between 65 and 200 fathoms.

Efimenco, N. M., 1957, *World Political Geography*, New York: T. Y. Crowell. 'That portion of the sea bed off the coast, extending outward to a depth of about 600 feet.' (p. 711)

*Comment*. The definition and delimitation of the continental shelf assume increasing importance in international law, in connection with ownership of minerals such as oil which may lie under the shelf.

**Continental shoulder**
The boundary or edge of the continental shelf.

**Continental slope**
The slope, usually steep or abrupt, between the edge of the continental shelf and the deep ocean, the bathyal zone (*q.v.*).

Holmes, 1944. 'The surface of the crust reaches very different levels in different places ... there are two dominant levels: the continental platform and the oceanic or deep-sea platform. The slope connecting them, which is actually quite gentle, is called the continental slope.' (p. 10)

Dietz, R. S., 1952, Geomorphic Evolution of Continental Terrace, *Bull. Ass. Am. Petr. Geol.*, 36, pp. 1802–1819. 'From the shoreline, the shelf typically slopes out to a depth of about 65 fathoms where an abrupt increase in declivity or "shelf-break" is present. This marks the top of the irregular and comparatively steep slope which descends to the abyssal ocean floor at a depth of 2000 fathoms or more.' (p. 1803)

*Comment*. Mill uses Continental talus (obsolete?).

**Continental terrace**
Dietz, R. S., 1952, Geomorphic Evolution of Continental Terrace (Continental shelf and slope). *Bull. Ass. Am. Petr. Geol.*, 36, pp. 1802–1819. 'The single geomorphic unit comprising the continental shelf and the continental slope has been termed the continental terrace (Daly, R. A., *The Floor of the Oceans*, Univ. N. Carolina Press, 1942).' (p. 1804)

Thornbury, 1954. 'Seaward from the *marine-cut terrace* there is usually a *marine-built terrace* or what Johnson (1919) called the *continental terrace*, which consists of materials removed in the cutting of the marine-cut terrace. The abrasion platform and marine-built terrace together constitute what is commonly called the *continental shelf*.' (p. 434)

**Continentality**
*O.E.D. Suppl.* The conditions of being continental as distinguished from oceanic; spec. in Meteorol., the qualities possessed by or typical of a continental climate.

*Met. Gloss.*, 1944. 'In meteorology, a measure of the extent to which the climate of any place is influenced by its distance from the sea.... Measures of meteorological continentality have been derived by Brunt and others, based on either the mean daily range of temperature or the mean annual range in relation to the latitude.

**Continuum** (Latin; *pl.* **continua**)
*O.E.D.* A continuous thing, quantity or substance; a continuous series of elements passing into each other.

*Comment*. A word frequently used in the 17th century onwards, recently revived in geographical writings to describe conditions where clear boundaries are absent. See urban-rural continuum.

**Contour**
*O.E.D.* 1. The outline of any figure: a. introduced as a term of Painting and Sculpture; *spec*. the line separating the differently coloured parts of a design. c. gen.; especially frequent as applied to the outline of a coast, mountain pass, or other topographical feature. 4. Comb. Contour line, a line representing the horizontal contour of the earth's surface at a given elevation. The contour line of a mountain at a given height represents the edge of a horizontal plane cutting the mountain at that height. A series of such lines at successive elevations laid down on a map shows the elevations and depressions of the surface. A map in which this is done is a contour map.

Ansted, D. T., 1844, *Geology*. 'contour-lines; by which is meant lines of equal altitude above a certain standard level' (p. 238). The earliest use quoted by *O.E.D.*

Mill, *Dict.* Contour line. A line drawn through all points having the same elevation above the sea-level as on the surface which is being mapped. See Isohypse.

Webster. Contour line. An imaginary line connecting the points on a land surface that have the same elevation; also, the line representing this on a map or chart. Contour map. A map showing the configuration of a surface by means of contour lines drawn at regular intervals of elevation (contour intervals), as one for twenty feet.

Monkhouse and Wilkinson, 1952. Use contour-line or contour indifferently 'Contourlines or contours (sometimes known as isohypses) are drawn on a map through all points which are at the same height above, or depth below, a chosen datum.' 'These interpolated contour-lines are commonly known as form lines, while the surveyed contours....'

Robinson, 1953. 'A contour is an isometric line of equal elevation above sea level' (p. 198) (G.R.C.).

Beaman, W. B., 1928, *Topo. Instr. of the U.S. Geol. Surv.* 'A contour line may be variously defined as representing (a) an imaginary line on the ground every point of which is at the same height above sea level (b) a level or grade line (c) a line of constant elevation (d) a coast line or other shore line of level water (e) an assumed shore line resulting from assumed rising of a body of level water' (p. 164).

Johnstone, J., 1919, *Bathymetric contours* (used in index—adj. dropped in text). 'A contour line on a chart is usually defined as "a line joining all points where the depth is the same" but this is not quite the same. The contour passes as nearly as possible to *all* the points where depth of (say) 50 fathoms are marked, but, as a rule, it would go near points where (say) 49 or 51 fathoms are recorded ... contour lines are generally approximations...' (p. 46).

Willis, B. and R., 1934. '*Structure contours* ... sometimes called *subsurface contours*, to distinguish them from topographic contours which are drawn on the surface of the ground' (p. 352).

*Comment.* Many authors insist that the contour is the imaginary line on the earth or its representation on a map and argue that contour-line is tautologous.

The word contour affords a good example of different use in everyday speech (*sensu lato* as in *O.E.D.*) and in geographical literature (*sensu stricto*). Geographical definitions often suggest (as Mill and Robinson) that contours refer only to land but the general practice is to include submarine contours (as in Monkhouse and Wilkinson). Geologists have extended the use of the word to various underground surfaces, *e.g.* the surface of the palaeozoic platform which underlies the later rocks in south-eastern England. A predecessor of the word standard in the 18th century would seem to be 'curve' (specifically Halleyan curve). Many cartographers are careful to draw a distinction between 'contours' which are based on instrumental survey and 'form-lines' which are sketched in from general observations. On British Ordnance Survey maps contours are usually given in feet above O.D. (Ordnance Datum, *q.v.*).

**Contour interval**

On a contour map, the vertical distance between the contours or contour lines used. The contour interval need not be constant, *e.g.* the contours may be shown for every 50 feet (a 50-foot interval) up to 500 feet, above that at every 100 feet (a 100-foot interval) and·so on. See also Hinks, 1942, p. 25.

**Contour-ploughing, contour cultivation**

The system of cultivation which involves ploughing approximately following the contours of the fields rather than parallel to their sides, the object being to reduce soil erosion and gullying. Widely used in the drier parts of the United States.

**Contraposed shoreline** (C. H. Clapp, 1913)

Clapp, C. H., 1913, Contraposed Shorelines, *Jour. of Geol.*, **21**, 537–540. 'When a shore-line which has been cut in a soft mantle covering hard rocks is, through the complete retrogression of the mantle, placed against the hard rocks, it changes radically since it becomes dependent on the character of the hard rock surface which was covered by the soft mantle.' Footnote: 'Acknowledgement is due to Professors W. M. Davis and D. W. Johnson who suggested that I devise a name for the type of shoreline described.... As superimposed has recently and conveniently been contracted to superposed, contraposed is preferable to contraimposed.' (p. 537)

Comment. Similar use in: D. W. Johnson, *Shore Processes*..., 1919. C. A. Cotton, *Geomorphology of New Zealand*, 1922 ('retrogradation of a coast of emergence.') *Geomorphology*, 1945 ('Those shorelines on which the surface of the undermass has been resurrected from beneath marine cover'), Wooldridge and Morgan, 1937.

**Conurbation** (Patrick Geddes, 1915)

Geddes, P., 1915, *Cities in Evolution*, London: Williams and Norgate. 'Some name, then, for these city-regions, these town aggregates, is wanted. Constellations we cannot call them; conglomeration is, alas! nearer the mark at present, but it may sound unappreciative; what of "conurbations"? That may perhaps serve as the necessary word, as an expression of this new form of population-grouping.' His conurbations include 'Lancaston' (Manchester and Liverpool); 'Waleston' (South Wales); West Riding; South Riding; 'Midlandton'; Tyne-Wear-Tees, Clyde-Forth (becoming). (p. 34 et seq.)

Fawcett, C. B., 1932, *Geog. Jour.*, 79. '... I use it in the strict sense of a continuously urban area: *i.e.* a conurbation is an area occupied by a continuous series of dwellings, factories, and other buildings, harbour and docks, urban parks and playing fields, etc., which are not separated from each other by rural land; though in many cases in this country such an urban area includes enclaves of rural land which is still in agricultural occupation. In Great Britain the larger of these conurbations have usually been formed by the growth and expansion of a number of neighbouring towns, which have spread out towards each other until they have reached practical coalescence.' (pp. 100-101)

Census, 1951, Prelim. Report. In addition to the obvious case of London, five conurbations in England were delineated with the aid of the Interdepartmental Committee on Social and Economic Research and the Central Statistical Office. They are: S.E. Lancs.; W. Midlands; W. Yorks.; Merseyside; Tyneside. p. xxii. Clydeside is similarly delineated in Scotland.

Barlow Report, 1940, Cmd. 6153. (Refers to Geddes and quotes Fawcett and comments:) 'This definition seems to place too much emphasis on bricks and mortar as constituting the link: while in many cases it may be an adequate definition, in others a better test would seem to be how far out from a given centre does industry or the industrial population look to that centre as essential to its life and as the focus of its business activities.' (p. 6)

Dickinson, R. E., 1947, *City, Region and Regionalism*, London: Kegan Paul. '... defines the urban agglomeration that extends beyond administrative boundaries, and was in fact first defined by Geddes as a group of two or more contiguous administrative units that were urbanised.' (p. 168) (Dickinson uses Urban Tract (built up area), City Settlement Area (inc. rural-urban fringe) and City Trade or circulation Area).

Taylor, G., 1949, *Urban Geography*, London: Methuen. '... a conurbation more strictly means the welding together of several more or less equally important towns.' (thus Manchester and Leeds are better examples).

Smailes, A. E., 1953, *The Geography of Towns*. 'In the massing of people into conurbations, we may distinguish between the process of accretion, peripheral growth around a nucleus, and the process of agglomeration, coalescence of neighbouring but originally separate nuclei as they have grown and the interstices have become filled in. According to the relative prominence of one or other process conurbations show either a uninuclear or polynucleate structure.' (p. 113) Similarly, '... great tracts of built-up country' ... (p. 36).

Comment. It is clear that Fawcett changed the original definition of Geddes and that currently both concepts are used. T. W. Freeman, 1959, *The Conurbations of Great Britain*, Manchester University Press, greatly extends the use of the term by including 'minor conurbations' with populations down to 50,000.

**Convection**

*O.E.D.* Physics. The action of carrying; conveyance; spec. the transportation of heat or electricity by the movement of a heated or electrified substance, as in the ascension of heated air or water.

*Webster.* 2. *Meteorol.* The upward or downward movement, mechanically or thermally produced, of a limited portion of the atmosphere. Convection essential to

the formation of many clouds, esp. of the cumulus types.
*Met. Gloss.*, 1944. In convection heat is carried from one place to another by the bodily transfer of the matter containing it....

**Convection current** (geology and oceanography)
Holmes, 1944. 'A hot current ascends near the middle, and turning along at the surface, it sweeps ... to the edges, where the current descends.... When the earth was molten it would cool off by similar circulations. Convection currents would rise in certain places, spread out horizontally and then turn down again.' (p. 20)
'At least in the early stages of the earth's history cooling would be brought about by convection in the substratum. In this process ... currents of relatively hot and light material ascend in certain places, so carrying heat up to the base of the crust, through which some of it escapes by conduction. Towards the top, the currents spread out in all directions from each centre, until they encounter similar currents from neighbouring centres and turn downwards. The descending currents consist of somewhat cooled and slightly heavier material. The driving force arises from the difference in density between the central and marginal columns. This kind of circulation continues until the temperature falls nearly to the freezing point of the material concerned.' (p. 408)
Holmes then proceeds to elaborate his theory of sub-crustal convection currents, stimulated by radio-active break-down.
Johnstone, J., refers to convection currents in the ocean due to temperature differences (p. 171).

**Convectional rain**
*Met. Gloss.*, 1944. Caused by the heating of the surface layers of the atmosphere which expand and rise, giving place to denser cool air. The warm air is frequently heavily charged with moisture taken up from the ground or vegetation. As it rises this moist air is cooled and its moisture condenses to fall as rain.

**Conventional name**
Aurousseau, M., 1957, *The Rendering of Geographical Names*, London: Hutchinson. 'Many places in parts of the world where English is not spoken are known to us by names which ... are not used at the places to which we apply them ... the Crusaders brought us a name for Cairo from the Venetians ... we call the capital of Denmark, Copenhagen ... the Danish name København is particularly hard for us to pronounce but its German form Copenhagen we can manage with ease, though we pronounce it imperfectly ... such a name, taken into English from a source foreign to the place to which it refers, or from a former state of the language spoken there, is a "*conventional*" *name*, or, more precisely, a "conventionally English name".' (p. 20)
*Comment.* Other nations use conventional names *e.g.* Londres, Londen, etc. for London. There is at present a strong international movement to give the local name priority with conventional names in brackets, *e.g.* Roma (Rome), Napoli (Naples), Wien (Vienna), but this still leaves the difficulty of varied transliterations from non-Roman scripts.

**Conventional sign**
Hinks, A. R., *Maps and Survey*, Cambridge: C.U.P., 1913. '... it is necessary to adopt a carefully considered scheme of conventions, so that the character of every line and the style of every letter may convey a definite meaning.' (p. 5)
*Comment.* A 'Standard Symbol' (U.S.) used on a map (*i.e.* a cartographic symbol) based upon a convention in the sense of 'a rule or practice generally accepted, arbitrarily or artificially determined'. (see *O.E.D.*)

**Convergence**
*O.E.D.* The action or fact of converging; movement directed toward or terminating in the same point.
*Webster*. 2. *Anthropol.* The development of similarities between cultures by accidental changes; convergent evolution. 3. *Biol.* The development or possession of similar characters by animals or plants of different groups, due usually to similarity in habits or environment. The resemblance in form of body of the whales and fishes is an example.
*Comment.* This word has many special applications in scientific literature notably in evolution biology convergent forms (morphological) as response to similar habitat (*e.g.* wings for flight in insects, reptiles, birds, mammals, fish); in meteorology (see ITCZ), in oceanography (see

below) in geology (see below). Note also convergence of meridians in surveying.

Sverdrup, H. U., *et al.*, 1942. 'Sinking of surface water is not limited to regions in which water of particularly high density is formed, but occurs also wherever converging currents (convergences) are present, the sinking water spreading at intermediate depths according to its density.... The most conspicuous convergence is the Antarctic convergence, which can be traced all around the Antarctic Continent' (p. 139).

Willis, B. and R., 1934. Thickening or thinning of formation is known as *convergence* (p. 68).

**Convulsionism**—see Catastrophism

**Coolie, cooly**

*O.E.D.* The name given by Europeans in India and China to a native hired labourer or burden-carrier; also used in other countries where these men are employed as cheap labourers (origin uncertain: now found in most Indian vernaculars).

**Coombe-rock**—see Combe-rock

**Co-ordinate**

*O.E.D.* Co-ordinate B.2. Math. Each of a system of two or more magnitudes used to define the position of a point, line, or plane, by reference to a fixed system of lines, points, etc. (usually in *pl.*).

Close, C. F., 1913, *Text Book of Topographical and Geographical Surveying*, 2nd ed. rev. E. W. Cox, London: H.M.S.O. Distinguishes between Geographical co-ordinates (latitude and longitude), Rectangular co-ordinates and Astronomical co-ordinates. (W.M.)

**Cop**

*O.E.D.* 1. The top or summit of anything, esp. of a hill.

Mill, *Dict.* A little round-topped hill (Central and N. England), *e.g.* Mow Cop.

**Coppice**

*O.E.D.* 1. A small wood or thicket consisting of underwood and small trees grown for the purpose of periodical cutting.

Moore, 1949. Coppice or Copse. A small wood, or plantation of trees, which are periodically cut before growing into large timber.

*Comment.* Although derived from the same root there is in current use a distinction between copse, a small wood, and coppice woodland, subjected to periodic cutting or 'coppicing' whereby a tree which has been cut to within a few inches of the ground then sends up a number of shoots, each of which after a few years can be cut to provide fencing posts, poles, etc. Oak, hazel, sweet chestnut and other trees are used in this way.

**Coppice-with-standards**

A coppice in which certain trees are left to grow to full size (standards). See Tansley, 1939, 181.

**Coprolite**

*Webster. Paleontol.* Fossil dung or excrement. Such remains often give valuable information as to the food and habits of extinct animals.

Himus, 1954. The fossilized excrements of fishes, reptiles, and mammals. Generally composed of calcium phosphate, hence the term is loosely used to include phosphatic nodules of any kind.

**Copse**—see Coppice

**Coral, coral reef**

*O.E.D.* 1. A hard calcareous substance consisting of the continuous skeleton secreted by many tribes of marine coelenterate polyps for their support and habitation. Found, according to the habits of the species, in single specimens growing plant-like on the sea-bottom, or in extensive accumulations, sometimes many miles in extent, called *coral-reefs*.
a. historically... the beautiful Red Coral....
b. afterwards extended to other kinds....
2.a. A particular species of the preceding, or of the colonial zoophyte of which it is the skeleton; also, a single polypary or polypidom in its natural condition. (= Corallum)

Mill, *Dict.* Coral Reef. A reef composed of the skeletons of madreporic corals and other lime-secreting organisms.

*Webster. Phys. Geog.* A reef, often of great extent, made up chiefly of fragments of corals, coral sands, and the solid limestone resulting from their consolidation. It is called a *fringing reef* when it borders the land, and a *barrier reef* when separated from the shore. See also *atoll*.

*Adm. Gloss.*, 1953. Coral: hard calcareous substance secreted by many tribes of marine polyps for support, habitation, etc. It may be found either dead or alive.

Coral reefs. Reefs, often of large extent,

composed chiefly of coral and its derivatives.
*Comment.* See also biostrome, bioherm. Note coral-sand derived mainly from mechanical destruction of coral and coral reefs, similarly coral-mud. Coral Rag is a subdivision of the English Corallian (Jurassic) made up largely of rolled and broken corals forming a loose, rubbly limestone.

**Cordillera** (Spanish)
*O.E.D.* (Sp. = mountain-chain, 'the running along of a rocke in great length' (Minsheu in 1599).
A mountain chain or ridge, one of a series of parallel ridges; in pl. applied originally by the Spaniards to the parallel chains of the Andes in South America (las Cordilleras de los Andes). Subsequently extended to the continuation of the same system through Central America and Mexico.
Some geographers in the U.S. have proposed to transfer the name to the more or less parallel chains of the Rocky Mountains and Sierra Nevada, with their intervening ridges and tablelands, termed by them the Cordilleran region; but this is not approved of by European geographers. (Comment: the American terminology is now accepted—L.D.S.)
*Webster.* A mountain range or system, orig. one of the Andes; sometimes, the main mountain axis of a continent. Thus, the western *cordillera* of North America (called also the Cordilleras) includes the Rocky Mountains, Sierra Nevada, Coast and Cascade ranges.
*Dict. Am.* A range or system of mountains, often *pl.* 1808 Pike, *Sources Mississippi* (1810) 1. App. 56. 'But we must cross (what is commonly termed) the Rocky Mountains, or a spur of the Cordeliers, previous to our finding the waters, whose currents run westward, and pay tribute to the western ocean.
Mill, *Dict.* A great mountain mass, a group of sierras (*q.v.*) (Spanish America).
Knox, 1904. (Sp.) A chain or ridge of mountains, a long elevated and straight tract of land.
(U.S.A.) A group of mountain ranges, including the valleys, plains, rivers, lakes, etc.; its composite ranges may have various trends, but the cordillera will have one general direction.
(S. America). A chain of mountains with distinct summits, but closely connected like the links of a chain or the strands of a rope.
Holmes, 1944. 'A mountain system is the whole series of ranges belonging to an orogenic belt. The term cordillera is sometimes used for a broad assemblage of ranges—such as that of western North America—belonging to more than one system.' (p. 378)

**Core** (of city, etc.)
*Cf.* core area.
Smailes, A. E., 1953, *The Geography of Towns*, London: Hutchinson. The functionally specialized centre of a town including shops, offices, banks, public services, nursing homes, hotels, administrative offices (examples from 'City Centre' of St. Albans) = 'downtown' district. (pp. 88–89; 91)
Dickinson, R. E., 1951, *The West European City*, London: Routledge. The central functional district, empty of residences. 'Central business core.' (pp. 5, 27)
*Comment.* See note under core area.

**Core area**
*Cf.* Core of cities or of states; nuclear area.
*O.E.D.* 13. The innermost part, very centre, or 'heart'.
Pearcy, G. E. and Russell, H. F., 1948, *World Political Geography*, New York: Crowell. (G. T. Renner) 'The core area of a state is usually the nucleus or cradle region in which the nation was germinated and nurtured.' (p. 19) (A. R. Hall) 'Such a state expands outward from a central core area. The core area is usually the original seat of the state and contains a population that may be more homogeneous culturally than other areas in the country.' (p. 517)
... 'Beyond is frequently found a zone of more recently acquired territories, called the frontier.'
Hartshorne, 1939. 'An incomplete study may limit itself to the "cores" or "hearts" of regions, although this involves the questionable assumption that such areas are geographically more important than the areas of less distinctive character.' (p. 305)
Ogilvie, A. G., 1957, *Europe and its Borderlands*, Edinburgh: Nelson. 'The place where each nation was formed, its core or nuclear area, has been plotted....' (p. 255)
*Comment.* This term and the quotations above have been included because of the frequent use of the term in geographical

literature but some members of the Committee deprecate this specialized use of a common word.

**Core-frame concept**
Horwood, E. M. and Boyce, R., 1959, *Studies of the Central Business District and Urban Freeway Development*, Seattle: University of Washington Press. '... different functional, geographical, and historical attributes are ascribed to the core and frame respectively. The C.B.D. (central business district) frame is a uniform area.... When all of the properties of the C.B.D. are considered as an organizing spatial concept, the C.B.D. frame is one of the major functional components of the city ... there is a considerable difference in the characteristics ascribed to the C.B.D. core and C.B.D. frame, despite their both being part of the C.B.D.' (pp. 19–21)
See also Central Business District.

**Core-periphery model (Centre-periphery model)**
Friedmann, J., 1973, *Urbanization, Planning and National Developments*, Beverly Hills and London: Sage Publication. '... a simple core-periphery model in which peripheral areas are defined in terms of their dependency relationships to relevant core regions. Urbanization processes are consequently viewed as spatial processes connecting core and peripheral areas in ways agreeable to systematic study.' (p. 18)
Darwent, D. F., 1969, Growth Poles and Growth Centres in Regional Planning: a Review, *Environment and Planning*, 1.1. 'By dealing with the whole of economic space in a given region, rather than particular points or area of it, and by defining subregions of the periphery in terms of the problems for which solutions are sought, it is a valuable step towards the prescription of policies for the distribution of economic and social development given a set of goals.' (p. 6)
See Friedmann, J., 1966, *Regional Policy: A Case Study of Venezuela*, Cambridge, Mass: M.I.T. Press.
See also Growth pole.

**Coriolis (wrongly Corioli's) force**
Sverdrup, H. U., *et al.*, 1942. 'The force per unit mass that must be introduced in order to obtain the correct equation is called the deflecting force of the earth's rotation, or Coriolis's force, after the French physicist who first made the transformation from a fixed to a rotating system' (p. 433).
Kuenen, 1955. 'Allowance has to be made also for the influence (Coriolis force) of the earth's rotation, which tends to deflect a current towards the right in the Northern Hemisphere' (p. 49).
*Comment.* This force underlies the simple principle stated in Buys Ballot's and Ferrel's law (A.A.M.). Named from G. G. de Coriolis (1792–1843).
See Buys Ballot's law.

**Corn**
*Webster.* 3. Collectively, the seeds of any of the cereal grasses used for food; grain, as wheat, rye, barley, oats, maize, etc. *Corn* is often specifically used for the important cereal crop of a given region; thus in England it is so used of wheat, and in Scotland and Ireland of oats; and in the United States and Australia the word is restricted to Indian corn, or maize, the other cereals being there collectively called *grain*, and separately *wheat, rye, oats,* etc.

**Corn Belt**
See Belt, and comments under cotton belt.

**Cornbrash**—see Brash

**Cornice**
*Webster. Mountaineering.* An overhanging formation of snow or rock, usually on a ridge or at the top of a couloir.
*Ice Gloss.*, 1958. An overhanging accumulation of *ice* and wind-blown *snow* on the edge of a ridge or cliff face.

**Corniche (French)**
An architectural term (*cf.* cornice) hence *fig.* 1 'sentier étroit qui borde le précipice au flanc d'une montagne (Hatzfeld, A., et Darmesteter, A., *Dict. gén. de la Langue Française*, c. 1908). La route de la Corniche—from Nice to Genoa high up on the hills overlooking the Mediterranean. From this Harrap's French-English Dictionary gives 'coast-road'. In 1959–60 term applied in Egypt to new scenic road along the east bank of the Nile, north and south of Cairo.
*Comment.* A very interesting example of evolution in meaning.

**Corral (Spanish)**
*O.E.D.* An enclosure or pen for horses, cattle etc., a fold; a stockade. Chiefly in Spanish America and United States. *cf.* kraal.
*Dict. Am.* An enclosure for horses, cattle, etc. 1829.

*Webster.* A pen or enclosure for confining or capturing animals; also, an enclosure, made with wagons, in an encampment, as a place of defense and security.

**Corrasion** (see also Corrosion)

*O.E.D.* Corrasion *obs. rare.* 1. The action of scraping together: see Corrade, ex. 1611.

*O.E.D. Suppl.* 2. Geol. The wearing away of stones, etc. by mutual friction or by running water. 1877. G. K. Gilbert, *Geol. Henry Mts.*, 101. In corrasion the agents of disintegration are solution and mechanical wear. 1895. Dana, *Man. Geol.* (ed. 4), 168. 1905. Chamberlin and Salisbury, *Geol.* I, 108. The more active and tangible processes by which surface rocks are broken up, such as wave wear, river wear and glacial wear, are processes of corrasion.

*Webster.* The scouring action of wind-borne sand.

Mill, *Dict.* Corrasion: 1. Sometimes merely a synonym for corrosion.
2. Sometimes used for the work of wind-blown sand in wearing down and polishing rock surfaces. This term is too ill defined to make its retention desirable, though Salisbury suggests it for the picking up of rock material loosened in any way.

Gilbert, 1877. (see above, and add:) '... in all fields of rapid corrasion the part played by solution is so small that it may be disregarded.' (p. 101) 'Streams of clear water corrade their beds by solution. Muddy streams act partly by solution, but chiefly by attrition.... Corrasion is performed by solution, and by mechanical wear.... Corrasion is distinguished from weathering chiefly by including mechanical wear among its agencies....' (p. 102) (Referring to stream erosion).

Powell, 1875. '... the power of the water is constantly exerted in corrasion.' (p. 179)

Cotton, 1922. A river is capable of doing erosive work of two kinds—corrasion (or cutting) and transportation. 'The corrasion is of two kinds—namely, chemical and mechanical; and the material transported is carried in two ways, corresponding respectively to these—namely, in solution and in suspension.' (p. 38)

Geomorphology, 1945 (of river erosion). 'The attack on rocks is of two kinds—namely, chemical corrosion and mechanical corrasion....' (p. 35)

Wooldridge and Morgan, 1937. 'The work of vertical or lateral cutting performed by a river in virtue of the abrasive power of its load is termed corrasion. This process should be distinguished from corrosion—a purely chemical process.' (p. 153)

Lobeck, 1939. (of erosion by young streams). 'This is accomplished by (a) corrasion, or scraping and scratching away the bedrock; (b) impact, or the effect of definite blows on the bed of the stream by large boulders; (c) quarrying, due to the lifting effect of the water as it pushes into the cracks of rocks; and (d) solution.' (p. 193)

Strahler, A. N., *Physical Geography*, New York: Wiley, 1951 (of streams) 'There are three kinds of erosion: (1) Hydraulic action, in which the force of the moving current alone breaks away the particles.... (2) Corrasion, in which rock particles carried by the current strike against the channel floor and sides without sufficient impact to break out new particles.... (3) Corrosion, the chemical action or simple solution....' (p. 145)

Holmes, 1944 (of river erosion:)
(a) '*Corrosion* is the solvent and chemical action of the water of the stream on the materials with which it comes into contact.'
(b) *Hydraulic action* is the mechanical loosening and removal of materials by water alone.
(c) *Corrasion* is the wearing away of the sides and floor with the aid of the boulders, pebbles, sand and silt which are being transported....
(d) *Attrition* is the wear and tear suffered by the transported materials themselves.... (pp. 150–151)
(of marine erosion)
'The sea operates as an agent of erosion in four different ways:
(a) *hydraulic action*
(b) by *corrasion*, when waves, armed with rock fragments, hurl them against the cliffs and, co-operating with currents, drag them to and fro across the rocks of the foreshore.
(c) by *attrition*, as the fragments or "tools" are themselves worn down by impact and friction; and
(d) by *corrosion*, i.e. solvent and chemical action,...' (p. 286).

Rice, 1941. Corrasion is a term first used by Powell to designate the erosion effected by running water. Running water erodes its bed chiefly by the mechanical wear or abrasion of the rock waste which it carries against the bottom and sides of its bed, by removal of joint blocks and by solution. Hence corrasion must be defined as consisting of these various processes (Bryan). (Bryan, K., *U.S. Geol. Survey Bull*, 730, pp. 20–90. Gloss. 86–90, 1922.)

Comment. See also the long discussion in Penck (trans. Czech and Boswell) 1953. 'We are giving the name *corrasion* to the freeing of loosened rock fragments from their place of origin.' (p. 112)

**Corrasion-valley** (W. Penck)

Penck, W. (trans. H. Czech and K. C. Boswell), 1953. 'Increased denudation beneath a corrading "mass-stream" must excavate its sub-stratum into an elongated hollow, dig a furrow. I propose the term corrasion valleys for valleys with such an origin.' (p. 112)

**Corridor**

*O.E.D. Suppl.* A strip of the territory of a state running through another territory and so contrived as to give access to a certain part, *e.g.* the sea.

*Webster* (*often cap*). *Polit. Geog.* A narrow strip of land across territory previously foreign, joining a country to its seaport; as, the Polish *Corridor* (across Germany to Danzig); also, such a strip connecting two countries or two parts of country.

Pounds, N. J. G., 1959, A Free and Secure Access to the Sea, *A.A.A.G.*, **49**, 256–268. 'A corridor reaching to the sea coast implies a transfer of sovereignty over an area of land in order to give to an inland state an assured outlet to the ocean' (p. 258).

Haushofer, A., 1933, Was ist ein Korridor, in *Deutschland und der Korridor*, Berlin, 202–220.

Comment. Also extended in use to an 'air corridor' giving right of access by air.

**Corrie** (Scottish)

*O.E.D.* (Gaelic coire (pronounced ko.re) cauldron, kettle; hence whirlpool and circular hollow).

The name given in the Scottish Highlands to a more or less circular hollow on a mountain side, surrounded with steep slopes or precipices except at the lowest part, whence a stream usually flows. 1795. *Stat. Acc. Scot.* XVI, 104. 'The Corries or Curries of Balglass. They are semi-circular excavations hollowed out in that ridge of hills.'

Mill, *Dict.* Corrie, Coire. An amphitheatre shaped notch in a mountainside, with steep walls on three sides and a floor which may either slope outwards, be flat, or be hollow. In all cases the corrie floor has not a slope continuous to that of the main valley below. Corrie is used by some writers as an equivalent of cirque (*q.v.*), but the term is best restricted to the form described above, which has many local names (*cf.* Cros in Dauphigny, Kahr or Kar in Tirol, and both in Norway).

Wooldridge and Morgan, 1937. 'At first sight a typical corrie bears some resemblance to the valley-head amphitheatre produced by normal erosion.... The slopes of a corrie are much steeper, and the rock basin in the floor constitutes another departure from the form of a normal valley-head, though admittedly it is not always present.' (p. 376)

'Between the convergent arêtes are remarkable hollows, open on one side, but bounded elsewhere by an almost semi-circular or sub-rectangular line of cliffs. These have been given a distinctive name in most European languages (corrie, in Scotland; cwm, in Wales; cirque, in France), ...' (p. 375)

von Engeln, 1942. 'The glacier cirque is so distinctive as a geomorphic form that it is identified by a specific word in many languages, thus kar, corrie, cwm, botn, caldare, oule, zanoga. These foreign words are occasionally used in English....' (p. 447)

Comment. In geographical literature cirque, corrie and cwm are commonly used as alternative names for the landform of glacial origin described above. In each case this is a restricted meaning of the original word. 'Glacial cirque' or 'glacier cirque' indicates the character of this restriction.

**Corrom**

Charlesworth, 1957. 'Streams have built watershed deltas or corroms in the passes to give rise to lakelets, *e.g.* in the north-west Highlands' (p. 406).

See also Bailey, E. B. and Kendall, P. F., 1907, Glaciation of East Lothian South of the Garleton Hills, *Trans. Roy. Soc. Edin.*, **46**, 1–31 (p. 25).

**Corrosion** (see also Corrasion)

*O.E.D.* Geol. The solution of a mass of rock

or mineral by water; the eating away by fusion and absorption of a mineral in its magma.
1897. G.J., X, 502. 1903. Geikie, Text-bk. Geol. (ed. 4), 141.
Note also l.c. The gradual wasting action of water, currents, etc.; erosion. Obs. example 1781.
*Webster.* Action or effect of corroding, or of corrosive agents; the process of corrosive change. *Corrosion* is now generally used of a gradual wearing away or disintegration by a chemical process, as in the rusting of iron; but it formerly sometimes included *erosion*, which is a mechanical process.
Mill, *Dict.* Corrosion. 1. The wearing of a stream—or glacier-bed—by friction of transported material (Penck). 2. Also used for the above wearing whether due to the water itself or to its transported material. Corrasion has sometimes been used in the above senses, and corrosion has been restricted by some to (3) the solution of rock material by water. The second meaning is the most general one.
Rice, 1941. 'Corrosion literally means gnawing away, but in this country it is generally restricted to the gnawing effected by chemical agencies—in other words, it means the solution of rocks and other material, especially in contradistinction to corrasion (Bryan). The modification of phenocrysts or xenoliths, etc., by the solvent action upon them of the residual magma in which they are contained (Holmes).'
Lake, P. and Rastall, R. H., 1910, *A Textbook of Geology*, London: Arnold. '... actual corrosion or wearing away of the rocks by the dynamical effect of moving sand.' (p. 73)
Guilcher, A., 1958. Coastal Corrosion forms in Limestones around the Bay of Biscay, *Scot. Geog. Mag.*, **74**, 137–149. 'Corrosion of the rocks ... creates ... a particular morphology which does not derive from the mechanical action of the waves.' (p. 137)
*Comment.* Should be restricted to chemical action.

**Cors** (mutated form, *gors*: Welsh)
*Lit.* bog: has been used in geographical-botanical works for a sphagnum bog.

**Cosmography**
*O.E.D.* Adaption of Greek, 'description of the world'.
1. The science which describes and maps the general features of the universe (both the heavens and the earth), without encroaching on the special provinces of astronomy or geography. But formerly often = geography in its present sense, or specifically as including hydrography.
2. A description or representation of the universe or of the earth in its general features.
Mill, *Dict.* The science of the universe, anciently equivalent to physiography.
*Comment.* A term in little use at the present time, which has had a long and complex history. In general its introduction and use implied belief in an ordered universe or cosmos (in contrast to chaos). So also many other related terms: cosmogeny, cosmogony, cosmology, cosmometry.

**Cosmopolitan**
*O.E.D.* 1. Belonging to all parts of the world; not restricted to any one country or its inhabitants.
2. Free from national limitations or attachments.

**Cosmopolite**
*O.E.D.* 1. A 'citizen of the world'; one who regards or treats the whole world as his country; one who has no national attachments or prejudices.
2. A plant or animal at home in all parts of the world.

**Costa** (Spanish)
*L.D.G.* The Spanish word for coast, now applied especially to the stretches of Spanish coasts along the Mediterranean which have been developed as holiday resorts, *e.g.* Costa Brava (north of Barcelona), Costa del Sol etc.

**Cost space (Time space)**
Represents human spatial structure by relative space as opposed to absolute space.
Abler, Adams and Gould, 1971. 'Attempts to explain human spatial behaviour are an increasingly important concern of geographers, and such attempts will be more successful if they are cast in relative spatial contexts. People shipping goods or taking trips between towns *A*, *B* and *C* are not as much concerned with absolute distance as they are cost and time; they make their decisions in cost and time space, not absolute space.' (p. 75)
*Comment.* Some geographers prefer to use 'cost distance' and 'time distance' as being more straightforward, 'cost space'

and 'time space' encompassing too many dimensions.

**Côte (French)**
1. An escarpment or cuesta, e.g. Côte d'Or in Burgundy.
2. Coast, e.g. Côte d'Ivoire, Ivory Coast; Côte d'Azur (French Riviera).

**Cottar, Cotter (Scottish); cottier (Irish)**
*O.E.D. a.* A peasant who occupies a cothouse or cottage belonging to a farm (sometimes with a plot of land attached) for which he has (or had) to give or provide labour on the farm, at a fixed rate, when required. *b.* A peasant especially in the Highlands who occupies a cottage and rents a small plot of land under the form of tenure similar to that of the Irish cottier.

*Cottier tenure.* The main feature of this system was the letting of the land annually in small portions directly to labourers, the rent being fixed not by private agreement but by public competition; legal and political changes have rendered this practice obsolete.

**Cotton Belt**
*O.E.D. Suppl.* Cotton Belt, U.S., the area in which cotton is grown. (quote: 1871).

Baker, O. E., 1927, *Econ. Geog.*, 3, p. 65. The Cotton and Corn Belts are simply two of the several 'belts' or agricultural regions into which North America has been divided for descriptive purposes. See Belt, region (agricultural).

**Couchant**—see Levant and couchant

**Coulée, coulee**
*O.E.D.* French, coulée, flow. Sense 2 appears to have arisen among French trappers in the Oregon region.
1. Geol. A stream of lava, whether molten or consolidated into rock; a lava flow.
2. In the western region of Canada and the United States. A deep ravine or gulch scooped out by heavy rain or melting snow, but dry in summer.

*Webster.* 1. *Geol.* A stream or sheet of lava. 2. Any of a number of steep-walled, trenchlike valleys, as the Grand Coulee, formerly occupied by the Columbia River. Western U.S.

*Dict. Am.* A small stream, or the bed of such a stream when dry. 1807 in *Amer. State P., Pub. Lands* (1832), I, 313. 'Bounded in front by the river Detroit, and in rear by a *coulee* or a small run.'

Thornbury, 1954. Glacial: Abandoned sluiceways which temporarily carried glacial melt waters. (p. 398) Volcanic: 'Individual lava flows from the crater or flank of a cone may form tongue-like extensions down the cone called coulees. (p. 496)

Putnam, W. C., 1938, The Mono Craters California, *Geog. Rev.*, 28. 'The use of the term "coulee" for a lava flow has become less familiar than its meaning as a steep-sided stream valley, such as the Grand Coulee, in Washington. It seems desirable to revive the first usage for these steep-fronted obsidian flows.' (Footnote, p. 78)

*Comment.* The accent is usually dropped in American literature.

**Coulisse (French)**
Scrivenor, J. B., 1921, The Physical Geography of the Southern Part of the Malay Peninsula, *Geog. Rev.*, 11, 351–371. This French term was introduced by Scrivenor to describe the mountain ranges which, arranged *en echelon*, make up the mountain divide of Malaya. See Stamp, L. D., 1950, *Asia*, 10th Edition, pp. 411–412 and Fig. 214. It has not been generally adopted; in French the word, *inter alia*, means the side scenes on a stage, the wings.

**Couloir (French)**
*O.E.D.* French: passage, lobby, the steep incline down which timber is precipitated on a mountain side: from late Latin, *colare*, to flow. A steep gorge or gully on a mountain side: first used in reference to the Alps.

Mill, *Dict.* A vertical cleft wider than a chimney.

**Counterdrift (D. Rigby Childs, 1963)**
Childs, D. R., *Local Govt. Chron.* 'Counterdrift is a method of planning which sets out to allow people who are living on the fringe of large areas of growth to participate in the benefits that accrue' (p. 13).

**Counter-trade wind**—see Anti-trade wind

**Country belt**
Osborn, F. J., 1946, *Green-Belt Cities*, London: Faber and Faber. 'Country Belt, Agricultural Belt, Rural Belt. These terms are synonymous. They describe a stretch of countryside around and between towns, separating each from the others, and predominantly permanent farmland and parkland, whether or not such land is in the ownership of a town authority.' (p. 182)

See also Belt, greenbelt.

**Country rock**
Holmes, 1944. 'The form and size of an igneous mass as a whole, and its relations

to the adjacent rocks (often referred to as the "country rocks") are called its *mode of occurrence*' (p. 71).

Geikie, 1912. 'The rocks traversed by a lode are known as the *country or country-rock*' (p. 244).

## County

*O.E.D.* From Anglo-French, counté (in Laws of Wm. I) derived from Latin, comitatus. The Latin word had primarily the sense of 'a body of companions, a companionship', subsequently 'an escort or retinue'; when *comes* became a designation of a state officer, *comitatus* followed as the name of his office, and when the *conte* became a territorial lord, the *conté* became his territory—the stage at which the word entered English.
1. The domain or territory of a count. *Obs.*
2. One of the territorial divisions of Great Britain and Ireland, formed as the result of a variety of historical events, and serving as the most important divisional unit in the country for administrative, judicial, and political purposes.
b. The status of county was also given at various times to a number of cities and towns in England and Ireland, with a certain portion of adjoining territory... more exactly called *corporate counties* or *counties corporate*.
c. By the Local Government Act of 1888 ... boroughs of above 50,000 inhabitants are made administrative counties under the name of *county boroughs*....
3. Introduced into most of the British colonies as the name of the administrative divisions; in the United States, the political and administrative division next below the State, into which all the States of the Union are divided—except South Carolina ... and Louisiana....

Mill, *Dict.* A subdivision of a state or province having its own courts of justice and the administration of its own local affairs.

*Comment.* With the separation of numerous County Boroughs which for most local government purposes are outside the jurisdiction of the old historical counties within which they lie it has become usual to refer to geographical counties for the old units—which would be better described as the historical counties. Some of these are divided for administrative purposes; each of the three Ridings of Yorkshire has a separate County Council, as have East and West Sussex, East and West Suffolk and others. See Fawcett, C. B., 1960, *Provinces of England*, London: Hutchinson, 42–5.

## Covariance

Cole and King, 1968. 'The relationship between two variables, given by $\Sigma\, xy/n$ where $x$ and $y$ are the differences between each variate and the mean of its set, and $n$ is the number of pairs.' (p. 650)

See *Biometrics*, 1967, **13**, 261–405. Series of papers on covariance analysis. An example of its use in geographical literature is provided in Taaffee, E. J., 1958, A Map Analysis of United States Airline Competition, Pt. II: Competition and Growth, *The Journal of Air Law and Commerce*, **25**, 402–427.

## Cove

*O.E.D.* 1. In O.E. A small chamber, inner chamber, bed-chamber, cell, etc.; ...
2. A hollow or recess in a rock, a cave, cavern.
3. A recess with precipitous sides in the steep flank of a mountain (common in the English Lake District, where small lateral valleys often end in 'coves'.) b. In some parts of U.S. = gap, pass.
4. A sheltered recess in a coast; a small bay, creek, or inlet where boats may shelter.

Webster. 5. *Phys. Geog.* A basin or hollow where the surface of the land has caved in as from solution of underlying rock.

Mill, *Dict.* 1. A small inlet creek or bay. 2. A hollow or recess in a rock; cave. 3. A recess with precipitous sides in the steep flank of a mountain. 4. Gap, pass (U.S.A.).

*Adm. Gloss.*, 1953. A small indentation in a coast (usually a cliffy one), frequently with a restricted entrance and often circular or semi-circular in shape.

Bencker, 1942. Very small gulf in the sand or gravel beach.

*Comment.* The famous example on the south coast of England which seems to have inspired the *Adm. Gloss.* definition is Lulworth Cove. *Webster's* sense is not used in British writings.

## Cover crop

Webster. *Agric.* A catch crop, as rye or clover, planted, esp. in orchards, to protect the soil in winter, to fix nitrogen in the soil, etc. See *catch crop.*

A crop grown primarily to protect cleared land from the danger of soil erosion. The practice of planting a cover crop is especially important in tropical forest country liable to torrential downpours. So long as the forest cover is intact, danger of soil erosion is small but as soon as it is cleared—for example for rubber planting —the danger is very great. A successful cover crop must be quick growing; when it has served its purpose it is commonly hoed in or ploughed in, so that leguminous plants able to enrich the soil are favoured. (L.D.S.)

**Covert**
A man-made spinney or small piece of woodland or scrub to 'cover' (*i.e.* serve as refuges for) foxes in hunting country (designated fox-covert) or game birds and animals. (England: Midlands). Also called locally 'thorns'. (I.L.A.T.)

**Crab hole** (Australia)
A term used in Australian pedological literature for hollows in clay plains not unlike shallow graves about 8 feet long (E.S.H.).

Wadham, S., Wilson, R. K. and Wood, J., *Land Utilization in Australia*, 3rd ed., 1957. 'Soil areas known locally as "crab holes" or "puff-banks"... the whole countryside may be thrown into a series of pits or troughs each of which is two or three feet below the adjacent banks. The soil in the pit is different from that of the bank' (p. 18).

**Cradle**
Sauer, C. O., 1952, *Agricultural Origins and Dispersals*, New York: Amer. Geog. Soc. 'As the cradle of earliest agriculture, I have proposed Southeastern Asia' (p. 24). Also referred to, and shown on the map, as the 'hearth'.

**Crag**
O.E.D. 1. A steep or precipitous rugged rock.
2. A detached or projecting rough piece of rock.
3. A local name for deposits of shelly sand found in Norfolk, Suffolk, and Essex, and used for manure; applied in geology to the Pliocene and Miocene strata to which these deposits belong, called in order of age, the Coralline Crag, Red Crag and Mammaliferous or Norwich Crag. (It is doubtful whether this is the same word; the connexion is not obvious.) Examples 1735–1885.

**Crag and tail**
O.E.D. 1.b. *Crag and Tail* (*Geol.*): see quot. 1865. Page, *Handbook Geol. Terms*, Crag and Tail (properly 'craig and tail') applied to a form of Secondary hills common in Britain, where a bold precipitous front is exposed to the west or north-west, and a sloping declivity towards the east. The phenomenon... is evidently the result of the currents of the Drift epoch.

Mill, *Dict*. Crag-and-tail. A hill or crag showing an abrupt and often precipitous face on one side, and sloping away probably to the low ground in the opposite direction. In a glacial-formed crag-and-tail the steep face is referred to as the stoss-side, and the lower the lee-side.

Cotton, 1945. 'These rock drumlins grade into crag-and-tail forms. The latter consist of ice-smoothed knobs of solid rock which give place leeward to tapering streamlined tails of ground moraine instead of terminating in plucked surfaces like typical roches moutonnées.' (p. 343)

Wooldridge and Morgan, 1937. As Cotton, but examples Edinburgh Castle Rock.

Holmes, 1944. 'Highly resistant obstructions, such as old volcanic plugs, that lay in the path of the ice, are responsible for an erosion feature known as *crag and tail*. The crag boldly faces the direction from which the ice came, while the tail (bedrock with or without a covering of boulder clay) is a gentle slope on the sheltered side, where the softer sediments were protected by the obstruction from the full vigours of ice erosion.' (pp. 217–218)

*Comment*. As indicated by Holmes, the 'tail' is not necessarily of moraine. Page is wrong in naming a direction; it depended upon the movement of ice.

**Crater**
*Cf*. Caldera.
O.E.D. 2. A bowl- or funnel-shaped hollow at the summit or on the side of a volcano, from which the eruption takes place; the mouth of a volcano.
4. *Mil*. The excavation or cavity formed by the explosion of a mine; the funnel.

Mill, *Dict*. The bowl-like upper portion of the orifice of a volcano through which lava or other volcanic products are ejected.

*Webster*. 4. *Geol*. a. The depression above or around the orifice of a volcano. The crater may be (1) a funnel-shaped pit, maintained by successive explosions at the top of a built-up cone; (2) the result of a single great explosion, in which the whole top of the volcano has been blown away; or (3) produced by subsidence or collapse consequent upon the withdrawal of underlying lava. b. The flaring or bowl-shaped opening of a geyser. c. A depression formed by the impact of a meteor.

*Adm. Gloss.*, 1953. A bowl-shaped cavity. In particular, at the summit or on the side of a Volcano.

*Comment.* Mill has also 'Craterlet. A small rounded depression or hole in the Earth's surface caused by an earthquake.' An early attempt was made by von Buch, quoted by de la Beche, 1853, to separate 'craters of elevation' from 'craters of eruption'. Though most frequently used of volcanic craters, the word crater is applied to other depression, especially meteorite craters, caused by the fall of large meteorites on the earth's surface (see also 'bays' *q.v.* of Carolina coast)—see Lobeck, 1939—and hollows on the surface of glaciers (Charlesworth, 1957, p. 61). Some writers use 'erosion craters' for eroded domes; for such D. H. K. Amiran has proposed the term makhtesh (*q.v.*).

**Crater basin**
*Webster. Geol.* A depression containing craters.

**Crater lake**
*Mill, Dict.* A body of water collected in the crater of an extinct volcano.

**Craton, cratonic**
*Hills, E. S.*, 1953, *Structural Geology*, 3rd ed. 'Resistant blocks or cratons' (p. 44). 'Cratons are relatively stable blocks that comprise the major portion of continents and perhaps also of the ocean basins.... The larger cratons have cores or nuclei of Pre-Cambrian rocks which form broadly arched plateaux and are therefore known as *shields* ... may be covered in places by little-disturbed younger rocks, thus constituting *tables*. Smaller areas of Pre-Cambrian rocks are termed *blocks*. The deformed rocks of older orogens, buttressed with igneous intrusions are welded to the nuclei and act as part of the cratonic regions for younger orogens. The terms resistant block (or mass), stable block or rigid block are approximately synonymous with craton....' (p. 46)

*Comment.* This term is specially used by Hills; Kober's *kratogen* has the same meaning.

**Creek**
*O.E.D.* 1. A narrow recess or inlet in the coast-line of the sea, or the tidal estuary of a river; an armlet of the sea which runs inland in a comparatively narrow channel and offers facilities for harbouring and unloading smaller ships.
b. A small port or harbour; an inlet within the limits of a haven or port. c. In the Customs administration of Great Britain, an inlet, etc., not of sufficient importance to be a separate Customs station, but included within the jurisdiction of another port station.
d. Applied to any similar opening on the shore of a lake.
2. As part of a river or river-system.
a. An inlet or short arm of a river, such as runs up into the widened mouth of a ditch or small stream, or fills any short ravine or cutting that joins the river (This is merely an occasional extension of sense 1.)
b. In U.S. and British Colonies: a branch of a main river, a tributary river; a rivulet, brook, small stream, or run.
3.a. Applied more widely and loosely to any narrow arm or corner of the sea. Obs.
b. A narrow corner of land running out from the main area; a narrow plain or recess running in between mountains. *cf.* Cove.

*Webster.* 1. A small inlet or bay, narrower and extending farther into the land than a cove; a narrow recess in the shore of the sea, a river, or a lake, 2. Hence: *Chiefly U.S.* a. The estuary of a small river or a brook, emptying on a low coast or into the lower reaches of a wide river, together with the upper course of the small river or brook to its source. b. A stream of moderate size constituting the principal, or some comparable, affluent of an inlet or bay. c. A flowing rivulet or stream of water, normally smaller than a river and larger than a brook or run in the same general locality; a lowland or valley watercourse of medium size. Such creeks may range in length from insignificant runs to rivers. This usage now prevails in the United States throughout the Middle Atlantic and South Atlantic States, the Mississippi Valley, and the Pacific slope; in Alaska, and in Canada, Australia, and certain British colonies. The term *creek* failed of general acceptance in New England, where, as in the British Isles, streams of this order are classed as *rivers*, and smaller streams as *brooks*. A *creek*, in this sense, may ordinarily be distinguished from a river in not being navigable for commercial purposes. d. *Southwestern U.S.*, and *Australia*. A long, shallow stream of intermittent flow; an arroyo. 3. A small port or harbor; an inlet within the limits, or included within the jurisdiction, of a port or haven. *Eng. Dict. Am.* 1. A stream larger than a brook but smaller than a river, orig. a tributary to a larger stream or body of water.

Mill, *Dict.* 1. A narrow arm of the sea. 2. A relatively small river (America and Australia).

*Adm. Gloss.*, 1953. A comparatively narrow inlet, of fresh or salt-water, which is tidal throughout its whole course.

Bencker, 1942. Small narrow bay generally forming a natural port at the mouth of a river at tide-water and serviceable as an anchorage for vessels.

**Creep**

*O.E.D. Suppl.* 7. Geol. A gradual movement of disintegrated rock due to atmospheric changes, water, etc.; the slow displacement of strata or the earth's crust by expansion or contraction, or under compressive strain; more explicitly continental, crust, tangential creep.

Webster. 7. Geol. A gradual movement, usually downhill, of loose rock, soil, or clay, due to alternate freezing and thawing, wetting and drying, or other causes.

Mill, *Dict.* An extremely slow movement of the Earth's surface or of loose material on it. In the Rhymney Valley (Wales) a creep of 6 to 10 feet in fifty years has been noticed.

Rice, 1941. (of soil, or rock). The slow movement of soil or rock waste down the slope from which these materials have been derived by weathering. Creep is primarily due to gravity and is facilitated by the presence of water, alternate wetting and drying, freezing and thawing, growth and decay of roots, and the work of burrowing animals (p. 91). Creeping (England): the settling or natural subsidence of the surface caused by extensive underground workings.

Balch, H. E., 1947, *Caves of Mendip*. Creeps are tiny passages where to carry one's pack is impossible and everything has to be passed from hand to hand (p. 6).

Johnstone, 1923. '... the creep of water along the sea bottom from the temperate regions towards the equatorial zone is slower still' (p. 254).

Salisbury, 1923. 'Since variations in saltiness are being produced all the time, motion due to unequal density is constant. Movements brought about in this way are usually very slow and may be called creep' (p. 495).

*Comment.* Commonly used is soil-creep, the gradual movement downhill of soil.

**Crenulate shoreline** (D. W. Johnson, 1919)

*O.E.D.* Crenulate (from Latin crēna, incision, notch). Zool. and Bot. Having the edge divided into minute rounded teeth; finely notched or scalloped: said of a leaf, a shell, etc. (No geographical example.)

Johnson, 1919. 'The hills and valleys of the land may have been well graded and characterized by smooth, flowing contours, in which case the initial shoreline must be composed of well rounded curves. But early in the youth of the shoreline the curves will be changed to sharply and irregularly crenulate lines by differential wave erosion.... We may call a shoreline of this character a crenulate shoreline.' (p. 278)

Similar usage in Cotton, *Geomorphology of New Zealand*, 1922, p. 404. Wooldridge and Morgan, 1937, p. 349.

**Creole** (French), **Criolle** (Spanish)

*O.E.D.* In the West Indies and other parts of America, Mauritius, etc. *orig.* A person born and naturalized in the country but of European (usually French or Spanish) or of African Negro race; the name having no connotation of colour; later, a descendant of European settlers, born and naturalized and more or less modified by climatic conditions (*Creole White*) or a negro born in the West Indies not in Africa (*Creole Negro*).

Webster. 1. (*usually cap.*) Orig. as used esp. in Spanish America and the southern French colonies, a person of European descent but born and bred in the colony 2. (usually cap.) Hence: a. A person of French or Spanish descent born and raised in a colonial or remote region, esp. an intertropical region, the environment and culture of which have entailed a characteristic adaptation of the national type. b. In the United States, a white person descended from the French or Spanish settlers of Louisiana and the Gulf States, and preserving their characteristic speech and culture. c. The French patois spoken in Louisiana. 3. A negro born in America—more properly, *creole negro*. 4. A person of mixed Creole and negro blood speaking a dialect of French or Spanish; a half-breed. b. In Alaska, a person of mixed Russian and Eskimo or Indian descent. 5. (*usually cap.*) In Haiti, and some other West Indian Islands, the language of the people. It is a degenerate French with admixture of native West African and Carib, and sometimes, Spanish words.

*Dict. Am.* A person of mixed blood, esp. a white person descended from French or Spanish settlers in Louisiana and the Southwest. 1792. J. Pope. *Tour S. & W.* 22. 'He is a Creole of French extraction.' In Alaska, one born of Russian and native Indian parents. 1867.

Mill, *Dict.* Creole (French), criolle (Spanish). Natives in Spanish America of any colour but not of indigenous blood.

*Comment.* The word has come to be used of the peoples in other parts of the world, with an increasing tendency to think of mixed blood being involved.

## Cretaceous

Although this word from the Latin *creta* (meaning chalk) is literally 'of the nature of chalk; chalky' (*O.E.D.*) it is rarely if ever used in that sense. It is reserved as the name of the geological system or period, the youngest of the Mesozoic, which includes the great thickness of chalk in its upper part. When an adjective is required 'chalky' (as in chalky boulder clay) or simply 'chalk' (as in Chalk-marl or Chalk-rock) is used.

**Creu** (French)—see Karst terminology

## Crevasse (1) and (2) (French)

*O.E.D.* (From French, crevasse, in both Switzerland and Louisiana).

1. A fissure or chasm in the ice of a glacier, usually of great depth, and sometimes of great width. (Quotations 1823, 1872.)

b. trans. Any similar deep crack or chasm.

2. U.S. A breach in the bank of a river, canal, etc.; used esp. of a breach in the levée or artificial bank of the lower Mississippi. (See *O.E.D. Suppl.* for quotation, 1812.)

Mill, *Dict.* 1. A fissure in a glacier, either running transversely, due to a more or less rapid fall in the bed of the glacier, or diagonally at the sides (diagonal or side crevasse due to the slower movement of the ice at the sides than in the middle of the glacier). 2. In the United States (a) a breach in a levee or river embankment; (b) a fissure in a plain.

*Ice Gloss.*, 1958. (Sense 1.) A fissure formed in a *glacier*. Transverse crevasses are found where a glacier falls over a step; longitudinal crevasses develop where the ice has been able to spread laterally. Crevasses are often hidden by *snow bridges.*

Thornbury, 1954. (Sense 2.) '*Crevasses* or breaks through natural levees' (p. 170).

## Crevasse filling

Thornbury, 1954. 'One special variety of kame is that commonly designated as a *crevasse filling* ... ridge-like ... resemble eskers ... usually smaller.' (pp. 394–395)

**Crib** (Welsh)

Crest or summit.

**Crib** (New Zealand)—see Bach.

## Croft

*O.E.D.* 1. A piece of enclosed ground, used for tillage or pasture: in most localities a small piece of arable land adjacent to a house (examples from 969 A.D.). Note: 1880, *W. Cornwall Gloss.* Croft, an enclosed common not yet cultivated.

2. A small agricultural holding worked by a peasant tenant; *esp.* that of a Crofter in the Highlands and Islands of Scotland. 1851. 2nd Rep. *Relief of Destitute Highlands,* 1850, 42. 'The crofting system was first introduced to the arable part of the small farms previously held in common being divided among the joint tenants in separate crofts, the pasture remaining in common.'

## Crofter

*O.E.D.* One who rents and cultivates a croft or small holding; *esp.* in the Highlands and Islands of Scotland, one of the joint tenants of a divided farm (who often combines the tillage of a small croft with fishing or other vocation).

## Crofting

*O.E.D.* The practice or system of croft-tenancy.

Stamp, 1948. 'This system is one of ancient hereditary tenure of small patches of cultivated land combined with rights of common grazing on the steep hillsides. It is little more than subsistence farming helped out by part time occupation in fishing, domestic spinning and weaving and in some of the occupations provided by the need for catering for the summer and autumn sporting and holiday trade' (p. 322).

*Comment.* Crofting may be summarized as a system of small hereditary tenant holdings consisting of arable land held with a right of pasturage in common with others.

Because of the economic difficulties of such a system of subsistence farming on poor land and the resulting low standard of living the system has been exhaustively examined and described in several government reports and has been legally defined for purposes of Acts of Parliament. See especially F. Fraser Darling, 1958.

**Crofting counties**
The seven officially recognized 'crofting counties' are Inverness, Ross, Argyll, Orkney, Shetland, Sutherland and Caithness.

**Crossing**
Fisk, 1947. 'The Mississippi channel throughout its length, consists of deep water pools separated by comparatively shallow water areas. This relationship is best developed in bends, where deep "bendways" on alternate sides of the stream are separated by shallower water "crossings"' (p. 9).

*Dict. Am.* a. A place where a river or other body of water is crossed, in full crossing place. 1753. b. A place in a river where steamboats seeking the safest channel cross from one side to the other. 1875.

*Comment.* 'An American term used for the shallower part of a stream channel between the deeper pools of meander bends. Note also the usage of crossing as equivalent to ferry or even bridging point—possibly derived from the frequent association of crossing points at the shallower "inflexions"' (G.T.W.).

**Crotovine**—see Krotovina

**Crumb structure**
L.D.G. An important physical characteristic of good soils where the constituent particles are aggregated into 'crumbs' which permit the percolation of air and water between them.

**Crush breccia**
Himus, 1954. A rock consisting of angular fragments, often cemented, which were produced as the result of crushing and shearing stresses during folding or faulting.

**Crush conglomerate**
Himus, 1954. A rock produced during faulting and folding, much as in *Crush Breccia*, but the fragments in which are rounded.

**Crust**
*O.E.D.* (From Latin, crusta, hard surface, rind, shell, incrustation).
  4.a. The upper or surface layer of the ground. *Obs.*, having passed into b. Geol. The outer portion of the earth; that part of the body of the earth accessible to investigation.

*Webster. Geol.* The exterior portion of the earth, formerly universally supposed to enclose a molten interior. The term is still used to designate the relatively cool outer part of the globe as distinguished from the unknown hotter part within.

Mill, *Dict.* The outer layer of the solid earth; the lithosphere.

Geikie, J., 1898. *Crust of the Earth: the outer portion of the earth which is accessible to geological investigation.* (p. 300)

Bucher, W. H., 1933, *The Deformation of the Earth's Crust*, Princeton: Princeton U.P. 'The "crust" is the outermost shell of the earth, which on the whole possesses sufficient strength to offer resistance to deformation and to transmit long-continued stresses within certain limits.' (This rests on the sub-crustal asthenosphere. W.M.) (This definition is followed by A. L. Du Toit, 1937.) (p. 39)

Holmes, 1944. 'The Lithosphere is the outer solid shell or crust of the earth.... It is usual to regard the crust as a heterogeneous shell, possibly about 30 miles thick, in which the rocks at any given level are not everywhere the same. Beneath the crust, in what may be called the substratum, the material at any given level appears to be practically uniform, at least in those physical properties which can be tested. Some authors use the term "lithosphere" to include both crust and substratum.' (p. 9)

Rice, 1941. 1. Hard external covering of anything. An incrustation.
2. The lithosphere, or solid external portion of the earth....
3. (Salop). Fine grained sandstone. (p. 93)

Umbgrove, 1947. 'The earth's crust, composed of 40 to 80 km. of crystalline rocks. This solid, elastic crust, presumably rests on amorphous foundations, *i.e.* rocks which, as a result of high temperature at this depth, must be above their melting point....' (p. 11)

Daly, 1940. Claims crust is synonymous with lithosphere only if it is proved that the 'layer of compensation is strong because crystalline'. (p. 11). He objects to use of 'crust' as it implies a theory unacceptable to some findings of physical geology—also because it is used in various other meanings (p. 15).

Bucher, 1933. 'Opinion 3. The "crust" is the

outermost shell of the earth, which on the whole possesses sufficient strength to offer resistance to deformation and to transmit long-continued stresses within certain limits' (p. 39).
*Comment.* Note also other uses of crust, *e.g.* the self-explanatory term 'breakable crust' in ice terminology.
**Crust-block**
Mill, *Dict.* A mass of the Earth's crust separated by lines of fracture, and either elevated or tilted relative to the rest of the crust.
*Cf.* block mountain.
**Crustification**
Rice, 1941. The English equivalent of a term suggested by Posepry for those deposits of minerals and ores that are in layers or crusts and that, therefore, have been distinctively deposited from solution (Fay) (p. 98).
**Cryergic**
Not in *O.E.D.* or *Webster* but currently used by physical geographers to denote phenomena due to cold conditions.
**Cryoconite** (A. E. Nordenskjöld, 1870)
Gajda, R. T., *The Canadian Geographer*, **12**, 1958, 35. 'The term "cryoconite" was introduced by A. E. Nordenskjöld, who, during his first expedition to the Greenland Ice Cap, observed numerous water-filled cylindrical holes sunk into the ice. The bottoms of these holes were covered with dust a few millimetres thick. This powdery dust he called "cryoconite" from the Greek *kruos* (ice) and *konis* (dust). Consequently, the forms which resulted from differential melting under cryoconite on the ice were called cryoconite holes, and the process of melting and deepening of the holes was called cryoconite or indirect ablation.'
Cryoconite stripes, cryoconite cover, cryoconite holes, cryoconite ablation etc. *Op. cit.*
**Cryogenics**
Science of extreme cold.
**Cryology**
Seligman, G., 1947, *Jour. Glaciology*, **1**. 'Shortly before the war this new word for the study of glaciology was coined in Central Europe... in America "cryology" is coming into fashion to describe the study of refrigeration... it is to be hoped that "cryology" so far as the scientific study of ice is concerned, will not be heard of again.' (p. 35)

*Comment.* At the International Association of Scientific Hydrology in Zürich, Meinzer referred to four divisions of hydrology—potamology, limnology, hydrology (subterranean water) and cryology (the scientific study of ice and snow) (G.T.W.).
**Cryopedology** (Kirk Bryan, 1946)
Bryan, K., 1946, Cryopedology... *Am. Jour. Sci.*, **244**. 'The science of intensive frost action and permanently frozen ground including studies of the processes and their occurrence and also the engineering devices which may be invented to avoid or overcome difficulties induced by them.' (pp. 639–640)
See also Congelifraction.
**Cryoplanation** (Kirk Bryan, 1946)
Bryan, K., 1946, Cryopedology... *Am. Jour. Sci.*, **244**, Cryoplanation: 'Land reduction by the processes of intensive frost-action, *i.e.* congeliturbation including solifluction and accompanying processes of translation of congelifracts. Includes the work of rivers and streams in transporting materials delivered by the above processes.' (p. 640)
**Cryostatic hypothesis** (A. L. Washburn, 1947)
Mackay, J. R., *The Canadian Geographer*, **3**, 1953: 'In the cryostatic hypothesis, debris is thought to be squeezed between downward freezing ground and the permafrost table. Upward injection would take place in areas of easiest relief' (p. 36). See summary of his earlier work by A. L. Washburn, 1956, Classification of patterned ground and review of suggested origins, *Bull. Geol. Soc. Amer.*, **67**, 622–637.
**Cryoturbation**
Swayne, 1956. Soil disturbance, erosion and deposition due to extreme cold.
*Comment.* Kirk Bryan's term covers all phases of weathering under very cold conditions.
**Cryptphyte**
Raunkiaer, 1934. 'A plant... characterized by having its buds concealed in the ground or at the bottom of the water' (p. 17). It is convenient to divide cryptophytes into geophytes, helophytes and hydrophytes.
**Crystalline**
*O.E.D.* of rocks: composed of crystals or crystalline particles; opposed to amorphous.
*Comment.* Hence applied to (a) igneous rocks which solidify on the cooling of a molten magma and, unless the cooling is so rapid

that they are glassy, develop a crystalline structure: the slower the cooling the more completely crystalline (*e.g.* granite) and (b) to metamorphic rocks of varied origin which have become crystalline, *e.g.* gneiss. Some definitions restrict the term crystalline to (a) but actually crystalline rock is a very convenient term because it enables ancient metamorphosed sediments and also igneous rocks which occur intimately associated in the pre-Cambrian to be included together independently of origin.

**Crystocrene**
Muller, 1947. Surface masses of ice formed each winter by the overflow of springs. (G.T.W.)

**Crystosphene**
*Webster.* A buried sheet of ice under the tundra of northern America, formed by the freezing of spring water which rises from the rock beneath alluvial deposits, or under swamps, and spreads laterally at the zone of freezing.
Muller, 1947. Mass or sheet of ice developed by a wedging growth between beds of other material. (G.T.W.)

**Cuesta** (Spanish)
*Cf.* scarp, escarpment.
*Webster.* A sloping plain, esp. one with the upper end at the crest of a cliff; a hill or ridge with one face steep and the opposite sloping. *Southwestern U.S.*
Mill, *Dict.* A land form whose sides slope in opposite directions from a crest of relatively uniform level, one slope being much steeper than the other. The term 'scarped ridge' has also been used for this form. See Côte.
Knox, 1904. Cuesta (Sp.), rising ground, eminence.
Cuesta (U.S.A.), an ascending slope, a tilted plain or Mesa (*q.v.*) top.
Davis, W. M. and Snyder, M. H., 1898, *Physical Geography.* 'An upland of this kind may be called a *cuesta*, following a name of Spanish origin used in New Mexico for low ridges of steep descent on one side and gentle slope on the other.' (p. 133, *fn.*) See also Davis, 1899, The Drainage of Cuestas, *Proc. Geol. Assoc.*, **16**, 75–93.
Lobeck, 1939. 'The ridges or uplands between the lowlands are termed cuestas. Each cuesta has a steep *inface* and a gentle *back slope*, down the dip of the beds.' (p. 451)
Cotton, 1941. 'Homoclinal ridges grade into *cuestas*, which are developed on escarpment-forming strata of very gentle inclination ... present greater contrast between escarpment and dip slopes than is found in typical homoclinal ridges.' (p. 94)
Committee, List 2. See Scarp (= scarped ridge).
*Comment.* This common Spanish word has been given a special meaning in English. It refers to a whole land form and thus avoids the ambiguity of 'escarpment' which may refer only to the steep face (scarp face or scarp slope) of a scarped ridge, the other slope being the dip slope. Not recorded in *O.E.D.*, it is now very widely used.

**Culm, Culm Measures**
*O.E.D.* Culm: coal-dust, small or refuse coal, slack; applied esp. to the slack of anthracite from the Welsh collieries; by extension to varieties of anthracite and hence to its principal current use. *Culm* or *Culm Measures*: a series of shales, sandstones, etc. containing, in places, thin beds of impure anthracite which represent the Carboniferous series in North Devon; also to strata supposed to be analogues of these elsewhere.

**Cultigen**
*Webster.* A plant race or form, as the cabbage, which has arisen or is known only in cultivation; also, a cultural variety.
Sauer, C. O., 1952, *Agricultural Origins and Dispersals*, New York: Amer. Geog. Soc. 'such man-made plants, or cultigens' (p. 25).

**Cultural geography**
Equated by Schmidt with human geography (*fide* Hartshorne, 1939, p. 90).
Geography which is primarily concerned with human culture; for discussion of its scope see Hartshorne, 1939. Also discusses 'culturogeographic regions'.

**Cultural landscape**—see Landscape

**Culture World**
Russell, R. J. and Kniffen, F. B., 1951 *Culture Worlds.* 'The type of change varies in accordance with the culture group concerned. Each culture world, realm, or region tends to assume its own landscape peculiaries.... The most logical introduction to regional geography is through the medium of these culture worlds.'

**Cupola**
Himus, 1954. A dome or boss-like protrusion from a batholith in the roof of which it forms a conspicuous irregularity.

**Curragh** (Irish and I. of Man)
Wet land, ill-drained land.

**Current**
*O.E.D.* 1. That which runs or flows, a stream; spec. a portion of a body of water, or of air, etc. moving in a definite direction.
2. The action or condition of flowing; flow, flux (of a river, etc.); usually in reference to its force or velocity.
Mill, *Dict.* Defined movement of water in the ocean in a definite direction with a perceptible velocity.
*Met. Gloss.*, 1944. Current, Ocean. A general movement of a permanent or semi-permanent nature of the surface water of the ocean. The term must not be used to denote tidal streams, which change direction and velocity hour by hour. A drift current is a drift of the surface water, which is dragged along by a wind blowing over it.
*Comment.* Note also vertical currents in the ocean and convection currents (*q.v.*) within the earth.

**Current bedding**—see False bedding

**Cusec**
An abbreviation of cubic feet per second, used commonly as a measure of river-flow.

**Cuspate foreland**
Johnson, 1919. 'A group of forms in which the shoreline is systematically prograded by wave and current action, and an appreciable area of more or less continuous dry land added to that previously existing. .... We will follow Gulliver's suggestive terminology and speak of these features as forelands. They may have a variety of forms, but where most typically developed are more or less triangular in shape with the apex of the triangle pointing out into the water; they are then called cuspate forelands.' (p. 319)
Cotton, 1922. 'Where conflicting currents meet—eddies, generally, of ocean or tidal currents—progradation commonly takes place, and a projecting foreland is built out, either as a local incident in an otherwise retrograded coast or as a salient of a continuous foreland. Though sometimes rounded at the end, such a salient is bounded typically by the two curves followed by the littoral currents, tangential to the general line of the coast some distance away from the projection at either side, and sweeping out to intersect each other in a sharp cusp at its extremity. Such a cuspate foreland may form part of a barrier or may spring from the main shoreline.' (p. 443) Similarly in ed. 1945.
Wooldridge and Morgan, 1937. 'Progradation of such cuspate bars converts them into cuspate forelands. . . .' (p. 337)
Shepard, F. P., 1952, Revised Nomenclature for Depositional Coastal Features, *Bull. Ass. Am. Petr. Geol.*, **36**, pp. 1902–1912. 'The largest of the cusps occur along the open coasts as capes, measuring in some places miles across their apices and extending miles seaward from the rest of the coast. These capes have been called "cuspate forelands" (Johnson, 1919, p. 324) and this name seems appropriate for such features as Cape Hatteras, Cape Fear and Cape Canaveral. The cuspate forelands contain a series of ridges with troughs which may be above water as in Cape Canaveral or submerged as in Cape Hatteras.' (p. 1911)

**Cuspate spit**
Shepard, F. P., 1952, Revised Nomenclature for Coastal Features, *Bull. Ass. Am. Petr. Geol.*, **36**, pp. 1902–1912. '... prominent points or horns found extending into the bays and lagoons inside many barriers. These may be referred to as "cuspate spits". Like giant cusps they have shoals extending out from the cusp point although these shoals occur independently. . . . The cuspate spits have much more prominent points than giant cusps and distances between points are ordinarily a mile or more. These spits are more common on the barrier side of a lagoon, but are also found on the mainland side and in some places two points are found *vis-à-vis* with or without a connecting shoal.' (p. 1911)

**Cutan** (R. Brewer, 1960)
Brewer, R., 1960, Cutans: their definition, recognition, and interpretation. *Jour. Soil Science*, **11**, No. 2, 280–292. 'The term cutans is proposed for a broad group of pedological features, including so-called 'clay-skins', associated with the surfaces of the skeleton grains, peds, and various kinds of voids within soil materials. The chief differentiating characteristics—kind of surface ... mineralogical nature of the material ... fabric of the cutans themselves'. (p. 280)

**Cutcha**—see Kucha
**Cut-off, Cut-off Lake**—see Ox-bow
See also Chute (for Fisk's chute cut-off).
**Cuvette** (French)
Wills, L. J., 1929, *The Physiographical Evolution of Britain*, London: Arnold.

'Cuvette is a convenient term for a basin in which sedimentation is going on (German, Sammelmulde), as distinct from a tectonic basin due to folding of pre-existing rocks, and not necessarily basin shaped as far as the present structure is concerned, e.g. the Anglo-Parisian cuvette was the region in which the Lower Cainozoic rocks of Britain and North France accumulated It is now folded into a number of basins—Paris, Hampshire and London Basins. (p. 79 footnote)

**CVP-Index** (S. S. Paterson, 1956)

Paterson, S. S., 1956, *The Forest Area of the World and its Potential Productivity*, Goteborg. '... the "plant-growth ability" of a climate... because of the intimate connection between climate, vegetation and productivity, the index has been called the CVP-index after the initials of the three words' (p. 68).

See review in *Geog. Jour.*, **124**, 1958, 405. 'An index of climate, vegetation and productivity (the CVP Index) is constructed from the temperature of the warmest month, annual temperature range, precipitation, evapo-transpiration, length of the growing season... and... used to draw "climato-isophytes".'

**Cwm** (Welsh)

*Lit.* valley, combe; has been used in geographical literature in the specialized sense of a glacial valley identical with cirque (French) or corrie (Scottish).

Mill, *Dict.* Cwm. Used for cirques and corries and also for narrow deep valleys in a mountain region (Wales).

North, F. J., Campbell, B. and Scott, R., 1949, *Snowdonia*, London: Collins. 'A cwm (using the word in its geographical sense) differs from a normal valley-head in having much steeper sides and a rounded or roughly quadrangular outline... they are valley-heads which once harboured small glaciers.' (p. 98)

Note: *cwms* used as plural, although Welsh form is *cymau* or *cymoedd*. Cwm is equated with corrie and cirque, 'although cwm is also loosely used for valleys in general'. (p. 97)

See cirque, corrie.

**Cycle, cyclic**

*O.E.D.* A recurrent period of a definite number of years adapted for purposes of chronology. A period in which a certain round of events or phenomena is completed, occurring in the same order in succeeding periods of the same length.

*Comment.* Many natural phenomena recur in nature in cycles, though not usually in periods of the 'same length'. The cycle of erosion or denudation and cycle of sedimentation are separately listed, also Brückner's cycle. A cycle of igneous activity is also very familiar to geologists, also cycle of weathering, cycle of evolution etc.

**Cycle of erosion, denudation etc.**

Davis, 1909 (paper published in 1902). 'The period of time during which an uplifted land mass undergoes its transformations by the processes of land sculpture, ending in a low, featureless plain, has been called a geographical cycle, or, as Lawson phrases it, a "geomorphic cycle".' (p. 408) (Hayes is quoted as using 'cycles of gradation', 'gradation period', and 'cycle'. The references are to Hayes, C. W., *U.S. Geol. Survey*, XIX, *Ann. Rep.*, Pt. II, 1899, 1–58 and Lawson, A. C., *Bull. Dept. Geol., Univ. Cal.*, I, 1894, 241–271.) (p. 408) Note that Davis continues with a distinction between denudation, operative at an early stage in the cycle, and degradation, at the later stages, suggesting that Davis would not approve of 'Cycle of Denudation'. (W.M.)

*Webster.* Erosion cycle. *Geol.* The succession of stages through which a newly uplifted land mass must pass before it is worn down to a peneplain or a surface near sea level. In the juvenile stages the surface is sharply cut by canyons; in the mature stage, it may disappear and the topography be characterized by high steep hills and fairly open valleys; and in the old-age stages the land is so worn down that the streams meander sluggishly across a lowland.

Mill, *Dict.* Cycle of Erosion. The time necessary for the development of a base-level throughout a drainage basin.

Cycle of Denudation. The varying phases of a land form from its initiation to its disappearance.

Salisbury, 1907. 'The time necessary for the development of a base-level throughout a drainage basin is a cycle of erosion.' (p. 153)

Cotton, 1922. 'The period occupied by the whole series of changes in relief produced by erosion following uplift of a surface of any form above sea-level is called a cycle

of erosion, or geographical cycle.' (p. 45)
Wooldridge and Morgan, 1937. 'The Cycle of erosion. 'A landscape has a definite life history of gradual changes ... *initial* forms pass through *sequential* forms to an *ultimate* form, we may broadly group the many successive stages into the major stages of *youth, maturity* and *old age*, the possibility of eventual obliteration of relief or planation is the first principle of the cycle of erosion ... landscape is a function of *structure, process* and *stage*. (p. 114)
Holmes, 1944. 'The whole sequence of changes passed through during this long evolution is called a cycle of erosion' (p. 144). (The preceding description is of a newly emergent land being eroded to a peneplain. W.M.)
Hayes, *op. cit.*, 1899 (see Davis, 1909). 'The term "gradation period" is employed for the entire time during which the base-level remains in one position; that is, the interval between two elevations of the earth's surface of sufficient magnitude to produce a marked change in the position of sea level.' (p. 22)
Lobeck, 1939. 'The cycle of erosion, sometimes called the geographical or geomorphic cycle, concerns the larger land masses rather than the streams ... the cycle of erosion refers to the stages through which a land mass passes from the time of its uplift until peneplanation.' (p. 163)
Committee, List 2. 'The hypothetical sequence of changes or stages through which a land surface would pass in its reduction to base level by the processes of erosion. The period of time taken by the process of reduction.'
Rice, 1941. The interval during which a land surface newly uplifted, either as plain or mountain, is worn down to the level of the sea (base level). (Grabau, 1906.)
*Comment.* See Cycle, cycle of sedimentation. Both Mill and Salisbury misuse the term 'base-level' (*q.v.*) in their definitions (A.A.M.).

**Cycle of Sedimentation, Cycle of Deposition**
Stamp, L. D., 1921, On Cycles of Sedimentation in the Eocene, *Geol. Mag.*, **58**. 'A "Cycle of Sedimentation" comprises the deposits of a complete oscillation of the basin, each oscillation including a positive phase of marine invasion and a negative phase of regression.' (p. 109)

**Cyclone**
*Webster*. 4. *Meteorol.* A storm or system of winds, often violent in the tropics and moderately elsewhere, with abundant precipitation and usually a diameter of 50 to 900 miles. It moves with a velocity of 20 to 30 miles an hour and is characterized by winds rotating, often at the rate of 90 to 130 miles an hour, clockwise in the Southern Hemisphere, counterclockwise in the Northern, about a calm center of low atmospheric pressure. Called also *hurricane* in the West Indies and *typhoon* or *baguio* in the Philippine Islands and the China Sea. The *extratropical cyclone* is a low-pressure storm of the middle latitudes, frequently 1,500 miles in diameter, advancing at the rate of 20 to 30 miles an hour, the wind circulating spirally inward at a velocity of 9 to 15 miles an hour, clockwise in the Southern Hemisphere, counterclockwise in the Northern, about a center of low barometric pressure, yet not attaining the center, owing to the forward movement of the cyclone. Usually the storm is accompanied by extensive clouds and precipitation. It commonly contains a wind-shift line that extends equatorward for hundreds of miles from the center of low pressure and divides the warmer humid winds of the forward portion from the cooler dry winds of the rear. The term *cyclone* should not be applied to the tornado, waterspout, or twister. *Cf.* Anticyclone.
Hare, 1953. 'Cyclone—see depression: the large, travelling, low-pressure systems called depressions are largely confined to middle and high latitudes ...' (p. 70). 'Tropical revolving storms ... in the south Pacific "cyclones".' (p. 110)
*Comment.* Temperate cyclones and tropical cyclones are of very different character and origin; the modern tendency is to restrict the term to the violent tropical storms and to refer to the temperate cyclones as depressions, lows, or cyclonic disturbances. See Miller, 1953, p. 26 *et seq.*

**Cyclothem** (Wanless and Waller)
*Webster. Geol.* A stratigraphic unit consisting of a series of beds deposited during a single sedimentary cycle.
Wills, L. J., 1956, *Concealed Coalfields*, London: Blackie. 'This term for a unit of cyclic deposition appears better than

"sedimentary cycle", "cycle", "rhythm" etc. For a discussion of them, see Edwards and Stubblefield, 1948, Wills, 1948 . . .' (p. 17). References are to *Quart. Jour. Geol. Soc.*, **103**, 219 and Wills' *Palaeogeography of the Midlands.*

# D

**Dahabiya** (Egypt: Arabic) also dahabeeyah, dahabiah and other spellings
*Webster*. Dahabeah. A long, light-draft houseboat, lateen-rigged, and now often propelled wholly or partly by engines, used on the Nile.
The large sailing-boat used on the Nile.

**Dahanah** (Arabia: Arabic)
Stamp, L. D., 1959, *Asia*, London: Methuen, 10th ed. (of the deserts of Arabia) 'dahanah is comparatively hard gravel plain covered at intervals with sand belts of varying width' (p. 145).

**Dala**
Mill, *Dict*. A basin depression in a sandy desert much smaller than a shott and possessing pasturage (N. Africa).

**Dak-bungalow** (properly dāk, formerly written dawk)
An Anglo-Indian word meaning a house for the use of travellers on a dāk route. The word dāk meant post or transport by relays of men or horses stationed at intervals. Under the British administration dāk bungalows were built at approximately 15-mile intervals along main travelling routes. The expression dāk bungalow has survived into the motor age.

**Dale** (mainly northern England)
*O.E.D.* A valley. In the northern counties of England the usual name of a river valley between its enclosing ranges of hills or high land. In geographical names, *e.g* Clydesdale, Annandale, Borrowdale, Dovedale it extends from Lanarkshire to Derby. shire and even farther south, but as an to-appellative it is more or less confined the district from Cumberland to Yorkshire. In literary English, chiefly poetical and in the phrases, hill and dale, dale and down.
Note also dale-head, dale-end etc. See also Hill-and-dale.

**Dalesman** (northern England)
*O.E.D.* A native or inhabitant of a dale, especially of the dales of Cumberland, Westmorland, Yorkshire and adjacent northern counties of England.

**Dallol** (Nigeria)
A flat-bottomed valley of width up to several miles, with steep sides 50–200 feet high—a form similar to that of a wadi. The name is applied to features on the S.E. margin of the Sahara (J.C.P.).

**Dam**—see Barrage, Beaver dam

**Dāman, dāmān** (Indo-Pakistan: Urdu-Hindi)
Talus slope.
Spate, 1954. 'the daman-i-koh—the skirts of the hills—the sharp break of slope at the detrital pediment' (pp. 149–150).

**Dambo** (Bantu)
River flood-plain in Northern Rhodesia, swampy during the wet season, but dry for the greater part of the year, with a vegetation of long grass (P.S.).

**Dans** (Afrikaans)
Mill, *Dict*. A broad shallow valley, *e.g*. Leeuwens Dans (S. Africa).

**Daryākhurdi** (Pakistan: Sindhi)
The alluvial lands on either side of the lower Indus, within the protecting bunds and flooded annually. The word is by origin Persian, 'darya-khurdi' meaning eaten by water.

**Dasht** (Iran)
Fisher, W. B., 1950, *The Middle East*, London: Methuen. 'firm sandy or stony stretches (*dasht*) in which sand dunes (*rig*) may occur' (p. 267).
A flat floor of a valley, normally dry, but with some water in spring. A narrower valley with periodic water is called a *darreh* (*cf*. wadi and nālā). (Information from Professor I. Shams, Teheran University.)

**Dasymetric** (A. G. Ogilvie, 1952)
Ogilvie, A. G. (1952). Report of the Commission for the Study of Population Problems, *International Geographical Congress*, Washington, 1952, p. 7. 'Dasymetric-Choropleth maps. It appears necessary to use this rather clumsy expression to designate maps in which the aim is to make density measurable while also showing the limits of areas of given density more realistically ... B. Semenov-Tian-Shansky seems to have invented the term.' [In Russian as Dazimetricheskaya and French

as dasymétrique in 1923.] Not in *O.E.D.*; dasymeter has quite a different meaning. See also *Geog. Rev.*, **16**, 1926, 341–3.

*Comment.* When densities, *e.g.* of population, are shown by administrative units the sharp contrast between one area and another does not correspond with reality which is usually a gradation. Thus dasymetric plotting involves an areal redistribution of densities in light of geographical knowledge.

### Dasymetric (mapping) technique

International Cartographic Association, Commission II, 1973, *Multilingual Dictionary of Technical Terms in Cartography*, Wiesbaden: Franz Steiner Verlag G.M.B.H. 'The technique of plotting densities after they have been areally redistributed in the light of geographic knowledge.' (p. 122)

### Date Line—see International Date Line

### Datum, Datum level, Ordnance Datum

*O.E.D.* datum (*pl.* data) Latin; A thing given or granted; something known or assumed as fact, and made the basis of reasoning or calculation.

*Webster.* 2. *datum point*, a point assumed or used as a basis of reckoning, adjustment, or the like—*datum line*, a horizontal line from which heights and depths are reckoned, as in a railroad plan.—*datum plane* or *level*, a plane or level assumed or used as a base of reckoning. 3. The low-water mark of all tides, assumed as a basis of reckoning but not admitting rigorous scientific determination.

Mill, *Dict.* Datum level. The zero with reference to which the altitudes of land surfaces are determined.

Hence *Ordnance Datum* (O.D.) from which heights on British official (Ordnance Survey) maps are calculated is mean sea level at Newlyn, Cornwall.

### Day degree

*Webster. Meteorol.* One degree Fahrenheit or Centigrade as specified, above or below 42°F. (temperature at which vegetation commences) for a period of 24 hours, or its equivalent (as 2° for 12 hours, 4° for 6 hours, etc.)

See definition given under accumulated temperature.

### Daylight Saving Time—see under Standard Time

### Dead cave

Muma and Muma, 1944. 'A cave wherein the formations are dry or dead; one in which excavation and deposition have finished.' (G.T.W.)

*Comment.* The contrast is with 'active cave'.

### Dead ground

The area which is invisible from a particular point in the field owing to the form of the surface. (G.R.C.)

### Dead ice

Flint, 1947. 'Here the ice is stagnant (dead) and devoid of further flow.' (p. 30)

Wooldridge and Morgan, 1937. 'In Spitzbergen and elsewhere, the ice has sometimes advanced over the low ground but there has been no correspondingly rapid retreat. It has simply been left as "dead ice", decaying by melting very slowly and without the production of large quantities of water.' (p. 381)

No reference in *Adm. Gloss.*, 1953; Wright, 1914; Bencker, Hyd. Reg., 1931; Geikie, J., 1898; Davis, 1909; Salisbury, 1907; Cotton, 1922, 1942; *U.S. Ice Terms*, 1952; *Ice Gloss.*, 1956, 1958.

### Dead reckoning

*O.E.D.* The estimation of a ship's position from the distance run by the log and the courses steered by the compass, with corrections for current, leeway etc., but without astronomical observations.

### Dead valley

According to Moore, 1949, the term used by French geographers for a dry valley, and hence sometimes translated accordingly.

### Death rate—see Birth rate

### Debacle, débâcle (French)

*O.E.D.* (French, from débâcler, to unbar, remove a bar)

1. A breaking up of ice in a river; in Geol. a sudden deluge or violent rush of water, which breaks down opposing barriers, and carries before it blocks of stone and other debris.

*Comment.* This word has been so widely used in a figurative sense that its use as a scientific term is rare.

### Débris (French: often anglicized to debris)

*O.E.D.* In geology applied to any accumulation of loose material arising from the waste of rocks; also to drifted accumulation of vegetable or animal matter (Page).

*Comment.* This common word is now more frequently used than detritus (*q.v.*).

**Debris-cone**
*O.E.D. Suppl.* A cone formed by the accumulation of volcanic ejecta and debris (Dana, 1892).

**Debris glacier**
*Webster. Geol.* A glacier composed of ice that has fallen in fragments from a larger and higher glacier.

**Deciduous**
*O.E.D.* 1. Falling down or off, *obs.*
2b. *Bot.* of a tree or shrub: that sheds its leaves every year; opposed to evergreen.
*Comment.* Though certain conifers, notably larch, are deciduous, the word is commonly used of broad-leaved trees, especially those of mid-latitudes or the temperate zones, which shed their leaves in the 'fall' (so called because of the leaf fall) or autumn and which consequently have a winter resting period. In the monsoon forests of India and the Far East the trees are also deciduous but the hot dry season is the resting season and the trees are protected against excessive transpiration.

**Decke** (German)—see Nappe

**Deckenschotter** (German; *pl.* Same)
Sheet gravel, notably in the northern Alpine Foreland corresponding to Günz and Mindel glaciations (K.A.S.).

**Declination**
*O.E.D. Astron.* The angular distance of a heavenly body (north or south) from the celestial equator, measured on a meridian passing through the body: corresponding to terrestrial *latitude.* Formerly also the angular distance from the ecliptic.
Of the magnetic needle: the deviation from the true north and south line, *esp.* the angular measure of this deviation, also called *variation.*
*Comment.* Magnetic declination, due to the irregularities of the earth's magnetic field, must be distinguished from magnetic deviation (*q.v.*). (E.G.R.T.)

**Deep**
*O.E.D.* 2. The deep part of the sea, or of a lake or river (opposed to shallow); deep water, a deep place.
4. A deep place in the earth, etc.; a deep pit, cavity, valley; an abyss; a depression in a surface.
Mill, *Dict.* A depression in the sea-floor of limited extent and great depth. Also used popularly of slight depressions on fishing-grounds, *e.g.* Lynn deeps.
Bencker, 1942. 'Area of relatively small dimensions and the deepest zone in a submarine trough or basin.'
Sverdrup *et al.*, 1946, *The Oceans.* 'The deeps by definition exceed 6,000 m.' (p. 20)

**Deer forest**—see Forest

**Deferred junction**
Wooldridge and Morgan, 1937. 'The common phenomenon of "deferred tributary junctions" on flood-plains. A tributary course is prolonged downstream parallel with the main river, its ultimate confluence occurring on the convex side of a major meander' (p. 173).

**Defile**
*O.E.D.* (Formerly defilé (French 17th century) with the military meaning given by Mill.)
Mill, *Dict.* A narrow pass or gorge between mountains; (literally) a road traversable only by files of men.

**Deflation** (geomorphology)
*O.E.D. Suppl.* 2. *Geol.* The removal of solid particles by the wind, leaving the rock exposed to the weather.
1898. J. Geikie, *Earth Sculpture,* 20. The transporting action of the wind, or 'deflation' as it is termed, goes on without ceasing. 1910. Lake and Rastall, *Text-bk. Geol.* 73. Erosion by wind divides itself naturally into two parts—removal of material or deflation, which of course comes under the heading of transport, and actual corrosion or wearing away of the rocks by the dynamical effect of moving sand.
Mill, *Dict.* Aeolian transport of material (Walther).
Wooldridge and Morgan, 1937. 'The term "deflation" (Latin *deflare,* to blow away) was applied by Richthofen to the process of winnowing the finer material from the desert floor.' (p. 298) Similarly Cotton, *Geomorphology,* 1945; Lobeck, *Geomorphology,* 1939, p. 378.
*Comment.* This much used word appears in geographical literature both as a specialized physical term and in economic geography borrowed from the economists. Reference should be made to works on economics for the special meanings relative to currency.

**Deforest, Deforestation**
*O.E.D.* 1. Law. To reduce from the legal position of forest to that of ordinary land.
2. gen. To clear or strip of forests or trees.

*Comment.* See Afforestation. The legal sense is used in works on historical geography.

**Deformation**
Webster. 4. *Geol.* a. The process whereby rocks are folded, faulted, sheared, compressed, or the like by earth stresses, as in the growth of mountain ranges. b. The result of the process.

**Deformational eustatism**—see eustatism

**Dega** (Ethiopia)
Fitzgerald, W., *Africa*, 7th ed., 1950. 'On the Ethiopian massif, where contrasts of altitude are marked, three zones are distinguished by the inhabitants (a) the *kolla*, extending up to approximately 5,500 feet, (b) the *woina-dega*, between the upward limit of the *kolla* and 8,000 feet, (c) above 8,000 feet the *dega* zone' (pp. 454–455).
Stamp, L. D., *Africa*, 1953. 'From the point of view of agricultural production the Ethiopian highlands may be divided into three zones. The lowest, called *Kolla*, comprising the lower slopes of the plateau itself and the deeper valleys up to 5,000 or 6,000 feet is forested . . . frost unknown. . . . Above is the *Voina Dega* or wine highlands, extending up to 8,000 or 9,000 feet. . . . Above rises the *Dega* proper, extending to 14,000 feet. The degas are actually equivalent to Alpine pastures of grassland with bushes' (pp. 357–358).
*Comment.* Since V has only recently been introduced, the spelling *Woina* is preferable (L.D.S.).

**Deglaciation**
L.D.G. The withdrawal of an ice sheet from an area. Some authorities restrict the term to the past (*e.g.* after the Great Ice Age) using **deglacierization** when referring to the present. See Glaciation.

**Degradation, degrade**
*O.E.D.* 6. *Geol.* The disintegration and wearing down of the surface of rocks, cliffs, strata, etc., by atmospheric and aqueous action.
1799. Kirwan, *Geol. Ess.*, 327. Those of siliceous shistus are most subject to this degradation and decomposition.
1802. Playfair.
Mill, *Dict.* The wasting or wearing down of the land by epigenetic agents; denudation or erosion. See Gradation.
Oldham, 1879. 'Degradation—a gradual wearing away, step by step. . . . The degrading agents are (1) atmospheric, (2) fluviatile, (3) marine.

Geikie, J., 1898. *Gloss.* 'The wasting or wearing down of the land by epigene agents.'
Davis, 1909, contrasts with denudation: '"Degradation", on the other hand, is more appropriately associated with those leisurely processes, characteristic of the later stages of the cycle, in which a graded slope is reduced to fainter and fainter declivity, although maintaining its graded condition all the while.' (p. 408)
Cotton, 1922 (also ed. of 1945). 'When, owing to excess of transporting-power over waste-supply, a stream cuts downward to establish or maintain grade, it is said to degrade; and the process is termed degradation.' (p. 61)
Wooldridge and Morgan, 1937. Similar to Cotton.
Rice, 1941. The general lowering of the surface of the land by erosive processes, especially by the removal of material by erosion and the transportation by flowing water.
*Degrade*—to wear down to the condition of grade, as a river.
Committee, List 2. 'The process of lowering a surface by erosion and the removal of rock waste.'
*Comment.* A word of general application during the 19th c. as defined by the Committee. See Powell, 1875; Gilbert, 1877; Lyell, 1832; also Campbell, *A.A.A.G.* 8, 1928, 35. The use by Davis and Cotton is a modern specialized one.

**Degradation** (Soil science), **degraded**
Jacks, 1954. 'Change in soil due to increased leaching'. *e.g.* Degraded chernozem 'dark well-drained soil of grassland-forest transition, with a grayish $A_2$ horizon.'

**Degree**
L.D.G. One degree is 1/360th part of a circle; a right angle is 90 degrees (90°). Used in latitude and longitude each degree is divided into 60 minutes (') and each minute into 60 seconds (''). A degree of latitude is 60 geographical miles, roughly 111 km (69 statute miles). See also Temperature.

**Dejection, cone of**—see cone of dejection

**Dell**
*O.E.D.* A deep natural hollow or vale of no great extent, the sides usually clothed with trees or foliage.
*Comment.* A literary word best avoided in scientific writing.

### Delta
*O.E.D. Hist.* (*The Delta*). The tract of alluvial land enclosed and traversed by the diverging mouths of the Nile; so called from the triangular figure of the tract enclosed between the two main branches and the coastline. b. *Geog.* The more or less triangular tract of alluvial land formed a the mouth of a river.

Mill, *Dict.* A more or less triangular terminal flood-plain of a river extended beyond the general trend of the coast, across which the river usually but not always reaches the sea by two or more distributaries.

Strickland, C., 1940, *Deltaic Formation*, Calcutta: Longmans. Distinguishes 'delta' from 'paradelta' (*q.v.*).

Thornbury, 1954. 'Not all rivers build deltas, nor are all deltas shaped like the Greek letter from which Herodotus in the fifth century B.C. coined the name for the deltaic plain at the mouth of the Nile ... at least four forms ... true delta form; arcuate or fan-shaped (Rhine); digitate form (Mississippi) and the estuarine form (Susquehanna)' (p. 172).

### Delta structure
Thornbury, 1954. 'What has come to be called delta structure is produced by the three sets of beds often observable in a delta. *Bottom-set beds* consist of the finer materials carried farthest seaward, *Fore-set beds* are somewhat coarser and they represent the advancing front of the delta and the greater part of its bulk. *Top-set beds* lie above the fore-set beds and are in reality a continuation of the alluvial plain of which the delta is the terminal portion.' (p. 172)

### Deltalogy, deltology
The study of deltas and delta formation.
*Comment.* Though not in standard works, the meaning of the word is obvious. It is used in *The Professional Geographer*, 7, 1955, p. 15.

### Demb lands (Indo-Pakistan: Kashmiri)
Spate, 1954. 'new land on the shallow lake-margins [made] by planting willows in the water and filling up the compartments so formed with boat-loads of lake-mud and weeds.' (p. 376)

### Demersal
*O.E.D. Supp.* Of fishes, living near the bottom [of the sea].

### Demesne
Royal Commission on Common Land, 1955–1958. *Report*. 'That part of a manor which the lord did not grant out but normally retained for his own occupation and use or that of his servants, as distinct from the manorial land farmed by the villagers.' (p. 273)

### Demoiselle (French)
Lobeck, 1939. Earth pillar capped by a large boulder or fragment, weathered from volcanic breccia, glacial till, etc. Earth pillar. (p. 81)
See also Hoodoo.

### Demography
The study of human population.

### Demopleth—see Choropleth

### Den (Scotland, local)
*O.E.D.* A deep hollow between hills; a dingle—especially if wooded.
*Comment.* Apparently used only in the Lowlands, particularly around Dundee and Glasgow. *Cf.* English dean or dene. (C.J.R.) The Old English denn (a pasture, especially a swine-pasture) shortened to den is very common in place names in the weald of Kent and Sussex.

### Dendritic drainage plan
Holmes, 1944. 'Where the rocks have no conspicuous grain and offer nearly uniform resistence to erosion, the headward growth of a tributary is governed primarily by the initial regional slope, with modifications controlled by haphazard irregularities of surface and structure ... such streams are said to be *insequent*. ... As each insequent develops its own valley, it receives in turn a second generation of tributaries. The branching drainage pattern so established is tree-like in plan and is described as *dendritic*. (pp. 173–174)
Wooldridge and Morgan, 1937, p. 189. Similar.

### Dendrochronology (Bryant Bannister, 1929)
Bannister, B., 1963, *Dendrochronology* in *Science and Archaeology*, Ch. 17, New York. 'The term dendrochronology refers both to the method of employing tree-rings as a measurement of time, wherein the principal application is to archaeology, and to the process of inferring past environmental conditions that existed when the rings were being formed, mainly applicable to climatology.' (p. 161)

### Dene, dean
*O.E.D.* 1. Dean, dene: A vale (*e.g. Taunton Dean*); now, usually, the deep, narrow, and wooded vale of a rivulet.
2. Dene: A bare sandy tract by the sea; a

low sand-hill; as in the *Denes* north and south of Yarmouth, *Deneside* there, the *Den* at Exmouth, Teignmouth, etc.

**Dene-hole, Dane-hole**

*O.E.D.* The name applied to a class of ancient excavations, found chiefly in Essex and Kent in England, and in the Valley of the Somme in France, consisting of a narrow cylindrical shaft sunk through the superincumbent strata to the chalk, often at a depth of 60 or 80 feet, and there widening out horizontally into one or more chambers. Their age and purpose have been the theme of much discussion.

*Comment.* Popularly supposed for long to be either hiding places for plunder used by the Danish invaders, or alternatively hiding places from the Danes, they are more likely to have been mines for flints from the chalk. A series of papers in the *Trans. Croydon Nat. Hist. Soc.* should be consulted.

**Denn**

Not in *O.E.D.*

Ekwall, E., 1940, *The Concise Oxford Dictionary of Place Names*, Oxford. 'a pasture, especially a swine-pasture' (p. XV).

Stamp, 1948. 'The upland settlements had rights of feeding swine in sections or "denns" of the great Wealden oak forests, hence such names as Tenterden and Biddenden' (p. 46).

**Density**

*Webster.* Quality or state of being dense; hence the quantity of anything per unit of volume or area; as *density* of population.

*Density of population.* The average number of persons per unit of area, usually the square mile in English-speaking countries.

*Comment.* Density=specific gravity hence density current, density flow etc.

**Denudation**

*O.E.D.* (L. denudare, to make naked, lay bare)

2. *Geol.* The laying bare of an underlying rock or formation through the wearing away or *erosion* of that which lies above it, by the action of water, ice, or other natural agency.

Mill, *Dict.* The uncovering of deeper rocks by any agency, and, even more generally the wearing down of the land.

Similarly Cotton, 1945, p. 10 and Strahler, 1951, p. 111.

Davis, 1909, distinguishes between denudation and degradation: '"Denudation" might be used advisedly as the name of those active processes, chiefly operative in the youth and maturity of a cycle, by which rock structures are laid bare literally denuded, because their waste is removed as fast as it is formed.' (p. 408)

Holmes, 1928. Quotes Poullet-Scrope, 1825. The sum of the processes that result in the wearing down of the surface of the earth. The term is wider in its scope than erosion, the restriction proposed by Lyell (limiting it to the action of running water) not having been generally adopted.

Campbell, M. R., 1928, Geographic Terminology, *A.A.A.G.*, **18**. In the revised edition of Lyell, 1833, denudation was substituted for degradation as a general term for weathering and removal. This usage is attacked and the following is quoted from Jukes-Brown, *Handbook of Physical Geology*, 1884: 'To denude is to uncover or lay bare a surface, and though such denudation necessarily involves the removal of matter, it is not this matter, but the underlying surface which is denuded; consequently, it is incorrect to speak of denudation in the sense of wearing away and removing material without reference to the uncovering of any particular stratum.' (p. 36)

*Comment.* The term is commonly used widely and loosely (Mill, Holmes) and not with the limitation suggested by Campbell.

In the past the term has had an almost catastrophic meaning—see Farey, 1811, *Present State of the Agriculture of Derbyshire*, where he often refers to the 'Great Denudation'. (G.T.W.)

**Denudation chronology**

The term now generally used by geomorphologists to describe studies in the sequence of events leading to the evolution of an existing physical landscape.

See Sparks, B. W. The Denudation Chronology of the Dip-Slope of the South Downs, *Proc. Geol. Assoc.*, **60**, 165–215 (1949).

**Denudation mountain**—see circumdenudation

**Departure**

*O.E.D. Navigation* a. The distance (reckoned in nautical miles) by which a ship in sailing departs or moves east or west from a given meridian; change of longitude.

b. The bearing of an object on the coast, taken at the commencement of a voyage, from which the dead reckoning begins.

Admiralty *Navigation Manual*. 'Departure

**Depasture**

*O.E.D.* 1. Of cattle : To consume the produce of (land) by grazing upon it; to use for pasturage. 2. To graze. 3. To put (cattle) to graze. 4. Of land : To furnish pasturage to (cattle).

**Deposit**

*O.E.D.* In *Geol.* any mass of material deposited by aqueous agency or precipitated from solution by chemical action.

*Comment.* This definition is too narrow since it would not include the important class of aeolian deposits. See deposition.

**Deposition**

*O.E.D.* The action of depositing, laying down, or placing in a more or less permanent or final position. No specific geological reference.

Holmes, 1944. 'Leading geological processes fall into two contrasted groups. The first group—denudation and deposition—includes the processes which act on the crust at or very near its surface, as a result of the movements and chemical activities of air, water, ice, and living organisms.... The second group—earth movements, igneous activity and metamorphism—are essentially of internal origin.' (p. 30) Holmes then gives the following classification of geological processes:
I. Processes of External Origin.
1. *Denudation* (Weathering, Erosion and Transport)
2. *Deposition*
  (a) of the debris transported mechanically (*e.g.* sand and mud)
  (b) of the materials transported in solution:
     (i) by evaporation and chemical precipitation (*e.g.* rock salt)
     (ii) by the intervention of living organisms (*e.g.* coral limestone)
  (c) of organic matter, largely the remains of vegetation (*e.g.* peat) (p. 31)

Fay, 1920. 'The precipitation of mineral matter from solution, as the deposition of agate, vein quartz, etc.'

**Depot** (anglicized from the French depôt)
Used by early traders in Canada and the U.S. to denote a place, usually at a junction of routes, for loading and unloading of goods. With the construction of railways, depot became the usual American term for a railway station.

**Depressed area**—see Development area

**Depression**

Mill, *Dict.* 1. Any hollow or relatively sunken area. It would be more convenient to use only the term 'hollow' in this wide sense, restricting the term 'depression' to (2) a hollow in which the surface of the Earth lies below sea level. (3) A spot or area in which the atmospheric pressure is below that of surrounding regions.

*Comment.* Meteorologists have succeeded in giving the word depression a specialized meaning equivalent to an atmospheric low and the old 'cyclone' has been replaced for mid-latitudes. Geographers and geologists have been less successful and the word if used has no precise meaning. In geographical writings the economic usages may be met (*e.g.* trade depression).

**Desert**

*O.E.D.* 1. An uninhabited and uncultivated tract of country; a wilderness; *a.* now conceived as a desolate, barren region, waterless and treeless, and with but scanty growth of herbage; *e.g.* the *Desert of Sahara*... etc. Examples 1225–1856. b. formerly applied more widely to any wild, uninhabited region, including forest land. *Obs.* Examples 1398–1834.

*Webster.* 1. A tract which, though it may capable of sustaining a population has been left unoccupied and uncultivated; a deserted region. 2. An arid region in which the vegetation is especially adapted to scanty rainfall, with long intervals of heat and drought, or more rarely, is entirely lacking; a more or less barren tract incapable of supporting any considerable population without an artificial water supply. Rock disintegration in deserts predominates over rock decay, causing the long slopes of stony detritus extending far up the mountain sides, the undrained basins, the salt pans or playas, and the areas of shifting sand so common to deserts. Desert rainfall is usually less than ten inches annually.

Mill, *Dict.* A tract of land nearly destitute of vegetation on account of scanty rainfall or low temperature.

*Met. Gloss.*, 1944. 'A region in which high temperature and small rainfall cause the evaporation to exceed the precipitation, and consequently there is insufficient moisture to support vegetation.'

Shantz and Marbut, 1923, *Vegetation and Soils of Africa*. 'The desert is a waste of sand, rock, or soil surface, devoid or apparently devoid of vegetation throughout the year.' (p. 82)
(Yet note the vegetation categories of Desert Shrub-Desert Grass; Desert Shrub; Salt Desert Shrub.)

Trewartha, G. T., 1937, *An Introduction to Weather and Climate*, New York: McGraw-Hill. (Summary of Köppen) Boundary of arid or desert climate is where rainfall is one half of the dry-humid boundary (see 'Dry') (p. 225).

Giaever, J., 1954, *The White Desert*, London: Chatto and Windus. (title refers to Antarctica).

Huntington, E. rev. Shaw, W. B., 1951, *Human Geography*, New York: Wiley. As 'cold deserts' includes Antarctica and N. Canada (2M sq. miles) (p. 487).

Carpenter, 1938. The climatic community which originates when, because of too great a drought or cold, climatic conditions are hostile to all vegetation (quoting Schimper).

*Comment.* Desert remains a difficult word to define and apply exactly. It is normally equated with aridity and heat but not to the extent of eliminating vegetation. Indeed some writers have distinguished between 'tame' deserts (with some vegetation serving as sparse grazing) and 'true' deserts. See the summary statement in Stamp, *Asia*, 1st ed., 1929 and also the classification of deserts.

With Mill many writers extend 'desert' to the cold wastes of the Arctic and Antarctic and in recent years Pierre Gourou has revived the old usage (*O.E.D.*b.) by referring to the 'forest deserts' of equatorial Africa where the forests are almost devoid of people. Statistically defined limits have been proposed by Köppen, de Martonne, Thornthwaite and others—see Miller, 1953.

See also Arid, ecumene.

### Desert-pavement, desert-mosaic

Holmes, 1944. 'Where the bedrock of the desert floor is exposed to blown sand it may be smoothed or pitted or furrowed. . . . Where pebbles have become sufficiently concentrated by removal of finer material they become closely packed, and in time their upper surfaces are ground flat. In this way mosaic-like tracts of *desert pavement* are developed.' (p. 258) Hence also desert-mosaic.

### Desert varnish

*O.E.D. Suppl.* A film of iron oxide or quartz on rocks polished by wind-blown sand.

Holmes, 1944 (of weathering in the desert) 'By evaporation minute quantities of dissolved matter are brought to the surface. The loose salts are blown away, but oxide of iron, accompanied by traces of manganese and other similar oxides, form a red, brown or black film which is firmly retained. The surfaces of long-exposed rocks and pebbles thus acquire a characteristic coat of "desert varnish".' (p. 270)

*Comment.* The distinction of desert varnish from polishing due to wind blowing sand against the rock is often difficult. Mill has only Desert Polish. No reference in Wooldridge and Morgan, 1937; Cotton, 1922, 1941; 1942; Moore, 1949. The term desert-patina is also used. Manganese oxide is often a major constituent.

### Desire lines

Abler, Adams and Gould, 1971. 'A desire line map depicts the straight line connections between origins and destinations for one class of trip. The desire line is thus the shortest line between origin and destination. On a map a desire line presents visually the basis for travel behaviour by expressing how a person would like to go if such a way were available. A desire line map provides a strong impression of the geographic pattern of travel demand.' (p. 211). See pp. 212–213 for illustrations.

### Desquamation

*S.O.E.D.* The removal of scales or any scaly crust; a coming off in scales; exfoliation.

Mill, *Dict.* Weathering of rocks by detachment of scaly fragments from the surface.

*Comment.* An old word; not in Wooldridge and Morgan, 1937; Lobeck, 1939; Holmes, 1944; Thornbury, 1953. Apparently obsolescent, but common use in French.

### Determinism, Determinist

*O.E.D.* 1. The philosophical doctrine that human action is not free but necessarily determined by motives, which are regarded as external forces acting upon the will.

2. gen. The doctrine that everything that

happens is determined by a necessary chain of causation.

Taylor, 1951. 'Determinist (in Geography). One who believes in dominant environmental control' (p. 612).

Wooldridge, S. W. and East, W. G., 1951, *The Spirit and Purpose of Geography*, London: Hutchinson. An attempt to deduce the human resultant from the physical causes (p. 32).

Martefiore, A. C. and Williams, W. M., 1955, Determinism and Possibilism, *Geog. Studies*, 2. 'Determinists hold that the environment can be modified only to a very limited degree and that its fundamental features are always the most important factors in explaining or in formulating laws about human phenomena' (p. 1).

**Stop-and-Go Determinism** (T. Griffith Taylor)

Taylor, T. G., *Australia*. 'The best economic programme for a country to follow has in large part been determined by Nature, and it is the geographer's duty to interpret this programme. Man is able to accelerate, slow or stop the progress of a country's development. But he should not, if he is wise, depart from the directions as indicated by the natural environment. He is like the traffic controller in a large city, who alters the *rate* but not the direction of progress; and perhaps the phrase "Stop-and-Go Determinism" expresses succinctly the writer's geographical philosophy' (p. 445).

**Detrition**

O.E.D. The action of wearing away by rubbing. See Detritus. Rare.

**Detritus, detrital deposits**

O.E.D. From L. detritus, rubbing away.

Matter produced by the detrition or wearing away of exposed surfaces, especially the gravel, sand, clay, or other material eroded and washed away by aqueous agency; a mass or formation of this nature.

Waste or disintegrated material of any kind; debris.

An accumulation of debris of any sort.

Mill, *Dict*. Material removed by disintegration and other processes from the surface of rocks.

Geikie, J., 1898. Any accumulation of materials formed by the breaking-up and wearing-away of minerals and rocks.

Rice, 1943. Matter worn from rocks by mechanical means. A general term applicable to several grades and types (p. 143).

*Comment*. There is a general tendency to use debris.

**Detritus: mountain-top detritus**

Ives, J. D., 1958. 'The term "mountain-top detritus" is used to designate a specific type of boulder field formed by the disintegration of bed-rock in situ by frost-shattering.' *The Canadian Geographer*, 12, 1958, 25.

**Deuterozoic**

The younger Palaeozoic systems—Devonian, Carboniferous and Permian. *Obsolete or obsolescent (cf.* Proterozoic).

**Development area**

In Britain in the inter-War years, certain areas, mainly the older industrial areas in certain of the coalfields, suffered severely from unemployment consequent upon the disappearance of some basic heavy industries. They were 'depressed' or 'distressed areas' but these designations were resented by the inhabitants, and they became for purposes of ameliorative legislation 'special areas'. Again this term was resented, hence the official term 'development areas'.

**Deviation**

In Haggett, 1965, Areal differentiation is given as the traditional view of geography from which there have been three 'deviations'—the landscape school, the ecological school and the locational school.

**Deviation (magnetic)**

The apparent declination or variation of the magnetic needle due to local conditions, *e.g.* the presence of magnetic iron ore, iron objects etc. (E.G.R.T.)

**Dew**

O.E.D. The moisture deposited in minute drops upon any cool surface by the condensation of the vapour in the atmosphere; formed after a hot day during or towards night, and plentiful in the early morning.

**Dew-point**

O.E.D. That point of atmospheric temperature at which dew begins to be deposited.

Mill, *Dict*. The lowest temperature to which air may be cooled without causing condensation of atmospheric moisture; below this temperature dew begins to be deposited. It depends on pressure and humidity.

## Dew-pond, dew pond

*O.E.D. Suppl.* A shallow pond, usually of artificial construction, fed by condensation of water from the atmosphere, occurring on downs where there is no adequate water supply from springs or surface drainage.

*Comment.* The *O.E.D.* gives the explanation which for long was accepted but is now known to be incorrect. Hollows in the downs were lined with well-puddled clay which prevented loss of water draining into them but there is no evidence of any appreciable augmentation of the water from dew.

## Dhānd (Pakistan: Sindhi)
Salt-lake or alkaline-lake in Sind.

## Dhāyā (Indo-Pakistan: Panjabi)
A Panjabi word for Khadar, but Spate, 1954, says 'broad floodplains of new alluvium (Khadar) bounded by deep bluffs (dhaya) which may be 20 or more feet high'. (pp. 462–463)

## Dhōrō dhōrū, (Pakistan: Sindhi)
A dry water channel.

Spate, 1954. 'long narrow depressions, apparently fragments of old drainage systems, now disrupted by shifts of course, desiccation and sand encroachments' (in Sind) (p. 454).

## Dhow (?Arabic)
A comprehensive term covering a wide variety of locally constructed wooden ships in the Arab and Moslem world.

*Webster.* An Arab lateen-rigged vessel of the Indian Ocean. It usually has a long overhang forward and a high poop and an open waist. Dhows have been notorious in the slave trade of the east coast of Africa.

Typically an Arab vessel with a single mast and sail and in the case of ocean-going craft with a burden of from 200 to 300 tons. Taking advantage of the monsoons these ships have from time immemorial provided communications around the periphery of the Arabian Sea; and they still frequent the harbours of East Africa. (S.J.K.B.)

## Diabase
A name formerly given to various dark-coloured basic igneous rocks, especially those partly metamorphosed and in which some of the original minerals had been altered. As the science of petrology has progressed the rocks have been separated into different types and the word 'diabase' is better avoided. Many diabases are now known to be dolerites.

## Diachronism (W. B. Wright, 1926)
Wills, L. J., 1929, 'A lithological unit that is of varying age in different areas and one therefore that transgresses the palaeontological zones.' (p. 140) See Report, Brit. Assoc., Oxford, 1926, 354–355.

## Diaclase, diaclasis
*Webster. Geol.* diaclase—a break or fracture; diaclasis, breaking or refraction.
*Comment.* Rarely used.

## Diaclinal stream (J. W. Powell, 1875)
*Webster.* Diaclinal. *Geol. & Phys. Geog.* Crossing a fold; as a *diaclinal* river.
Mill, *Dict.* Applied to streams which are in part cataclinal and in part anaclinal, and to regions which possess both these types of streams (Powell).
Powell, 1875. Diaclinal valleys are those having a direction at right angles to the strike and which pass through a fold.
*Comment.* Now rarely used.

## Diagenesis (K. W. von Gümbel in German 1888, A. E. Fersman in English, 1922)
Read, H. H., and Watson, J., 1962, *Introduction to Geology* 1. London. 'Diagenesis comprises all those changes that take place in a sediment near the earth's surface at low temperature and pressure and without crustal movement being directly involved. It continues the history of the sediment immediately after its deposition and with increasing temperature and pressure it passes into metamorphism.'
See also Taylor J. H., 1964, Some Aspects of Diagenesis, *Adv. of Science*, **20**, 417–436.

## Diara—see Char

## Diastrophism, diastrophic
*O.E.D. Geol.* (Greek, distortion, dislocation). A general term for the action of the forces which have disturbed and dislocated the earth's crust, and produced the greater inequalities of its surface.
*Webster. Geol.* The process or processes by which the earth's crust is deformed, producing continents and ocean basins, plateaus and mountains, folds of strata. and faults; also, the results of these processes. See *epeirogeny, orogeny*.
Lobeck, 1939. 'The relief features of the second order have been brought into being by the action of internal forces beneath the

earth's crust. The general term diastrophism comprehends all such movements of deformation' (p. 7). (The relief features of the first order are the continents and the ocean basins.)

Thornbury, 1953. Diastrophic processes are usually classed as two types, *orogenic* (mountain-building with deformation) and *epeirogenic* (regional uplift without important deformation).' (pp. 50–51)

**Diastrophic eustatism**—see eustatism, rejuvenation

**Die-back**

*O.E.D.* The fact of dying back; the term for a disease affecting orange-trees in Florida, etc., in which the tree dies from the top downwards.

*Webster. Plant Pathol.* A diseased condition in woody plants, in which the terminal twigs are blighted either by parasites or by other agencies, as winter injury.

*Dict. Am.* A disease affecting trees causing them to die from the top downward. 1886

*Comment.* This term has been extended and is now commonly used to describe the phenomenon of considerable tracts of vegetation, especially forests, so affected by some change in physical conditions that a large proportion of the individuals, or the individuals of one species, die off at the same time. In many cases the phenomenon is unexplained; in certain examples it is believed to be due to a secular climatic change.

**Differential disequilibrium**

Harvey, D., 1973, *Social Justice and the City*, London: Edward Arnold. '. . . the speed with which different parts of an urban system can adjust to the changes occurring within it. . . . It is therefore misleading to think of adjustment in the urban systems as a homogeneous process proceeding at a uniform rate. This varying speed of adjustment means that there are substantial differentials in the disequilibrium in the urban system at any one point in time. . . . Certain groups, particularly those with financial resources and education, are able to adapt far more rapidly to a change in the urban system. . . . Any urban system is in a permanent state of differential disequilibrium (by which I mean that different parts of it are approaching equilibrium at different rates).' (p. 56)

**Diffluence, glacial**

*O.E.D.* diffluence: 1. The action or fact of flowing apart or abroad; dispersion by flowing.

Cotton, 1942. Diversion of part of a glacier into a distributary channel. (p. 234)

*Comment.* More generally applied to a valley glacier overflowing by a col into an adjoining valley. (G.T.W.)

**Diffusion**

*O.E.D.* 2. The action of spreading abroad; the condition of being widely spread, dispersion through a space or over a surface; wide and general distribution. 3. Spreading abroad, dispersion, dissemination (of abstract things, as knowledge) 1750, Johnson, *Rambler* No. 101, P 2. 'The writer . . . receives little advantage from the diffusion of his name.'

A general term that gained recognition in in geography through Hägerstrand, T., 1967, *Innovation and Diffusion as a Spatial Process*, Chicago: University of Chicago Press. 'The spread of ideas and/or artifacts from a given point or set of points over space. This spread usually takes place over a surface or along the lines of a network or from centre to centre.'

Morrill, R. L., 1970, *The Spatial Organization of Society*, Belmont, California: Wadsworth Publishing Co. 'The process of the gradual spread over space of people or ideas from critical centres of origin.' (p. 244)

See also Neighbourhood effect.

**Digitate margin**

The seaward extension of a delta into finger-like forms; a bird's-foot delta. See Delta.

**Dike, Dyke**

*O.E.D.* (The spelling dyke is very frequent, but not etymological)

I 1. An excavation narrow in proportion to its length, a long and narrow hollow dug out of the ground; a ditch, trench or fosse. *Obs.*

2. Such a hollow dug out to hold or conduct water; a ditch.

b. extended to any water-course or channel, including those of natural formation. On the Humber, a navigable channel, as *Goole Dike, Doncaster Dike*, etc. (A local use).

3. A small pond or pool. *dial.*

II An embankment, wall causeway.
5. 'A bank formed by throwing the earth out of the ditch' (Bosworth).
6.b. A low wall or fence of turf or stone serving as a division or enclosure. Now the regular sense in Scotland.
7. A ridge, embankment, long mound, or dam, thrown up to resist the encroachments of the sea, or to prevent low-lying lands from being flooded by seas, rivers, or streams.
9. *Mining (Northumb.)* A fissure in a stratum, filled up with deposited or intrusive rock; a fault.
b. Hence, in *Geol.* A mass of mineral matter, usually igneous rock, filling up a fissure in the original stratum, and sometimes rising from these like a mound or wall, when they have been worn down by denudation.

Holmes, 1944. 'One of the commonest signs of former igneous activity is provided by the wall-like intrusions called *dykes*. Here the magma has ascended through approximately vertical fissures, forcing the walls apart as it rose and so on cooling, became a vertical sheet of rock ... cutting across the "bedding" planes (*transgressive* or *discordant*).' (p. 82)

### Dike rocks (more often Dyke rocks)
Those which occur as in 9b above, hypabyssal rocks.
Sandstone dykes arise when a crack has become filled with sand, afterwards consolidating to sandstone.

### Dike (Dyke) phase
Himus, 1954. The closing episode in a volcanic cycle, characterized by the injection of minor intrusions, especially dykes.

### Dike (Dyke) swarm
Himus, 1954. A set of dykes of the same age, generally with a common trend over a wide area. Occasionally, the dykes of a swarm radiate outwards from a common centre.

### Dilatancy
King, C. A. M., 1959, *Beaches and Coasts*, London: Arnold. 'The term applied to sediments which, when wet and shaken in this state, will exude water' (p. 4).

### Diluvion
Lambourn, G. E., 1918, *Malda*, Bengal District Gazeteers. 'Alluvion and diluvion are perpetually taking place on the Malda bank, which is throughout of sand, offering little resistance to the changes of current.' (p. 3)
O'Malley, L. S. S., 1923, *Pabna*, Bengal District Gazeteers. 'Alluvion and diluvion are constantly taking place along the courses of the principal rivers ... especially the Padma and the Jamuna, the river channels perpetually swinging from side to side of their sandy beds ... the surface of the land is thus subject to constant changes which naturally give rise to innumerable disputes. ... Alluvion and diluvion are the causes of a large proportion of the litigation in the district.' (p. 6)
*Comment*. Alluvion (*q.v.*) is an old and well-established word, (though one to be avoided) but diluvion seems to be confined to these Indian publications. It appears to mean loss of land by river erosion after flooding. (G.T.W.)

### Diluvium, diluvial deposits
*O.E.D.* (fr. Latin, diluvium, flood, inundation, deluge) A term applied to superficial deposits which appear not to have been formed by the ordinary slow operations of water, but to be due to some extraordinary action on a vast scale; such as were at first attributed to the Noachian or Universal deluge, whence the name; the chief of these deposits were those of the Northern Drift or Boulder formation at the close of the Tertiary Period, to which the name continued to be applied after the theory of their origin was given up; it is now generally 'applied to all masses apparently the result of powerful aqueous agency'.
1819. J. Hodgson in Raine, Mem., 1857, 265. 'The cliffs are very white, excepting where they are tarnished by diluvium falling from the top of the cliff.'
1823. W. Buckland, Reliq. Diluv., 2 'The term diluvium ... I apply to those extensive and general deposits of superficial loam and gravel, which appear to have been produced by the last great convulsion that has affected our planet.'
Mill, *Dict*. 1. An almost obsolete term almost synonymous with 'glacial drift'.
2. Used by some German geologists and geographers to designate the older or Pleistocene Quaternary deposits, as distinguished from recent deposits which are termed 'alluvium'.
Flint, 1947, points out German division

Quaternary into Diluvium (=Pleistocene) and Alluvium (=Recent) and states term derived from Mantell (Sussex, 1822) where Diluvium=superficial deposits laid down by agencies no longer operative, such as the biblical flood; *cf.* alluvium, sediments laid down by agencies still in force, such as existing streams. (p. 199)

Cizzarz and Jones, German-English Geol. Term, 1931, equate German Diluvium with English Pleistocene.

Davies, G. M. A French-English Vocab. in Geol., 1932, translates diluvium as river drift, a usage not justified by Hatzfeld and Darmesteter.

Rice, 1941. The older usage included all drift of the Glacial Period. The term is now used only of the continent of Europe, and refers to the drift deposited prior to the waning of the ice-sheet (Antevs and MacClintock) (also quote Holmes, similar to Mill, 2).

*Comment.* A term which is best avoided except in a historical discussion. See also Proluvial, Alluvial.

**Dimple**

Hare, F. K., 1947. The Geomorphology of a part of the Middle Thames, *Proc. Geol. Assoc.*, **58**. '"Dimples" on bare Chalk surfaces.'

'These are neat little depressions about 5-15 yards across and up to 25 ft. in depth. Usually circular, they have the regular form characteristic of all Chalk terrains. They occur mainly on steep slopes, chiefly valley sides. They are probably abandoned swallets of the valley side class and date from a wetter period when surface water was more abundant.' (p. 327)

*Comment.* May be compared with a small doline.

**Dingle**

*O.E.D.* A deep dell or hollow; now usually applied (app. after Milton) to one that is closely wooded or shaded with trees; but, according to Ray and in modern Yorks. dialect, the name of a deep narrow cleft between hills.

Mill, *Dict.* Small well-wooded narrow valley (S. W. England).

*Comment.* Rare except in place names, *e.g.* Marrington Dingle, a famous geological section along a stream course in Shropshire, where the word is common. A literary word, obsolescent and best avoided in scientific writing.

**Diorite**

An intermediate (silica 55-65 per cent) plutonic rock, usually coarsely crystalline. It normally consists of plagioclase felspar and a dark mineral, hornblende or augite, and biotite mica. Free quartz is typically absent; with presence of quartz (quartz-diorite) and diminution in quantity of dark minerals it passes into granite.

**Dip**

*O.E.D.* 3. *Astron* and *Surveying*. The angular distance of the visible horizon below the horizontal plane through the observer's eye; the apparent depression of the horizon due to the observer's elevation, which has to be allowed for in taking the altitude of a heavenly body.

4. The downward inclination of the magnetic needle at any particular place; the angle which the direction of the needle makes with the horizon.

5. Downward slope of a surface; *esp.* in *Mining* and *Geol.* the downward slope of a stratum or vein: estimated, as to direction, by the point of the compass towards which the line of greatest slope tends, and as to magnitude, by its angle of inclination to the horizon.

*Webster. Geol.* The angle which a stratum, sheet, vein, fissure, fault, or similar geological feature makes with a horizontal plane, as measured in a plane normal to the strike.

*Comment.* Best restricted to sedimentary rocks or those interbedded with them and to denote the inclination of the strata measured from the horizontal. There is an unfortunate tendency to careless use among geographers to denote the inclination of land surfaces. This is definitely wrong. When beds are exposed as in a quarry face or cliff the 'apparent dip' *i.e.* that seen by the observer, may not be the 'true dip' which is the maximum. At right angles to the true dip is the *strike* which by definition must be horizontal. From dip one gets such combinations as dip-fault, a fault roughly parallel to the dip, and dip-stream (*q.v.*). Despite Webster dip should not be used of a fissure or fault: the term is then hade (*q.v.*).

**Dip slope**

*O.E.D. Suppl.* The surface slope of ground when parallel to the dip of the strata over which it lies.

Mill, *Dict.* The inclined surface of the out-

crop of a succession of strata facing towards the dip. It usually possesses a corresponding escarpment which faces in a direction opposite to that of the dip.
*Comment.* The *O.E.D.* is not quite correct: it should read roughly or approximately parallel to the dip since there is rarely exact parallelism. Mill's definition is confirmed. Cotton and others use the term 'back slope'.

**Dip stream**
Mill, *Dict.* A consequent river (*q.v.*).
*Comment.* A stream running parallel to the dip. Mill's definition is not reversible, *i.e.* not all consequent streams are dip streams.

**Dirichlet polygons (first-order)**
Haggett, 1972. 'These polygons, named after G. L. Dirichlet, a German mathematician of the last century, have the geometric property of containing within them areas that are nearer to the point around which they are constructed than any other points. Meteorological agencies originally used Dirichlet polygons to determine the average distribution of rainfall around individual recording stations.' (pp. 366–367)
*Comment.* Also known as Thiessen polygons and first-order Brillouin regions.

**Dirt cone**
Krenek, L. O., 1958. The Formation of Dirt Cones on Mount Ruapehu, New Zealand *Jour. of Glac.*, **3**, 313–319. 'All consisted of an ice core covered by a layer of black ash 5 to 7 cm. thick.' (p. 312)

**Discharge**
Gilbert, G. K., 1914. The Transportation of Debris by Running Water. *U.S.G.S. Prof. Paper*, **86**. 'The quantity of water passing through any cross-section of a stream in unit time is the *discharge* of the stream at that point. It is measured in cubic feet per second.' (p. 35)
*Comment.* C.f.s. or cusecs in English units or cubic metres per second or litres per second in metric units. The measurement of discharge constitutes river-gauging.

**Disconformity**
*Webster.* Want of conformity.
*Comment.* Occasionally used as less definite in meaning than unconformity.

**Discontinuous distribution**
Applied to individual species of plants or animals, or of specialized communities which occur in more or less isolated locali-

ties or in districts far removed from one another.
*Webster. Biol.* Occurrence of like or related organisms in widely separated areas with no closely related forms in intervening areas.

**Discordant coast**
Mill, *Dict.* A coast-line cutting across the end of the ranges of folded mountains and the mouths of the intervening valleys.
*Comment.* Wooldridge and Morgan use 'transverse coasts, cutting across the structural grain.' See also Atlantic type of coastline. No reference in Geikie, J., 1898; Salisbury, 1907; Davis, 1909; Johnson, 1919; Cotton, 1922, 1941, 1942; Holmes, 1944.

**Discordant junction**
(*cf.* accordant junction).
Cotton, 1922 [of young valleys] '... the tributary streams sometimes failing to deepen their valleys as rapidly as the main, so that the junctions remain discordant or "hanging".' (p. 49) Similar usage for glacial hanging valleys (p. 289).
No reference in Geikie, J., 1898; Salisbury, 1907; Davis, 1909; Holmes, 1944; Wooldridge and Morgan, 1937; Moore, 1949.
*Comment.* Not in common use but unequivocal (G.T.W.).

**Dismembered river(s)**
Mill, *Dict.* The original tributaries left as independent rivers entering an arm of the sea when the lower part of the river system is submerged.

**Dispersed city**
Philbrick, A. K., 1961, *Analyses of the Geographical Patterns of Gross Land Use and Changes in Numbers of Structures in Relation to Major Highways in the Lower Half of the Lower Peninsula of Michigan*, East Lansing, Michigan State University, used the term more or less synonymously with the phrase 'urban sprawl'.
Burton, I., 1963, A restatement of the dispersed city hypothesis, *Ann. Assoc. Am. Geographers*, **53**, notes that Beimfohr, O. W., 1953, 'used the term with respect to southern Illinois, however, and as subsequently developed it refers to a group of politically discrete cities which, although separated by tracts of agricultural land, function together economically as a single urban unit'. But, 'although Beimfohr is credited with the

introduction of the term into literature, it was in common use in southern Illinois prior to 1950'.

See Beimfohr, O. W., 1953, Some Factors in the Industrial Potential of Southern Illinois, *Trans. Illinois State Acad. Sci.*, 46, 97–103; and Stafford Jr., H. A., 1962, The Dispersed City, *Professional Geographer*, 14.4, July, 8–10.

**Dispersed settlement**
Applied to the pattern of rural settlement when most of the people live in isolated farmhouses and cottages so that groups (hamlets or villages) tend to be absent. The contrasts are with nucleated settlement. See also Hamlet.

**Disphotic**—see under Pelagic

**Displacement Theory**—see Continental drift

**Dissection, dissected**
Mill, *Dict*. In rainy countries plateaus are attacked by streams which cut deeply into them, so that they become 'dissected plateaus'.

Cotton, 1922. 'The gradual etching of the land by the action of streams is termed dissection.' (p. 64) Discusses mature and youthful dissection.

Moore, 1949. Dissected plateau. A plateau into which a number of valleys have been carved by erosion; its origin as a plateau is patent, however, when the tops of the mountains and ridges are seen to be level against the skyline, showing that they once formed part of a continuous surface.

Rice, 1941. 'Dissected. Cut by erosion into hills and valleys or into flat interstream areas and valleys, as dissected plateau or pediment.'

*Comment*. Dissection normally conveys the idea of a fairly close network of valleys—often of a dendritic pattern. (G.T.W.)

**Distributary**
*O.E.D.* B.1. Something whose function is to distribute; applied to branch canals distributing water from a main one.
2. A river branch which flows away from the main stream without returning to it, as in a delta.

*Comment*. Also used of ice streams leading from an ice sheet or ice cap (G.T.W.).

**Divagation, divagating meander**
Mill, *Dict*. Divagation (of rivers). The lateral shifting of a river's course in the regions of deposition of alluvium.

*Divagating meanders*. Meanders which because of their occurrence on a surface approaching the condition of a peneplain are liable to variations from time to time.

**Divide**
*Webster*. 2. A dividing ridge or section of high ground between two basins of drainage; a watershed.

Swayne, 1956. The line of separation between two adjacent drainage basins. *Continental divide*. The line separating the drainage to separate oceans which runs through the heart of a continent.

*Comment*. Mill equates 'divide' with 'watershed' (*q.v.*). Also used for ground-water divide (G.T.W.). See also Continental divide.

**Do** (Indo-Pakistan: Urdu-Hindi; Hindustani)
Two, hence do-fasli harsala, land yielding two crops a year; do-fasli-do-sala, a two-year rotation; contrast with ek- (one) fasli-harsala, land yielding one crop a year.

**Doab** (Indo-Pakistan; Persian and Urdu)
*O.E.D. Lit*. 'two waters'. Used in India of the tongue of land between the Ganges and Jumna, and of similar tracts in the Punjab, etc. The 'tongue' or tract of land between two confluent rivers.

Mill, *Dict*. A series of rounded hills occurring in plain land (India). The ridges left between consequent valleys as they develop on a slightly inclined plain of alluvium or soft recent deposits (W. M. Davis). Penck has suggested the name 'Riedel' for such valleys.

*Comment*. Mill seems to have been misled. Riedel (German) = interfluve; Doab is used simply to denote the land between two rivers in the Indo-Gangetic Plain of northern India. It is commonly restricted to the plains portion, of alluvium and with very little relief, and thus differs from interfluve. The north Indian doabs are named by taking appropriate letters from the names of the bordering rivers, *e.g.* Bari doab between the Bias and Ravi, Rech doab between the Ravi and Chenab. 'The Doab' is the Jamuna-Ganga (Jumna-Ganges) doab. The word is now widely used in geographical descriptions in many parts of the world.

**Dod, dodd**
*O.E.D.* In the north of England and South of Scotland a frequent term for a rounded summit or eminence, either as a separate hill, or more frequently a lower summit or distinct shoulder or boss of a hill. Rarely applied to a lower buttress when

not rounded, as Skiddaw Dod. Usually forming part of a proper name, like the equivalent Welsh Moel (Foel), but also an appellative.

**Dogger**

L.D.G. A large concretion or mass of consolidated material found in certain sedimentary rocks notably Jurassic sandstones in Dorset and Yorkshire.

**Dokeph** (T. Griffith Taylor, c. 1920)

Taylor, 1951. 'A narrow-headed race' (p. 612). Abbreviated from dolichocephalic (for person or race). Contrast Brakeph.

**Doldrums**

The belt of calms or light winds in the equatorial region of the oceans. The evolution of this word, originally slang for a dull fellow, is discussed in O.E.D.

**Dole**

O.E.D. A portion of an undivided or common field.

**Dolerite**

The typical hypabyssal basic igneous rock (silica 45–55 per cent) corresponding in composition to the lava basalt and the plutonic gabbro. It is intermediate in coarseness of crystalline structure between basalt and gabbro and is typically found in dykes, sills and laccolites.

**Dolina, doline**

O.E.D. no reference.

*Webster. Geol.* A sink.

Mill, *Dict.* Dolina (Italian or Slavonic), or Doline (French and German). A basin or closed depression in an area of limestone, usually flat-bottomed and round or oval. Rock-walled dolinas have steep rocky sides and varied vegetation at the bottom, in some cases including large trees. Funnel-shaped dolinas have gently sloping sides covered with vegetation, and are usually cultivated at the bottom. The larger rock-walled form may be half a mile in diameter, and as much as 600 feet deep; the smallest dolinas are sometimes swallow-holes (*q.v.*). Alluvial dolinas are depressions of similar form in alluvial soil, being the loamy bottoms of larger dolinas. The term was first applied in Carinthia.

Knox, 1904. 'Dolina (S. Slav), a valley (Dol.).'

Geikie, J., 1898, *Gloss.* 'Dolina (It.): name given to the funnel shaped cavities which communicate with the underground drainage system in limestone regions. Similar cavities are known in this country as sinks and swallow-holes.'

Similarly Cotton, 1922; Wooldridge and Morgan, 1937; Strahler, 1951.

Wray, D. A., 1922, The Karstlands of Western Yugoslavia, *Geol. Mag.*, 59. 'The funnel-shaped hollows which are so frequently met with on the surface of the karst are termed ponors. Many of these are largely filled with red earth, formed by the decomposition of the limestone, and are then known as dolinas...' (p. 406).

von Engeln, 1942. 'The doline, of which the swallow hole and sink hole are variants, is the initial and fundamental unit of karst topography. The type doline has the form of a funnel top and occurs in the Adriatic karst.... Dolines range in size and form from mere chimneylike shafts, called jamas,...a number of small steep-sided dolines may in time coalesce... when this has happened on a large scale the resulting extensive depression is a *uvala*.' (pp. 572–575)

*Comment.* This is an interesting example of a very common word in the language from which it is borrowed having been given a specialized meaning in technical literature. In Slovenia the Logarska dolina means simply the Logar Valley—a broad open normal mountain valley. Those who studied karst phenomena in Carinthia learnt that the local name for the specialized type of valley, described by Mill above, was dolina and it has accordingly become restricted, but differently interpreted—compare Wray and von Engeln.

**Dollar Curtain** (L. Dudley Stamp, 1952)

Stamp, L. D., *Land for Tomorrow*, New York and Indiana University, 1952. 'Man-made barriers ... there is the territory of the United States with Alaska, the 3,600,000 square miles within the Dollar Curtain ... from certain points of view as effective a barrier as the Iron Curtain. In broad general terms Canada lies within the Dollar Curtain but with closer relationship, economically as well as politically, with the outside world.' (p. 204)

*Comment.* In the years following the Second World War dollars were unobtainable by those living in the Sterling area, except for strictly business purposes approved by the Bank of England. It was accordingly impossible to visit relatives or friends in the United States or Canada unless all costs were paid by them. Similar currency restrictions apply to other 'man-made' barriers.

**Dolomite**
Strictly a mineral with the formula $CaCO_3 Mg_2CO_3$, i.e. equal molecules of Calcium carbonate and Magnesium carbonate, but applied to a rock consisting predominantly of this mineral. Hence *dolomitic limestone*, a limestone with some dolomite, and *dolomitization*, the conversion of limestone to dolomite.

**Dolostone** (R. R. Shrock)
*A.G.I.* 'A term proposed by R. R. Shrock for a sedimentary rock composed of fragmental, concretionary, or precipitated dolomite of organic or inorganic origin.'

**Dome**
*O.E.D.* (Derived via French from Italian, duomo, house, house of God (and thus of the distinguishing church dome) 4. transf. b. The convex rounded summit of a mountain, a wave, etc. In U.S., frequently entering into the names of rounded mountain peaks.
*Webster. 9. Geol.* a. An anticline broad in comparison with its length and consequently approximately circular or elliptical in plan. A dome may be many miles in diameter, as the Ozark dome of the United States, or small and steep-sided, as some of the salt domes of Louisiana. In oil-producing regions, it often indicates the presence of oil. b. A rock mass of domical form, as the granite domes of the Yosemite. c. A rounded snow peak.
*Dict. Am.* (*cap.*). Used in the names of certain mountain peaks in allusion to their rounded tops. 1833.
Mill, *Dict.* 1. A mass of high land with a relatively rounded top and steep sides. 2. See Laccolith. 3. An outcrop formed of strata dipping in all directions away from a central point. *cf.* pericline.
Knox, 1904. Dome (Eng.), as a minor form of sub-oceanic relief, a single elevation or submarine mountain of small area, but rising with a steep angle to a depth more than 200 metres from the surface. (This is not in *Adm. Gloss.*, 1953; Bencker, 1942; and seems to be obsolete or erroneous.)
Holmes, 1928. This term is used not only to connote certain crystallographic, structural, and geographical phenomena, but also to describe at least two different types of modes of occurrence of igneous rocks: It is applied (a) to stocks whose sides slope away quaquaversally at low and gradually increasing angles beneath the invaded formations; and (b) to rounded extrusions of highly viscous lava squeezed out from a volcano, and congealed above and around the orifice instead of flowing away in streams. Portions of older lavas or ejectamenta may be elevated by the pressure of the new lava rising from beneath. The second type of dome is usually distinguished as a volcanic dome (*q.v.*).
Holmes, 1944, discusses salt, tectonic and volcanic domes (pp. 349–350, 73, 456–457).
Hatch and Wells, 1937. 'The roofs of batholiths and laccoliths are similar... in the absence of definite evidence as to the character of the lower parts of the intrusion, the non-commital term *Dome* is frequently used.' (p. 9)
*Comment.* Loosely used for almost any dome-shaped mass of rock, or a rock structure, or a surface feature.

**Dome volcano**
Mill, *Dict.* A volcano composed of very viscid lava, which on eruption heaps itself up dome-shaped above the vent, leaving no crater.
Cotton, 1944. Volcano forms are divided into 'domes' or 'cones', being generally composed of lava or cinders, etc., respectively. There is no requirement of viscosity, *e.g.* the Hawaii domes. (p. 71)
Holmes, 1928. 'Rounded extrusions of highly viscous lava squeezed out from a volcano, and congealed above and around the orifice instead of flowing away in streams Portions of older lavas or ejectamenta may be elevated by the pressure of the lava rising from beneath... usually distinguished as a volcanic dome.'
*Comment.* The Hawaiian volcanoes are more usually called 'shield volcanoes' (G.T.W.).

**Domestic**
*O.E.D.* 1521. ad L. domesticus fr. domus = home.
3. Of or pertaining to one's own country or nation; internal, inland, home, 1545; indigenous, home-grown, home-made.
*Comment.* Hence Domestic market—the home market.
Domestic port—port serving internal or coastwise trade as opposed to international.
Domestic trade—internal as opposed to international trade.
*Domestic animal:* living under the care of man, in or near his habitations; tame not wild. (*O.E.D.*)

**Donga** (Bantu)
*O.E.D.* S. Africa (native name). A channel or gully formed by the action of water; a ravine or watercourse with steep sides.
Mill, *Dict.* 1. A small ravine or wash-out caused by floods in soft ground. 2. A gulley or dry watercourse with steep sides (South Africa), synonymous with the Eastern terms 'wadi' and 'nullah'.
*Comment.* Now used especially of gullies formed as a result of soil erosion. (P.S., G.T.W.)

**Dorbank** (Afrikaans; *lit.* dry layer)
Wellington, J. H., 1955, *Southern Africa*, I. 'In the Little Karoo ... a peculiar concretion of lime and silica, known locally as "dorbank" occurs under the surface layer of sandy loam.' (p. 322)

**Dore**
*O.E.D.* no reference.
Mill, *Dict.* An opening or fissure between rocks, often a widened joint (Cumbrian Lakes).
*Comment.* Very localized usage, not recommended for general use. (G.T.W.)

**Dormant**
Literally sleeping, applied to volcanoes which, though not actively erupting, are regarded as liable to erupt.

**Dorp** (Afrikaans; Nederlands: dorp; German Dorf)
The word has undergone a change of meaning in Afrikaans. In Afrikaans Capetown (Kaapstad) was originally the only place indicated as *stad* (town, city). All the rest used to be called *dorp* till far into the 19th century. Conservative people persisted even in the early 20th century in calling Pretoria and Johannesburg '*dorps*'. This was natural, because the smaller centres, though rural in appearance, are urban in their functions; they have hardly any active farmers among their inhabitants. For this reason people saw no real difference between the smaller and the larger centres. In South African English dorp is correctly translated as town (never village), when attention is focused on its urban functions. The term dorp is sometimes used untranslated, when it is desired to stress the rural appearance. (P.S.)

**Dot-map**
A map showing distribution by means of dots, usually of a uniform size and each representing a specific number of the objects concerned. Sometimes referred to as the absolute method of representing distribution.

**Doup**
*O.E.D.* 1. A rounded cavity or hollow bottom Obsolete.
Mill, *Dict.* A hemispherical depression or cavity in a rock or hillside (N. England).

**Down, downs, downland**
*O.E.D.* (O.E., dun, hill).
1. A Hill, 1653, obsolete.
2. An open expanse of elevated land; spec. in pl., the treeless undulating chalk upland of the south and south-east of England serving chiefly for pasturage; applied to similar tracts in the colonies, etc.
3. A sand-hill, Dune, 1523.
4. The Downs: the part of the sea within the Goodwin Sands off the east coast of Kent, a famous rendezvous for ships 1460.
5. Applied to a breed of sheep raised on the chalk downs of England, *e.g.* South downs.
Mill, *Dict.* A rounded hill summit sparsely covered with soil (S. England, especially Chalk Country).
*Comment.* The definition given by Mill is too narrow: normally used as *O.E.D.*2 *O.E.D.*3 equivalent to sand-dune is an old usage, now obsolete. Downs and downland convey the idea of open, rolling, treeless grassland often on limestone, especially chalk, but not necessarily so. In New Zealand applied to undulating rounded terrain developed mainly on Tertiary rocks flanking the foothills of the Southern Alps.

**Downtown**—see Central Business District

**Dråg** (Swedish)
Soak; a narrow strip of fen running between mosses, or across a moss. (Droge; Rüllen in German) (E.K.).

**Drain**
*O.E.D.* 1. A channel by which liquid is drained or gradually carried off; *esp.* an artificial conduit or channel for carrying off water, sewage, etc. In the Fen districts, including wide canal-like navigable channels. Elsewhere, applied chiefly to covered sewage drains or field drains.
c. Applied to a natural water-course which drains a tract of country. Examples 1700–1876 and (in Suppl.) of American use, 1816–1836.
Mill, *Dict.* Drains (special use). Wide canal-like navigable channels (Fen district).
*Comment.* See also Drawn. In water-meadows

the water is led on by drains and taken off by drawns.

**Drainage area**
Mill, *Dict*. The whole land having a common outlet for its surface water.

**Drainage pattern, system, network** etc.
Commonly used terms which are self-explanatory.

**Draw**—see Blind creek

**Drawn**
Moon, H. P. and Green, F. H. W., 1940, Water Meadows in Southern England, Append. II, *Report of the Land Utilisation Survey, Hampshire*. 'These drawns or drains are the beginning of a maze of channels which drain the water away from the various flooded meadows.' (p. 374)

**Dreikanter** (German)
Lobeck, 1939. 'Wind-faceted pebbles or ventifacts or dreikanter, three corners.' (p. 379)
*Comment.* Widely used in England. Dreikanter are usually polished by wind action and commonly indicate formation under desert conditions. See also Ventifact.

**Driblet cone**
*O.E.D. Suppl.* A cone produced by the successive ejections of small quantities of lava (Dana, *Man. Geol.* (ed. 4), 1895, 271).
Mill, *Dict*. Small volcanic cones formed of lava sent so short a distance into the air that it falls round the crater still more or less fluid, and builds highly porous crater walls by partial adhesion.
Lobeck, 1939. 'Where gases sputter out through the side of the dome, a *spatter* or *driblet cone* may be built up 10 to 12 feet above the ground.' (p. 675)
*Comment.* 'Spatter cone' is more usual than 'driblet cone'.

**Drif** (Afrikaans), **drift**
Ford in river (*pl.*: driwwe; *antiq.*: drift, driften).
*O.E.D.* 17. *S. Africa*, A passage of a river; a ford (from 1849).
*Comment.* Used in many parts of Africa to indicate a ford or a sudden dip in a road over which water may flow at times (hence roadside notices 'beware of drifts').

**Drift** (Geological)
*O.E.D.* 10. *Geol*. A term applied (*a*) to any superficial deposit caused by a current of water or air; also (*b*) *spec*. (*the Drift*) to Pleistocene deposits of glacial and fluvio-glacial detritus, also known as *boulder-clay*, and *till*; diluvium. (examples 1839–1892).

Also: 15. *Mining*. A passage 'driven' or excavated horizontally; *esp.* one driven in the direction of a mineral vein. 1653.
Webster. 19. *Geol*. a. Rock material of any sort deposited in one place after having been moved from another; as river *drift*. b. Specif. a deposit of earth, sand, gravel, and boulders transported by glaciers (*glacial drift*) or by running water emanating from glaciers (*fluvioglacial drift*) and distributed chiefly over large portions of North America and Europe, esp. in the higher latitudes. 23. *Mining*. A nearly horizontal mine passageway driven on, or parallel to, the course of a vein or rock stratum. Also loosely applied to the main exploratory workings of a given level.
Mill, *Dict*. a. A general term for superficial unconsolidated deposits, clay and gravel, which have been transported; especially used with reference to that which has been transported by ice.
Fay, 1920. 1. Horizontal passage under ground. A drift follows the vein. . . .
2. In coalmining, a gangway or entry above water level, driven from the surface in the stream.
6. Any rock material such as boulders, till, gravel, sand, or clay, transported by a glacier and deposited by or from the ice or by or in water derived from the melting of the ice. Generally used of the glacial deposits of the Pleistocene epoch. Detrital deposits (La Forge)
Rice, 1941. Quotes Antevs and MacClintock, 1934 for a similar glacial connection. (W.M.)
Stamp, L. D., 1923, *An Introduction to Stratigraphy*, London: Murby. 'The term "Drift" is often loosely but conveniently used to include all superficial deposits.' (p. 329)
Himus, 1954. A generic term for the superficial, as opposed to the *solid*, formations of the earth's crust. It includes all the various deposits of boulder-clay, outwash gravel and sand of Quaternary age. Much is of fluvio-glacial origin.
Thornbury, 1954. 'The name *drift* has come down from the days when what we now know to be glacial deposits were thought to represent materials rafted by icebergs during the Noachian deluge. The word survives . . . convenient . . . if we want to be non-committal and merely indicates that materials of glacial origin are present

without saying whether deposited by ice or water' (p. 376).

**Comment.** British Geological Survey Maps, Drift edition, show Glacial and post-glacial deposits (*i.e.* post-Pliocene) and also Plateau gravels, which may be Pliocene. The contrast is with 'Solid edition' excluding superficial deposits. The *O.E.D.* 10 (b) is confused. See also Continental drift. As a mining term a drift is not necessarily horizontal but may be at almost any angle except vertical, when shaft would be used.

**Drift (2)**
*Webster.* 25. *Phys, Geog.* One of the slower movements of oceanic circulation; a general tendency of the surface water, subject to diversion or reversal by the wind; as the easterly *drift* of the North Pacific.

Moore, 1949. A slow movement of surface water at sea, on a lake etc., caused by the wind; a similar movement of sand, etc.

**Drift, sand and snow**
*O.E.D.* 10. *Geol.* A term applied (c) to any superficial deposit caused by a current of water or air....

*Webster.* 7. A mass of matter, as of snow or sand, that has been driven or forced onward together in a body or piled together in a heap, esp. by the wind.

Bagnold, R. A., 1941, *The Physics of Blown Sand and Desert Dunes*, London: Methuen. 'A drift ... is formed in the lee of a gap between two obstacles and is due to "funnelling", or the concentration of the sand stream on the windward side from a broad front to a narrower one.' (p. 191)

Thornbury, 1954. More commonly applied to thin deposits of sand, called by Bagnold 'sandsheets' (p. 307).

**Comment.** The sand drift as used by Bagnold may be compared with snow drifts, *i.e.* snow drifted and piled up by wind.

**Drift-ice**
*O.E.D.* Drifting or drifted ice; esp. detached pieces of ice drifting with the wind or ocean currents.

Mill, *Dict.* Portions of icebergs or floes carried by currents into the open sea beyond the limit of a pack.

*U.S. Ice Terms*, 1952. 1. Floating ice. 2. Any ice that has drifted from its place of origin. Note: Floating ice is 'A general term applied to all types of ice (other than icebergs and other land ice) floating in the water'.

*Adm. Gloss.*, 1953. Loose, very open pack-ice, water predominating over ice. Unattached pieces of floating ice; navigable with ease.

**Driftless, drift-free**
*O.E.D.* 2. *Geol.* Free from drift. J. Geikie, 1873, *Great Ice Age.* 'The "driftless region" of Wisconsin, Iowa, and Minnesota.' (p. 465)

**Driftway, Drift-way**
*O.E.D.* A lane or road along which cattle are driven; a drove-road (sometimes shortened locally to drift).

**Drought, absolute**
*O.E.D. cf.* Absolute—II 6. Pure and simple, mere; in the strictest sense.
Drought or Drouth. 2. *Spec.* Dryness of the weather or climate, lack of rain (the current sense). First ref. c. 1200. 3. Dry or parched land, desert. Obs. rare. 1000 A.D.; 1671 Milton P.R. III, 274.

*Met. Gloss.*, 1944. 'A period of at least 15 consecutive days, to none of which is credited ·01 in. of rain or more.' (introduced in British Rainfall, 1887, p. 21)

**Drove-road, drove-way**
*O.E.D.* An ancient road or track along which there is a free right of way for cattle, but which is not 'made' or kept in repair by any authority.
See also drift-way; *cf.* also 'drover'—one who drives droves of cattle, sheep etc., to distant markets; hence a dealer in cattle.

**Drowned valley**
Mill, *Dict.* Drowned valley or submarine valley. River or ice valleys which have been submerged by the advance of the sea or a lake.

Swayne, 1956. A valley which owing to a change of sea-level has become partially submerged.

**Drubbel** (German; *pl.* same: *lit.* small nucleation)
Introduced by W. Müller-Wille to describe an ancient type of settlement in north-west Germany which, associated with a strip field (Langstreifenflur) was the ancestor of a number of different types of settlement which developed subsequently. An alternative is 'Urweiler' (aboriginal, ancestral or primeval) hamlet (K.A.S.).

**Drumlin** (Irish; *dim.* of drum which is rarely used)
*O.E.D.* Drum. A ridge or 'rigg', a long narrow hill often separating two parallel

valleys: a frequent element in Scottish and Irish geographical proper names. Hence Geol. A term for a long narrow ridge of 'drift' or diluvial formation, usually ascribed to glacial action. (Examples, 1725–1882, inc. Geikie.)

Mill, *Dict.* 'Drumlin, Drum. A ridge or bank of boulder clay having a lenticular or wedge-like form, and having its long axis longitudinal with the valley.

Geikie, J., 1898, *Gloss.* 'Drum, Drumlin (Ir. and Gael. druman, the back, a ridge); a ridge or bank of boulder-clay alone or of "rock" and boulder clay. Ridges of this kind often occur numerously. There seem to be two varieties (a) long parallel ridges or banks, and (b) short lenticular hillocks; the former usually consist of glacial accumulations alone; the latter not infrequently contain a core or nucleus of solid rock, or they may show solid rock at one end and glacial materials at the other.'

Knox, 1904. 'Drumlin (U.S.A.), a smooth oval or elongated hill or ridge, composed chiefly of glacial detritus.'

Cotton, 1922. 'Drumlins are elliptical or somewhat elongated low hills of subdued outline either formed by deposition under the margin of an icesheet by which they were moulded into shape, or carved out of an earlier thick drift deposit by the ice re-advancing over it.' (This expanded in ed. of 1945, and reference to 'rock drumlins' which grade into 'crag and tail' forms.) (pp. 323–324)

Wooldridge and Morgan, 1937. 'They are composed entirely of boulder clay'. (Footnoted: 'The term has been extended by some authors to cover the case of "rock-drumlins", smoothed mounds of rock with or without a thin veneer of boulder-clay.') (p. 384)

Lobeck, *Geomorphology*, 1939. 'Drumlins are smooth, oval hills composed mainly of till but sometimes including lenslike masses of gravel and sand.... Their stoss end, facing the glacier, is usually blunter and steeper than the tail or lee side.... Some drumlinlike hills have rock coves with a thin veneer of till. These are called *rocdrumlins*.' (p 306)

*Comment.* The meaning of drumlin has undergone an evolution. Knox, Cotton, Wooldridge and Lobeck give the modern usage; the *O.E.D.* is out of date.

**Drumlin Field**—see Basket-of-eggs topography

**Drumlinoid** (J. B. Tyrrell, 1906)
*Webster. Geol.* A hill of drift resembling a drumlin.
Tyrrell, J. B., Report on the Dubawni, Ferguson and Kazan Rivers, *Geol. Surv. Can. Ann. Report*, 9, 1906—described rather than defined—see Dean, W. G., The Drumlinoid Landforms of the Barren Grounds, N.W.T., *The Canadian Geographer*, 3, 1953, 19–30: 'Drumlin-like landforms ... the most characteristic features of glacial origin in the Barren Grounds ... every gradation from faint streak-like patterns to "true" drumlins ... their most common form is that of a parallel series of low spindle-shaped ridges' (p. 19).

**Dry**
Mill, *Dict.* A joint in rock determining a line of easy weathering (Scotland).
*Cf. O.E.D.* Masonry: 'A fissure in a stone, intersecting it at various angles to its bed, and rendering it unfit to support a load.' (Ogilvie).

**Dry** (climate)
*O.E.D.* 1c. Of a season or climate; free from or deficient in rain; having scanty rainfall; not rainy.
*Met. Gloss.*, 1944. 'In meteorology we may take it that air referred to as "dry" has at least a sufficiently low Relative Humidity for Evaporation to be taking place actively from earth, rock, etc., as well as from vegetation.' (In Beaufort notation, rel. hum. less than 60%—for which the letter is 'y'.)

Trewartha, G. T., 1937, *An Introduction to Weather and Climate*, New York: McGraw Hill. 'The essential feature of a dry climate is that evaporation shall exceed precipitation. In other words, during a normal year more water could be evaporated than actually falls as precipitation.... permanent streams cannot originate in these areas.' (p. 225)

On p. 359 the boundary is given (from Köppen) as where:

$r = 0.44t - 8.5$ (where rainfall evenly distributed)

$r = 0.44t - 3$ (where rainfall max. in summer)

$r = 0.44t - 14$ (where rainfall max. in winter)

These are the *B* climates of Köppen.

Thornthwaite, C. W., 1948, *Geog. Rev.*, **38**. When ratio $(100s - 60d)/n$ is negative. This boundary is between climates $C_1$ and

$C_2$ (s = water surplus; d = water deficiency: n = water need).
See also Arid, aridity index.

## Dry delta
A synonym for alluvial cone or fan (q.v.) (G.T.W.).

## Dry farming
O.E.D. Suppl. Dry farming U.S., farming without a good supply of water. 1908 Sci. Amer. 22 Aug., 120 'Dry farming' consists in so preparing the soil in semi-arid regions that it will catch what little annual rainfall there is, and store it within reach of the roots of the plants to be grown.

Webster. Agric. Production of crops without irrigation in regions of low or otherwise unfavorable rainfall, principally by tillage methods conserving soil moisture and by the use of drought-enduring or drought-evading crops.

Dict. Am. In semiarid regions, a method of farming whereby the rainfall is conserved to the utmost.

Mill, Dict. Such treatment of the soil that the water which falls during all parts of the year is retained in the soil and subsoil till required at the growing season (Salisbury).

Widtsoe, J. A., 1921, Dry Farming, New York: Macmillan. 'Dry-farming, as at present understood, is the profitable production of useful crops, without irrigation, on lands that receive annually a rainfall of 20 inches or less.' Rainfall of 20–30 inches may still require dry-farming, according to local conditions, etc.

Russell, E. J., 1950, Soil Conditions and Plant Growth, 8th ed. (rev. E. W. Russell), London: Longmans. 'Dry farming is farming under conditions of water shortage when irrigation cannot be used....' (p. 390)

Comment. Dry farming implies a specialized technological treatment of land to overcome the short supply of water. One much-used method is to crop the land only once in two years and, by breaking up or pulverizing the surface, to conserve at least a part of the moisture received in one year to add to that received in the next. To prevent loss of moisture by evaporation another method is to protect the surface by a mulch.

**Dry gap**—see Wind gap

## Dry point settlement
Aurousseau, M., 1920, Geog. Rev., 10. Contrasts 'wet point' and 'dry point' villages. Examples of the latter are S. of Minsk; Nile delta (mounds to avoid flooding); Flanders dune belt (on dykes or canal banks); villages on alluvial cones in Alpine valleys; Vallais, Rhône valley.

Comment. A settlement which has sought a location free from flooding in a region of wet soils or liable to floods, e.g. patches of gravel terrace in a clay vale.

## Dry valley
Himus, 1954. A valley which, though originally carved out by running water, is now streamless. Drying may be caused by a fall in the water-table, river-capture or by a change in the climate. The dry valleys which are common in the Chalk of Southern England are believed to date mainly from the close of the Pleistocene glaciations.

Wooldridge and Morgan, 1937, discuss the origin of the dry valleys of the Chalk and quote the theory of their origin developed by C. C. Fagg, Trans. Croydon Nat. Hist. Soc., 9, 1923 (pp. 283–287). Fagg acknowledged the hypothesis had been previously developed by Chandler, R. H., 1909 (Geol. Mag., 46, 538–539).

See also Sparks, B. W. and Lewis, W. V., 1957 (Proc. Geol. Assoc., 68, 26–38).

## Dry wash
Webster. 2. A gravelly or stony stream bed, normally dry, but occasionally filled by a torrent of water. Western U.S. Dict. Am. 1872.

## Dualism (Dual economies)
Thoman, R. S., 1962, The Geography of Economic Activity: An Introductory World Survey, New York: McGraw-Hill. 'This is the existence, resulting from invasion or indigenous growth, of an affluent and economically and socially advanced controlling group side by side with an amorphous mass of underdevelopment... the elite tend to be located at the central city, with branches in the principal market towns, while the underdeveloped people are scattered over the remaining inhabited portions of the country.' (p. 48)

**Duār** (Indo-Pakistan: Bengali)—see Terai

## Dub
O.E.D. A muddy or stagnant pool; a small pool of rain water in a road; a puddle (chiefly Sc.).

Mill, *Dict.* A deep pool of still water in the course of a swift stream (N. England).
*Comment.* Also used in plural for a limestone resurgence (G.T.W.).

**Duff** (United States; soil science)
*Webster.* The partly decayed vegetable matter on the forest floor.
Jacks, 1954. 'Surface horizon of partly decomposed organic remains resting on mineral soil (= raw humus of U.K.).'

**Dumb-bell, dumb-bell island**
Schofield, W., 1920. Dumb-bell islands and Peninsulas on the Coast of South China. *Proc. Liverpool Geol. Soc.*, **13**, 45–51.
'A dumb-bell may be defined as two areas connected by a relatively narrow isthmus of sand which is never below high water mark in any part of its length. The size of the areas of land may vary to any extent, but the highest points in both must rise higher above sea-level than any part of the isthmus.' *Cf.* tombolo.

**Dūn** (Indo-Pakistan: Urdu-Hindi)
Longitudinal intermontane valley, especially in the Siwaliks; also used in place names. Originally a word of the Pahari dialect.

**Dune**
*O.E.D.* A mound, ridge, or hill of drifted sand on the sea coast (or, rarely, on the border of a lake or river); applied esp. to the great sand-hills on the coast of France and the Netherlands. In earlier English use, down occurs.
*Webster.* A hill or ridge of sand piled up by the wind. Dunes are common along shores, along some river valleys, and generally where there is dry surface sand during some part of the year.
Mill, *Dict.* A hill of blown sand.
Bagnold, R. A., 1941, *The Physics of Blown Sand and Desert Dunes*, London: Methuen. Distinguishes dunes from sand shadows and sand drifts which are dependent for their continued existence on the presence of an obstacle, and cannot move away from it. 'Unlike shadows and drifts, dunes can exist independently of any fixed surface feature, and do, in fact, reach their most perfect development on flat featureless country. Although capable of movement from place to place, they are able to retain their own characteristic shape.' (p. 188)
*Comment.* Physiographically it is important to distinguish coastal from desert dunes. See also note under down, also barchan, seif dune and dune (subaqueous).

**Dune (subaqueous)** (G. K. Gilbert, 1914)
Gilbert, G. K., 1914, The Transportation of Debris by Running Water, *U.S.G.S. Prof. Paper*, 86. 'In another experiment a bed of sand was prepared with the surface level and smooth. Over this a deep stream of water was run with a current so gentle that the bed was not disturbed. The strength of the current was gradually increased until a few grains of sand began to move and then was kept steady. Soon it was seen that the feeble traction did not affect the whole bed, but only certain tracts and after a time a regular pattern developed and the bed exhibited a system of waves and hollows.... A current ... follows the rising slope and crest of the wave ... and then shoots free, to reach the bed again.... [On the rising slope] there is traction, the material being derived from the slope.... [Beyond the crest.] The debris being abandoned by the current is dumped and it slides by gravity down the [steeper] slope. So the upstream face of the wave is eroded and the downstream face built out, with the result that the wave as a surface form, travels downstream. As this is precisely what takes place when a sand hill travels under the influence of the wind, the name of the eolian hill has been borrowed and the waves are called *dunes*.' (pp. 30b–31a)

*Comment.* The term does not appear to have been generally adopted, though within its context it serves a useful purpose for distinguishing between the two types of debris waves or ripples. The term ripple is usually associated with small scale features but dunes can obviously be of large dimensions. (G.T.W.) The term 'sand-wave' is commonly used in the United States and Britain.

**Duricrust**
Woolnough, W. G., 1930, *Geol. Mag.*, **67**. Relating to Australia on surfaces believed to originate from the Miocene. Concentrations in the upper part of the soil profile of aluminous, siliceous, ferruginous, calcareous or magnesian materials thought to have been formed under a semi-arid climate with alternating wet and dry seasons by means of capillarity, which brought into the upper soil during the dry season minerals which had been dissolved during the wet season. It is only formed in areas of highly perfect peneplanation

almost devoid of topographic relief, where drainage cannot destroy the effects of chemical weathering.
See discussion in Thornbury, 1954, p. 83.

### Dust

Mill, *Dict*. Any substance comminuted or pulverized, especially earth or other solid matter in a minute and fine state of subdivision, so that the particles are small and light enough to be easily raised and carried in a cloud by the wind. Cosmic dust has a meteoric origin.

### Dust bowl

*Dict. Am*. A region, esp. in the western part of the U.S., subject to dust storms and drought. 1936.

Holmes, 1944. 'The most serious effects of wind deflation—from the human point of view—are experienced in semi-arid regions like the Great Plains west of the Mississippi where originally an unbroken cover of grass stabilised the ground, but long-continued ploughing and over-exploitation ... exposed it as a loose powder to the driving force of the wind. This national menace become critical during a period of severe droughts, culminating in 1934–35, when great dust storms originating in the "Dust Bowl" of Kansas swept over the States towards the Atlantic.' (p. 253)

*Comment*. Since commonly applied to similar conditions and areas elsewhere.

### Dust devil

Miller, 1953. 'Hot deserts (Saharan). The heat gives rise to strong convection currents resulting in strong, though variable, winds which whip up the dust and bear it along in clouds. These may be only local swirls (*dust devils*), but sometimes they are connected with cyclonic storms passing to the north and are on a much greater scale. (pp. 256–257)

*Comment*. A common phenomenon is most arid lands and the term dust-devil or simply 'devil' is widely used.

### Dust storm

Mill, *Dict*. A whirlwind passing over a dry or sandy district and carrying up the dust into the air, *e.g.* brickfielder and simoon (*q.v.*).

*Webster. Meteorol*. A violent dust-laden whirlwind moving across an arid region. The air is very hot, excessively dry, and attended by high electrical tension. When very violent, a dust storm is commonly called a *simoom*, or, in India, a *devil*, *shaitan*, or *peesash*.

Swayne, 1956. 'A type of storm occurring in arid regions due to a very turbulent wind passing over dusty soil. A storm is often preceded by a "wall" of dust which may extend up to 10,000 feet. It differs entirely from a sandstorm, which lifts and carries sand for a short distance but rarely to a greater height than 50–100 feet.'

*Comment*. See also dust bowl, dust devil, brickfielder, simoon. It seems the current practice to use dust storm as the broad general term covering dust-laden winds in general, not necessarily a 'whirlwind'.

### Duty of water

*Webster*. Duty. 6. *Agric*. The quantity of irrigation water required to mature a given area of a given crop, expressed in acre inches or acre feet per acre—called specif. *duty of water*.

### Duty of water traction (G. K. Gilbert, 1914)

Gilbert, G. K., 1914. The Transportation of Debris by Running Water, *U.S.G.S. Prof. Paper*, 86. 'Capacity varies with discharge but is not proportional to it. The load which may be borne by a unit of discharge varies with the discharge and also with the other conditions. It is the capacity per unit discharge, or the quotient of capacity by discharge, and will be called the *duty*.' (p. 36a) 'The duty of water traction, as defined by the units adopted for this paper, is the capacity in grams per second for each cubic foot per second of discharge.' (p. 74b)

### Dy (Swedish)

Swedish folk name; proposed by H. v. Post, 1862, for muddy acid AC soils, biologically extraordinarily inert, occurring at the bottom of brown waters which consist to a great extent of an amorphous precipitation of humus gels. (E.K.). AC refers to the existence of these horizons in the soil profile.

Jacks, 1954. 'Muddy material formed of plant residues and deposited from nutrient-poor water.'

### Dyke—see Dike

### Dynamic equilibrium

Chorley and Kennedy, 1971. 'A circumstance in which fluctuations are balanced about a constantly-changing system condition which has a trajectory of unrepeated "average" states through time.' (p. 348)

### Dynamo-metamorphism, dynamic metamorphism

*Webster. Geol*. Metamorphism in which

mechanical energy, as exerted in pressure and movement, is the principal agent—contrasted with *pyrometamorphism, hydrometamorphism*.

Mill, *Dict*. Alteration of rocks by action of high pressure without great rise of temperature (*Alfred Harker*).

*Comment*. The contrast is with Contact-metamorphism. See Metamorphism. The *O.E.D.* gives the puzzling definition of dynamo-metamorphism as 'the transformation of energy from one mode of action to another'.

**Dyngja** (*pl*. dyngjur)

Mill, *Dict*. A gently-sloped volcano formed by successive outpourings of fluid lava unaccompanied by accumulations due to violent ejection (Iceland).

**Dysgeogenous**

*Webster*. Not easily decomposing into soil.

Mill, *Dict*. The type of rocks which yield by weathering but a small amount of detritus (*Thurmann*) *cf*. eugeogenous.

# E

**Eagre, eigre**
*O.E.D.* (of unknown etymology). A tidal wave of unusual height, caused by the rushing of the tide up a narrowing estuary; Bore. Chiefly with reference to the Humber (and Trent) and the Severn.
Mill, *Dict.* The bore of the Humber or Severn.

**Earth**
Rice, 1943. 1. The solid matter of the globe in distinction from water and air. The ground. The firm land of the earth's surface. 2. Loose material of the earth's surface; the disintegrated particles of solid matter in distinction from rock, soil. 3. In chemistry—a name formerly given to certain inodorous, dry and uninflammable substances which are metallic oxides but were formerly regarded as elementary bodies. 4. Term used for soft shaley ground when sinking through coal measures.
*Comment.* Commonly used also for the planet which we inhabit.

**Earthflow**
*Webster.* A landslide consisting of unconsolidated surface material that moves down a slope when saturated with water. Cf. Mudflow.

**Earth-movement**
*Webster. Geol.* Differential movement of the earth's crust; elevation or subsidence of the land; diastrophism; faulting; folding; mountain building or orogeny.
Wooldridge and Morgan, 1937. 'Earth-movements of two distinct types are taking place within the range of our observation. Slow, or secular, movements of upheaval and depression are evidenced on many coastlines, while sudden movements, sometimes producing visible features at the earth's surface, are responsible for earthquakes (p. 54) ... we may adopt the following broad classification of earth-movements:
Epeirogenetic (radial forces)
Orogenetic (tangential forces)
  a Movements of contraction (forces of compression)
  b Movements of extension (forces of tension)' (p. 64).
See Orogenesis, orogeny, epeirogenetic, fault, etc.

**Earth pillar**
*Webster.* Earth pillar or pyramid. *Geol.* A high pillar of earth capped by a stone, formed by the erosion of surrounding material.
Mill, *Dict.* A mass of softer material capped by a harder rock which shields the underlying portions from the effects of denudation, though all the surrounding material may have gone. If rain is the chief agent of removal the block acts as an umbrella, and the pillar widens out towards the base all round. If a stream does most of the work it will eat more and more into the pillar, which will therefore narrow down towards the base. They are commonest in boulder clay. In the French Alps these pillars are called demoiselles, cheminées de fées (fairy chimneys), penitents.
*Comment.* Mill's definition is accurate, except the reference to boulder clay. They are particularly characteristic of badlands where hard sandstone blocks form the capping to unconsolidated sands. Miniature examples are often seen in course of formation in rainstorms. Miniature forms are also occasionally found in caves where roof-drips form the denuding agent. See also demoiselle, hoodoo, penitent.

**Earthquake**
*O.E.D.* 1. A shaking of the ground; usually *spec.* a convulsion of the earth's surface produced by volcanic or similar forces within the crust.
Mill, *Dict.* A series of elastic and quasi-elastic waves often modified by gravity, usually originating by sudden geological adjustments in or beneath the crust of the Earth, and then propagated from this origin to varying distances in all directions through and over the surface of the globe. A few originate by explosions at volcanic foci (Milne).
*Webster.* Earthquakes are divisible into two classes, volcanic and tectonic, most great destructive earthquakes being tectonic and caused by faulting.
Wooldridge and Morgan, 1937. 'Many

important shocks are demonstrably related to lines of pre-existing fracture in the crust. A small movement along such a fracture ... sets up secondary wave-like disturbances which travel through the body of the earth as well as along its surface.' (p. 58) '... The larger shocks ... frequently produce actual topographic features which indicate both vertical and horizontal movement.' (p. 60)

Comment. The O.E.D. definition tends to perpetuate the association in the popular mind of volcanoes and earthquakes. In fact the most severe earthquakes occur in association with fault lines and where there are no volcanoes to act as safety valves.

**Earth's crust**—see Crust

**Ebb, ebb-tide**

O.E.D. Ebb: to flow back or recede, as the water of the sea or a tidal river; *sub.* the return of tide-water towards the sea.

Ebb-tide: the falling tide; contrast flood-tide.

**Ebb channel**

Mill, *Dict*. The channel in a river estuary in which the tide flows out most strongly; found on the side of the estuary towards which the movement of the ebb tide is directed in accordance with Baer's law (Krümmel).

No reference in *Adm. Gloss.*, 1953.

**Éboulis** (French)

A scree (Jura etc.).

**Ecliptic** (*lit.*, of or pertaining to an eclipse)

O.E.D. 1. The great circle of the celestial sphere which is the apparent orbit of the sun. So called because eclipses can happen only when the moon is on or very near this line.

2. The great circle on the terrestrial sphere which at any given moment lies in the plane of the celestial ecliptic.

Admiralty *Navigation Manual.* 'The apparent path of the Sun in the Celestial Sphere is called the *ecliptic*. It is a great circle, and it makes an angle of 23° 27' with the celestial equator.'

**Ecoclimate**

Climate considered in relation to plant and animal life.

**Ecology**

O.E.D. Œcology. The science of the economy of animals and plants; that branch of biology which deals with the relations of living organisms to their surroundings, their habits and modes of life, etc. 1879 tr. Haeckel's Evol. Man, 1, 114. 'All the various relations of animals and plants to one another and to the outer world, with which the Oekology of organisms has to do ...' (*Note:* Haeckel spells œcology in 1873.)

The supplement contains 'ecology' as the more usual form, with examples from 1896.

*Chambers's Encyclopedia* (G. Salt). 'Ecology can be briefly defined as the study of the relations between living organisms and their environment.... The environment ... is ... the whole association of factors, inert and living, of which the organism is a part.'

Pearse, A. S., 1939, *Animal Ecology*, New York: McGraw-Hill. 'Ecology is the branch of biological science that deals with relations of organisms and environments. ... Ecology is often divided into autecology and synecology....' (p. 1)

Tansley, A. G., 1946, *Introduction to Plant Ecology*, London: Allen and Unwin. 'In its widest meaning ecology is the study of plants and animals as they exist in their natural homes; or better, perhaps, the study of their household affairs, ...' (p. 15)

Clements, F. E. and Shelford, V. E., 1939, *Bio-ecology.* New York: Wiley. Proposes the term bio-ecology because of confusion with biology, sociology, geography, or geobotany and obscurity of subdivisions, autecology, synecology, insect ecology, human ecology. (p. 1) Human ecology should be included.

See also Carpenter, 1938; Shelford, *The Science of Communities*, 1929; Haeckel, *The Science of Relationships of organisms to environment*, 1866.

Comment. At least one school of geographers has defined geography as human ecology (see H. H. Barrows, Geography as Human Ecology, *Ann. Assoc. Amer. Geog.*, 13, 1923, 1–14). Geographers tend to use 'human ecology' as the relationship between people and place, sociologists as association of people with people. See James, P. E., and Jones, C. F., *American Geography*, Syracuse: Syracuse U.P., 1954, p. 28.

Without qualification 'ecology' has tended to become restricted to plant ecology. This is the sphere of the *Journal of Ecology* since the appearance of the *Journal of Animal Ecology* as a separate publication, but both are issued by the Ecological Society in Britain.

Ecologists have introduced a large number of special terms into their studies. There is first the simple idea of the plant communities—the association (*q.v.*) and the consociation (*q.v.*) which have developed in harmony with the environment. The idea of development is fundamental and the development sequence is a *sere*: in water the sequence is a *hydrosere*, under dry conditions inland a *xerosere*, on a rock surface a *lithosere*. A group of plants representing a stage in development is a *seral community*: where this is kept more or less permanent as a result of human activity it may be termed a *subclimax* or *subsere*; when definitely due to human interference the term *plagiosere* or *plagioclimax* is used (*e.g.* much grassland in farmed country). Theoretically the final stage in development is the *climax* (*q.v.*) which will be primarily determined by climate unless soil is the more important (*edaphic climax*). More complex is the concept of the *ecosystem* (*q.v.*), *ecotope* and *ecotone* (*q.v.*).

For ecological terms see Ecoclimate, ecosystem, biosphere, climax, sere. It has only been possible to include in this glossary some of the ecological terms most commonly used in geographical works. See Dansereau, P., 1957, *Biogeography: an Ecological Perspective*, New York: Ronald Press, for a very wide range of terms.

**Economancy**
Zimmermann, E. W., 1951. *World Resources and Industries*, 2nd ed., New York: Harper. 'Economancy is the by-product economy that results incidentally from conservation.' (p. 807) Other terms relative to conservation are defined on the same page.

**Economic geography**
Weber, A., 1929, *Theory of the Location of Industries* (Trans. C. J. Friedrich). Chicago: U. of Ch. Press. '... the theoretical consideration of the distribution of economic processes over a given area.' (p. 1)

Smith, W., 1948, *An Economic Geography of Great Britain*, London: Methuen. '... the analysis of the distribution forms or patterns of economic life.' (p. xii)

Brown, R. N. R., 1946, *The Principles of Economic Geography*, 5th ed., London: Pitman. 'Economic geography is that aspect of the subject which deals with the influence of the environments, inorganic and organic, on the economic activities of man.' (p. ix)

James, P. E. and Jones, C. F. (ed.) 1954, *American Geography*, Syracuse: Syracuse U.P. 'Economic geography has to do with similarities and differences from place to place in the ways people make a living. The economic geographer is concerned with economic processes especially as manifested in particular places modified by the phenomena with which they are associated.' (p. 243)

Committee, List 1. 1. The branch of geography dealing with the production, distribution, exchange and consumption of wealth.

2. The study of economic factors affecting the areal differentiation of the Earth's surface.

*Comment.* Economic Geography and Commercial Geography are often used interchangeably especially in the titles of textbooks and examination syllabuses. In many, indeed most, such cases nothing more is implied than geography with an emphasis on production and trade.

**Economic man**
*The McGraw-Hill Dictionary of Modern Economics*, 1973, 2nd ed., New York: McGraw-Hill. 'The concept of man motivated solely by economic reasons, credited by the economists of the classical school. Economic man was considered representative of the average man, with all the characteristics of the human race, but average only in an economic sense. All his characteristics were economic.... They failed to realize, however, that besides profit and other pecuniary interests, man desires leisure, independence, security, and other non-monetary benefits.' (pp. 189–190)

See Noyes, C. R., 1948, *Economic Man in Relation to his Natural Environment*, New York: Columbia University Press.

**Economic rent (Land rent)**
Thünen, J. H. Von, 1966 edn, Hall, P. (ed.), *Von Thünen's Isolated State*, Oxford: Pergamon Press. 'That part of the total (gross) product of land which remains as a surplus after deduction of all costs, including interest on invested capital. Equivalent to "economic rent" in English classical economics.' (p. li)

**Economy**
O.E.D. 2. The administration of the concerns and resources of any community or establishment with a view to orderly con-

duct and productiveness; the art or science of such administration. Frequently specialized by the use of adjectives as Domestic, Naval, Rural, etc. ...
b. esp. Management of money, or of the finances.
3. Political economy. 8. The organization, internal constitution, apportionment of functions, of any complex unity.
Wooldridge and Morgan, 1937. Sub-heading of 'The general economy of erosion', containing an analysis of the load of a river and its work (p. 152). (Used in *O.E.D.* Sense 8.)
Committee, List 4. The whole economic structure and function of a country, region or society. An economy may be described as agricultural, primitive, or industrial according to the activities which preponderate.

### Ecosphere
Cole, L., 1958, The Ecosphere, *Scientific American*, 198 (4). 'It is intended to combine two concepts: the "biosphere" and the "ecosystem" ... "biosphere" as the collective totality of living creatures on earth ... "ecosystem" means a self-sustaining community of organisms—plants as well as animals—taken together with its inorganic environment.' 'Now all these are interdependent ... a self-sustaining community must contain not just plants, animals and nitrogen-fixers but also decomposers which can free the chemicals bound in protoplasm.' (pp. 83–85) '... each ecosystem is a composite of the community and the features of the inorganic environment that govern the availability of energy and essential conditions that the community members must tolerate.' (p. 85)
*Comment.* Cole introduced the term to refer to the whole terrestrial ecosystem.

### Ecosystem (A. G. Tansley, 1935), ecotope
Tansley, A. G., 1935, 'The Use and Abuse of Certain Vegetational Concepts and Terms', *Ecology*, 16, 284–307
Tansley, 1939. 'A unit of vegetation considered as a system includes not only the plants of which it is composed, but the animals habitually associated with them, and also all the physical and chemical components of the immediate environment or habitat which together form a recognisable self-contained entity. Such a system may be called an *ecosystem*, because it is determined by the particular portion, which we may call an *ecotope* (Greek τοπος, a place), of the physical world that forms a home (οίκος) for the organisms which inhabit it. A prisere is the gradual development of such a system, the climax represents the position of relative equilibrium which it ultimately attains, and a subsere is the redevelopment of the same type of system after partial destruction.' (p. 228)
Fosberg, F. R., 1957. *Abs. of Papers, Ninth Pacific Science Congress*, Bangkok. 'An ecosystem is a segment of nature with its included organisms and their environment. Such a segment may be of any magnitude convenient to facilitate study or understanding.' (p. 259)
See also full discussion in Fosberg, F. R., 1963, *Man's Place in the Island Ecosystem*, Honolulu: Bishop Museum Press, where the term is compared with microcosm (Forbes) and biosystem (Thienemann). Recent introductions have included ecosystematic and ecosystematics (the study of ecosystems).

### Ecotone
Dansereau, P., 1957, *Biogeography*, New York: Ronald Press. 'In the contact between two climaxes, there is an area where small changes in climate produce a contamination of one zone by the other. These transition zones are called ecotones.' (p. 178)
Tansley, 1939. 'Between two adjacent communities there is very commonly a "tension belt" or *ecotone* (Greek τονος, strain).' (p. 215)

### Ecotope—see Ecosystem
### Ecumene—see Œcumene
### Ecumenopolis
Doxiadis, C. A. and Papatoannou, J. G., 1974, *Ecumenopolis: the Inevitable City of the Future*, Athens Publishing Centre. 'The coming city that will together with the corresponding open land which is indispensable for Anthropos, cover the entire earth as a continuous system forming a universal settlement. It is derived from the Greek words ecumene, that is the total inhabitable area of the world, and polis, or city, in the broadest sense of the word. Term coined by C. A. Doxiadis and first used in October, 1961, issue of *Ekistics*.' (p. 436, glossary)
*Comment.* Attempt to redefine the expanding scale of urban areas. Previous defini-

tions like 'metropolitan region', 'megalopolis' and so on had a smaller conception of urban life.
See also Conurbation, Megalopolis, and Œkumene.

**Edaphic**
*O.E.D. Suppl.* Pertaining to, produced or influenced by, the soil.
Mill, *Dict.* Relating to the soil, due to, or having characteristics due to the nature of the soil (Schimper).
Tansley, 1939. 'The arresting factors in both these types of subclimax are *edaphic factors, i.e.* dependent on *soil* (p. 222). . . . Various climaxes determined by edaphic, physiographic and biotic factors are normally dominated by life-forms simpler than the climatic climax (p. 225). . . . A. F. W. Schimper pointed out that the great plant formations were determined by climatic factors (climatic formations) and that the vegetation so determined was sorted out into smaller units by the influence of different soils (edaphic formations). In trying to classify British vegetation on natural lines C. E. Moss based his scheme on the view that the climate was so uniform as to justify classification according to the soil types . . . these primary edaphic divisions he called "plant formations" . . . corresponded with Schimper's "edaphic formations".' (p. 236)
*Comment.* Also used by Chamberlain and Salisbury, *Geology*, 2, 343, of the seabottom. See also Epedaphic.

**Edaphon**
*Webster. Biol.* The animal and plant life present in soils.
Jacks, 1954. 'The whole living community of the soil.'

**Edge**
*O.E.D.* 6. The crest of a sharply pointed ridge; frequent in topographical names, as Swirrel Edge, Striding Edge. More frequently, however, names of this kind denote escarpments terminating a plateau . . . *e.g.* Millstone Edge, Bamford Edge. In Scotland edge usually denotes merely a ridge, watershed.
10. The line which forms the boundary of any surface; a border, verge. By extension, that portion of the surface of any object, or of a country, district, etc., adjacent to its boundary. In geographical sense formerly often used where frontier or boundary would now be preferred.

Mill, *Dict.* A sharply-scarped ridge, the outstanding terminal line of hard uplifted strata. It has an abrupt slope (scarp slope) on the side on which the strata in question have been broken or worn away, and a gentler slope (dip slope) in the direction in which the strata dip down.
Knox, 1904. (Anglo-Saxon ecg), a sharp point, a narrow part rising from a broader; the highest part of a moorish and elevated tract of ground of considerable extent, generally that which lies between the streams, *e.g.* Axe Edge, Ipstone Edge, Claverton-Edge.
*Comment.* Mill's definition is that of a cuesta. Edge is used loosely rather than with his precise meaning; *O.E.D.* is satisfactory. A topographic term and one not generally used in geographical literature in any specific sense.

**Effective precipitation**
Jacks, 1954. 'That part of total precipitation which is of use to plants.'

**Efficiency of a stream** (G. K. Gilbert, 1914)
Gilbert, G. K., 1914. The Transportation of Debris by Running Water., *U.S.G.S. Prof. Paper*, 86. 'The load which may be borne by a unit of discharge on a unit slope varies with all the conditions of transportation. It is the capacity per unit discharge and unit slope, or the quotient of capacity by the product of discharge and slope and will be called the *efficiency*. It is a measure of the stream's potential work of transportation in relation to its potential energy.' (p. 36a) 'The measure of efficiency is $E = C/QS$, where C is the capacity for stream traction in grams per second; Q is the discharge in cubic feet per second and S is the slope of channel bed per cent.' (p. 75)
See also Capacity, load.

**Efflorescence**
*O.E.D.* (Latin, ex, out + florescere, to blossom) 4. Chem. The process of efflorescing, in various senses; also the powdery deposit which is the result of this process. Effloresce. 2.c. of the ground, a wall, etc. To become covered with a powdery crust of saline particles left by evaporation from a solution which has been drawn to the surface by capillary attraction.
Mill, *Dict.* The surface crust of previously dissolved material left after the evaporation of water which has arisen from the soil in arid regions.

### Effluent

*O.E.D.* B.*a.* A stream flowing from a larger stream, lake or reservoir.
*b.* The outflow from a sewage tank, or from land after irrigation or earth-filtration of sewerage.

Mill, *Dict.* A stream issuing from a lake.

Rice, 1941. (1) A stream whose upper surface stands lower than the water table in the locality through which it flows and which is not separated from the water table by ny impervious bed. (2) A stream which flows out of another stream or out of a lake. (p. 116)

*Comment.* Also used of waste products from a factory (C.J.R.).

### Effluent cave (G. T. Warwick, 1953)

Warwick, G. T., 1953, *British Caving.* Effluent cave: introduced in lieu of gallicized 'Cave of Debouchure', *i.e.* 'a cave entered at the lower end from which a stream issues or is known to have issued'. (p. 438)

See Cave.

### Egocentrism

Tuan, Y. F., 1974, *Topophilia*, Englewood Cliffs, New Jersey: Prentice-Hall. 'Since consciousness lies in the individual, an egocentric structuring of the world is inescapable; and the fact that self-consciousness enables a person to view himself as an object among objects does not negate the ultimate seating of that view of the individual. Egocentrism is the habit or ordering the world so that its components diminish rapidly in value away from self.' (p. 30)

Gould, P. and White, R., *Mental Maps*, Harmondsworth: Penguin Books. '... see directions in relation to their own position at the moment. These people seem to be able to navigate much more surely ...' (p. 28)

*Comment.* Ideas about egocentrism and ethnocentrism introduced by Trowbridge, C. C., 1913, On Fundamental Methods of Orientation and Imaginary Maps, *Science*, **38**, 9 December, 888–897.

See also Ethnocentrism.

### Einzelhof (German; *pl.* Einzelhöfe: *lit.* isolated farm)

The farm land is not necessarily consolidated (K.A.S.).

### Eis (German)

Ice. Some writers prefer to retain the German in such compounds as Eisblink, instead of using what is really a partial translation in 'ice blink' (*q.v.*).

### Eiscir (Irish)—see Esker

### Ek fasli harsala—see Do

### Ekistics

The science of human settlements.

*Comment.* A large part of the *Report of Proceedings* of the Town and Country Planning Summer School, Southampton, 1959 (London: Town Planning Institute) is devoted to the new science of Ekistics. 'Ekistics is derived from the Greek word EKOS or habitat and from the verb EKO meaning to settle down ... the science of human settlements ... the settlement itself, its evolution and its formation.' (p. 119) (C. A. Doxiadis, but it is not stated who introduced the term.)

### Elbasin (T. Griffith Taylor, 1940)

Taylor, 1951. 'An elevated basin, often wrongly called a plateau (*e.g.* upper Yukon R.).' (p. 613) In MS. communication Griffith Taylor instances also the Swiss Plateau and the British Columbia Plateau but regrets that the term has not been generally adopted.

### Electoral Geography

Prescott, J. R. V., 1959, The Function and Methods of Electoral Geography, *A.A.A.G.*, 49, 296–304. 'The study of election statistics has some value for political or social geography.'

### Elevation

The same as altitude (*q.v.*).

### Eluvial horizon (Soil science)

Jacks, 1954. 'Layer from which material has been removed in solution or in water suspension and in which silt- and sand-sized particles have become concentrated.'

Eluvial horizon is broadly the A-horizon.

### Eluviation (Soil science)

The process by which material is removed in solution or in water suspension from the upper horizon or horizons of a soil. *Cf.* downward leaching.

Rice, 1943. 'The process in soil formation of removal by either mechanical or chemical means. By this process, clayey material is characteristically removed from the top-soil horizon.' (p. 114)

### Eluvium, Eluvial

*O.E.D.* Geol. (mod. Latin, e, out+luere, to wash, on the analogy of Alluvium). A term proposed for accumulations of debris either produced in situ by atmospheric agencies, or carried by wind-drift.

*Webster.* Geol. Rock debris produced by the

weathering and disintegration of rock in situ.

Mill, *Dict.* Products of weathering lying on the original place of deposit (Supan).

Geikie, A., 1885, *Text-book of Geology*, 2nd ed., London: Macmillan. 'For atmospheric accumulations of this nature, Trautschold has proposed the name eluvium. They originate in situ, or at least only by wind-drift, whereas alluvium requires the operation of water, and consists of materials brought from a greater or less distance.' (p. 308)

Holmes, 1928. 'A general term, used more particularly by Continental and American writers, for residual deposits.' (p. 89)

Rice, 1943. *Eluvial.* Formed by the rotting of rock in place to a greater or less depth. *Eluvium.* General term, used more particularly by American and Continental writers, for residual deposits. (p. 114)

*Comment.* A term which has tended to go out of use. In speech the distinction between eluvium and alluvium is dangerously slight.

**Emigrant**—see Migrant

**Emir, emirate**

*Webster.* From Arabic amīr, commander. An independent chieftain or ruler of a province.

*Comment.* Formerly used especially in West Africa: emirate, the territory ruled by an emir.

**Empolder**

To reclaim land by the creation of polders (*q.v.*).

**Emposieu** (French)—See Karst terminology.

**Enclave** (French)

*O.E.D.* (French, enclaver, to enclose, shut in.) A portion of territory entirely surrounded by foreign dominions. Also figuratively.

Mill, *Dict.* A small district entirely surrounded by foreign territory.

*Comment.* Also used for outlying portions of counties, *e.g.* Dudley in Worcestershire and for small ethnic groups surrounded by a dominant nationality (G.T.W.). See also Exclave.

**Enclosure**—see Inclosure.

**Endogenetic** (also **endogenic**)

*Webster.* Endogenetic: *Geol.* Designating rocks formed by solidification from fusion precipitation from solution, or sublimation—opposed to *clastic.*

*Webster.* Endogenic: 2. Pertaining to, having to do with, or designating the processes of metamorphism, etc., taking place within the earth—opposed to *exogenic.*

Wooldridge and Morgan, 1937. '. . . internal uplifting, distorting or disrupting forces, which may be grouped as "endogenetic".' (p. 54)

Thornbury, 1954. 'The term *process* applies to the many physical and chemical ways by which the earth's surface undergoes modification. Some processes, such as diastrophism and vulcanism, originate from forces within the earth's crust and have been designated by Penck as *endogenic*, whereas others, such as weathering, mass-wasting and erosion, result from external forces and have been called *exogenic* in nature. In general, the endogenic processes tend to build up or restore areas which have been worn down by the exogenic processes.' (p. 19)

Himus, 1954. Endogenetic. A term applied to geological processes which originate within the earth, and to rocks, ore-deposits and land-forms resulting therefrom. Contrasted with *Exogenetic.*

*Comment.* Endogenetic is the usual English form, endogenic the American. As stated by Thornbury the meaning of endogenic and exogenic is quite clear, but actually there has been a great deal of confusion between these words and the ambiguous endogenous and exogenous which have quite different meanings. Notice that *Webster* distinguishes between endogenic and endogenetic.

**Endogenous**

*O.E.D. Suppl. Geol.* Formed within a mass of rock, or within the earth's surface; spec. applied to intrusive rock changed by contact with surrounding rocks.

*Webster.* 3. *Geol.* Designating or pertaining to intrusive igneous rocks. *Rare.*

Mill, *Dict.* Proceeding from the interior of the Earth.

*Comment.* See also Endogenic, endogenetic and contrast exogenous (with quotation from Dana), exogenic. Both endogenous and exogenous (quite different from endogenic and exogenic) are ill-defined and have been used in different senses.

**Endoreic**—see Areic

**Endrumpf** (German)

Penck, W., Morphological Analysis of Land Forms (Trans. H. Czech and K. C. Boswell). London: Macmillan, 1953. 'The final surface of truncation, the end-peneplane or Davisian peneplain.' (p. 144)

*Comment.* Although this is a quotation from Penck, the endrumpf is the final result of Penck's theoretical cycle and should have characters different from the Davisian peneplain. (C.T.S.) This German term, not in general use, corresponds rather to an extended use of the pediplain of L. C. King (G.T.W.). See also Fischer, 1950, p. 13.

**Englacial drift, river, moraine**

*O.E.D. Suppl.* Embedded in or passing over the surface of a glacier.

*Webster.* Englacial: *Phys. Geog.* Embedded in a glacier; as *englacial* drift; also, traversing the body of a glacier; as, an *englacial* stream.

Mill, *Dict.* Englacial Drift. Debris carried enclosed in the mass of a glacier. Its position may be due to upward movement of some of the ice from the bottom of the glacier, or to the fact that the glacier has passed over some considerable elevation in its path.

Englacial River. A river running in a mass of land-ice.

*Comment.* A river passing over the surface of a glacier may be distinguished as superglacial. (Salisbury, R. D., Geol. Survey, New Jersey, 87, 1891.)

**Engrafted rivers**

Mill, *Dict.* Rivers, separate on the older land, which may unite or engraft on their way to the sea, as when a submarine coastal shelf is uplifted and laid bare as a coastal plain.

No reference in Geikie, J., 1898; Salisbury, 1907; Davis, 1909; Wooldridge and Morgan, 1937; Cotton, 1922, 1941, 1942; Holmes, 1944.

*Comment.* To be regarded as obsolete.

**Ennoyage** (French)—see Inverted relief.

**Enters**

Dwerryhouse, A. R., 1905, *Proc. Yorks. Geol. Soc.*, 15. 'The terms "enters", "shake-hole", "water swallow" are used indifferently by the O.S. map to indicate smaller openings of a similar nature or even those funnel-shaped hollows, otherwise known as swallow holes' (p. 238).

**Entrenched meander**—see Intrenched meander

**Entrepôt** (French), entrepot

*O.E.D.* (French, after Latin, interponere inter, between + ponere, to place).
1. Temporary deposit of goods, provisions etc.; chiefly *concr.* a storehouse or assemblage of storehouses for temporary deposit
2. A commercial centre; a place to which goods are brought for distribution to various parts of the world.
3. A mart or place where goods are received and deposited, free of duty, for exportation to another port or country.

Mill, *Dict.* A commercial town or district of importance for trade between other countries, and itself not necessarily either a great producing or consuming centre; = staple-land, staple-town (Eng.); = Stapelländr, Stapelstad (Ger.).

*Comment.* The correct form is entrepôt, but the word is commonly anglicized to entrepot and frequently, used as an adjective especially 'an entrepot port', 'entrepot-trade'.

**Envirium** (*pl.* enviria), envirial (Patrick Horsbrugh, c. 1963)

A term introduced in lectures and apparently generally accepted by Patrick Horsbrugh, Professor of Architecture successively at the Universities of Nebraska and Texas, to denote a specific environment (or environments).

**Environment**

*O.E.D.* That which environs; the objects or the region surrounding anything. Especially the conditions under which any person or thing lives or is developed; the sum-total of influences which modify and determine the development of life and character.

See Environmentalism and determinism.

**Environmentalism**

The philosophical concept which stresses the influence of the environment on the life and activities of man. In its extreme form it becomes Environmental Determinism (*q.v.*).

Tatham, G., in Taylor, 1951. In his article *Environmentalism and Possibilism* George Tatham discusses the origin and development of the philosophical concepts involved.

Haggett, 1972. 'Late nineteenth and early twentieth century geography saw the high point of environmentalism, in which extreme claims were made for the direct role of the natural world in many aspects of man's affairs, from the bedroom (e.g.

climatic rhythms and conception times) to the battlefield (e.g. terrain and tactics).' (p. 44)

*Comment.* Not in *O.E.D.* or *Suppl.* Tatham discusses but does not define environmentalism which seems to have crept into use without being defined, but see environment. See also R. S. Platt, 1948, *Amer. Jour. Soc.*, 53, 351–358.

**Eohypses,** eoisohypses (Swedish eohyps, pl. eohypser; or eoisohyps, pl. eoisohypser). (From Greek, eos, dawn and isohypse, contour line.)
A term introduced by Gerard De Geer (1912) for reconstructed (imaginary) contour lines. (E.K.)

**Eolian**—see Aeolian

**Eolithic**
The dawn of the Stone Age, pre-Palaeolithic, with 'eoliths' or pebbles crudely chipped possibly by human agency; see Palaeolithic.

**Eolomotion** (R. C. Kerr and J. O. Nigra, 1952)
Kerr, R. C. and Nigra, J. O., 1952, Eolian Sand Control, *Bull. Amer. Ass. Petroleum Geol.*, 36. 'Eolomotion is a relatively slow down-wind movement or down-hill creep as a consequence directly or indirectly of wind action on surface rock particles. This overall surface movement is caused (1) by the pushing effect of the wind on the exposed face of each grain, (2) by the impingement of rebounding grains on other grains with sufficient force to overcome inertia, and (3) by the effect of gravity, in which case the sheer angle or angle of repose of a particle is disturbed by eolian undermining resulting in tractional creep' (pp. 1541–1542).

**Epedaphic** (I. B. Balfour, 1902)
Balfour, I. B., 1902, *Encycl. Brit.*, 25, 430. 'The varying climatic or environmental conditions to which Angiosperms may be exposed in their wide distribution, including those of the soil, edaphic, those of the atmosphere, epedaphic, and those of water, aquatic.'
*Webster.* Pert. to, or depending upon, atmospheric conditions.
*Comment.* Edaphic has been generally adopted but not epedaphic.

**Epeirogenetic, Epeirogenic, Epeirogeny, Epeirogenesis**
*O.E.D. Suppl.* Epeirogenic. Geol. (from Greek mainland + -genic) Of or pertaining to the formation of continents. Also Epeirogenetic in the same sense. Epeirogenesis, epeirogeny: the formation of continents.
1890. G. K. Gilbert, *Lake Bonneville.* Th process of mountain formation is *orogeny*, the process of continent formation is *epeirogeny*, and the two collectively are diastrophism.
Mill, *Dict.* Epeirogenic Movement—the type of earth-movements which result in such changes of level, over large areas, as are indicated, for example, by raised beaches or coast-lines (Gilbert).
Wooldridge and Morgan, 1937. '... the result of radial movements, due to forces acting vertically, which have been styled epeirogenetic (continent building or plateau building).' (p. 62)
Baulig, H., 1935, The Changing Sea Level, *Inst. Brit. Geog. Pub.* No. 3. London: Philip. '... uplifts or sinkings over large areas, accompanied by very gentle warping or folding, so gentle indeed as to be hardly perceptible in short distances: such movements are known as continent-making or *epeiorogenic* movements.' (p. 1) See also Eustatic.

**Epicentre, epicenter, epicentrum**
*O.E.D.* gives Epicentrum. The point over the centre applied in *Seismology* to the outbreaking point of earthquake shocks.
*Webster.* Epicentre: *Seismol.* The portion of the earth's surface directly above the focus of an earthquake; a seismic vertical.
Wooldridge and Morgan, 1937. The point or small area immediately above the focus of an earthquake (p. 58).
*Comment.* Epicentrum has given place to epicentre. The *O.E.D.* is not very clear.

**Epicontinental sea**
Mill, *Dict.* Expanses of salt water on the surface of the continental shelf.
Chamberlin, T. C. and Salisbury, R. D., 1904, *Geology*, Vol. 1. New York: Holt. 'Those shallow portions of the sea which lie upon the continental shelf, and those portions which extend into the interior of the continent with like shallow depths, may be called epicontinental seas, for they really lie upon the continent, or at least upon the continental platform.' (p. 11)

**Epicycle**
*O.E.D.* 1. A small circle, having its centre on the circumference of a greater circle. Chiefly Astron.

Wooldridge and Morgan, 1937. '... there are practical, if not logical, advantages in regarding small changes of base-level as initiating sub-cycles or epi-cycles, which are essentially subdivisions of a major cycle of erosion.' (p. 218)

Cotton, 1945. 'epicycle', similar to Wooldridge. (p. 240)

No reference in Davis, 1909, who regarded any such change as initiating a new cycle. The term 'hemi-cycle' has also been proposed (G.T.W.).

### Epigenesis

Yates, R. A., 1957, in Bowen, E. G., *Wales*, London: Methuen. 'Epigenesis and superimposition are often synonymous, being used to indicate any incongruence between drainage and tectonic structure. They involve the inheritance of a drainage pattern developed on an overlying cover by an exhumed old land.' (p. 28)

See also Epigenetic.

### Epigenetic (1) (von Richthofen, 1886)

*O.E.D. Suppl.* 2. Geol. and Phys. Geog. A term applied to those rivers whose courses have been determined by the slope of a once overlying series of strata, now removed by erosion so as to disclose rock-structures of another arrangement; superposed; inherited. (*Cent. Dict. Suppl.* 1909) Note in *O.E.D.* first meaning 'of or pertaining to, or of the nature of, epigenesis.' *i.e.* the biological theory that '(the germ is brought into existence (by successive accretions) and not merely developed, in the process of reproduction'.

Mill, *Dict.* Similar to *O.E.D.* 2. (v. Richthofen).

Wooldridge and Morgan, 1937. 'superimposed or epigenetic, ...'

Cotton, 1941. 'This term (superposed) as adapted from Maw and used by Powell (1875) was originally "superimposed". It was shortened to "superposed" by McGee. Richthofen termed the valleys of such rivers "epigenetic" (1886).' (p. 51 *fn.*)

### Epigenetic (2)

Holmes, 1928. A term now generally applied to ore-deposits of later origin than the rocks among which they occur; contrasting them with those that are contemporaneous with the enclosing rocks (syngenetic).

### Epigenic

*O.E.D. Suppl.* Originating above the surface of the earth.

Geikie, J., 1898, *Earth Sculpture*, Gloss. 'Epigene: applied to the action of all the geological agents of change operating at or upon the earth's surface; also to all accumulations formed by the action of those agents.'

*Comment.* Rarely used; the confusion with epigenetic is dangerous but see D. L. Linton, 1958, quoted under hypogenic. Contrast tectogene, hypogenic.

### Epiglacial bench

Cotton, 1942. '... terraces cut by the lateral erosion of supra-glacial water streams originating on large glaciers and also actual gutters developed on valley sides where such melt-water streams have become superposed on them.' (p. 292)

### Epoch

Derived from the Greek for a check or stop, this word has been extended to mean an event of importance or the time of such event and then to a period of time. Some geologists give it the restricted meaning quoted under Geological Time (*q.v.*) but others give it different though specific meanings (see *Webster*).

### Equator

*O.E.D.* 2. Geog. A great circle of the earth, in the plane of the celestial equator, and equidistant from the two poles.

Mill, *Dict.* The great circle of the Earth perpendicular to the axis of rotation or line of the poles. The maximum great circle of the Earth's surface.

*Note: cf.* rainfall and thermal 'equator'

### Equatorial

Of or pertaining to the Equator

### Equatorial climate

The type of climate found as a belt a few degrees wide on each side of the equator on low ground.

### Equatorial current, Equatorial countercurrent

*Webster.* Equatorial current. The surface current moving westward in the oceans near the equator. Equatorial counter-current. The surface current moving eastward in a few places near the equator.

Mill, *Dict.* The movement of the surface waters of tropical seas north-east to south-west in the northern hemisphere, south-east to north-west in the southern. Where these two streams come into opposition (near the equator) the resultant movements are either east to west (equatorial current) or west to east (counter equatorial current). The general result is that there are north and south equatorial currents with a counter equatorial current between them.

### Equatorial forest
Mill, *Dict*. The impenetrable tangle of vegetation, much of it of great size, which covers such equatorial regions as receive more than 60 inches of rain per annum and have no dry season. It is rich in lianas and epiphytes; characteristic of Amazon valley, Congo region and parts of East India archipelago.

### Equatorial Front—see Intertropical Convergence Zone

### Equilibrium
*O.E.D.* The condition of equal balance between opposing forces.
*Webster.* A state of balance between opposing forces or actions, either *static* or *dynamic*.
*Comment.* This is a good example of a common word which has been given a variety of restricted meanings in geographical writings. In physical geography Gilbert, 1877, (pp. 113, 115) uses it for equality of action between capacity for corrasion and resistance; W. M. Davis (*Geog. Jour.* 5, 1895, p. 130) says 'streams ... having attained a profile of equilibrium ... may be said to be graded', i.e. their capacity to do work is just equal to the work they have to do. In talking of social equilibrium (Odum, H. W., and Moore, H. E., *American Regionalism*, 1938) and economic equilibrium (Baker, J. N. L., *History of Geography*, 1963, p. 267, Smailes, A. E., *Soc. Rev.*, 36, 1946, 18) geographers sometimes use it in a static sense of no change, sometimes in a dynamic sense of maintaining a balance. There is room for much research into past usage (note by Mary Marshall).

### Equilibrium line (Glaciology)—see Firn line

### Equinoctial
*O.E.D.* 1. Pertaining to a state of equal day and night.
2. Pertaining to the period or point of the equinox. Happening at or near to the time of the equinox; said *esp.* of the 'gales' prevailing about the time of the autumnal equinox.

### Equinox (lit. equal night)
*O.E.D.* One of the two periods in the year when the days and nights are equal in length all over the earth, owing to the sun's crossing the equator. (March 21 and September 23)

### Equiplanation (D. D. Cairnes, 1912)
Cairnes, D. D., Differential erosion and equiplanation in portions of Yukon and Alaska, *Geol. Soc. of Amer. Bull.* 23. '... all physiographic processes which tend to reduce the relief of a region, and to cause the topography to become more and more plainlike in contour, without involving any loss or gain of material to the area affected.' (pp. 333–348)
*Comment.* Refers to conditions in a frost climate as in Alaska. See also cryoplanation. (G.T.W.) See also Thornbury, 1954, p. 412.

### Equiplanation terrace—see Altiplanation

### Era—see Geological time

### Erg (North Africa: Arabic)
Mill, *Dict.* (*pl.* Areg). Tracts of sandy desert (Sahara). Distinguished from the stone or rock deserts ('hamada').
*Webster.* Geog. A desert region of shifting sand; specif.: (*cap.*) a. The *Iguidi Erg*, south of Morocco. b. The *Western Erg*, south of Algeria. c. The *Eastern Erg*, adjoining Tunisia.
Holmes, 1944. 'As a result of wind erosion, transport, and deposition three distinctive types of desert surface are produced:
1. The rocky desert (the *hammada* of the Sahara), with a surface of bedrock kept dusted by deflation and smoothed by abrasion;
2. the stony desert, with a surface of gravel (the *reg* of the Algerian Sahara), or of pebbles (the *serir* of Libya and Egypt); and
3. the sandy desert (the *erg* of the Sahara)' (p. 255).
See also Hamada. The local vocabulary of the people of the Sahara is rich in terms distinguishing sub-types of desert within the three main groups.

### Erosion, erode
*O.E.D.* 1. The action or process of eroding; the state or fact of being eroded.
Erode: (from Latin via French, 'out' + 'to gnaw')
2. Geol. of the action of currents, glaciers, etc.: a. to wear away; to eat out. b. to form (a channel etc.) by gradual wearing away. a. 1830. Lyell; 1871. Tyndall; b. 1830. Lyell; 1859. Burton; 1872. Symonds.
*Webster.* 3. Geol. The general process of the wearing away of rocks at the earth's surface by natural agencies.
Mill, *Dict.* 1. All processes of wearing down of land, including weathering (Gilbert). 2. In general use in Britain. The widespread lowering of the land by wind, rain, and weather and by rivers and glaciers acting laterally (Gregory). 3. The wearing of rock by pot-hole formation (Geinitz).
Cotton, 1941. 'Most English and American

writers use "erosion" as an inclusive term, for which "denudation" is a synonym. Some geomorphologists of the German school attempt to distinguish between "denudation", defined as "degradation by surficially extensive mass movements" (C. Sauer), and "linear erosion" (corrasion).' (p. 13 *fn.*)

Wooldridge and Morgan, 1937. Rockwastage or denudation is divided into three distinct phases: (1) weathering (2) transport (3) erosion. Erosion appears to be confined to the work of abrasion or corrasion. (pp. 146–153)

Cotton, Geomorphology. 1945. Erosion = denudation. (p. 10)

Davis, W. M. and Snyder, W. H., 1898, *Physical Geography*, Boston: Ginn. 'The general process of wasting and washing, by which the surface structures are slowly worn away and the deeper and deeper structures of the earth's crust are attacked, is called denudation, or erosion.' (p. 105)

Holmes, 1944. 'It is convenient to regard weathering as rock decay by agents involving little or no transport of the resulting products, and erosion as land destruction by agents which simultaneously remove the debris. Both sets of processes cooperate in wearing away the land surface, and their combined effects are described by the term denudation.' (p. 24)

*Comment.* The restricted use of Wooldridge and Morgan and of Holmes is far from being generally accepted. In this narrower sense corrasion is a better term. (G.T.W.) See also Cycle of erosion.

### Erosion platform
A platform or relatively level surface formed by erosion in contrast to a platform or terrace formed by deposition or aggradation.

### Erratic blocks, erratics
*O.E.D.* Stray masses of rock foreign to the surrounding strata, that have been transported from the original site, apparently by glacial action.

Mill, *Dict.* Transported fragments of rock at a distance from their original bed. The term is specially used for large boulders deposited by an ice-sheet or glacier which has since disappeared.

*Comment.* The *O.E.D.* definition should not be used because the word 'stray' fails to suggest the common and regular nature of the phenomena and the word 'apparently' suggests an unnecessary doubt. In fact erratic blocks enable the former course of glaciers and ice sheets to be traced with precision back to the source of the rocks. In the case of valley glaciers the erratics are often found perched precariously on the valley sides; hence the term perched blocks. Far-travelled or 'foreign' boulders may be further distinguished as 'exotic'. (L.D.S. G.T.W.)

### Escarpment
*O.E.D.* 2. Geol. 'The abrupt face or cliff of a ridge or hill range' (Page). 1845. Darwin, Voy. Nat. viii (1852), 165. 'The view is generally bounded by the escarpment of another plain.'

Mill, *Dict.* An inland cliff, produced by the erosion of inclined hard strata, the worn edge of which crowns the cliff.

Rice, 1943. 'A more or less continuous line of cliffs or steep slopes facing in one direction and due to erosion or faulting. The abbreviated form "scarp" is sometimes used as a synonym, but is more commonly limited to cliffs produced by faulting' (p. 124).

Committee, List 2. A general term for a continuous scarp face.

*Comment.* Though commonly used by British geographers as equivalent to cuesta this usage is indefensible.

### Esgair (Welsh)
*Lit.* long ridge; geographers normally use the Irish spelling of the same word esker (*q.v.*), but the Welsh word would never be used for a glacial land form. (E.G.B.)

### Esker (formerly also eskir, escar, eskar from old Irish eiscir)
*O.E.D.* 'The name given in Ireland to the elongated and often flat-topped mounds of post glacial gravel which occur abundantly in the great river-valleys of that country.' (Quoted from Page, 1865, *Handbook of Geology*.)

Geikie, A., 1882, Text-Book of Geology, VI, V, par. 1, 892. 'Ridges known in Scotland as kames, in Ireland as eskers, and in Scandinavia as ösar.'

Mill, *Dict.* 'Esker (Ireland), Kame (Scotland), Åsar (Sweden)—A sinuous ridge or mound formed by glacial action of stratified falsebedded sand or gravel often receiving tributary ridges.'

Geikie, J., 1898, Gloss. 'Ridges of gravel and sand which appear to have been formed in tunnels underneath the great glaciers and ice-sheets of former times; same as the Swedish osar.'

Wooldridge and Morgan, 1937. 'Eskers (comparable if not identical with the Swedish "ösar") are long sinuous ridges with steep sides, composed of gravel and sand' (p. 387—describes also 'beaded esker' with diagrams).

Holmes, 1942. 'Drift-covered regions are diversified by mounds (*kames*), long winding ridges (*eskers*) and relatively short and straight ridges (*crevasse infillings*). All of these are built of crudely bedded gravel and sand, showing that they are features for which glacial streams were responsible ... most eskers are regarded as the infillings of the tunnels of unusually long subglacial streams ... eskers characteristically disregard the underlying topography which they cross like long railway embankments ... their courses, though winding, are generally aligned more or less at right-angles to the receding ice front' (pp. 233–235).

Flint, 1947, p. 150. 'A long, narrow ice-contact ridge commonly sinuous and composed chiefly of stratified drift.'

Cotton, 1945, p. 345. 'Most of the "eskers" of Ireland are not true eskers in technical sense.'

Comment. Esker is now used in the restricted sense of Wooldridge and Morgan and of Holmes, hence the comment of Cotton. The definitions of Page in *O.E.D.*, A. Geikie and Mill are therefore no longer acceptable.

**Estancia** (South America: Spanish)
Literally a station: a cattle ranch in South America comparable with a cattle station in Australia.

**Estovers**
Royal Commission on Common Land, 1955–1958. *Report*. 'The word derives from the Norman-French "estouffer"—to furnish. Hence, the common right of cutting and taking tree loppings, or gorse, furze, bushes, or underwood, heather or fern, of a common for fuel to burn in the commoner's house or for the repair of the house and farm buildings, hedges, fences and farm implements. The Early English equivalent was "bote" hence fire-bote, house-bote, plough-bote, cart-bote, and hey- or hedge-bote.' (p. 273)

**Estuarine**
Of or pertaining to an estuary: applied especially to deposits laid down in brackish water or to the environment in biological studies.

**Estuary**
*O.E.D.* 1. gen. A tidal opening, an inlet or creek through which the tide enters; an arm of the sea indenting the land. Rare in mod. use. 2. spec. The tidal mouth of a great river, where the tide meets the current of fresh water.

*Webster.* 1a. A passage, as the mouth of a river or lake, where the tide meets the river current; more commonly, an arm of the sea at the lower end of a river; a firth. b. In physical geography, a drowned river mouth, caused by the sinking of the land near the coast.

Mill, *Dict.* The widening-out lower portion of a tidal river.

*Comment.* According to physical circumstances a river on reaching the sea may broaden out and flow into its estuary, or divide and deposit material to form a delta. Thus the mouth of a river may be either estuarine or deltaic in character.

**Étang** (French)
In French, an expanse of still water, natural or artificial; étang salé, one which communicates with the sea.

A pool or lagoon caused by the ponding back of water draining from the land by beach material thrown up by the sea. Lobeck, 1939, lists in index and shows on a map of the Landes coast of south-western France but does not describe.

**Etchplain** (E. J. Wayland, 1934, L. C. King, 1947)
King, L. C., 1947, *Proc. Geol. Soc. S. Afr.*, 50, xxvi. A relatively inextensive erosional platform.

Thornbury, 1954. 'Wayland described five topographic levels in Uganda ... the surfaces below the uppermost he attributed to a process he called etching.' See *Geol. Surv. Uganda, Ann. Rept. Bull.*, 1934 77–79.

**Etesian wind**
*Webster.* Etesian. Periodical; annual—applied to winds in the Mediterranean region, esp. those which blow from the north in the Aegean Sea and the Levant during summer.

Mill, *Dict.* Originally applied to the prevailing northerly summer winds of Greece, now extended to similar winds prevailing for a shorter or longer period of the summer in other parts of the Mediterranean. Described at length in Miller, 1953, 171–172, 'in the eastern Mediterranean reach great force and constancy ... on land clouds of dust ... force and dryness sufficient to prohibit tree-growth in exposed places. ...' Called *meltemi* by the Turks.

**Ethnobotany**
The relationship between wild or primitive

man and wild plants. Used in this sense in various UNESCO publications (*e.g.* NS/173, 1961), urging the necessity for studies in view of the rapidity of change taking place in the environment of the few remaining groups of primitive man.

### Ethnocentrism (Collective egocentrism)
Tuan, Y. F., 1974, *Topophilia*, Englewood Cliffs, New Jersey: Prentice-Hall. 'Individuals are members of groups, and all have learned—though in varying degree—to differentiate between "we" and "they", between real people and people less real, between home ground and alien territory. "We" are at the center. Human beings lose human attributes in proportion as they are removed from the center.' (pp. 30-31)
See also Egocentrism.

### Ethnography
*O.E.D.* The scientific description of nations or races of men, with their customs, habits, and points of difference.

### Ethnology
*O.E.D.* The science which treats of races and peoples, and of their relations to one another, their distinctive physical and other characteristics etc.
*Comment.* The general tendency is to drop ethnography and ethnology in favour of anthropology.

### Ethology
The biology of behaviour of animals.
But see long history of varied use in *O.E.D.*, including the science of character-formation (J. S. Mill).
*O.E.D.* Suppl. 4. The branch of Natural History which deals with an animal's actions and habits, its reaction to its environment.

### Etiology, aetiology, etiologist
*O.E.D.* Etiologist: one who studies the science of causes.
*Webster.* d. the science, doctrine, or demonstration of causes; *esp.*, the investigation of the causes of any disease.

### Euclidean space
James, G. and James, R. C., 1949, *Mathematics Dictionary*, New York: P. Van Nostrand. 'Euclidean space (1) Ordinary three dimensional space (2) A space consisting of all sets (points) of $n$ real numbers $(x_1, x_2, \ldots, x_n)$, where the distance $p(x, y)$ between

$$x = (x_1 \ldots x_n) \quad \text{and} \quad y = (y_1 \ldots y_n)$$

is defined as $p(x, y)$

$$= \left[ \sum_{i=1}^{n} (x_i - y_1)2 \right]^{1/2}$$

This is an $n$-dimensional space. An infinite dimensional Euclidean space is the (real) Hilbert space of all infinite sequences of real numbers $(x_1, x_2 \ldots)$ for which $\sum_{i=1}^{\infty} x_i^2$ is finite, with $p(x, y)$ defined as

$$\left[ \sum_{i=1}^{\infty} (x_i = y_i)2 \right]^{1/2}$$

Based on the geometry of Eukleides the Greek geometer.

### Eugeogenous
Mill, *Dict.* Rocks which yield by weathering a large amount of detritus (*Thurmann*) *cf.* dysgeogenous.

### Euryhaline—see Stenohaline

### Eurythermic—see Stenothermic

### Eustatism, eustasy, eustatic
*O.E.D. Suppl.* Eustatic. Geol. and Physiogr. Of a land area: not subject to depression or elevation; thoroughly established. 1906. tr. E. Suess., Face of Earth II, 538. (No reference to Eustatism.)
Mill, *Dict.* Eustatic Movements. Changes affecting the level of the strand taking place in the same direction (either positive or negative) so as to bring about an approximately equal difference of level over the whole globe. (Suess)
Cotton, 1922. 'Actual rise and fall of the ocean-surface must be regarded as a probable cause of some movements of the strand-line (and such probable movements are sometimes distinguished as "eustatic").' (p. 214)
Similarly Wooldridge and Morgan, 1937; Holmes, 1944.
Baulig, H., 1935, The Changing Sea Level, *Inst. Brit. Geog.*, Publ. 3. 'Movements of the land may take place along more or less sharply defined lines, and result in fractures, folds, and pronounced distortions of the earth's crust ... called mountain-building or *orogenic* movements. But they may also consist of uplifts or sinkings over large areas, accompanied by warping or folding scarcely perceptible ... known as continent-making or *epeirogenic* movements. ... Again, movements of the sea level may result from variations either in the liquid mass of the oceans or in the capacity of their basins ... such movements are simultaneous and uniform the

world over... due to the restoration of hydrostatic equilibrium, Ed. Suess has called them *eustatic*.' (pp. 4–5) After discussing *glacio-eustatism* (*q.v.*) 'a quite different sort of eustatic movement, resulting from changes in the capacity of the oceanic basins, might be called *deformational* or *diastrophic*.' (p. 4) Swayne, 1956. Eustasy: the theory that world-wide changes in ocean-level derive either from changes in the capacity of the ocean basins or from variations in the amount of ocean water.

*Comment*. The *O.E.D.* definition (also given by Fay, 1920 and Rice, 1941) is utterly different from normal usage. Eustatic has also been used for movements resulting in vertical displacement in contrast to tilting, warping or folding but the correct term is epeirogenic. See also rejuvenation.

### Eutrophication (of lakes)
A change to higher productivity status, often caused by mild pollution (E. B. Worthington, communicated in MS.).

### Eutrophy, eutrophic (lakes)
'Lakes termed eutrophic are those whose waters are relatively rich in plant nutrients (such as phosphates and nitrates), usually have much rapidly decaying organic mud on the lake bottom, and usually show reduced oxygen tensions in their waters in summer... usually shallow with gently sloping banks.' W. D. R. Hunter, *The Glasgow Region*, 1958, 106. (Contrast oligotrophy.)

### Euxinic
Applied to conditions such as those found in the Black Sea (Euxine Sea of the Greeks), water stagnant at depth.

### Evapotranspiration
Thornthwaite, C. W., 1948, An approach towards a Rational Classification of Climate, *Geog. Rev.*, 38. 'The combined evaporation from the soil surface and transpiration from plants, called "evapotranspiration", represents the transport of water from the earth back to the atmosphere....' (p. 55)

*Comment*. The measurement of evapotranspiration is now recognized as of great importance in climatological studies as something different from the measurement of evaporation from water or earth surfaces. Other workers contend that the measurement of evaporation and of transpiration (from plants) should be kept quite distinct.

### Everglade (United States)
*O.E.D.* A marshy tract of land mostly under water and covered in places with tall grass; chiefly in plural as the name of a large swampy region of South Florida.
*Webster*. A swamp or low tract of land inundated with water and interspersed with hummocks, or small islands, and patches of high grass.—the Everglades. A great tract of this nature in Florida.
*Dict. Am.* A low, marshy region, usu. under water and overgrown with tall grass, cane, etc. 1823. Tanner. *Map Florida* "Extensive inundated region... generally called the Everglades."
*Comment*. Presumably from ever and glade (*q.v.*).

### Evorsion
Mill, *Dict*. Erosion by vortices or eddies in a river.
Ängeby, O., 1951. Pothole erosion in Recent Waterfalls, *Lund Studies in Geography*, Ser. A, No. 2. Gives an illustration of what he calls 'evorsion marks', apparently the same as the features called 'flutes' by American writers and 'scallops' by British speleologists. Evorsion, apparently borrowed from the German, is used as synonymous with pothole erosion of rivers (G.T.W.).
Not in *O.E.D.*, Wooldridge and Morgan, Holmes, von Engeln, Lobeck, Thornbury.

### Evorsion hollow
Pothole in a stream-bed (G.T.W.).

### Exceptionalism
Schaefer, F. K., 1953. Exceptionalism in Geography: A Methodological Examination, *A.A.A.G.*, 43, 226–249.
Hartshorne, R., 1955. 'Exceptionalism in Geography Re-examined.' *A.A.A.G.*, 45, 205–244.
Haggett, 1965. 'Exceptionalist traditions in geography... the general law of geography that all its areas are unique (Hartshorne, 1939, p. 468)... the fact that one can do little with the unique except contemplate its uniqueness, has led to the present unsatisfactory position.' (pp. 2–3)
*Comment*. In the first two lengthy papers certain ideas are put forward and also refuted; no succinct definition of the word emerges. Haggett expresses the viewpoint of the modern school of quantifiers which has led them to seek theoretical 'models' from which deviations can be measured.

## Exclave (orig. French)
*Webster.* A portion of a country which is separated from the main part and surrounded by politically alien territory. The same territory is an *enclave* in respect to the country to which it is politically attached (*i.e.* in which it is physically located).

Van Valkenburg, S., 1939, *Elements of Political Geography*, New York: Prentice-Hall, 1939. 'Small disconnected sections in a foreign territory are enclaves of the surrounding area and exclaves of the country to which they politically belong.' (p. 111)

Zimmermann, E. W., 1951, *World Resources and Industries*, 2nd ed., New York: Harper. 'An economic exclave may be defined as a splinter of one economy lying inside another economy.' (*e.g.* Some oilfields of Venezuela ... may be viewed as splinters or exclaves of the economy of the United States.) p. 129

Robinson, G. W. S., 1959, Exclaves, *A.A.A.G.*, 49, 283–295, separates Normal, Pene-, Quasi-, Virtual, and Temporary Exclaves. Pene-exclaves 'are parts of the territory of one country that can be approached conveniently—in particular by wheeled traffic—only through the territory of another country'. (p. 283) Quasi-exclaves 'are those exclaves which for one reason or another do not in fact function as exclaves today'. (p. 283)

## Exfoliation
*O.E.D.* 3.b. of minerals, metals, rocks, etc.: To split into laminae, come off in layers or scales.

Mill, *Dict.* The separation of scales parallel to the surface of rock by erosion characteristic of the granitic rocks of the east coast of Brazil.

Holmes, 1944. In desert and semi-arid regions and in monsoon lands with a marked dry season a characteristic effect on the outlines of upstanding hillocks and peaks, especially where they are made of crystalline rocks, is produced by *exfoliation*, the peeling off of curved shells of heated rock ... rounded outlines are developed and the hills become dome-shaped ... well seen in *inselbergs* of Africa' (pp. 114–115).

See also Insolation, Onion weathering.

## Exhumation, exhume, exhumed
*S.O.E.D.* Exhume: to dig out or remove (something buried) from beneath the ground; to unearth. Hence exhumation: the action or process of removing a body from beneath the ground.

*Comment.* Widely used in geographical literature in the sense of uncovering by erosion previously existing features, such as mountains, peneplained surfaces and plateaus which had become buried by later deposits. Hence exhumed landscapes etc.

## Exile
*S.O.E.D.* A banished person; one compelled by circumstances to reside away from his native land.

*Comment.* In current usage the distinction between exile, expatriate and refugee has become important. An exile is *compelled*, an expatriate *chooses* to live away from his native land, a refugee does so through fear.

## Exogenetic, Exogenic
*O.E.D.* Exogenetic: that arises from without.

Thornbury, 1954. '... weathering, mass-wasting and erosion result from external forces and have been called exogenic in nature.'

Wooldridge and Morgan, 1937. 'external denuding agencies grouped together by some writers as "exogenetic"' (p. 54).

Himus, 1954. Exogenetic. Geological processes which originate at or near the surface of the earth, such as denudation and deposition and land forms arising therefrom.

*Comment.* Exogenetic is the usual English form, Exogenic (not in *O.E.D.*) the American.

## Exogenous
*O.E.D.* Suppl. Geol. 'of external sea-border origin: said of the stratigraphical growth of continents' (Funk's Standard Dict. 1895). Also applied by von Humboldt to extrusive volcanic rocks changed by contact with surrounding rocks.

Dana, J. D., 1890, *Bull. Geol. Soc. Amer.*, I. 'The growth of the continent, so far as through marine waters, may be said to have been endogenous. It began to be exogenous on the Atlantic side in the Cretaceous era.' (p. 48)

Mill, *Dict.* Proceeding from or acting on the outside of the crust of the Earth.

See Endogenous.

## Exonym (M. Aurousseau, 1957)
Aurousseau, M., 1957, *The Rendering of Geographical Names*, London: Hutchinson. 'Name used in English for other parts of the world and geographical features out-

**Exoreic**—see Areic

**Exosphere**
The upper Atmosphere is regarded as the atmosphere above the Tropopause and below the exosphere. *Weather*, April 1960, p. 146A.

**Expatriate**
Not in *O.E.D.* as substantive; as verb to withdraw from one's native country; to renounce one's allegiance (*S.O.E.D.*).
*Comment.* As substantive, now widely used for one who voluntarily, especially for personal reasons, chooses to live away from his native country. The contrast is with exile and refugee (*q.v.*).

**Exponential curve**
A curve which may be assumed by a surface in accordance with a particular mathematical formula. See Krumbein, W. C., 1937, Sediments and exponential curves. *Jour. Geol.*, **45**, 577.

**Exposure (1)**—climatological
*O.E.D.* 1. The action of exposing; the fact or state of being exposed. 3. The manner or degree in which anything is exposed; *esp.* situation with regard to sun and wind; aspect 1664.
*Webster.* 2b. Position as to points of compass, or to influences of a climate, etc.; as, a southern exposure.
*Met. Gloss.*, 1939. 'In meteorology, the method of presentation of an instrument to that element which it is designed to measure or record, or the situation of the station with regard to the phenomena there to be observed.'
Hann, J., 1903, *Handbook of Climatology* (tr. R. de C. Ward), 'the surface with the south-western exposure has the highest temperature in winter' (p. 240).
Peattie, R., 1936, *Mountain Geography*. 'In considering land values of mountain fields a concern of the first importance is exposure to the sun' (p. 88). 'the law of exposure' (p. 90) (see under Adret).
Miller, 1953. 'The exposure may cause curious features in the daily march of temperatures; an east-facing slope will have warm mornings and cool afternoons' (p. 270). (This passage is indexed under Aspect, Climatic effect of.)
Brooks, C. E. P., 1954, *The English Climate*. 'Superimposed on the general climate of a district are local variations due to exposure and aspect ...' (p. 77) (exposure being the nature and degree of exposure to sun, wind and weather; aspect being the direction and angle of slope).
See also Aspect, Adret.

**Exposure (2)**—geological
Rice, 1943. In geology, the condition or fact of being exposed to view, either naturally or artificially, hence also that part of a rock bed or formation which is exposed; an outcrop (p. 128).

**Exsudation, exudation**
*O.E.D.* Exudation: The process of exuding; the giving off or oozing out (of moisture) in the manner of sweat.
Thornbury, 1954. 'The scaling off of rock surfaces through growth of salines by capillary action has been called exsudation.' (p. 39)
*Comment. O.E.D.* gives exsudation as an obsolete form of exudation. Not widely used and of doubtful value as a term.

**Extensive (agriculture, farming)**
*O.E.D. Suppl. Econ.* Applied to methods of cultivation in which a relatively small crop is obtained from a large area at the minimum of attention and expense; opposed to intensive.
*Webster.* 4. *Agric.* Designating, or pertaining to, any system of farming in which large areas are used with a minimum of labor and outlay—opposed to *intensive*.
James, P. E., 1949, *A Geography of Man*, Boston: Ginn. '... where the expenditure of labor and capital per unit of land is relatively small, the economy is said to be extensive.' (p. 60)
Venn, J. A., 1933, *The Foundations of Agricultural Economics*, Cambridge: C.U.P. 'The earliest forms of intensive cultivation, as opposed to extensive—which implied a constant moving on of nomadic or other tribes....' (p. 31)
Committee, List III. Extensive agriculture. 1. Loosely used of farming in very large units. 2. (in economic geography) farming in which the amount of capital and labour applied to a given area is relatively small.
*Comment.* See also intensive agriculture, where the *O.E.D.* definition is criticized. Venn seems to confuse extensive with shifting agriculture.

**Exterritoriality**

Fishel, W. R., 1952, *The End of Extraterritoriality in China*. University of California Press, C.U.P. 'The immunities accorded a diplomatic envoy and his suite in accordance with international law' (p. 2)—not to be confused with extraterritoriality (*q.v.*).

**Extractive industry**—see Industry, extractive

**Extraterritoriality**

*Webster.* A quality, state, or privilege of general or partial exemption from the application of local law or jurisdiction of local tribunals, as in the case of diplomatic agents and their dwellings, public ships, or subjects of states which by long assumption or by treaty exercise jurisdiction over their own subjects in certain Oriental countries.

Fishel, W. R., 1952, *The End of Extraterritoriality in China*, University of California Press, C.U.P. 'The extension of jurisdiction beyond the borders of the state. It embodies certain rights, privileges and immunities which are enjoyed by the citizens, subjects, or protégés of one state within the boundaries of another, and which exempt them from local territorial jurisdiction and place them under the laws and judicial administration of their own state'. (p. 2)

*Comment.* In its modern form introduced in China in 1843. Abolished by successive events between 1917 (the entry of China into World War I against Germany and Austria-Hungary) and 1943 when Britain and the United States relinquished their rights. Extraterritoriality exists, however in modified forms elsewhere, *e.g.* to some extent United States and other forces stationed in foreign countries.

**Extrusive rock**

Himus, 1954. Rocks resulting from the consolidation of magma at the surface of the ground, in contrast with *intrusive* rocks, which have consolidated below the surface. Also known as *Lavas*.

**Eyot** (see also Ait)

Eyot, as -ey, survives in many place names with the meaning of isle.

# F

**Fabric analysis**
The statistical analysis of fabric diagrams which are maps of either two or three dimensional orientation of a measurable axis in a suitable coordinate system. Used in the study of Till (*q.v.*).

**Facies mapping**
Forgotson Jr., J. M., 1960, Review and Classification of Quantitative Mapping Techniques, *Bull. Am. Assoc. Petr. Geol.*, **44**.1. 'Facies maps illustrate the areal variation of lithological character, faunal content, geochemical attributes, or any other compositional variable that can be quantitatively expressed. Facies maps may be divided into two sub-classes according to the number of variables considered. A. Univariate. Percentage maps and isolith maps belong to this class. B. Multivariate. Ratio maps describe the relationship between two variables. Combined ratio maps, entropy-function maps, and distance-distribution maps describe the relationship among three variables.' (p. 85)
See Krumbein, W. C., 1952, Principles of Facies Map Interpretation, *Jour. Sed. Petrology*, **22**.4, 200–211.

**Facet**
*O.E.D.* (from French, a little face)
1. 'One of the sides of a body that has numerous faces;...' (originally applied to cut gems, but extended).
Davis, 1909. Applied to faceted spurs along the front of a tilted block. (p. 746)
Cotton, 1922. 1. As Davis (fault scarps). (p. 160)
2. Triangular shaped cliffs between deeply cut valleys. (p. 248)
3. Also refers to 'faceted pebbles' caused by wind corrasion or on sea-beach (similar usage in edition of 1945).
Wooldridge and Morgan, 1937. Any part of an historic (or present) surface, *e.g.* a 'flat' or 'slope'. 'In the ultimate analysis, indeed, it is the variously inclined facets of intersecting surfaces which must form the units of geographical study.' In example, contrast the northern, steeply sloping facet of the Chalk of East Kent with the southern, hill-top area. (p. 258)
Holmes, 1944. 1. 'Faceted spurs' caused by valley glaciation. (p. 220)
2. Faceted pebbles caused by wind erosion: dreikanter or ventifacts. (p. 258) See Ventifact.
Flint, 1947. 1. Faceted spurs, in glaciated valleys.
2. Faceted pebbles from glacial 'pavement'.
No reference in Mill, *Dict.*; Geikie, 1898; Lyell, 1830; Powell, 1875; Gilbert, 1877.
*Comment.* The extension of meaning in the sense of Wooldridge and Morgan above, by analogy with the facets of a cut gemstone, is one for which Wooldridge is essentially responsible. It is now widely used by British geomorphologists.

**Factorial ecology**
Berry and Horton, 1970. 'The basic purpose of factor analysis as applied in urban ecology is to reduce a matrix of $n$ tracts by $m$ variables to one of the $n$ tracts and $r$ factors, where the number of significant (i.e. of practicable, interpretable importance) factors, $r$, is less than $m$. The $r$ factors summarize the common patterns of variablity in the data and make possible more concise statements about the population under consideration. The mathematical dimensions isolated are an objective outcome of the analysis, their interpretation depending upon the nature of the variables used in the analysis and the body of concepts or theory that is brought to bear. The theory provides the investigator with a set of expectations regarding the factor structure which can be compared to the actual set of factors produced.' (p. 316)

**Fadama** (West Africa; Hausa)
A floodplain, in a wide, flattish river valley, subject to annual inundation, commonly occurring in the Sudan and Guinea Savanna zones, with a characteristic vegetation of grasses, sedges, and the tree species *Mitragyna inermis* and *Borassus* (Fan Palm). (J.C.P.)

**Fairway**
*O.E.D.* A navigable channel in a river or between rocks, sand banks, etc.; the usual course or passage of a vessel on the sea or in entering and leaving a harbour.
*Adm. Gloss.*, 1953. The main navigable channel, often buoyed, in a river, or running through or into a harbour. The usual course followed by vessels entering and leaving harbour.
*Comment.* The words 'on the sea' in the *O.E.D.* seem to extend the meaning too widely and are better omitted.

**Fako** (West Africa: Hausa)
Land, in the Sudan and Sahel zones, from which the surface soil has been eroded to leave an infertile subsoil which supports little vegetation even in the rains, so that it is never farmed. The surface is churned to a slippery layer by cattle, but never cracks in the dry season (J.C.P.).

**Falaise** (French—*lit.* a cliff)
von Engeln, 1942. 'The low cliffs which re-establish contact of the open sea with the oldland on an emergent coast are designated by the French term *falaises*, of which the coast of Normandy is the type example.' (p. 541)

**Fall line, fall zone**
*O.E.D. Suppl.* Fall-line, an imaginary line drawn through a number of rivers where they make a sudden descent, as at the edge of a plateau; specifically, such a line in the eastern United States marking the western limit of the Atlantic coastal plain. Example: 1902. Lord Avebury, *Scenery of England*, 481.
*Webster.* A line characterized by numerous waterfalls; the edge of a plateau, in passing which the streams make a sudden descent; specif., *Geol.* the boundary between the ancient crystalline rocks of the Appalachian piedmont and the younger, softer rocks of the Atlantic coastal plain.
*Dict. Am.* The boundary between the older Appalachian uplands and the younger sea plain. 1882 *Nation* 13 July 33/1 'It is here, at the "fall line" that most of the available water-powers are to be found. . . . Above the fall line the currents of the streams are much more rapid.'
Lobeck, 1939. 'The zone where the streams pass from their rocky channels in the oldland to their more gentle courses in the coastal plain is called the *fall line* or *fall zone*.' (p. 451)

**Fallow**
*O.E.D.* Ground that is well ploughed and harrowed, but left uncropped for a whole year or more; called also *summer fallow* as that season is chosen for the sake of killing the weeds.
*Webster.* 2. Land ordinarily used for crop production when allowed to lie idle either in a tilled or untilled condition during the whole or the greater portion of the growing season. 3. The plowing or tilling of land, without sowing it for a season; the state or period of being fallow; as, summer *fallow* is a method of destroying weeds.
Watson, J. A. S., 1938, *The Farming Year*, London: Longmans. Land left without a crop during the growing season in order that weeds may be killed and the land very thoroughly tilled. A *bare fallow* is left without a crop for a whole season. A *partial fallow* may be taken in the first half of the summer (for instance before a late crop of turnips) or in the late summer (after a crop of hay).
*Comment.* These definitions, especially the *O.E.D.*, based on British practice, are much too narrow. Land lying fallow or uncropped is being allowed to rest; fallow land is arable land lying uncropped and may so remain for one season (current fallows of India) or more. In 'bush fallowing' (shifting cultivation, land rotation, swidden farming) weeds and scrub are allowed to invade the fallow for ten years or more until the time comes to clear and cultivate it again.

**Fallow, Green**
*O.E.D.* 2. Green, cropped, or bastard fallow: one from which a green crop is taken.
*Encyc. Britt.* 14th ed. 'A "green fallow" is land planted with turnips, potatoes, or some similar crop in rows, the space between which may be cleared of weeds by hoeing.' 'The "bastard fallow" is a modification of the bare fallow, effected by the growth of rye, vetches, or some other rapidly growing crop, sown in autumn and fed off in spring, the land then undergoing the processes of ploughing, grubbing and harrowing usual in the bare fallow.'

**False bedding**
*O.E.D.* False-bedded. Geol. Quotations include: 1876. Page, Adv. Text Bk. Geol. v, 91. Sandstones are said to be false-

bedded when their strata are crossed obliquely by numerous laminae.
1876. Woodward, Geol. (1887), 13. False-bedding ... is a feature produced in shallow water by currents and tidal action, whereby beds are heaped up in irregular layers without any approach to horizontality or continuity.
Mill, *Dict.* False Bedding. See Current Bedding.
Current Bedding of Strata, especially sandstones. Deposits in which the laminae are parallel to each other for short distances, but inclined at varying angles and in different directions to the general layering of the mass. This indicates deposit in the course of a current of variable direction.
Fay, 1920. False bedding. Current bedding. Laminations in sandstone parallel to each other for a short distance, but oblique to the general stratification; caused by frequent changes in the currents by which the sediment was carried along and deposited (Power). Also Cross bedding.
Fay, 1920. Cross-bedding. Lamination, in sedimentary rocks, confined to single beds and inclined to the general stratification (La Forge). Caused by swift, local currents, deltas, or swirling wind-gusts, and especially characteristic of sandstones, both aqueous and eolian (Kemp).
*Comment.* From a careful study of false bedding it is often possible to deduce much information regarding current direction and conditions of deposition.

**Fan** (alluvial fan, alluvial cone, rock fan)
*Webster.* The alluvial deposit of a stream where it issues from a gorge upon an open plain.
Holmes, 1944. 'When a heavily laden stream, flowing swiftly through a ravine or canyon, emerges at the base of a slope, its velocity is suddenly checked by the abrupt change of gradient, and a large part of its load of sediment is therefore dropped. The obstructed stream—as in delta formation—divides into branching distributaries, and the heap of debris spreads out as an *alluvial fan*. If the circumstances are such—arid or semi-arid conditions are specially favourable—that most of the water sinks into the porous deposit, practically the whole of the load is dropped and the structure rapidly gains height and becomes an *alluvial cone*. (p. 201)
Thornbury, 1954. Quoting Johnson, D. W., 1932, Rock fans of arid regions, *Amer. Jour. Sci.*, 223, 389–416, says 'he believed that the pediments evolve from features that he called *rock fans*. Johnson applied this name to fan-shaped rock surfaces which apex where mountain streams debouch upon a piedmont slope.' (p. 287)
See alluvial fan, alluvial cone.

**Fan-foiding, fan-structure**
*Webster.* Fan fold. *Geol.* A fold in which the anticlines are broader at the summit than at the base; called also *Alpine structure*, because typical of the Alps.
Swayne, 1956. A belt of folded rocks raised in the centre and consisting of a number of individual subsidiary folds with axes dipping inwards from both sides.
Wooldridge and Morgan, 1937. 'Before the recognition of recumbent folds or nappes the Alps were usually interpreted as showing "fan-folding" (Fig. 25).' (p. 68)
*Comment.* In cross section a belt of country exhibiting fan-structure suggests an open fan, hence the term. Such a structure is essentially a symmetrical anticlinorium (*q.v.*). Webster's definition reflects the old erroneous idea of Alpine structure.

**Fanglomerate**
*Webster. Geol.* The material of an alluvial fan in which the rock fragments are only slightly waterworn.
The consolidated material of a rock fan. See Fan.

**Farm**
*O.E.D.* 5. Originally a tract of land held on lease for the purpose of cultivation; in modern use often applied without respect to the nature of the tenure. Sometimes qualified by sb. prefixed, as dairy-, grass-, poultry-farm.
6. A farm-house. Examples 1596–1600.
*Webster.* 6. Orig., a piece of land held under lease for cultivation; hence, any tract of land (whether consisting of one or more parcels) devoted to agricultural purposes, generally under the management of a tenant or the owner; any parcel or group of parcels of land cultivated as a unit.
7. Hence, a plot or tract of land devoted to the raising of domestic or other animals.
Ministry of Agriculture and Fisheries, 1946, *National Farm Survey of England and Wales: A Summary Report.* 'The Survey ... confined to agricultural holdings of 5 acres and above ... could not be confined to holdings which were also farms—

farms, that is, in the sense of having sufficient capital resources (both landlords' and tenants') to provide the occupier with a main occupation and a chief source of livelihood from farming. There are a substantial number of agricultural holdings, mainly small in size, which do not conform to this definition of a farm ... sometimes difficult to decide when to use the word "holding" and when the word "farm". "Holding" is the more inclusive.'

Comment. In general use a farm is a vague term for a holding of agricultural land, varying enormously in size. The special definition quoted above was worked out for a specific purpose. On the other hand a large area of hill grazing which would be called a farm in Wales or Scotland would most likely be termed a ranch in America. See also Farmer.

**Farm, ladder** (E. E. Evans, 1958)
A farm with a succession of small ribboned fields. 'They are the "striped farms" or, as I prefer to call them, ladder farms.' (E. E. Evans, 1958, in *S.G.M.*, 74, 190.)

**Farmer**
O.E.D. 3. spec. One who rents land for the purpose of cultivation; = tenant farmer. Now chiefly as a contextual application of 5.

5. One who cultivates a farm, whether as tenant or owner; one who 'farms' land or makes agriculture his occupation.

Ministry of Agriculture and Fisheries, 1946. *The National Farm Survey of England and Wales: A Summary Report.* 'The various types of occupier ... are:
  (i) Full-time *farmers* in occupation of *farms*.
  (ii) Part-time *farmers* in occupation of *farms*.
  (iii) 'Regular' spare-time *occupiers* in occupation of *holdings*.
  (iv) 'Occasional' spare-time *occupiers* in occupation of *farms* (run by an agent).
  (v) *Occupiers* of residential *holdings*.
  (vi) Hobby *farmers* in occupation of *farms*.
  (vii) *Occupiers* of accommodation *holdings*.
  (viii) *Occupiers* of institutional *farms* or *holdings*.' (pp. 9–10)

Comment. The Farm Survey distinguished for the first time in statistical form, full-time, part-time, spare-time and hobby farmers. Other terms used of British farming include 'cheque book farmers' (who expect to lose rather than gain money by their farming enterprise), whilst a distinction is sometimes drawn between the 'gentleman farmer' and the 'dirty-boot farmer'. American usage refers to 'sidewalk farmers'. In general use, however, 'farmer' as with 'farm' is used very loosely though in many countries 'farming' implies cropping as opposed to ranching which is simply use of grazing lands.

**Farmscape**
Coleman, Alice, 1969, A Geographical Model for Land Use Analysis, *Geography*, 54, 48. 'Farmscape is the area dominated by viable agriculture: crops, pasture and orchards. Farmland is clearly the chief element in farmscape but it is not the only element nor is it restricted to farmscape. ... farmscape may include many non-agricultural uses as long as these are neither of the type nor extent to preclude agricultural dominance. Agricultural villages, for example, are definitely an element of farmscape and not of townscape.'

Comment. Farmscape has subsequently been defined more precisely by means of a pattern-recognition technique (as yet unpublished) which allows its boundaries to be mapped to the nearest millimetre at 1:100,000 scale. The farmland which defines farmscape is improved land and not rough grazings.

**Fast ice**—see Ice, sundry terms
Webster. Sea ice fastened to the shore. *Arctic.*

**Fault**
O.E.D. 9. Geol. and Mining. A dislocation of break in continuity of the strata or vein. Cf. F. Faille. (quots. 1796–1883).

Webster. 6. Geol. & Mining. A fracture in the earth's crust accompanied by a displacement of one side of the fracture with respect to the other and in a direction parallel to the fracture. Webster p. 924 then defines fault surface, fault plane, vertical or inclined fault, strike fault, dip fault, oblique fault, cross faults, step faults, distributive faults, normal faults, everse fault, thrust fault, overthrust fault, horizontal fault, rotary or pivotal fault.

Mill, *Dict.* A fracture of rocks where those on one side have been displaced relatively to those on the other side. Normal Fault—a fault which hades to the downthrow side. Reversed Fault or Thrust Fault—a fault which hades to the upthrow side. Hade—the inclination of the direction of downthrow or upthrow of a fault to the vertical.

Holmes, 1944. 'A fault is a fracture surface against which the rocks have been relatively displaced. Vertical displacements up to thousands of feet and horizontal movements up to many miles are well known . . . three types occur most commonly . . . *normal faults* . . . the fault plane . . . generally inclined at an angle between 45° and the vertical; the beds abutting against the fault on its upper face or "hanging wall" are displaced downwards relative to those against the lower face or "footwall". The terms "downthrow" and "upthrow" for the two sides are of course purely relative . . . stress appears to have been a tension, involving stretching instead of compression. *Reverse* or *thrust faults*—when fracture inclined at an angle inclined between 45° and the horizontal the fault is described as an *overthrust* . . . may be driven forward many miles . . . such far-travelled rock-sheets are called *nappes* . . . along the thrust-plane a hard streaky or banded rock *mylonite* may be produced. Tear-faults . . . movement is predominantly horizontal the fracture being vertical or nearly so.

*Comment.* Note also *dip-faults* approximately parallel to the dip of the rocks; *strike-faults* or *longitudinal* faults approximately parallel to the strike; oblique faults with planes between dip and strike; also groups of faults forming a complex fault or fault zone, step faults, trough faults (forming a graben), ridge faults (forming a horst), also imbricate structure.

**Fault apron**
Hills, E. S., 1940. *The Physiography of Victoria.* 'The [fault] scarps suffer rapid dissection, much of the rock waste being dropped along their bases as *fault aprons* formed by numerous coalescing alluvial cones' (p. 155).

**Fault block**
A block of country, often of considerable size, bounded by faults, hence block-mountains in contrast to fold mountains. See also Horst.

**Fault breccia, fault zone**
*Webster.* A rock composed of angular fragments that have resulted from movement along a fault; crush breccia.

Holmes, 1944. 'Sometimes, instead of a single fracture, there are two or more, forming a strip consisting of a sheet of crushed rock of variable thickness. This is distinguished as a *fault zone,* and the shattered material within is called a *fault breccia.*' (p. 79)

**Fault-line scarp**
*Webster.* A fault-line scarp results when a hard rock, brought against a soft rock by faulting, is, through erosion of the soft rock, left standing as a cliff.

A fault frequently brings into contact groups of rocks of differing hardness and resistance to weathering so that the line of the fault is often marked, through differential weathering, by an abrupt scarp but the scarp is not necessarily the original footwall of the fault as in a true fault scarp (*q.v.*).

**Fault scarp**
*Webster.* A high steep face of rock . . . the direct expression of the relative uplift along one side of the fault.

Holmes, 1944. 'In faults recently active the footwall is exposed at the surface as a fault scarp' (p. 79).

See Fault, scarp, escarpment.

**Fauna**
The animal life, collectively of any given age or region. *Cf.* flora.

**Fell**
O.E.D. 2. A wild, elevated stretch of waste or pasture land; a moorland ridge, down. Now chiefly in the north of England and parts of Scotland.

Mill, *Dict.* 1. Bare and uncultivated mountain or hill.

2. Any high-lying sheep pasture (Cumbrian Lakes and N. England).

*Comment.* In *A Survey of the Agricultural and Waste Lands of Wales,* R. G. Stapledon and W. Davies drew a distinction between heather moor and heather fell by saying 'The typical "fell" is rocky and steep, often strewn with large boulders and in places showing small precipices of bare rock.' Although this distinction was preserved by Stapledon in later works it does not coincide with popular usage; in northern England any open hill-side or mountain is designated fell *e.g.* Cross Fell. See the discussion in Stamp, 1948, 153; also Tansley, 1939.

**Fellah** (Arabic; *pl.* fellaheen, fellahs)
*O.E.D.* A peasant in Arabic-speaking countries; in English applied *esp.* to those of Egypt.

**Felsenmeer** (German; *pl.* Felsenmeere)
Cotton, 1941. 'The boulders of a *felsenmeer*, which is a field of angular blocks on a flat-topped mountain in a temperate region, or perhaps on a lowland in the frigid zone (*e.g.* on Bear Island in the Arctic Ocean), are subject to creep of the nature of solifluction down very gentle slopes' (p. 12).
'A veneer of loose blocks' (p. 15).
Fischer, 1950. 'Boulder field, rock-block field, rock river.' (p. 16)

**Felucca** (prob. of Arabic origin)
*O.E.D.* A small vessel propelled by oars or lateen sails, or both, used, chiefly in the Mediterranean, for coasting voyages.
*Comment.* Webster derives from Italian feluca. Now most commonly used for the characteristic sailing craft of the Nile, usually single-masted.

**Fen**
*O.E.D.* 1. Low land covered wholly or partially with shallow water, or subject to fresh inundations; a tract of such land, a marsh.
Mill, *Dict.* A very damp area with reeds and rushes and other water-loving vegetation, the remains of which decay under the water and form peat.
Tansley, 1939. 'In this book the term *marsh* is applied to the "soil-vegetation type" in which the soil is waterlogged, the summer water level being close to, or conforming with, but not normally much above, the ground level, and in which the soil has an inorganic (mineral) basis; *fen* to a corresponding type (whose vegetation is closely similar) in which the soil is organic (peat) but is somewhat or decidedly alkaline, nearly neutral, or somewhat but not extremely, acid. *Bog*, on the other hand (bearing a radically different vegetation), forms peat which is extremely acid. *Swamp* is used for the type in which the normal summer water level is above the soil surface.' (p. 634)

**Fenland**
This word is sometimes used for a tract of land occupied by fen but, spelt with a capital, is used as a regional name for the country around the Wash in eastern England formerly occupied by fen, but now drained and cultivated and known also as 'The Fens'.

**Fenêtre** (French; *lit.* window), **Fenster** (German)
von Engeln, 1942, 'a fenster or window i[s] an erosional opening [in the upper limb o[f] an overfold or nappe] comparable to a breached anticline that... the fenster ha[s] a floor of rocks as young or younger tha[n] those of its rim' (p. 332–333).
Lobeck, 1939, 'windows or fensters are produced by the erosion of plateau areas so that the youngest strata of an underlying nappe are exposed at the bottom of the valley... the frame of the window is made up of older strata belonging to an upper nappe which was thrust over the lower one.' (p. 605, also diagram)
Thornbury, 1954, 'In the early stages of dissection of a nappe, erosion may penetrate the overthrust mass and locally expose patches of the younger rocks below it. Such exposures are called *fensters* o[r] *windows*.' (p. 273)
*Comment.* Some famous early examples were described in French, such as the Fenêtre de Theux, and writers in English prefer to retain the French term fenêtre.

**Feng-shui** (Chinese: in Cantonese fung-shui)
*Lit.* Wind water but used as the Spirit of Wind and Water of a locality, as interpreted in geomancy, particularly important in the siting of habitations, hence often the reason for a location. See Tregear, T. R., *Land Use in Hong Kong*, 1958, pp. 53, 72.

**Fenster** (German; *lit.* window)—see Fenêtre

**Feral**
*O.E.D.* Of an animal; wild, untamed. Of a plant, also (*rarely*) of ground: uncultivated. Now often applied to animals or plants that have lapsed into a wild from a domesticated condition.
*Comment.* Should be reserved for the second sense of *O.E.D.* Also used by Cotton of landscape in contrast to 'subdued' of Davis (see *Landscape*, 2nd ed., 1948, p. 215).

**Ferik** (Sudan; Arabic)
Huts occupied for part or the whole of the year by a group of nomad or semi-nomad pastoralists whose animals (usually cattle or camels) form a common herd, in the northern Sudan. (J.H.G.L.)

**Ferling** (Agricultural History)—see under Yardland

**Fermtoun** (Lowland Scots: farm-town, farm hamlet).
A hamlet consisting of a small group of

farms, comparable with a *baile* ('bal') in the Highlands. Where a church is present it becomes a kirktoun or *clachan* (*q.v.*). According to Arthur Geddes the term covers farm steadings (dwellings plus barns and cattlesheds), often with the dwellings of cottars (landless men, usually with grass for a cow). With depopulation in the Highlands, *baile* came to mean a single family farm, but not a croft, which is invariably a small holding. (C.J.R.)

### Fetch
Mill, *Dict.* Fetch of Waves. The distance that waves have travelled before they attack a coast-line (America).
*Adm. Gloss.*, 1953. The distance a sea wave has travelled to the observer from the locality of the wind that caused it from the nearest weather shore.

### Ffridd, ffrydd (Welsh; *pl.* ffriddoedd)
*Report Royal Commission on Common Land, 1955–58*, H.M.S.O., 1958, Cmd. 462. 'Welsh equivalent to "in-by land", the term widely used in the North of England for the fenced-in land nearest the homestead' (p. 274). Rough grazing (enclosed from mountain), sheep-walk, wood.

### Fiard (J. W. Gregory, 1913)—see Fjärd

### Field
*O.E.D.* Ground; a piece of ground. 1 Open land as opposed to woodland. 4 Land or a piece of land appropriated to pasture or tillage, usually parted off by hedges, fences etc.
See also Open field.

### Field capacity (Soil Science)
The amount of water held in a soil after excess water has drained away by gravity and downward movement has practically ceased.

### Field system
*Webster.* The prevailing system of husbandry in medieval times in England and parts of western Europe, whereby the arable land of a village unit was composed of unenclosed strips held by the different owners or cultivators subject to use as a common for pasture during a certain period of each year.
*Comment.* Where the common arable was divided into two, one cultivated and one lying fallow, the system is known as the Two-Field; the more usual division into three, one lying fallow, is the Three-Field system.
See also Open field.

### Figure of the Earth
Bowie, W., 1931, Physics of the Earth, II. The Figure of the Earth, *Bull. Nat. Res. Council*, 78, 103–115. '*Figure of the Earth*—the defining elements of the mathematical surface which approximates the geoidal surface. The figure of the earth has been proved to be approximately an oblate spheroid.' (p. 113)

### Filter mapping
Chorley, R. J. and Haggett, P., 1965, Trend Surface Mapping in Geographical Research, *Trans. Inst. Brit. Geog.*, No. 37. 'Where the areal array is in rectangular grid form, Nettleton (1954) has shown that many of the results hinge on their spacing, for the grid acts like an electric filter which will pass components of certain frequencies while excluding others. Indeed, Holloway (1958) views all smoothing attempts as forms of filtering.' (p. 51)

### Fine-textured topography—see Coarse-textured topography

### Finger lake
A long narrow lake. Some writers limit the term to those occupying rock basins.

### Fiord, Fjord (Norwegian; the anglicized form like the old Norwegian is fiord)
*O.E.D.* A long narrow arm of the sea, running up between high banks or cliffs, as on the coast of Norway.
Mill, *Dict.* 'Fiord or Fjord. A long deep arm of the sea occupying a portion of a channel having high steep walls, a bottom made uneven by basins and sills, and sidestreams entering from high-level valleys by cascades or steep rapids. The term is usually confined to such valleys where the sides rise more or less steeply to considerable heights, but is extended by Penck to the inlets of the low flat Swedish coast (see Fjärd).'
Knox, 1904. 'Fjord (Da., Nor.), firth or long narrow inlet. *cf.* Fjärd,
Wooldridge and Morgan, 1937. 'Fjord coasts resemble ria coasts, but have the typical form of glaciated troughs, with steep parallel walls, truncated spurs, hanging valleys and irregular rock-floors. The floor is often deeper than the sea-floor off-shore from which it is separated by a submerged "sill".' (pp. 354–355)
Gregory, J. W., *The Nature and Origin of Fiords*, London: Murray, 1913. Gives a long discussion, stresses special features due to crushing of rock along fracture lines and its consequent easy removal by

agents of denudations: glaciation not essential.

*Comment.* Another case of a widely used Danish-Norwegian word applied to almost any sea inlet (contrast shallow, low-shored Limfiord in Denmark with Hardanger Fiord in Norway) which has been given a special meaning in geographical literature in English—usually in the sense of Wooldridge and Morgan.

**Firiki** (West Africa; Kanuri), firki

Very flat land in the Chad Basin, often stretching for many miles, with a fine-particled clay soil which is completely impervious. In the rains the firiki is flooded and the soil becomes a soft, sticky mass; at the beginning of the dry season crops are planted for harvesting after about four months. The surface cracks badly by the end of the dry season (J.C.P.).

Harrison-Church, 1957. '*Grey Earth*, known as Firki in Northern Nigeria, occurs in badly drained depressions' (p. 86).

**Firka** (Indo-Pakistan: Hindi)

A sub-division of a thana (*q.v.*).

**Firn** (German)

*O.E.D.* (German, firn, firne, *lit.* 'last year's' (snow))

A name given to snow above the glaciers which is partly consolidated by alternate thawing and freezing, but has not yet become glacier-ice.

*Ice Gloss.*, 1958. *Old snow* which has been transformed into a dense material. Firn is characterized by the fact that (a) the particles are to some extent joined together, but that (b) the air interstices still communicate with each other. (a) distinguishes it from snow, and (b) from ice.

See also nevé (French), but the reasons for preferring the German term have been discussed in the *Journal of Glaciology*.

**Firnification**

The process by which snow is transformed into ice.

**Firn line**

Llibuotry, L., 1958, *Jour. Glaciology*, 3, 'the highest level reached by the snow line during the year; equilibrium line is the line where the net ablation, allowing for refreezing, equals the accumulation, ... it would seem that the firn line is rather lower than the equilibrium line of the smallest glaciers.' (p. 265)

**Firth** (Scottish)

*O.E.D.* A Scottish word, apparently from Old Norse. An arm of the sea; an estuary of a river.

Mill, *Dict.* 1. Portion of a river entrance of considerable depth, in which there is little or no tidal change in salinity. 2. Also a fiord. 3. The lower part of an estuary of gradually increasing depth towards the sea (Mill).

*Comment.* An essentially Scottish word with no precise meaning. Pentland Firth is a strait, the Firth of Forth, Firth of Tay, and Solway Firth are the seaward extensions of estuaries, the Firth of Lorne is comparable with a fiord.

**Fjäll, fjell** (Swedish)

In the Swedish sense of the word, fjäll (fjell) is a mountain rising above the tree limit. Flat undissected areas are a characteristic feature of the Scandinavian fjells. Translated as 'mountain' the word fjäll (fjell) would loose a part of its sense. In the form fjeld it was formerly sometimes used in international literature. In Northern Britain and Scotland the old Scandinavian word fell corresponds to the Norwegian fjeld, in modern orthography fjell, and the Swedish fjell, nowadays written fjäll. (E.K.)

**Fjärd**, *pl.* fjärdar (Swedish), fiard

Large continuous water areas surrounded by skerryguard islands. Coastal areas, where such dominate, are termed fjärdkuster (fjard-coasts), sing. fjärdkust. (E.K.)

Mill, *Dict.* 'Fjärde. Specifically applied to the numerous small narrow inlets of the low flat coast of Sweden.'

Knox, 1904. 'Fjärd (Sw.), firth or long narrow inlet, bay.'

Note: an identical definition is given for Fjord (Da., Nor.) and a Fjördr (Ice.) is given as a fiord.

Gregory, J. W., The Nature and Origin of Fiords, London: Murray, 1913. Among types of drowned valleys allied to fiords, those of Sweden are exampled, incorrectly named 'fjärden' which Gregory makes into 'fiard' in the singular (fiards in *pl.*). 'The essential difference between fiords and fiards is that the latter occur in coasts where the land is low.' (p. 67)

*cf.* rias they contain 'basins' and a 'threshold'. (p. 5)

Also refers to fiards in Dalmatia, Eastern U.S.A., N.W. Australia, Tsusima, Korea, Japan.

Shackleton, M. R., *Europe*, 4th ed., London: Longmans, 1950. (of Fjords) 'Other sea lochs believed to be of similar origin but with lower banks have been termed "fjards", such as those occurring on the southern coast of Norway, from Stavanger eastwards, and round the coasts of Sweden.' pp. 228–229

Comment. The above quotations show how far usage in English has departed from the Swedish definition given by Professor Kant. For this Gregory is responsible and fiard is better dropped.

**Fjeld** (Norwegian)
*O.E.D.* (Norwegian, field) An elevated rocky plateau, almost devoid of vegetation.

Comment. A Norwegian word with a wide meaning which when used in literature in English has been given the restricted meaning indicated by *O.E.D.* Such large areas as Dovre Fjeld and Hardanger Fjeld in Norway are above the tree line; hence their distinctive character. *Cf.* fjäll (Swedish) and fell.

**Fjord** (Norwegian)—see Fiord

**Fjord topography** (Swedish fjordtopografi, German fjordtopographie)
Introduced by G. De Geer, 1912, 1913. (E.K.)

**Flark** (Swedish)
A limited wet area (or feature) occupied by a sparse, feebly peat-forming (*i.e.* mud bottom) fen vegetation alternating with damming drier areas (or features). The Swedish word flark is also used in other languages. The Finnish word rimpi used in German is equivalent to flark. (E.K.)

**Flash (1)**
*O.E.D.* 1. A pool, a marshy place. Obsolete except locally.
2. (Quot.) 1888. Gresley, *Gloss. Coal Mining*; Flash (Cheshire), a subsidence of the surface due to the working of rock salt and pumping of brine.
Mill, *Dict.* 1. A pool or marsh. 2. Specially used for a subsidence due to working of rock-salt and pumping of brine (Cheshire). See also Swag, Pitfall.

**Flash (2)**
A sudden rise of water in a river. *Cf.* flush.

**Flat**
*O.E.D.* 5. A piece of level ground; a level expanse; a stretch of country without hills, a plain; the low ground through which a river flows.
b. A tract of low-lying marshy land; a swamp.
c. Australian (quot.) 1869. R. B. Smith, Goldfields Victoria, 611, Flat, a low even tract of land, generally occurring where creeks unite, over which are spread many strata of sand and gravel, with the usual rich auriferous drift immediately overlying the bed rock.
6. Chiefly pl. A nearly level tract, over which the tide flows, or which is covered by shallow water; a shallow, shoal.
Mill, *Dict.* A stretch of land without marked hollows or elevations.
Sparks, B. W., 1949, *Proc. Geol. Assoc.*, **60**. 'an area which visibly possesses a lower degree of slope than its surroundings' (p. 169).
(See also Sweeting, M. M., *Trans. Inst. Brit. Geog.*, **21**, 1955, 37.)
Brown, E. H., 1950, Erosion Surfaces in North Cardiganshire, *Trans. Inst. Brit. Geog.*, **16**, 51–66. 'The spur profiles are seen to consist, not of simple smooth slopes, but of a series of flats and breaks in slope.' (p. 63)
See also Hare, F. K., *Proc. Geol. Assoc.*, 1947; Linton, D. L., *London Essays in Geography*; also alluvial flats. A very large number of uses are given in Rice, 1941.

**Flatiron**
Lobeck, 1939. 'Other hogbacks are triangular masses, adhering to the ends of the spurs or rising up to the crest of the range, to form so-called *flatirons*.' (p. 511)

**Flatt**
Alternative name in some places for 'furlong', a unit in the open-field system. See Orwin, C. S. and C. S., 1954, *The Open Fields*, Oxford Univ. Press, 2nd ed., pp. 35 and 96. See shott (2).

**Fleet**
*S.O.E.D.* A place where water flows; a creek, inlet, run of water.
Comment. Now obsolete or occasionally used locally but common in place names, *e.g.* Fleet St., London; Northfleet, Kent.

**Floe**—see Ice floe

**Flood plain, flood-plain, floodplain**
Gilbert, 1877. 'Whenever the load reduces the downward corrasion to little or nothing, lateral corrasion becomes relatively and actually of importance.... As an effect of momentum the current is always swiftest along the outside of a curve of the channel, and it is there that the wearing is performed; whilst at the inner side of the curve the current is so slow that part of the load is deposited. In

this way the width of the channel remains the same while its position is shifted, and every part of the valley which it has crossed in its shiftings comes to be covered by a deposit which does not rise above the highest level of the water. The surface of this deposit is hence appropriately called the flood-plain of the stream.' (pp. 126–127) Mill, *Dict.* Flood-plain or Valley-flat. The general name for an alluvial plain bordering a river, the surface of which is formed by the accumulation of deposits during floods. Moore, 1949. A plain, bordering a river, which has been formed from deposits of sediments carried down by the river.... Thornbury, 1954. '...*floodplain*, the deposit of alluvium which covers a valley flat.... along many valleys a veneer of alluvium over a bedrock floor... initial floodplain is different in many respects from the final floodplain of old age.' (pp. 130–131) *Comment.* May also include deposits at the heads of drowned valleys. Note also 'fossil flood plain' and 'occasional flood plain'. (G.T.W.)

**Flood-Plain streams**
Melton, F. A., 1936, An Empirical Classification of Flood-Plain Streams, *Geog. Rev.*, **26**, 593–609. In this paper an elaborate terminology is introduced. The streams are first divided into single-crest and double-crest streams. Extensive and special use is made of the term 'bar' (*q.v.*); scroll, bar, and lacine meander plains are described. The more important terms have been included in this glossary.

**Flood-tide**
The rising tide; contrast ebb, ebb-tide (*q.v.*).

**Flora**
The plant life, collectively, of a given age or region. *Cf.* fauna.

**Flow meadow**
Water meadow (*q.v.*).

**Flume**
*O.E.D.* 3. *U.S.*, etc. An artificial channel for a stream of water to be applied to some industrial use.
*Webster.* 2. *U.S.* a. A ravine or gorge with a stream running through it. b. An inclined channel for conveying water from a distance to be utilized for power, transportation, irrigation, etc. as in placer mining, logging, etc. c. A channel for admitting water to a water turbine.
*Dict. Am.* 1. A narrow channel for water maintained to provide power, float logs convey water to a placer mine, etc. 1748 2. A narrow defile or ravine worn or cu out by a stream. Also (*cap.*) as a prope name. 1784.

**Flurosion** (W. S. Glock, 1928)
Glock, W. S., 1928, *Amer. Jour. Sci.*, **15** 'Proposed as a term to designate the work of transportation and destruction carrie on by streams' (p. 477). Not generall adopted.

**Flush**
*O.E.D.* A sudden increase in the volume of a stream; a rush of water coming dow suddenly.
Mill, *Dict.* A state of flood during which th water does not quite overflow its bank (N. England).

**Flute, fluting**
Fluting and grooving refer to small-scal ridges and depressions caused by differen tial erosion, especially by wave action, o an exposed rock-surface with marked jointing.

**Fluvial, fluviatile**
Mill, *Dict.* (1) Of or pertaining to a river o rivers. (2) Found or living in a river.
*Comment.* Geologists nearly always us 'fluviatile' when referring to deposits lai down in or by rivers. Note also 'fluvio marine'.

**Fluvioglacial, fluvio-glacial**
Produced by, or due to, the action of stream of water derived from the melting o glacier ice. Applied especially to th deposits of the outwash fan.

**Fluviraption** (C. A. Malott, 1928)
Thornbury, 1954. 'Hydraulic action, th sweeping away of loose material by movin water... (from *fluvius*, river and *rapere* to seize).' (p. 47) See Malott, C. A., 1928 An Analysis of Erosion, *Proc. Indian Acad. Sci.*, **37**, pp. 153–163.

**Fly** (Norwegian)
The moderately steep slope lying between a scarp face and the vidda below it (esp. East Norway) (Strøm, 1945, *Norsk Geologisk Tidsskrift*). (P.J.W.)

**Flysch** (German)
Mill, *Dict.* Sandstone and some shale etc., of Eocene and Oligocene age, stretching from southwest Switzerland into the Carpathians, and of great importance in the building of the Alps.
Shackleton, 1958. '...the Alpides.... As the folds rose from beneath the sea,

shallow-water deposits were laid down, on the outer sides of the ranges, and these consist usually of sandstone and clays known as Flysch.' (p. 8)
*Comment.* Primarily a formation name; Cretaceo-Tertiary in age.

### Fog
*O.E.D.* 2. Thick mist or watery vapour suspended in the atmosphere at or near the earth's surface; an obscured condition of the atmosphere due to the presence of dense vapour.
3. trans. a. Any abnormal darkened state of the atmosphere.
b. Any substance diffused through the atmosphere, so as to cause darkness; a dark mass (of smoke).
Mill, *Dict.* Water existing as minute droplets in the atmosphere. A cloud on or near the surface of land or water; its formation is facilitated by the solid particles of carbon, etc., in a smoky atmosphere.
*Met. Gloss.*, 1944. Fog is defined as obscurity in the surface layers of the atmosphere caused by particles of condensed moisture or of smoke held in suspension in the air. In international meteorological practice the term 'fog' is limited to a condition of atmospheric obscurity in which objects at a distance of one kilometre are not visible.

### Fog drip, fog precipitation
Moisture deposited from fog or cloud. See Parsons, J. J., 1960, 'Fog Drip' from Coastal Stratus, with special reference to California, *Weather*, 15, 58–62 (with bibliography); also Byers, H. R., 1930, Summer Sea Fogs of the Central California Coast, *Univ. Calif. Publ. Geog.*, 3, 291–338.

### Foggara (Arabic; Sahara)
The term used, especially in Mauritania, for gently inclined underground channels bringing water for irrigation from aquifers near the base of mountains. Comparable with the Karez of Baluchistan (*q.v.*).

### Föhn (German)
Mill, *Dict.* 'Föhn, Foehn, Fön. A descending warm and relatively dry wind on the lee side of mountains; originally a local name for such winds occurring in northern Alpine valleys, especially from Geneva to Vorarlberg, where in winter they are remarkable for the rapidity with which they cause the snow to disappear.'
*Met. Gloss.*, 1944. 'Föhn. A warm, dry wind which blows down the slopes on the leeward side of a ridge of mountains... likely to occur wherever cyclonic systems pass over mountainous regions;... Examples of Alps, E. Rocky Mts. (Chinook) and Greenland coast. See also Berg wind.

### Fordes
Charlesworth, J. K., *The Quaternary Era* I, 1957, 353. 'Relatively shallow submerged valleys which branch at their heads into several subaerial valleys containing a number of lakes separated by shallows.' Introduced into English terminology by J. W. Gregory, *Origin and Nature of Fiords.*

### Föhrde (German; *pl.* Föhrden), Förde, Förden
A long and narrow inlet of the sea (*cf.* firth) in boulder clay country or surrounded by terminal moraines; a local name for the inlets in eastern Schleswig-Holstein (K.A.S.).
Shackleton, M. R., 1958. 'West of the Oder ... two types of coast are distinguished, the *bodden* and *förden* ... the *förden* coast lies at the south eastern side of the Jutland peninsula ... the *förden* are believed to be drowned valleys, formed originally by rivers flowing beneath the ice-sheet (*cf.* the *fiords* of Danish Jutland). They provided very good harbours until the days of really big ships.' (pp. 250–251) Shackleton uses italics but no capital letter.

### Fold (1)
*O.E.D.* A pen or enclosure for domestic animals esp. sheep.

### Fold (2), folding, folded rocks
Fold: the general term for a flexure of the rocks of the earth's surface produced by compression in the course of *orogenesis* or mountain-building The folds may be a simple arch or *anticline*, a hollow or *syncline*; either may be symmetrical or asymmetrical, the axis of an asymmetrical anticline may be so inclined as to result in an *overfold* or *recumbent fold*. A very large anticline is a *geanticline* or *geoanticline*, a large syncline a *geosyncline*, a complex anticline is an *anticlinorium* and a complex syncline a *synclinorium*. A fold with one limb is a *monocline* (English sense), repeated tight folds from *isoclinal* strata. All the words in italics are separately listed.

### Folded mountains, fold mountains, fold-mountains
Mill, *Dict.* Fold-mountain. Mountains the ridges of which are due primarily to up-

folds, *i.e.* anticlines, and the valleys, *i.e.* synclines, to downfolds.

Geikie, J., 1898. Mountains which have resulted from the flexuring of the earth's crust are termed folded or flexured mountains.

*Comment.* Mill is wrong in his restricted usage: in the older fold mountains synclines frequently coincide with the ranges and anticlines with the valleys (see inverted relief). The contrast is between fold-mountains and block-mountains due primarily to faulting.

**Foldage, foldcourse, foldland**
*O.E.D.* The practice of feeding sheep in movable folds (foldage). Fold-course, foldland: land to which pertained the right of foldage; the right itself; hence a sheep-walk.

Fold course. An area comprising, usually, open field arable land, permanent pasture (heath, marsh), and perhaps closes, and within which a flock of sheep was confined, using the different types of feed when each was available. Norfolk 15th to 18th centuries (I.L.A.T.).

*Comment.* See Allison, K. J., 1957, The Sheep-Corn Husbandry of Norfolk in the Sixteenth and Seventeenth Centuries, *Agric. Hist. Rev.*, 5, 12–30; Simpson, A., 1958, The East Anglian Foldcourse: some queries, *Ibid.*, 6, 87–96. See also discussion in Denham, D. R., 1958, *Origins of Ownership*, London: Allen and Unwin (includes also leanland and sokeland).

**Fontein** (Afrikaans)
Fountain, water-spring, source of river. The Afrikaans form is not used by itself in South African English. As a part of a proper name however it is left untranslated, *e.g.* Bloemfontein (*i.e.* Flower Fountain).

**Footrail** (North Staffordshire coalfield)
An adit mine.

**Force** (northern England)
*O.E.D.* A name in the north of England for a waterfall or cascade. From Old Norse, *fors*; *cf.* Swedish fors (*q.v.*). Formerly spelt forse.

**Foredeep**
*Webster. Geol.* A deep depression in the ocean bottom fronting a mountainous land area, as the Tuscarora deep, off the coast of Japan.

Mill, *Dict.* Elongated depressions indicating the subsidence of a mountain foreland beneath the folded mountains. All the great marine abysses are foredeeps in this sense.

von Engeln, 1942. '... relatively narrow steep-sided troughs, located near continents or associated with island arcs. These are referred to as *foredeeps* and probably result either from graben faulting, or are geotectoclines.' (p. 46)

*Comment.* Also used where the foredeeps have been infilled with sediment, *e.g.* the Himalayan foredeep is now the Indo-Gangetic Plain of Northern India-Pakistan.

**Foreland** (in physical geography)
*O.E.D.* 1. A cape, headland or promontory.
2. A strip of land in front of something.
a. Examples include space between canal bank and drainage ditch or river.
b. (In fortification = berm (between wall and moat)).
3. Land or territory lying in front. *e.g.* 1870. Daily Tel., 22 Sept., Alsace and Lorraine ... will form a German foreland.

Mill, *Dict.* 1. A strip of low, flat land built by waves and currents at the foot of a cliff.
2. The region in front of the line of advance of the main folds of a mountain chain. It is characterized by folds more or less parallel to the main ones, *e.g.* Alpine Foreland, often called Fore-Alp and contrasted with Middle Alp (5,300 to 9,000 feet) and High Alp (over 9,000 feet).

Johnson, 1919. Used only in connection with 'cuspate foreland' (*q.v.*) in which foreland may be used in sense of *O.E.D.* 1 rather than Mill 1.

Cotton, 1945. 'Prograded areas are forelands, and if continuous for some distance alongshore are strand plains.' (p. 431) (Similarly in edition of 1922.) Also 'cuspate foreland'. (pp. 485–486)

Wooldridge and Morgan, 1937, (of Alpine folding) (Tectonic). 'The African "hinterland" is believed to have moved northward towards the European "foreland", and the amount of crustal shortening achieved ...' (p. 76). Used in connection with 'cuspate foreland'. (pp. 337, 344, 346)

Billings, M., 1952, *Structural Geology*, New York: Prentice-Hall. 'The region in front of the overthrusts is often called the foreland.... The rocks of the foreland are essentially where they were deposited and are said to be autochthonous.' (p. 181)

*Comment.* A word with many and varied

meanings: still commonly used in all the many senses illustrated above, and in others. Structural geologists (see also Willis, Hills) give it the most precise definition.

**Foreland** (in economic geography)
Weigand, Guido G., 1955. A term proposed by Weigand as a contrast to hinterland in the geographical study of ports. He defines it thus (*A.A.A.G.*, **45**, 303) 'denotes not maritime space extending from the port but the seaward trading areas which are connected with the port through maritime organization'.
*Comment.* The wisdom of Weigand in using a word with several existing meanings in a new specialized sense is doubtful.

**Fore-set Beds**—see Delta structure, bottom-set beds

**Foreshore**
*O.E.D.* The fore part of the shore; that part which lies between the high- and low-water marks; occasionally the ground lying between the edge of the water and the land which is cultivated or built upon.
Mill, *Dict*. Loosely applied in British Isles to various parts of the region between low-water mark and the margin of occupation.
Wooldridge and Morgan, 1937, 'the region extending between the ordinary tidal limits.' (p. 322) (following D. W. Johnson).
*Comment.* The descriptive terminology of the 'coast' or 'shore' is in a state of some confusion. (Wooldridge and Morgan; D. W. Johnson, *Shore Processes and Shoreline Development*, New York, 1919) as the above quotations indicate.

**Forest**
*O.E.D.* The 'outside' wood (*i.e.* that lying outside the walls of the park, not fenced in, from foris, out of doors).
1. An extensive tract of land covered with trees and undergrowth, sometimes intermingled with pasture. Also, the trees collectively of a 'forest.'
b. In Great Britain, the name of several districts formerly covered with trees, but now brought more or less under cultivation, always with some proper name attached, as Ashdown, Ettrick, Sherwood, Wychwood Forests.
2. Law. A woodland district, usually belonging to the king, set apart for hunting wild beasts and game, etc.; having special laws and officers of its own.
3. A wild uncultivated waste, a wilderness.
Mill, *Dict.* As *O.E.D.* 1 and 1b.
*Webster*. In popular usage, a dense growth of trees and underbrush covering a large tract of land; technically, an extensive plant society of shrubs and trees with a closed canopy, and having the quality of self-perpetuation (in which case it is a *climax forest*) or of development into a climax.
Darby, H. C. (ed.), 1936, *An Historical Geography of England*, Cambridge: C.U.P. '... "forest" land—that is, land outside (foris) the common law and subject to a special law, whose object was the preservation of the king's hunting. The word "forest" was then a legal and not a geographical term.' (p. 173)
Stamp, 1948. '"Deer Forests" derive from an older meaning of "unenclosed hunting ground" and were defined by a Departmental Committee, 1922 (Cmd. 1636) as "an area from which the stock of sheep or cattle has been wholly or partly removed in the interests of deer-stalking, or which is derelict for want of a pastoral tenant and used for this purpose. ...' (p. 164)
Stroud, 1952. '"Forrest" is a place priviledged by royall authority or by prescription, for the peaceable abiding and nourishment of the Beasts or birds of the forrest, for disport of the King.' (Termes de la Ley, 1641) (A subject may hold a forest by grant from the Crown).
FAO, 1960, *World Forest Inventory*. Rome. 'All lands bearing vegetative associations dominated by trees of any size, exploited or not, capable of producing wood or of exerting an influence on the local climate or on the water regime, or providing shelter for livestock and wild life.' It includes bamboo; wattle; savanna types with average density of tree cover at least 0·05 per hectare. *Accessible forests* are within reach of exploitation by existing transportation systems. The economic definition of accessibility adopted in the 1953 inventory was abandoned in 1958 since (1) it was not found easy to decide whether a forest was within reach of economic management or exploitation and (2) the area to be included would vary with the price of wood. *Productive forests* include 'all forest land which is now producing or is capable of producing usable crops of wood or other forest products' but do not include forests in which there

is merely occasional fuelwood cutting. The shift of emphasis to accessibility in an economic sense will be noted. See also *accessibility*. (C.J.R.)

Sears, P. B., 1956 in *A World Geography of Forest Resources*, New York: Amer. Geog. Soc., 'Roughly the great vegetation types can be divided into forest, grassland, scrub, and desert, in descending order of available moisture necessary for their maintenance. (p. 3)

Paterson, S. S., 1956. 'Vegetal associations dominated by trees of size capable of producing timber or other forest products or influencing water regime.' *The Forest Area of the World and its potential productivity*. Göteborg.

Comment. In present day English speech forest is used to mean an extensive area occupied mainly by trees and there is often the implication of large trees in contrast to woodland where the trees are smaller or where there is much undergrowth or scrub. Such a distinction between forest and woodland cannot be maintained. Because the historical meanings of the word cannot be eliminated, the context should make clear in what sense the word is being used. Note the use of forest-tree, *O.E.D.*, any tree of large growth, fitted to belong to a forest. This links with the simple statement given above of Sears as a botanist. Contrast the economic definition of Paterson, a Swede writing in English, with which may be compared the old definition of Zon and Sparhawk 'land covered with woody growth of economic importance'. For some of the complications of the former use of the term in a legal sense see Bazeley, M. L., 1921, The extent of the English Forest in the thirteenth century, *Trans. Roy. Hist. Soc.*, **4**, 4th Ser., 140–172, London; Petit-Dutaillis, C., 1915, *Studies and Notes supplementary to Stubbs' Constitutional History*, 2, 147–251, Manchester (L.D.S., H.C.D., C.J.R.).

**Forestal**
*O.E.D.* Of or pertaining to a forest.
Comment. See also agro-forestal as defining a type of land use.

**Forestry**
*O.E.D. Suppl.* 2. The science and art of forming and cultivating forests, management of growing timber.
Forestry conveys the idea of economic development which 'woodland management' need not.

**Formation**—see Geological Time

**Formkreis, Formenkreis** (German; pl. Formenkreise: *lit.* circle of forms)
Related land forms which owe their existence to the same action, *e.g.* of running water, ice action etc., hence 'glazialer Formenkreis' (K.A.S.).

Thornbury, 1954. Formkreisen or 'morphogenetic regions'. 'The concept of a morphogenetic region is that under a certain set of climatic conditions, particular geomorphic processes will predominate and hence will give to the landscape of the region characteristics that will set it off from those of other areas developed under different climatic conditions.' (p. 63) See Büdel, J., *Geol. Rundschau*, **34**, 1944, 482–519, and *Erdkunde*, **2**, 1948, 25–53, who uses the correct Formenkreis.

**Form-line**—see Contour

**Form Ratio** (of a Stream) (G. K. Gilbert, 1914)
Gilbert, G. K., 1914, *The Transportation of Debris by Running Water*, U.S.G.S. Prof. Paper 86. 'Of the variable factors which in combination produce the multifarious channel forms of natural streams, the laboratory dealt exclusively with but a single one, the relation of depth to width. The relation is a simple ratio and either of the two terms might be the divisor. The width has been chosen because, as in the case of fineness, its selection conduces to symmetry in formulation. The ratio *depth of stream/width of stream* will be called the *form ratio*' (pp. 35b–36a).

Woodford, A. O., 1951, Stream Gradients and Monterey Sea Valley, *Bull. Geol. Soc. Amer.*, **62**, 799–852. 'A second measure of the form of cross section . . . is the *form ratio* depth/width (Gilbert 1914). The range recorded from 1:16 to 1:160 is thought to be representative. Since form ratios are usually such small fractions, hydraulic radii are usually identical with mean depths' (pp. 804b and 806a).

**Förna** (Swedish)
Litter; leaves and other undecomposed residues lying on the soil. The Swedish word förna is also used in other languages. (E.K.)

**Fors** (Swedish; *pl.* forsar)
A waterfall of low inclination in a river bed; rapids, cataract. (E.K.) *cf.* force.

**Fosse** (1), also **fossa**
*O.E.D.* (Latin, fossa, ditch) 1. An excavation narrow in proportion to its length; a canal, ditch, or trench.

**Mill**, *Dict*. 1. A swallow-hole (Champagne). 2. A trench used for protection and for drainage. 3. An ocean deep.

*Comment*. In French an ocean deep: in English the word is now generally used in such special cases as the Fosse Way (a Roman road). The use of fossa to indicate a great rift valley should be noted in the *Fossa Magna* of Japan—'a great depressed zone, the Fossa Magna of Naumann' (Trewartha, 1945, p. 11). Also foss, a waterfall (northern England) *cf.* fors, force.

**Fosse (2)**
Thornbury, 1954. 'Depressions between a glacier and the sides of its trough are called *fosses*, and they owe their existence to the more rapid rate of melting which takes place here because of the added effect of heat absorbed or reflected from the valley sides. Fosses may be the sites of short lakes, or they may have stream courses through them.' (p. 378) See kame terrace.

**Fossil**
*O.E.D.* 1. In early use: any rock, mineral or mineral substance dug out of the earth. *Obsolete*.
2. Now only in restricted sense: anything found in the strata of the earth, which is recognisable as the remains of a plant or animal of a former geological period, or as showing vestiges of the animal or vegetable life of such a period.

Himus, 1954. The remains and traces of animals and plants which are found naturally embedded in rocks. They comprise not only the actual remains of organisms, such as bones, shells of molluscs, tests of echinoderms, twigs, bark, leaves and seeds of plants, but moulds and casts thereof, and even impressions left by soft-bodied animals, such as worms and jelly-fish, and the footprints of birds, reptiles and other animals.

Twayne, 1956 adds 'The term is sometimes applied to anything buried beneath the earth's surface by natural causes or geological agencies (*e.g.* a fossil cliff).' Also *fossil erosion surface* (exhumed erosion surface); *fossil water* or connate water, water entrapped in sediments during their deposition.

**Founderous, foundrous**
*O.E.D.* Causing or likely to cause to founder miry, full of ruts and holes.

*Comment*. An old term formerly much used when roads were frequently in a state of disrepair. It survives mainly in legal documents.

**Frazil ice**—see Ice, sundry terms
**Freeboard**
A narrow irregular belt of land marking the boundary between medieval parishes and counties affording access to the arable fields on either side. The origin of many of the winding lanes of the Midlands of England after inclosure. (H.C.D.)

**Free face** (Alan Wood, 1942)—see waning slope
**Free port**
*Webster. Com.* An enclosed, guarded port, or section of a port, where goods may be received and shipped free of customs duty and of most customs regulations.

A port free from customs dues and similar restrictions. For obvious reasons free ports often have a large entrepôt trade; even in free ports or ports which are free for the greater proportion of merchandise there may be restrictions and dues payable on certain articles, notable alcoholic liquors. Hong Kong and Singapore are good examples of free ports. (L.D.S.)

**Freshet**
*S.O.E.D.* A flood or overflowing of a river caused by heavy rains or melted snow.

**Friagem (Brazil)**—see Surazo
**Frigid Zone**—see Zone
**Fringing reef**—see Barrier reef
**Front (Meteorology), frontal surface, frontal zone**
Hare, 1953. 'If we recognize the existence of airmasses, characterized by homogeneity over wide areas, we must also give consideration to the boundaries which separate the airmasses... often separated by a sloping boundary surface which we call a *frontal surface*. The line along which this surface intersects with the ground surface is termed a '*front*'. Sometimes airmasses are separated by a zone too wide to be regarded strictly as a front; we then use the term 'frontal zone'. (p. 64)

**Frontal Apron**—see Valley train, outwash plain
**Frontier**
*O.E.D.* 4. sing. and pl. The part of a country which fronts or faces another country; the marches; the border or extremity conterminous with that of another.

b. U.S. 'That part of a country which forms the border of its settled or inhabited regions: as (before the settlement of the Pacific coast), the western frontier of the United States.' (Cent. Dict.)

5. A fortress on the frontier, a frontier town; examples 1604–1796.

Mill, *Dict.* The country contiguous to the boundary-line between two territories.

Webster. 1. That part of a country which fronts or faces another country. 2. The border or advance region of settlement and civilization. Chiefly *U.S.*

*Dict. Am.* A region in what is now the U.S., newly or sparsely settled and immediately adjoining the wilderness or unoccupied territory. 1671. 1893. Turner. *Signif. of Frontier in Amer. Hist.* 3. 'What is the frontier?... In the census reports it is treated as the margin of that settlement which has a density of two or more to the square mile.'

Boggs, S. W., 1940, *International Boundaries*, New York: Columbia U.P. 'The term "boundary" denotes a line, such as may be defined from point to point in treaty, arbitral award, or boundary commission report. A "frontier" is more properly a region or zone, having width as well as length.' (p. 22)

Fawcett, C. B., 1918, *Frontiers*, Oxford: Clarendon. '... we may start by regarding a political frontier as an area of separation between two areas of more or less homogeneous, and usually denser, population.' (p. 15) (The 'Zonal Character of Frontiers' is given a complete chapter.)

Billington, R. A., 1950, *Westward Expansion*, 'In this geographical sense, the frontier has been usually defined as an area containing not less than two nor more than six inhabitants to the square mile. Census Bureau statisticians have adopted this definition in tracing the frontier's advance from the records of each population poll since the first tabulation of 1790.' (United States.) (p. 3) Billington emphasizes the several zones involved: fur traders, cattlemen, miners, pioneer farmers, equipped farmers, urban frontier.

*Comment.* In some British literature 'The Frontier' is the North-West Frontier of India where lies the North-West Frontier Province which in 1947 became part of Western Pakistan. Considerable numbers of British military forces were always engaged in guarding this frontier hence it became The Frontier, *par excellence*.

*American Comment.* The most influential definition in American historiography is that of F. J. Turner, who spoke of the frontier as 'the meeting point between savagery and civilization' and the 'hither edge of free land' (*The Frontier in American History*, New York: Henry Holt, 1920), p. 3. See also Fulmer Mood: Notes on the History of the Word *Frontier*, *Agricultural History*, 22, 1948, 78–83; and Ladis K. D. Kristof: The Nature of Boundaries and Frontiers, *A.A.A.G.*, 49, 1959, 269–282 (M.W.M.).

**Frontogenesis** (Meteorology)

*Webster. Meteorol.* The bringing together of two masses or currents of air which differ and commonly so react upon each other as to induce cloud and precipitation.

Hare, 1953. 'The term we use for the processes which lead to the formation or intensification of fronts.' (p. 65)

**Frost**

*O.E.D.* 1. The act or state of freezing or becoming frozen; the temperature of the atmosphere when it is below the freezing point of water; extreme cold. Often used with qualifying adjective as *hard, sharp*, etc. *Black frost*: frost not accompanied by rime, as opposed to white frost.
b. Viewed as an agent which penetrates and freezes the contained moisture of a porous substance, especially the ground. 2. Frozen dew or vapour. More fully *hoar* (frost), *rime* or *white frost*.

Mill, *Dict.* Similar to *O.E.D.* except that Mill equates rime with frozen dew and hoar frost with frozen vapour.

**Frost climate(s)**

A term often applied to Köppen's E F climates 'characterized by eternal frost' (Haurwitz and Austin, 1944).

**Frost heaving, frost riving, frost shattering, frost wedging**

Some of the many terms used in describing frost action characteristic of high latitudes and high altitudes.

See Felsenmeer, stone stripe, stone polygon, etc.

**Frost-line, frost line**

Mill, *Dict.* The limit of frost. *Cf.* snow-line. Similarly *O.E.D.*

*Comment.* In tropical countries there is an altitudinal limit below which frost never occurs; normally beyond the tropics the limit is at sea-level and so the frost-line becomes the limit towards the equator beyond which frosts never occur. As with snow-line, frost-line is sometimes used with reference to seasonal limits.

**Frost smoke**

*Ice Gloss.*, 1956. Fog-like clouds, due to the contact of cold air with relatively warm sea water, which appear over newly formed *leads* or *pools*, or leeward of the *ice edge* [in winter].

**Fu, -fu (Chinese)**
*Webster.* A department or prefecture in China comprising *hsien*; also, the chief city of a department.
As suffix to place name means city or town, e.g. Yunnan-fu, but sometimes an integral part of the place name.

**Fula (Sudan; Arabic)**
A shallow pool formed during the rainy season in the northern Sudan, and an important source of water for pastoral tribes until about the middle of the dry season. (J.H.G.L.)

**Fulgurite**
A vitrified sandtube, caused by fusing of mineral grains where lightning strikes the ground. First observed by Hermann in Silesia in 1711; found especially on exposed mountains and deserts. See also Australite.

**Fulji (Northern Arabia: Arabic)**
Bagnold, R. A., 1941, *The Physics of Blown Sand and Desert Dunes*, London, Methuen (rep. 1954). 'interdune hollows are known in northern Arabia as *fuljis*. They occur wherever a string of barchans are pressing closely on one another' (p. 214).

**Full**
*O.E.D.* Kent: A ridge of shingle or sand pushed or cast up by the tide. Examples: 1864–1907 (a feature of the foreshore)
Mill, *Dict.* Full (of sandbanks). The rounded or more swelling part of a bank. Also a ridge of shingle upon the seashore above the reach of ordinary tides.
Johnson, 1919. 'Beach ridges have long been recognised as representing successive positions of an advancing shoreline, and are known to the English as "fulls", while the depressions between them are known as "swales", "slashes," or "furrows".' (p. 404) Similarly in Wooldridge and Morgan, 1937.

**Fumarole (Italian)**
*O.E.D.* Also fumarol, fumerole. A hole or vent through which vapour issues from a volcano; a smoke-hole.
Mill, *Dict.* An opening of the Earth's crust emitting aqueous or other vapour.
Fay, 1920. Similarly, and *cf.* Solfatara, Mofette, Soffioni.
Himus, 1954. One of the manifestations of dying or extinct volcanic activity. A hole in the earth's crust from which steam (sometimes charged with boric acid) and gases, such as carbon dioxide, are emitted under pressure.

**Functional analysis**
Harvey, D., 1969, *Explanation in Geography*, London: Edward Arnold. 'Functional analysis attempts to analyse phenomena in terms of the role they play within a particular organization. Towns may be analysed in terms of the function they perform within an economy, rivers may be analysed in terms of their role in denudation, and so on.' (p. 81)

**Functional region**
Hartshorne, R., 1959, *Perspective on the Nature of Geography*, Chicago: Rand McNally. 'Regions of functional organization may be determined by a single feature, as in the case of a river system, but in most cases one is concerned with different kinds of features in interrelation —for example the integration of different types of production through transport routes and the media of trade. In any case, the unity of an area involved is a reality based upon dynamic connections among phenomena at different places. The functional region is therefore not a descriptive generalization of character, but rather the expression of a theory of process—relationships, a generalization in the logical sense.' (p. 136)
Morrill, R. L., 1970, *The Spatial Organization of Society*, Belmont, California: Wadsworth Publishing Co. 'Area under the economic and social domination of a centre; nodal region is a better term.' (p. 242)
See also Nodal region.

**Functional (rurban) fringe**
Coleman, Alice, 1976, Approaches to a Typology of Environmental Planning, *Urban Prospects*, Ottawa: Department of State, Urban Affairs. '... an area that is completely dominated by settlement, with no remaining fragmented farmland or run-down rough vegetation, so that the element of conflict has been removed, although the actual types of settlement concerned are those which would have tended to disrupt or dislocate the fabric of townscape if they had been located within it. They may include barriers to communication or large parcels which

coarsen the fine-textured scale of townscape and common examples are railroads and freight yards, freeways and spaghetti junctions, pipelines, energy transmission lines, quarries, sewerage plants, refuse dumps, golf courses and other large recreational spaces. Built-up and open land uses are still intermixed but in a form of spatial organisation where they can co-exist peacefully. This often happens where both components are in the same ownership in the form of institutions with large grounds, such as educational and research establishments, experimental farms ... power stations ... airfields, etc.'

**Fundament**

*O.E.D.* Geographical use is derived from an obsolete use: The foundation or base of a wall, building, etc.

James, P. E., 1934, *A.A.A.G.*, **14**. 'By an extension of its dictionary meaning, fundament is used to indicate the foundation on which the works of man have been built. Fundament may be defined as the face of the earth as it existed before the entrance of man into the scene.' (p. 80) The first use is credited to J. B. Leighly, The towns of Mälardalen in Sweden, *Univ. Calif. Publ. in Geog.*, **3**, 1928, 1–134, 'The forces which condition and shape a cultural landscape are many and of varied origin, each fluctuating in intensity through time. Their combination at a given place and moment of time is probably absolutely unique. The NATURAL FUNDAMENT which they modify, on which they erect their proper structures, is similarly varied from place to place, itself changing through natural processes.' (p. 3)

*Comment.* The usage of Leighly and James is not widely current, probably because of the common and old established geological term 'fundamental complex' (*q.v.*). But see A. H. Meyer, 1926, *Papers Michigan Acad. Sci. Arts and Lett.*, **21**, 360–396; also 1954, *A.A.A.G.*, **44**, 245–274. Leighly has informed the Committee that he hopes the term will be forgotten.

**Fundamental complex**

In geology the rocks of the 'original' crust of the earth formerly applied to the great areas of pre-Cambrian crystalline rocks. It is still used although it is now recognized that probably no part represents the 'original' crust of the earth. Basal complex is now more commonly used. (G.T.W.)

**Furlong**

*O.E.D.* 1. Originally the length of a furrow in the common field. One eighth of an English mile.

Orwin, C. S. and C. S., 1954, *The Open Fields*, Oxford Univ. Press, Compact block of strips forming a unit in openfields. (p. 35) See Shott (2).

**Furrow**

*O.E.D.* 1. A narrow trench made in the earth with a plough. (See also furlong.)

# G

**Gabbro**
Mill, *Dict*. A group of basic holocrystalline igneous rocks, usually coarsely crystalline. The term gabbroid is often used to denote the broad group (see intermediate), gabbro being more precisely defined. The granites are the corresponding acid rocks, the diorites and syenites the intermediate. A gabbro consists essentially of a basic plagioclase felspar, usually labradorite, a ferromagnesian mineral (commonly augite) and accessory minerals.

**Gait**—see Cattlegate

**Gale**
*O.E.D.* A wind of considerable strength.
Mill, *Dict*. In popular literary use, a wind not tempestuous, but stronger than a breeze.
A wind of force 8 on the Beaufort Scale (*q.v.*) or roughly 35 miles per hour.

**Galeria** (Italian and Spanish)
James, P. E., *Latin America*. New York: Odyssey Press, 1942. 'In both wet and dry savannas, as in the prairies, the banks of the streams are usually covered with a dense ribbon of forest, known as galeria—literally a "tunnel forest", since the narrower streams flow beneath an arch of foliage.' (p. 2)
*Comment*. Commonly corrupted to Gallery forest (*q.v.*).

**Gallery forest**
*Webster*. A fringing forest.
Mill, *Dict*. Forest occurring along both banks of a river in what is otherwise a region of open country. (The view that most of the Central African rainforest was 'gallery forests' was advanced by E. de Wildman in 1913 and 1926. See Richards, P. W., *The Tropical Rain Forest*. Cambridge: C.U.P., 1952, pp. 11–12.)
*Comment*. Apparently a corruption, probably through ignorance, of the Italian or Spanish *galeria* with the meaning of tunnel, since travellers by boat appear to be travelling through a tunnel of trees. Not in *O.E.D.*

**Gangue**
Mill, *Dict*. The material surrounding or accompanying a metallic ore in a lode or vein.
*Comment*. The worthless vein matter in which the valuable mineral occurs, hence gangue-mineral, of which a very common example is quartz.

**Gap**
Mill, *Dict*. Gap or Notch. A cutting through a ridge, either indenting only the upper part of the ridge, as in a wind-gap, or penetrating to the base and having water flowing through it (water gap).
See also Gate.

**Gap town**
A town situated in, or at the entrance to, a gap. Since gaps frequently afford natural lines of communication such a site is a very natural one to be chosen for a town. Some so-called gap towns were genetically settlements at river crossings. See also gate; but the American gate city is used in a different sense.

**Garden**
*O.E.D.* An enclosed piece of ground devoted to the cultivation of flowers, fruit and vegetables.
*Webster*. 2. Hence, a rich, well-cultivated spot or tract of country.

**Garden City** (A. T. Stewart, 1869; Ebenezer Howard, 1898)
*Enc. of Soc. Sc.*, 1931. 'The term itself was originally used in the United States by A. T. Stewart, who founded Garden City, Long Island, in 1869.'
*Webster*. 2. A real-estate development planned to combine the advantages of city and country life, as by providing open spaces, intervals between buildings, and garden plots.
*Barlow Report*, 1940 (summarizing the Report of the Dept. Committee of 1935). '... new self-contained towns, separated from any existing urban unit by a protecting belt of open land, which were to offer properly planned facilities for industry, residence of inhabitants of various grades of society, and cultural and recreational opportunities.' (para. 271)
Osborn, F. J., 1946, *Green-Belt Cities*.

London: Faber and Faber. 'Howard was unconscious of the Long Island use of the name when he adopted it. Its world wide currency is due to his book; and the term as descriptive of a type of urban settlement should only be used in the sense which he gave to it. A short definition was adopted, in consultation with him, by the Garden Cities and Town Planning Association in 1919: "A Garden City is a Town designed for healthy living and industry; of a size that makes possible a full measure of social life, but not larger; surrounded by a rural belt; the whole of the land being in public ownership or held in trust for the community."'

Stamp, L. D., 1955, *Man and the Land*. London: Collins. 'Ebenezer Howard published his scheme for a Garden City in *Tomorrow* (1898), and the first Garden City, carrying out his ideas of a self-contained industrial-residential unit with adequate living space, was initiated by private enterprise at Letchworth in 1903, followed in 1920 by Welwyn Garden City. A housing density of ten units per acre, or 35 persons per acre, accepted for Garden City standards, means that the same population occupies, for residential purposes alone, at least twenty times the area it did in the back to back slums. Despite vigorous advocacy by the Garden Cities Association (later renamed the Town and Country Planning Association), Howard's ideas were slow to gain recognition until adopted in general for the new towns of the post-Second World War.' (p. 236)

### Garden Suburb, Garden Village

Osborn, F. J., 1946, *Green-Belt Cities*. London: Faber and Faber. 'Garden Suburb, Garden Village. In these combinations the word Garden connotes simply a well-planned open layout. It is misleading, though good authorities have been guilty of the practice, to describe a Garden Suburb as a suburb "laid out on Garden City lines".'

Stamp, L. D., 1955, *Man and the Land*, London: Collins. 'George Cadbury laid the foundation of Bournville (1879) and the "factory in a garden" idea. Private enterprise launched Hampstead Garden Suburb in 1906.' (p. 235)

### Garigue, garrigue (French)

*O.E.D. Suppl.* Garrigue, garigue (Fr.). In the south of France, uncultivated land of a calcareous soil overgrown with scrub-oak and pine.

Mill, *Dict*. Garigue or Garrigue. Dry wastes with short bushes of prunus and a variety of frequently odoriferous xerophytes. They occur on limestone rock (Causses, S. France), where they as it were replace the maquis or macchia.

Kuchler, A. W., 1947, Localizing Vegetation Terms, *A.A.A.G.*, 37, 'A Provençal term for low, rather stunted evergreen shrub vegetation occurring where there is a calcareous substratum. It is anthropogenic in that under the influence of human settlement (including the effects of browsing by goats and of fire) the original forest of tall oaks (*Quercus ilex*) gradually gives way to the scrub oak (*Q. coccifera*) known in Provençal as *garroulia*, hence garigue.'

Comment. Degenerate forest found on calcareous soils in Mediterranean lands (E.D.L.).

### Garúa

Mill, *Dict*. A dense mist that prevails on the Pacific slope of the Peruvian Andes at the height of 1,000 to 3,000 feet or more above sea-level.

James, P. E., 1959, *Latin America*, 3rd ed., New York: Odyssey. 'Where the cloud rests against the slopes of the Coastal Range or the lower foothills of the Andes, the heavy mist, known as *garúa*, applies a soaking moisture to the land. The dense growth of quick-flowering plants and grasses which appears with the garúa is described in Peru as *loma*.' (p. 191) '2,600 to 4,600 feet above sea-level.' (pp. 191–2)

### Gat

*O.E.D.* Gat, also gate. An opening between sandbanks: a channel, strait: in Kent, an opening, natural or artificial, in the cliffs serving as a landing place.

Mill, *Dict*. (1) A strait connecting a lagoon of a coastal plain with the sea. Low German coast; thence sometimes extended to a narrow sea. 'Grau' is the term used in S. France (2) A swallow-hole (S. Africa).

*Adm. Gloss.*, 1953. A swashway or natural channel through shoals.

Comment. In Kentish place names it has usually become 'gate', *e.g.* Margate, Ramsgate, Sandgate.

## Gate (1)

*O.E.D.* 4. tranf. An entrance into a country through mountains; a mountain pass. Also *plural*.

*O.E.D. Suppl.* 13. Gate City, U.S., a city placed at the entrance to a district (applied spec. to Atlanta, Keokuk, etc.).

Knox, 1904. 'Gate (England), a passage, road, street, from A-S. geat, an opening, gap, e.g. Reigate.'

Taylor, 1951. 'A broad low gap between highlands.' (p. 614)

*Comment.* In geographical literature 'gate' is often used loosely to indicate either an entrance into a country (as in the title of the book *The Hampshire Gate* by H. G. Dent, Benn, 1924) or as alternative to gap (as in the Midland Gate between the southern end of the Pennines and the Wrekin, Shropshire). Gate is also used in the normal English sense of a gate through city walls but confusion results because in many parts of Britain formerly under Scandinavian influence town streets are known as gates, the equivalent of the modern Swedish 'gata'. A notable example is the city of York. Frequently in geographical literature a 'gate' implies something wider than a 'gap' (Burgundian Gate, Moravian Gate) but this is not always the case (Iron Gates on the Danube).

## Gate (2)—see Cattlegate

## Gateway

*Webster.* 4. *Transportation.* Any one of a limited number of points by which the traffic of a defined region can enter.

*Dict. Am.* A place serving as the entrance *of* or *to* a specified region. 1884.

## Gathering ground

The area over which water is collected from precipitation, combined sometimes with percolation from springs; a term commonly used in connection with the supply of water to a reservoir for urban water supply or other purposes. The same as catchment area, which is usually applied to a natural basin. See comments under watershed.

## Gauthānā (Indo-Pakistan: Marathi)

Spate, 1954. 'the village common, an essential part of its economy, the centre of all harvesting' (Dharwar) (p. 173).

## Gazetteer

*O.E.D.* A geographical index or dictionary.

*Comment.* The O.E.D. definition is scarcely adequate. A gazetteer is better described as a catalogue or list of places and named geographical features with references to their position, with or without brief descriptions or notes. When issued in conjunction with an atlas there are references to the position on specific maps.

## Geanticline, geoanticline; geanticlinal, geoanticlinal

*O.E.D.* geanticlinal: A. of the nature of a general upward flexure of the earth's crust. B. A general upward flexure of the earth's crust.

Mill, *Dict.* Geanticline. See geoanticline. Geoanticline or geoanticlinal = anticlinoria (1) A ridge-shaped uplift of a portion of the Earth's crust, with many minor foldings on its flanks. (2) Anticlines in which the formative movement has been comparatively free from the irregularities causing deformation (Dana).

Geikie, J., 1898. 'A broad or regional arching or bending up of the crust—thus a geanticline may be composed of strata showing all kinds of geologic structure. It is simply a bulging or swelling of the crust which affects a wide region.'

Rice, 1943. 'Geanticline—a large, broad and usually very gentle anticline, commonly many miles in width.' (p. 148)

Himus, 1954. Geanticline. A major geological structure, anticlinal in nature, resulting from the operation of mountain-building movements on the site of a former geosyncline (*q.v.*).

*Comment.* These selected references show a remarkable range and variation in definition. Geoanticline is the more usual form and may be taken to mean simply an anticline on a large scale (as in *O.E.D.* for geanticlinal). In this Geikie, 1898, is to be followed. Mill is definitely wrong according to modern usage in equating geoanticline with anticlinorium. Whilst it is true that some geoanticlines (*e.g.* The Weald) have replaced former geosynclines, this is not necessarily the case and by including this condition in the definition Himus goes too far. (L.D.S.)

## Geest (German)

*O.E.D.* (From German, geest, dry or sandy soil, opposed to marshland). Old alluvial matter on the surface of the land; coarse drift or gravel.

Mill, *Dict.* A dry, sandy and barren zone

inland from the coast marshes of the Low Countries.

Shackleton, 1958. 'Wide, infertile, dry, sandy regions called *geest* are intersected by the marshes of the river valleys and fringed by the marshes of the coast.' (West of the Elbe; p. 251.)

**Geltozem**

Yellow soil developed under Mediterranean conditions.

**Gemma** (Sudan; Arabic)

A shallow well dug at the foot of rocky hills, especially in Blue Nile Province, of the Sudan, and which, exceptionally, yield water throughout the dry season. The word is also applied to shallow wells close to the Nile. (J.H.G.L.)

**General land use model**

Coleman, Alice, 1969, A Geographical Model for Land Use Analysis, *Geography*, 54, 43–46. '... a *general land use model* designed to direct and streamline the interpretation of (land use) maps.' '... consists of three relatively stable and necessary scape territories, *wildscape*, *farmscape* and *townscape*, together with the relatively unstable and unnecessary fringe territories between them: the *marginal fringe* and the *rurban (rural-urban)* fringe.' '... the model shall be hierarchical. Its concepts must be structured into different levels so that particular ranges of detail can be instantly dismissed or instantly called into play merely by specifying the level of organisation ... major divisions into which partial models can be fitted at lower levels, e.g. models which refer to urban areas only or to agriculture only.' '... that land use territories should be synonymous with planning territories and that classification of an area as a Type A territory should automatically imply that it needs Type A planning treatment ...'

**Generic region**—see Regional systems.

**Gentrification**

Glass, R., 1964, *London: Aspects of Change*, London: MacGibbon and Kee. '... many of the working class quarters of London have been invaded by the middle classes—upper and lower. ... The current social status and value of such dwellings are frequently in inverse relation to their size, and in any case enormously inflated by comparison with previous levels in their neighbourhoods. Once this process of "gentrification" starts in a district, it goes on rapidly until all or most of the original working class occupiers are displaced, and the social character of the district is changed.' (pp. xvii–xix)

**Genya** (Japanese)

Pastures, plains.

**Gendarme** (French)

*Webster.* 3. A pinnacle of rock on a ridge.

Cotton, 1942. 'Sharp rock pinnacles surmounting serrated arêtes.' (p. 192)

**Geo** (Northern Scotland and Faroes)

*Webster.* A deep, narrow coastal inlet; creek. *Local Scot.*

A narrow coastal cleft or harbour (E.E.E.).

A long narrow inlet of the sea (in northern Scotland especially in well-jointed Old Red Sandstone of Caithness and Orkney). (A.C.O'D.)

*Comment.* The word has been adopted by geomorphologists to describe coastal clefts, often marking joints, faults or dikes from which material has been removed by wave action.

**Geo-, ge-**

*O.E.D.* Representing Greek γεω combining form of γη earth; in compounds formed in Greek itself and in many of modern formation.

It usually implies relating to the earth as a whole, as in geology but sometimes on a world or large scale as in geosyncline. Before a vowel the prefix is usually ge- as in geoid.

**Geoanthropolitics** (S. B. Jones, 1959)

Jones, S. B., 1959. Boundary Concepts in the Setting of Place and Time, *A.A.A.G.*, 49, 241–255. 'I begin with the boundary concepts of people we call "primitive", whose political systems we loosely call "tribal". One might call this the study of anthropological political geography, though I am tempted by "geoanthropolitics".' (p. 202)

**Geoanticline**—see Geanticline

**Geobiocenosis**

A term first used by a Russian school of ecologists, using the terms geocenosis for the physical habitat and biocenosis for the biome, the two uniting to form the geobiocenosis.

See Sukachev, V. N., 1944, On the Principles of Genetic Classification in Biocenology *Zhür. Obschei. Biol.* (Journal of General

Biology), **5**, 213–227; Sukachev, V. N., 1950, Biogeozönose, *Bol'sheia Sovetskaia Entsiklopediia*, **5**, 180–181; and Blydenstein, J., 1961, The Russian School of Phytocenology, *Ecology*, **42**, 575–577. See also Biocoenosis, biocenosis; and Ecosystem.

### Geochronology, geochronological
Swayne, 1956. The study of the time-measurement of geological periods. Not in *O.E.D.* See also Absolute age.

### Geo-Code
Gould, S. W., 1968, *Geo-Code*, Vol. 1, West Edition, New Haven: The Gould Fund. This world-wide reference system, mainly of use to biologists, 'uses the 80 column card in the most economical way possible to store information in a hierarchical form about localities. Four letters, attached to a placename can define its county, state, country, continent or hemisphere—and, if these placenames are attached to species, information about the specimens available in the museum from any subdivision of the earth's surface, large or small, can be recovered with the minimum of sorting either by punched-card machines or computers.' (Foreword.) See also Georef system.

### Geocol
Taylor, 1951. 'Geocol—see Gate' (p. 614).

### Geocratic (T. Griffith Taylor, 1936)
Taylor, 1951. 'Control of man by Nature.' (p. 614) See comment under WE-ocratic.

### Geocratic (2)
Mill, *Dict.* Geocratic Earth-movements.—Movements which add to the amount of land surface exposed above water. The opposite is Hydrocratic. Both appear to be obsolete; not in *O.E.D.*

### Geodesy, geodetic
*O.E.D.* a. Land surveying, the measuring of land (obs.).

b. In modern use: That branch of applied mathematics which determines the figures and areas of large portions of the earth's surface, and the figure of the earth as a whole.

Bowie, W., 1931, *Physics of the Earth*, Bull. Nat. Res. Council, **78**, 103–115. 'That branch of science which deals with the determination of the shape and size of the earth, and with surveys in which the shape and size of the earth must be taken into consideration.'

### Geodetic position
Bowie, W., 1931, *Physics of the Earth*, II, Bull., Nat. Res. Council, **78**, 103–115. The latitude and longitude of a point on the earth's surface referred to an adopted spheroid of reference. Geodetic positions are determined by triangulation from an initial station whose position has been arbitrarily adopted after a consideration of astronomical data. (p. 113)

### Geodynamics
Scheidegger, A. E., 1958, *Principles of Geodynamics*, Berlin: Springer-Verlag. 'The theory of the deformation of continuous matter is the basic mechanical background of geodynamics.' (Preface)

*Webster.* geodynamic. Of, pert. to, or noting the forces or processes within the earth.

### Geo-econometrics (W. Warntz, 1959)
'combined are ideas and methods from Geography, Economics and Statistics' (p. 5). Warntz, William, *Towards a Geography of Price: A Study in Geo-econometrics*, Philadelphia: University of Pennsylvania Press, 1959.

### Geographer
The following definition was prepared officially for 1951 Canadian Census Classification Professional Service Occupations by N. L. Nicholson, Director of the Geographical Branch, Ottawa.

*Geographer:* A geographer is a professional man who studies the nature and uses of areas of the earth's surface, relating and interpreting the interactions of physical and cultural phenomena. He conducts research on the physical aspects of an area or region and its human activities, making direct observation within the area under study and incorporating available knowledge from related fields, such as physics, geology, oceanography, meteorology, biology, demography, anthropology, history and economics. He may act as an adviser or consultant to governments and international organizations on such matters as the economic exploitation of a region.

### Geographic, geographical
*O.E.D.* Geographic = geographical. Now somewhat rare.

*Comment.* Broadly -ic is American and never used by British writers; even in America geographic is more restricted in use than

**Geographical mile**
Theoretically equal to one minute (1′) of latitude but, owing to the fact the earth is a geoid, a sphere with flattening at the poles, a minute of latitude varies slightly in length so that a geographical or nautical mile has been standardized at 6080 feet. See Mile.

**Geographical momentum**
Committee, List IV. The tendency of places with established installations and services to maintain or increase their importance after the conditions originally determining their establishment have appreciably altered.
See also Industrial inertia, industrial momentum.

**Geographical name**
Aurousseau, M., 1957, *The Rendering of Geographical Names*, London: Hutchinson. 'The proper name, or geographical expression by which a particular geographical entity is known.' (p. 4)
See also Toponymy.

**Geographical region, geographic region**
Crowe, P. R., 1938, *Scot. Geog. Mag.*, **54**, 1–18, suggests three types:
1. 'The world has been sub-divided on the basis of groups of factors closely related genetically....'
2. '... individual areas have been called regions on the grounds of concomitant distribution of a wide range of factors.'
3. '... regions have been held to exist in Nature, the geographer's task being only to examine them.'
For a general *critique* see G. H. T. Kimble, 1951, The Inadequacy of the Regional Concept, in *London Essays in Geography*, London: Longmans, in which he claims (p. 151) there are 'no less than one hundred definitions of it (*i.e.* the region) in our geographical literature.' But see D. L. Linton, The Delimitation of Morphological Regions in the same volume. See also Hartshorne, 1939, especially pp. 312–313; also the entries under region, morphological region, in this glossary.

**Geography**
*O.E.D.* The science which has for its object the description of the earth's surface, treating of its form and physical features, its natural and political divisions, the climate, productions, population, etc., of the various countries. It is frequently divided into *mathematical, physical*, and *political geography*.
*Webster*. 1. The science of the earth and its life; esp., the description of land, sea, air, and the distribution of plant and animal life, including man and his industries, with reference to the mutual relations of these diverse elements. 2. A treatise of this science. 3. The natural features, collectively, of an area.
*American College Dictionary*, 1947, New York: Random House, 1947. 1. The study of the areal differentiation of the earth surface, as shown in the character, arrangement, and interrelations over the world of elements such as climate, relief, soil, vegetation, population, land use, industries, or states, and of the unit areas formed by the complex of these individual elements. 2. the topographical features of a region. [Hartshorne]
Committee, List I. The science that describes the *Earth's surface*, with particular reference to the differentiation and relationship of areas.
Mill, *Dict*. The science of the distribution and interaction of phenomena on the Earth's surface.
Hare, F. K., 1964, *Canadian Geographer*, III, 3. 'The scientific study of a highly evolved complex—the man-land relationship'. (p. 113)
*Comment*. The meaning of geography has undergone a progressive evolution, discussed at length in Hartshorne, 1939 and later editions. Brief definitions range from the old 'a description of the earth and its inhabitants' through the definitions like that of the *O.E.D.* which refer to the 'science' to the succinct 'human ecology' or, from a different point of view, 'the science of distributions'. It is still difficult to find a brief definition which the majority would accept. See also Wooldridge and East; also the definitions under the several branches of geography.

**Geoid**
*O.E.D.* A geometrical solid, nearly identical with the terrestrial spheroid, but having the surface at every point perpendicular to the direction of gravity. (Quot. Merriman, 1881)
Bowie, W., 1931, *Physics of the Earth*, Bull. Nat. Res. Council, **78**, 103–115. 'An equi-

potential surface coinciding with the surface of the waters of the ocean, if they were free from the periodic disturbing effects of the sun and moon, the wind, the varying barometric pressure, and differences of temperature and density of the water. Under land areas the geoid surface is that which would coincide with the water-surfaces in narrow sea-canals if they were extended inland through the continents.' (p. 113)

Moore, 1949. A term sometimes used to signify the shape of the *Earth*; the latter is often taken to be an oblate spheroid (a sphere flattened at the poles), but, in view of certain variations, the term geoid, which simply means 'earth-shaped body', has been introduced.

**Geoisotherm, geo-isotherm**
Himus, 1954. Surfaces in the lithosphere below ground-level on which the temperature at all points is the same.

**Geologic, geological**
O.E.D. Of, pertaining to, or derived from geology. There is now a slight distinction in usage between *geologic* and *geological*; the former tends to be used only as an epithet of things forming part of the subject-matter of the science; we may say a *geologic epoch*, but hardly a *geologic student*, a *geologic theory*.
*Comment.* The O.E.D. note refers to American usage; English usage is almost invariable geological.

**Geological record**
Mill, *Dict.* The history based on inferences from a study of the succession of stratified rocks.
Holmes, 1944. 'To a geologist a rock is a page of the earth's autobiography with a story to unfold, if only he can read the language in which the record is written .. stratigraphy ... historical geology.' (p. 6)

**Geological time**
Himus, 1954. 'Geological time is divided into a number of *eras*, each represented by a group of rocks.' Himus goes on to explain that the eras are divided into periods and that during a geological period a *system* of strata was deposited. Periods are further divided into epochs during which *series* of rocks were deposited; epochs are divided into ages during which *formations* were laid down.
Stamp, 1923. 'Geological time is divided into five great Eras, and into a number of Periods. The rocks deposited during a Period constitute a System. A System of rocks usually comprises several Groups or Series or Stages. The smallest division of geological time is a *hemera*, and the rocks deposited during a hemera constitute a zone.' (pp. 5–6)
*Comment.* The hierarchy of time-periods given by Himus is not always fully observed. The Eras long recognized are Eozoic or Pre-Cambrian or Archaean; Palaeozoic or Primary (comprising Cambrian, Ordovician, Silurian or Gotlandian, Carboniferous and Permian); Mesozoic or Secondary (comprising Trias, Rhaetic, Jurassic, Cretaceous); Kainozoic or Tertiary (comprising Eocene, Oligocene, Miocene and Pliocene); Quaternary (Pleistocene and Holocene). Several eras are now distinguished though with greater difficulty in the vast time-span of the pre-Cambrian and there are considerable variations in the nomenclature and of the periods used.

**Geology**
O.E.D. The science which treats of the earth in general. *Obsolete.*
2. The science which has for its object the investigation of the earth's crust, of the strata which enter into its composition with their mutual relations, and of the successive changes to which their present condition and positions are due.
Holmes, 1944. 'Modern geology has for its aim the deciphering of the whole evolution of the earth and its inhabitants from the time of the earliest records that can be recognized in the rocks right down to the present day.' (p. 5)
*Comment.* The inclusion by Holmes of all 'inhabitants' refers of course to fossils not to human beings.

**Geomorphology, geomorphological**
O.E.D. The theory of the conformation of the earth. (quot.: Pop. Sc. Monthly, XLVIII, Apr. 1896, p. 815. The new phase of geography, which is sometimes known as physiography, and later, as geomorphology.)
Mill, *Dict.* Geomorphology or Geomorphogeny. The science of landforms.
Gregory, J. W., 1899. The Plan of the Earth in H. R. Mill, *The International Geography*, London: Newnes. '... veiled by the great variety in topographical details there is some underlying symmetry in continental

form, the discovery of which is the main problem of geomorphology.' (p. 37)
Committee, List 1. The science (or genetic study) of land forms.
*Comment.* The derivation from the Greek is self-explanatory: ge, the earth; morphe, form; logos study. It is difficult to find any justification for the *O.E.D.* definition, and the quotation from Popular Science Monthly is nonsense. Geomorphogeny (Mill) is no longer used. See also Rice, 1943, 151.

**Geonomics**
Renner, G. T. (see under Geopolitics). 'Geonomics is an abbreviated term for economic geography.' (p. 3)
*Comment.* Not generally adopted. Purists point out that if such a word is needed it should be geoeconomics but that is rather cumbersome.

**Geopacifics** (T. Griffith Taylor, 1947)
Taylor, T. Griffith, 1947, *Our Evolving Civilization—an Introduction to Geopacifics*, London: Methuen. Geopacifics is an attempt to base the teachings of freedom and humanity upon real geographical deductions; in a sense it is humanized geopolitics.
Taylor, 1951. 'Study of geography to promote peace.' (p. 614, but see also pages 587–608) Pronounced geopácifics.

**Geophysics, geophysical**
*Webster. Geol.* The physics of the earth, or the science treating of the agencies which modify the earth, including dynamical geology and physical geography; esp. the causes of movements and warpings of the surface of the lithosphere.
The science concerned with the physics of the earth's crust and the earth's interior; the application of the techniques of the physical sciences to the investigation of the earth. (L.D.S.)
*O.E.D.* (given only under geo- as a minor entry and quoting only the *Century Dict.*). The physics of the earth.
*Comment.* In recent years geophysical surveying has become a very complex technique of investigating the earth's crust, especially in the search for economic minerals.

**Geopolitics**
The usual translation of *Geopolitik, q.v.*
Renner, G. T., 1951, *Political Geography and its point of view. World Political Geography*, ed. Pearcey, G. E. and Fifield, R. H., New York: Crowell. 'Geopolitics may be regarded as a shortened designation for political geography; just as geonomics is an abbreviated term for economic geography.' (p. 3)
Houston, J. M., 1953, *A Social Geography of Europe*, London: Duckworth. 'The study of geographical factors in politics. As in the case of Nazi politics, it can be only too readily subversive propaganda. (p. 245)
Wooldridge, S. W. and East, W. G., 1951, *The Spirit and Purpose of Geography*, London: Hutchinson. '... properly conceived and soberly studied, geopolitic may fittingly be regarded as an extension or application of political geography to the external geographical relationships of States.' (p. 123)
Taylor, 1951. 'Study of Geography to promote conquest.' (p. 614, but see also pp. 587–608)
Taylor, T. Griffith, 1942, *Canada's Role in Geopolitics*, Contemporary Affairs, Toronto. 'Haushofer and his school seem to imply that Geopolitics necessarily includes discussions of world domination and of racial superiority. To the present writer these are arbitrary and unnecessary extensions of the term Geopolitics.' (p. 1
Efimenco, N. M. (ed.), 1957, *World Political Geography*, New York: T. Y. Crowell. 'The study of political areas, particularly from the point of view of national self-interest.' (p. 713)
*Comment.* Griffith Taylor's earlier attempt to use Geopolitics in its natural meaning ('a study in situation and status') was negatived by the later identification of *Geopolitik* with the Nazi philosophy hence his later rather naïve definition of 1951 and his introduction of 'Geopacifics' as a term.

**Geopolitik** (German)
Hartshorne, 1939. 'The application of geography to politics . . . one's estimate of its value and importance will depend on the value that one assigns to the political purpose it is designed to serve. . . . produced the concept of "*Lebensraum*".' ... (p. 404)
Kish, G., 1942, Political Geography into Geopolitics, *Geog. Rev.*, **32**. Kjellén was the first to use the word "Geopolitik" (*Ymer*, **19**, 1899, 1900, 283–331) as 'the science of the state as a realm in space.' (pp. 632–645)
Translates Haushofer and others, *Bausteine zur Geopolitik.* 'Geopolitik is the science of the determination and conditioning of political developments by the earth.' (p. 27)

Van Valkenburg, S., in Taylor, 1951. 'Haushofer himself defined Geopolitik as the science of the political life in the natural environment, which tries to understand the political life in its close relation to the earth... in *Bausteine zur Geopolitik* defined as the study of the geographic foundation of political events....' (pp. 106–107)

For the history of *Geopolitik* and how it became a tool of the Nazi philosophy see Van Valkenburg, op. cit. sup. and Hartshorne, R., 1935, *Amer. Pol. Sci. Rev.*, **29**.

**Georef system**
Dickinson, G. C., 1969, *Maps and Air Photographs*, London: Edward Arnold. '...the World Geographic Reference System, more commonly known as Georef, which simplifies positional references based on latitude and longitude by omitting the tedious north, south, east and west degrees, minutes and seconds of the old description.' (p. 110)

Haggett, 1972. 'The system divides the global surface into 288 segments from the intersection of 15° meridianal and latitudinal bands, each defined by an index letter. By taking the letter of the meridianal first, geographers can uniquely define each segment by a two-letter code ...' (p. 89)

See also Geo-Code.

**Geosophy**
Wright, J. K., 1962, *Geog. Jour.*, **128**, 'the study of the nature and expression of geographical ideas' (p. 73).

**Geostrategy** (G. B. Cressey, 1944)
Cressey, G. B., 1944, *Asia's Lands and Peoples*, New York: McGraw-Hill. 'Geostrategy considers the significance of the environment as applied to the understanding of a problem of economic or political welfare, primarily but not necessarily of international scope.' (p. 8)

**Geostrophic**
*Webster*. Of or pertaining to deflective force due to the rotation of the earth.

**Geosyncline, geosynclinal** (J. D. Dana, 1873).
*O.E.D.* A. adj. Forming a large depression in the surface of the earth, from the lowest point of which there is a gradual rise to either side, even although the continuity of this is broken by smaller depressions. The opposite of geanticlinal.
B. sb. A geosynclinal dip or depression in the earth's surface.
Mill, *Dict.* 1. A trough-shaped depression of a portion of the Earth's crust on a very large scale, with many minor foldings on its flanks. 2. Synclines in which the formative movement has been comparatively free from the irregularities causing deformation. (Dana.)
Geikie, J., 1898. '...a wide or broad region of depression; a sinking of the earth's crust as a whole.'
Holmes, 1944. 'Geosynclines. By the work of several generations of geologists it has been firmly established that the orogenic belts of each geological era originated in long downwarps of the crust in which extraordinarily thick deposits of sedimentary rocks accumulated during the era that preceded the orogenic revolution. Such elongated belts of long continued subsidence and sedimentation were called geosynclines by Dana in 1873.' (pp. 379–380)
Himus, 1954. An elongated downwarping of the crust of the earth, forming a deep trough in which a great thickness of sediments accumulated, especially in the central zone. The floor of the trough is thus synclinal in form, but ultimately it may become completely filled with sediment, even in the centre. Compression of the sediments within the geosyncline involving intense folding and thrusting, is believed to be the prime cause of mountain-building.
*Comment*. There has been considerable confusion in the use of this term. It has been used in the *O.E.D.* sense as the simple opposite of geanticline. Mill in his sense 1 confuses it with synclinorium. Whilst many geologists follow Holmes and Himus in maintaining a link with orogenesis, others do not. See also E. S. Hills, chapter on Geosynclines and note the series of terms which has been coined such as auto-geosyncline, para-geosyncline etc.

**Geosystem**
Stoddart, D. R. in Chorley and Haggett, 1967. '... the fundamental concept of system in geography is central to the development of the subject as a nomothetic science. The study of these geosystems may now replace that of ecosystems in geography, which remains concerned with precisely the same body of data, which intuitively grasped in all its complexity, led to the use of simplistic organic analogies.... System analysis at last provides geography with a unifying methodology.' (pp. 537–538)

**Geotectocline**
von Engeln, 1942. 'The discovery by gravity measurements of great deficiencies in the vicinity of island arcs afforded F. A. Vening Meinesz (1934) and H. H. Hess (1938) a basis on which to construct a theory of the origin and manner of such uplifts. It is inferred that over the negative strip the sialic crust is bent vertically downwards... this crustal downbuckle, as a structure, is called a *tectogene* and the basin between the two limbs of the downbend a geotectocline.' (pp. 44–45)

**Gerf** (Sudan; Arabic)
The land uncovered by the Nile after the annual flood in the northern Sudan, and cultivated by *seluka*. (J.H.G.L.)

**Gestalt theory**
Kirk, W., 1963, Problems of Geography, *Geography*, **48**, 357–371. 'The term gestalt can be translated as configuration, pattern, structure, form. Gestalt theory arose from pioneer studies of Max Wertheimer fifty years ago on the perception of stroboscopic motion.' (p. 371)
See Lowenthal, D., 1961, Geography, Experience and Imagination, *Ann. Assoc. Am. Geographers*, **51**, 241–260, for the increasing importance of perception in geographical studies.
See also Behavioural and Phenomenal environments.

**Gewann** (German, *pl.* Gewanne)
Fischer, 1950. 'A division of the communal lands of a village community composed throughout of lands of about the same quality.... The resulting form of the farms, composed of scattered, non-contiguous plots, may survive." (p. 21)

**Gewanndorf** (German)
A nucleated village with strip-fields (C.T.).

**Geyser** (Icelandic)
*O.E.D.* (Icelandic, Geysir, proper name of a particular hot spring in Iceland. Related to geysa, to gush). 1. An intermittent hot spring throwing up water etc., in a fountain-like column.
Mill, *Dict*. A hot spring the water of which is shot upwards in columns more or less explosively at intervals; usually the water is projected from a cylindrical tube opening at the bottom of a basin with gentle slopes leading up to and down on the other side from the rim, and composed of sinter deposited from the geyser water. Jaggar distinguishes two types of geysers: (1) the flowing geyser, in which the hot water is constantly overflowing under the pressure of cold water pouring in from higher levels so that eruptions take place only in the season in which the inflow is diminished and (2) the standing geyser, which undergoes periodic eruption.
Holmes, 1944. 'Geysers are hot springs from which a column of hot water and steam is explosively discharged at intervals.' (p. 138)

**Geyserite** (Delamétherie, 1812)
Holmes, 1944. 'The waters [of hot springs and geysers] are highly charged with mineral matter of considerable variety... rich in calcium carbonate deposited as *travertine*, silica in solution deposited as *siliceous sinter* or *geyserite*.' (p. 138)

**Gezira** (Arabic)
A tract of land partly or wholly enclosed by water, either rivers or the sea. The Arabic term for Arabia is *Gezirat al Arab* (=the arab island). The area between the Blue Nile, White Nile and the Sobat R. in the Sudan is called the *gezira*, although the term nowadays is often applied to the much smaller area north of Sennar which is irrigated (J.H.G.L.).

**Ghat, ghaut** (Anglo-Indian; Hindi is ghāt)
*O.E.D.* The senses are here placed in the order of their occurrence in English. The order of development, however, is as follows:
1. A path of descent to a river; hence a landing-stage, a quay, the place of a ferry.
2. A path down from a mountain; a mountain pass.
3. In *plural* the name erroneously given by Europeans to the mountain ranges parallel to the east and west coasts of India.
1. *The Ghauts:* the name applied by Europeans to two chains of mountains along the eastern and western sides of southern Hindostan, known as the Eastern and Western Ghauts.
2. A mountain pass or defile.
3. A passage or flight of steps leading to the river-side; hence, *generally*, a landing-place, the place of a ford or ferry.
*Comments.* The spelling now used is ghat, strictly ghāt; ghaut should never be used. The Western and Eastern Ghats are respectively the western and eastern edges of the great Indian plateau; it is usually presumed that the word ghat was transferred from the passes (*e.g.* Palghat) to the ranges in each case. The Ghats would

not now be called 'chains' (*q.v.*). In the other sense of the word the famous bathing-ghats of Benares (Varanasi or Banaras should be noted.

**Gher** (Indo-Pakistan)
Spate, 1954. '"sweet water" lands above the areas of salt marsh on the west coast' (p. 599).

**Ghetto**
*Webster.* 1. The quarter of a town or city to which Jews were restricted for residence, esp. in Italy; a Jewry. 2. A quarter of a city where members of a racial group are segregated.

**Ghyben-Herzberg Relation**
Todd, D. K., 1959, *Ground Water Hydrology*, New York: Wiley. 'Along the (north) European coast, salt water occurred underground, not at sea level but, rather, at a depth below sea level of about forty times the height of the fresh water above sea level. This distribution was attributed to a hydrostatic equilibrium existing between the two fluids of different densities. The equation derived to explain the phenomenon is generally referred to as the Ghyben-Herzberg relation after its originators' (p. 278).

**Ghyll**—see Gill

**Giant cusp**
Shepard, F. P., 1952, Revised Nomenclature for Coastal Features, *Bull. Ass. Am. Petr. Geol.*, 36, Pt. 2, pp. 1902–1912. 'Along other beaches slightly protruding cuspate points are more widely spaced, commonly 1,000–1,500 feet between points. These may be called "giant cusps". The distances separating these cusps do not vary according to wave heights so far as could be determined. The giant cusps have submarine ridges continuing seawards beyond them as transverse bars. Along one or both sides of these bars there are deep channels. Ordinarily beach cusps are poorly developed between giant cusps.' (pp. 1909–1911)

**Giant's Kettle**—see Kettle

**Gibber** (Australian)
*O.E.D.* Aboriginal Australian; a large stone, a boulder. Used in

*Gibber plain.* A level surface in arid parts of Australia with numerous pebbles; a type of gravel desert (E.S.H.).

Cotton, 1942. 'Boulder pavements of Australian stoney deserts, forming "gibber plains".' (p. 10)

**Gibli**
Local name for the sirocco (*q.v.*) in Libya.

**Gilgai** (Australian)
'A feature, characteristic only of certain regions [of the Murrumbidgee basin, New South Wales] is gilgai formation which shows up as a series of low puff-like mounds on the surface of the plain. Gilgais are developed as a result of fragments of the A horizon falling down cracks into the B horizon. When the soil becomes saturated, the B horizon is forced to expand. The A horizon is pushed up into a series of mounds which are continually subject to erosion, until finally the B is exposed at the surface. Gilgais impart a very desolate appearance to the landscape.' T. Langford-Smith (unpublished thesis, A.N.U., 1958).

**Gill, Ghyll**
Topley, W., 1872, The Weald, *Mem. Geol. Surv.* 'steep-sided wooded valleys are called "gills", a word, which in the south-east of England, appears to be confined to the Hastings Beds country' (p. 245).

Wooldridge, S. W. and Goldring, F., 1953, *The Weald*, London, Collins, 'The Central Weald remains a very hilly country cut up by steep-sided "ghylls"' (p. 89).

*O.E.D.* The spelling ghyll, often used in guide books to the Lake District, seems to have been introduced by Wordsworth. 1. A deep rocky cleft or ravine, usually wooded and forming the course of a stream. 2. A narrow stream, a brook, a rivulet.

*Comment.* Gill is used for streams which are in shallow open valleys in north-west Yorkshire. They may end in pot-holes to which the word 'gill' became transferred *e.g.* when Gaping Gill Hole by the dropping of 'hole' became known as Gaping Gill. The spelling 'ghyll' is being dropped on Ordnance Survey maps and should not be used. See also Ekwall. (G.T.W.)

**Gipfelflur** (German: W. Penck)
Penck, W., *Morphological Analysis of Land Forms* (Trans. H. Czech and K. C. Boswell), London: Macmillan, 1955. 'A further feature that is independent of glacial remodelling is the uniformity of the summit levels. This proves to be independent not only of the folded structure but also, within wide limits, of the nature of the rock. A. Penck has termed this the *gipfelflur* and has shown that, so far as it is a matter of Alpine conditions, it cannot well be the heritage of an hypothetical

peneplane... shows a notable connection with the distribution of slope form. (p. 231) 'One can only fitly speak of a gipfelflur when the mountain masses have been broken up into peaks and sharp edges and when slopes of maximum gradient meet in such edges.' (p. 264) There is a long discussion; see also A. Penck, Die Gipfelflur der Alpen. *Sitz.-Ber. preuss. Akad. Wiss.*, Berlin, 1919, 17, p. 256.
See also Peak-plain, summit-plain, accordance of summit levels.

**Gipsy, gypsy, gip**
Intermittent spring or bourne.

**Girt**—see Bayou

**Gja, Gia**
Mill, *Dict*. Gja = geodha in *Gaelic*. A narrow opening in sea cliff into which the waves rush, usually a weathered-out dyke (Skye and west Highlands).
Swayne, 1956 adds A fissure from which volcanic eruptions take place.
Not in *O.E.D.*

**Glacial**
*O.E.D.* Characterized by the presence of ice.
*Comment.* The general tendency is to use the word glacial very loosely as connected in some way with ice, glaciers, the ice age etc., as in glacial deposits, glacial lakes, glacial valleys, troughs and cols.

**Glacial age, epoch, era, period**—see Ice age

**Glacial boulder**
Mill, *Dict*. A boulder which has been transported by a glacier. See also Erratic.

**Glacial control theory**
Swayne, 1956. A theory concerning the formation of coral-reef atolls. It assumes that during the Great Ice Age the sea-level was lowered by about 300 feet and the coral reefs then existing were destroyed and filled in by debris leaving smooth surfaces thus providing ideal platforms on which new coral colonies were formed as the sea-level rose again.

**Glaciation** (*cf.* glacierization)
*O.E.D.* 2. Geol. The condition of being covered by an ice sheet or by glaciers; glacial action or its result. Examples 1863. Lyell—1881.
Mill, *Dict*. 1. Occupation of an area by ice in the form of glaciers, ice-sheets, or floating ice.
2. The action of a glacier or ice-sheet on the rocks over which it has passed.
Flint, 1947. 'By glaciation is usually meant the alteration of any part of the Earth's surface (usually by means of erosion or deposition) in consequence of glacier ice passing over it.' 'By deglaciation is meant the uncovering of any area as a result of glacier shrinkage.' (p. 64)
Wright, C. S. and Priestley, R. E., 1922 *Glaciology*, London: Harrison. '"Glaciation"—the erosive action exercised by Land-Ice upon the land over which it flows.' (p. 134)
*Comment.* Glaciation is also used as equivalent to ice age to denote a period of time 'the first glaciation', 'the Würm glaciation', etc. An anonymous reviewer in *Ice* 5, 1960, remarks that a writer unfortunately makes 'no distinction between glacierization (present ice cover) and glaciation (evidence of past ice cover)' (p. 12). Under 'Glaciated' *Ice Gloss.*, 1958 gives 'land covered in the past by any form of glacier is said to be glaciated.'

**Glacier** (originally French)
*O.E.D.* A large accumulation or river of ice in a high mountain valley, formed by the gradual descent and consolidation of the snow that falls on the higher ground. The resulting mass is often many miles in length, and continues to move slowly downward until it reaches a point where the temperature is high enough to melt the ice as fast as it descends.
*Ice Gloss.*, 1956. A mass of snow and ice continuously moving from higher to lower ground or, if afloat, continuously spreading. The principal forms of glaciers are: *inland ice sheets, ice shelves*, ice caps, *ice piedmonts*, cirque glaciers, and various types of mountain (valley) glaciers.
Mill, *Dict*. A river of solid ice, descending from its source in the high *névé* of a snowfield and carving a broad steep-sided valley on its way.
*Comment.* The above quotations from Mill and *O.E.D.* and the *Ice Glossary* are sufficient to indicate the recent wide extension of meaning from what would now be called a valley-glacier or Alpine glacier to ice sheets in general.

**Glacier breeze**
Moore, 1949. A cold breeze of *katabatic* type blowing down the course of a glacier, caused by the cooling of the air on contact with the ice.

**Glacier-cone**
Mill, *Dict*. A steep block of ice rising above the surface of a glacier formed in the same

way as a glacier-table, but having as a protective covering, not a stone or rock, but accumulations of sand and mud weathered out from the glacier. Now rarely used. See Dirt cone.

**Glacier funnel**
Mill, *Dict*. (1) A large glacier mill. (2) a superficial hollow in a glacier arising from the subsidence of the ice above a glacier cavern.

**Glacier gate**
Mill, *Dict*. The wide vaulted opening above the ground moraine at the lower end of a glacier. Rare; *cf.* glacier snout.

**Glacier ice**
*Ice Gloss.*, 1958. Any *ice* originating from a glacier, whether on land or floating in the sea as *icebergs*.

**Glacier mice**
See Polster.

**Glacier milk**
Mill, *Dict*. The turbid water of a stream issuing from under a glacier; so called on account of its colour, which is like that of dirty milk but slightly greenish.

**Glacier mill**—see Moulin.

**Glacier snout**
Moore, 1949. The arch at the end of a *Glacier* from which flows a stream formed by the melting ice.

**Glacier table**
Mill, *Dict*. A block of stone supported on a pedestal of ice on the surface of a glacier, the stone serving to protect the pedestal from melting.

**Glacier tongue**
*Ice Gloss.*, 1956. An extension of a *glacier*, projecting seaward and usually afloat.

**Glacière** (French)
Ice-cave.

**Glacieret** (T. Griffith Taylor)
Taylor, 1951. 'Small glaciers, may develop from snow-drifts.' (p. 614)

**Glacierization, glacierisation** (*cf.* glaciation)
*O.E.D.* Conversion into glacier. 1850. *Westm. Rev.*, Oct. 267. 'A general glacierization (*vergletscherung*) of the whole island is a thing not to be thought of.
Wright, C. S. and Priestley, R. E., 1922 *Glaciology, British (Terra Nova) Antarctic Expedition*, 1910–1913. London: Harrison and Sons. '"Glacierisation"—the inundation of land by ice (German Vereisung).' (p. 134)
*Comment*. The term was spelt by Wright and Priestley, 'glacierisation'. Its use is supported by the editors in *Jour. of Glaciology*, **2**, 1954, p. 378 and by the following correspondents on pp. 507–509; F. Debenham; R. S. Taylor; C. Swithinbank; H. Thompson; N. E. Odell; B. Roberts; and D. L. Linton. Debenham reports it introduced by Griffith Taylor. Opposed by Flint.
Also 'deglacierization', H. Thompson, *Jour. of Glaciology*, **2**, 1954, 507. Under the heading Ice covered, *Ice Gloss.*, 1958 notes 'the alternative form "glacierized" has not yet found general favour.'

**Glacification**
*O.E.D.* a. The action of converting into ice. b. The action of covering with ice.
1860. Tyndall, Glac. II, v, 252. The second great agent in the process of glacification, namely pressure.
1875. tr. Schmidt's Desc. and Darw. 63 The diluvial period ... includes, both in Europe and America, a repeated glacification of countries, and vast portions of the world.
*Comment*. Obviously ambiguous and may be regarded as obsolete.

**Glacio-eustatism** (H. Baulig, 1935)
Baulig, H., The Changing Sea Level, *Inst. Brit. Geog.*, *Publ.* 3, London: Philip. After discussing changes in sea-level due to waxing and waning of ice-sheets, Baulig says '...this kind of phenomena, for which the name *glacio-eustatism* would seem to be appropriate.' (p. 4)
See also Eustatism, rejuvenation.

**Glaciology, glaciological**
The study of ice and ice action in all its forms.
In Britain a small group formed a permanent 'Committee for the study of Snow and Ice'. When the widespread interest in its field of work became apparent a permanent society was formed. There was much discussion as to a suitable title: that chosen was the British Glaciological Society, and its publication is the *Journal of Glaciology*.

**Glacis**
*O.E.D.* A gently sloping bank. In modern use probably transferred from sense 2: a parapet extending in a long slope so that every part shall be swept by the fire of the ramparts.
Mill, *Dict*. Glacia or glacis: an easy slope or bank less steep than a talus. '
*Comment*. This old word may have been derived from mediaeval latin relative to fortifications but the French of the 16th century seems to have meant 'a place made slippery by wet and then frozen'. It is best avoided.

**Glaçon**—see Ice terminology

**Glade**

*O.E.D.* A clear open space or passage in a wood or forest, whether natural or produced by the cutting down of trees.

*Dict. Am.* 1. A marshy tract of low ground covered with grass 1644. 2. In New England, an opening in the ice of rivers or lakes, or a place left unfrozen. 1698; also Webster 1828. 3. mountain meadows in the Alleghanies. 1788. 4. *N. Eng.* A long, narrow tract or way covered with ice. 1806.

*Comment.* The concept of an open or sunny place was variously interpreted when the word was taken to America. Webster (1828–32) notes its use in New England for an opening in the ice of rivers or lakes; Morse (1796) refers to glades of rich swamp; Bartlett (1859) to tracts of land in the South covered with water and grass whence presumably Everglade—a permanent glade.

**Glam** (Malay)

Inland freshwater swamp.

**Glaze**

*Ice Gloss.*, 1958. Glaze (clear ice). A generally homogeneous and transparent deposit of *ice* formed by the freezing of super-cooled drizzle droplets or raindrops on objects the surface temperature of which is below or slightly above 0°C. (32°F.). It may also be produced by the freezing of non-supercooled drizzle droplets or raindrops immediately after impact with surfaces the temperature of which is well below 0°C. (32°F.).

**Glazed frost**

*cf.* Glaze. U.S. A coating or covering of ice; also a stretch of ice (*O.E.D.*).

Mill, *Dict.* Glazed Frost (Fr.) Verglas, (Ger.) Glatteis. The coating of ice produced either by a frost setting in after a partial thaw, or by rain falling during hard frost and congealing as it falls; frequently mistaken for hoar-frost. It has been known to add over 900 grains weight to the weight of a lime twig of 8 grains.

*Met. Gloss.*, 1944. When rain falls with the air temperature below the freezing point a layer of smooth ice, which may attain considerable thickness, is formed upon all objects exposed to it. This is known as glazed frost.... A phenomenon similar to glazed frost may also result if sleet, formed by the passage of snow through an upper stratum of air above the freezing point, occurs during frost.

*Comment.* Because of the weight of the ice produced, a glazed frost can often do great deal of damage. Well establishe trees can be completely destroyed, bough breaking off till nothing but a stump is left (*Cf.* Mill, above)

**Glebe**

*O.E.D.* The soil of the earth, regarded as the source of vegetable products. Now onl poetical or rhetorical.

A portion of land assigned to a clergyman as part of his benefice.

*Comment.* In the latter and current sens found frequently in works dealing with land ownership and historical geography

**Glen** (Scottish)

*O.E.D.* A mountain-valley, usually narrow and forming the course of a stream. A first applied to the narrow valleys of the mountainous districts in Scotland and Ireland, but now extended to similar places in other countries.

Mill, *Dict.* Glen (Gaelic), Glean (Irish) Glyn (Welsh), a contracted valley, usually wooded and with a stream.

Steep-sided (usually glaciated) valley characterized by a narrow floor. (A.C.O'D.)

*Comment.* Often contrasted with a broad open and cultivated valley or strath (*q.v.*).

**Gley, glei** (Soil science)

Jacks, 1954. 'Yellow and gray mottling in the soil produced by partial oxidation and reduction of iron caused by intermittent waterlogging.'

Hence gleying, the development of this structure.

**Glint**—see Klint

**Glint-line, glint line**

Mill, *Dict.* Used in Russia for the escarpment of almost horizontal Palaeozoic rocks which rise above the Baltic Shield; employed by Suess for all such erosion escarpments, more particularly for the homologous arc which surrounds the Hudson Shield. The glint-line runs through a succession of great lakes in both Europe and America.

*Comment.* Not in *O.E.D.*; Holmes, 1944; Lobeck, 1939; von Engeln, 1942; Thornbury, 1954 and other standard works. Apparently fallen into disuse. Mill also gives 'glint-lakes'—lakes along the glint-line.

**Glitter**

A scree (Northumberland).

**Globe, global**
*O.E.D.* 1. A body having the form of a sphere.
2. The earth.
3. A spherical structure on whose surface is depicted the geographical configuration of the earth (*terrestrial globe*).
*O.E.D. Suppl.* Global: pertaining to or embracing the totality of a group of items, categories or the like.
*Comment.* The modern use of global is with reference to the earth as a whole *e.g.* global war, global geography and in this sense "the globe" is frequently used as meaning "the earth" or "the world".

**Gloup, gloap** (Scottish)—see **Blow-hole**

**Glyders, glydrs**
, scree (North Wales).

**Glyptolith**—see **Ventifact**

**G.M.T.**—see **Greenwich Mean Time**

**Gneiss**
Himus, 1954. A foliated, coarsely crystalline rock generally of ... granitic or dioritic composition. Highly micaceous layers alternate with bands or lenticles or 'eyes' that are granular and more or less like granite in composition. Gneisses may be the result of metamorphism on igneous rocks, giving *orthogneisses,* or on sediments, when the products are called *paragneisses.*

**Gneissose structure**
Himus, 1954. A structure shown by coarsely crystalline rocks, having foliation which is discontinuous. . . .

**Gobi** (as term Berkey and Morris, 1924)
Stamp, L. D., *Asia*, 10th ed. 1959. 'The great basin of the Gobi contains many minor basins, which Berkey and Morris have called "talas" ... each tala has its own local interior drainage and is bounded by inconspicuous warp divides or mountain ranges. Within each tala are still smaller basins which contain late Mesozoic or Tertiary sediments. These Berkey and Morris have termed "gobis"' (p. 597).
See Berkey and Morris, Basin Structures in Mongolia, *Bull. Amer. Mus. Nat. Hist.*, **51**, 1924, 103–127; *idem., Geology of Mongolia*, 1927.

**Gombolola** (Uganda; Luganda language)
An administrative division, smaller than a *saza.* The alternative term, sub-county, is sometimes used. (S.J.K.B.)

**Gondwanaland, Gondwana Land** (Suess)
Mill, *Dict.* The ancient continent (late Palaeozoic) which is supposed to have included south and central Africa, Madagascar and the Indian peninsula, the lofty tablelands of which may not have been covered by sea since Carboniferous times. It may also have included the eastern part of South America, and is characterized by an absence of folding of any recent date. It is named from Gondwána (India) where are found late Palaeozoic strata with a fern flora found also in South Africa.
Himus, 1954, adds 'separation of these fragments may have resulted either by sinking of the floor of what are now oceanic areas, or, according to the theory of continental drift, by their drifting apart at a comparatively late stage of the earth's history. There is strong evidence that during the Carboniferous period there was heavy glaciation of the areas mentioned.'
*Comment.* Also the land of the Gond tribe in India.

**Gore**
*O.E.D.* A triangular piece of land; one of the triangular pieces that form the surface of a globe.
Triangular piece of ploughland found in corners where ploughland area is not rectangular—sometimes called 'gore acres'. See Orwin, C. S. and C. S., 1954, *The Open Fields*, Oxford Univ. Press, pp. 28, 35 and 36.
*Comment.* The *O.E.D.* definition is puzzling and seems to combine the use by historians quoted from Orwin with an entirely different meaning in cartography for the map of a narrow strip of the earth's surface stretching from pole to pole and bounded by two meridians not many degrees apart. Such gores can be pasted on the surface of a sphere to make a globe.

**Gorge**
*O.E.D.* 7. A narrow opening between hills; a ravine with rocky walls, especially one that gives passage to a stream.
Mill, *Dict.* A small canyon.
Swayne, 1956. A valley deep in proportion to its width, usually with precipitous sides.

**Gorich** (Pakistan: Baluchi)
The dry, steady, powerful north-west wind in Baluchistan.

**Gouging**—see **Plucking**

**Goz** (Sudan: Arabic)
Grove, A. T. (1958), *Geog. Jour.*, **124**, 1958. 'In the Sudan Republic some 200,000

square kilometres of Darfur and Kordofan are covered by sand a few decimetres to tens of metres in thickness. The name "*goz*" refers to thick dune-like accumulations of the sand.' (p. 531)

Bagnold, R. A., 1941, *The Physics of Blown Sand and Desert Dunes*, London: Methuen (rep. 1954). 'under all wind conditions the grassy surface acts as a continuous deposition area, and we get great undulating tracts of accumulated sand such as the gozes of Kordofan, which are devoid of steep-sided dunes' (p. 184). 'gentle large-scale undulations apparently associated with slight present, past, or periodic rainfall and its resulting light intermittent vegetation' (p. 189).

**Graben** (German: *lit*. ditch)

*O.E.D. Suppl.* A rift-valley.

*Webster. Geol.* A depressed tract bounded on at least two sides by faults and generally of considerable length as compared with its width.

Wooldridge and Morgan, 1937. Graben = a rift = a rift valley (pp. 96, 103).

Similarly Cotton, 1946, etc. Not in Mill.

Knox, 1904. 'Graben (Ger.), a ditch, canal; a term now generally applied to the "rift" valleys. For use as a form of sub-oceanic relief *see* Trench.'

Bencker, 1942. Graben is given as the German for Trench = 'A long narrow oceanic depression with relatively steep sides' (p. 62).

See also Taphrogeosyncline

**Gradation**

*Webster*. 11. *Phys. Geog.* A bringing to a uniform or nearly uniform low grade or slope. The *gradation* of the land by streams resulting in the formation of plains, is a twofold process, involving: (1) *degradation* of tracts originally above final grade; (2) *aggradation* of tracts originally below it. If, after gradation, a change in level is produced by crustal movement, *regradation* may occur.

Mill, *Dict*. The process of bringing the surface of the lithosphere to a common level.

**Grade, graded** (1)

*O.E.D.* 10. U.S. In a road, railway, etc. Amount of inclination to the horizontal; rate of ascent or descent; = Gradient.

11. of a surface; Degree of altitude; level. rare. At grade (U.S.): on the same level. [Note: grade crossing = level crossing.]

*Webster*. 4. In a railroad or highway: a. The rate of ascent or descent; gradient; deviation from a level surface to an inclin plane; stated as so many feet per m (2) as one foot rise or fall in so many horizontal distance, (3) as so much in hundred feet, or (4) as a percentage horizontal distance. *U.S.*

*At grade* a. *U.S.* On the same level; sa of the crossing of a railroad with anoth railroad or highway, when they are on t same level at the point of crossin b. *Phys. Geog.* At such a level, with rel tion to slope, that no erosion or depositi is effected; said of a stream bed that h been so established.

Mill, *Dict*. 1. To fix the levels and gradien of any means of communication. 2. Gr dients (N. America).

Graded. Produced by the action of denuding agent, as a graded plain, a grad river-bed.

Committee, List 2. 'A surface or featu (*e.g.* a river) may be said to be at grade, graded, when neither degradation n aggradation is taking place.'

Woodford, A. O., 1951, Stream Gradien and Monterey Sea Valley, *Bull. Geol. So Amer.*, **62**, 799–852. 'W. M. Davis (190 introduced the term "grade" for th condition of "balance between erosion an deposition, attained by mature rivers" . . Kesseli (*Jour. Geol.*, **49**, 1941, 587) attacke the whole concept of the graded river . . Mackin (1948) set Kesseli right. . . .'

Mackin, J. H., 1948. Concept of the Grade Stream, *Bull. Geol. Soc. Amer.*, **59**, 463 512. 'A graded stream is one in which, ove a period of years, slope is delicately adjus ted to provide . . . just the velocity require for transportation of the load supplie from the drainage basin.' (p. 471)

**Grade** (2)

*O.E.D.* A class of things, constituted b having the same quality or value.

Hence applied 'to those grains in any detrita sediment which are of the same order o size.' (Himus, 1954)

*Comment*. Used especially in the mechanica analysis of sediments and soils. For soil the agreed international standards are:

| Stones | diameter above 2 mm |
| Coarse sand | ,, | 0·2–2 mm |
| Fine sand | ,, | 0·02–0·2 mm |
| Silt | ,, | 0·002–0·02 mm |
| Clay | ,, | below 0·002 mm |

(See Russell, E. J., 1957, *World of the Soil* London: Collins, p. 7; Stamp, 1948, p. 287.)

For other conventional classifications of grade-size, see especially British Standard Code (1947), Wentworth, and Inman, 1952; also King, C. A. M., 1959, *Beaches and Coasts*, London: Arnold.

**Graded sediment**

Himus, 1954. A general name for loose or cemented detrital sediments in which the grains lie mainly within the limits of a single grade.

**Grain (of a country)**

The general trend of the physical features, mountain ranges, valleys, rivers etc., or of the underlying geological structure.

Wooldridge and Morgan, 1937. 'Von Richthofen distinguished:
(a) longitudinal coasts, parallel with the structural grain,
(b) transverse coasts, cutting across the structural grain. . . .' (p. 352) 'The regional "grain" of the mainland can only affect coastline or shoreline detail . . .' (p. 354).

*Comment:* Some writers insist on the maintenance of a distinction between structural or tectonic and relief or topographical trend-lines and grain. Grain in this sense not in *O.E.D.*

**Graffito** (Italian; *pl.* graffiti)

*O.E.D.* A drawing or writing scratched on a wall or other surface; a scribbling on an ancient wall, as those at Pompeii and Rome. Also, a method of decoration in which designs are produced by scratches through a superficial layer of plaster, glazing etc., revealing a ground of different colour.

*Comment.* A term much used by Egyptologists.

**Gran Chaco**—see Chaco

**Grange**

*O.E.D.* 1. A repository for grain, a granary. 2. An establishment where farming is carried on. Now, a country house with farm buildings attached, usually the residence of a gentleman-farmer. 3. An outlying farm house with barns etc. attached to a monastery or a feudal lord.

**Granite**

*O.E.D.* 1. A granular crystalline rock consisting essentially of quartz, orthoclase-feldspar, and mica, much used in building.

Holmes, 1928. 'A phanerocrystalline rock, consisting essentially of quartz and alkalifelspars with any of the following: biotite, muscovite and amphiboles and pyroxenes (including soda varieties in the alkaligranites).'

Himus, 1954. A coarsely-crystalline acid plutonic rock.

*Comment.* Strict definitions of granite are applied by petrologists but the word is used in ordinary speech more loosely. Essentially the plutonic rocks (*q.v.*) comprise the acid granites, very widely distributed in the earth's crust, the intermediate diorite and syenite and the basic gabbro.

**Grao** (Spanish)

The escape-channel of a littoral marsh across a tombolo, corresponding to a '*crau*' or '*grau*' in French Languedoc (P.D.).

**Graphicacy, graphicate** (W. G. V. Balchin and Alice M. Coleman, 1965)

Balchin, W. G. V., and Coleman, Alice M., Graphicacy should be the fourth ace in the pack, *Times Educational Supplement*, 5 Nov., 1965, p. 947, 'the historian may resort to a picture or the geographer to a map or photograph to communicate the likeness of something remote in time or place. Graphs, photographs, cartography, the graphic arts. . . . The syllable "graph" which is common to all these names for visual aids can be used as a root to coin the word "graphicate" by analogy with literate, numerate and articulate.

It can be said that art is a form of self-expression and communication, a skill that involves both an expresser and an expressee. Great art includes both. Graphicacy, however, includes only the skill aspect. It is the communication of relationships that cannot be successfully communicated by words or mathematical notation alone.

Graphicacy is concerned especially, but not wholly, with spatial relationships as in maps. It also uses spatial relationships, usually in two dimensions but sometimes in three, to portray other kinds of relationships, such as time scales, rates of change, derivatives, abstractions etc. Spatial relationships and graphs may suggest geometry and other branches of mathematics, but mathematics cannot absorb the whole of graphicacy, which is more the province of geography than of anything else. Just as geography plays a role as a bridge subject between the arts and the sciences, so graphicacy spills over into literacy on the one hand and numeracy on the other without being more than marginally absorbed in either. It is inter-disciplinary without losing its own identity as a distinct medium of communication.'

Also ingraphicate—not versed in graphicacy.

**Grass, grassland**

*O.E.D.* 1. Strictly, a member of the *Gramineae*
2. Loosely, herbaceous plants on which sheep, cattle and other herbaceous (herb-eating) animals feed; herbage.
3. Land on which grass and other herbaceous plants are grown for hay or pasture (usually grasses grown as cereals for their seeds are excluded).

Grassland: land covered mainly with grass (see quotation under forest).

**Graticule**

Mill, *Dict*. Alternative and more satisfactory name for the network of meridians and parallels on a map.

*Comment*. Not to be confused with grid (*q.v.*).

**Grau** (French)—see Gat

**Gravel**

*O.E.D.* 2. A material consisting of coarse sand and water-worn stones of various sizes, often with a slight intermixture of clay, much used for laying roads and paths.

Glentworth, 1954. Mineral fragments between 2 mm. and 50 mm. in diameter; fine medium and coarse. (p. 163)

Himus, 1954. A loose detrital sediment in which the predominant grain size is from 2 mm. to 10 mm. If the deposit is of more coarsely graded detritus, it is known as a pebble-bed or a boulder-bed. The term is, however, used to cover a wider range of sizes than that defined. In common parlance, the name gravel is generally applied to a river deposit which contains a considerable proportion of more or less rounded pebbly material.

*Comment*. The definition of Himus is of the common usage, that by the pedologist (Glentworth) to denote a certain size in the constituents of a soil is a special one. The modern use of gravel (contrast *O.E.D.*) is in making concrete, and commercial usage of the term covers also such deposits as the Bunter Pebble Beds if suitable for this purpose.

**Gravel train**

A valley train (*q.v.*) where the dominant material is gravel

**Graviplanation** (Alice Coleman, 1952)

Coleman, A., 1952, Selenomorphology, *Jour. of Geol.*, 60. 'Though the moon's gravitational force is lower than the earth's, its morphological influence is vastly greater as it appears to be almost the sole agent transport and deposition... condition come as near as may be envisaged "equiplanation" as conceived by D. Cairnes (1912).... However Cairne concept embraces processes which are n operative on the moon, and regional lo and gain would be expected in the la stages of lunar evolution. Hence the ter "equiplanation" is not employed her but the self-explanatory "graviplanation is substituted.' (pp. 454–455) See Equ planation.

**Gravitative transfer**—see Mass-wasting

**Gravity**

*O.E.D.* The attractive force by which bodies tend to move towards the cent of the earth.

*Comment*. For connection between the varie meanings and use of gravity and grav tation, the *O.E.D.* should be consulted. adjectives 'gravity' and 'gravitationa are often used as alternatives.

**Gravity anomaly**

Variations in the force of gravity may be du to the existence of rocks of different der sities or magnetic characters in the earth crust and so the detection of gravi anomalies plays an important part in gec physical surveying and in the determinatio of the structure of the crust. See als Anomaly.

**Gravity model**

Haggett, 1972. 'As early as 1850 observer of social interaction had noted that flow of migrants appeared to be directl related to the size of the groups involve and inversely proportional to the distanc separating them. By 1885... Ravenstei had incorporated similar ideas into laws o migration. Although the specific term gravity model does not appear until the 1920s, it is clear that nineteenth century workers were drawing on the relationship formalized by Sir Isaac Newton in his lav of universal gravitation (1687), which states that two bodies in the universe attract each other in proportion to the product of their masses and inversely a the square of their distance.... Gravi tational concepts were specifically intro-duced by W. J. Reilly in 1929 to study trade areas.... Expanded by researchers concerned with predicting flows in applied fields like highway design or retail marketing studies.' (p. 329)

Wilson, A. G., 1972, *Papers in Urban and Regional Analysis*, London: Pion Ltd., reformulated the original concept of Newton in terms of statistical mechanics: '... the force between two masses becomes the number of trips (interaction) between "masses" of population, jobs or other measures of trip productions and attractions ... the population is now treated as a collection of individuals, and trips are obtained by estimating "means" over this population, or, equivalently by seeking a maximum probability solution.' (p. 15)

See also Reilly's law of retail gravitation.

### Gravity slope
Thornbury, 1954. 'According to Penck, there are two basic elements in any valley side slope formed under declining development or uplift—the upper, relatively steep *bösche* or *steilwand* or what Meyerhoff (1940) has called the *gravity slope*; and the lower, more gentle *haldenhang* or *wash slope* of Meyerhoff.' (p. 200)

### Gravity or gravitational slumping or sliding
The downward slipping of masses of rocks or sediment due to gravity. See slump; see also Bull, A. J., 1950, Gravitational Sliding, *Proc. Geol. Assoc.*, 61, 198–201.

### Gravity or gravitational water
Swayne, 1956 (1) water in the soil seeping towards the water-table. It moves by the force of gravity and is not retained by the soil.
(2) Sometimes used as a synonym for vadose water.

### Gray-brown podzolic soil
FAO, 1963. '*Gray-brown podzolic soils* have A, B and C-horizons, the important characteristic being the marked illuviation of clay into the B-horizon as expressed by the presence of clay coatings on ped surfaces or clay linings to pores and cavities .... The B-horizon frequently contains small rounded ferruginous concretions. The A-horizon contains an A2-horizon which, however, is not strongly bleached; the thin A0-horizon has a mull-humus form; the A1-horizon may only be weakly developed.'

### Graywacke, greywacke, grauwacke
Himus, 1954. Felspathic or tuffaceous grit or coarse sandstone, generally dark coloured, strongly cemented, often with an argillaceous cement. Characteristically developed among the older formations.

Mill, *Dict.* Represents the muddy sand of Palaeozoic sea-shores. An old and loose term.

### Grazing, graze
Webster. A pasture, growing grass.
*O.E.D.* Graze: to feed on growing grass or other herbage.
*Comment*. An important distinction is often made between land used for grazing and what may be precisely similar land but where the herbage is cut for hay or drying.

### Great Circle
Any hypothetical circle on the earth's surface whose plane passes through the earth's centre and which thus divides the earth into two hemispheres (*q.v.*); each meridian is half of a great circle. The shortest distance between any two points on the earth's surface is along the arc of the great circle on which they lie. See also Small Circle.

### Greenbelt, Green belt
Osborn, F. J., 1946, *Green-Belt Cities*, London: Faber and Faber. 'Originally used by Unwin as a further synonym for Country Belt, this term has also been applied, thus far confusingly, to a narrow strip of parkland more or less encircling part of a built-up metropolitan or large urban area. Park Belt is a better name for such a strip' (p. 182).

Report of the Committee on Land Utilisation in Rural Areas (Scott Report), 1942, London: H.M.S.O., Cmd. 6378. 'Although the term "green belt" is of comparatively recent introduction it seized the public imagination and has become not only widely used but still more widely misused ... we conceive the green belt to be a tract of ordinary country of varying width, round a town ... where the normal occupations of farming and forestry should be continued ... in essence just a tract of the countryside' (p. 71 *fn.*).

*Comment*. There is an extensive literature on the meaning and function of green belts. Note also the green wedge—'wedges' of green penetrating towards the heart of the town itself (Scott Report, p. 71 *fn.*) See also belt, country belt.

### Green lane
Unmetalled road still bearing a turf or grass surface, but definitely a right of way. Often identical with drift way (*q.v.*).

**Green village**

Thorpe, H., 1949, The Green Villages of County Durham, *Trans. Inst. Brit. Geog.*, **15**, 153–180. 'Nucleated settlements in which the homesteads cluster around an open grassy space or village-green'. (p. 155)

Thorpe, H., *Bull. Soc. belge d'Etudes Géog.*, **30**, 1961. 'The green village in which the homesteads cluster around an open space or Village-green.' (p. 93)

**Greenwich Mean Time (GMT, G.M.T.)**

The local mean time of the Meridian of Greenwich (Longitude 0°) which is used as the Standard Time for the United Kingdom and which is used as a standard from which time in nearly every country of the world is reckoned. Nearly all countries of the world choose their standard times so that they shall be an exact number of hours or at least half-hours ahead of or behind Greenwich time. Thus Eastern Standard Time (New York) is five hours back; Indian Standard Time $5\frac{1}{2}$ hours ahead.

**Gregale**

Kendrew, 1953. 'The Gregale (the "wind from Greece") of Malta and its neighbourhood is a strong NE wind, generally between a large anticyclone in the Balkans and a depression over north Africa.' (p. 359)

**Greywether, graywether**

Himus, 1954. Grey rounded blocks of sand stone or quartzite remaining as residual boulders when less resistant material has been eroded away. So-called from their resemblance, at a distance, to grazing sheep (wethers). Also known as sarsen stones (*q.v.*).

**Grid**

*O.E.D. Suppl.* A network of lines used *e.g.* for finding places on a map.

*Comment.* The network usually consists of two sets of parallel lines at right angles, the lines being such a distance apart as to represent a fixed distance, *e.g.* a kilometer or 10 kilometers. By numbering the lines, usually eastwards and northwards from a fixed point to the south-west of the area to be covered, the position of any point may be given by a series of numbers. British Ordnance Survey maps are now overprinted with the 'National Grid' with a reference point to the south-west of the British Isles. A grid should not be confused with the graticule. *Grid north* is the direction of the roughly north-south grid lines but coincides with true north only along the meridian of origin.

**Griffograph (L. D. Stamp, 1959)**

Stamp, L. D., 1959, *Geog. Jour.*, **125** (in review of the autobiography of Griffith Taylor). '... the highly characteristic Griffographs, if one may coin a word for those sketch block-diagrams which so often convey the story as nothing else could do....' (p. 122)

**Grike, gryke**

*O.E.D. Suppl.* A crack or slit in rock; a ravine in a hill-side.

Mill, *Dict.* A cranny or fissure; used chiefly for the widened joints which traverse the limestone clints (Cumbrian Lakes and Yorkshire).

Wray, D. A., *Geol. Mag.*, 1922. 'In the Craven district of Yorkshire ... the floor of the Carboniferous Limestone to the north of Settle is cut up by irregular widened joints known as *grikes*.' (p. 394)

Kendal and Wroot, 1924, *Geology of Yorkshire*. '... the distinctive type of scenery known as "clints", "helks" ... or "grykes".' (p. 62)

*Comment.* The term 'gríc' is used in the west Balkans for the same feature. Grike is normally used only of limestone (G.T.W.).

**Grit**

Himus, 1954. An ambiguous term which has been applied to (i) coarse-grained sandstone, (ii) sandstones, whether coarse or fine, made up of angular grains, (iii) sandstones made up of grains of conspicuously unequal sizes and (iv) stratigraphically in the names of such formations as Millstone Grit etc.

Also gritstone. Note that grit is also used for small particles of stone or rock, especially if angular.

**Grotto**

*O.E.D.* 1. A cave or cavern, esp. one which is picturesque, or which forms an agreeable retreat.

Mill, *Dict.* Loosely used for small caves; strictly applied to the larger cavities produced in limestone regions by the solvent action of underground streams and percolating water.

Rice, 1943. 'A small cavern or cavern-like apartment or retreat, natural or artificial; especially a cavern having some attractive features, as beautiful stalactite formation or rockwork.' (p. 163)

**Ground frost**
*O.E.D. Suppl.* A frost which occurs on the surface of the earth but not in the circumambient air.

Wayne, 1954. A frost occurring when the reading of the thermometer on the grass falls to 30·4°F. or below.

*Comment.* A light frost does not affect plant tissues, but when the temperature drops below 30·4°F. the effect on some plants is very marked.

**Ground ice**
Mill, *Dict.* Ice formed on the bed of a river, lake, or shallow sea while the water as a whole remains unfrozen.

*U.S. Ice Terms*, 1952. 1. Bottom ice. Ice formed on the bed of a river, lake or shallow sea. 2. Fossil ice. Ice associated with permafrost.

*Comment.* Used by many authors as synonymous with anchor-ice (see Ice, sundry terms). Not in *Ice Gloss.*, 1956, 1958.

**Ground water, groundwater**
Normally used for all water found under the surface of the ground but not including underground streams (see Thornbury, 1954, 47).

*Webster.* Water within the earth such as supplies wells and springs, specifically that below the unsaturated zone of percolation and above the region where all openings are closed by pressure. Its upper surface (the water table) may coincide with the surface or lie deep below.

Meinzer, *Outline of Ground-Water Hydrology.* Exact synonym 'phreatic water'; 'underground water' and 'subterranean water' are sometimes used as synonymous with ground water, sometimes as including *all* water found below the surface.

**Group** (geology)—see Geological Time

**Grove**
*O.E.D.* A small wood; a group of trees affording shade or forming avenues or walks, occurring naturally or planted for a special purpose. Groves were commonly planted ... in honour of deities to serve as places of worship or for the reception of images.

*Comment.* The groves mentioned in the latter part of the *O.E.D.* definition are often distinguished as 'sacred groves' (*e.g.* those associated with temples in China). By a rather doubtful analogy the term 'grove belt' has been applied in Canada to the transitional belt between the prairies and the coniferous forest, where small patches of woodland resemble groves.

**Growing season**
Moore, 1949. That part of the year in a specific region when the growth of the indigenous vegetation is made possible by the favourable combination of temperature and rainfall. In general its duration decreases with distance from the equator: in the *Equatorial Forest*, for instance, the growing season is continuous throughout the year, whilst in the *Tundra* it lasts only for two or three months. Many cultivated plants demand growing seasons of special character.

*Comment.* The crucial temperature for many plants is 42° or 43°F., below which vegetative growth does not take place and the plant is dormant.

**Growler**
*Webster.* 5. A very small iceberg or mass of floe ice large enough to be a menace to ships.

*Ice Gloss.*, 1956. A piece of ice almost awash, smaller than a *bergy bit* (*q.v.*).

**Growth pole (growth centre)**
Darwent, D. F., 1969, Growth Poles and Growth Centres in Regional Planning—a Review, *Environment and Planning*, 1.1. 'The term "growth pole" was introduced into economic literature in 1949 by Francois Perroux (1950), since when it has become associated with an enormous variety of indistinct and ill-defined concepts and notions' '... within these poles that growth and change is initiated, while the connections between the poles, in terms of flows of inputs and outputs, transmit the forces generated. The poles are therefore best regarded simply as sectors of an economy represented by an input-output matrix in which growth effects can be transmitted across the rows and columns.' 'Perroux defines growth poles only and specifically in relation to abstract economic space and not in relation to geonomic (or geographic) space, which he dismisses in 1950 as banal.' (p. 6)

See Perroux, F., 1964, La Notion du Développement, *L'Economie de XX Siècle*, Paris:

Presse Universitaire de France (p. 143).
Isard, W. and Schooler, E. W., 1959, Industrial Complex Analysis: Agglomeration, Economies and Regional Development, *Jour. Reg. Sci.*, Spring, discusses growth poles in relation to geographic space.

**Groyne, groin**
*O.E.D.* A framework of timber, or now sometimes a low broad wall of concrete or masonry run out into the sea, for the purpose of arresting the washed-up sand and shingle and thus raising a barrier against the encroachment of the sea.

King, C. A. M., 1959, *Beaches and Coasts* London: Arnold. 'The aim of a groyne, usually built approximately at right angles to the beach, is to trap beach material moving alongshore to build up the level of the beach.' (p. 313)

**Grumusol** (Soil science)
A general American term for self-mulching black tropical and sub-tropical clays such as regur and tirs. (G.V.J.)

FAO, 1963. '*Grumusols* are AC soils of heavy texture. The clay fraction consists of expanding clay minerals which causes the soil to crack badly during the alternating wet and dry seasons and the soils show evidence of strong churning action. The A1-horizon is thick and the soils may contain calcium carbonate.'

**G-Scale** (Haggett, Chorley and Stoddart, 1965)
Haggett, P., Chorley, R. J. and Stoddart, D. R., 1965, Scale Standards in Geographical Research: A New Measure of Area Magnitude, *Nature*, **205**, 844–847. 'We propose ... that the standard geographical measurement shall be the Earth's surface area (Ga), and that scale of measurement (the G-scale) shall be derived by successive subdivisions of this standard area by the power of ten.' (p. 846)

| G-value | Earth's surface | Area in sq. miles |
|---|---|---|
| 0 | Ga | $1\cdot968 \times 10^8$ |
| 1 | $Ga(10)^{-1}$ | $1\cdot968 \times 10^7$ |
| 2 | $Ga(10)^{-2}$ | $1\cdot968 \times 10^6$ |
|   |   | etc. |

According to Haggett, 1965, it has four characteristics. 'First, it uses a natural standard, the surface of the earth, rather than existing arbitrary standards. Second, through its logarithmic nature it reduces a very wide range of values to a simple scale. ... Third, it allows ready comparison of the relative size of areas, in that regions which are different in area by factor of one hundred are two units apart and so on. Fourth, it simplifies the present confusion of conventional standards and substitutes a simpler scale.' (pp. 5–8)

**Guadi**—see Wadi

**Guano**
A deposit of excrement which can be used as a fertiliser. It is a Spanish word (*huano* in South America) and is applied especially to the thick layers of excreta of sea-bird found in many islands of the dry Peruvian and Chilean coasts, where there is little or no rain to wash away the material. The term is also applied to similar deposits elsewhere *e.g.* bat-guano in the limestone caves of Malaysia. (L.D.S.)

**Gulch**
*O.E.D.* United States: a narrow and deep ravine, with steep sides, marking the course of a torrent; especially one containing a deposit of gold.

**Gulf** (maritime)
*O.E.D.* 1. A portion of the sea partially enclosed by a more or less extensive sweep of the coast; often taking its name from the adjoining land.
The distinction between *gulf* and *bay* is not always clearly marked, but in general a *bay* is wider in proportion to its amount of recession than a *gulf*; the latter term is applied to long land-locked portions of sea opening through a strait, which are never called *bays*.

Mill, *Dict.* Applied to very varied forms of inlets of the sea, but on the whole it is given to the largest of these inlets, *e.g.* Gulf of Guinea, Persia, Mexico, whereas the term 'bay' is applied to only a few of these larger forms and to many smaller ones. A gulf may contain more than one bay, *e.g.* the Gulf of Guinea with the Bays of Benin and Biafra.

**Gulf** (2)
*O.E.D.* A deep hollow, chasm, abyss.
*Comment.* There are various specialized meanings associated with this general one. Note especially Muma and Muma, 1944, a steep-walled sink hole with a flat alluvial floor in which an underground stream either sinks or rises (p. 60) (G.T.W.).

**Gull**
*O.E.D.* A breach or fissure made by a torrent;

a gully, chasm; a channel made by a stream. *Obs. exc. dial.*

Himus, 1954. Fissures or joints which are steeply inclined and trend parallel with the surface contours, associated with the formation of cambers in rocks. The gulls are usually filled with material which has fallen in from above.

*Comment.* The varied meanings of this dialect word suggest it should be avoided.

### Gully, gullying

*O.E.D.* 2. A channel or ravine worn in the earth by the action of water, esp. in a mountain or hill side.

*Webster.* 1a. A miniature valley or gorge excavated by running water, but through which water commonly runs only after rains.

Moore, 1949. '... it is smaller than a ravine....'

*Comment.* The use of the term 'gullying' or 'gully erosion' should be noted for the type of soil erosion or erosion of soft rocks by heavy downpours, especially in the tropics, which produce gullies. The term 'gully gravure' has also been used to indicate the type of erosion surface scored by numerous gullies (Kirk Bryan, 1940, *Jour. Geomorphology*, 3, 89). Spelling gulley also used.

### Gumbo

*Webster.* A soup thickened with the mucilaginous pods of the okra; specif. applied, esp. in the western U.S., to a class of peculiar, fine-grained, silty soils, which when saturated with water become impervious and soapy or waxy and sticky.

*Dict. Am.* (a native name for okra in the Bantu or central or northern Angola, Africa). 1. The okra plant or its pods. Also short for gumbo soup, any one of various thick, palatable soups in which okra pods are used. 4. Applied, esp. in the western states, to any one of various types of soil suggestive in some way of mucilaginous gumbo soup. 1881. 1894 *Cent. Mag.* Jan. 453/1. 'Gumbo is ... the clay of Northern Wyoming. When wet, it is the blackest, stickiest ... mud that exists on earth.' 1888. Roosevelt in *Cent. Mag.* March 659/1 'gumbo clay, which rain turns into slippery glue.'

### Gumbotil

*Webster.* A dark, leached, non-laminated, very sticky clay, formed by the weathering of glacial till.

Lobeck, 1939. 'In the regions of older drift ... the glacial deposits are profoundly weathered. Even the boulders have decayed and the mass of till, known as *gumbotil*, has gained a uniformity of character. The soil profiles indicate a very long period of adjustment to weathering conditions, whereas the recent glacial soils are immature.' (p. 314)

*Comment.* Also used to refer to fossilised soils beneath later tills. See Flint and Charlesworth (G.T.W.).

### Gums, Gum trees (Australia)

A general name applied to the varied species of *Eucalyptus* in Australia. This usage is mentioned because of confusion with the many other usual meanings of gum.

### Gumland

In New Zealand is properly land from which fossil Kauri gum has been obtained but commonly applied to land which may or may not have yielded gum, now covered with scrub.

### Gun (Japanese)

County.

### Gutter

*O.E.D.* 1. A watercourse, natural or artificial; in later use, a small brook or channel. b. A furrow or track made by running water. c. Austral. gold mining. The lower part of the channel of an old river of the Tertiary period containing auriferous deposits.

Mill, *Dict.* 1. Natural or artificial water-channels.
2. Shallow steep-sided valleys which drain marshy uplands, usually marking areas where the drainage is just becoming rejuvenated.
3. The lower and auriferous part of the channel of an old river of the Tertiary period (Australia).
4. A submarine valley-like depression on a continental shelf.

*Comment.* Applied mainly to the artificial channel made by the side of a road to take away surplus rain water or to channels of metal to take away rainwater from roofs of buildings. As a technical term is obsolescent and is best avoided. No reference in Lyell, 1850; Geikie, *Earth Sculpt.*, 1898; Davis, Essays, 1909; Wooldridge and Morgan, 1937; Cotton; von Engeln, 1942.

**Guyot** (H. H. Hess, 1946)

Hess, H. H., 1946, Drowned ancient islands of the Pacific Basin, *Amer. J. Sci.*, **244**, 772–791.

Thornbury, 1954. 'Lesser topographic features on the deep-sea floor are volcanic islands, seamounts and guyots.... Hess gave the name *guyot* to the larger flat-topped forms.' (p. 479)

*Comment.* Named in honour of a distinguished Swiss-American scientist. See Seamount.

**Gwaun** (Welsh)

Moor, mountain pasture.

**Gwely** (Welsh; *lit.* family)

A social-anthropological term covering an extended family claiming descent from a common ancestor (E.G.B.). See also Tyddyn.

**Gyttja** (Swedish; *pl.* gyttjor)

Swedish folkname, introduced by H. v. Post (1862) as a term for humic soils of sedimentary origin in which the humus consists mainly of plant and animal residues precipitated from standing water; active, muddy, grey brown to blackish AC soils, rich in organisms occurring in waters with adequate nutrients and oxygen and containing great quantities of organic food. (E.K.)

Jacks, 1954. 'Name for humic soils of sedimentary origin in which the humus consists mainly of plant and animal residues precipitated from standing water.'

# H

**Haar**
Mill, *Dict.* A sea fog accompanying an easterly drift of air on the east coast of Britain.
*Comment.* Especially in summer from Lincolnshire northwards, notably the Firth of Forth; a cold sea-fog.

**Habitant** (French; *lit.* an inhabitant)
*O.E.D.* Pronounced abitan; a native of Canada (also of Louisiana) of French descent; one of the race of original French colonists, chiefly small farmers or yeomen.
*Comment.* Originally the peasant settlers as opposed to the nobility; now applied especially to the rural population of the province of Quebec. See Lower, A. R. M., 1946, *Colony to Nation: A History of Canada,* Toronto: Longmans.

**Habitat**
*O.E.D.* The locality in which a plant or animal naturally grows or lives; habitation. Sometimes applied to the *geographical area* over which it extends, or the special *locality* to which it is confined; sometimes restricted to the particular *station* or spot in which a specimen is found; but chiefly used to indicate the kind of locality, as the sea-shore, rocky cliffs, chalk hills, or the like.
b. Hence generally: Dwelling place; habitation.
*Webster.* In *Biol. habitat* refers especially to the kind of environment in which a plant or animal is normally found, as opp. to *range,* which denotes the geographical region throughout which it is distributed. *Locality* indicates a circumscribed area of its occurrence within the range; and *station,* in its usual restricted sense, refers to the exact spot at which a given specimen or species is found.
Mill, *Dict.* (As *O.E.D.*).
Yapp, R. H., 1922, The Concept of Habitat. *Jour. of Ecology*, 10. '... from the point of view of plant ecology: a habitat may be described as the place of abode of a plant, a plant community, or in some cases even a group or a succession of related plant communities, together with all factors operative within the abode, but external to the plants themselves.' (p. 12) This is said to correspond more or less closely to environment, milieu or Standort. (W.M.)
Mackinder, H. J., 1931, The Human Habitat, *Scot. Geog. Mag.*, **47**, 321-335. Does not define the habitat; discusses the hydrosphere principally in its effect upon humanity. (W.M.)
Forde, C. D., 1934, *Habitat, Economy and Society,* London: Methuen. 'Habitat' is applied to humans, with reference to their cultural groups. Appears to be covered by brief descriptions of the local physical geography, or natural environment; *e.g.* see index under 'habitat'. (W.M.)
*Comment.* Frequently used by geographers as synonym for environment, *e.g.* E. Huntington, 1928, *The Human Habitat,* London: Chapman and Hall; E. C. Semple, 1913, *Influences of Geographical Environment,* London: Constable.

**Habūb, Haboob** (Sudan; Arabic *lit.* to blow violently)
A local strong wind, accompanied by thick dust and, in the rainy season, preceding a thunder-shower in the northern Sudan. (J.H.G.L.)

**Hachure** (French, but naturalized)
Moore, 1949. Hachures: short lines of shading drawn on a map to represent differences in the slope of the ground. If the slope is steep the lines or hachures are thick and close together; if the slope is gentle, they are thinner and farther apart ... the hachures are drawn from the highest point of a mountain to the foot, and should be perpendicular to the contour lines. Unlike the contour lines, they give no indication of the actual height of land above sea level.

**Hacienda** (Spanish)
*O.E.D.* 'In Spain, and existing or former Sp. colonies: an estate or "plantation" with a dwelling house upon it; a farming, stockraising, mining or manufacturing establishment in the country; sometimes a country-house.'

*Comment.* Geographers have tended to give this word specialized meanings varying from country to country. In the Philippines for example 'the estate or hacienda system' (Dobby) of sugar-cane growing is in contrast to small-holdings; in Chile it was 'the estate of vast size' (James); in Mexico 'large, privately-owned feudal estate' (James) and is associated mainly in Latin America with extensive ranching.

**Hade**
*Webster. Geol.* The angle made by a fault plane or a vein with the vertical. Contrast dip (*q.v.*).

**Haff** (German; *pl.* Haffe)
A coastal lagoon of fresh water due to its being fed by a river; separated from the open sea by a Nehrung (*q.v.*) and linked with it by a narrow channel. The classic haffe which gave rise to the term are on the Baltic coast of East Prussia, *e.g.* Kurisches Haff, Frisches Haff (K.A.S.).

**Hafir** (Sudan; Arabic, *pl.* hafayrer, *lit.* a pit)
Applied in the northern Sudan to a reservoir for watering stock made by deepening a *fula*, or, more recently, by mechanical excavation. (J.H.G.L.)
See Lebon, J. H. G., Current Developments in the Economy of the Central Sudan, *I.G.U. Report of a Symposium at Makerere*, 1955, 57–66.

**Hafod, hafoty** (Welsh: *lit.* summer dwelling, shieling)
*Report Royal Commission on Common Land*, 1955–58, H.M.S.O., 1958, Cmd. 462. 'The upland pastures in Wales to which transhumance took place in the summer months' (p. 274).

**Hag, hagg**—see Peat hag

**Ha-ha**
*O.E.D.* A boundary to a garden, pleasure-ground, or park, of such a kind as not to interrupt the view from within, and not to be seen till closely approached; consisting of a trench, the inner side of which is perpendicular and faced with stone, the outer sloping and turfed; a sunk fence.

**Hail**
*Ice Gloss.*, 1958. Precipitation of small balls or pieces of *ice* (hailstones) with a diameter ranging from 5 to 50 mm., or sometimes more, falling either separately or agglomerated into irregular lumps.

**Hained cow pasture**
A cow pasture from which the cattle have been temporarily excluded in order to allow the grass to grow (Midlands of England) (I.L.A.T.).

**Haldenhang** (German)
Penck, W., *Morphological Analysis of Land Forms* (Trans. H. Czech and K. C. Boswell). London: Macmillan, 1953. '... basal slope (the less steep slope found at the foot of a rock wall, usually beneath an accumulation of talus.... It is the top part of the Fusshang or foot-slope, which includes all the slopes of diminishing gradient).' (p. 419)
Fischer, 1950. 'Haldenhang = under-talus rock slope of degradation (Penck).' (p. 24)
Wood, A., 1942, The Development of Hillside Slopes, *Proc. Geol. Assoc.*, 53, 128–140. 'The constant slope and the waning slope appear to correspond to the "Steilwand" and "Haldenhang" of W. Penck (1924) and to the "gravity slope" and "wash slope" of Meyerhoff (1940).'
See also *A.A.A.G.*, 30, 1940; Wooldridge, S. W., 1958, *Trans. Inst. Brit. Geog.*, 25. 'A toe-slope or haldenhang' (p. 32).

**Half-year Close**
An enclosure opened for sheep pasturage during the winter half year (Eastern England) (I.L.A.T.).

**Half-year Land**—see Lammas Land

**Hällanalys** (Swedish)
Analysis of the glacial surfaces of outcropping rocks (E. Ljungner). (E.K.)

**Halomorphic soil** (Saline soil)
Jacks, 1954. 'Soil whose properties have been determined by the presence of salts.'

**Halophyte**
*O.E.D. Suppl.* A plant which grows in soil impregnated with salt, as on the seashore or in the sea.
Tansley, 1939. 'Very specialized *halophytic* or salt-marsh vegetation' (p. 99).... the master factor is the salt water which bathes the whole plant body during the periodic immersions and forms the soil solution (though varying in concentration) at all times ... the plants are said to be *halophytes* or *halophilous*.' (p. 819)
*Comment.* A plant growing in a salt-marsh or other saline environment such as a salt desert but not in the sea.

**Ham**
*O.E.D.* (1) A plot of pasture-ground; in some places especially meadow-land; in others an enclosed plot, a close.
(2) The Old English *ham* = home.
*Comment.* In the first sense it is applied to

some well-known tracts of rich pasture land in southern England.
In the second sense it was applied to a wide variety of settlements from a town to a single homestead and is extremely common as part of a place name in all those areas of England which were under Anglo-Saxon influence. It is obsolete as a term but survives in the diminutive form hamlet (*q.v.*).

**Hamada, Hamāda, Hammada** (North Africa; Arabic)
Mill, *Dict*. The stony uplands of the desert which have been stripped of their sand and dust by air-currents (Sahara). The Arabic term is Nijd. *Cf.* Reg.
Knox, 1904. Hamāda (Arab., N. Africa), plateau with rocky soil; calcareous waterless plateau.
Holmes, 1944. 'The rocky desert (the *hammada* of the Sahara), with a surface of bedrock kept dusted by deflation and smoothed by abrasion.' (p. 255)
Fitzgerald, W., *Africa*, London: Methuen, 1934. 'Rocky wastes with the bare exposure of fissured rocks as dominant features of the scene, form the "hamada" type of the Sahara.' (p. 60)
See also Erg. Knox's introduction of 'calcareous' is unwarranted.

**Hamlet**
O.E.D. A group of houses or small village in the country; esp. a village without a church, included in the parish belonging to another village or town. (In some of the United States, the official designation of an incorporated place smaller than a village.)
Mill, *Dict*. Similar to *O.E.D.*
Stroud, 1952. 'The distinction seems to be that a vill has a constable and a hamlet has none.'
*Comment*. It is doubtful whether 'hamlet' should be given any more precise meaning than that of a settlement of several dwellings but too small to be called a village. The adjective 'hamletted' has been used to denote a settlement pattern between one of nucleated (villages) and dispersed (isolated dwellings) in which habitations are grouped in small units.

**Hamūn** (Pakistan: Pashto)
A playa or lake (Baluchistan).

**Hanger**
O.E.D. A wood on the side of a steep hill or bank.
Mill, *Dict*. A steep and wooded declivity; the wooded part of an escarpment (Sussex and South Downs).
*Comment*. Mill seems to have transferred the meaning from the wood to the slope on which it grows. On steep chalk slopes in southern England the tree commonly found is the beech, hence beech-hangers. Some writers state that the word refers to the way in which the woodland appears to hang precariously on the steep slopes.

**Hanging valley**
Mill, *Dict*. Tributary valleys the lower ends of which are well above the bed of the main valley because this has been deepened (usually by glaciers) and the tributaries have not been so deepened.
Rice, 1941. 'A valley, the floor of which is notably higher than the level of the valley or shore to which it leads.'
*Comment*. The earlier discussions included notably discordant valleys both involving glacial action and the earliest stages of normal river-erosion (*e.g.* Davis, 1909, p. 632, and still in Cotton, 1941, p. 29). Later discussion generally ignores 'normal' discordant junctions and provides alternative theories for their presence in glaciated areas (*e.g.* Wooldridge and Morgan, 1937, pp. 371–372). Hanging valleys of sea cliffs are usually mentioned. A 'review of previous writings on hanging valleys' is provided in Davis, 1909, pp. 677–687. (W.M.)

**Haor, haur** (Indo-Pakistan: Bengali) (more correctly hrād)
A bhil or jhil; especially in the Sylhet plain; depressed water-filled areas (P. Sengupta, *The Indian Jute Belt*, Bombay, 1959, p. 150).

**Harbour**
O.E.D. 3. A place of shelter for ships; spec. where they may lie close to and sheltered by the shore or by works extended from it; a haven, a port.
*Adm. Gloss.*, 1953. A stretch of water where vessels can anchor, secure to buoys or lie alongside, wharves, piers, etc., to obtain protection from sea and swell. The protection may be afforded by natural features or by artificial works.
Island Harbour: a harbour formed, or mainly protected, by islands.
Morgan, F. W., 1952, *Ports and Harbours*, London: Hutchinson. A port has two main sections, the harbour and the port proper. A good harbour or anchorage has

deep water, shelter, extent, good holding ground, etc. 'The port, or port works ... comprise the installations at which ships are dealt with. ....'
*Comment.* Also used in place names, especially for farms in sheltered situations, *e.g.* Cold Harbour (G.T.W.).

**Hardebank**—see Blue ground

**Hardeveld** (Afrikaans)
Veld where a thin layer of soil is underlain by hard gravel or rock so that no ploughing would be possible. The name cannot be applied to bare rock, as the term veld always denotes some form of vegetation. The term is most frequently used on the border of sandy regions like those of the west coast or the Kalahari. Here the name is often found in a favourable sense, because the sandveld, in its natural condition, was difficult for traffic, while the hardeveld was much easier for draught animals. (P.S.)

**Hard-pan, hard pan**
*O.E.D.* U.S. 1. A firm subsoil of clayey, sandy, or gravelly detritus; also, hard unbroken ground.
*Webster. Chiefly U.S.* A cemented or compacted layer in soils (often containing some proportion of clay) through which it is difficult to dig or excavate. It results from the accumulation of cementing material and may be caused by continuous plowing at the same depth.
*Soils and Men*, 1938. 'An indurated (hardened) or cemented soil horizon. The soil may have any texture and is compacted or cemented by iron oxide, organic material, silica, calcium carbonate, or other substances.' (pp. 1169)
Jacks, 1954. 'Indurated or cemented layer of soil.' The general term including: *clay pan*, 'dense subsoil horizon formed by washing down of clay or by synthesis of clay'; *iron pan*, 'layer, usually of sand cemented with iron oxides'; *lime pan*, 'thick layer of calcium carbonate'.
*Comment.* Now used in the special sense quoted, either as 'a hardpan', the layer, or 'hardpan' the material. Hardpan may occur at varying depths below the surface, or at the surface itself; in the latter case especially when overlying unconsolidated material has been removed by weathering, Mill says boulder-clay is called hard pan in the United States; if this was the case it is obsolete.

See also Nazzaz, Calcareous crust, Caliche, Carapace latéritique, and Laterite.

**Hariq** (Sudan; Arabic, *lit.* a conflagration)
An area of land rotation in which wild grasses are fired just before sowing the crop, usually *dura*, cotton or *simsim*. (J.H.G.L.)

**Harmattan, hamattan**
*O.E.D.* (From harmata, the name in the Fanti or Tshi lang. of W. Africa).
A dry parching land-wind, which blows during December, January, and February, on the coast of Upper Guinea in Africa; it obscures the air with a red dust-fog.
Mill, *Dict.* Harmattan or Hamattan. A dry, dust-laden, parching but cool wind, which blows at intervals, chiefly from November to February, from the interior of Guinea to the Atlantic Ocean. *cf.* Khamsin.
*Met. Gloss.*, 1944. A dry wind blowing from a north-east or sometimes easterly direction over north-west Africa. Its average southern limit is about 5°N. latitude in January and 18°N. in July. Beyond its surface limit it continues southward as an upper current above the south-west monsoon.

**Harrah** (Arabia: Arabic)
Stamp, L. D., 1959, *Asia*, London: Methuen, 10th ed. (of the deserts of Arabia) 'tracts of rough lava surface which cuts the feet of man and animals to pieces' (p. 145).

**Harratin** (Sahara; Arabic, Harāthīn)
Freed slaves; negroid agricultural serfs in the Sahara (M.W.M.).

**Ha-ta** (Japanese)
Ta or Suiden means paddy field: dry cultivation of rice.

**Haufendorf** (German; *pl.* Haufendörfer: *lit.* heap, thrown together, village)
A nucleated village of irregular ground plan (K.A.S.).

**Haughland, haugh** (Scottish)
*Webster.* A low-lying meadow by the side of a river. *Scot. and N. of Eng.*
Flood plain of a river. (A.C.O'D.)

**Havsband** (Swedish)
The seaward skerries. The outermost part of the skerry guard with its mainly bare skerries and smallest rock islets. See also skärgård (skerry guard) (E.K.).

**Hawaiian volcano**—see Strombolian, also shield

**Hayes**
Land won from the forest or woodland; synonymous with 'assart' used in old

deeds in Leicestershire; in neighbouring areas haggs, or hagges and in Derbyshire heage used with the same meaning (I.L.A.T.).

**Haystack Hill** (Puerto Rico)—see Hum.

## Haze

*O.E.D.* 1. An obscuration of the atmosphere near the surface of the earth, caused by an infinite number of minute particles of vapour, etc. in the air. In 18th c. applied to a thick fog or hoar-frost; but now usually to a thin misty appearance, which often makes distant objects indistinct, and often arises from heat (heat-haze).

*Webster*. Light vapor or smoke in the air which more or less impedes vision, with little or no dampness. Haze, mist, fog. *Haze* suggests a lack of transparency in the air, commonly due to dust, but often in slight measure, to shimmering caused by heat. *Mist* always suggests moisture. *Fog* is thick or dense *mist*.

*Met. Gloss.*, 1944. When visibility is more than one and less than two kilometres and the obscuration is due to solid matter such as dust or smoke.

## Head, headland (1)

*O.E.D.* 16. spec. The source of a river or stream. Now chiefly in Fountain-head, *q.v.*
17. A body of water kept at a height for supplying a mill, etc.; the height of such a body of water, or the force of its fall (estimated in terms of the pressure on unit of area). Sometimes the bank or dam by which such water is kept up. c. A high tidal wave, usually in an estuary; = Bore, Eagre.
22. A projecting point of the coast, esp. when of considerable height; a cape, headland, promontory. Now usually in place-names. b. A projecting point of a rock or sandbank.

Mill, *Dict*. 1. Upper end of a slope. 2. Source of a river. 3. Bore or Eagre. 4. A projecting part of a coast or a sandbank.

Wooldridge and Morgan, 1937. 'The "Head" of South-west England is a rubble-drift made of local rocks, mantling the slopes and filling the valley-bottoms.' (p. 403)

*Adm. Gloss.*, 1953. A comparatively high promontory with a steep face. An unnamed Head is usually described as a headland. The inner part of a bay, creek, etc., *e.g.* The head of the bay. Head Sea. A sea coming from the direction to which the ship is steering (heading).

Moore, 1949. Head of navigation. The farthest point up a river to be reached by vessels for the purposes of trade.

Fay, 1920. 13. The attitude or direction of the set of parallel planes in a massive crystalline rock along which fracture is most difficult. It is normal to the direction of the strongest cohesion. 16. A layer of angular debris of adjacent strata, which generally overlies the raised beaches of England (Standard Dictionary, 20th century edition, 1910).

*Comment*. All the uses above are current. Only the context can reveal the meaning intended. The rubble-drift mentioned by Wooldridge and Morgan has been recognized as a periglacial deposit and precisely defined by S. E. Hollingworth *et al. Geol. Mag.*, 1940. 'Head' is now accepted as an official term by the Geological Survey.

## Headland (2)

The unploughed strips between the ploughlands of open fields in medieval England on which the ploughs were turned and which formed winding routeways between the fields.

## Head-Dyke

Robertson, I. M. L., 1949, 'The Head-Dyke: A Fundamental Line in Scottish Geography', *Scot. Geog. Mag.*, 65, 6–19. 'A dry stone wall built across the head of every farm before the era of general enclosure to secure the effectual separation of hill pasture from the arable and meadow grounds ... not often the present day boundary of agriculturally exploited land, as the crops-and-grass land seldom now extends as far' (pp. 6–7).

## Head-pond (U.S.)

Area reserved for waters lying above the control dam from which water is used for power-generation.

## Head-race

The channel through which water is led to a water mill or generating station.

## Headward erosion

Mill, *Dict*. Headward Erosion of Rivers. The action of a stream in extending its valley farther up-stream; thereby inflecting the watershed.

Cotton, 1922. 'They (Insequent Streams rapidly eat their way back into the interfluves by headward erosion.' (p. 72)

Wooldridge and Morgan, 1937. '... in all cases the graded curve will be first attained near the mouth or base-level point, for here the amount of necessary down-cutting is least. With the progress of time it proceeds backwards, *i.e.* headwards, from the lowest point. The grading of a stream course thus involves headward erosion. (p. 162)

Holmes, 1944. Uses similarly to Cotton (p. 153) but also (p. 219) to erosion in a corrie by frost sapping and the influence of the bergschrund. (W.M.)

Fay, 1920. Headwater erosion. The extension of a stream valley by erosion of the upland at its head (*Webster*).

### Headwater, head-water, head-waters

*O.E.D.* Plural the streams from the sources of a river.

*Comment.* Better the source or sources and upper parts of a river, especially a large river. See above, headwater erosion = headward erosion.

### Heaf

*O.E.D.* Accustomed pasture ground (of sheep).

*Comment.* Although this is a dialect word from the north of England, it is widely used in works on agriculture and agricultural geography. Flocks of many breeds of mountain sheep become so closely attached to a particular tract of mountain grazing—their own 'heaf' that they pine and lose condition if attempts are made to move them. If the farm changes hands the flocks ('hefted' flocks) are accordingly sold with the farm.

### Hearth

Sauer, C. O., 1952, *Agricultural Origins and Dispersals*, New York: Amer. Geog. Soc. 'I like the combination of nature and culture in north-western South America for locating there the first hearth of agriculture ... it seems to me there is a case for one basic hearth' (p. 43). Elsewhere Sauer uses the more common 'cradle'.

### Heartland (H. J. Mackinder, 1904)

Mackinder, H. J., 1904, The Geographical Pivot of History, *Geog. Jour.*, **23**, 421–437. The term heart-land is only used in a descriptive sense of the 'Pivot Area' which is the area inaccessible to sea-power. As illustrated it appears to include the areas of inland drainage and that drained to the Arctic Ocean excluding the Barents Sea. (W.M.)

Mackinder, H. J., 1919, *Democratic Ideals and Reality*, London: Constable. pp.95–97. The area of Asia inaccessible to navigation from the ocean because it drains inland or to the Inaccessible Coast, beset with ice. 'The Heartland is the region to which, under modern conditions, sea-power can be refused access, though the western part of it lies without the region of Arctic and Continental drainage.' (p. 141)

Fawcett, C. B., 1947, Marginal and Interior Lands of the Old World. *Geography*, **32**, 1–12. Fawcett contrasts these two definitions which result in a 'Physical Heartland' and a 'strategic Heartland'. (W.M.)

Mackinder, H. J., 1943, The Round World and the Winning of the Peace, *Foreign Affairs*, **21**, 595–605. The area east of the Yenisei, called Lenaland, is excluded from the Heartland apparently because of its small population and little development. (W.M.)

Whittlesey, D., 1939, 1944, *The Earth and the State*, New York: Holt, p. 84. 'Continental Heartlands' are discussed with reference to the peripheral setting of Australia, Africa and South America, contrasted with Asia and N. America. The interior of continents, removed from coastal (political) influence. (W.M.)

*Comment.* The terminology developed by Mackinder in his classic essay quoted above and expanded in *Democratic Ideals and Reality* involved the concept of the World Island (called by Fawcett, 1947, the Mainland) of Eurasia-Africa, the core of which is the Heartland.

### Heath, heathland, heather moor

*O.E.D.* 1. Open uncultivated ground; an extensive tract of waste land; a wilderness; now chiefly applied to a bare, more or less flat, tract of land, naturally clothed with low herbage and dwarf shrubs, esp. with the shrubby plants known as heath, heather or ling.

2. A name given to plants and shrubs found upon heaths or in open or waste places. b. The ordinary name for undershrubs of the Linnaean genus Erica.... By botanical writers sometimes limited to the modern genus Erica, sometimes extended to other cognate genera of Ericaceae.

Mill, *Dict.* (Follows *O.E.D.* and Robert Smith, see below).

Smith, Robert, 1900, Botanical Survey of Scotland, 1. Edinburgh District. *Scot. Geog. Mag.*, **16**, 404. Divides the heather association into three types: Heath, dominated by Calluna-Erica, D.C.; Heather-moor in which Calluna is still dominant but peat is abundant and moisture loving plants; Sphagnum-moor or Moss-moor, where living sphagnum is an important constituent. (W.M.)

Tansley, 1939. 'This is an oceanic and sub-oceanic plant formation dominated by dwarf ericaceous undershrubs, mainly the common ling or heather (*Calluna vulgaris*) but also by other species of the heath family.' (p. 199)

'It has been usual to separate "heaths" from "heather moors" (German, *Heide* and *Heidemoor*), the former on sandy or gravelly soil with a minimum of peaty humus, the latter on deep acid peat, and most commonly in Britain only at higher altitudes.' (p. 743)

*Comment.* The common heather or ling was formerly included in the genus *Erica* (the heaths) but since its reclassification as *Calluna vulgaris* botanists commonly distinguish between true heaths (*Erica* spp.) and heather.

### Heat-island
In the modern study of urban climates, it is found that the air over any large town has a higher average temperature than that over the surrounding country, constituting a 'heat-island'. See T. J. Chandler, *Geog. Jour.*, **128**, 1962, 279–302.

### Heave
O.E.D. A horizontal displacement or dislocation of a vein or stratum at a fault.
*Comment.* A miner's term not in scientific use (but used by Geikie, J., 1912, 159); see Fault. See also Frost-heave.

### Heaviside Layer
O.E.D. Supp. The name of Oliver Heaviside used attrib. to denote a layer of the atmosphere which reflects back wireless waves. Heaviside (1850–1925) first suggested the existence of the layer now bearing his name, otherwise the lower part of the ionosphere above the stratosphere and about 18 miles above the earth's surface.

### Heavy industry—see Industry

### Heide (German)
Heathland characterized by heather (*Calluna vulgaris*) in northern Germany; used in English mainly in the combination Steppenheide (*q.v.*).

### Helm wind
Mill, *Dict.* A strong wind that rushes down from the summits of hills in the north of England, especially from Cross Fell, and causes a stationary isolated helm-shaped cloud to form over the mountain summit; due to continuous condensation on the windward side and continuous evaporation on the leeward. *Cf.* banner cloud.

### Hemantic Season (Indo-Pakistan: Bengal)
Johnson, B. L. C., Crop-Association Regions in East Pakistan, *Geography*, **43**, 1958. 'In Bengal the agricultural year is divided into three parts: the *rabi* or dry season; the *bhadoi*, or rainy season, which corresponds most closely with the *kharif* season of northern India and West Pakistan; and the *hemantic* season, the principal cropping season of East Pakistan, extending from the latter part of the rains into the cool dry season' (p. 87).

### Hemera—see Zone (geological)

### Hemicryptophyte
Raunkiaer, 1934. A plant which 'has the surviving buds actually in the soil surface, protected by the soil itself and by the dry dead portions of the plant' (p. 17).

### Hemisphere
Half a sphere. It is common practice to divide the earth into the Eastern Hemisphere (the Old World) and Western Hemisphere (the New World) using the meridians of 20°W and 160°E. See also Land Hemisphere, Water Hemisphere

### Hendref, Hendre (Welsh)
*Lit.* old home; winter dwelling, permanent home. Contrast hafod.

### Hercynian
O.E.D. Applied by and after the ancient writers to the wooded mountain-system of Middle Germany, or to portions of it; esp. in more recent times to the Erzgebirge, whence Hercynian gneiss.

Mill, *Dict.* According to Marcel Bertrand, applied to the folds, mainly of Carboniferous date, in North Europe. According to L. von Buch, lines of structure in North Central Europe having a north-westerly direction. They are also called Asiatic fractures or Karpinsky's lines. Suess prefers not to use the term. Some of the north-western lines of structure, at any rate, are of Tertiary age.

Holmes, 1944. 'Geosyncline... across Europe from South-west Ireland to the Sea of Azov... became transformed into orogenic belts during Carboniferous and early Permian times. This orogenesis and the structures and mountains that resulted are known by the name *Hercynian*, after the Harz Mountains' (p. 386).

East, W. G., 1935. *An Historical Geography of Europe*, London: Methuen. 'The Hercynian forest, as Caesar described it, could be crossed in nine days by a fast runner, although its eastern end lay more than a sixty-days' march distant.' (p. 51)

*Comment.* The modern usage is that of Holmes. The *O.E.D.* reference to Erzgebirge should surely be to the Harz whilst Mill's comment is only of historic interest. The Hercynian earth movements were widespread and are alternatively known as Armorican (*q.v.*). They correspond to the younger Appalachian movements in America.

**Heteroclinal**
Swayne, 1956. Heteroclinal fold: a fold one side of which slopes at a steeper angle than the other.

**Heteropic**
Himus, 1954. A term applied to two formations deposited contemporaneously but of different facies. *Cf.* isopic.

**Hide** (Agricultural History)
Butlin, 1961. 'In Domesday Book the majority of hides were reckoned at four yardlands...not an exact area...some equate it with Bede's 'land of one family' which either actually supported one family or for fiscal purposes was deemed capable of doing so.' See also Yardland.

**Hierarchy, urban**—see Urban hierarchy

**Highland, Highlands**
*O.E.D.* 1. High or elevated land; a lofty headland or cliff.
b. The mountainous or elevated part of any country; occas. also in the names of geographical districts. 2. spec. (now always pl.). The mountainous district of Scotland which lies north and west of a line drawn from the Firth of Clyde through Crieff to Blairgowrie and thence north and northwest to Nairn on the Moray Firth; the territory formerly occupied by the Celtic clans.

Mill, *Dict.* Any elevated land. In particular a dissected mountainous region composed of old folded rocks, and possessing a fairly uniform summit level.

*Comment.* In district names 'Highlands' is commoner than 'Highland', *e.g.* the Kenya White Highlands. For small areas 'heights' is sometimes used. Geographically the 'Highlands of Scotland' is the name applied to the whole massif north of the Highland Boundary Faults which run from the Firth of Clyde to Stonehaven. The term 'Highlander' may be applied to any inhabitant of highlands but where unqualified may be taken as referring to those people of Scotland native to the Highlands, especially the area defined in *O.E.D.* and/or belonging to one of the Highland clans.

**'High Rise' estate**
Estate of tall buildings.

**High seas**
Mill, *Dict.* The open sea beyond territorial waters (*q.v.*).
Nearly always in plural.

**High water, High Water Mark**
*O.E.D.* High water. The state of the tide when the surface of the water is highest, the time when the tide is at the full.
High-water mark. a. lit. the mark left by the tide at high water, the line or level then touched; esp. the highest line ever so touched. Also, by extension, the highest line touched by a flooded river or lake.

Mill, *Dict.* In the sea, high tide. In rivers and lakes, the highest level reached by their waters.

*Adm. Gloss.*, 1953. High water. The highest level of the sea in any one tide oscillation.

*Comment.* The level of high water varies through the year, hence the abbreviations H.W.O.S.T. (High Water Ordinary Spring Tides) and H.W.O.N.T. (High Water Ordinary Neap Tides) used on British Ordnance Maps. See also Mean Sea Level, Ordnance Datum.

**Highway**
*O.E.D.* A public road open to all passengers, a high road; esp. a main or principal road forming the direct or ordinary route between one town or city and another, as distinguished from a local, branch, or cross road leading to smaller places off the main road, or connecting two main roads.

*Comment.* The phrase 'The King's Highway' denotes not only a highway open to all but one where, in the past, travellers might enjoy some measure of special protection when the 'King's peace' did not necessarily apply to the whole country. With the advent of fast automobile traffic changes in the law were necessitated in Britain. The highways there have been classified into A-roads (all numbered but including trunk roads maintained centrally, and ordinary A-roads maintained by the counties as highway authorities), second class or B-roads, all numbered, maintained by the counties, unclassified or C-roads (not numbered) and private roads. Superimposed on the old network are the motorways (M1, M2, M3, M4 etc.) with restricted access and restricted users only possible under new legislation. Following on the development of the *Autobahnen* (*q.v.*) in Germany, *autostrada* in Italy, Europe has now a growing international system of E-routes. In North America various terms have been introduced for modern-type highways: a dual highway is one in which a central barrier fence or strip with grass, trees or shrubs separates the opposing streams of traffic (dual carriageway in Britain); an expressway is a public, toll-free, controlled-access highway (*e.g.* in Ontario); a parkway is a controlled-access highway with landscaped margins; a turnpike is a toll-paying controlled-access highway; similar to the last is the thru-way of New York State.

### Hill

*O.E.D.* 1. A natural elevation of the earth's surface rising more or less steeply above the level of the surrounding land. Formerly the general term, including what are now called mountains; after the introduction of the latter word, gradually restricted to heights of less elevations;... A more or less rounded and less rugged outline is also usually connoted by the name. In Great Britain heights under 2,000 feet are generally called hills;... The plural hills is often applied to a region of hills or highland; esp. to the highlands of northern and interior India.

3. A heap or mound of earth, sand, or other material, raised or formed by human or other agency.

Mill, *Dict.* Strictly an elevation of less than 1,000 feet; generally any slightly elevated ground.

Geikie, J., 1898. '... the term hill is properly restricted to more or less abrupt elevations of less than 1,000 feet, all the altitudes exceeding this being mountains.' (p. 271)

*Comment.* Various attempts have been made to define more precisely 'hill' and 'mountain' and to distinguish between them but usage remains loose.

### Hill-and-dale

A literary or poetic phrase (see Dale) which has been applied also in recent years to the alternating ridges and hollows of waste rock or overburden left after the quarrying of ironstone as in the English Midlands (L.D.S.).

### Hill-billy

*Dict. Am.* A southern backwoodsman or mountaineer. 1900. *N. Y. Jrnl.* 23 April 2/5. 'A Hill-Billie is a free and untrammelled white citizen of Alabama, who lives in the hills, has no means to speak of, dresses as he can, talks as he pleases, drinks whiskey when he gets it, and fires off his revolver as the fancy takes him.'

An American slang term which has passed into general literature to denote a dweller in a hill region badly served by communications, who continues to live under primitive conditions closely comparable with those prevailing in the past or pioneer days.

### Hill farming

Farming carried out in hill country, but the term has been given specialized and legal meanings in Britain because of a system of grants to those who qualify as 'hill farmers'.

See especially *Report of the Committee on Hill Sheep Farming in Scotland*, H.M.S.O., Cmd. 6494, 1944; W. Ellison, 1953, *Marginal Land in Britain*, London: Bles.

### Hill-island

A term used to translate the Danish *bakkeöer* which are glacial moraines, mainly sand, rising as mature hills from an outwash plain of a later glacial period. Vary greatly in size (P. R. Barham in MS.).

### Hillock

A small hill.

### Hill-shading

Swayne, 1956. A method of indicating relief on maps by shading normally the slopes facing south and east, *i.e.* it is assumed that the area mapped is illuminated from the north-west. The method is often used in association with contours.

### Hill-station, hill station

A mountain-resort in the tropics where elevation results in a lowering of temperature. The term is associated especially with the period of British administration in India when hill-stations were established on the healthy heights of the Himalayas at 5,000 feet and over (Mussoorie, Naini Tal, Simla, Darjeeling) to which wives and families of the white administrators repaired to escape the heat of the plains and to which in some cases the Government itself moved for a period. In this sense the hill-station of Simla was the hot-weather capital of India. In early days reference was always to going to 'the hills', hence 'the hills' as referring to hill stations or resorts in the tropics; and 'hill-stations' were established in other tropical countries such as Maymyo in Burma, Garoet in Java, and many others. See Spencer, J. E. and Thomas, W. L., 1948, The Hill Stations and Summer Resorts of the Orient, *Geog. Rev.*, 38, 637–651.

## Hill village

Sylvester, D., 1947, The Hill Villages of England and Wales, *Geog. Jour.*, 110. '... the determining criterion has been that the ground should slope downwards from the site through a total angle of more than 180 degrees' (p. 76). Relative elevation is required and thus both high site-types, *e.g.* hill-top, and lowland-elevation types, *e.g.* marsh island sites, are included. Village is widely defined to include hamlets and small market towns and includes non-agricultural settlements (W.M.).

## Hinderland (G. G. Chisholm, 1889)

Chisholm in his *Handbook of Commercial Geography* (London: Longmans, First Edition, 1889) used this translation of the German *Hinterland* with reference to ports and Chisholm himself continued to use it (see Ninth Edition, 1922). It did not however gain acceptance and was dropped when others revised the *Handbook*—see footnote p. 97 in the Twelfth Edition, 1932.

## Hinge-line

*Webster. Geol.* Any imaginary line on the earth's surface which can be regarded as a boundary between a stable region and one undergoing upward or downward movement.

Dury, 1959. '... covered with hinged plates ... land-areas which were pushed down by the weight of ice-caps, and which are now rising, are moving precisely as if they were so hinged—indeed the term *hinge-line* is applied to those lines where the rate of uplift changes sharply.' (p. 122)

## Hinterland (German)

*O.E.D.* The district behind that lying along the coast (or along the shore of a river); the 'back country'.

Sargent, A. J., 1938, *Seaports and Hinterlands*. London: Black. 'A port, essentially, is a transit area, a gateway through which goods and people move from and to the sea, by way of rail, inland waterway, or sometimes by road. The region to and from which this movement is directed is commonly and somewhat vaguely described as the hinterland.' (p. 3) 'By limiting the hinterland to an area of which the greater part or a substantial part of the trade passes through a single port we approach nearer to reality.' (p. 15)

Mill, *Dict.* Hinderland, Hinterland. 1. The region the seaborne trade of which belongs to a particular seaport or seaboard. 2. Country lying inland from the district bordering a coast, or more generally, lying beyond a given district.

Morgan, F. W., 1949, *Trans. Inst. Brit. Geog.* 'What is the problem of seaport hinterlands? It is the problem of delimiting in terms as far as possible measurable, the distribution and origins of the commodities which pass in and out of a port.' (p. 46)

Green, F. H. W., 1950, Urban Hinterlands ... *Geog. Jour.*, 116. '... urban spheres of influence or hinterlands, ...' (p. 64).

Van Cleef, E., 1937, *Trade Centers and Trade Routes*, New York: Appleton-Century. Hinterland is used for inland cities (*e.g.* Chicago) but not defined. It is the 'region about' a centre. Note: 'A trade center plus its continuous hinterland may be referred to as an urban region.' (pp. 30–31) Distinguishes between a continuous and a discontinuous hinterland and a division known as the umland. 'The immediately contiguous territory within the continuous hinterland which in some instances contributes to the formation of the metropolitan city has been termed by the Germans the "Umland" or country about.' (p. 34)

Van Cleef, E., 1941, *Geog. Rev.*, 31. 'Hinterland (Continuous). The area adjacent to a trade center (extending to and including its satellites) within which economic and some cultural activities are focussed largely on the primary center. ... Umland: The area contiguous to a trade center (extending to and including its suburbs or "urblets") whose total economic and cultural activities are essentially one with those of the primary center. ... In both definitions "trade center" is used as synonymous with "city", "town", "village", or any other settlement term given a specific political connotation. It is used in a geographical sense to serve an all-inclusive purpose. (p. 308) Van Cleef points out that A. Allix coined the word in 1914 for the 'hinterland' of centre other than a port (Also *Geog. Rev.* 1922). He suggests that in

the sense of environs the word has an historical German use dating to 1437.
Taylor, 1951. 'Regions behind the coast.' 'Umland: area affected by a city's trade.' (pp. 615, 621)
Comment. The foregoing quotations illustrate the remarkable evolution in meaning of the German word 'hinterland' since it was introduced into English. From the early vague meaning of back-country it became essentially associated with ports to be later extended to spheres of influence of inland towns, the spheres defined in various ways. The word may be regarded as naturalized in English but is now being used in differing senses and is partly giving way to umland. See Umland, urban hinterland, hinderland.

**Hinterland** (tectonic)
Wooldridge and Morgan, 1937, pp. 76–78. Africa is regarded as a 'hinterland' moving northwards to the European 'foreland', during the Alpine orogenesis. (W.M.)
Comment. Not found in the many works in geology and geography investigated, but used in the above sense by Argand, *La Tectonique de l'Asie* (1922).

**Hirsel** (Scottish)
Robertson, J. C., 1964, Sheep and Cattle Systems of Hill Land Use, *Adv. of Science*, 21. 'These "hirsels" are natural divisions of land, say a glen with the hills rising above it, and they depend on a sheep's "hefting" instinct whereby it tends always to remain on or near the ground on which it was reared'. (p. 174)

**Hirst**—see Hurst

**Histogram**
*O.E.D. Suppl.* A form of graph employed in statistics.
*Webster. Statistics.* A graphical representation of a frequency distribution by means of rectangles (*frequency rectangles*) whose widths are the class intervals and whose heights are the corresponding frequencies.
Christodoulou, D., 1959, *The Evolution of the Land Use Pattern in Cyprus*, London: World Land Use Survey. 'Frequency distribution histograms for Nicosia give an idea of the incidence of rainfall and the amounts involved.' (p. 21)

**Historical geography**
Committee, List I, The geography (physical and human) of historical periods.
Swayne, 1956 adds 'In the study of international relations the term is restricted in meaning to political geography.'
Hartshorne, 1939. 'Historical geography is not a branch of geography comparable to economic or political geography. Neither is it the geography of history or the history of geography. It is rather another geography, complete in itself, with all its branches.' (pp. 184–185)
Comment. The term has been used in a variety of senses. One usage has been concerned with the influence of geographical conditions upon historical events; thus the object of George Adam Smith's *The historical geography of the Holy Land* (London, 1st ed. 1894) was 'to discover from "the lie of the land" why the history took certain lines'. Another usage has been concerned with changes in political boundaries and with the extent of administrative divisions at different times. Such was E. A. Freeman's *The historical geography of Europe* (London, 1st ed. 1881), and L. Mirot's *Manuel de géographie historique de la France* (Paris, 1st ed. 1929). A third usage has been devoted to the study of geographical changes through time, and a variety of papers along these lines have been entitled 'the historical geography' of a town or of an industry. The term has also been used to denote the history of geographical exploration and geographical thought; this is the usage encountered, for example, in some bibliographies.
Another meaning of the term which has gained general acceptance among geographers is that it implies the reconstruction of the geography of a past period. Phrases used in this connection are 'retrospective geography' and 'cross-sections' of the geography of the past. In this sense we can speak of the geography of France in 1500 or of Sicily in 1250. It follows from this, as Alfred Hettner wrote, that 'an historical geography of any region is theoretically possible for every period of its history, and is to be written separately for each period; there is not one but many historical geographies' (*Die Geographie, ihre Geschichte, ihr Wesen und ihre Methoden*, p. 151, Breslau, 1927). Examples of this approach are provided by A. H. Clark, *Three centuries and the island* (Toronto, 1959), W. G. East, *An Historical Geography of Europe* (London, 1935) and

H. C. Darby (ed.), *An historical geography of England before A.D. 1800.* (Cambridge, 1st ed. 1936). The method has been carried to its logical conclusion in Ralph H. Brown's *Mirror for Americans: Likeness of the Eastern Seaboard, 1810* (American Geographical Society, 1943); this was written from the viewpoint of an imaginary author of 1810, with the understanding and sources that were available at the time. Although there is wide agreement that 'historical geography' is logically concerned only with the geographies of past periods, one must recognize that such a view reflects but one aspect of the field of enquiry that lies in the borderland between geography and history. For a discussion of the variety of viewpoints see: Clark, A. H., 1954, 'Historical Geography' in James and Jones *American Geography: Inventory and Prospect*, Syracuse (pp. 70–105); Darby, H. C., 1953, On the relations of geography and history, *Trans. Inst. Brit. Geog.*, **19**, 1–11; Dion, R., 1949, La géographie humaine retrospective, *Cahiers Internat. de Sociologie*, **6**, 3–27, Paris; Hartshorne, R., 1939, 175–188; Gilbert, E. W., 1932, What is historical geography? *Scot. Geog. Mag.*, **48**, 132; Wooldridge, S. W. and East, W. G., 1951, *The Spirit and Purpose of Geography*, London; What is historical geography, *Geography*, **17**, 1932, 39–45. (H.C.D.) See also Sequent occupance.

## Historicism
*Comment.* Not in *Webster, O.E.D.* For this philosophical approach to history and geography see *E.B.*, 1946, **20**, 59a.

## History of geography
Committee, List I. The history of geographical knowledge and ideas.
*Comment.* Hartshorne, 1939, observes 'British geographers have occasionally used the term "historical geography" to refer to the history of geography as a science.' (p. 184 *fn.*) See Historical geography.

## Hithe, hythe
*O.E.D.* A port or haven; especially a small haven or landing place on a river. Now obsolete except in historical use, and in place-names.

## Hoarfrost
*Webster.* A silvery-white deposit of ice needles formed by direct condensation at temperatures below freezing, due to nocturnal radiation. Hoarfrost forms during still, clear nights, is small in amount, needlelike in texture, the "needles" approximately perpendicular to the objects on which they occur, and most abundant along their edges. It is confused by some with *rime*.

*Ice Gloss.*, 1958. A deposit of *ice* having a crystalline appearance, generally assuming the form of scales, needles, feathers or fans; produced in a manner similar to dew (*i.e.* by condensation of water vapour from the air), but at a temperature below 0°C. (32°F.).

## Hoe
*O.E.D.* 'A projecting ridge of land, a promontary' (Sweet); 'originally a point of land, formed like a heel and stretching into the sea' (Kemble); a height ending abruptly, or steeply; *cf.* Heugh. Now only in the names of particular places....
*Comment.* Obsolete except in place names, *e.g.* Plymouth Hoe, Ivinghoe.

## Hogback, Hog-back, Hog's back
*O.E.D.* Hogback, hog-back, also hog's back. 2. Something shaped like a hog's back.
a. A sharply crested hill-ridge, steep on each side and sloping gradually at each end; a steep ridge of upheaval.
Salisbury, 1907. 'An elongate narrow ridge, due to the isolation of a tilted layer of resistant rock, is sometimes called a "hogback".'
Fay, 1920. 1. (Eng.) A sharp rise in the floor of a coal seam (Gresley)
2. A ridge formed by the outcropping edge of tilted strata; hence any ridge with a sharp summit and steeply sloping sides, as an esker (*Webster*). Called also Horseback.
Wooldridge and Morgan, 1937. 'With a high dip the inclination of the dip-slope may become equal to, or greater than, that of the scarp-slope, and in this way more or less symmetrical "hog-backs" arise.' (p. 202)
Rice, 1941. Hog-backs. Sharp ridges caused by unequal erosion on alternating hard and soft layers of steeply inclined rock. Sometimes applied to ridges formed by hard dikes. A hog-back differs from a cuesta in having the rocks more nearly vertical and the two slopes of the ridge more nearly equal.

## Holarctic
A zoogeographical region comprising the Palaearctic and the Nearctic together,

*i.e.* the whole of the cold and temperate lands of the Northern Hemisphere.

**Holding (agricultural)**
*O.E.D.* 1.b. spec. The tenure or occupation of land.
3. Land held by legal right, esp. of a superior; a tenement.

Min. of Ag. and Fisheries, National Farm Survey of England and Wales (1941-1943). A Summary Report, London: H.M.S.O., 1946. 'For the purposes of the Farm Survey it ("an agricultural holding") comprises any area of 5 acres and above used for the growing of crops (including grass), which is being farmed separately, that is to say, as a self-contained unit.' 'For the Farm Survey, holdings which were under the same occupancy and day-to-day management and had a common source of labour, machinery and other permanent equipment were regarded as forming a single unit....' (p. 6) This is similar to that used in the collection of the annual agricultural statistics.

Stroud, 1952. 'An agricultural holding is defined by s. 1 (1) of the Agricultural Holdings Act, 1948 (11 and 12 Geo. 6, c. 63), as being the aggregate of the agricultural land comprised in a contract of tenancy, not being a contract under which the said land is let to the tenant during his continuance in any office, appointment or employment held under the landlord.'

*Comment.* Normally an agricultural holding is the same as a farm but in British practice an area of a few acres would not be dignified by the title of 'farm'; it might however be a 'small-holding' owned, tenanted or cultivated as a unit. On the other hand several farms may be combined and worked as one 'holding'. See Farm.

**Holism** (J. C. Smuts)
*O.E.D. Suppl.* A term coined by General J. C. Smuts to designate the tendency in nature to produce wholes (*i.e.* bodies or organisms) from the ordered grouping of unit structures.

*Comment.* Several writers have referred to the 'holistic' view of geography.

**Holisopic**—see Isopic

**Holm**
*O.E.D.* 2. A small island, an islet; esp. in a river, estuary or lake, or near the mainland (Frequent in place names as *Steep Holm(e)* in the Severn ... but, as a living word, applied only to the small grassy islets in Orkney and Shetland....)
3. A piece of flat low-lying ground by a river or stream, submerged or surrounded in time of flood. In living use in the south of Scotland (howm) and north of England, and extending far south in place names; 'a flat pasture in Romney Marsh (Kent) is yet called the *Holmes*' (Way).

Mill, *Dict.* 1. An island. 2. Low flat land, more or less surrounded by water, but affording good landing-places for small boats (N. England). 3. A sand-dune (Germany).

**Holokarst**
Lit. whole or entire Karst.
This little-used term was introduced by Cjivić in 1918 for 'le Karst complet dans lequel toutes les formes karstiques sont parfaitement developées ainsi que l'hydrographie souterraine.' See also Cotton, 1941, p. 448.

**Home range** (Scottish)
Hunter, R. F., 1964, Social behaviour in hill sheep, *Adv. of Science*, **21**. 'The basic unit within a flock of hill-sheep is the family (dam/daughter) unit, but family groups, in association with other unrelated sheep, form higher groups, which have been called home range groups ... at dusk they draw together and settle down for the night in the same lairs or night resting areas.' (p. 170)

**Homestead**
*O.E.D.* (O.E. ham, home + stede, place)
1. *gen.* The place of one's dwelling or home: a. The place (town, village, etc.) in which one's dwelling is. *Obs.* b. A home or dwelling.
2. A house with its dependent buildings and offices; esp. a farm-stead.
3. *U.S.* A lot of land adequate for the residence and maintainance of a family; 'a farm occupied by the owner and his family'; esp. the lot of 160 acres granted to a settler by the Homestead Act of Congress, 1862.
Also gives the U.S. verb. Homestead: to take up and occupy as a homestead.
*O.E.D. Suppl.* applies also to similar settlement in Canada. Hence homesteading, the act of taking up or occupying a homestead.
*Webster.* 2. The home place; a home and the enclosure or ground immediately connected with it. 3. The residence of a squatter situated on his run or station.

*Australia.* 4. *Law.* The land and buildings thereon occupied by the owner as a home for himself and his family, if any, and more or less protected by law from the claims of creditors. *Homestead law.* b. U.S. Any of several acts of Congress, esp. the first (Homestead Act), passed in 1862, authorizing the sale of public lands, in parcels of 160 acres each (*homesteads*), to settlers.

*Dict. Am.* 1. A place where a family makes its home, including a plot of land, the house, and other buildings. 1683. 2. A tract of land taken up from the public domain by a settler under the homestead laws; the developed farmstead established on such land. Ordinarily 160 acres each, these tracts were as large as 640 acres in mountainous or semiarid regions.

*Comment.* Used by many writers in human geography as a convenient term to denote a rural settlement type of dispersed habitations or farms in contradistinction to a village or hamlet. (W.G.E.)

**Homoclimes** (T. Griffith Taylor, 1951)

Taylor, 1951. Places having similar climates (p. 615).

Miller, 1953. 'Stations having similar climographs are described as "homoclimes". Alice Springs is the homoclime of Biskra (Algeria), Perth of Cape Town etc.' (p. 15) See Climograph.

**Homoclinal ridge**

Cotton, 1941. 'Moderately inclined strata now outcrop as *homoclinal* ridges bounded by *escarpments* and *dip slopes*... homoclinal ridges on steeply dipping strata grade into *hogbacks*... into *cuestas* on strata of very gentle inclination.' (pp. 91–94)

**Homoclinal shifting**

Cotton, 1922. The migration of a homoclinal ridge in the direction of the dip due to differential rates of erosion between the dip and the obsequent slope (p. 97). This is called monoclinal shifting in Gilbert, 1877, p. 140 (W.M.).

**Homocline** (R. A. Daly, 1916), **homoclinal strata**

*Webster. Geol.* A structure, as one limb of an anticline or syncline, part of a monocline, or a tilted fault block, in which the strata dip uniformly in one direction.

Not in *O.E.D.*, Mill, Geikie, 1898, Holmes, 1944, Himus, 1954, Monkhouse, 1954.

Cotton, 1941, *Landscape*, C.U.P. 'The term "homocline" was introduced by R. A. Daly (Homocline and Monocline, *Bull. Geol. Soc. Amer.*, **27**, 1916, 89–92) to replace the ambiguous "monocline" where it is used to signify a succession of strata dipping continuously in one direction. The synonym "unicline", being a hybrid, is less acceptable' (p. 80 *fn.*).

*Comment.* Monocline as used by British authors means a fold with one limb, but it has long been used by American writers for evenly dipping strata (see quotation from Gilbert, 1877 under homoclinal shifting). Monocline is still used in this sense by von Engeln, 1942, but in both senses by Lobeck, 1939.

**Homologue**

That which is the same or corresponding in type to an original, *e.g.* a climatic homologue or homoclime.

**Homotaxis** (T. H. Huxley, 1862)

*Webster.* Similarity in arrangement; esp. *Geol.*, similarity in fossils and in order of arrangement of stratified deposits which are not necessarily contemporaneous.

Hawkes, L., 1957, Progress in Geology, *Quart. Jour. Geol. Soc.*, **103**, 309–312. 'Signifying a similarity of position in a sequence which did not imply similarity of age. The name did not catch on... but ... progress... has emphasized the importance of Huxley's thesis.' (p. 313) See Huxley, T. H., *Q.J.G.S.*, **18**, xl-liv.

**Hoodoo** (American)

*O.E.D. Suppl.* 'A fantastic rock pinnacle or column of rock, formed by erosion... occurring in the Western United States.'

Cotton, 1941. 'The picturesque minor surface-relief forms termed *hoodoo columns* and *earth pillars* are slender residual columns of unconsolidated sediment... capped by slabs or boulders of resistant material' (p. 15).

*Comment.* Used in popular rather than scientific literature since 1879 presumably from the fancied resemblance of some earth pillars to embodied evil spirits. There are well known groups in the Banff and Jasper National Parks of Canada, also called 'demoiselles' (L.D.S.).

**Hook**

*O.E.D.* 9. A sharp bend or angle in the course or length of anything; esp. a bend in a river (now in proper names).

11. A projecting corner, point, or spit of land (app. from Dutch, hoek, as in Hoek van Holland).

Mill, *Dict.* A recurved spit resembling a fish-hook.
See Spit.
## Hope
*O.E.D.* A small enclosed valley, especially 'a smaller opening branching out from the main dale and running up to the mountain ranges; the upland part of a mountain valley'; a blind valley. Chiefly in the south of Scotland and the north-east of England where it enters largely into local nomenclature.
## Horizon, geographical
*O.E.D.* 1. The boundary-line of that part of the earth's surface visible from a given point of view; the line at which the earth and sky appear to meet. In strict use, the circle bounding that part of the earth's surface which would be visible if no irregularities or obstructions were present (called the *apparent, natural, sensible, physical,* or *visible horizon*), being the circle of contact with the earth's surface of a cone whose vertex is at the observer's eye. On the open sea or a great plain these coincide.
## Horizon, geological
*O.E.D.* A plane or level of stratification assumed to have been once horizontal and continuous; a stratum or set of strata characterized by a particular fossil or group of fossils.
Rice, 1941. A particular stratigraphic position, as the horizon of a certain fossil, meaning the particular place in the stratigraphic column where that fossil occurs, even though in widely separated localities (Cox-Dake-Muilenburg).
## Horizon, Soil
U.S. Dept. of Agriculture, *Soils and Men, Yearbook of Agriculture,* 1938. 'Horizon, Soil. A layer of soil approximately parallel to the land surface with more or less well defined characteristics that have been produced through the operation of soil-building processes' (p. 1169).
## Horizontal equivalent
Monkhouse and Wilkinson, 1952. 'If two points on a hill-side are projected on to a horizontal plane, as they are on a map, the distance between them is known as the Horizontal Equivalent (H.E.).' (p. 74) See Vertical interval.
## Horn
A pyramidal peak; see Matterhorn, also col.
## Horse latitudes
*O.E.D.* Origin of the name uncertain. The belt of calms and light airs which borders the northern edge of the N.E. trade-winds. Mill, *Dict.* Areas of high barometric pressure and calms.
*Met. Gloss.*, 1944. The belts of calms, light winds and fine, clear weather between the Trade Wind belts and the prevailing westerly winds of higher latitudes. The belts move north and south after the sun in a similar way to the Doldrums.
## Horst (German)
*cf.* Block Mountain.
*O.E.D. Suppl.* Geol. (German = heap, mass, cluster, sandbank, etc.) A term introduced by Suess to denote tracts of the earth's surface which have resisted lateral thrust and against which surrounding areas have been pressed and dislocated; also, an area that has become raised above surrounding areas which have been depressed by faulting.
1893. Geikie, *Text-bk. Geol.* (ed. 3), 1071. Suess has pointed out various areas of the earth's surface, named by him 'Horsts', which seem to have served this purpose (as buttresses) in the general rupture and subsidence of the terrestrial crust.
Mill, *Dict.* Crust-block isolated and left outstanding by subsidence of surrounding parts (Suess). Usually a large area, a small mass being often described as a block-mountain. 'Horst' is an old German word for a wooded hill or rise or a heap of earth, *e.g.* a dry spot in a wet moor: English equivalents of O. Ger. Haust are hurst or hirst = a wood or grove.
Suess, E., 1904, *The Face of the Earth,* Vol. 1 (Trans. by H. B. C. Sollas), Oxford: Clarendon. '... the sheets of Mesozoic deposits have sunk down, and out of this general subsidence these mountains rise as horsts, ...' (pp. 202–203)
Geikie, J., 1898. 'Horste: name given by German geologists to isolated mountains severed by dislocations from rock-masses with which they were formerly continuous, but which have since subsided to a lower level. Rumpfgebirge (lit. rump-mountain) is another name for this type of mountain.' (p. 303)
Cotton, 1941. 'Such blocks may sink or rise uniformly (movement relative to adjacent blocks only being taken into account), becoming in the one case a trough, or Graben, or in the other an uplifted block, or Horst, bounded on two long sides by fault scarps.' (pp. 240–241)
Lobeck, 1939. 'If a block is raised between

two faults, such an elevated mass is called a horst.' (p. 61)

*Comment.* There is confusion in common usage between the structural and relief forms. A structural horst or faulted inlier (*q.v.*) may have been denuded to the level of the surrounding country. It is also not clear whether a block between two faults is properly a horst or whether there must be faults on all sides. Mill's suggestion that a horst is large, a block-mountain small is not in accord with current usage. (G.T.W., L.D.S.)

### Horticulture

*O.E.D.* Adaption of Latin type, *horticultura*, cultivation of a garden. The cultivation of a garden; the art or science of cultivating or managing gardens, including the growing of flowers, fruits and vegetables.

The Committee, List 3. The cultivation of vegetables, fruit or flower crops, often on very small plots, with a higher intensiveness than in field cultivation; sometimes used of nursery gardening.

*Comment.* The meaning of horticulture has moved away from the cultivation of gardens to cover intensive agriculture in general. The University of London grants a degree of B.Sc. (Horticulture) covering both the 'Principles' and 'Practice' of Horticulture. The latter covers both gardens and market gardens and also glasshouse cultivation.

### Hosier bails

A portable milking apparatus that became widely used in Britain during the Second World War which obviates the cows being brought back to the farm for milking.

### House-bote—see Estover

### Howe (Eastern Scottish)

*O.E.D.* A hollow place or depression; *esp.* a hollow on the surface of the earth, a basin or valley.

Miller, Ronald, *The Glasgow Region*, 1958, 1. 'Glasgow ... lies on the floor of what would be called in east Scotland a Howe— a hollow place—ringed around by bleak plateaus.'

*Comment.* Best known examples are the Howe o' the Mearns, the Howe of Fife, and the Howe of the Merse. (C.J.R.)

### Hoyt's sectoral model—see Sector model.

### Hpoongyi, hpongyi (Burmese)

A Buddhist monk; hpoongyi-kyaung = monastery commonly found in every Burmese village of any size and providing the village school and a shelter for travellers. (L.D.S.)

### -hsien (Chinese)

As suffix to place name means county town

### Huasipungo

From Quechua words meaning "housedoor", word peculiar to Highland Ecuador. The plot of land which a landless agricultural labourer is allowed to cultivate in return for three to five days' work for his master's estate.

Linke, L., 1960, *Ecuador: Country of Contrasts.* 3rd ed. revised. London: O.U.P. 'The landless Indian ... is attached to a hacienda and in exchange for his work is given the usufruct of a small plot called *huasipungo* (pp. 60–61).

### Huerta (Spanish)

From Latin *hortus*, a garden.

Shackleton, M. R., 1958, *Europe*, 6th ed. London: Longmans. 'In the north from the Ebro delta to Cape de la Nao in Valencia, the irrigated districts known as *vegas* and *huertas* are practically continuous along the coast' (p. 102).

Fisher, W. B. and Bowen-Jones, H., 1958, *Spain*, London: Christophers. 'Originally developed by the Moslems, the "huertas" are small, highly cultivated plots which depend on irrigation water brought by an intricate system of channels, aqueducts and lifts. The Arab name for this kind of land utilisation is a *vega*. *Vega* is still used in Andalusia and parts of Valencia' (p. 52).

Chisholm, G. G., 1956, *Handbook of Commercial Geography*, 15th ed., London: Longmans. 'Irrigation schemes dating back to the Moors, or even to the Romans, have made famous the *huertas* (horticultural lands) of Valencia and Murcia and the *vega* (agricultural plain) of Granada. A distinction is usually made between the *vegas* which yield only one crop a year, and the *huertas* which yield two or more' (p. 471).

*Comment.* In other parts of the world *vega* seems to have been used with quite a different meaning: an extensive fertile and grass covered plain (*O.E.D.*) and in Cuba for a tobacco field. *Huerta* seems to be used mainly in Spain. See Corominas, J., 1954, *Diccionario Critico Etimologico de la Lengua Castellana*, Bern: Francke.

### Hum (Southern Slav.)

Thornbury, 1944. 'In an advanced stage of solutional destruction of a limestone upland, residual hills analogous to monadnocks in the fluvial cycle often remain. In the karst area of Jugoslavia, where such hills are rather common, they are known as *hums*. Similar features are known as *pepino hills* or *haystack hills* in Puerto Rico, as *mogotes* in Cuba, and as *buttes temoines* in the Causse region of France.' (p. 334)

von Engeln, 1942. 'Residual conical summits, resembling haystacks.' (p. 569—same equivalents)

Commission on Karst Phenomena, I.G.U., 1956. Hums = Karstinselberge.

## Human geography

Committee, List 1. The geographical study (the complement of physical geography) of those features, objects and phenomena of the Earth's surface which relate directly, or are due, to man and his activities.

Lebon, J. H. G., 1952, *An Introduction to Human Geography*, London: Hutchinson. 'Human geography... seeks to elucidate ... the general problem of man's relationship with his environment... this distinction is not one that is generally accepted among geographers....' (p. ix)

Huntington, E. and Cushing, S. W., 1921, *Principles of Human Geography*, New York: Wiley. '... Human Geography may be defined as the study of the relation of geographical environment to human activities.' (p. 1)

Newbigin, M. I., 1922, Human Geography: First Principles and Some Applications, *Scot. Geog. Mag.*, **38**. '... human geography is the biology of man....' (p. 221)

Roxby, P. M., 1930, The Scope and Aims of Human Geography, *Scot. Geog. Mag.*, **46** 'I believe that in essence human geography consists of the study of (a) the adjustment of human groups to their physical environment, including the analysis of their regional experience, and of (b) interregional relations as conditioned by the several adjustments and geographical orientation of the groups living within the respective regions.' (pp. 282–283)

See also the discussion in Hartshorne, 1939, who, writing of the 18th century, observes 'very few writers of this period made the distinction which is most familiar to us, between a geography of natural, or non-human features, and a human geography...' (p. 43).

*Comment.* 'Human Geography' seems first to have been used by H. J. Mackinder in his Presidential Address to Section E of the British Association at Ipswich in September 1895 (*Geog. Jour.*, **6**, 1895) 'The facts of human geography... are the resultant for the moment of the conflict of two elements, the dynamic and the genetic' (p. 375). He does not use it in *The Scope and Methods of Geography* (1887) and H. R. Mill is still using 'Anthropo-geography' (*q.v.*) in his *International Geography* (1899). A. J. Herbertson uses the term in the preface to the first edition of *Man and his Work* (1899) (Mary Marshall in MS.).

## Humic acid

Jacks, 1954. Organic matter extractable from soil by alkali and precipitated by acid.

## Humic coal—see Coal

## Humid Tropicality (B. J. Garnier, 1960)

The climatic condition relative to a standard period of time (*e.g.* a month) in which relative humidity exceeds 65 per cent, pressure 20 mb, mean temperature exceeds 68°F (20°C). To this may be added rainfall exceeding or equalling evaporation for the period in question, approximating to 3 inches or 75 mm. per month. (Definition communicated to the meeting of the Humid Tropics Commission of the International Geographical Union, Stockholm, 1960.) The 'Humid Tropics' must have conditions of humid tropicality for a minimum of nine months of the year.

## Humidity, absolute

*Met. Gloss.*, 1944. ... is the mass of aqueous vapour per unit volume of air, and is usually expressed in grammes per cubic metre of air. The term is sometimes incorrectly applied to the pressure of the aqueous vapour in the air.

## Humidity, relative

*Webster.* The ratio of the quantity of vapor actually present to the greatest amount possible at the given temperature... saturation is expressed by *humidity* 100.

*Comment.* A RH of 60 may be considered as approximately separating 'dry' from 'moist' atmospheres.

## Humification (Soil science)

Jacks, 1954. 'Transformation of organic matter into humus.'

## Hummocked ice, hummocking—see Ice, sundry terms

## Humus

*O.E.D.* (L. = mould, ground, soil). Vegetable mould; the dark-brown or black substance

resulting from the slow decomposition and oxidation of organic matter on or near the surface of the earth, which, with the products of the decomposition of various rocks, forms the soil in which plants grow.
Mill, *Dict.* Similar to *O.E.D.*
Robinson, G. W., *Soils.* 3rd ed., London: Murby, 1949. 'Whilst many writers use humus as synonymous with organic matter, others have applied the term to a definite fraction, for example, the so-called *matière noire* of Grandeau, soluble in 4% ammonia.... In the opinion of the present writer the name humus might be conveniently retained for the organic matter of the soil that has been decomposed and has lost its original structure.' (p. 31)
Jacks, 1954. The amorphous (colloidal) organic matter of soil (U.K.). The organic complex of the soil which is more or less resistant to microbial decomposition (U.S.)
See also Mor, mull, humification.
*Comment.* Note also derivatives, especially 'humic acid'.

### Hundred

*O.E.D.* 5. In England (and subsq. in Ireland): A sub-division of a county or shire, having its own court; also formerly applied to the court itself.
Most of the English counties were divided into hundreds; but in some counties wapentakes, and in others wards, appear as divisions of a similar kind. The origin of the division into hundreds, which appears already in O.E. times is exceedingly obscure, and very diverse opinions have been given as to its origin.
b. A division of a county in the British American colonies or provinces of Virginia, Maryland, Delaware, and Pennsylvania, which still exists in the State of Delaware.
Mill, *Dict.* As *O.E.D.*
*Comment.* An oft-repeated assertion which may or may not be well founded is that the hundred was originally the area inhabited by a hundred families, or equivalent to 100 hides. See Stenton, F. M., 1946, *Anglo-Saxon England*, 2nd ed.

### Huronian

From Lake Huron; a pre-Cambrian orogenesis responsible for part of the folding seen in the rocks of the Canadian shield; also applied to one of the main rock systems of the Shield; a division of pre-Cambrian time.

### Hurricane

*O.E.D.* (Spanish, *huracan*, from the Carib word given by Oviedo as *huracan*....)
1. A name given primarily to the violent wind storms of the West Indies, which are cyclones of diameter of from 50 to 1,000 miles, wherein the air moves with a velocity of from 80 to 130 miles an hour round a central calm space, which with the whole system advances in a straight or curved track; hence, any storm or tempest in which the wind blows with terrific force.
*Met. Gloss.*, 1944. (As *O.E.D.*).... In the Beaufort scale of wind force the name hurricane is given to a wind force 12, and its velocity equivalent is put at a mean velocity exceeding 34 m./sec. or 75 mi./hr....
Miller, 1953. 'During the season of calms are apt to occur the violent storms known as hurricanes, typhoons or tropical cyclones, a feature of the trade-wind belt ... irregular and infrequent... resulting in considerable loss of property and life' (p. 120). 'The West Indian hurricane and the Chinese typhoon are fairly frequent visitors to the extra-tropical coasts of the United States and China and Japan.' (p. 181)

### Hurst, hirst

*O.E.D.* An eminence, hillock, knoll, or bank, especially one of a sandy nature.
A grove of trees; a copse; a wood; a wooded eminence.
*Comment.* Very common in place names but almost obsolete as a term.

### Husbandry, husbandman

*O.E.D.* 2. The business or occupation of a husbandman or farmer; tillage or cultivation of the soil (including also the rearing of live stock and poultry, and sometimes extended to that of bees, silkworms, etc.); agriculture, farming.
*Comment.* Though husbandman is archaic and almost obsolete, husbandry survives in 'animal husbandry' and in such phrases as 'according to the rules of good husbandry', *i.e.* good farming practice. When the word is used at the present time the tendency is to use it for pastoral farming (*i.e.* animal husbandry) as distinct from crop farming or general farming. The *Agriculture Act, 1947*, makes great use of the term, without precisely defining what is meant. Part II of the Act states '... this Part of this Act shall have effect for the

purpose of securing that ... occupiers of agricultural land fulfil their responsibilities to farm the land in accordance with the rules of good husbandry' (p. 6). See also R. Trow-Smith, 1951, *English Husbandry*, London: Faber.

**H.W.O.S.T., H.W.O.N.T.**—see High water

**Hydraulic**

*O.E.D.* Pertaining or relating to water as conveyed through pipes or channels, especially by mechanical means.

Applied to various mechanical contrivances operated by water-power, or in which water is conveyed through pipes.

Wittfogel, Karl A., 1956, *The Hydraulic Civilizations* in *Man's Role in Changing the Face of the Earth*, University of Chicago Press; *Idem.*, 1957, *Oriental Despotism*, Yale University Press. In his studies of the development of large scale irrigation and human civilizations based thereon, Wittfogel has introduced a whole new terminology including such expressions as hydraulic agriculture, hydraulic revolution, hydraulic civilization together with words like agrohydraulic, hydroagriculture and many others. His basic ideas may be illustrated from two quotations. 'There are at least two major types of rural-urban agrarian civilizations—hydraulic and non-hydraulic.' (1956, p. 152) 'The emergence of big productive water-works (for irrigation) was frequently accompanied by the emergence of big productive water-works (for flood control). I suggest this type of agrarian economy be called "hydraulic agriculture" to distinguish it from rainfall farming and hydroagriculture.' (p. 153)

**Hydraulic limestone**

A limestone containing some silica and alumina and yielding a quicklime or cement (hydraulic cement) that will set or harden under water.

**Hydraulic radius**

Rubey, W. H., 1938, The force required to move particles on a stream bed. *U.S.G.S. Prof. Paper 189E*, pp. 121–141. 'The hydraulic radius of a stream is defined as the area of cross-section divided by the length of the wetted perimeter. In a wide shallow stream in which the height of the channel side-walls is negligible in comparison with channel width, the hydraulic radius is essentially equal to the mean depth.' (p. 124)

Mill, *Dict.*, gives *hydraulic mean depth* or *hydraulic radius* as derived from Penck.

**Hydrocratic**—see Geocratic (2)

**Hydroelectricity, hydro-electricity**

*O.E.D. Suppl.* Electricity produced by the application of the motive power of water.

*Comment.* This meaning of the word has completely replaced the older meaning which included 'the electricity of the galvanic battery' (*O.E.D.*). Used in contradistinction to carbo-electricity—electricity produced by the utilization of coal and, presumably, other carbonaceous materials.

**Hydrogeology**

*O.E.D.* That part of geology which treats of the relations of water on or below the surface of the earth.

**Hydrographer**

*O.E.D.* One skilled or practised in hydrography; specially one whose business it is to make hydrographic surveys and to construct charts of the sea, its currents, etc., as the *Hydrographer to the Admiralty*.

**Hydrography**

*O.E.D.* 1. The science which has for its object the description of the waters of the earth's surface, the sea, lakes, rivers, etc., comprising the study and mapping of their forms and physical features, of the contour of the sea-bottom, shallows, etc., and of winds, tides, currents, and the like. (In earlier use, including the principles of Navigation.) Also a treatise on this science, a scientific description of the waters of the earth. (Quote 1559)
2. The subject-matter of this science, the hydrographical features of the globe or part of it; the distribution of water on the earth's surface.

Mill, *Dict.* 1. Marine surveying. 2. The study of the physical conditions of sea-water. 3. The study of the river and lake systems of a region.

*Adm. Gloss.*, 1953. Is the science and art of measuring the oceans, seas, rivers, and other waters with their marginal land areas, inclusive of all the fundamental elements which have to be known for the safe navigation of such areas, and the publication of such information in a form suitable for the use of navigators.

*Comment.* Despite some overlap and change in meaning the general tendency is to use hydrology for the science of water, especially with an emphasis on the waters

## Hydrolaccolith

Kuenen, 1955. 'Hummock caused when ground-water under artesian pressure is checked by formation of layers of ice ... Siberia is the best known scene of these strange icy blisters.' (p. 220)

*Comment.* In section these masses of ice, with a water core, do resemble laccoliths. Compare pingo, but hydrolaccoliths develop only in the presence of permafrost; the pingos in East Greenland but not those in the Mackenzie delta are hydrolaccoliths. (J. B. Bird in MS.)

## Hydrology

*O.E.D.* The science which treats of water, its properties and laws, its distribution over the earth's surface etc. (Quote: 1762)

*O.E.D. Suppl.* The science of treatment by baths and waters. (Quote: 1913) but this would seem to be 'medical hydrology'.

*Webster.* The science treating of water, its properties, phenomena, and distribution over the earth's surface. The term is used specifically in the United States Geological Survey with reference to underground water sources, as distinguished from *hydrography*, which is applied to surface water supplies and sources.

Moore, 1949. 'The study of water, especially in relation to its occurrence in streams, lakes, wells etc., and as snow, and including its discovery, uses, control and conservation.'

*Comment.* Hydrology and hydrologist have replaced hydrography and hydrographer for reasons given under the latter when the reference is to the science which studies water.

The Hydrological Group of the Institution of Civil Engineers set up in collaboration with the R.G.S. and other bodies in 1963 accepted a definition proposed by P. O. Wolf: 'The science dealing with the occurrence and movement of water and ice on or under the earth's surface from the moment of precipitation to the moment of its return to the atmosphere or discharge into the ocean.' (*Proc. Inst. Civil Eng.*, March 1964)

## Hydrophyte

*Webster. Biogeog.* A plant which grows in water or in saturated soil. *Cf.* xerophyte, mesophyte.

*O.E.D.* An aquatic plant.

See also Hygrophyte, hygrophilous.

## Hydroplastic

Kuenen, 1955. 'If strata of varying composition and, above all, containing different percentages of water, alternate in a thawing zone (even if they lie horizontally) deformations are bound to occur. Heavily soaked strata are liable to become so "hydroplastic" that they yield to every change of pressure.... This explains how the soil, after thousands of seasons, gives the impression of having been thoroughly kneaded.' (pp. 220-221)

## Hydrosphere

*O.E.D.* The waters of the earth's surface collectively.

b. By some used to designate the moisture contained in the air enveloping the earth's surface (Cent. Dict.).

Mill, *Dict.* The waters of the Earth.

Moore, 1949. The comparatively shallow layer of water which covers over two-thirds of the earth's surface, and forms the oceans, seas, and lakes; ...

*Comment.* Hydrosphere is the 'water sphere' in comparison and contrast with lithosphere and atmosphere. Because of the necessity of treating the water-cycle as a whole it should not properly be restricted as suggested by Moore's definition.

## Hydrothermal

*O.E.D.* Of or relating to heated water; specially applied to the action of heated water in bringing about changes in the earth's crust.

Himus, 1954. This term may be applied to magmatic emanations which are rich in water, or to the processes in which they are concerned, and to the rocks, ore-deposits and springs which they produce.

## Hyetograph

*O.E.D.* A chart showing the rainfall (*Syd. Soc. Lex.*, 1886).

Mill, *Dict.* A self-recording rain-gauge.

*Met. Gloss.*, 1944. A pattern of self-recording rain gauge in which the recording pen is actuated by a series of stops attached to a vertical float rod.

*Comment. O.E.D.* definition is almost obsolete.

## Hygrophilous, hygrophyte

Dansereau, P., 1957, *Biogeography*, New

York: Ronald Press. 'Xerophilous, mesophilous and hygrophilous reflect three degrees of water requirements. Hygrophilous plants are "moisture-loving".' (p. 206)

*Comment.* In view of the range of meaning which has been given to hygrophyte (which some botanists equate with hydrophyte and so with an aquatic plant) reference should be made to botanical works. The modern tendency would seem to be to agree with the simple statement of Dansereau quoted above. In this case the trees of the tropical rain forests are described as hygrophilous but they are far from being aquatic.

**Hylea, hylaea, hileia**

*Int. Geog. Congress*, Brazil, 1956, Excursion Guidebook No. 8, Amazonia. 'Impressed by the extreme floristic exuberance, density and richness of the equatorial forest, Humboldt gave the regions in which it is dominant the name *hylaea* (from the Greek hylé: dense forest), a designation later extended to the actual forest of the equatorial zone... one must distinguish an *American hylea*, an *African hylea* and an *Asian hylea*' (p. 47). Compare selva which James, 1959, continues to use in common with most British and American authors but UNESCO founded a Hylea Institute.

**Hypabyssal rock**

Himus, 1954. Minor intrusions such as sills and dykes and the rocks of which they are made.

*Comment.* Hypabyssal rocks thus form a group intermediate between plutonic and volcanic.

**Hypogene, hypogenic**

*Webster. Geol.* a. Formed or crystallized at depths beneath the earth's surface; also, plutonic—said of granite, gneiss, and other rocks; as, *hypogene* action. Opposed to *epigene*. b. Formed by generally ascending solutions; said of ore deposits. Opposed to *supergene*.

Himus, 1954. The term as originally proposed by Lyell included all the plutonic and metamorphic rocks, that is, rocks formed *within* the earth. It was later applied by A. Geikie to geological processes originating within the earth, and therefore, if applied in this sense, the hypogene rocks would include the effusive rocks which were excluded by Lyell.

Mill, *Dict.* Hypogenic geology: The study of changes taking place in the interior of the earth.

*Comment.* Not in Geikie, J., Holmes, 1944 or other standard works. A term best avoided, but see D. L. Linton, 1958, The Everlasting Hills, *Adv. of Science*, **14**. 'The growth and decay of mountain ranges—their erection by the hypogenic, and destruction by epigenic forces acting on the earth's crust.'

**Hypsographic curve, Hypsometric curve**

Monkhouse and Wilkinson, 1953. 'A hypsographic (sometimes known as a hypsometric) curve is used to indicate the proportion of the area of the surface at various elevations above or depths below a given datum.' (p. 86)

See also Wooldridge and Morgan, 1937. pp. 38–39; Holmes, 1944, p. 10.

Clarke, J. I., 1966, Morphometry from Maps, in Dury, G. H. (ed.), *Essays in Geomorphology*, London: Heinemann. 'This is an ogive, or cumulative-frequency curve, which represents the absolute or relative areas of land above or below each contour.'

*Comment.* Hypsometric analysis of the earth as a whole dates from the last two decades of the nineteenth century. See Lapparent, A. de, 1883, *Traité de Géologie*, Paris: Masson, and Murray, J., 1888, On the Height of the Land and the Depth of the Ocean, *Scottish Geog. Mag.*, **4**, 1–41. Important modifications made by Kossina, E., 1933, Die Erdoberfläche, *Handbuch der Geophysik*, Bd. 2, 875, and Orlicz, M., 1931-5, *Krzywe hipsograficzne lądów*, Lwów: Przegląd Kartograficzny.

See also Clinographic curve for a more sensitive method.

'Hypsometry' is used on some maps to mean the measurement and delineation of relief. See also below, hypsography.

**Hypsography, Hypsometry**

*O.E.D.* That department of geography which deals with the comparative altitude of places, or parts of the earth's surface.

*Webster. Geog.* a. Topographic relief. b. The observation or description of topographic relief. c. The parts of a map, collectively, which represent topographic relief. d.

Hypsometry, or the measurement of heights.

Clarke, J. I., 1966, Morphometry from Maps, in Dury, G. H. (ed.), *Essays in Geomorphology*, London: Heinemann. 'Hypsometry is the measurement of the inter-relationships of area and altitude.' (p. 237)

*Comment.* Now rarely used except in connection with the hypsometric curve, or hypsographic curve.

**Hythe**—see Hithe

**Hythergraph** (T. Griffith Taylor, 1915)

Taylor, 1951. 'Twelve-sided graph of temperature and rainfall.' (p. 615)

Moore, 1949. A graphical representation of the differentiation between various types of climate. Mean monthly temperatures are plotted as ordinates against the mean monthly rainfalls as abscissae and a closed, twelve-sided polygon, the hythergraph, is obtained. This reveals the type of climate at a glance.

See Climograph for further details.

# I

**Ice**
*Gloss.*, 1956, 1958. The solid form of water, in nature formed either by (a) the freezing of water as in the case of river or sea ice, (b) the condensation of atmospheric water vapour direct into *ice crystals*, (c) the compaction of *snow* with or without the motion of a *glacier*, or (d) the impregnation of porous snow masses with water which subsequently freezes.

**Ice crystal.** A single ice particle with regular structure.

**Ice prisms.** A fall of unbranched *ice crystals*, in the form of needles, columns, or plates, often so tiny that they seem to be suspended in the air. These crystals may fall from a cloud or from a cloudless sky. They are visible mainly when they glitter in the sunshine (diamond dust); they may then produce a luminous pillar or other halo phenomena.

See also Ice terminology and separate entries.

**Ice age**
*O.E.D.* The glacial period.
*Webster.* Glacial epoch. *Geol.* Any of those parts of geological time, from Pre-Cambrian time onward in both the Northern and Southern hemisphere, during which a much larger portion of the earth was covered by glaciers than at present. Specif. (*cap.*), the latest of the glacial epochs, that of the Quaternary period, known as the *Pleistocene epoch* during which Canada, northern and northeastern United States, northern and northwestern Europe, and northern Asia, together with most high mountain regions in the Northern Hemisphere were largely covered with ice.
Mill, *Dict.* Ice Age or Glacial Epoch. Usually used with reference to the period of extensive glaciation in Pleistocene times, though there have been other ice ages.
Moore, 1949. A geological period in which ice sheets and glaciers covered large areas of the continents, reaching the sea in places and lowering the temperature of the oceans.
*Comment.* When used without qualification the Ice Age signifies the latest of the geological ice ages corresponding broadly with the Pleistocene Period and better termed the Great Ice Age. Some writers contend that the term suggests that the Great Ice Age was monoglacial rather than polyglacial in character and accordingly advocate the dropping of the term but it is too widely used for that to happen.

**Ice apron**
*Ice Gloss.*, 1958. A thin mass of snow and ice adhering to a mountain side.

**Ice-barchan**
Bagnold, R. A., 1941, *The Physics of Blown Sand and Desert Dunes*, London: Methuen (rep. 1954). 'Moss has described in detail the formation of small barchan dunes of very cold ice crystals' (p. 221) (see Barchan). R. Moss, The Physics of an Ice-Cap, *Geog. Jour.*, 92, 1938, 211–231, describes rather than defines an ice-barchan which is simply a crescentic dune or barchan composed of ice crystals instead of sand. Similarly snow-barchan.

**Ice barrier**
The edge of the Antarctic ice-field, applied specifically to the Ross Ice Barrier which is the edge of the great Ross Ice Shelf.
*Comment.* Now obsolete and replaced by 'ice front' or 'ice wall'. (T. E. Armstrong in MS.)

**Iceberg**
*O.E.D.* 1. An Arctic glacier, which comes close to the coast and is seen from the sea as a hill or 'hummock'. Obs. (quotes, 1774–1821)
2. A detached portion of an Arctic glacier carried out to sea; a huge floating mass of ice, often rising to a great height above the water. Formerly also called ice-island, also island or shoal of ice.
Mill, *Dict.* A mass of land-ice, broken from a glacier and floating in the sea.
*Ice Gloss.*, 1956. 'A large mass of floating or stranded ice broken away from a *glacier* or from an *ice-shelf*. Often of considerable height (in any case more than 5 m. above the level of the sea).'
*Comment.* Arctic and Antarctic icebergs are commonly different in form, those from

the Antarctic which have broken off from the Ross ice shelf are flat-topped (tabular berg), but there is no justification for the *O.E.D.* to restrict icebergs to the Arctic.

### Ice blink
*Ice Gloss.*, 1956. Yellowish white luminous glare in the sky produced by the reflexion on the clouds of *pack ice* or an *ice sheet* which may be beyond the range of vision. See also Eisblink, isblink.

### Ice cake
*U.S. Ice Term.*, 1952. A general term, like floe, used in reference to flat fragments of ice. The dimensions of a cake are not fixed and the term is used when no specific size is intended.
*Ice Gloss.*, 1956. A *floe* smaller than 10 m. across.

### Ice cap
*O.E.D.* 1. A permanent cap or covering of ice over a tract of country, such as exists on high mountains, and on a large scale at either pole.
Mill, *Dict.* Ice-cap. A continuous covering of ice, névé or snow, such as occurs in polar lands.
Flint, 1947. 'Small ice sheets and sometimes large ones as well, are referred to as ice caps.' (p. 13)
*Ice Gloss.*, 1958. 'A dome-shaped *glacier* usually covering a highland area. Ice caps are considerably smaller in extent than *ice-sheets*.'
*Comment.* The distinction of size between an ice sheet (larger) and an ice cap (smaller) is maintained by Holmes, 1944; *Adm. Gloss.*, 1953; Cotton, 1945; also *Webster's Dictionary*.
See also Ice-sheet, inland ice.

### Ice cluster
*Ice Gloss.*, 1956. A concentration of sea ice covering hundreds of square kilometres, which is found in the same region every summer.

### Ice cover
*Ice Gloss.*, 1956. The amount of sea ice encountered; measured in tenths of the visible surface of the sea covered with ice.

### Ice-column, ice-pillar
Mill, *Dict.* A column or pillar of ice with a boulder capping it.

### Ice covered
*Ice Gloss.*, 1958. Land overlaid at present by an extensive *glacier* is said to be ice-covered. The alternative term 'glacierized'

has not yet found general favour. Contra glaciated.

### Ice-dam Lake
Mill, *Dict.* Lake formed by a barrier of ice stretching across a valley mouth.
Now called an ice-dammed lake.

### Ice edge
*Ice Gloss.*, 1956. The boundary at any given time between the open sea and sea ice of any kind, whether drifting or fast. When wind or swell compact the ice edge it may be called a compacted ice edge; when dispersed, it may be called an open ice edge.

### Ice fall
*O.E.D.* 1. A cataract of ice; a steep part of glacier resembling a frozen waterfall.
2. The fall of a mass of ice, from an ice-cliff or iceberg.
Mill, *Dict.* A fall of heavy masses of ice detached from a glacier, and caused by an abrupt change of slope in its bed.
*Ice Gloss.*, 1958. A heavily *crevassed* area in a *glacier* at a point of steep descent.

### Ice field, Ice-field
*O.E.D.* A wide flat expanse of ice, esp. of marine ice in the Polar regions.
Mill, *Dict.* The mass due to accumulation of snow to great depths on a surface and its subsequent conversion into ice.
*U.S. Ice Term.*, 1952. The largest of sea ice areas. An ice field is so called because of its size only (more than 5 miles across). The effect of pressure, erosion, or age have no part in the definition.
*Adm. Gloss.*, 1953. An area of pack-ice consisting of any size of floe of such extent that its limits cannot be seen from the masthead. When, by air observation or otherwise, the full extent of an ice field is known, the following terms may be used:
Large (over 20 kms. across), Medium (15–20 kms. across), or Small (10–15 kms. across).
*Comment.* Despite both the British and American official definitions the term ice field is applied to large stretches of land ice, notably the Columbia Ice-field of Canada. 'Large glaciers are conspicuous the most extensive being in the Columbia Ice-field ... and in the Reef Ice-field' (Putnam, D. F., and others, *Canadian Regions*, Toronto: Dent, 1952, p. 424). Because of its ambiguity, the term is omitted entirely in *Ice Gloss.*, 1956, 1958
See also Ice terminology, Dunbar.

## Ice floe

*O.E.D.* Ice-floe. A large sheet of floating ice: sometimes several miles in extent.

Floe. A sheet of floating ice, of greater or less extent; a detached portion of a field of ice.

*U.S. Ice Term.*, 1952. The term floe is used for referring to fragments of ice (other than icebergs and other land ice) with no specific size intended. However, unlike the term cake, when floe is used with such qualifying terms as small, medium, or giant, a rather definite size is implied. Terms for describing floes of specific size are: Brash, block, small floe, medium floe, giant floe, and ice field. Another distinction is that a floe may consist of a single unbroken fragment of ice or many consolidated fragments. Whereas cake implies a single unbroken fragment of ice. Floe is also used with such qualifying terms as heavy and light, but these terms imply thickness rather than a real limit.

*Ice Gloss.*, 1956. Floe: A piece of sea ice other than *fast-ice*, large or small. Light floes are anything up to a metre in thickness. Floes of greater thickness, both level and hummocked, are called heavy floes. Floes over 10 km. across are described as vast; those between 1 and 10 km. across as large; those between 200 and 1,000 metres across as medium; between 10 and 200 metres across as small. Floes less than 10 metres across are called *Ice Cakes*.

## Ice fog

*Ice Gloss.*, 1958. A suspension of numerous minute *ice crystals* in the air, reducing visibility at the earth's surface. The crystals often glitter in the sunshine. Ice fog produces optical phenomena such as luminous pillars, small haloes, etc.

## Icefoot, ice foot

*Ice Gloss.*, 1956. A narrow strip of ice attached to the coast, unmoved by tides and remaining after the *fast ice* has broken free.

*Webster.* a. A wall or belt of ice that forms along the shore in arctic regions between high and low watermarks as a result of the rise and fall of the tides. b. The ice at the front of a glacier.

## Ice front

*Ice Gloss.*, 1956. The vertical cliff forming the seaward face of an *ice shelf*, varying in height from 2 to 50 m. (*cf. Ice wall*).

Comment. The authors of *Ice Gloss.* would now add after ice shelf 'or other glacier which enters water'. An ice front is afloat.

## Ice island

*Ice Gloss.*, 1956. A rare form of tabular berg found in the Arctic Ocean... regularly undulating surface... originate from *ice shelves* in northern Ellesmere Island and northern Greenland.

## Ice jam

Mill, *Dict.* A temporary accumulation of broken-up river-ice which has been carried downstream and then stopped by some obstruction in the river.

*U.S. Ice Term.*, 1952. 1. An accumulation of broken river ice, caught in a narrow channel. *cf.* ice gorge, debacle (also *Adm. Gloss.*, 1953).
2. Fields of lake ice thawed loose from the shores in early spring may be blown against the shore exerting great pressures. Also, masses of broken-up ice may drift with the wind and produce jams on and against the shore.

*Ice Gloss.*, 1958. An accumulation of broken river or sea ice caught in a narrow channel.

## Ice limit

*Ice Gloss.*, 1956. The average position of the *ice edge* in any given month or period based on observations over a number of years.

## Ice piedmont

*Ice Gloss.*, 1956. Ice covering a coastal strip of low-lying land backed by mountains. The surface of an ice piedmont slopes gently seawards and may be anything from about 50 m. to 50 km. wide, fringing long stretches of coastline with ice cliffs. Ice piedmonts frequently merge into *ice shelves*.

## Iceport

*Ice Gloss.*, 1958. An embayment in an *ice front*, often of a temporary nature, where ships can moor alongside and unload directly onto the *ice shelf*.

## Ice rind

*Ice Gloss.*, 1956. A thin, elastic, shining crust of ice, formed by the freezing of *sludge* on a quiet sea surface. Thickness less than 5 cm.

## Ice pedestal—see Sérac

## Ice rise

*Ice Gloss.*, 1956. A mass of ice resting on rock and surrounded either by an *ice shelf*, or partly by an *ice shelf* and partly by sea and/or ice-free land. No rock is exposed and there may be none above sea-level. Ice rises often have a dome shaped surface. The largest known is about 100 km. across.

*Comment.* The authors of *Ice Gloss.* note that an ice rise may include much snow and occasionally rock may appear.

## Ice sheet

*O.E.D.* Ice-sheet. A sheet or layer of ice covering an extensive tract of land; spec. that supposed to have covered a great part of the northern hemisphere during the glacial period.

Mill, *Dict.* Ice-sheet = Ice-cap (*q.v.*).

Flint, 1947. 'Ice sheets ... are not confined to valleys but are broad, cake-like ice masses usually (though not invariably) occupying highlands.'

*Ice Gloss.*, 1956. A continuous mass of ice and snow of considerable thickness and large area. Ice sheets may be resting on rock (see *inland ice sheet*) or floating (see *ice shelf*). Ice sheets of less than about 50,000 sq. km. resting on rock are called ice caps. But see Ice cap.

## Ice shelf

*Ice Gloss.*, 1956. A floating *ice sheet* of considerable thickness. Ice shelves are usually of great horizontal extent and have a level or gently undulating surface. They are nourished by local snow accumulation and often also by the seaward extension of land glaciers. Limited areas may be aground. The initial stage is called *bay ice* until the surface is more than about 2 m. above sea-level. The seaward edge is termed an *ice front*.

## Ice storm

Webster. *Meteorol.* A storm in which falling rain freezes as soon as it touches any object.

*Dict. Am.* A storm in which rain freezes on the object it falls upon. 1877 Mark Twain. 1886 Geikie. *Outline Geol.* 50.

## Ice stream

*Ice Gloss.*, 1958. Part of an *ice sheet* in which the *ice* flows more rapidly and not necessarily in the same direction as the surrounding ice. The margins are sometimes clearly marked by a change in direction of the surface slope, but may be indistinct.

## Ice wall

*Ice Gloss.*, 1956. An ice cliff forming the seaward margin of an *inland ice sheet*, *ice piedmont*, or *ice rise*. The rock basement may be at or below sea-level (*cf. Ice front*).

*Comment.* In contrast to an ice front, an ice wall is aground, not afloat.

## Ice, sundry terms

*Ice Gloss.*, 1956, 1958 defines the following:
*Anchor ice*—submerged ice which attached to the bottom.

*Bay ice*—Fast ice of more than on winter's growth, which may be nourished also by surface layers of snow. Thickness of ice and snow up to about 2 m. above sea-level. When bay ice becomes thicker than this it is called an *ice-shelf*.

*Belt*—a long area of *pack ice* from a few kilometres to more than a 100 km. in width.

*Brash ice*—small fragments of floating ice not more than 2 m. across, the wreckage of other forms of ice.

*Fast ice*—sea ice of greatly varying width which remains fast along the coast, where it is attached to the shore, to an *ice wall* to an *ice front*, or over shoals, generally in the position where originally formed. Fast ice may extend 400 km. from the coast.

*Frazil ice*—fine spicules or plates of ice in suspension in water. The first stage of freezing, giving an oily or opaque appearance to the surface of the water (*cf.* sludge).

*Hummocked ice*—sea ice piled haphazardly one piece over another and which may be weathered.

*Hummocking*—process of pressure formation by which *level ice* becomes broken up into humps and ridges. When the floes rotate in the process it is termed 'screwing'.

*Level ice*—sea ice with a level surface which has never been hummocked.

*New ice*—a general and indefinite term for ice newly formed. It will be in the form of crystals, *sludge*, *ice rind*, or *pancake ice* of thickness less than 5 cm.

*Pack ice*—any area of sea ice other than *fast ice*, no matter what form it takes or how it is disposed. Pack ice cover may be reported in eighths, or may be described as *very open pack ice* (1/8th to 3/8ths) *open pack ice* (3/8ths to 6/8ths), *close pack ice* (6/8ths to 7/8ths) and *very close pack ice* (7/8ths to 8/8ths with little, if any, water visible). *Note.* The authors of *Ice Gloss* now suggest measurement in tenths (1/10 to 3/10; 4/10–6/10; 7/10 to 9/10) and would add that in consolidated pack ice the floes are frozen together (10/10ths).

*Pancake ice*—pieces of newly formed sea ice usually approximately circular in shape with raised rims, and from one-third to 2 m. in diameter.

*Patch*—a collection of *pack ice*, whose limits can be seen from a masthead or even lower.

*Polar ice*—extremely heavy sea ice, up to 3 m. or more in thickness, of more than one winter's growth. Heavily hummocked, and may be ultimately reduced by weathering to a more or less even surface.

*Pressure ice*—a general term for sea ice which has been squeezed, and in places forced upwards when it can be described as *rafted ice*, *hummocked ice*, or *pressure ridge*.

*Pressure ridge*—a ridge or wall of *hummocked ice* where one floe has been pressed against another.

*Rafting*—the over-riding of one *floe* on another, the mildest form of pressure. Hence 'rafted ice'.

*Ram*—the underwater projection of ice from an *ice front*, *ice wall*, *iceberg* or *floe*.

*Rotten ice*—sea ice which has become honeycombed in the course of melting, and which is in an advanced state of disintegration.

*Sludge*—a later stage of freezing than *frazil ice*, when the spicules and plates of ice coagulate to form a thick soupy layer on the surface. Sludge reflects little light, giving the sea a matt appearance.

*Strip*—a long narrow area of *pack ice*, about 1 km. or less in width, usually composed of small fragments detached from the main mass of ice, and run together under the influence of wind, swell or current.

*Tide crack*—the fissure at the line of junction between an immovable *icefoot* or *ice wall* and *fast ice*, the latter being subject to the rise and fall of the tide.

*Tongue*—a projection of the *ice edge* up to several kilometres in length, caused by wind and current.

*Weathered ice*—hummocked polar ice subjected to weathering which has given the hummocks and *pressure ridges* a rounded form. If the weathering continues the surface may become more or less even.

*Winter ice*—more or less unbroken, level sea ice of not more than one winter's growth originating from *young ice*. Thickness from 15 cm. to 2 m.

*Young ice*—newly formed *level ice* in the transition stage of development from the *ice rind* or *pancake ice* to *winter ice*. Thickness from 5 to 15 cm.

See also separate entries for Dead ice, Drift-ice, Ground ice.

## Ice Terminology

In view of the need for an agreed terminology for any system of ice reporting and ice recording, the Scott Polar Research Institute of Cambridge undertook the preparation of an illustrated Ice Glossary. Pending its completion and appearance as a Special Publication by the Institute, a short version has been published in the *Polar Record*: a first part dealing with sea-ice in Vol. 8, No. 52, January 1956, pp. 4–12, and a second part dealing with land ice in Vol. 9, No. 59, May 1958, pp. 90–96, both by Terence Armstrong and Brian Roberts. These papers are referred to as *Ice Gloss.*, 1956 and *Ice Gloss.*, 1958. The first is a development of an Admiralty Glossary (Glossary of terms used on Admiralty charts and in associated publications, 1953, Part 5, Ice and Snow terms: London, Hydrographic Dept. Prof. Paper No. 11 2nd Edition) and the Ice Glossary is very close to that prepared by the World Meteorological Organization (W.M.O.).

In 1955 in the *Geography of the Northlands*, edited by G. H. T. Kimble and D. Good (Amer. Geog. Soc., Spec. Pub. No. 32) there was reproduced with slight modifications the classification given in both the British and the United States *Arctic Pilots*, used and enlarged by Koch in his account of the Greenland ice (1945). It is in the chapter by M. J. Dunbar and is referred to as Dunbar, 1955. Because it differs considerably from the Ice Glossary, Dunbar is quoted below in full.

All the main terms given by Armstrong and Roberts have been included in this Glossary. In addition to the words listed under *ice* the following are included:

*Sea ice*: bergy bit, bight, calving, floe, frost smoke, growler, lead and polynya,

*Land Ice*: avalanche, bergschrund, crust, cirque glacier, cornice, crevasse, crust, drifting snow, firn, glaciated, glacier, glacier tongue, inland ice sheet, glacier ice, glacierized, glaze (clear ice), hail, hoarfrost, icicle, icing, moraine, nunatak, piedmont glacier, randkluft (bergschrund), rime, ripple marks, skavler, barchan, snow line valley glacier.

In addition Armstrong and Roberts have given specialized meanings in ice terminology to such words or expressions as crack, pool, puddle, shore lead, tabular berg, water sky, blow hole, blowing snow, breakable crust, depth of snow, freezing drizzle, freezing rain, new snow, old snow, powder snow, slush, snow, snow barchan, snow bridge, snow cornice, snow cover, snow drift, snowflake, snow patch, strand crack, unbreakable crust, and whiteout.

Certain ice terms, not used by Armstrong and Roberts, will also be found in the Glossary—notably ice-field, sastrugi, penitents, pocket penitent.

Ice terminology according to Dunbar, following Koch, in Kimble and Good, *Geography of the Northlands*, 1955

Glacier ice: Iceberg—floating ice detached from a glacier.
    Calf ice—smaller pieces of glacier ice either direct from the glacier or from the breakdown of icebergs.
    Bergy bits—pieces of glacier ice up to 3 ft in diameter.

Fast ice: Ice foot—a belt of sea ice adhering to the shore, unaffected by the tide.
    Fiord ice—ice formed in sheltered places and melting where it is formed, more or less.
    Sikussak—fiord ice more than 10 yr. old, with rough surface.

Drift ice: Land ice—ice formed in the mouths of large fiords and other inlets which drifts out to sea before melting.
    Pack ice—more or less hummocky ice, formed at sea, usually with pressure ridges.

Conventional names are used to describe various types of drifting ice, as follows:

Field—an area of floating ice, more than 5 km. in diameter.

Floe—an area of floating ice, more than 1 km. in diameter.

Small floe—an area of floating ice, more than 200 m. in diameter.

Glaçon—floe less than 200 m. in diameter.

Growler—a piece of hummocky ice, rising at least 1·5 m. above the surface of the sea.

Cake—a relatively flat piece of ice smaller than a small floe.

Cake ice—accumulation of cakes.

Bit—a piece of ice less than ½ m. in diameter.

Brash ice—accumulation of bits, melting

Sludge or slush—new sea ice in the initial stages of formation, when it is of a gluey consistency with a peculiar steel gray or leaden color.

(pp. 54–55).

**Icicle**
*Ice Gloss.*, 1958. Hanging spike of clear ice formed by the freezing of dripping water

**Icing**
*Ice Gloss.*, 1958. The accumulation of a deposit of *ice* on exposed objects. The ice may be dense and clear (transparent) or white and opaque, or anything in between. Icing may be produced by the deposition of water vapour as frost or by the freezing on impact of droplets suspended in the air. Hence de-icing (removal), anti-icing (prevention).

**Iconic models**
Iconic models are the first stage of abstraction in model building. They are simple scale representations of reality.
*O.E.D.* Iconic. Of or pertaining to an icon image, figure, or representation.
See Haggett, 1972, pp. 14–16.
See also Analogue model and Symbolic model.

**Idd** (Sudan; Arabic)
A place in the bed of an intermittent stream where water may be obtained in shallow wells for most if not all of the dry season in the northern Sudan. (J.H.G.L.)

**-ides**
A suffix frequently added to names of major mountain ranges or chains, *e.g.* Alpides, Altaides, Caledonides. The purpose is to indicate the broad geological or structural units of which the present geographical units form part.

**Idiographic approach**
Keeble, D. E., in Chorley and Haggett, 1967. '... geography's traditional preoccupation with the individuality and uniqueness of different countries and areas.' (p. 243)
*Comment.* The difference between nomothetic and idiographic clearly put in Bunge, W., 1964, Geographical Dialects, *Professional Geographer*, **16**.4, 28–29. See Wooldridge, S. W. and East, W. G., 1951, *The Spirit and Purpose of Geography*, London: Hutchinson University Library, p. 145.

## Igneous

*O.E.D.* Of, pertaining to, or of the nature of fire, fiery. Produced by volcanic agency (opposed to Aqueous).

Lobeck, 1939. 'Igneous rocks result from the cooling of a molten mass' (p. 41).

*Comment.* Although this word dates from 18th century arguments regarding the origin of the rocks of the earth's crust, it has persisted in use as the general term for rocks formerly in a molten condition, originating from a magma, whether plutonic (deep-seated) or volcanic or between the two.

## Illuvial horizon (Soil science)

Jacks, 1954. 'Horizon that has received material in solution or suspension from the upper part of the soil.'

## Illuviation (Soil science)

The process by which material removed in solution or in suspension from the upper part of the soil is deposited at a lower horizon, usually the B horizon. Hard-pan (*q.v.*) is formed in this way.

## Imbricate, imbricate structure, imbrication

*O.E.D.* Imbricate: 4b. overlapping like tiles (but not given in the special geological sense).

Himus, 1954. Imbricate structure: a structure resulting from the intense pressures which obtain in mountain building, as a result of which individual blocks of rock are thrust over each other, like a pack of fallen cards.

See also Wooldridge and Morgan, 1939, p. 66.

## Immature soil

*Soils and Men*, 1938. 'A young or imperfectly developed soil.' (p. 1170)

Robinson, G. W., 1949, *Soils*, 3rd ed., London: Murby. 'Profiles which have not attained full development are termed immature or underdeveloped.' Immature soils occur on recently laid down deposits or where erosion keeps pace with profile development. (p. 70)

**Immature town**—see Urban hierarchy
**Immigrant**—see Migrant

## Impermeable

*O.E.D.* That does not permit the passage of water or other fluid.

Moore, 1949. Impermeable rocks: rocks which, being non-porous or practically so, do not allow water, *e.g.* rain water, to soak into them. Granite provides an example; this rock may be pervious, however, owing to joints and fissures. Some geographers assume the term to be synonymous with *Impervious rocks* (*q.v.*).

## Impervious

*O.E.D.* Through which there is no way; not to be passed through or penetrated; impenetrable, impermeable, impassable.

Moore, 1949. Impervious rocks: rocks which do not allow water, *e.g.* rain water, to pass through them freely; they may be porous, like clay, or practically non-porous, like unfissured granite. Some geographers use the term *Impermeable rocks*.

*Comment.* Whilst some writers are very insistent on the distinction between permeable and pervious or between impermeable and impervious, others draw no distinction at all. Permeability should not, however, be confused with porosity.

## Improved land

*O.E.D.* Improve: 2b. To turn *land* to profit; to inclose and cultivate (waste land); hence to make land more valuable or better by such means, and so, in later use, merged in sense 5. *Obs.*

3. To enhance in monetary value; to raise the price or amount of. *Obs.* As said of lands and rents, app. connected with senses 1 and 2b, land that was 'emprowed' or inclosed and cultivated being enhanced in value or in rent.

5. To advance or raise to a better quality or condition; to bring into a more profitable or desirable state; to increase the value or excellence of; to make better; to better, ameliorate. (The prevailing modern sense, in which 2b is now merged.)

*Webster.* Improve. 5. Specifically *U.S.* a. To enhance in value by bringing under cultivation or reclaiming for agriculture or stock raising; as, to *improve* virgin land; *improved* farms.

*Dict. Am.* Land in use, under cultivation, fenced and provided with farm buildings, 1643.

Stroud, 1952. 'Barren heath or waste land is not "improved and converted into arable-ground, or meadow" (s. 5 (2 & 3 Edw. 6, c. 13)), by merely turning cattle thereon; the phrase connotes some act of cultivation (Ross v. Smith, 1B & Ad. 907).'

*Comment.* In some countries an official distinction is drawn for statistical purposes between 'unimproved' and 'improved' farm land (*e.g.* United States). The two categories would correspond to 'rough grazing' (*i.e.* range land) and 'crops and (improved) grass' in British statistics.

**In-by land**
Report Royal Commission on Common Land, 1955–58, H.M.S.O., 1958, Cmd. 462. 'The term is widely used in the north of England and derives from the Scandinavian word *by* for a farm. Hence it means the fenced-in land nearest the homestead' (p. 274).

**Inch** (Scottish; also northern Ireland)
*O.E.D.* From Gaelic, *innis*, island, land by a river.
A small island (frequent in the names of small islands belonging to Scotland).
b. Applied locally to a meadow by a river (as the Inches of Perth); also to a piece of rising ground in the midst of a plain.
Mill, *Dict.* 1. An island, usually small and rocky (Scotland).
2. An alluvial flat beside a river (Scotland) which may become an island during a moderate flood. It is thus the Gaelic equivalent for *Ing* in some place names. Innis, ennis, ynys are other Celtic forms.

**Incised meander**
Thornbury, 1954. 'Incised or inclosed meanders are usually considered evidence of rejuvenation. Unfortunately there is confusion in the usage of terms intended to describe valley meanders which are entrenched in bedrock.... Five terms, incised, entrenched, intrenched, inclosed and ingrown have been applied to them with varying meaning. It seems to be more common practice to use incised and inclosed as generic terms to include any meanders inclosed by rock walls. We shall therefore consider *incised meanders* and *inclosed meanders* as synonymous group names for all meanders set down in bed rock. Two types are generally recognised (1) *entrenched* or *intrenched meanders* (the difference is only in spelling) which show little or no contrast between the slopes of the two valley sides, and (2) *ingrown meanders* (Rich, 1914) which exhibit pronounced asymmetry of cross profile.' (p. 145)
*Comment.* For discussion of this complex problem of nomenclature see Dury, 1959, 93–95; Davis, 1909 (1896), Wooldridge and Morgan, 1937, Cotton, 1922, 1945. See also Intrenched meander.

**Incision**
Mill, *Dict.* The deepening of its channel by river.

**Inclosure**
Report Royal Commission on Common Land, 1955–58, H.M.S.O., 1958, Cmd. 462. 'The process by which common right over a piece of land are extinguished and the land turned into "ordinary freehold". (p. 274)
*Comment.* The Royal Commission was careful to draw the distinction between 'enclosure' (the simple act of surrounding land with a fence so as to cut it off from other land) and 'inclosure' as a legal term. Strictly therefore one should refer to 'Inclosure Awards' or 'Inclosure Acts' but the more normal spelling enclosure is commonly used in both senses.

**Incompetence, incompetent**—see Competence

**Inconsequent drainage**
Gilbert, 1877. Drainage which is not consequent upon the structure. It may be antecedent or superimposed. (pp. 143–144)
Mill, *Dict.* Anomalous river courses, *i.e.* river courses which do not seem to be conditioned by the present structure of the river beds (Gilbert).
*Comment.* No reference in Salisbury, 1907; Davis, 1909; Cotton, 1944; Strahler, 1951. Not in general use; replaced by Insequent (*q.v.*).

**Incrop** (P. L. Collinson and R. E. Elliott, 1953)
Wills, L. J., 1956, *Concealed Coalfields*, London: Blackie. 'The term "incrop" has recently come into use for the area where a formation comes to a concealed outcrop below an unconformable cover. It represents an outcrop on an ancient land-surface.' (p. 59) See *Trans. Inst. Min. Eng.*, 112, 895.

**Indaing** (Burmese)
Literally a woodland of In or Eng (*Dipterocarpus tuberculatus*) which is usually codominant with ingyin (*Pentacme suavis*) to form an open dry forest on the margins of the Burmese Dry Belt with an annual rainfall of about 30 inches. (L.D.S.)

**Indelta** (T. Griffith Taylor, 1940)
Taylor, 1951. 'Inland area where a river subdivides (*e.g.* mid-Murray)' (p. 615). Now in common use especially in Australia.

**Index of dissimilarity**
Duncan, O. D. and Duncan, B., 1955,

Residential Distribution and Occupational Stratification, *Am. Jour. Sociology*, 60.5, 493–503. 'The spatial "distance" between occupation groups, or more precisely the difference between their areal distributions, is measured by the index of dissimilarity. To compute this index, one calculates for each occupation group the percentage of all workers in that group residing in each area unit. The index of dissimilarity between two occupation groups is then one-half the sum of the absolute values of the differences between the respective distributions, taken area by area.' (p. 494)
See also Index of segregation.

**Index of segregation**
Duncan, O. D. and Duncan, B., 1955, Residential Distribution and Occupational Stratification, *Am. Jour. Sociology*, 60.5, 493–503. 'When the index of dissimilarity is computed between one occupation group and all other occupations combined (i.e. total employed males except those in the given occupation group), it is referred to as an index of segregation.' (p. 494)
See also Index of dissimilarity.

**Indian Summer**
*Webster*. a. A period of warm or mild weather, late in autumn or in early winter, usually characterized by a clear cloudless sky and by a hazy or smoky appearance of the atmosphere, especially near the horizon. The name is of American origin, the reason for it being uncertain; it is now used also in England.
*Dict. Am.* A period of mild, warm weather in late October or early November, after the first frosts of autumn. 1778. Crèvecoeur *Sk. 18th Cent. Amer.* (1925) 41 'a short interval of smoke and mildness, called the Indian Summer.' Many theories, none of them convincing, have been advanced to account for this expression. For the most competent discussion of the term see Albert Matthews in the *Monthly Weather Review* for Jan. and Feb., 1902.
Mill, *Dict*. A period of calm, dry, mild weather, with hazy atmosphere occurring in the late autumn in the northern United States. The name is generally attributed to the fact that the region in which the meteorological conditions in question were originally noticed was still occupied by Indians.

**Induration**
Himus, 1954. The hardening by heat of a compact rock.
*Webster*. Indurated rock. *Petrog.* A rock hardened by the action of heat, pressure, or cementation.

**Industrial archaeology**
This term seems first to have been used in Britain by Michael Rix in 1955. It covers the study of the remains of former industrial enterprise, notably of the early part of the Industrial Revolution. Some may be removed to museums or recorded by photography, others must be preserved in situ as industrial monuments.

**Industrial complex**
Chardonnet, J., 1953, *Les grands types de complexes industriels*, defines an industrial complex as having one or more basic industries plus a certain diversity of other industries, technically and economically interdependent and generally well endowed with a network of transportation, commercial and financial facilities. It is distinguished from an industrial centre by its greater size and diversity and from an industrial region by being concentrated in a relatively restricted area. (pp. 9–17) Chardonnet distinguishes complexes primarily based on various sources of energy, a raw material base (particularly iron ore), various transportation facilities and population. (C.J.R.)

**Industrial inertia**
Committee, List 4. The tendency of industries or of firms to remain in a given location or site after the conditions originally determining that location or site have appreciably altered.

**Industrial momentum**
Committee, List 4. Geographical momentum. The tendency of places with established installations and services to maintain or increase their importance after the conditions originally determining their establishment have appreciably altered.
*Comment*. The Committee defined *geographical momentum*, something wider than industrial momentum which may be defined as the tendency of an industry in a given locality to increase its importance after the conditions originally favouring its establishment have appreciably altered. *Cf.* industrial inertia.

## Industrialisation, Industrialization

*O.E.D. Suppl.* The process of industrializing or fact of being industrialized; also, the conversion of an organization into an industry.

Wythe, G., 1945, *Industry in Latin America*, New York: Columbia U.P. Suggested criteria for judging the extent of industrialization are the number of persons employed in manufacturing industries and the value of the output (p. 11) (W.M.).

## Industry

Committee. 1. In the widest sense, work performed for economic gain.
2. In a narrower sense, mining, manufacture or handicrafts (as distinct from agriculture, commerce and personal service).

*O.E.D.* 4. Systematic work or labour; habitual employment in some useful work, now especially in the productive arts or manufactures. (This, with 5, is the prevalent sense.)
5. A particular form or branch of productive labour; a trade or manufacture.

General Register Office, Census, 1951, *Classification of Industries*, London: H.M.S.O., 1952, pp. iii–iv. For Census purposes, men and women in their working capacity are classified separately by their occupation and the industry in which they are employed. In each 'industry' is grouped together all persons, whatever their occupation, who contribute their labour to a particular service or product, the occupation of any person being the kind of work which he or she performs. (W.M.)

Florence, P. S., 1929, *The Statistical Method in Economics and Political Science*, London: Kegan Paul. '... any kind of transaction for exchange (production, distribution, services, etc.) usually specialized in by a group of firms who do not usually perform other transactions' (p. 356).

Sloan, H. S. and Zurcher, A. J., 1949, *A Dictionary of Economics*, New York: Barnes and Noble. 'Productive enterprise, especially manufacturing or certain service enterprises such as transportation and communications, which employs relatively large amounts of capital and labor.' 'The word industry is generally used and thought of as referring exclusively to manufacturing industry where a more or less transportable raw material is converted into a more or less transportable product.' (Considers the addition of other, wider, senses.) (p. 337)

*Comment.* Studies in 'industry' are sometimes bounded by the activities measured by the Census of Production. *e.g.* P.E.P. Location of Industry in Great Britain, 1939. The census for 1949 covered undertakings in the field of industrial production, including building and contracting, public utilities, and mines and quarries. It included textile converting, laundry work, dyeing and dry cleaning, wig-making, and tea blending and coffee roasting. The scrap metal trade was included in 1948.

## Industry, basic

Royal Commission on the Distribution of the Industrial Population (Barlow Report, 1940) Cmd. 6153. 'The industries which, for the purposes of exchange, send their products to places outside the area in which they are situated, may be termed "basic" industries.' (para. 65, p. 28)

Committee, List 4. A term variously used in the sense of heavy industry of national economic importance, or of industries fundamental to other industries, *e.g.* sulphuric acid, iron and steel.

Ministry of Labour Gazette, 62, 2, Feb. 1954. Employment per industry is grouped under three headings: Basic Industries, Manufacturing Industries and an uncharacterized group. Basic industries are Coal Mining; Other Mining and Quarrying; Gas, Electricity and Water; Transport and Communication; Agriculture and Fishing.

## Industry: classification

Many different classifications have been proposed for industries. For the common distinction into primary, secondary and tertiary industry see below. A rough two-fold distinction of manufacturing industry into 'heavy' and 'light' is also discussed below with the concept also of 'basic' industry and 'local' industry. When the *Committee on Land Utilisation in Rural Areas* was considering the impact of industrial development on the countryside, it was compelled to prepare a provisional classification of industry which is detailed in the *Report* (London: H.M.S.O., 1942) and the main classes with their self-explanatory titles are:

*Extractive industries* (a) of minerals in short supply—location fixed by the occurrence of the minerals;
(b) of minerals—clay, limestone etc.—in

abundant supply and so susceptible of planned location.

*Heavy industries*—largely fixed by heavy and bulky character of raw materials and finished products and much concerned with transport costs.

*Linked industries*

*Mobile or footloose industries*—broadly those generally called 'light'.

*Servicing industries*

This classification tended to emphasize the limited number of industries which could be located at will. For its particular purpose the Scott Report regarded agriculture and forestry apart from industry. *Cf.* above, industry, Committee 2.

**Industry, extractive**

Committee, List 4. Those *primary* industries in which non-replaceable materials are removed in their natural state, *e.g.* mining and quarrying.

*O.E.D.* Pol. Econ. *Extractive Industry*: an industry (*e.g.* agriculture, mining, fisheries, etc.) that is concerned with obtaining natural productions.

*Comment.* The *O.E.D.* definition is definitely incompatible with current use and it is not clear that it is justified from the authorities quoted. Extractive industry is often equated with 'robber economy' and contrasted with agriculture and forestry which should maintain the means of production.

**Industry, heavy**

Committee, List 4. Loosely used for those secondary industries in which large weights of materials are handled. More precise tests have been suggested based on the following criteria: 1. weight of material per operative; 2. value of a given weight of product; 3. cost of materials as a proportion of gross value of output; 4. proportion of men to total workers employed; 5. horse-power capacity in use per operative.

Jones, J. H., 1940, Memorandum on the Location of Industry in *Report of the Royal Commission on the Distribution of the Industrial Population*, Appendix II, London: H.M.S.O. 'The "heavy" basic industries are those basic industries in which the number of insured females was less than one-tenth of the total number of insured persons at July, 1937. In the "intermediate" basic industries the propor-tion of females exceeds one-tenth but is less than one-third; and in the "light" basic industries the proportion of females exceeds one-third.' (p. 275)

**Industry, light**

Committee, List 4. Secondary industries which do not come under the definition of heavy industry (*q.v.*).

*Comment.* In town planning the term tends to be used for an industry which can be located in a residential area without damage to amenity.

**Industry, local**

Jones, J. H., 1940, Memorandum on the Location of Industry, in *Report of the Royal Commission on the Distribution of the Industrial Population*, Appendix II. From the 45 industries of the Ministry of Labour 7 are selected that are mainly local industries: 1. distributive trades; 2. building; 3. gas, water and electricity supply; 4. tramway and omnibus service; 5. road transport (other than tramway and omnibus service); 6. laundries, job dyeing and dry cleaning; 7. bread, biscuits, cakes, etc. This group is then contrasted with the 'basic industries'. (pp. 271–272)

*Comment.* It is difficult to substantiate this use of 'local'. The industries concerned belong to the 'servicing industries' of the Scott Report quoted above under Industry, classification.

**Industry, primary**

Committee, List 4. Activities concerned with collecting or making available materials provided by nature, *e.g.* agriculture, fishing, forestry, hunting, mining.

Clark, C., 1940, *The Conditions of Economic Progress*, London: Macmillan. '... agriculture, livestock farming of all kinds, hunting and trapping, fisheries and forestry, though in certain cases it would be desirable to separate the statistics of forestry and fishing.' (p. 337) In subsequent editions Clark included mining as 'primary'.

**Industry, secondary**

Committee, List 4. Industry which transforms the material provided by primary industry into commodities more directly useful to men, *e.g.* manufacture, construction, energy generation.

Clark, C., 1940, *The Conditions of Economic Progress*, London: Macmillan. 'Secondary industry is defined to cover manufacturing production, building and public works construction, mining and electric power production. Mining and electric power production are in certain countries included with primary production, on the grounds that they represent the exploitation of natural resources.' (p. 337)

**Industry, tertiary**
Committee, List 4. Industry, in the sense of Industry (1), other than primary or secondary industries, *e.g.* transport, trade, finance, communications, the professions, administration, personal and other services. (Industry, 1 = 'work performed for economic gain'.)

Clark, C., 1940, *The Conditions of Economic Progress*, London: Macmillan. Tertiary industry seems to be equated with services, including 'Commerce and distribution, transport, public administration, domestic, personal and professional services.' (p. 338)

**Industry, quaternary**
From tertiary industry quaternary can be identified.

Abler, Adams and Gould, 1971. 'Quaternary enterprises specialize in the assembly, processing, and transmission of information and the control of other enterprises.' ... 'The fundamental inputs and products in quaternary activity are information ...' (pp. 307, 553)

**Inface**
The scarp face of a cuesta, in contrast to the back slope. See quotation from Lobeck under cuesta.

**Infantile land forms**
Cotton, 1941. 'The theoretical distinction between "infantile" forms developing on a peneplain as it is slowly uplifted and the "senile" forms it exhibited before uplift was first made by Walther Penck.' (p. 191—see the discussion on pp. 191–193 and elsewhere)

**Infantile town**—see Urban hierarchy

**Infield, in-field**
*O.E.D.* The land of a farm which lies around or near the homestead, as opposed to the outlying parts, which are usually on higher ground and may consist of moorland; hence arable land as opposed to pasture; land regularly manured and cropped.

Comment. The *O.E.D.* definition is incomplete as the quoted examples themselves show. In Scotland where the system was characteristic the outfield was cropped in places after the manner of systems of shifting cultivation or land rotation.

**Infield-outfield** (infield and outfield)
*O.E.D.* A system of husbandry which confines manuring and tillage to the infield land.

Stamp, 1948. 'In the Celtic system there was a group of small enclosed fields near the farmstead which were kept in more or less constant cultivation and which received all the manure produced by the farm animals which were also kept mainly in the enclosures. This constituted the "infield". In addition tracts of the nearby hillside or other open land were cleared of the scrub or rough herbage by which they were naturally clothed, cultivated (often with one grain crop such as oats) for as many years as they would produce even a meagre yield and then abandoned in favour of new tracts. The area so cropped constituted the "out-field" and the system is comparable to that well known in many parts of the tropics today as the system of shifting agriculture' (p. 46).

**Infield-Outfield.** System whereby the arable field of a township is continuously cultivated as 'infield' while portions of (usually) heathland are periodically cultivated as the 'outfield'; the latter composed of 'brecks'. Norfolk 16th to 18th centuries (I.L.A.T.).

**Influent**
A stream flowing *in*, especially into a cave—contrast effluent.

**Information field**—*see* Mean information field and Private information field.

**Ingraphicate**—See graphicate.

**Ingrown meander**—See incised meander, intrenched meander.

**Ings** (Yorkshire)
Holliday, R. and Townsend, W. N., 1959, *Soils in York: A Survey, 1959* (British Association). 'Some of the alluvial soils alongside rivers subject to regular flooding in both summer and winter are referred to as "Ings" land—good examples may be seen alongside the River Derwent' (p. 51). *Cf.* innings.

**Inherited valley**
Mill, *Dict.* Valleys cut by an old and obsolete drainage system which have been occupied subsequently by a new drainage system.
See also Superimposed.

## Inland Basin

Mill, *Dict*. A depression entirely surrounded by higher land *i.e.* having no communication with the ocean.

## Inland ice

Mill, *Dict*. An ice-cap of very great extent, as in Greenland.

Flint, 1947. An extensive ice sheet which buries mountains beneath itself, is called by some, especially in Scandinavia, inland ice.

No reference in Wright, 1914; *U.S. Ice Terms*, 1952; Daly, 1934; Wooldridge and Morgan, 1937.

*Ice Gloss.*, 1956. Inland Ice Sheet. An *ice sheet* of considerable thickness and more than about 50,000 sq. km. in area, resting on rock. Inland ice sheets near sea level may merge into *ice shelves*.

*Comment*. Sometimes used in the German form *Inlandeis* (G.T.W.).

## Inland sea

Mill, *Dict*. A large isolated portion of the hydrosphere entirely surrounded by land.

## Inlet

*O.E.D.* A narrow opening by which the water penetrates into the land; a small arm of the sea, an indentation in the sea-coast or the bank of a lake or river; a creek.

## Inlier

*O.E.D.* (Given but not defined except by reference to quotations from Page, 1865).

Mill, *Dict*. A mass of older stratified rocks surrounded by newer strata.

Page, 1865. A term introduced by Mr. Drew, of the Geological Survey, to express the converse of 'outlier', which see. 'It means', says Mr. Drew, 'a space occupied by one formation which is completely surrounded by another that rests upon it'.

Geikie, J., 1912. 'An *Inlier* is the converse of an outlier, and consists of rocks which are surrounded on all sides by rocks which are geologically younger. The rocks may belong to the same series as those by which it is surrounded or they may be overlaid discordantly by the latter.... Faulting also sometimes accounts for the presence of an inlier forming elevated ground ... this is the structure of the "Horst mountains" of German geologists.' (pp. 183–184)

## Inning, innings

*O.E.D.* The action of taking in, inclosing etc.; especially the reclaiming of marsh or flooded land.

Innings: Lands taken in or reclaimed.

## Innovation waves

Abler, Adams and Gould, 1971. 'Thinking of diffusion processes in terms of waves is a common verbal and conceptual analogy which originated in an early work entitled *The Propagation of Innovation Waves* by Törsten Hägerstrand.... One of his early studies in Sweden investigated the diffusion of automobile ownership which seemed to sweep like a wave across the southern province of Skane.' (p. 346)

*Comment*. Like all analogies it must not be used too literally. The diffusion wave concept appears to work best when the process involves individual decisions to adopt an innovation. See Abler, Adams and Gould, 1971, pp. 395–397, and Hägerstrand, T., 1952, *The Propagation of Innovation Waves, Lund Studies in Geography*, Series B, No. 4, Lund: Gleerup.

See also Diffusion.

## Input-output analysis

Keeble, D. E. in Chorley and Haggett, 1967. '... input-output models explicitly recognize that changes in production in one sector inevitably affect production in many other sectors as well—by means of altered demands from outputs from these other sectors as inputs to the original sector.' (p. 255)

Abler, Adams and Gould, 1971. 'Input-output analysis set the foundation for ... how the specialized industrial sectors of the economy interact.' (p. 204)

See Holland, S., 1976, *Capital Versus the Regions*, London: Macmillan Press for discussion on the limitations of input-output analysis. (pp. 25–26, 185–194, 228–231)

## Inselberg (German; plural Inselberge = island mountains)

Wooldridge and Morgan, 1937. 'As long ago as 1904 Passarge, in his studies of arid South Africa, noted the existence of vast flat surfaces studded with sharply rising residual hills, termed in German literature, "Inselberge".'

*Comment*. Similar descriptions and discussions of steep sided, isolated, residual hills in Africa, Australia, Arabia are contained in Holmes, 1944; Cotton, 1945. Their formation is characteristic of hot dry lands. Now in general use as a descriptive term implying little more than an isolated hill of circumdenudation. Fischer and Elliott, 1950, go so far as to equate with outlier, monadnock and bornhardt.

**Inselberg-and-pediment landscape**
Dury, 1959. 'Inselberg-and-pediment landscapes have been studied by Lester King and others, chiefly in regions where the climate includes one wet season and one dry season.' (p. 64) See Pediment.

**Insequent stream, drainage** (W. M. Davis, 1894)
*O.E.D.* Insequent. *Obs. rare.* Following on, succeeding; subsequent.
Davis, 1907 (1894). '...regions of horizontal structure have no normal subsequent streams. All the branch streams are either perpetuated consequent streams, or else they are developed under accidental controls, of which no definite account can be given. It is to these self-guided streams that I have applied the term "insequent".' (p. 174)
Similarly Cotton, 1922; Wooldridge and Morgan, 1937.
Committee, List 2. Insequent drainage. A drainage system developed under accidental controls of which no definite account can be given. Reference is made to W. M. Davis, The Geographical Cycle, *Geog. Jour.*, **14**, 1899, 490.
Fay, 1920. In geology, developed on the present surface but not consequent on nor controlled by the structure; said of streams, drainage and dissection of a certain type.
Rice, 1941. Streams beginning as runs over the edges of a cliff, where they cut a gully which gradually becomes elongated (Grabau).
*Comment.* Wooldridge and Morgan, 1937, point out that the distinction between subsequent and insequent has greater practical than theoretical validity (p. 191).
See also Dendritic drainage plan.

**Inshore**
Johnson, 1919. Used as equivalent to the 'shoreface' of Johnson which lies between low tide shoreline and the beginning of the more nearly horizontal surface of the offshore. It is a narrow zone over which the beach sands and gravels actively oscillate with changing wave conditions. (p. 161)

*Adm. Gloss.*, 1953. Close to the shore. Used sometimes to indicate shorewards of a position in contrast to seawards of it Inshore water. A strip of open water to seaward of the ice foot or the land, formed by the melting of the fast-ice near shore due to the movement of the ice and thaw
*Comment.* No reference in Wooldridge and Morgan, 1937; Cotton, 1945; Steers, 1946 Moore, 1949. An indefinite term with no precise meaning.

**Insolation**
*O.E.D.* The action of placing in the sun; exposure to the sun's rays; sometimes (in modern use) the action or effect of the sun's rays on a body exposed to them.
Mill, *Dict.* The amount of solar radiant energy which reaches the actual surface of the Earth in any locality.
*Met. Gloss.*, 1944. A term applied to the solar radiation received by terrestrial or planetary objects (Willis Moore).
*Comment.* For the year as a whole insolation is greatest at the equator. It decreases at first slowly, then more rapidly, then slowly again towards the poles. It shows least variation throughout the year at the equator, and most at the poles. Insolation is thus an important factor in climate; it plays a significant if disputed role in atmospheric weathering, whilst the intimate relationship between relief and insolation plays an important role in human geography. See Adret, etc. For full discussion see Garnett, A., Insolation and Relief, *Publ. Inst. Brit. Geog.*, **5** and Peattie, R., 1936, *Mountain Geography*.

**Installed capacity**
Committee, List 4. The total potential capacity of a plant or machine, as distinct from the utilized capacity. The percentage relationship between utilized and installed capacity is the capacity factor (or plant factor) when average load is taken.
*Comment.* Used especially of water power and sometimes contrasted with potential capacity, *e.g.* of a waterfall.

**Insular Climate**
Mill, *Dict.* A climate with comparatively little range between summer and winter temperatures, characteristic of many islands (compare oceanic climate, contrast continental climate).

**Intake**
*O.E.D.* 2. A piece of land taken in from a moorland, common etc.; an enclosure.

3. The place where water is taken into a channel or pipe from a river or other body of water, to drive a mill, or supply a canal, waterworks etc.

**Intensive (agriculture)**
O.E.D. Intensive. 5. Econ. Applied to methods of cultivation, fishery, etc., which increase the productiveness of a given area: opposed to extensive in which the area of production is extended.
Committee, List 3. Farming in which large amounts of capital or labour are applied to a given area.
Moore, 1949. Intensive Cultivation. The system of farming by which the land, being too valuable to be allowed to lie fallow at any period, is kept continually under cultivation by means of a Rotation of Crops and manuring. Intensive cultivation is practised in countries such as Denmark and England, where land is relatively scarce, and high crop yields are sought and obtained.
*Comment.* It is the application of capital and/or labour which should be emphasized. 'Extensive' does not necessarily mean extending the area, whilst crop rotation is usual in many types of extensive farming (C.J.R.).

**Interactance, interactance hypothesis**
Mackay, J. Ross, 1958. The Interactance Hypothesis and Boundaries in Canada: A Preliminary Study, *The Canadian Geographer*, **11**, 1–8. 'The interactance hypothesis has been defined most fully by S. C. Dodd whose model is shown in equation:

$$I = \frac{KTP_A P_B A_A A_B}{D}$$

K is a constant. T is the time element, such as a day or a week over which interactions are measured. $P_A$ and $P_B$ are the populations of the two interacting groups. $A_A$ and $A_B$ are the specific indices of per capita activity of the populations. D is a space dimension.
Dodd, S. C., 1950, The Interactance Hypothesis: A Gravity Model Fitting Physical Masses and Human Groups, *Amer. Soc. Rev.*, **15**, 245–256.

**Intercision**
Lobeck, 1939. 'Capture by sidewise swinging of mature streams ... mature meandering Huron River impinged against the course of Oak Run and captured it. This type of capture has been termed *capture by stream intercision*.' (p. 201)

**Intercolline**
Mill, *Dict.* Intercollines. Hollows lying between the conical hillocks made up of accumulations from volcanic eruptions.
*Comment.* Appears to be obsolete.

**Intercommoning**
Parishes that had a shortage of common pasture often shared a tract of land with a neighbouring parish or parishes (I.L.A.T.).

**Interfluve**
Webster. The district between adjacent streams flowing in the same direction.
Wooldridge and Morgan, 1937. 'The inter-stream areas or *interfluves*. ...' (p. 175)
Swayne, 1956. 'Higher land initially separating two river valleys.'
Cotton, 1945. 'Interfluves (spaces between rivers), which are termed doabs if flat.' (p. 60)
*Comment.* This word, not in *O.E.D.* or *Suppl.*, seems to have slipped into common use without being precisely defined. It means literally the area between rivers (Latin *fluvius*) and there seems no reason to restrict the meaning as Swayne has done. Himus, 1954, goes further and gives 'a ridge separating two parallel valleys'.

**Interface**
O.E.D. A surface lying between two portions of matter or space, and forming their common boundary ... a face of separation, plane or curved, between two contiguous portions of the same substance.
*Comment.* Has been used in recent years of the boundaries between related disciplines especially if ill-defined, *e.g.* between geography and related studies.

**Interglacial**
Webster. *Geol.* Occurring between two glacial epochs; as, an interglacial climate.
Mill, *Dict.* Interglacial Period—A period characterized by a relatively warm climate intervening between two periods of glacial cold. Similar use by most writers.
Dury, 1959. 'The Ice Age is divided into a number of glacials and interglacials. During interglacial times, the climate was at least as genial as it is to-day. When attempts are made to draw up a scheme of successive glacials and interglacials, difficulties arise.' (p. 155)
*Comment.* Dury's use of interglacial as a noun meaning interglacial period may lead to confusion because interglacial is also commonly applied by geologists to deposits formed during the period.

**Interior basin, drainage** etc.—see Internal basin, drainage etc.

**Interlobate moraine**
Swayne, 1956. A moraine formed between two tongues of a single glacier.
Thornbury, 1954. '*Interlobate moraines* form in the reentrant between two ice lobes.' (p. 389)

**Interlocking spurs**—see Meander

**Intermediate rocks**
Himus, 1954. Igneous rocks containing between 66 and 55 per cent of silica and so intermediate in chemical composition between the acid (granitic) and the basic (gabbroid) rocks. Include syenite and trachyte (sub-acid), diorite and andesite (sub-basic).
The plutonic rocks are syenite and diorite, the volcanic are trachyte and andesite.

**Intermont, intermontane, intermountain**
*O.E.D. Suppl.* Situated between mountains. The most common form is intermontane and is given by *Webster* as early as 1828. Its use spread from America and is now usual. Mill, *Dict.*, gives 'Intermont—a hollow between mountains' but this appears to be unusual.

**Internal drainage**
Mill, *Dict.* A drainage system the waters of which do not reach the ocean. See inland basin.

**Internal migration**
Movement of people within a country, *e.g.* in search of employment.

**International Date Line**
Swayne, 1956. An imaginary line, internationally accepted as a date line, following closely the 180° meridian, adjustments being for convenience' sake so that Alaska and the Aleutian Islands lie to the east of the line and some of the South Sea Islands to the west of it.

**International Map**
Unless otherwise qualified refers to the International Map of the World on the scale of 1:1,000,000 which was planned under the auspices of the International Geographical Union and carried out by very many of the countries of the world.

**Interstadial**
Between stages; used especially of the period or deposits formed during the period between two stages in glacial retreat. See Stadial moraine

**Intertropical**
Mill, *Dict.* Of or pertaining to regions between the tropics.

**Inter-Tropical (Intertropical) Convergence Zone (ITCZ), Intertropical Front (ITF)**
Hare, 1953. '... The trade winds meet along a fairly definite "front". '... This fundamental atmospheric boundary has become known as the *intertropical front* (abbreviated ITF) or sometimes as the equatorial front. Plainly the intertropical front must generally be the axis of the doldrum belt ... the word "front" is something of a misnomer and it has been suggested that the term *"intertropical convergence zone"* should replace the older form.' (p. 108)
Comment. ITCZ is now the usual form.

**Intervening opportunity**
Stouffer, S. A., 1940, Intervening Opportunities: A Theory Relating Mobility and Distance, *Am. Soc. Rev.*, 5.6, December, 845–867. '... introduces the concept of intervening opportunities. It proposes that the number of persons going a given distance is directly proportional to the number of opportunities at that distance and inversely proportional to the number of intervening opportunities.' (p. 846)

**Intratelluric water**—see Juvenile water

**Intrenched meander**
(*cf.* Incised meander).
Cotton, 1922. Winding streams which, because of uplift, have cut down into their valleys, create entrenched meanders. (p. 225)
Wooldridge and Morgan, 1937 (of incised meanders). 'Rapid incision gives rise to "intrenched meanders", of more or less symmetrical cross-section, while if the incision is less rapid, affording time for lateral shift, "ingrown meanders" result.' (p. 225)

**Intrusion**
*Webster.* 3. *Geol.* The forcible entry of molten rock or magma into or between other rock formations; the intruded mass.
Himus, 1954. Bodies of igneous rocks of varying size and structure which, in the condition of molten magma, were intruded into pre-existing rocks. Intrusive igneous rocks are thus contrasted with extrusive igneous rocks or volcanic rocks. See dyke sill, laccolith, phacolith etc.

**Intrusive igneous rock**
Webster. Intrusive 4. *Petrog.* a. That has been forced, while in a plastic or liquid state, into the cavities or cracks or between the layers of other rocks. According to its shape, the intrusive mass is called a boss, sheet or sill, dike, etc. b. Plutonic.
Mill, *Dict.* A rock formed by the consolidation of molten material which has been forced into cracks or between the layers of rock masses.

**Invagination**
Mill, *Dict.* Limited local subsidence of a volcano or surrounding country.

**Inversion (geological)**
In regions of intense folding, strata may actually be inverted (as on the lower limb of an overfold) and the observed sequence is upside down (Geikie, J., 1898, p. 303).

**Inversion (of temperature)**
Mill, *Dict.* A phenomenon sometimes observed in winter in narrow valleys or in hollows or occasionally on plains near isolated peaks. The temperature falls as one proceeds down into the valley. This occurs during winter anticyclonic calms, when the cold layers of air simply flow to the lower level. This inversion of temperature is one reason why villages in regions subject to it often avoid the lowest land.
*Met. Gloss.*, 1944. Inversion. An abbreviation for 'inversion of temperature-gradient'. The temperature of the air generally gets lower with increasing height, but occasionally the reverse is the case, and when the temperature increases with height there is said to be an 'inversion'. ... The term 'counter-lapse' has been suggested as an alternative to inversion.

**Inverted relief**
Mill, *Dict.* Inversion of Relief. An extreme case of ennoyage (*q.v.*); the synclines stand out and the anticlines are subordinate. The former are called perched synclines, and they abound in highly dissected regions (*e.g.* Alps, Wales, etc.).
Cotton, 1922. Streams cut ravines in the sides of dormant volcanoes which will be occupied by the next lava-flows, forcing the streams to flow on their former divides, thus giving rise to 'inversion of topography' (pp. 354–356) (W.M.). Stated more generally with reference to lava-buried landscapes in 1945 (pp. 366–367).
Wooldridge and Morgan, 1937. Landscape where ridges and high ground are synclines and anticlines form low ground. Attributed to successful competition of subsequent with consequent valleys and structural instability of the anticlines enabling more rapid weathering (pp. 203–204) (W.M.).

**Invierno**
Mill, *Dict.* The rainy season in intertropical America. *Cf.* Verano.

**Invisible exports**
Moore, 1949. Items, representing services rendered to a foreign country which do not involve actual transfer of goods. Examples are the trade conducted for foreign countries by national shipping, expenditure in the country by foreign tourists, the interest on national capital invested in foreign countries.
*Invisible imports* are similar movements in the reverse direction.

**Ionosphere**
About 18 miles from the earth the stratosphere gives place to the ionosphere so called because free ions or electrons are present. Several layers D, E, and F are distinguished according to the concentration of electrons and it is these layers which reflect radio waves back to the earth and so control wireless reception. The ionosphere extends at least for several hundred miles.
See also Heaviside layer.

**Iron Curtain** (Winston Churchill, March 1946)
In a speech at Fulton, Missouri, in the presence of President Truman of the United States Winston Churchill referred to 'an iron curtain which had come down in Europe'—separating the Communist countries of the East from the West. Behind the Iron Curtain are U.S.S.R., Poland, Czechoslovakia, Hungary, Roumania and Bulgaria. The position of Yugoslavia and Albania is less certain; the former Russian zone of East Germany, now the Deutsche Demokratische Republik (D.D.R.), is also included, and in some contexts, the Chinese People's Republic. See also Bamboo Curtain.

**Iron pan**
Himus, 1954. A hard layer often occurring in sands and gravels a short distance below the surface, resulting from the leaching of iron from the upper layers and its

subsequent deposition as hydrated oxide of iron at a lower level. See also Hard pan.

**Irrigation**
*O.E.D.* The action of supplying land with water by means of channels or streams; the distribution of water over the surface of the ground, in order to promote the growth and productiveness of plants.

**Isarithm, isarithmic** (maps)
Used by A. G. Ogilvie, *Report of the Commission for the Study of Population Problems*, International Geographical Congress, Washington, 1952, p. 7, as equivalent to isopleth for maps of population density. Not in *O.E.D.*

**Isblink** (Danish)
Mill, *Dict*. The seaward splayed-out end of a stream of inland ice (Greenland).

**Ishinna** (Swedish)
Ice-film, ice-scum. The initial freezing in calm and even in windy frosty weather taking the form of thin oil- or film-like ice on the sea surface. (E.K.)

**Island**
*O.E.D.* A piece of land completely surrounded by water. Formerly used less definitely, including a peninsula or a place insulated at high water or during floods, or begirt by marshes, a usage which survives in particular instances.
*Transf.* an elevated piece of land surrounded by marsh or 'intervale' land; a piece of woodland surrounded by prairie or flat open country; a block of buildings etc.
*Comment.* Some writers would extend the definition by saying 'a piece of land of less than continental size' (*e.g.* Swayne, 1956) but Mackinder coined the phrase 'world-island' for Eurasia-Africa. By analogy, the use of the word island is now commonly extended to many uses, *e.g.* an island site—a building site surrounded by roads; a recreational island in a sea of houses, etc.

**Island arc**
Steers, J. A., 1932, *The Unstable Earth*, London: Methuen. The arclike disposition of island chains is common in the Pacific. Suess showed they tend to have many points in common including an outer Tertiary zone, an inner cordillera of folded Paleozoic rocks within which is a volcanic zone, the outer (convex) side of the Tertiary zone flanked by a foredeep (W.M.) (pp. 49–53).

**Isle**
*O.E.D.* A portion of land entirely surrounded by water; an island. Now more usually applied to an island of smaller size, except in established appellations, as 'the British Isles'.
*Comment.* The word isle has long been favoured in poetic and romantic literature; in scientific writing it is normal and preferable to use 'island' independently of the size of the area concerned.

**Islet**
*O.E.D.* A little island, an eyot or ait. See also Inch, innis.

**Iso-**
*O.E.D.* Iso-, before a vowel sometimes is-, combining form of Greek ισος equal, used in numerous terms, nearly all scientific the second element being properly and usually of Greek origin, rarely of Latin (the proper prefix in the latter case being *Equi-*). Many recent words of this class are terms of Physical Geography, Meteorology etc. formed on the analogy of *isotherm isothere, isochimenal*, the French originals of which were introduced by A. von Humboldt in 1817.
*Comment.* The terms introduced by Humboldt referred to lines on maps and it is in this sense that most of the words listed below are used by geographers. As noted by Monkhouse and Wilkinson, 1952, 'to cover all lines representing constant values on maps the terms isopleth, isarithm, isoline, isobase, isogram, isontic line and isometric line have been used at various times' (p. 29). The most favoured general term is isopleth (but see below) though purists may protest that this has been used in different senses and refers to a measure not to a line on a map. Griffith Taylor, 1951, claims to have introduced isopleth to geographical literature in 1910 meaning 'a line of equal abundance', Moore, 1949, defines isopleth as 'a line on a map drawn through places having the same value of a certain element'. The word isoline may be self-evident in its meaning but it is unfortunately a bastard word. Isogram was proposed by F. Galton in 1889 (*Nature*, 31 Oct., p. 651) ... 'isobars, isotherms, and other contour lines ... (to which the general name *isograms* might well be given)' ... but has not been generally adopted.

New terms are constantly being introduced whilst others fall into disuse. The following have been selected as the more important in current literature.
For a brief definition of many of the terms in German, see Gulley, J. L. M. and Sinnhuber, K. A., 1961, Isokartographie: eine terminologische Studie, *Kartographische Nachrichten*, **4**, 89–99.

*Isabnormal line.* Formerly used for isanomalous line.

*Isallobar.* Line on a map joining places with equal pressure-differences in a given interval (Mill, *Dict.*). Line on a weather chart through places at which the same change of pressure or *Pressure Tendency* has taken place during some period of time (Moore, 1949).

*Isanakatabar.* Line on a map joining places with equal pressure amplitudes (Mill, *Dict.* quoting Shaw).

*Isanomal.* 'isopleth of anomalies' (Monkhouse and Wilkinson, 1952, p. 125) = isabnormal, also iso-abnormal (*O.E.D.*).

*Isanomalous line.* Line on a map joining a continuous series of stations for which the difference between the average temperature of all stations in its latitude is equal. There will be lines of positive anomaly, as in the N. Atlantic, and negative anomaly, N.E. Asia (Mill, *Dict.* quoting Dove). Line joining places having equal departures from the normal in some meteorological element (Moore, 1949).

*Isarithm.* Any line representing constant value on maps (Monkhouse and Wilkinson, 1952, p. 29). 'Among German and Scandinavian geographers "Isarithmen" is in common use...'. *Geog. Rev.*, **20**, 1930, 341) (compare Isopleth).

*Isoamplitude*, line of. Line on a map joining a continuous series of stations with the same amplitude of variation of the character studied. They may be drawn for variations of temperature, barometric pressure, etc. (Mill, *Dict.*).

*Isobar.* 'Lines on a chart joining places of equal barometric pressure' (*Met. Gloss.*, 1944).

*Isobath.* Line on a map representing lines at the bottom of the ocean or any body of water every point of which is at the same depth (Mill, *Dict.*). Submarine contour; equal depths below sea-level.

*Isobathytherm* (also *isothermobath*). A line connecting points having the same temperature in a vertical section of any part of the sea (*O.E.D.*).

*Isobront.* Line on map joining places experiencing a thunderstorm at the same instant (Mill, *Dict.*).

*Isocheim.* A line drawn on a map through points having the same mean winter temperature (Swayne, 1956). See *Webster*, 1864.

*Isochimenal line.* An isocheim (*O.E.D.*, from Humboldt, 1817).

*Isochrone.* Originated by Francis Galton, see On the Construction of Isochronic Passage-Charts, *Proc. R.G.S.*, 3 (N.S.), 1881, 657–658. An isochrone bounds the distance that can be traversed in equal time from a common starting point. 'Isopleths which join places having the same travelling time to the centre of the city are sometimes called isochrones' (Monkhouse and Wilkinson, 1952, p. 314). Also adj. sense, commonly of pendulum, as quote in *O.E.D.* 'The degrees of the meridian, and the lengths of an isochrone pendulum, will always increase together' (1762 tr. Busching's Syst. Geog. 1. Pref. 35). *Isoclinal line.* A line drawn on a map through points at which the dip of the magnetic needle is the same (Swayne, 1956).

*Isocline, isoclinal.* See below, separate entry.

*Isocryme.* A line on a map connecting places at which the temperature is the same during a specified coldest part (*e.g.* the coldest 30 consecutive days) of the year (*O.E.D.*). A line drawn on a chart through points on the surface of the ocean having the same temperature during the coldest season of the year (Swayne, 1956).

*Isodynamic line* (A. von Humboldt). A line drawn on a map through points where intensity of the terrestrial magnetism is the same (Swayne, 1956).

*Isogeotherm.* A surface (usually imaginary) connecting points in the interior of the Earth having the same temperature (Mill, *Dict.*). A line drawn on a map through subterranean points having the same mean temperature (Swayne, 1956).

*Isogonal, isogonic line.* (A. von Humboldt). Applied to a line on a map connecting parts of the earth's surface where the magnetic declination, or variation from the true

*Isogone.* Line of equal magnetic variation (Atlas of Canada, 1959).
*Isogram.* See above (Galton).
*Isohaline.* A line on a map joining points in the oceans which have equal Salinity (Moore, 1949). Same as isohalsine of which it is the more modern form (*O.E.D. Suppl.*).
*Isohalsine.* A line on a map or chart connecting points at which the waters of the sea have an equal degree of saltness.
*Isohel.* Line showing equal amounts of sunshine (*Met. Gloss.*, 1944).
*Isohyet.* Line showing equal amounts of rainfall (*Met. Gloss.*, 1944).
*Isohydrics.* Equal hydrogen-ion concentration (Carpenter, 1938).
*Isohyomene.* Equal wet months.
*Isohypse* (also *Isohyp*). Contour line or contour (Monkhouse and Wilkinson, 1952, p. 57).
*Isoikete.* Line of equal degree of habitability, derived from Econographs. (From G. Taylor, The Distribution of White Settlement, *Geog. Rev.*, 12, 1922, pp. 375–402)
*Isokeph.* A line charting cranial variation (Taylor, 1951, introduced by Griffith Taylor).
*Isokeraunic.* Having equal frequency or intensity of thunderstorm phenomena (Swayne, 1956).
*Isokrymene.* Line on a map indicating a continuous series of marine stations with the same minimum temperature. Prepared by Dana in connection with his study of distribution of corals (Mill, *Dict.*).
*Isoline.* As isarithm, etc. See above (Monkhouse and Wilkinson, 1952, p. 29).
*Isomer.* A line joining places where the average monthly rainfall expressed as a percentage of the annual average is equal (*Met. Gloss.*, 1944).
*Isomesic.* Applied by Mojsisovics to sediments formed under the same conditions, *e.g.* all formed in a certain sea or lake etc. (Mill, *Dict.*). Opposite is heteropic. Swayne gives isomeisic.
*Isometric.* Applied to a method of projection in which the plane of projection is equally inclined to the three principal axes of the object, so that all dimensions parallel to those axes are represented in their actual proportions (*O.E.D.*). Gives a bird's eye view combining the advantages of a ground plan and elevation (Swayne 1956)—hence used in block diagrams.
*Isometric line.* 'A line that represents constant value or intensity pertaining to every point through which it passes (Wright, J. K., *Geog. Rev.*, 34, 1944, p. 653).
*Isoneph.* Line showing equal amounts of cloudiness (*Met. Gloss.*, 1944).
*Isonif.* Equal snow.
*Isonoet.* See below, separate entry.
*Isonotide* (P. Hirth). A 'courbe d'aridité or line of equal rain-factors.
*Isontic line.* As general term = isogram etc. proposed by A. C. Lane in *Science*, 68 No. 1750.
*Isopachyte, Isopach* (L. J. Wills, 1956). A line drawn on a map connecting all points at which a selected geological bed has the same thickness (Himus, 1954).
*Isophaenomenal.* Connecting places at which phenomena of any kind are equal (*O.E.D.*).
*Isophene, isophane.* Line on a map joining places of equal seasonal phenomena (in botany floral isophenes: equal flowering dates).
*Isophode.* A cost-contour, *i.e.* a contour of transport costs.
*Isophotic.* Of or pertaining to the emission of an equal quantity of light (*O.E.D.*).
*Isophyte.* Equal height of vegetation (Carpenter, 1938).
*Isophytochrone.* See below, separate entry.
*Isopic.* Applied by Mojsisovics to formations possessing the same fauna and flora though occurring in different provinces or in the same province at different times. If the similarity extends also to lithology and general characters, *holisopic* is used (Mill, *Dict.*). Applied to two contemporaneous formations which are of the same facies (Himus, 1954).
*Isopleth.* See above.
*Isopore.* Line of equal annual change in magnetic variation. (Atlas of Canada, 1959, plate 18.)
*Isopotential* (level). Surface to which artesian water can rise (Taylor, 1951).
*Isopract.* A special chart to demonstrate populations etc. (Taylor, 1951).
*Isopycnic.* Of equal density.
*Isorad.* Line showing equal amounts of radiation from bedrocks (A. Segall, *Science*, 140, 1963, 1337); line joining points of equal radioactivity drawn from Geiger data.

*Isoryme.* Line showing equal frost (Monkhouse and Wilkinson, 1952, p. 119).

*Isoseismal.* Line of equal seismic activity or of a given phase of an earthquake wave (Mill, *Dict.*). A line joining places which have suffered an equal intensity of shock from an earthquake (Moore, 1949).

*Isoseismic* (or *co-seismic*) line. Line round an epicentre enclosing an isoseismal area in which the mechanical effects of an earthquake are approximately equal (Mill, *Dict.*, quoting Milne); also used as equivalent to isoseist.

*Isoseist.* Line joining a continuous series of stations which feel a given phase of an earthquake wave at the same instant (Mill, *Dict.*).

*Isostade.* Line of equal significant dates, *e.g.* foundation of towns etc. (Monkhouse and Wilkinson, 1952, p. 314).

*Isostasy, isostatic theory.* See below.

*Isotachic map.* Showing the distance that can be travelled in 24 hours, *i.e.* a 'travel-speed' map (Boggs, S. W., *A.A.A.G.*, 31, 1941, 119-128).

Abler, Adams and Gould, 1971. 'Within each city neighbourhood a specified maximum rate of travel is possible. A rate of travel or isotachic map is derived from isochrone maps. Each isochrone map is drawn for a specific point. By measuring rates of travel speed at different places on the isochrone map, we can assemble observations for each neighbourhood and create the isotachic map ... one isotachic or rate of travel map underlies an infinite number of isochrone maps of an area.' (pp. 284-285)

*Isotalantose.* Line on a map connecting places with an equal range of temperature between the mean of the hottest and that of the coldest months of the year (Mill, *Dict.*, quoting von Kerner).

*Isoterp.* Line indicating equal comfort (for human beings) (Taylor, 1951, introduced by Griffith Taylor).

*Isothere.* A line drawn on a map through points having the same mean summer temperature (Swayne, 1956).

*Isotherm.* A line joining places along which the temperature of the air or of the sea is the same. When the places are at different altitudes a correction may be necessary on account of the general upward decrease of temperature (*Met. Gloss.*, 1944).

*Isothermal zone* or *layer.* Now more commonly known as stratosphere (*q.v.*).

*Isothermic gradient.* The slope of an isothermic surface (Mill, *Dict.*).

*Isothermic surface.* An imaginary surface connecting points all having the same temperature at the same time (Mill, *Dict.*).

*Isothermombrose.* A line drawn on a map through points having the same rainfall during the summer (Swayne, 1956).

*Isotype.* A type or form of animal or plant common to different countries or regions (*O.E.D.*).

*Isovol.* A line drawn on a map through points where coal has the same ratio of fixed to volatile carbon (Swayne, 1956).

*Isoxeromene.* A line on a map connecting points of equal atmospheric aridity (R. H. Fuson, 1963, *Professional Geographer*, 15, 3, p. 4).

**Isobase** or isoanabase (Swedish isobas, *pl.* isobaser, isoanabas, *pl.* isoanabaser, fr. Greek, isos, equal and basein, rise). A topographic or imaginary contour line on a map, drawn through a series of points of equal elevation in a topographic surface of line, formerly level, but at present deformed; lines of equal land uplift. Introduced by G. De Geer, 1890. (E.K.)

*O.E.D.* Line drawn through areas formerly of equal elevation which have undergone deformation.

Mill, *Dict.* Line on a map of an elevated or depressed region, indicating a continuous series of points which have been equally elevated or depressed by some Earth movements, *e.g.* old shore-lines at an equal elevation above the present sea-level.

Himus, 1954. Line drawn through places where equal depression of the land took place in glacial times, as a result of loading with ice.

**Isoclinal folding, isocline**

Himus, 1954. A fold which has been partially overturned so that both limbs dip in the same direction and at approximately the same angle.

Swayne, 1956. Isoclinal: dipping at the same angle or in the same direction; isocline: a fold in which the strata have been subject to unilateral thrust so that the limbs dip in the same direction.

But also A. von Humboldt *isocline* applied to magnetic declination.

**Isoclinal ridge**
Mill, *Dict*. A ridge composed of strata having the same dip throughout.

**Isoclinal valley**
Mill, *Dict*. A valley in which the strata on both sides have the same dip.

**Isodapanes**
Weber, A., 1929, introduced this term in analogy to the 'isotherm' to denote 'curves of equal transportation cost'. (p. 103)

Isard, W., 1956, *Location and Space Economy*, New York: Wiley. '... isodapanes as used by Palander and Hoover, is a locus of points at each of which the location of the production process would incur the same over-all (combined) transport costs in the movement of both raw materials and finished product.' (p. 122)

Haggett, 1972. 'Lines connecting points with equal transportation costs are termed isodapanes and, as with isotims, may be regarded as contours showing the cost terrain of a particular region.' (p. 195)

*Comment*. Iso (equal) dapane (expense, cost). See Weber, A., 1929, *Theory of the Location of Industries*, Chicago: University of Chicago Press, App II, p. 240; Hoover, E. M., 1948, *Location of Economic Activity*, New York: McGraw-Hill, and Palander, T., 1935, *Beiträge zur Standartstheorie*, Uppsala.

**Isonoet, isonoetic** (Peter Scott, 1957)
'The pattern presented by the distribution of the percentages [of children of low intelligence, with I.Q. ratings D and E] was found to be so significant that it was possible to interpolate isopleths, for which the writer suggests the term "isonoets" (Greek, *noeticos*, intellectual).' Peter Scott, An Isonoetic Map of Tasmania, *Geog. Rev.*, 47, 1957, 311–329, (p. 317).

**Isopach**
Wills, L. J., 1956. *Concealed Coalfields*, London: Blackie. 'An *isopach* is a contour of equal thickness or, if measured from the present surface, a contour of equal depth.' (p. 6)

*Comment*. This definition by Wills involves an unusual use of 'contour', where isopleth would be more usual, and a redefinition of the older word 'isopachyte'.

**Isophytochrone**
Paterson, S. S., 1956, *The Forest Area of the World and its Potential Productivity*, Goteborg. 'Lines connecting places with equally long growing season' (p. 8).

**Isopleth**
*O.E.D. Suppl*. Meteorol. (from Greek, fullness), a line plotting the distribution of a given phenomenon.

Mill, *Dict*. Strictly an Isogram; in practice a line drawn on a map separating the region of values above a certain fixed limit from that of those below, *e.g.* Isobars are Isopleths of Pressure.

*Met. Gloss.*, 1944 (of the use of isopleth = isogram, as a generic term). 'The use of isogram, however, seems advisable, in order to avoid confusion, as isopleth is often used with a more specific meaning, viz. a line showing the variation of an element with regard to two co-ordinates, such as the time of the year (months) and the time of day (hour).'

*Comment*. See above under *Iso-*. Plethron was also a Greek measure of distance and area.

**Isostasy** (C. E. Dutton, 1889)
*O.E.D*. Equilibrium or stability due to equality of pressure.

Mill, *Dict*. Isostasy, Isostatic Theory. The theory that the Earth's crust is brought to a state of equilibrium only when the elevation of any portion of the Earth's surface is dependent solely upon its weight (Gregory).

Steers, J. A., *The Unstable Earth*, London: Methuen, 1932. The doctrine of isostasy states that 'wherever equilibrium exists on the earth's surface, equal mass must underlie equal surface areas.' (p. 71)

Dutton, C. E., 1892 (read 1889). On some of the greater problems of Physical Geology, *Bull. Phil. Soc. Washington*, 40, 51–54.
'If the earth were composed of homogeneous matter its normal figure of equilibrium without strain would be a true spheroid of revolution; but if heterogeneous, if some parts were denser or lighter than others, its normal figure would no longer be spheroidal. Where the lighter matter was accumulated there would be a tendency to bulge, and where the denser material existed there would be a tendency to flatten or depress the surface. For this condition of equilibrium of figure, to which gravitation tends to reduce a planetary body, irrespective of whether it be homogeneous or not, I propose the name

*isostasy*. I would have preferred the word *isobary* but it is preoccupied. We may also use the corresponding adjective, isostatic. An isostatic earth, composed of homogeneous matter and without rotation, would be truly spherical. If slowly rotating, it would be a spheroid of two axes. If rotating rapidly within a certain limit, it might be a spheroid of three axes.

**Isostatic adjustment, compensation and equilibrium**

Bowie, W., 1931, *Figure of the Earth*, Ch. VII, Nat. Res. Bull., **78**, 102–115.

*Isostatic Adjustment*—The movement, mostly vertical of sections of the earth's crust and the movement, mostly horizontal of subcrustal material necessary to balance the loading and unloading of the earth's surface by erosion and sedimentation. (p. 113)

*Isostatic Compensation*—A deficiency of mass under land areas and an excess of mass under oceans. The isostatic compensation is in the crust of the earth below sea-level and is assumed to be equal to the mass of the topographic features. (p. 113)

*Isostatic Equilibrium*—The condition of rest which the outer materials of the earth tend to acquire. The equilibrium is between the crust and the sub-crustal material and not within the crust itself. It is based on the idea that the mass of each unit section of the earth's crust exerts the same pressure on the subcrustal material. (p. 113)

Davies, A. Morley, 1918, The Problem of the Himalaya and the Gangetic Trough, *G.J.*, **51**, 175–183. 'The adjustment of the material towards this condition, which is produced in nature by the stresses due to gravity, may be called the *isostatic adjustment*.... The compensation of the excess of matter at the surface (continents) by the defect of density below, and of the surface defect of matter (oceans) by excess of density below, may be called *isostatic compensation*' (p. 177).

These ideas are derived from J. F. Hayford, *The Figure of the Earth and Isostasy*, U.S. Coast and Geodetic Survey (Washington, 1909) and Supplementary Investigation (1910).

**Isostatic anomaly**

Daly, R. A., 1949, *Strength and Structure of the Earth*, New York: Prentice-Hall. Gravity anomaly which is due to horizontal and vertical variations of density below the surface of the geoid. The remainder is due to height relative to the geoid and the mass between the station and the geoid (p. 120) (W.M.).

**Isotims**

Isotims are a series of equally spaced, concentric and circular contours that describes the locus of points about each source (i.e. a resource supply site) where delivery or procurement costs are equal. Transport costs per unit weight are assumed to be equal.
See Haggett, 1972, p. 195.
See also Isodapanes.

**Isotope**

Certain elements, such as lead, uranium and carbon, occur in different forms or isotopes which can be distinguished by physical or chemical means from one another. Because one isotope changes very slowly but at known rates into another, the proportion of one to the other gives a measure of the age of the deposit which can be expressed in years.

Holmes, A., Isotopes in the Service of Geology (Review of K. Ramkama's *Isotope Geology*, New York: McGraw-Hill, 1954), *Nature*, **176** (No. 4492), 1955, p. 1038. 'The investigation of geological problems by means of research on the occurrence and natural history of isotopes and especially the variations of their relative abundances.'

**Isotropic (planes)**

*O.E.D.* Exhibiting equal physical properties or actions (*e.g.* refraction of light, elasticity conduction of heat or electricity) in all directions. 1879. Rutley, *Stud. Rocks*, ix. 79. To distinguish singly-refracting or isotropic from doubly-refracting or anisotropic minerals.

Chorley and Kennedy, 1971. 'Exhibiting equal properties or actions in all directions. When the properties or actions are unequal the medium is termed anisotropic.' (p. 351)

**Isthmus**

*O.E.D.* From Latin, isthmus, from Greek, neck, narrow passage, a neck of land between two seas, specifically the Isthmus of Corinth connecting the Peloponnesus with northern Greece.

1. Geog. A narrow portion of land, enclosed on each side by water, and connecting two larger bodies of land; a neck of land.

Mill, *Dict*. A narrow strip of land connecting two larger masses of land bordered by water.

**ITCZ**—see Inter-tropical Convergence Zone

# J

**Jama** (Serbo-Croat; *pron.* yama)
A pothole in Karst country. The Commission on Karst Phenomena of the International Geographical Union in its *Report* presented to the Rio congress in 1956 defines *jamas* as 'cavités dont le fond n'est pas apparent. Le terme peut être conservé internationalement ... les uns aboutissent à un réseau de cavités souterraines; les autres se terminent assez rapidement. On peut appeler les premiers des *abîmes*; les seconds des *avens*.' (pp. 34–35)

**Jebel, jabal, djebel** (Arabic)
Mountain.

**Jessero**
Mill, *Dict.* Marshy or lake-filled dolina (Carniola).
*Comment.* As a term this appears to have been dropped (L.D.S.).

**Jet-stream**
*Met. Gloss*, 1939. 'A jet stream is a more or less horizontal, flattened, tubular current in the vicinity of the tropopause, with its axis on a line of maximum wind speed and characterised by high wind speeds and strong transverse wind shears.
Generally speaking a jet stream is a few thousand kilometres in length, a few hundred kilometres in width and a few kilometres in depth.'

**Jetty**
Taylor, 1951. 'Silt built out by rivers in lakes etc.' (p. 616)
*Comment.* Not in *O.E.D.* or standard works; in view of confusion with the ordinary meanings of the word where human agency is involved this use is not to be recommended.

**Jhil, jheel** (Indo-Pakistan: Bengali; *jhīl* in Urdu-Hindi)
Same as bhil, bil, or bheel; a lake, swamp, or marsh.

**Jhum**—see Shifting cultivation

**Jilla, zila, jila** (Indo-Pakistan: Urdu; Hindi; Bengali)
A district.

**Joint**
*O.E.D.* 5. Geol. A crack or fissure intersecting a mass of rock; usually occurring in sets of parallel planes, dividing the mass into more or less regular blocks (quotations from 1602).
Mill, *Dict.* Divisional planes in rocks, along which they may be divided into blocks.
Himus, 1954. Vertical, inclined, or horizontal planes of division which are found in almost all rocks; produced by tension or torsion. In stratified rocks, the joints are more or less perpendicular to the planes of bedding.
Lyell, C., 1833, *Principles of Geology*, Glossary. 'The partings which divide columnar basalt into prisms are joints.'
Holmes, 1947. 'The stronger beds in a folded series of strata become ruptured when they are bent, the cracks being known as joints. Unfolded rocks are also commonly divided into blocks by jointing ... in addition to the bedding planes the rock is traversed by fractures that are generally approximately at right angles to the stratification, and therefore nearly vertical when the beds are flat. Joints frequently occur in sets consisting of two series of parallel joints one series approximating in trend to the dip, the other to the strike. Such joints are of great assistance to the quarryman in extracting roughly rectangular blocks, especially the "master joints" which are often remarkably persistent and strongly developed. In most rocks there are subordinate joints which cut across the main sets. Joints may be due either to shearing under compression or to tearing apart under tension. At the surface joints of all kinds are very susceptible to the attack of weathering agents. In igneous rocks tensile stresses are set up by contraction during cooling ... in granite three series of joints ... polygonal cracks of basalt' (pp. 75–76).

**Joint valleys**
Thornbury, 1954. 'Some valley courses or portions of valley courses are controlled by major joint systems and may be classed as *joint valleys*' (p. 114).

**Jökla mýs** (Icelandic; J. Eythórsson, 1951 *Lit.* glacier mice). See Polster

**Jökulá** (Icelandic)
River from a glacier.
**Jökull** (Icelandic; *pl.* jöklar), **jokul**
*O.E.D.* In reference to Iceland; a mountain permanently covered with snow and ice; a snow-mountain.
Knox, 1904. An ice-covered mountain or plateau.
G.S.G.S., 1944. 'Ice cap, permanent land ice.'
Taylor, 1951. 'A small icecap' (p. 616).

**Joran**
Mill, *Dict*. The cold, dry wind blowing at night from the Jura towards the Lake of Geneva.

**Junction** (of rivers)—see Accordant junction, Discordant junction

**Jungle**
*O.E.D.* (adapted from Hindi and Marathi, *jangal*, desert, waste, forest, Sanskrit *jangala*, dry, dry ground, desert)
In India, originally, as a native word, waste or uncultivated ground (='forest' in the original sense); then such land overgrown with underwood, long grass, etc.; hence, in Anglo-Indian use, a. Land overgrown with underwood, long grass or tangled vegetation; also the luxuriant and often almost impenetrable growth of vegetation covering such a tract. b. A particular tract or piece of land so covered; *esp*: as the dwelling-place of wild beasts. c. Extended to similar tracts in other lands, especially tropical.
Mill, *Dict*. Wild or uncultivated; applied in Hindi to wild animals or plants and to all waste land. In English the term is used mainly for tracts covered with wild vegetation, forest, scrub or grass.
Küchler, A. W., Localizing Vegetation Terms. *A.A.A.G.*, **38**, 1947, 208. 'The term Jungle is not included because almost every author attaches a different meaning to it. Vestal has shown that the term Jungle does not now imply a type of vegetation, nor is it a localizing vegetation term.' (See: Vestal, A. G., Use of Terms Relating to Vegetation, *Science*, **100**, 1944, 100.)
*Comment*. This is a word to be avoided in scientific literature. Particularly to be deprecated is the phrase 'jungle-forests' for equatorial or hot wet forests. Such adjectives as 'jungley' and the pseudo-Hindi 'jungli' should also be avoided, since they have been employed in a derogatory sense as implying uncivilized, boorish, primitive as in 'jungle manners' or 'junglibusti'—like a jungle village. (L.D.S.)

**Junk**
*O.E.D.* A word of oriental origin, now adopted in most European languages. A name for the common type of native sailing vessel in the Chinese seas. Early writers applied it still more widely.

**Juvenile relief, juvenile topography**
In addition to the specialized use of 'juvenile' in juvenile water, it is also used in the normal sense of youthful or immature especially of a landscape in the early stages of a cycle of erosion. Somewhat naïvely defined by Taylor, 1951, 'juvenile topography: where all valleys are narrow'. (p. 616)

**Juvenile town**—see Urban hierarchy

**Juvenile water** (*lit*. young water)
Holmes, 1944. 'Part of the water which ascends from the depths by way of volcanoes reaches the surface for the first time; such water is called *juvenile* water to distinguish it from the *meteoric* water already present in the hydrosphere and atmosphere' (p. 23).
Wooldridge and Morgan, 1937. 'Groundwater ... must be supplemented locally by such other sources as: (a) 'intratelluric' or juvenile, water, set free in the crystallization of igneous rocks ...' (p. 268).
Himus, 1954. 'Juvenile. A term applied to water and other volatile materials that are known to be magmatic emanations of primary endogenetic origin' (p. 79).

# K

**Kaatinga**—see caatinga
**Kabouk** (Ceylon: Sinhalese)
Laterite, especially as there used as a building material and road stone.
**Kachchi** (Indo-Pakistan: Panjabi; Katchchi in Urdu)
Flood-plain: the area actually flooded by the river (same as sailābā).
**Kachha**—see Kucha
**Kaffir, kafir** (Afrikaans)
Kaffer was the term applied by the Arabs on the east coast of Africa to the non-Mohammedan natives. In Arabic the word means infidel, and has a contemptuous connotation, but the Portuguese and after them the Dutch and the English took it over as the ordinary name for the south-eastern Bantu tribes. For three centuries it has been used in this way without any bad intention, and during the 19th century it was quite common in scientific publications. In the present century, however, the rise of a class of Bantu intellectuals has brought a renewed loss of status for the word. It is again being felt as a term of contempt, and is now usually avoided by well-meaning white people, who prefer to speak of Bantu or native (Afrikaans: naturel). Lately the word African has become popular in English, but it has the disadvantage of being easily confused with Afrikaans and Afrikaner. (P.S.)
*Webster*. 4. Any of certain grain sorghums derived from *Sorghum vulgare*, having stout, short-jointed, semijuicy stalks and erect heads. They are cultivated for grain and forage in southwestern U.S. and other regions of limited rainfall. Called erroneously kaffir corn.
**Kaffir corn** (South Africa)
South African variety of *Sorghum*, which used to be a very important article of diet for the Bantu, but has largely been replaced by maize. It is still important for the brewing of Kaffir beer, a thick liquid with a low alcoholic content. (P.S.)
**Kaffir farming** (South Africa)
This does not mean farming by Kaffirs, but a form of land use in which a white landowner seeks his profit not through his own farming operations, but by means of keeping a large number of Kaffir or Bantu tenants. This practice is condemned by public opinion and made difficult by law, but it has not been quite eradicated. (P.S.)
**Kainga** (New Zealand)—see pa
**Kair, Kair farming** (Russian)
Kovda, V. A., 1961 in *A History of Land Use in Arid Regions*, Paris: UNESCO, p. 196. 'The narrow strips of non-saline alluvial soil with fresh groundwaters in the vicinity are called *kairs*. Agricultural crops growing on these kairs use the fresh underground water filtering into the flood terraces from the river-beds.'
*Comment.* A local term even in the U.S.S.R. used of the Amu Darya and other rivers crossing the deserts of central Asia.
**Kallar** (Indo-Pakistan: Punjabi)
Saline soil.
Spate, 1954. 'useless expanses of alkaline efflorescence—*reh* or kallar...' (p. 463).
**Kame** (originally Scottish)
*O.E.D.* Northern and Scottish form of Comb in various senses, especially that of a steep and sharp hill ridge; hence in Geology one of the elongated mounds of post-glacial gravel, found at the lower end of the great valleys in Scotland and elsewhere throughout the world; an esker or osar. (*sic.*, should be ås).
Mill, *Dict.* A sharp and steep hill ridge. Geol., an esker (*q.v.*) (Scotland and U.S.A.).
Geikie, J., 1898. Ridges and mounds of gravel and sand generally, but now and again of rude rock-rubbish. They are of glacial and fluvio-glacial origin, having been accumulated, in many cases, along the terminal margins of large glaciers and ice-sheets.
Wooldridge and Morgan, 1937, 'The term "kame" has been applied to a variety of features, but in many cases the mounds or ridges so named appear to lie along the line of a former ice-front and to represent the confluent fans or deltas of a series of closely spaced sub-glacial streams... many kames simulate terminal or stadial

moraines... however a kame consists of sand and gravel while a moraine consists of unassorted rock-debris.' (p. 389)

Rice, 1941, summarizes loose uses and also lists: kame, crevasse-filling; kame, frontal; kame, moraine; kame, moulin; kame terrace.

Thornbury, 1954, 'The term *kame* was introduced into geologic literature by Jamieson in 1874 and has been applied to so many different types of glacial and glacio-fluviatile deposits that it has been argued by some that it is so ambiguous that it should be abandoned... it is still useful when we do not want to be or cannot be specific about the origin of a particular deposit of sand and gravel. It is difficult to define a kame but almost everyone agrees that it is a mound or hummock composed usually of poorly assorted water-laid materials.' (p. 379)

Comment. This is a word which has undergone remarkable changes in meaning. The original is a Scottish variant of the word *comb* common in place names and signifying a long steep sided ridge (comparable with a cock's comb). It is probably impossible now to endow it with any more precise meaning than that suggested by Thornbury. (L.D.S.)

See also Esker, Perforation deposit, Sag and swell topography

### Kame complex
Thornbury, 1954. 'Cook (1946) suggested ... *kame complex* for areas of sag and swell topography ... the term kame complex is a useful one if we restrict it to an assemblage of kames and do not apply it to *any* area of sag and swell topography' (pp. 378–379). See Perforation deposit.

### Kame moraine
Thornbury, 1954. 'Kames are sometimes so numerous in end moraines as to cause moraines to be designated *kame moraines*, but even here till is likely to be more abundant than water-laid materials.' (p. 379)

### Kame terrace
Thornbury, 1954. 'Kame terraces are fillings or partial fillings of depressions called *fosses* between a glacier and the sides of its trough.' (p. 378)

### Kampong (Malay)
Dobby, E. H. G., 1957, *Malayan J. of Tropical Geog.*, 10. 'Cluster of buildings making up a large homestead or a small hamlet, and including the surrounding mixed gardens' (p. vi).

'Kampong horticulture' is used for one of the types of land use on the detailed maps of Singapore, 1958 (Robert Ho).

Also used in Java for a village.

### Kanat—see Qanat, Kārez

### Kankar (Indo-Pakistan: Urdu-Hindi) also kunkur
Nodular concretions of limestone found in the older alluvium of the Indo-Gangetic plain.

### Kaoliang
A grain corn of north China. It is one of the small grains to which the term 'millet' is loosely applied in English. The enumeration and definition of the numerous 'millets' is beyond the scope of this Glossary.

### Kaolin
China clay: from Chinese *kao* high and *ling* hill from the name of the mountain in north China where it was originally obtained.

### Kaolinization, kaolinisation, kaolisation
The process by which granite is attacked and the felspars with other aluminium silicates altered by heated gases and waters to a soft white clay (china clay or kaolin). The details of the process and how far weathering plays a part have been matters of much discussion amongst mineralogists.

### Kar (German *pl.* Kare)
The German word for cirque (*q.v.*). See also Flint, 1947, p. 93, and Thornbury, 1954, p. 367. Not to be confused with Karre and Karrenfeld.

### Karaburan
Miller, A. A., 1953, *Climatology*, 8th Ed. 'Over these arid interiors... strong winds spring up by day, carrying clouds of dust and sand which darken the sky. The *Karaburan* is of this type, blowing strongly from the north-east in the Tarim basin.' (p. 260)

### Karewa (Indo-Pakistan: Kashmiri)
Spate, 1954. 'The word *Karewa* applies strictly to the level surface between the incised streams dissecting the terraces' [of the vale of Kashmir] (p. 372).

### Kārez (Pakistan: Baluchi)
Underground irrigation channel, tapping water from the foot of a hill range and bringing it out by gravity to a basin floor. See also Qanat and Foggara.

arling
aylor, T. Griffith, 1958, *Journeyman Taylor*. 'Snowdon is a good example of a karling, a dome into which cirques have nibbled deeply. Mount Anne in Tasmania is a perfect example' (p. 68).

aylor, 1951. 'A cluster of cirques.'

ot in *O.E.D.*, Holmes, Wooldridge and Morgan, von Engeln, Baulig or other standard works.

aroo, Karroo (Afrikaans)
lateau between the Swartberge (*antiq.*: Zwarte Bergen) and the Nuweveldberge (*antiq.*: Nieuweveld Bergen), covered with a semi-desert vegetation of small shrubs. The form with single 'r' is the accepted Afrikaans spelling; that with double 'rr' is often used in English still. The form Karru is found in some German books, but is not used in South Africa itself. The type of vegetation found in this region is also called Karroo; it extends, with some modifications, into the Little Karoo, south of the Swartberge, and the Northern Karoo, Upper Karoo, or Karroid Plateau, north of the Nuweveld Range. (P.S.)

Karre (German; usually in *pl.* Karren)
Equivalent of English clints and French lapiés. Channels up to a few inches in depth caused by chemical weathering on limestone surfaces (K.A.S.). See Karst terminology.

Karrenfeld (German)
An area dominated by *Karren*; a surface seamed with *Karren* (Mill).

Karroid vegetation (South Africa)
The vegetation of the country north of the Nuweveld Range till across the Orange River. The name is used in botanical literature and on maps, but not in everyday life. The farmer calls this whole area Karooveld, including even the south-western Free State. (P.S.)

Karst
*O.E.D. Suppl.* The name of a barren limestone plateau between Carniola and the Adriatic, marked by abrupt ridges, caverns, sinks and underground streams; used in *Phys. Geog.*, to designate a region or scenery of similar type. (First quotation supplied is 1894, *Geog. Jour.*, 3, 509, which is in a summary of a paper by K. Hassert.)

Mill, *Dict.* (Ital. Carso). A region composed of porous limestone and remarkable for surface and subterranean forms due to the action of water on the porous rock; primarily applied to the north-east side of the Adriatic.

Thornbury, 1954. 'The word *karst* is a comprehensive term applied to limestone or dolomite areas that possess a topography peculiar to and dependent upon underground solution and the diversion of surface waters to underground routes.' (p. 316)

Comment. Usage similar to *O.E.D. Suppl.* in Salisbury, 1907; Wooldridge and Morgan, 1937; Holmes, 1944; Cotton, 1945. Corresponding French term is 'causses' (Wooldridge and Morgan, 1937, p. 288). It is now normal to talk about 'karst' in any part of the world including the tropics. (L.D.S.) The word karst was derived originally from the Serbo-Croat *Kras*, meaning bleak waterless place. (G.T.W.)

Karstic
Of or pertaining to Karst country. Used especially in the phrase Karstic phenomena when translating the French 'phenomènes karstiques', but in English it is more usual to write simply karst phenomena, using karst as an adjective.

Karst terminology
The phenomena of limestone weathering in the Karst region of the Adriatic were originally described in detail towards the end of last century by Cvijić who introduced, naturally, a number of Serbo-Croat words into his descriptions (see the summary by E. M. Sanders, 1921, *Geog. Rev.*, 11). The position regarding terminology is summarized as follows by Wooldridge and Morgan, 1937, pp. 288–289. 'Local British equivalents for most of the terms exist, but they have never been systematically collected and defined. As a result British, French and German writings on the subject are exasperatingly studded with Serbian and other words which have taken on the status of technical terms... many of the words in question have much broader meanings than their technical use assumes and confusion inevitably results. ... Bare limestone surfaces commonly show a widening of joints by solution, or, in extreme cases, a complex fretting or fluting of the surface, which rises in minor ridges and pinnacles separated by deep and narrow clefts. Limestone pavements tending to this type are called "clints" or

"grykes" in the North of England. The French term for such areas of "solution fretwork" (often to be crossed on foot only with great difficulty) is "lapiés" while the Germans speak of them as "karren" or "schratten". The Serbian term for deeper fissures is "bogaz". Swallow-holes are characteristic of the karst regions. They are known as "dolines" in the type Karst, as "sotchs", "creux", or "emposieux" in various parts of France. Alternative English words are "swallet" and "sink", the latter being favoured in America. Vertical or highly inclined shafts, leading from swallow-holes, or direct from the surface to underground caves are "ponor" in the Serbian Karst and "avens" in France, while many other terms are used. Depressions larger than swallow-holes and arising often through their coalescence are termed "uvalas"; there is no English equivalent. Still larger depressions, some of rather problematical origin, are termed "poljes". Residual limestone masses or hummocks rising above polje floors are called "hums".'

The International Geographical Union set up a *Commission on Karst Phenomena* and in its *Report* of 1956 presented to the Rio Congress there is a section in French on the question of a vocabulary with recommendations for words to be used in French and German, but no reference to English use. However, different types of lapiés or karren are discussed, and the Report recommends the following terms should be retained internationally: dolina, jama, polje and kegelkarsts. Dolines are defined as 'pits at the bottom of which the soil can be cultivated' (contrast Wooldridge and Morgan) and are divided according to form. From the use which is made of them, the Commission evidently considered that other terms to be used include uvala, bogaz, cockpit, ponor, estavelle, abime, aven, hum, piton, tourelle (or their German equivalents). This list has been made the basis of words included in the present Glossary. For another review of the terminology see Sweeting, M. M., 1958, The Karstlands of Jamaica, *Geog. Jour.*, 124, 184–199.

**Kās** (*dim.*, Kassi) (Indo-Pakistan: Punjabi)
Ravines which carry water after the rains.

**Kasba** (Arabic, *pl.* ksabi)
A town or small city; in Algeria the citadel or fort in the town. (Knox)
*Comment.* The following definitions are taken from the Glossary of C. S. Coon, 1951, *Caravan: The Story of the Middle East*, New York: Henry Holt, p. 358.

*al-qasba* (Moroccan Arabic): 'the citadel or fortified refuge of a city; the old Muslim quarter of a modern mixed city'.

*qasr* (*pl.*: *qsar*): 'a fortress'—Either in urban or rural areas, the term is usually applied to rural areas (M.W.M.).

**Katabatic wind**
*O.E.D. Suppl.* Meteorology, katabatic. Of wind: caused by the local gravitation of cold air down a steep slope.

*Met. Gloss.*, 1944. Katabatic. An adjective applied to winds that flow down slopes that are cooled by radiation, the direction of flow being controlled almost entirely by orographical features . . .

*Comment.* Examples include the mistral of the Rhone valley and the bora of the Adriatic and the term is applied also to the cold winds flowing down from the ice-caps of Greenland and Antarctica.

In modern scientific literature the transliteration kata- is preferred to the older cata- from the Greek κατα meaning 'down'.

**Kataclastic structures**
Himus, 1954. 'Structures resulting in a rock from severe mechanical stresses; the constituent minerals generally are deformed and granulated.'

**Katamorphism**
Himus, 1954. 'The destructive processes of metamorphism, as opposed to the constructive processes, or *Anamorphism*.'

**Katohaline**
Mill, *Dict.* Increasing in salinity with depth (Krümmel).
*Comment.* Not in *O.E.D.* or standard works apparently defunct.

**Katothermal Lakes**
Mill, *Dict.* Lakes in which the temperature increases with the depth; a polar type of lake (Krümmel).
*Comment.* As katohaline.

**Kavir** (Iran)
Salt-marsh.

**Kayak**
An Eskimo canoe, usually of sealskin laced over a light frame.

**Kegelkarst** (German)
The term recommended for international use by the Commission on Karst Phenomena of the International Geographical Union in its *Report* (1956) to cover the types of karst country characterized by *coupoles, pitons* and *tourelles* (*q.v.*).

Sweeting, M. M., *Geog. Jour.*, **125**, 1959, gives reasons for using 'cone karst' as the nearest equivalent in English. (p. 290) She also equates with cockpit karst, *Ibid.*, **124**, 1958, 186.

### Keld
*O.E.D.* Now dial. a. A well, fountain, spring. b. A deep, still, smooth part of a river. (Frequent in place-names).

### Kerangas
Strictly a soil type (in Borneo), but coming to be used (wrongly in my opinion) for the types of vegetation (heath forest, "padang", etc.) which grow on it.' (P.W.R. in MS.)

### Kettle, kettle hole, kettle lake
*O.E.D.* Kettle: 4c. A deep circular hollow scoured out in a rocky river bed, or under a glacier, etc.; a pot-hole. 1893. Northumbld. Gloss., Kettle, a pot-hole or circular hole, scoured out in a rocky river bed by the swirling action of pebbles.

*Webster.* Kettle. 6. Geol. a. A pothole. b. A steep-sided hollow, without surface drainage, especially in a deposit of glacial drift.

*Mill, Dict.* Irregular ponds or hollows among boulder clay or morainic matter determined by irregular deposition of the drift.

Knox, 1904. (U.S.A.) A long-sided depression in sand or gravel; a hole in the bed of a stream.

Cotton, 1922. 'In these "glacial sand-plains" basin-shaped hollows, termed kettles, are rather common, which are due to great blocks of ice in the drift melting and allowing the surface to sink in.'

Wright, W. B., 1914. *The Quaternary Ice Age.* London: Macmillan (of the kettle-moraine of America). 'A characteristic of these belts is the presence of isolated basins, known as giant's kettles, containing a pond or marsh and without surface outlet. These are generally supposed to have been formed by the inclusion in the moraine of masses of ice, the subsequent melting of which has left a hollow . . . but it is clear also that the irregular dumping of ridges and mounds of debris will in itself inevitably produce many enclosed basins.' (p. 33)

Flint, 1947. 'A kettle is a depression that occurs in drift, usually stratified, and that has been made by the wasting away of a mass of ice that had lain wholly or partly buried in the drift. Although depressions of other kinds occur in drift, they are not kettles.' (p. 148) This restricted terminology appears to derive from Fuller, M. L., *The Geology of Long Island, New York*, U.S. Geol. Survey, Prof. Paper 82, 1914, pp. 38–44.

Holmes, 1944. As Flint, 1947. (p. 233)

Moore, 1949. As Flint, 1947, 'a hollow or depression in an outwash plain'. (p. 95)

Geikie, J., 1894. *The Great Ice Age*, 3rd ed., London: Stanford. '"Giants' Kettles" . . . are simply large pot-holes formed on the bed of a glacier by water plunging down through crevasses' (by moulins). 'Some of the examples occur in limestone, sandstone, and other rocks, while not a few appear in the diluvial deposits themselves.' (pp. 430–431)

Geikie, J., 1898, describes the Karst phenomena of 'kettle-valleys', trough-like or dish-shaped basins. (p. 217)

*Comment.* The above quotations make it clear that kettle-hole or kettle has, or has had, at least three meanings. The original, that recorded by the *O.E.D.*, is obsolescent in current geographical literature; the confusion with doline (Geikie, 1898, and repeated by Dicken, S. N., *J. Geol.*, **43**, 1935, 713) is to be avoided. See also Knob and kettle topography.

### Kettle-drift, Kettle-moraine
Wooldridge and Morgan, 1937. 'Mounds and ridges of gravelly drift are referred to in British glacial literature as eskers and kames, or, generally, as kettle-drift or kettle-moraine. They are regarded as formed by water or at beyond the margins of the ice.' (p. 387)

*Webster.* Kettle moraine. Geol. A terminal moraine the surface of which is marked by many kettle holes.

### Key, kay, cay (from Spanish, *cayo*, shoal or reef)
*O.E.D.* A low island, sand-bank or reef, such as those common in the West Indies or off the coast of Florida. *Cf.* the place name Key West. See also Cay, Sandkey.

### Key village
Adams, J. W. R., 1949, *Preliminary County Outline Plan, Explanatory Statement.* Maidstone: administration of the County of Kent. 'Minor centres, or key villages, where such facilities as primary schools and village halls would serve the smaller villages and hamlets in their vicinities. These minor centres would have to look to major centres for certain facilities.'

*Comment.* Such Key villages are a modern concept in rural planning and have also been called King villages.

**Keyline, Keypoint** (P. A. Yeomans, 1954)
P. A. Yeomans, *The Keyline Plan,* 1954. 'A line through this point, which may be either a true contour or a line with a slight grade, according to the circumstance of climate and landscape, which will affect the planning, is the Keyline of the Valley.'
P. A. Yeomans, *The Challenge of Landscape; The Development and Practice of Keyline,* Sydney, 1958. 'The point where the first higher slope meets the flatter lower slope [of a valley] named the Keypoint of the valley in my earlier book.'
*Comment.* P. A. Yeomans, an Australian engineer, developed a successful system of soil improvement and land management based essentially on a close study of geomorphology.

**Khad** (Indo-Pakistan: Panjabi)
A torrent in the hills.

**Khādar, khaddar, khuddar** (Indo-Pakistan: Urdu-Hindi)
New alluvium (in contrast to bangar or old alluvium); lowlying area of new alluvium liable to be flooded by river water.

**Khaderā, khuddera** (Indo-Pakistan: Panjabi)
Deep ravines, the result of destructive erosion by rain water.
Spate, 1954. 'Intricately dissected ravine lands (in Potwar), locally *khuddera*' (p. 448).

**Khāl** (Indo-Pakistan: Bengali)
Ahmad, 1958. 'Narrow natural channel of water' (p. 338). Sluggish creeks of the lower Ganges delta.

**Khamsin, khamseen** (Arabic, *lit.* fifty)
*O.E.D.* An oppressive hot wind from the south or south-east, which in Egypt blows at intervals for about 50 days in March, April and May, and fills the air with sand from the desert.
Hare, 1953. 'The Saharan type of cyclone are storms forming as frontal cyclones ... bring a wave of cT air out across the African coast ... the hot, dusty desert winds from the south-west which blow as soon as the warm front is passed are known as the Ghibli, the Scirocco in southern Italy and the Khamsin in Egypt and the Levant.' (p. 181)
*Comment.* Where the wind has crossed the Mediterranean it is often highly humid.

**Khari** (Indo-Pakistan: Bengali)
Ahmad, 1958. 'The surface [of the Ea Pakistan Barind] is cut by small stream of local origin which have deep channe and are known as *Kharis*' (p. 25).

**Kharif** (Indo-Pakistan: Urdu-Hindi; als Sudan; Arabic)
1. The rainy season of northern India.
2. Monsoon-planted and winter-harveste crop.
Spate, 1954. '... the autumn (*kharif*) an spring (*rabi*) harvests ... kharif is the mon soon crop, sown soon after the onset of th rains (June–July) and harvested in autum ...' (p. 204).
'Applied to the rainy season in the norther Sudan; Arabic for autumn' (J.H.G.L.).

**Khās Mahal** (Indo-Pakistan: Bengali)
Ahmad, 1958. 'Estate owned and manage direct by government' (p. 338).

**Khedive**
*O.E.D.* The title of the viceroy or ruler o Egypt, accorded to Ismail Pasha in 186' by the Turkish government, [at that tim in control of Egypt].

**Khirba** (Arabic: *pl.* Khirbab)
A ruin-site or temporarily occupied hamlet used in Israel. (D.H.K.A.)

**Khoai** (Indo-Pakistan: Bengali)
Spate, 1954. 'the lateritic areas (khoai [of West Bengal]' (p. 535).

**Khor** (Sudan: Arabic)
An intermittent stream. (J.H.G.L.)

**Khud Kasht** (Indo-Pakistan: Urdu)
Land cultivated by the owner himself.

**Khushkābā** (Indo-Pakistan: Panjabi)
Dry unirrigated land.

**Kibbutz** (*pl.* Kibbutzim) (Hebrew: Israel) also **Kibutz**
An enlarged version of the Kvutza (*q.v.*) but '*Kvutza* and *Kibbutz* are used interchangeably to describe Israel's collective or communal villages' A. Harman, *Israel Today*, 2 Agricultural Settlement, 1960, p. 8.

**Killas**
A Cornish miners' term which has passed into general use for clay-slates which have been altered by contact metamorphism with the granites into micaceous and schistose rocks. A general rather than a precise term, little used except relative to these rocks of Devon and Cornwall. (L.D.S.)

**Kimberlite**—see Blue ground

**-king** (Chinese)
As suffix to place name, or part of place

name, means capital. Thus Peking is northern capital, Nanking southern capital.

**King village**—see Key village

**Kink-band** (G. Voll, *Liv. & Manchester Geol. J.*, **2**, 1960, 503)

Anderson, T. B., Kink-bands and related geological structures. *Nature*, **202**, Apr. 18, 1964. 'Kink-bands are essentially small monoclinal or sigmoidal folds of a destructive morphology' (p. 272). See also E. S. Hills, 1963, p. 239.

**Kipuka** (Hawaiian)

An 'island' of old land, frequently with vegetation, left within a lava flow, a steptoe.

**Kirktoun** (Lowland Scots)

A hamlet or small village consisting of a group of farms with a church (kirk). See Fermtoun and Clachan.

*O.E.D.* gives kirk-town but quotes *Glasgow Herald* of 1864 'The word kirktoun ... applied to all collections of houses, not farmtouns, which surrounded parish kirks.'

**Kitchen-midden, kitchen midden**

*O.E.D.* A refuse heap of prehistoric date, consisting chiefly of the shells of edible molluscs and bones of animals, among which are often found stone implements and other relics of early man.

Trent, 1959. 'A mound containing the domestic refuse of early man. The term is obsolescent and the mounds are now generally known as shell mounds.' (p. 40)

**Kivas** (Hottentot)—see Straate

**Kiwi**

A flightless New Zealand bird; now used familiarly for a New Zealander.

**Kizdhi** (Pakistan: Baluchi)

Nomad's tent.

**Klint** (Swedish: *pl.* klintar)

Glint, a steep cliff, or a steep terrace, a steep edge of a plateau, *e.g.* the Baltic-Ladoga Limestone Cliff or klint extending from the coast of the Baltic to the south margin of Lake Ladoga.

In the countries surrounding the Baltic Sea klint also is used for a more or less vertical, free mountain-wall or abrasion precipice at least some yards high and a hundred or more long. (E.K.)

**Klint**

Thornbury, 1954. 'An exhumed bioherm or coral reef.' (p. 26)

*Comment.* Not to be confused with clint (*q.v.*).

**Klippe** (German: *pl.* Klippen)

*Webster. Geol.* A detached or isolated portion of an overthrust rock mass or nappe, owing its isolation to erosion; an outlier.

Lobeck, 1939. 'A *nappe outlier* or *Klippe* is a remnant of a higher nappe spared by erosion. In the field it is recognizable by the fact that older strata cap younger ones.' (p. 605)

Fischer, 1950. 'Klippe, (i) sea stack, (ii) nappe outlier.' (p. 29)

*Comment.* This is the specialized meaning in international literature of the German word which simply means crag or (granite) tor. Such is the simple meaning given in Czech and Boswell's translation of Walther Penck's *Morphological Analysis of Land Forms* (1953).

**Klong** (Thai)

The Thai word for the waterways, partly natural, partly artificial, which are particularly associated with Bangkok in international literature.

**Kloof** (Afrikaans)

Ravine, gorge; steep and short valley on a dissected mountain side. This original meaning has sometimes been watered down in proper names: *e.g.* Langkloof is a wide, open valley, nearly 200 km. long. (P.S.)

**Knick**

Tator, B. A., 1952, Pediment Characteristics and Terminology, *A.A.A.G.*, **40**. The angle between the erosion surface and adjacent higher terrain in a pediment area. It is a sharp angle, usually explained either by lateral corrosion or the contrast between hill wash and gravity transportation. It is called either the knick or knickpoint. (p. 304) (W.M.)

**Knick point, knick-point**

Committee, List 2. (German, Knick-punkt) A break of slope, particularly in a river profile.

Wooldridge and Morgan, 1937. 'A negative change of base-level gives rise to a complex chain of consequences. Rivers immediately begin to regrade their courses to the new sea-level, constructing a new curve, which, by headward erosion, progressively replaces the former curve. The junction of the two curves at any time is marked by a break of slope, which may be referred to as a "rejuvenation-head" or "knickpoint".' (p. 220)

Similarly Holmes, 1944, p. 195.
No reference in Fay, 1920; Rice, 1941; Cotton, 1945; Penck, W. (trans. Czech and Boswell), 1953. For an early American use see Knopf, E. B., 1924, *Bull Geol. Soc. Am.*, 35, 633–8.

**Knick-point, artificial**
In a lecture before the British Association at Glasgow in 1958, H. A. Moisley used the expression 'artificial knick-point' pointing out that if a river is dredged, as the Clyde was, up to a certain point such as a weir, an artificial knick-point is created. If the weir is removed this may work back upstream with serious consequences. (L.D.S.)

**Knickpunkt** (German: *pl.* Knickpunkte)
The form usually used by English writers is knickpoint, American writers more logically prefer nickpoint. But see Fischer, 1950.
Thornbury, 1954, uses the words with the following meanings:—
1. A sharp inflection in an interrupted river profile.
2. The sharp angle made by the *haldenhang* of Penck (wash slope of Meyerhoff) and the *steilwand* of Penck (gravity slope of Meyerhoff) (pp. 110, 290).

**Knob**
*O.E.D.* A prominent isolated rounded mound or hill; a knoll; a hill in general; especially in the United States.
Mill, *Dict.* A rounded hill or mountain summit.
*Comment.* Neither definition fits the current use in American terminology—see Knob and basin.

**Knob and basin topography, knob and kettle**
Thornbury, 1954. 'Slight oscillations of an ice front as it recedes may result in an irregular belt of knolls and basins, usually described as *knob and basin topography*. This type of end moraine is more commonly formed by ice caps than by ice streams.' (p. 374)
Lobeck, 1939, 'Terminal moraines grade from simple smooth ridges with very low slopes to the most complex aggregation of knobs and ridges interspersed with enclosed *kettles* or *pits*. This is sometimes termed *knob-and-basin* topography.' (p. 303)
von Engeln, 1942. 'Because the volume of material deposited varies from place to place in the moraine, and because buried ice blocks melt out after deposition cease the surface topography of end moraine is normally very irregular, having a *kno and kettle*, or *basin*, aspect.' (p. 488)

**Knoll, knowe, know, knowle**
*O.E.D.* A small hill or eminence of more o less rounded form; a hillock, a mound formerly also the rounded top of a moun tain or hill. Knowe and Know are given a north English and Scottish forms.
Ekwall notes derivation from Old Englis cnoll and the prevalence of the forr Knowle in place names. *Cf.* Knob.
Mill, *Dict.*, adds usually with trees.

**Knot**
A measure of speed used for ships. One kno is one nautical mile (see mile, geographica per hour. It is tautological and incorrect t refer to a speed of so many 'knots pe hour'.
Originally 'A piece of knotted string fastene to the log-line, one of a series fixed at suc] intervals that the number of them tha run out while the sand-glass is runnin; indicates the ship's speed in nautical mile per hour; hence each division so marke on the log-line, as a measure of the rat of motion' (*S.O.E.D.*).

**Knot** (2)
A complex of mountains, *e.g.* the Pami Knot, Armenian Knot.

**Koembang**—see Kumbang

**Kolkhoz** (Russian); contraction of Kollek tivnoe khoziaistvo)
A collective farm.
Mikhaylov, N., 1935, *Soviet Geography* London: Methuen. 'Socialist developmen of agriculture could only be achieved b; the consolidation of agriculture: by th creation of large-scale state agricultura enterprises—'sovkhozes'—and the amal gamation of the scattered households o the peasant toilers into large-scale collec tive farms—"kolkhozes".' (p. 115)

**Kolla** (Ethiopia)—see Dega

**Kona** (Hawaii)
The leeward side *i.e.* away from the Trade Winds.

**Kop, Koppie** (Afrikaans; *antiq.*: kopje; diminutive of Kop (*i.e.* head, German *Kopf*).
Isolated hill (German: *Inselberg*) or row o hills (*pl.* koppies) often composed of old volcanic rock, and characteristic of the interior regions of South Africa. The word Kop (without the diminutive ending) is

sometimes applied to mountains of large dimensions, especially as part of a proper name. (P.S.)
hill of any description if it stands out prominently. (J.H.W.)
fter the South African war the word was introduced into England and is occasionally found as part of the local name of a hill (G.T.W.)

**oum**—see Kum

**oup** (Afrikaans)—also **Goup** or **Gouph**
region in the driest part of the Great Karoo. (P.S.) See Wellington, J. H., 1955, *Southern Africa*, I, 117.

**oustar, Kustar** (Russian, Handicraftsman engaged in home industry)
oustar industries—the old peasant industries of Russia.
tamp, L. D., 1929, *Asia*, London: Methuen. 'The Russian immigrants into Siberia took with them their Koustar (peasant) industries and they found in their new homes an even greater need for these occupations in the long severe winter.' (p. 674) Spelling Kustar is now more usual.

**Kraal** (Afrikaans)
a) enclosure for cattle; (b) native village. In early colonial Dutch, both in the East Indies and South Africa, the word has been in use with the primary meaning of enclosure; derived from Portuguese corral. (P.S.)

**Kraaling** (South Africa)
Putting cattle or sheep into a kraal at night as protection against wild animals. The system has many disadvantages and has become obsolete by the gradual extermination of the larger carnivora, and by the erection of jackal-proof fencing as a protection against the smaller ones. (P.S.)

**Krans** (Afrikaans)
Precipitous rock-face on a mountain. Originally the word meant a wreath, crown, or garland (German: Kranz). (P.S.)
Cotton, 1941, *Landscape*, C.U.P., uses krantz and krantzes. (p. 93)

**Krasnozem**
Red soil developed zonally under Mediterranean conditions; not to be confused with azonal *terra rossa*.

**Kratogen**—see Craton

**Krotovina** (Russian: Soil Science). Also **crotovine**
Jacks, 1954. 'Filled-in animal burrow in soil.'

**Kuala** (Malay)
A confluence or estuary; common in place names as of the capital town Kuala Lumpur.

**Kucha, cutcha, kachha, kacha** and various anglicized spellings (Anglo-Indian from Hindi)
As opposed to pukka (built of stone or masonry, hence strong, good) of mud or earth (hence weak, temporary).
Kucha wells; wells not lined with masonry.

**Kum** (Turkestan: *lit.* sand), **Koum**
Applied to the sandy deserts of central Asia, e.g. Kizil Kum (Red Sand). There seems no justification for the spelling koum adopted in some English text-books. Knox gives kum only. Koum is French.
von Engeln, 1942. 'Wide regions, called erg in the Sahara, *koum* in Asia, are continuous dune tracts.' (p. 425)
*Comment*. *Kum* is the word for sand in many Turkish languages and also in Tadzhik. It does not occur as an independent word in Russian but only in place names. (C.D.H.)

**Kumatology** (Vaughan Cornish, 1899), also **Kumatologist**
Cornish, Vaughan, 1899, On Kumatology, *Geog. Jour.*, 13, 624–628. After the title—'The Study of the waves and wave-structures of the Atmosphere, Hydrosphere and Lithosphere,' later on the same page—'I think the time has come when it will be for the advantage of our science that there should be a distinctive word for the study of the waves and wave-structures of the Earth as a special branch of geography ... Greek for "a wave".' (p. 624)
*Comment*. The suggestion was not adopted and the word is not in *O.E.D.*

**Kumbang** (Java), **koembang**
Robequain, C., *Le Monde Malais* (translated Laborde, E. D., 1954). 'The southeast wind which, after crossing the Pembarisan Hills, Java, becomes a foehn known as the kumbang' (p. 187).

**Kumri**
Shifting cultivation in Kanara, India. See Spate, 1954, p. 624.

**Kunkur**—see Kankar

**Kursaī** (Pakistan: Pashto)
Shepherd's hut.

**K-value**
Garner, B. J., in Chorley and Haggett, 1967.

'Christaller used the term in his central place model. The K-value ... refers to the number of settlements at a given level in the hierarchy served by a central place at the next highest order in the system ... Christaller assumed that once the K-value was adopted in any region it would be fixed. Consequently, it would apply equally to the relationship between hamlets and villages, villages and towns, and so on up the central place hierarchy. Because of this the total number of settlements in any area should follow a regular progression. In the case where the K-value equals three, this would be one, three, nine, twenty-seven, eighty-one ... starting with the highest order place in the region.' (p. 308)

See Christaller, W., 1966, *Central Places in Southern Germany*, trans. Baskin, C. W., Englewood Cliffs, New Jersey: Prentice-Hall.

*Comment.* K often used to label a constant.
See also Central place theory.

**Kvutza, Kvutzah** (Hebrew, *lit.* group; Israel
'At Degania in 1909 a group of young men and women received from the Zionist Organisation a contract to farm a newly purchased area of land on their own financial responsibility as a collective group ... and developed a unique system of collective village living. They called their group a *kvutza* which means simply 'group' A. Harman, *Israel Today*, Agricultural Settlement, 1960, 7–8. The word has become almost obsolete and has been superseded by Kibbutz (*q.v.*). See also Moshav.

**Kwin** (Burmese)
A small division of land.

**Kyaung** (Burmese)—see Hpoongyi-kyaung

**Kyle** (Scottish; from Gaelic cael)
*O.E.D.* A narrow channel between two islands, or an island and the mainland (in the west of Scotland); a sound, a strait.

# L

Laagte, Laegte (Afrikaans)
Broad hollows between wide rises in a comparatively flat landscape. Less well defined than valleys. (P.S.)

Laccolite, laccolith (G. K. Gilbert, 1877)
O.E.D. A mass of igneous rock thrust up through the sedimentary beds, and giving a dome-like form to the overlying strata. Describing such a mass in the Henry Mountains, Utah, Gilbert, 1877, *Rep. Geol. Henry Mts.*, 2, says 'for this body the name *laccolite* will be used' (p. 19).
Holmes, 1944. 'Instead of spreading widely as a relatively thin sheet (*i.e.* a sill) an injected magma, especially if it is very viscous, may find it easier to arch up the overlying strata into a dome-like shape' (p. 86 with fig.).
Wooldridge and Morgan, 1937. 'Laccolites are in the nature of huge blisters, in which the covering sedimentary rocks have been thrown into a dome by the local accumulation of molten rock beneath them. They may be conceived as filled by a "pipe" from below, though this has rarely if ever been seen.... Both before and after the breaching of the sedimentary cover by erosion laccolites tend to form striking isolated conical hills. A beautiful example is afforded by the Traprain Law in Haddingtonshire' (pp. 109–110).
Comment. The *O.E.D.* definition is inadequate because it fails to convey the essential idea that the laccolite is an igneous mass *concordant* with the strata into which it is intruded and has a flat base. Gilbert's original word was laccolite but because -lite is the normal termination for a mineral name, J. D. Dana, 1879 (*Manual of Geology*, 3rd ed., 840), changed it to laccolith. Both forms are used. See also Sill, Phacolite, Lopolith, Batholith.

## Laccolith, Cedar-tree
A composite laccolith consisting of a series of intrusions, one above the other, each in form being comparable with a simple laccolith. So called because in section it recalls the horizontally spreading branches of a cedar tree.

## Lacustrine
O.E.D. Of or pertaining to a lake or lakes. Said especially of plants and animals inhabiting lakes and geologically of strata, etc., which originated by deposition at the bottom of lakes.
Himus, 1954, notes that many lacustrine deposits show seasonal banding or varves (*q.v.*).

## Lacustrine Plain
A plain which marks the site of a former lake.

## Ladang (Indonesia)
Robequain, C., *Le Monde Malais* (translated by Laborde, E. D., 1954). 'In Indonesia most of the cultivation is done on the *ladang* system ... based on burning the natural vegetation. The trees are cut down to within a foot or two of the ground—all except the biggest ... the wood and foliage are burnt before the first rains fall; seeds or slips are then planted in the ash-fertilized soil ... at the end of two or three years the patch is usually abandoned ... neither plough, nor irrigation, nor domestic animals are used.' (p. 94)

## Lag deposits (Happ, Rittenhouse and Dobson, 1940), lag gravel
Thornbury, 1954. 1. Coarse materials which have been sorted out and left behind on the bed of the stream. (p. 172)
2. Concentrations of pebbles and boulders left behind after deflation, also called desert pavement or armor. (pp. 302–303)
Comment. In the first usage above, lag deposits are one of six categories of alluvial deposits distinguished by Happ, Rittenhouse and Dobson. The second usage seems far from clear.

## Lag gravel
Cotton, 1942. 'Boulder pavement.' (p. 10)

## Lag mound (R. W. Packer, 1965)
A term introduced by R. W. Packer in a paper entitled Lag mound features on a dolostone pavement, *The Canadian Geographer*, 9, 1965, 138–143, but without definition. In reply to a specific enquiry the author replied: 'The term "lag" was used in the same sense as in "lag deposits"

and "lag gravels" to indicate what was left behind after some kind of sorting process.'
'*Lag Mounds*. Residual remnants of thin unconsolidated surficial materials on limestone or dolostone pavements, where the karst process is beginning. Solution and washing of loose material down fissures makes trench depressions leaving the appearance of mounds two to three feet high and the same size as the dolostone blocks.'

**Lagg** (Swedish: *pl.* laggar)
Marginal fen. The Swedish lagg is used as an international term; German *Lagg*. (E.K.)

**Lagoon**
O.E.D. (French, *lagune*; Latin, *lacuna*, pool)
1. An area of salt or brackish water separated from the sea by low sand banks esp. one of those in the neighbourhood of Venice.
2. The lake-like stretch of water enclosed in an atoll.
Rare: (Anglicized form of It. lagone. In Tuscany the basin of a hot spring from which borax is obtained.
Mill, *Dict*. 1. A sheet of water separated from the sea by a low ridge. 2. The sheet of water inside a coral atoll or fringing reef. 3. A pool into which the waters from Soffioni are discharged, and in which some of the materials they held in solution crystallize out (Tuscany).
*Adm. Gloss.*, 1953. An enclosed area of salt or brackish water separated from the open sea by some more or less effective, but not complete, obstacle, such as low sandbanks. The name most commonly used for the area of water enclosed by a barrier reef or atoll.
*Comment*. The first usage mentioned by O.E.D. corresponds broadly with the German *haff*. In the second usage the O.E.D. is incomplete because the commonest type of lagoon in the tropics is that between an off-shore coral reef and the shore.

**Lahar** (Indonesia)
Robequain, C., *Le Monde Malais* (translated by Laborde, E. D., 1954). 'A flow of mud, a landslide of volcanic material impregnated with water and moving down gentle slopes carrying blocks of stone at times larger than a native dwelling. The water may come from the heavy falls of rain which occur during eruptions . . . this makes a cold *lahar*. Or else it may come from the sudden emptying of a crater lake, giving rise to a hot *lahar*' (p. 24).
See also Cotton, C. A., 1944, *Volcanoes a Landscape Forms*, Christchurch, N.Z. p. 240.

**Lak** (Pakistan: Pashto)
A pass.

**Lake**
O.E.D. A large body of water entirely sur rounded by land; properly one sufficiently large to form a geographical feature, bu in recent use often applied to an orna mental water in a park etc.
Mill, *Dict*. An accumulation of water col lected in a depression in the earth'. surface.
*Comment*. The word lake is the broad genera term and is qualified in various ways. A very small natural lake is termed ir ordinary speech a pond or pool; a very large lake especially if containing salt water may have the proper name of 'sea' (*e.g.* Caspian Sea, Sea of Aral, Dead Sea). A lake may or may not have inflowing and outflowing rivers, a lake may not necessarily be a permanent body of water (*cf.* playa and *e.g.* Lake Eyre).
See also Llyn, Loch, Lough, Mere, Pond, Pool, Tank, Tarn, also Crater lake. The adjective applying to lakes is lacustrine; see also Limnology.

**Lake-dwelling**
A dwelling built upon piles driven into the bed of a shallow lake. Though common in many parts of the world right to the present day they were characteristic of certain periods of Neolithic times in Switzerland, France and elsewhere so that some authors talk of a Lacustrine period or Lacustrine civilization.

**Lake Rampart**
A conspicuous ridge or ridges found on lake shores and caused by the freezing of the lake in winter and consequent expansion of the ice which presses or shoves (*ice-shove*) against the lake shores and causes ridging of the lake-shore deposits.

**Lalang** (Malay)
Coarse grass (*Imperator cylindrica*) which grows up on deserted clearings; it occupies large areas and hence appears as a category on land-use maps.

**Lamination**
Himus, 1954. 'Stratification on a fine scale, each thin layer or *lamina* being a fraction of an inch thick. Typically exhibited by fine-grained sandstones and shales.'

**Lammas land**

Report, Royal Commission on Common Land, 1955–58. 'Also Half-Year Land. A class of commonable land, arable or meadow, held in severalty during part of the year, but, when the crop is gathered, thrown open to the severalty owners and other classes of commoners. The usual day for throwing open is Lammas day (1st August) or Old Lammas day (12th August).' (p. 274)

**Land**

*O.E.D.* The simple word: the solid portion of the earth's surface, as opposed to sea, water. A part of the earth's surface marked off by natural or political boundaries.

*Comment.* In estimating the land surface of the Earth commonly given as 55,786,000 square miles compared with a water surface of 141,050,000 square miles (total 196,836,000 square miles) (Whitaker, 1959, p. 195) a difficulty has recently arisen. Should the permanent solid ice of Antarctica be included as 'land'? Since it is a 'rock' at the temperatures prevailing, the general answer is yes. If the ice should disappear from the Antarctic continent the 'land' area would be much less. (L.D.S.) Among special uses of this common word should be noted (i) a German administrative Unit, (2) a strip of a ploughed open field divided by water furrows, (3) as a suffix in badlands, marshland etc.

**Land** (German; *pl.* Länder)

*L.D.G.* An important administrative unit in Germany which replaces the former kingdoms, grand-duchies, principalities and other units which made up the old Federation. When the Federal Republic of Germany (West Germany) became a sovereign independent country on 5 May 1955 it comprised 11 *Länder*. The Federation has a Basic Law and the constitutions of the *Länder* must conform to certain principles each being a republican, democratic and social state based on the rule of law. The *Länder* are Baden-Württemberg, Bavaria, Bremen, Greater Berlin (suspended), Hamburg, Hessen, Lower Saxony, North-Rhine-Westphalia, Rhineland-Palatinate, Saarland, and Schleswig-Holstein. The German Democratic Republic (East Germany) abandoned *Länder* in favour of 15 districts.

**Land, Land Units** and **Land Systems** (C. S. Christian, 1957)

Christian, C. S., 1957, *Abs. of Papers, Ninth Pacific Science Congress*, Bangkok. 'The word *land* is used to denote that complex of all factors of the land surface of importance to man's existence and success. The characteristics of any area of land may be regarded as the end products of land surface evolution governed by parent geological material, geomorphological processes, past and present climates, and time.... Where such parts of the land surface can be identified as having a similar genesis and can be described similarly in terms of the major observable inherent features, namely topography, soils, vegetation and climate, they are regarded as being members of the same *land unit*. Land units are naturally aggregated in recognisable and recurring landscape patterns and such patterns are referred to as *land systems*' (p. 257).

*Comment.* This abstract illustrates the difficulty inherent in giving specialized meanings to common words—in particular 'system' applied to land has usually quite a different meaning. The 'land unit' would seem to be the same as the minor natural region or 'stow' of Unstead, the 'land system' the 'tract' of Unstead. (L.D.S.)

**Land Breeze**—see Breezes, Land and Sea

**Land Classification**

*L.D.G.* Land classification is a very complex problem. Classification is needed as a basis for land planning and is designed in most cases to indicate the relative fertility of land for different types of farming or other land use. In the newer countries (notably the Americas) 'land capability' or 'potential land use' classes are favoured and a widely used American system has eight such classes. But potential is a matter of judgment and may be radically changed by technological progress—as seen with the addition of trace elements on much land, previously thought worthless, in Australia. A system devised by Dudley Stamp in Britain was into ten types of land (1–4 good; 5–6 medium; 7–10 poor) based on site and soil and history of land use. In Britain, as in much of Europe, land has been cultivated for two or three thousand years and the history of its use over that long period gives a guide to potential. Land of category 1 may be twenty times as productive as land of categories 8 and 9. Fully described in *The Land of Britain: its Use and Misuse* (Longmans). See also PPU, SNU.

**Land-forms, study of**—see Geomorphology

**Land Hemisphere**
It is possible to divide the earth into two halves or hemispheres in such a way that the bulk of the dry land is in one whilst the greater part of the surface of the other is occupied by oceans. If this is done the centre of the 'land hemisphere' lies approximately at Paris.

**Land Rotation**
Report of the Commission on World Land Use Survey. *International Geog. Union*, 1952. 'By land rotation we understand the system whereby cultivation is carried on for a few years and then the land allowed to rest perhaps for a considerable period before the scrub or grass which grows up is again cleared and the land recultivated. In such areas the farms or settlements from which cultivation takes place are fixed.'

*Comment.* This is one type of shifting cultivation (*q.v.*) in which there is a *regular* system of land management, hence the insistence of the Commission on its separate recognition, following the writings of Leo Waibel. To be distinguished from Crop rotation (*q.v.*).

**Landdrost** (Afrikaans)
Under the Dutch East Indian Company, title of an official controlling a rural district, both in an administrative and a judicial capacity. The office and the title were at first retained under British rule, but were afterwards abolished. They were re-established in the Transvaal and Orange Free States Republics. The house and office buildings of the landdrost were called drostdy. (P.S.)

**Lande** (French)
From Celtic landa, probably free, open land. Modern use: area in which only brushwood and wild plants—gorse, broom, heath—grow. (G.R.C.)

**Landes** (French)
Mill, *Dict.* The low-lying sandy plains bordered in the maritime parts by sand-dunes (S.W. France).

*Comment.* It is doubtful whether this word should be included as it is a district name rather than a term and is rarely if ever applied elsewhere, even to similar country.

**Landscape**
O.E.D. The word was introduced as a technical term of painters. From Dutch, landschap.
3. In generalized sense: inland natural scenery, or its representation in painting.

Hartshorne, 1939, pp. 149–174, devotes a chapter to 'Landschaft' and 'Landscape', discusses and dismisses such definition as (1) 'a piece of area having certain characteristics which our minds, if not actually in reality, set it off from other pieces of area', (2) 'the view of an area as seen in perspective', (3) 'the sum total of those things in an area that could produce "landscape sensations" in us if we placed ourselves in the different positions necessary to receive them.' He concludes 'The actual single and concrete reality which, we believe, underlies the thought of many who have used the term without attempting to define it, is the external visible surface of the earth' (p. 168).

Sauer, C., 1925, *The Morphology of Landscape*, Univ. of Calif. Pub. in Geog., No. 2, pp. 25–26. '... an area made up of a distinct association of forms, both physical and cultural.'

James, P. E., 1934, The Terminology of Regional Description, *A.A.A.G.*, 24, 79. '... a portion of territory which is found to exhibit essentially the same aspect after it has been examined from any necessary number of views.'

See also Broek, J. O. M., 1938, The Concept Landscape in Geography, *C.R. Cong. Int. Geog.*, Amsterdam, 2, 103–109.

**Landscape, cultural**
Hartshorne, *Nature of Geog.*, 1939 (after noting the usage of Sauer). 'Most American geographers who use the term "cultural landscape" mean simply the present landscape of any inhabited region. In this sense it is demanded only where we need to emphasize the present complete landscape in contrast to the "natural landscape".' (p. 170)

James, P. E., 1934, The Terminology of Regional Description. *A.A.A.G.*, 24. 'Many writers describe the aspect of the face of the earth which results from the presence of man as the cultural landscape —the natural landscape modified by man ... As a matter of fact the fundament ceases to exist and is replaced, after the arrival of man, by the cultural landscape. There is, after all, only one landscape.' (p. 80)

Bryan, P. W., 1933, *Man's Adaptation of Nature*. 'The concrete expression of the

process of adaptation takes the form of the cultural as opposed to the natural landscape, in other words, it is the natural landscape as modified by man.' (p. 14)
Sauer, C., 1925, Univ. of Calif. Pub. in Geog. No. 2, 1925, *The Morphology of Landscape*, pp. 20–53. 'The cultural landscape is the geographic area in the final meaning (Chore). Its forms are all the works of man that characterize the landscape.' (In a diagram these forms include those of Population, density, mobility; Housing, plan, structure; Communications, etc.) (p. 46)

In footnote, Sölch is given as proposing the term 'Chore' as equivalent to 'landscape' (an area made up of a distinct association of forms, both physical and cultural). (p. 26)

But cf. 'natural landscape': 'The thing to be known is the natural landscape. It becomes known through the totality of its forms.' (p. 41) (The forms include those of Climate; Land, surface, soil, drainage, mineral resources; Sea and coast; Vegetation.)

Also, 'The area prior to the introduction of man's activity' is called '... with reference to man, the original, natural landscape.' (p. 37)

Dickinson, R. E., 1939, Landscape and Society, *Scot. Geog. Mag.*, 55. '... the material features, in compact areal association, which are the result of the transformation of a natural landscape through human occupance, form the cultural landscape (Kulturlandschaft).' (p. 2)
See comment after Landscape, natural.

**Landscape, fossil**
Cotton, 1945. An erosion surface which has been buried under latter deposits. Similarly described but not termed a 'landscape' in 1922. (W.M.)
No reference in Geikie, J., 1898; Salisbury, 1907; Davis, 1909; Wright, 1914; Holmes, 1944.
*Comment.* A better term perhaps is 'exhumed landscape' since the character of the fossil landscape is not apparent until the superficial deposits have been removed and the old surface is exhumed though of course a fossil landscape can exist buried but not seen. (L.D.S.)

**Landscape, natural** (see also Landscape, cultural)
Hartshorne, 1939. 'For practical purposes it includes only the contrast between land and water, the relief, and the natural vegetation.' (p. 172)

Dickinson, R. E., 1939, Landscape and Society, *Scot. Geog. Mag.*, 55. 'The land forms and their plant cover build the natural landscape (Naturlandschaft).' (p. 2)
*Comment.* In most inhabited countries the natural landscape has been modified in a greater or less degree by man and become the cultural landscape. In this sense it is the cultural landscape which is the total landscape. It is perhaps of more practical use to refer to the 'natural and cultural elements' of the landscape. (G.T.W.)

**Landscape Architecture, Landscape Gardening**
Although the old profession of the landscape gardener and the newer profession of the landscape architect are well known, definition is difficult. The two differ mainly in scale. The landscape gardener seeks to lay out the surroundings of a house or other building in such a way as to create a landscape which will be a harmonious whole; the landscape architect is similarly concerned to create a landscape which will harmonize the works of man with the natural setting. The *Journal of the Institute of Landscape Architects* is a valuable guide to the range of work involved. (L.D.S.)

**Landschaft** (German: *pl.* Landschaften)
Landscape as usually understood conveys only part of the meaning. In German it is used in a great number of different ways, but in most cases it means a region seen and delimited under the aspect of appearance. (K.A.S.)
Fischer, 1950. i, landscape; ii, scenery; iii, region. (p. 32)
For a full discussion see Hartshorne, 1939, pp. 149–174.

**Landskap** (Norwegian)
A province; applied to the 25 provinces of Norway. In the publication *Norden* (1960) issued for the 20th International Geographical Congress it was accidentally translated 'landscape' so that 'landscape names' appears instead of 'province names' (see pp. 11–12).

**Landslide, Landslip**
*O.E.D.* Land-slide, U.S. = Landslip.
Landslip. The sliding down of a mass of land on a mountain or cliff-side; land which has so fallen.
Rice, 1943. 'Landslip—Portion of a hillside or sloping mass which becomes loosened or detached and slips down, a landslide.' (p. 212)
See also Mass-wasting; also Sharpe, C. F. S.,

1938, *Landslides and Related Phenomena*, Columbia University Press.

**Landslip Terrace**
Rice, 1943. 'A short, rough-surfaced terrace resulting from the slip of a segment of a hill.' (p. 212)

**Land Use, Land Utilization, Land Utilisation** also **Land Use Survey**
Literally the use which is made by man of the surface of the land but in sparsely populated areas including the natural or semi-natural vegetation. For details of the Land Utilisation Survey of Britain, carried out in the nineteen thirties, see Stamp, L. D., 1948, and later editions *The Land of Britain: its Use and Misuse*, London: Longmans; for the classification proposed for the World Land Use Survey see *Economic Geography*, **26**, 1950, 1–5; and *Reports* of the Commission on World Land Use Survey (International Geographical Union), 1952–1972. For details of Second Land Utilisation Survey of Britain see Coleman, Alice, 1968, *Land Use Survey Handbook*, 5th ed., King's College London: Second Land Utilisation Survey.
Fox, J. W., 1956, Land Use Survey: General Principles and a New Zealand Example, *Auckland University College Bulletin No. 49, Geog. Series, No. 1*. 'Land use, therefore, may be defined as the actual and specific use to which the land surface is put, both cultivated and non-cultivated land. It may be given over to crops, to grasses, to trees, to buildings—or it may be barren waste. These classes of land use are precise and definite, and their significance can be objectively assessed. Secondary aspects of land use derive from them—grasses may be utilized in several ways by the farmer, factories may produce different kinds of manufactured goods—but differentiation in terms of these characteristics falls within the sphere of agriculture and manufacturing industry respectively, not within that of land use *per se*.' (p. 8)
*Comment.* Some authors have attempted to draw a distinction between land utilization and land use, terms which are or have commonly been used interchangeably. A land use survey will also record the vegetational cover of land which is not directly used by man so that 'land cover' might be more appropriate. When the Commission for a World Land Use Survey of the International Geographical Union drew up it recommendations in 1949 (see *Economic Geography*, **26**, 1950, 1–5) we had in mind the desirability of using the simpler word *use* which would avoid the difficulty between the older English spelling *utilisation* (as in the official title of the British Survey of the 1930s) and the more usual American *utilization*. T. M. Burley has argued (Land Use or Land Utilization. *Professional Geographer*, **13**, 1961, 18–20) that Land Cover + Land Utilization = Land Use but it seems unlikely that this refinement will be generally accepted. (L.D.S.)

**Landward Population** (Scotland)
Stevens, A., The Distribution of the Rural Population of Great Britain, *Trans. Inst. Brit. Geog.*, **11**, 23–53. 'A term applied in Scotland to that section of the population which is not resident in a *burgh* ... rural and landward are not synonymous.' (p. 25)
*Comment.* In Scotland a number of ancient burghs or boroughs are only the size of villages; conversely some quite considerable settlements may not have the status of burgh. (A.C.O'D.)

**Lapiaz** (French)
*Lapié* is the individual form; *lapiaz* is the area—the two words comparable with *Karren* and *Karrenfeld* in German and, roughly, with clint or gryke and limestone pavements in English.

**Lapié** (French)
Thornbury, 1954. 'Karst Regions. Locally where relief is considerable, limestone surfaces are bare of terra rossa and there is exposed an etched, pitted, grooved, fluted and otherwise rugged surface to which the name lapiés is most commonly applied.' (p. 319)
*Comment.* Equivalent of *karren* in German; see Karst terminology. Lapiés are considered by Cvijić to be mainly due to rain water run-off, *i.e.* fluted channels with slopes determined by free-flowing rills. Similar forms found on volcanic and other rocks. Clints are more like solutional features, many having lapiés on them. There are also very small forms—microlapiés or German *rillensteine*—perhaps best called rock-rills. Occasionally marine lapiés have been described from the British Isles and continental Europe. (G.T.W.)

**Lapiésation**
Wray, D. A., *Geol. Mag.*, **59**, 1922. 'irregular

limestone surfaces are termed lapiés, while the weathering process is described as lapiésation.'

**Lapilli** (*pl.* of Italian *lapillo*)
*O.E.D.* Small stones or pebbles; now only specifically of the fragments of stone ejected from volcanoes.
Scrope, *Volcanoes*, 1857, used the singular form *lapillo*.

**Lapse Rate**
Hare, 1953. 'The rate of decrease of temperature with height is known as the *lapse rate of temperature*, or more often just the lapse rate ... usually expressed in Great Britain in degrees Fahrenheit per thousand feet ... 3·5° ... in certain cases the temperature actually rises with height ... an "inverted" lapse rate to which we assign a minus sign.' (p. 15)

**Laramide Revolution** or **Orogeny**
A period of earth-movement in the early Tertiary (*i.e.* earlier than the main Alpine orogeny) important in the development *inter alia* of the Rocky Mountains.

**Lateral moraine**—see Moraine

**Laterite** (F. Buchanan, 1807)
*O.E.D.* Min. (f.L., later, brick, + -ite). A red porous, ferruginous rock, forming the surface covering in some parts of India and south-western Asia. F. Buchanan, *Journey from Madras through Mysore*, etc., 1807, II, 460. 'In general the Laterite, or brickstone, comes very near the surface.'
Mill, *Dict*. An ochreous accumulation of disputed origin, chiefly found in tropical or subtropical regions. Some laterites appear to be due to decomposition of gneissose rocks, others are partly of volcanic origin.
Jacks, 1954. 'Buchanan's name for a red subsoil which hardens permanently on exposure or has already hardened under natural conditions.'
*Comment.* There is a huge literature on laterite which has been variously defined. So great has the confusion become that no definition is likely to command universal acceptance. It is probably best used in the loose general sense implied in Jacks's definition for a subsoil product of atmospheric weathering produced under a humid tropical climate with alternating wet and dry seasons. Under the conditions of formation it is mottled red and grey and quite soft (*cf.* gley); when exposed to the atmosphere it hardens and the grey sandy or clayey portions are washed out leaving a red rock hard enough to be used for building (especially as a foundation to wooden huts) or for farm roads. The grey sandy and clayey material being washed out the rock is full of holes. Writers who consider the solution should be a redescription of Buchanan's type-rock are hampered by the fact that it is not clear exactly where he saw his original 'brick stone'. Where surface material is washed away a hard lateritic crust (*carapace latéritique* of the French) is found at the surface. (L.D.S.) See also Laterite soil, Latosol.

**Laterite soils**
U.S. Dept. of Agriculture, *Yearbook of Agriculture*, 1938, Soils and Men. Washington: U.S. Govt. Printing Office, 1938, p. 1171. 'Laterite soils. The zonal group of soils having very thin organic and organic-mineral layers over reddish leached soil that rests upon highly weathered material, relatively rich in hydrous alumina or iron oxide, or both, and poor in silica; usually deep red in colour. Laterite soils are developed under the tropical forest in a hot, moist, or wet-dry climate with moderate to high rainfall.'
Robinson, G. W., *Soils, their Origin, Constitution and Classification*, 3rd ed., London: Murby, 1949, pp. 409–415. First used to describe a red soil in India by F. Buchanan, 1807, *Journey from Madras through ... Mysore, Canara, and Malabar* (p. 440), Martin, F. J., and Doyne, H. C., in *J. Agric. Sc.*, 1927, 17, defined it as a material in which the molecular ratio of silica to alumina in the clay is less than 1·33, whilst in lateritic soils the ratio is 1·35 to 2·0. Robinson favours usage of R. L. Pendleton, *Amer. Soil Survey Assoc.*, 1936, **17**, 102–138. Profiles characterized by the presence of concretionary material or crusts overlying mottled or vesicular horizons, developed originally as pseudo-illuvial deposits in the zone of a fluctuating water-table under peneplanic conditions, later uplift and dissection removing the ground water conditions with a new direction of profile development being induced. If it is merely intended to state that the soil is highly sesquioxidic, ferrallitic would be a better term than lateritic. (W.M.) See also Prescott, J. A. and Pendleton, R. L., 1952, *Laterite and Lateritic Soils*, Farnham Royal: Commonwealth Agri. Bur.

See also Latosol with Jacks's definition.

**Lateritization, Laterisation**

Thornbury, 1954. 'In temperate climates the residual products from the weathering of igneous rocks are clay minerals . . . under tropical climates final weathering products are hydrous oxides of such metals as aluminum, iron and manganese. This type of weathering has been called *lateritization*.' (p. 564)

Geikie, A., 1903. 'Laterite . . . the peculiar kind of alteration exemplified by this rock and by bauxite has been termed "Laterisation".' (p. 169)

*O.E.D. Supp.* Laterization. The hardening which takes place in laterite when it is quarried and dried (quotes Geikie).

See also Laterite.

**Lathe**

*O.E.D.* One of the administrative districts (now five in number) into which Kent is divided, each comprising several hundreds. For a discussion, see J. E. A. Jolliffe, *Pre-Feudal England: The Jutes*, pp. 39–72 (Oxford, 1933).

**Latifundia** (Latin plural; also anglicized as *latifunds*; sing. is *latifundium*).

Large estates.

*Comment.* This word has been used particularly in describing conditions in the Argentine and in Spain where the contrast is with intensively cultivated *huertas*. Thus Jones, 1930, says 'the establishment of the *latifundia* first began on a large scale during Rosas' first campaign' (p. 310) and 'the *latifundia* system' (p. 447), whilst W. B. Fisher and H. Bowen-Jones in *Spain* (London: Christophers, 1958) discuss the system at some length.

The comparable Italian term *latifondo* refers to a type of economic landscape, with extensive cereal cultivation and grazing and includes both large estates and peasant holdings. (C.J.R.)

**Latitude**

Mill, *Dict.* Distance north or south of the equator, measured as an angle with the centre of the earth in degrees, minutes and seconds. The equator itself is latitude 0°, the North Pole is latitude 90° N. and the South Pole 90° S. Hence low latitudes or, broadly, the Tropics between the Tropic of Cancer ($23\frac{1}{2}°$ N.) and the Tropic of Capricorn ($23\frac{1}{2}°$ S.); mid-latitudes ($23\frac{1}{2}°$–$66\frac{1}{2}°$ N. and S.) and high latitudes within the Arctic and Antarctic Circles ($66\frac{1}{2}°$ N. and S.). Mid-latitudes may, however, be given a rather more restricted meaning than this.

**Latitude, parallels of**

Lines of latitude are so called because they are parallel to one another and each is a Small Circle (*q.v.*) except the equator (0°) which is a Great Circle (*q.v.*).

**Latosol** (Soil science: C. E. Kellogg, 1948–9; Jacks, 1954. (Equated with laterite soil) 'Soil with thin $A_0$ and $A_1$ layers over reddish or red deeply weathered material which is low in silica and high in sesquioxides.'

*Comment.* Because of the innumerable interpretations which have been given to 'laterite' and consequently to 'lateritic soils', Kellogg proposed to use latosol for the latter, carefully redefined. See C. E. Kellogg and F. D. Davol, An Explanatory Study of Soil Groups in the Belgian Congo, *Pub. of I.N.E.A.C.*, 1949, Sci. Ser., 46, 73.

Not widely accepted outside U.S.A. The form latosol suggests 'broad soil'.

**Lava**

*O.E.D.* Italian, orig. 'a streame or gutter suddainly caused by raine' (Florio, 1611), Applied in the Neapolitan district to a lava-stream from Vesuvius whence . . . 2. The fluid or semi-fluid matter flowing from a volcano, 3. The substance that results from the cooling of the molten rock.

Himus, 1954. Molten rock or magma that issues from a volcanic vent or fissure and consolidates on the surface. Chemically, lavas vary widely in composition: they may be acid (with an excess of silica so that some crystallizes out as free quartz, $SiO_2$), intermediate, basic, or ultra-basic. Some lavas cool quickly and are vitreous or glassy, others finely crystalline; others cooling slowly exhibit a coarsely crystalline or holocrystalline character.

See also Volcanic rocks, Bombs.

**Lava tube**

A lava tube may be formed when a lava river solidifies where it touches the cold ground and at the same time crusts over at the surface. The fluid lava then drains out leaving a hollow tube which may be large enough for a man to walk through. Common in Hawaii.

**Law** (Scottish)

*O.E.D.* A hill, especially one more or less rounded or conical.

**Lawn**

*O.E.D.* 1. An open space between woods;

a glade, now archaic. 2. A portion of a garden or pleasure ground covered with grass which is kept closely mown.

Miller, A. A., 1954, The Mapping of Strip Lynchets, *Adv. of Sc.*, 11. Of lynchets (*q.v.*) 'in the Isle of Portland cultivated terraces of a similar form on the Portlandean (upper Jurassic) limestone are called "lawns"' (p. 277).

**Layer colouring, layer tinting**
A method employed on maps to emphasize differences using different colours or tints: most commonly applied to relief maps where land up to a certain height is given one colour, from that height to another limit a different colour or tint, and so on.

**Lazy-bed**
*S.O.E.D.* A bed for potato-growing, about six feet wide, with a trench on each side, from which earth is taken to cover the potatoes.

**Lea**
1. A tract of open ground, most commonly applied to grassland; now mainly in poetical or rhetorical sense.
2. An obsolescent spelling of ley or lay—see Ley farming.

*Comment.* A word to be avoided because of a multiplicity of meanings.

**Leaching**
The action of percolating water in removing soluble constituents: used especially of water percolating through soil. Hence to leach, leached layer, etc.

**Lead**
Among the innumerable meanings of the word 'lead' are several of geographical significance liable to be overlooked:
1. An artificial water-course, especially one leading to a mill; *cf.* leat.
2. A channel in an ice-field or sea-ice; *cf.* lane.
3. A lode (mining).
4. An old river-bed in which gold is found (Australia).

**Leanland**
See reference under foldage.

**Leasow**
Pasture or meadow land (laes, laeswe—old English), also individual allotment or field often arable (England: Midlands) (I.L.A.T.).

**Leat**
*O.E.D.* An open water-course to conduct water for household purposes, mills, mining works etc. *Cf.* flume.

**Lebensraum** (German: *pl.* Lebensräume: *lit.* living space, habitat)
The area occupied (area of distribution) of a living thing from plants to man. Used by Nazi geopoliticians in the sense of a claim to what they considered adequate living space. (K.A.S.)

Van Valkenburg, S., in Taylor, 1951. 'Kurt Vohwinkel (*Zeit. für Geopolitik*, 1939) distinguishes three kinds of German *Lebensraum*. The first kind is the real area occupied solidly by Germans; the second the area where besides Germans there are other people but the German cultural influence prevails; and the third is the one in which Germans are outnumbered by others but still because of their racial and cultural superiority have a right to dominate.' (p. 109)

**Legend** (of a map)
The explanation of the cartographic conventions used on a map; usually printed outside the frame.

**Leet** (Court-leet)
*O.E.D.* A special kind of court of record which the lords of certain manors were empowered by charter or prescription to hold annually or semi-annually.
See Manorial courts.

**Leeward**
*O.E.D.* Situated on the side turned away from the wind. Opposed to windward.
Also lee side: that side which is turned away from the wind (opposed to weather side) but lee-shore: a shore that the wind blows upon.

**Lembah** (Malay)
Dobby, E. H. G., 1957, *Malayan J. of Tropical Geog.*, 10. 'Low-lying land' (p. vi).

**Leste** (Portuguese)
Mill, *Dict.* The parching easterly to southerly wind blowing from the Sahara and experienced in Madeira; often dust-laden and heralding a depression.

**Levant**
Formerly the countries of the East; later the eastern part of the Mediterranean with its islands and the countries adjoining; now rarely used, being partly replaced by the equally ambiguous Middle East. Also used as a regional name for south eastern Spain.

**Levantine**
An inhabitant of the Levant, as defined above, but usually restricted to those of

European or mixed parentage who have no other country and have largely adopted the languages, customs etc. of the land in which they live.

**Levant and couchant, levancy and couchancy**
A legal phrase derived from medieval Latin through French meaning, literally, rising up and lying down; said of cattle. The name of the rule, dating back at least to the thirteenth century whereby the number of animals (especially cattle and sheep, but also horses) which a holder of common rights was entitled to graze on common land in summer was not allowed to exceed the number he could support on his farm through the winter. See *Report, Royal Commission on Common Land*, 1955–58.

**Levante, Levanter, Llevante, Llevantades** (Spanish)
Mill, *Dict*. A strong and raw easterly wind in the Mediterranean. Affects south-eastern Spain, the Strait of Gibraltar (causing a banner-cloud when moderate, dangerous currents and eddies when strong) and northern Algeria, especially in July to October.

Kendrew, 1953. '"Llevantades" when specially stormy.' (p. 350)

**Leveche** (Spanish)
Mill, *Dict*. A hot dry southerly or south-easterly wind originating in the Sahara and affecting south-eastern Spain.

Miller, 1953. 'On the advancing front the winds are southerly, coming from the deserts of north Africa and often excessively hot and dry, sometimes laden with red penetrating dust. This is the *Sirocco* of Algeria, the *Leveche* of Spain, the *Khamsin* of Egypt.' (p. 170)

**Levee, Levée**
*O.E.D.* U.S. 1. An embankment to prevent the overflow of a river. 2. A landing-place, pier, quay.

*Webster*. 1. An embankment to prevent inundation; as the *levees* along the Mississippi; also, a landing place, pier, or quay. *Southern & Western U.S.* 2. The very low ridge sometimes built up by streams on their flood plains, on either side of their channels. 3. *Irrigation*. A small continuous dike or ridge of earth for confining the checks of land to be flooded.

Mill, *Dict*. The raised banks artificially built to protect a flood-plain from floods; sometimes applied to the naturally aggraded banks of a stream in a flood-plain.

Holmes, 1944. 'Each time the stream overflows its banks the current is checked at the margin of the channel, and the coarsest part of the load is dropped there. Thus, a low bank or *levee* is built up on each side' (p. 167). 'To obtain increased protection (from floods) artificial levees are often built, but these provide only temporary security, since they accentuate the tendency of the river floor to rise.' (p. 168)

As an artificial embankment: see Cotton 1945, p. 192.

As a natural embankment: see Lake, P. 1952, *Physical Geography*, 3rd ed., Cambridge: C.U.P., p. 326.

Martonne, E. de, 1927, *A Shorter Physical Geography* (trans.), London: Christophers, p. 156 (levées = grinduf of Lower Danube).

*Comment*. Although ultimately of French origin, this word came into general use from Louisiana. There was originally an accent but it has long been used without—levee. See also Spill-bank.

**Level**
*O.E.D.* A level tract of land; a stretch of country approximately horizontal and unbroken by elevations; applied specifically (as a proper name) to certain large expanses of level country *e.g.* Bedford Level in the fen district of England.

*Comment*. The famous Bedford Level experiment to measure the curvature of the earth was however carried out by observing three stakes standing each the same height above a *water* surface; hence Mill, *Dict*., gives 'a surface of water concentric with the surface of the geoid and of similar curvature is level'.

Also used in many other senses including a nearly horizontal 'drift', passage, or gallery in a mine.

**Ley, lay, lea**
*O.E.D.* gives lea, ley, lay, but the usual spelling is now ley and lea is rarely if ever used. Land that has remained untilled for some time; arable land under grass; land laid down for pasture, pasture-land, grass-land.

*Comment*. The current usage is for land put down to grass or clover for a period of years. 'Short ley' applies to two, three or four years, 'Long ley' to longer periods. The character of the ley may be qualified by saying grass-ley, clover-ley, etc.

**Ley farming**
The system of farming in which grass-leys or clover-leys are an essential part of the land management.
Stamp, 1948. '... the normal practice is to plough up the old grass, grow normal field crops for three years and then to sow down to grass. The length of time the grass is left down ... will vary from three to anything up to twenty years, a common average being seven years' (p. 65).

**Liane, liana**
A climbing plant, used especially of those woody climbers which are a characteristic feature of tropical forests. The word is of French origin; the form liana seems to have resulted from a mistaken idea of a Spanish origin. (L.D.S.)

**Lido** (Italian)
Mill, *Dict.* See Barrier Beach—a beach of sand or silt in front of a shore lagoon. See Nehrung and Spit.
Not in *O.E.D.*, Moore, Swayne.
*Comment.* One of the best known examples is the lido which protects the lagoon of Venice. This has been converted into a famous bathing beach so that the word lido has now come to mean a bathing beach and is so used all over the English-speaking world, even of fresh water and artificial lake resorts. In this sense it has supplanted the French word *plage*. It is best avoided in its original sense.

**Life-forms of plants**
Raunkiaer, 1934 (first published 1903) distinguished five main types—phanerophytes, chamaephytes, hemicryptophytes, cryptophytes and therophytes *q.v.* See also Dansereau, P., Biogeography, New York 1957.

**Light Industry**—see **Industry, classification**

**Lignite**
*O.E.D.* A variety of brown coal bearing visible traces of its ligneous structure.
Himus, 1954. A variety of coal, generally post-Carboniferous in age, intermediate in properties between the peats and the bituminous coals. Distinguished from brown coal by its more evident vegetable origin.
*Comment.* Often used as synonymous with brown coal; defined by some writers by proportion of carbon, and in other ways.

**Liman** (Russian) 1.
*Webster.* A marsh, or marshy lake, at the mouth of a river; a lagoon.
Mill, *Dict.* A lagoon formed by the barring of the mouth of an estuary by sand (east coast of Black Sea) *cf.* Haff.
*Comment.* Russian authorities, quoted in Appendix II, derive from Greek *Limen* and define as 'broad freshwater bay of the sea' (Murzaev) especially of Black and Azov seas where spits block river mouths.

**Liman** (Russian) 2.
A valleyside cultivation terrace in arid U.S.S.R. (V. A. Kovda in MS., also in *History of Land Use in Arid Regions*, Paris: UNESCO, 1961, p. 194).

**Limes** (Latin: boundary; *pl.* limites)
A rare word in the sense of boundary, limit or end; met occasionally in works on historical geography meaning the limits of territory, especially of the Roman Empire. See Houston, J. M., 1953, *A Social Geography of Europe*, London, Duckworth, pp. 92–95.

**Limestone**
*O.E.D.* A rock which consists chiefly of carbonate of lime, and yields lime when burnt. (The crystalline variety of limestone is marble.)
*Comment.* Limestone is the broad general term; most definitions which attempt to go into greater detail than the *O.E.D.* fall into some error by failing to include the innumerable types (*e.g.*, Himus, Moore, Swayne). Types are distinguished by qualifying adjectives which may refer to mineralogical composition (*e.g.* dolomitic) lithology or texture (oolitic, pisolitic, earthy, crystalline, massive, etc.), origin (organic, coral, sedimentary, precipitated, shelly, etc.), geological age (Carboniferous, Jurassic, etc.) and other characters. In popular concept a limestone is relatively hard, hence such references as to 'chalk and limestone' despite the fact that chalk is a limestone but can be comparatively soft. (L.D.S.) Twenhofel, 1939, suggests limiting the term to rocks containing more than 50 per cent of calcium carbonate.

**Limnology**
*Webster.* The scientific study of fresh waters, especially that of ponds and lakes dealing with all physical, chemical, meteorological, and biological conditions in such a body of water.

**Limon** (French)
Mill, *Dict.* Limon or Lehm. Sandy clay soils of finely divided materials, resulting generally from the rearrangement of

particles by deposition from river-floods, etc. (France).

Shackleton, M. R., 1934, *Europe*, London: Longmans. '... superficial deposits of loam (limon). This is a finely grained deposit, believed to be residue from the erosion of now vanished Tertiary beds ...' (p. 129).

Tricart, J. L. F., 1952, *La Partie Orientale du Bassin de Paris*, T. II, Paris: Seder. pp. 252–253. Described as an important periglacial deposit, either a clay or loam, either true loess or resulting from decalcification. (W.M.)

Baulig, 1956, *par.* 105 equated with 'silt'; but *par.* 404 defined as 'sable très fin, argileux, mais non plastique'.

*Comment.* Limon is a widespread superficial deposit in northern France where it is spread like a blanket largely ignoring minor relief. This suggests an origin comparable with that of loess but formed under a more humid climate and it thus resembles certain brick-earths in southern Britain. On the other hand some resting on chalk and other limestones may be compared in origin with some of the clay-with-flints of southern England. It should certainly not be translated as silt and it is best to use limon. (L.D.S.) Plaisance and Cailleux distinguish between 'limons sub-aériens' and 'limons subaquatiques'. (p. 326)

**Line, The**

Sailors' and colloquial name for the Equator. Used especially in the phrase 'crossing the line'.

**Line-squall**

*O.E.D.* A squall, consisting of a violent straight blast of cold air with snow or rain, and occurring along the axis of a V-shaped depression.

Mill, *Dict.* The line of squalls which are associated with the trough of a cyclone or V-depression.

*Met. Gloss.*, 1944. Squalls may occur simultaneously along a line, sometimes 300 or 400 miles in length, advancing across the country, and to such phenomena the name 'line squall' is applied. A long description follows. It is related to the passage of a cold front. (W.M.)

Note: *O.E.D.*, line-storm, U.S., an equinoctial storm.

**Lines**

In a cantonment (*q.v.*) barracks and huts were arranged in lines and the whole was sometimes termed 'the lines'. In India homes for Europeans laid out in similar regular rows became the 'civil lines' (*q.v.*).

**Lingua franca** (Italian: *lit.* 'Frankish tongue')

*O.E.D.* A mixed language or jargon used in the Levant consisting largely of Italian words deprived of their inflexions. Also, transferred sense, any mixed jargon formed as a medium of intercourse between people speaking different languages.

*Comment.* The original sense is now almost forgotten and the term is used of a language, usually simplified but not necessarily mixed, which is widely used as a means of communication among people of different tongues. Thus Hindustani, a simplified form of Hindi, is the *lingua franca* of northern India, Tamil of much of southern India. Swahili is so used in East Africa, different forms of so-called pidgin-English (a Chinese corruption of Business English) in the East. It should be noted that *lingua franca* is Italian, not Latin, and that a Latin plural ('*linguae francae*') is incorrect.

**Links** (Scottish; always in plural)

*O.E.D.* Comparatively level or gently undulating sandy ground near the sea-shore, covered with turf, coarse grass etc. The ground on which golf is played often resembling such land.

Mill, *Dict.* Sand-dunes or hillocks of blown sand which have ceased to travel (S. Scotland).

Narrow coastal strip, frequently coinciding with the lowest raised beach, with accumulations of blown sand. (A.C.O'D.)

*Comment.* In Scotland the type of land described has been so often used as golf-links that the word 'links' has come to be regarded as almost synonymous with golf-course and is best avoided as a geographical term.

**Linn** (Scottish and northern English)

*O.E.D.* 1. A torrent running over rocks; a cascade, waterfall.

2. A pool, especially one into which a cataract falls.

3. A precipice, a ravine with precipitous sides.

*Comment.* Compare Welsh Llyn (Appendix II—Welsh words).

**Lis, liss** (Irish; Welsh, *llys*)

*O.E.D.* A circular enclosure having an earthen wall; often used as a fort.

**Lithology**
Literally the science of stones. The *O.E.D.* wrongly defines it as a department of mineralogy.
*Webster.* The study of rocks.
Himus, 1954. 'The character of a rock expressed in terms of its mineral composition, its structure, the grain-size and arrangement of its component parts; that is all those visible characters that in the aggregate impart individuality to the rock.'
*Comment.* Hence lithological, lithologic. The geographer is often far more concerned with the lithology of rocks than with their geological age and so is the soil scientist; a lithological map shows the textural character of rocks and so differs from an ordinary geological map which stresses age, though the two aspects may be combined.

**Lithosol** (United States: Soil Science; C. E. Kellogg)
Jacks, 1954. Thin stony soil, shallow over bedrock without a definite B horizon, due to relative youth.
Jacks equates with skeletal soil (U.K.) *q.v.* See also Thorp, J. and Smith, G. D., 1949, *Soil Sci.*, 67, 119.
FAO, 1963. '*Lithosols* have (A)C profiles and are characterized by stony, dominantly mineral, often thin (A)-horizons that merge into or rest on shattered hard rock at no great depth below the surface.'

**Lithosphere**
Mill, *Dict.* The solid crust of the earth.
Himus, 1954. The solid crust which envelops the central core of the earth. It includes the continuous *sima* and the discontinuous *sial.*
*Comment.* Used with Hydrosphere and Atmosphere. See also Barysphere, Moho and Psychosphere.

**Littoral (1)**
*O.E.D.* Of or pertaining to the shore; existing, taking place upon or adjacent to the shore. Applied loosely and generally to the seashore as well as to the shores of lakes and rivers.

**Littoral (2)**
A littoral district; the region of a country lying along the shore. In English adapted from Italian and French usage.

**Littoral zone**
Page, J., 1876, *Adv. Text-Bk. Geol.*, iii, 76. The Littoral [zone] lies between high and low water mark.
Mill, *Dict.* The land between high- and low-tide levels.
Sverdrup, H. U., Johnson, M. W. and Fleming, R. H., 1942, *The Oceans*, New York: Prentice-Hall. The littoral system is the upper portion of the Benthic or bottom terrain biotic environment. 'The eulittoral zone extends from high-tide level to a depth of about 40 to 60 m. The lower border is set roughly at the lowest limit at which the more abundant attached plants can grow. The sublittoral zone extends from this level to a depth of about 200 m. or the edge of the continental shelf' (pp. 275–276). The latter is also supposed, roughly, to be the depth separating the lighted from the dark portion of the sea. Some authors, *e.g.* Gislen, 1930, restrict eulittoral zone to the intertidal zone. (W.M.)
Hesse, R., 1937, *Ecological Animal Geography*, New York: Wiley. Equated with the shore zone or littoral region, as the area where green plants grow on the lake or sea floor, related to light conditions, may be to 30 m. in some lakes but usually 6–12 m. (p. 327) (W.M.).
Carpenter, 1938. Among other definitions are: 1. Of the seashore between the tide marks; 2. Pertaining to the shallower life zone near the shore of a body of water out to the usual limit of influence of wave action or tides and daylight on the bottom life (p. 160).
Holmes, 1944. *Littoral* deposits are those formed between high and low tides (p. 313).
*Comment.* The meaning given to the word depends upon the context. Some geologists, following Sverdrup, use 'littoral deposits' for all shallow water marine deposits. See also the definition given under Pelagic.

**Living space**
An English translation of the German *Lebensraum* but most authors prefer to use the German word when discussing its geopolitical implications. (L.D.S.)

**Liwa** (Iraq: Arabic)
A province.
*Statesman's Yearbook*, 1959. 'Each Liwa is administered by a Mutasarrif, and is subdivided into qadhas (under Qaimmaquams) and nahyahs (under Mudirs)' (p. 1130).

**Llan** (Welsh)
Church; enclosure. Originally an early religious enclosure and used in this sense in geographical literature. (E.G.B.)

**Llano** (Spanish: *lit.* a plain, level ground)
*O.E.D.* A level treeless plain or steppe in the northern parts of South America.
*Webster.* 1. An extensive plain with or without vegetation. *Sp. Amer.* 2. *pl.* The lowlands bordering the Orinoco River.
*Dict. Am.* An extensive treeless plain, steppe, or prairie. 1846. Sage *Scenes Rocky Mts.* xxxiii. 'the cheerless *llanos* of the Great American Desert'. b. Llano Estacado, the Staked Plain, a high arid plateau of forty thousand square miles, situated in Texas and New Mexico. 1834.
Mill, *Dict.* Level treeless plains or steppes in the northern parts of South America, *e.g.* Savannas of the Orinoco.
Küchler, A. W., 1947, Localizing Vegetation Terms, *A.A.A.G.*, 37. 'The Llanos extend from the Andes in the north to the Orinoco or slightly beyond, in the south, and from the dense forests of the Orinoco delta westward for about 1000 km. into eastern Colombia. They consist of large grass-covered plains, over which trees, often palms, are thinly scattered. In addition there are dense galeria forests along the permanent streams and occasional formations of shrub which at times become quite extensive.' '... the name Llanos has become inseparably linked with it [the region], even though in Spanish literature, the term llanos is frequently used to denote plains, especially grassy plains, anywhere.' (p. 200)
James, 1959. 'The Llanos consist of a vast almost featureless plain.' (p. 85)
*Comment.* The Spanish use the term for a plain, *e.g.* flat-floored basins in the Pyrenees. In South America the use is the same (see James above); only in international literature has the word come to mean the grassy vegetation (which James does not use).

**Load of a River**
Himus, 1954. The sum-total of the solid matter carried by a river, including (i) materials suspended in the water, mainly mud, silt and sand; (ii) larger, heavier material which is rolled along the bed of the river; and (iii) material in solution in the water. The maximum load of a river depends on its velocity and its volume, and on the size of the particles constituting the load. When the limit to the possible load has been reached, any further addition involves the dropping of an equivalent portion of the original load.

**Loadstone, lodestone**
Magnetic oxide of iron ($Fe_3O_4$); a piece of this mineral used as a magnet.

**Loam**
*O.E.D.* 1. Clay, clayey earth, mud; obsolete. Also loam pit=clay pit; obsolete. 2. Clay moistened with water so as to form a paste capable of being moulded into any shape; specifically a composition of moistened clay and sand with an admixture of horse-dung, chopped straw, or the like, used in making bricks and casting moulds, plastering walls, grafting, etc. 3. A soil of great fertility composed chiefly of clay and sand with an admixture of decomposed vegetable matter. It is called *clay loam* or *sandy loam* according as the clay or sand predominates.
Mill, *Dict.* An earthy mixture of sand, clay and organic matter.
Robinson, G. W., *Soils*, 3rd ed., London: Murby, 1949. A soil in which neither sand, silt nor clay comprise 50% by weight. (W.M.) (p. 23)
Jacks, 1954. Soil having clay and coarser particles in proportion which usually form a permeable, friable mixture (U.K.). Soil material containing 7–27% clay, 28–50% silt, and less than 52% sand (U.S.).
*Comment.* This is a very old word, as quotations in the *O.E.D.* show, but modern usage in scientific writing is as defined by Jacks. Organic matter is not an essential constituent and there is no reference to vegetable matter as a constituent of loam in standard works, *e.g. Soils and Men*, 1938; Stamp, *Land of Britain*, 1948; Robinson, 1949.

**Loam-terrain**
Wooldridge, S. W., and Linton, D. L., 1933. The Loam-Terrains of Southeast England and their Relation to its Early History, *Antiquity*, 1933; also in *The Geographer as Scientist*, London: Nelson, 1956. 'Extensive tracts of loamy or intermediate soils, which present optimum conditions for cultivation' (p. 224) '... a natural soil group, giving rise to highly distinctive and clearly bounded regions, which figure among the more important settlement areas of the country from Bronze Age times onward' (p. 225).

## Local climate
Kendrew, W. G., *Climatology*, Oxford: O.U.P., 1949. The temperature of the air is controlled by (a) the importation of air-masses and (b) local controls, which consist of the altitude of the sun, length of day and night, nature of the surface in respect of topography and type of surface (vegetation etc.). Reference to 'local climates' in the index however appears to apply to the section 'Topographical Control of Air Temperature', *i.e.* relief. (W.M.)

Brooks, C. F., 1931, How may one define and study local climates? *Comptes R. Cong. Inter. Geog.*, II., Paris: Armand Colin. 'Local climates may be defined as those which are appreciably different in one or more elements from others within a short distance, say, one kilometre or less. The contrast should be great enough to be obvious without instruments and to be consequential to man.' (p. 291)

No reference in *Met. Gloss.*, 1944; Moore, 1949.

See also Microclimate, mesoclimate.

## Local Time
At any given place it is twelve o'clock or midday local apparent time when the sun crosses the meridian and shadows are at their shortest.

## Locality
*O.E.D.* 4. The situation or position of an object; the place in which it is, or is to be found; especially geographical place or situation, *e.g.* of a plant or mineral.

b. A place or district, of undefined extent considered as the site occupied by certain persons or things, or as the scene of certain activities.

Mill, *Dict.* As *O.E.D.* 4 and 4b.

## Localization (localisation) of Industry
Committee, List 4. The concentration of an industry or trade in a certain district or districts.

## Localization, coefficient of (P. Sargent Florence, 1937)
Florence, P. S., 1937, Economic Research and Industrial Policy, *Econ. J.*, No. 188. 'A coefficient of localisation based on the mean deviation from unity of the industry's regional location factors' (pp. 621–641).

Florence, P. S., and Baldamus, W., 1948, *Investment, Location and Size of Plant*, Cambridge: C.U.P. Localization is 'the local concentration of that industry compared with the distribution of industries as a whole'. The formula for the coefficient of localization is: 'When workers are divided up region by region as percentages of the total in all regions, the coefficient is the sum (divided by 100) of the plus deviations of the regional percentages of workers in the particular industry from the corresponding regional percentages of workers in all industry.' Uniform dispersal gives a coefficient of 0, extreme differentiation a coefficient of 1 (pp. 34–35).

## Location factor; Location quotient (P. Sargent Florence, 1937, 1948)
Florence, P. S., 1937, Economic Research and Industrial Policy. *Econ. J.*, No. 188, 621–641. The location factor of any particular industry in any given area is provided by 'comparing the proportion of all occupied persons that were occupied in that industry in the given area with the corresponding proportion for the country as a whole'.

Florence, P. S., and Baldamus, W., 1948, *Investment, Location and Size of Plant* Cambridge: C.U.P. 'The location quotient for any industry in any region is obtained by dividing the percentage of workers in that industry found in the region by the percentage of total workers found there.' (p. 41)

See also Localization.

## Location of Industry
Committee, List 4. The areal distribution of industrial activities.

*O.E.D.* Location: The fact or condition of occupying a particular place.

## Location (least cost)
Chorley and Haggett, 1967. Hamilton, F. E. I. 'Weber stresses that within this heterogeneous environment, entrepreneurs will locate industries at the point of least cost in response to three general location factors—transport and labour (as intra-regional factors), and agglomeration or deglomeration (as intra-regional factors). He assumes that transport costs are a function of weight and distance (ton-miles).' (p. 370)

See Weber, A., 1909, *Uber den Standort der Industrien, I: Reine Theorie des Standorts*, Tübingen: and Weber, A., 1914, Industrielle Standortslehre, *Grandiss der Sozialökonomik*, 4.

**Location (minimax)**
Used by Greenhut to describe firms seeking a location with minimum cost but maximum profit. See Greenhut, M. L., 1952, Integrating Leading Theories of Plant Location, *Southern Economic Journal*, **18**, 526–538; and Greenhut, M. L., 1957, *Games, Capitalism and General Economic Theory*, The Manchester School of Economic and Social Studies, **25**, 61–68.

**Location (optimum)**
Morrill, R. L., 1970, *The Spatial Organization of Society*, Belmont, California: Wadsworth Publishing Co. '... the optimal site for a firm is that which is central, minimizes the costs of the necessary spatial relations, and, if selling prices vary for the product, maximizes the differences between costs and revenues. A classical theory of firm location was formulated by A. Weber in 1909.' (pp. 86–87)
See Weber, A., 1909, *Uber den Standort der Industrien, I: Reine Theorie des Standorts*, Tübingen.

**Location theory**
Berry and Horton, 1970. 'Location theory was developed out of the need to explain and predict the location of economic activities. . . . It provides either the medium for ex-post-factor rationalization of "survival of the fittest" within an "adoptive" economic system . . . or the basis for "adaptive" planning whereby businessmen can analyze and comprehend the economic system and make rational choices concerning location of their firms. This theory embraces three levels of observation: (1) location of a firm; (2) the location of groups of firms so they will be in a stable competitive situation in relation to each other; and (3) the location of sets of activities, such as different kinds of agricultural land use, in relation to each other so that all activities bear a stable competitive relationship.' (p. 95)
See translation of Weber, A., 1909, in Friedrich, C. J., 1929, *Alfred Weber's Theory of the Location of Industries*, Chicago: University of Chicago Press, for location theory in relation to theory of land rent. (pp. xi–xxi)

**Loch** (Scottish; Gaelic and Irish)
*O.E.D.* A lake; applied also to an arm of the sea, *esp.* when narrow or partially landlocked.

Mill, *Dict.* The Scottish term for 1. a lake and 2. a fiord.

**Loch or Loch-hole**
A Derbyshire miners' term for a cavity in a mineral vein, elsewhere called a vug.

**Lochan** (Scottish from Gaelic, diminutive of loch)
A small loch or lake.

**Lode, vein**
Himus, 1954. General terms for bodies of ore in forms having great depth or length but little thickness.
*Comment.* A lode is a mineral vein; a vein may be occupied solely by quartz or other non-metalliferous mineral.

**Lode**
Mill, *Dict.* A watercourse, usually partly artificial, and banked up above the surrounding country (Cambridge and East Anglia).
*Comment.* An old word now local, and best avoided in this sense.

**Lodestar, Loadstar**
*O.E.D.* A star that shows the way; *esp.* the pole star.

**Loess, löss** (German)
*O.E.D.* A deposit of fine yellowish-grey loam found in the valley of the Rhine and of other large rivers. From German dialect word *lösz*. Spelling loëss is erroneous.
*Webster. Geol.* An unstratified deposit of loam, ranging to clay at one extreme and to fine sand at the other, usually of a buff or yellowish-brown color, covering extensive areas in North America (especially in the Mississippi basin), Europe (especially north central Europe and Russia), and Asia (especially eastern China), and now generally believed to be an aeolian deposit chiefly. It is usually calcareous, and often contains shells, bones, and teeth of mammals, as well as concretions of calcium carbonate, and occasionally of iron oxide. It makes an excellent soil where adequately watered.
Mill, *Dict.* Loess. A fine yellowish or greyish-brown loamy earth with some sand, carbonate of lime, and easily soluble alkaline salts. It is of two kinds, (1) *land-loess*, which is unstratified or imperfectly stratified . . . (2) *Lake-loess*, originally deposited in salt-lakes; it is stratified and lacks the capillary structure.
Robinson, G. W., 1949, *Soils*. 3rd ed., London: Murby. 'Typically, it consists of a very uniform fine-grained material with

fine sand or silt as the dominant fraction. It is generally considered to be mainly material transported by wind during and after the Glacial Period. It is, in effect, wind-sorted morainic material.

Page, D., 1865, *Handbook of Geological Terms*, London: Blackwood. Loess or lehm. A German term for an ancient alluvial deposit of the Rhine replete with fresh-water shells of existing species.

Geikie, J., 1898. Primarily a flood-loam of glacial times, all 'more or less re-arranged and modified by subaerial action'. (pp. 192, 210)

Rice, 1941. Adopted from German dialect *löss*, a deposit of very fine yellowish grey loam in the valley of the Rhine. A widespread deposit of silt or marl. It is a buff or brownish coloured (when oxidised), porous, but coherent, deposit traversed by a network of narrow tubes representing the negatives of successive generations of grass roots. The comminution of the constituents is ascribed to the grinding action of glaciers; the fine grade and distribution to the action of the wind; and the accumulation in thick deposits to the grip of vegetation.

Russell, R. J., 1944, Lower Mississippi Valley Loess, *Bull. Geol. Soc. Amer.*, 55, 1–40. 'No one seems to have formulated a precise physical definition at an early date. A synthesis based on descriptions, however, indicates the intent to regard loess as homogeneous, unstratified, slightly indurated, porous, calcareous, sedimentary rock consisting predominantly of particles of coarse silt size, ordinarily yellowish or buff in colour, capable of splitting along vertical joints and tending to stand in vertical faces. Everyone to-day tends to regard such material as "typical" loess and it is doubtful if a better definition can be formulated.'

*Comment.* The theory of the subaerial origin of loess was developed by F. Richthofen in *Letter on the Provinces of Honan and Shansi*, Shanghai, 1870, pp. 9–10; also his *China*, Vol. 1, pp. 56–189, and *Geol. Mag.*, 1882, pp. 293–305. In modern literature it is applied to wind-borne material as defined by Robinson above and the *O.E.D.* definition like that of Page is out of date. It may well be argued that the work of Richthofen on the Chinese loess has led the great Asiatic deposits to be regarded as 'typical' and that the original loess of Germany, laid down under more humid conditions, is partly water-sorted and so no longer to be regarded as 'typical'. See also Loess-loam, Limon, Brick-earth.

**Loess loam (Loess-lehm; German)**
Zeuner, F. E., 1945, *The Pleistocene Period*. In discussing the loess at Achenheim he says 'Another "atypical" portion of the Middle Older Loess, called "humose loess-loam" by Schumacher has a hygrophilous forest fauna of molluscs associated with *Dicerorhinus merckii*, abundant red deer, roe deer but also the marmot. It is apparent from the nature of this deposit, which is often humose or loamy, and from its largely humid and temperate fauna, that the conditions of formation cannot have been those of the periglacial steppe. It cannot be decided as yet whether the middle Older Loess at Achenheim is a "secondary loess" re-deposited by hill wash in an interglacial climate (possibly in connection with tectonic disturbance, Wernert, 1936) or whether it is the marginal facies of an ordinary loess steppe, which in the climatically favoured Rhine valley, was bordered by forests along the mountains (Zeuner, 1937, p. 392). The presence of the Marmot supports the latter view. In any case the Middle Older Loess does not appear to represent a cold phase of great intensity' (p. 70).

Russell, R. J., 1944, Lower Mississippi Valley Loess, *Bull. Geol. Soc. Amer.*, 55, 1–40. 'While the initial phases of deloessification occur in an amazingly short time the completion of the process probably requires an indefinitely long period possibly of the order of that needed before a soil profile attains equilibrium. Eventually loess will probably alter completely into some type of brown loam.' Much of the European 'Lehm' appears to be of such origin (Mill, H. R., *Am. J. Sci.*, 3rd Ser. V., 49, 1895, p. 26). (G.T.W.)

**Logan Stone**
Mill, *Dict.* A rocking-stone (S.W. England).
*Comment.* Actually a well-known block of granite in the Land's End peninsula resulting from the atmospheric weathering which produces tors (*q.v.*) which is so balanced that it can be rocked with the hand; applied to similar balanced or 'logging stones' elsewhere.

**Loma** (Spanish America)
*Webster.* A broad-topped hill. *Southwestern U.S.*
*Dict. Am.* A small hill or elevation. Often in place-names. *S.W. U.S.*, 1849.
Mill, *Dict.* A hill or rising ground on a plain (Spanish America).
No reference in standard works, but given with the same meaning by Knox, 1904.

**Longitude**
*O.E.D.* Distance east or west on the earth's surface, measured by the angle which the meridian of a particular place makes with some standard meridian.
The meridian which is now almost universally used is that of Greenwich, London, which is considered as 0° and distance is recorded in degrees, minutes and seconds East and West Longitude. In time, fifteen degrees is equivalent to a difference of one hour in local time.

**Longitudinal Coast**
A coast running broadly parallel to the main tectonic structure or fold lines; also called a Pacific coast being found commonly around the Pacific Ocean. In contrast to Atlantic or Transverse coast.

**Longitudinal valley**
Mill, *Dict.* A valley whose sides are formed by parallel mountain or hill ranges.
Powell, 1875. '... having a direction the same as the strike.' (p. 160) May be anticlinal, synclinal or monoclinal, *q.v.*
*Comment.* It should be used in the sense given by Powell. Mill's definition is derived from an earlier use by W. D. Conybeare and W. Phillips, 1822, *Outlines of the Geology of England and Wales*, p. xxiv. According to current usage it is really only correct where the mountain or hill ranges are parallel to the strike. See also Linton, D. L., 1956, The Sussex Rivers, *Geography*, **41**, 233–247.

**Lōō** (Indo-Pakistan: Urdu), **Look** (Sindhi)
Hot dry wind; a heat-wave.
Spate, 1954. 'A very hot dust-laden wind which may blow for days on end' (p. 55).

**Loop District**
*Webster.* The Loop. That part of the business center of the city of Chicago surrounded by a 'loop' of elevated railways near the lake and bounded by Lake, Wells, and Van Buren streets and Wabash Avenue.
*Dict. Am.* Loop. Originally any completed turn in railroad or elevated tracks, then the territory inclosed by such tracks. Especially the territory in downtown Chicago bounded by the elevated railway; hence transferred (especially by Chicagoans), any business district. 1893.
The heart of the central business district of Chicago is approximately delimited by a loop of the elevated railway and hence is known familiarly as the Loop District. Being so well known, Americans not infrequently make reference in other cities to the heart of a CBD (*q.v.*) as the 'loop district'. (L.D.S.)

**Lopolith**
Holmes, 1944. 'Intrusions, which on the whole are concordant and have a saucer-like form, are distinguished as *lopoliths* (Gr. *lopas*, a shallow basin). The best known examples are of extraordinary dimensions, the largest, that of the Bushveld in South Africa, being nearly as extensive in area as Scotland (*cf.* Sill, Laccolith, Batholith).'
Himus, 1954. Similar: not in *O.E.D.*

**Lōrā** (Indo-Pakistan: Baluchi)
A hill torrent carrying rain water.
Spate, 1954. 'The bigger wadis or loras' (Baluchistan) (p. 425).

**Louderback** (W. M. Davis, 1930)
Davis, W. M., 1930. The Peacock Range, Arizona, *Bull. Geol. Soc. Amer.*, **41**.
Thornbury, 1954. 'The term louderback was proposed by Davis for displaced segments of a lava flow on two sides of a fault. If it can be established that the lava flow is of late geologic age, there is justification for assuming that associated scarps were produced by faulting.' (p. 258)
*Comment.* Named after the geologist G. D. Louderback who first described the phenomenon. An American term not in general use. (G.T.W.)

**Lough** (Irish)
*O.E.D.* A lake or arm of the sea; equivalent to the Scottish loch (which is the native Irish; lough is called 'Anglo-Irish').

**Lowland, lowlands**
Mill, *Dict.* Usually applied to land less than 600 feet above the sea level.
*Comment.* A word with no precise meaning: low and relatively level land at a lower elevation than adjoining districts.

**Lowry models**
Lowry, I. S., 1964, *A Model of Metropolis*, Santa Monica, California: Rand Corporation. 'This report describes a computer model of the spatial organization of human activities within a metropolitan area. The model is intended for eventual

use as (1) a device for evaluating the impact of public decisions (e.g. concerning urban renewal, tax policies, land-use controls, transportation investments) on metropolitan form; and (2) a device for predicting changes in metropolitan form which will follow over time as a consequence of currently visible or anticipated changes in key variables such as the pattern of "basic" employment, the efficiency of the transportation system, or the growth of population.' (p. v)

McLoughlin, J. B., 1969, *Urban and Regional Planning: a Systems Approach*, London: Faber and Faber. 'Lowry's model has become a classic demonstration of the potentialities of the equilibrium approach to urban metropolitan form.' (p. 246) 'Beginning, then, with a distribution of "basic" jobs by one mile square cells, the computer allocates around each cluster of jobs the appropriate populations to provide the labour force needed. But these residents call forth the retail activities to serve them, which are thus located so as to take maximum advantage of the market potentials offered. These retail activities of course represent employment opportunities in themselves and so in the next round or iteration further residents are distributed in order to provide the retail labour force. But this step "disturbs" the market potentials so that the pattern of retail activities has to be modified, which in turn will disturb the residential distribution ... and so on until an equilibrium position is reached.' (p. 247)

*Comment.* The Lowry model started with the *Economic Study of the Pittsburgh Region* in 1962 and developed within the Rand Corporation in 1963 and onwards. Lowry's model showed the potentialities of the equilibrium approach to urban metropolitan form.

**Loxodrome, Loxodromic curve**—see Rhumb line

**Lumb** (Yorkshire)
Used currently in the Sheffield area for the local steep-sided valleys, Not in *O.E.D.* or *Webster* in this sense.

**Lunette** (Australia; E. S. Hills, 1940)
Hills, E. S., 1940, The Lunette, a New Land Form of Aeolian Origin, *Australian Geographer*, 3, 3–21. 'Crescent-shaped formations of aeolian origin, found on the leeward side of lakes in Victoria, formed of black loam deposited from dust-laden winds travelling over the lakes' (T. Langford-Smith in MS.). See also Stephens, C. G., and Crocker, R. L., 1946, Composition and Genesis of Lunettes, *Trans. Roy. Soc. S. Aust.*, 70, 302–312.

**L.U.S., LUS**
Land Utilisation Survey (of Britain) which carried out a field to field survey of England, Wales and Scotland on the scale of six inches to one mile in the nineteen-thirties and published its findings as a series of one-inch maps with explanatory memoirs. The World Land Use Survey under the auspices of the International Union worked on comparable lines. The Second Land Utilisation Survey of Britain was inaugurated in 1960 by Alice Coleman of University of London, King's College. See Land Use.

**Lusaka** (Bantu)
A type of dense bush vegetation in Northern Rhodesia. Malusaka is the normal plural; the Bantu languages express the plural by means of a prefix. (P.S.)

**Lusuku** (Uganda; Luganda language)
A garden plot, averaging 1–2 acres and maintained year after year—characteristic of Uganda. The banana garden in the midst of which the individual homestead of rural Buganda is characteristically set. (S.J.K.B.)

**Lutite, lutyte**
Sediment or sedimentary rock which consists essentially of particles of clay-size (less than 0·002 mm diameter). A. W. Grabau proposed many types with names ending in -lutyte.

**Lynchet, linchet, linch** and other spellings
*O.E.D.* Linchet. dial. also Lynchet(t), Linchard, Linchet.
1. A strip of green land between two pieces of ploughed land.
2. A slope or terrace along the face of a chalk down.
(Also: Linch. A rising ground; a ridge, a ledge; esp. one on the side of a chalk down; an unploughed strip serving as a boundary between fields.)
Mill, *Dict.* Old terrace produced by the accumulation of earth along some artificial boundary constructed during cultivation (S. and E. England).
Crawford, O. G. S., 1953, *Archaeology in the Field*, London: Phoenix House. '... though artificial in the sense that they are caused by cultivation, they are produced by

natural agencies and not deliberately made' (pp. 88–89). (Produced by soil run-off accumulating at the downhill edge of a field. There is a negative lynchet on the upper side of the field and a positive lynchet on the lower side. W.M.)

Miller, A. A., *Chambers's Encyclopaedia*. 'Lynchet (locally also called Lynch or Rein) is a terraced feature on a hillside, ascribed to ancient cultivation.' Generally parallel to the contours but sometimes up and down and sometimes retained by a wall of rough stones. 'They represent a device, intentional or accidental, for conserving the soil that naturally moves downhill when disturbed and cleared of the protecting sod.'

Miller, A. A., 1954. The Mapping of Strip Lynchets, *Advancement of Science*, **11**, No. 43, 277–279. 'Lynchets are terrace-like features, commonest on steep chalk hill slopes in the south of England ... often occurring in "flights" of five, or more, when they are sometimes popularly called "shepherds' steps". The name, a dialect word in the first place, has many variants, linches, landshards, etc.; in the Isle of Portland ... are called "lawns".'

*Comment*. The modern tendency is to give a precise meaning to this old dialect word, sometimes particularizing 'strip-lynchet' as of definitely human origin. Contrast terracette.

### Lysimeter, lysimetry

*Webster*. A device for measuring the percolation of water through soils and determining the soluble constituents removed in the drainage.

Green, F. W., 1959, *Nature*, **184**. 'A lysimeter is an apparatus used for measuring the quantity or quality of water which has percolated through a container which is filled with soil or similar material ... installed to throw light on what happens in the field.' (p. 1186)

# M

**Maar** (German *pl.* Maare)
Lobeck, 1939. 'The Eifel region of western Germany contains a number of low craters called *maars*, often lake filled.' (p. 683)
Fischer, 1950. 'Explosion pit.' (p. 34)
*Comment.* The term seems originally to have referred to a crater lake but has sometimes been transferred to the crater itself. See Thornbury, 1954, 501 in contrast to Holmes, 1944, 452.

**Macadam, macadam-road, macadamized**
In the early part of the 19th century a Scottish road engineer J. L. McAdam discovered that certain types of stone, broken into angular pieces of nearly uniform size, would bind together to form a hard road surface under the pressure of ordinary wheeled traffic or better still under the heavy pressure of specially designed rollers. This became standard practice in road-making; the broken stone is known as macadam, the roads as macadam-roads or macadamized roads. The binding of the pieces of stone is assisted if they are coated with tar, whence tar-macadam.

**Macchia** (Italian)—see Maquis

**Machair** (Scottish)
Vince, S. W. E., 1944, *The Highlands of Scotland* in The Land of Britain, I. 'In the Uists there is a low plain along the western shores which is in part formed of the sandy accumulation known as machair' (p. 457).
Darling, F. F., 1947, *Natural History in the Highlands and Islands*, Collins New Naturalist Series, uses the term repeatedly without fully defining it and there are illustrations of machair in both colour and monotone. Essentially a belt of shelly sand with an interesting fauna and flora and affording useful light arable soils, hence 'machair soils' (L.D.S.). Not in *O.E.D.*

**Macro**—See Appendix I Greek words

**Macrogeography**
Stewart, J. Q. and Warntz, W., 1958), Macrogeography and Social Science *Geog. Rev.*, **48**, 1958, 167–184. Used in contrast to microgeography, this paper is devoted to defining the concept of macrogeography but the neat summary by O. M. Miller (p. 184) is open to much criticism:—

| Geography | Geography |
|---|---|
| In small degree | In large degree |
| Collects | Is thus |
| The facts that be | The gift to see |
| On land and sea | Not thee and me |
| Expects | But us |
| But little thought | Summed up with |
| Nor is it sought | grace |
| | In mass and space |

**Maelstrom** (Dutch)
Literally a stream which whirls round or millstream; probably introduced into English from the Dutch maps such as those in Mercator's Atlas of 1595. A famous whirlpool caused by tidal currents in the Lofoten Islands on the west coast of Norway, formerly supposed to suck in and destroy all vessels within a long radius. Hence any large whirlpool.

**Maerdref** (Welsh)
Hamlet attached to a chief's court; lord's demesne.

**Maestrale** (Italian)
Kendrew, 1953. In summer 'in the Adriatic the winds, blowing with fair constancy from N.W., usually light or moderate... called Maestrale' (p. 351). *Cf.* Mistral.

**Magma**
Swayne, 1956. The molten material in a state of fusion under the crust of the earth from which igneous rocks are formed.
Hence *magma basin*, a reservoir of such molten rock.
Such a mass of molten rock at very high temperatures would cause profound changes in the rocks with which it came in contact, a change known as *contact metamorphism* and whereby sedimentary and other rocks are converted into *metamorphic rocks* through a distance from the molten mass known as the *metamorphic aureole*. Such a magma would absorb into its own mass fragments, perhaps large masses, of adjoining rocks (*magmatic assimilation*) and the magma would penetrate into fissures, joints and cracks and replace fragments of the adjoining rocks by its own mass

(*magmatic stoping*). In the course of these changes and in the slow cooling of the mass differentiation takes place (*magmatic differentiation*) and a wide range of different igneous rocks may originate from one great magma basin. The main mass eventually solidifies as a *bathylith* or *boss* (*q.v.*) with associated smaller masses such as *laccoliths*, *phacoliths*, *sills* and *dykes* (*qq.v.*). There may be a connection with the earth's surface through *necks* or fissures with *volcanoes*. (L.D.S.) As noted by Holmes, 1944, 'magma is not merely molten rock, it differs ... in being more or less heavily charged with gases and volatile constituents' (p. 443).

**Magmatic water**—see Juvenile water

**Magnesian Limestone**

Himus, 1954. 'A limestone containing from about 5 to 15 per cent magnesium carbonate. Stratigraphically one of the major divisions of the Permian System in northeast England.'

**Magnetic bearing**—see Bearing

**Magnetic declination**—see Declination

**Magnetism, terrestrial**

O.E.D. The magnetic properties of the earth considered as a whole.

The North and South Magnetic Poles are not fixed and do not coincide with the geographical poles. For reasons not entirely clear a permanent magnetism was imparted to certain rocks in past geological times and much work now being carried out on palaeomagnetism is throwing light on the geographical conditions of past eras and especially on continental drift.

**Maidan** (Persian, *maidān*) and various spellings.

A word variously applied by Anglo-Indians. There is a large area of rolling plateau in peninsular India known as the Maidan, but the word has been used for open spaces in the heart of cities, *e.g.* Calcutta, and has since been applied to similar urban open spaces in various parts of the British Commonwealth. (L.D.S.)

**Mai yu** (Chinese)

Plum rain. *cf.* Bai-u of Japan

Cressey, G. B., 1955, *Land of the 500 Million*. 'When well-developed invading masses of Tropical or Equatorial air continue to move over stagnant or modified Polar air, high humidity and exceptionally heavy rains result. This is the case in the Yangtze Valley during June and July when the front is stationary ... this is known as the period of the mai yu or plum rains.' (p. 68)

**Maize rains**

Miller, 1953. 'East Africa. There is no real dry season here but two maxima occur (Nairobi and Entebbe), one in February to May, known as the "maize rains", the other in October to December, known as the "millet rains"' (p. 129).

**Makhtésh** (Hebrew: *pl.* makhteshim. D. H. K. Amiran, 1950–51)

Amiran, D. H. K., Geomorphology of the Central Negev Highlands, *Israel Exploration Journal*, 1, 1950–51, 107–120. 'In the area of the change of direction where both the vertical fold of the anticlines and the horizontal bend of the structure-lines combine to bring about strong tension effects erosion has opened huge hollows, somewhat resembling elongated meteorite craters, known locally as makhteshim (literally mortars). There is only one narrow V-shaped valley, breaching the elliptical walls.' Amiran notes they were wrongly called cirques by S. H. Shaw (*Southern Palestine, Geological Map on the scale of 1/250,000 with Explanatory Notes*, Jerusalem, 1947) and erosion-cirques by Y. L. Picard (*Structure and Evolution of Palestine*, Jerusalem, 1943).

**Maki-Hata** (Japanese: *maki* = pasture; *hata* = the fields)

Tanaka, Toyoji, I. G. U., *Regional Conference in Japan, Abstracts of Papers*, p. 62. 'Comparable to the three-field system in medieval Europe, maki-hata is one form of Japanese agriculture in feudal days which still exists in Oki-island in the Japan Sea ... the land is used alternately for pasturing and farming.'

**Malaysian**

A term used to include all Malays, Indonesians and Aborigines inhabiting Malaysia—the Malay peninsula and Indonesia or East Indian islands. Since the independence of Malaya (1957) Malayan has a political significance as a citizen of Malaya hence the adoption of Malaysian in a racial sense.

**Mallee scrub** (Australia)

O.E.D. Mallee (native Australian). Any one of several scrubby species of eucalyptus which flourish in the desert parts of South Australia and Victoria; esp. *Eucalyptus dumosa* and *E. oleosa*. Also the 'scrub' or thicket formed by these trees.

Mill, *Dict. Mallee Scrub.* Scrub rich in low eucalyptus bushes (S.E. Australia).

Kuchler, A. W., 1947, Localizing Vegetation Terms. *A.A.A.G.*, 37. 'Mallee derives its name from *Eucalyptus dumosa* which Australians also call Mallee. The Mallee is a densely growing evergreen shrub formation which, however, does not consist of pure stands of *E. dumosa*. ... largely restricted to the southern part of the continent ... especially the area of northwestern Victoria and the Eyre Peninsula of South Australia ... as well as large sections of Western Australia, south of Queen Victoria Springs and westward to the ocean.' (p. 207)

## Malnad

Spate, 1954. 'In Mysore there is a fundamental division between the high forested Malnad in the W and the more open "champaign" country of the Maidan (here = parkland) in the E' (p. 656).

*Comment.* Whereas 'maidan' (*q.v.*) has passed into general use in a wider sense, malnad has not.

## Malthusianism, Malthusianist

The body of doctrines derived from the writings of Thomas Robert Malthus (1766–1834); a follower of those doctrines.

Malthus, T. R., 1798, *An Essay on the Principle of Population as it affects the Future Improvement of Society.* 'In later editions the original pamphlet was enlarged into a book and modified. The fundamental thesis remained that population, if unchecked, increases in a geometrical ratio whilst subsistence increases only in arithmetical ratio; that population always increases up to the limits of the means of subsistence and is only checked by war, famine, pestilence and the influence of miseries derived from a consequent low standard of living.' (Stamp, L. D., 1953, *Our Undeveloped World*, 28; Stamp, L. D., 1952, *Land for Tomorrow*, 28.) See also *E.B.*, 14th ed. 1929, vol. 14, pp. 744–745.

*Comment.* French writers use this term for *any* form of restriction on production. A recent example is Dumont, R., 1957, *Types of Rural Economy* (English Edition (p. 12)). (C.J.R.)

## Malvernian, Malvernoid

From the Malvern Hills, a conspicuous ridge running north and south on the borders of Worcestershire and Herefordshire. The ridge results from intense folding of rocks of Carboniferous and older ages and is part of the Armorican orogeny (*q.v.*).

Stamp, L. D., 1923, *Introduction to Stratigraphy.* 'The Malvernian Folds—north and south folds found chiefly in the west of England' (p. 173) ... in contrast to the main Armorican folds which are east–west.

*Comment.* The use of the terms Malvernian and Malvernoid has been and still is a matter of much discussion among geologists. See especially Wills, L. J., 1948, *Palaeogeography of the Midlands*, Liverpool; Wills, L. J., 1951, *A Palaeogeographical Atlas*, London; Wills, L. J., 1956, *Concealed Coalfields*, London; Raw, F., 1952, Structure and Origin of the Malvern Hills, *Proc. Geol. Assoc.*, 63, 227. Wills, following Lapworth, distinguishes between Malvernian folds, initiated in pre-Cambrian times, and later Malvernoid folds following the same trend. See also Armorican and -oid in Appendix I, Greek roots.

## Mammillated surface

*O.E.D.* Mamillated. Also mammillated etc. Having rounded protuberances or projections; covered with mammiform excrescences.

Darwin, C., 1879, *Voyage of a Naturalist.* 'The mammillated country of Maldonado' (p. 46).

Geikie, A., 1865, *The Scenery of Scotland*, London: Macmillan. Of the surroundings of Loch Awa, 'the rocks are worn into smooth mammillated outlines, and covered with ruts and grooves that trend with the length of the valley. It is, in short, a rock basin of which all that can be seen is ice-worn; ...'

Cotton, 1945. 'Commonly the benches are short and ripple-like and very irregular in pattern—whalebacked mounds alternating with rock-rimmed hollows in some of which are tarns. This is the mammillated variety of glaciated surface, in which hollows appear to have been dug out on the weaker parts of the rock by selective glacial erosion.' (pp. 295–296)

Thornbury, 1954. 'Where the surface over which the ice caps moved was mountainous as in the Adirondacks ... the result was a general smoothing off or streamlining of the topography. The many granite bosses of New England exhibit this to a notable degree and form a special type of topography that is well described as a *mammillated surface*.' (p. 385)

No reference in Wooldridge and Morgan, 1937; Holmes, 1944; Flint, 1947.

*Comment.* The term has been applied to various scales of landform, *e.g.* a large-scale rounding due to periglacial conditions, a small-scale due to glacial gouging (G.T.W.).

## Mandated territory

*O.E.D.* Mandate: a command, order, injunction and other meanings.

With the creation of the League of Nations in 1919 after the First World War, the colonial possessions of Germany and Turkey, as the defeated nations, came under control of the League which granted mandates to Great Britain, France and others of the victorious nations to govern those territories until such time as they could become independent, in the meantime reporting to the League. The territories were grouped into three according to their capacity for self-government.

Class A:
    Iraq (Great Britain), mandate ended 1932
    Palestine and Transjordan (Great Britain), mandate ended 1948, now Israel and Jordan
    Syria (France), mandate ended 1936, now Syria and Lebanon.

Class B:
    Cameroons (France and Great Britain)
    Togoland (France and Great Britain)
    Tanganyika (Great Britain)
    Ruanda (Belgium)

Class C:
    South-West Africa (Union of South Africa)
    New Guinea and certain Pacific islands (Australia)
    Samoa (New Zealand)
    Certain Pacific islands (Japan)

The United Nations Organization which came into existence in 1945 after the Second World War took over the mandatory responsibilities of the old League and the remaining mandated territories became Trusteeships or Trustee Territories. (L.D.S.)

## Mango-showers

An expression commonly used in India especially by Europeans for the heavy showers, usually of thunderstorm origin, which occur before the breaking of the main monsoon—in March, April and May when the mangoes begin to ripen. (L.D.S.)

## Mangrove, Mangrove swamp

*O.E.D.* Mangrove. 1. Any tree or shrub of the genus *Rhizophora*, or the allied genus *Bruguiera* (N. O. *Rhizophoraceae*); esp. the Common Mangrove, *R. Mangle*. 2. Applied, on account of similarity of habit and appearance, to various other plants.

Mill, *Dict.* Mangrove Swamp. An area containing a group of xerophilous trees with deep green leaves arising from a tidal mud stretch which is covered by a tangle of branches and 'breathing roots'. Through the latter at low tide the plants obtain, whilst the tide is away, the oxygen required for their root systems.

Richards, P. W., 1952, *The Tropical Rain Forest*, Cambridge: C.U.P. 'The word mangrove is used both for an ecological group of species inhabiting tidal land in the tropics and for the plant communities composed of these species.' (p. 299)

## Manor

*O.E.D.* A unit of English territorial organization, originally of the nature of a feudal lordship. *Lord of the manor*, the person or corporation having the seignorial rights of a manor.

Mill, *Dict.* mod. It meant originally the lands held by a lord who kept for himself such parts of it (called demesne lands) as were necessary for his own use, and distributed the rest to tenants.

Report, Royal Commission on Common Land, 1955–58. 'A term denoting the extent of the land held by the lord of the manor together with in feudal times certain legal and administrative rights, dues and responsibilities' (p. 275). *Manorial waste*. 'Part of the demesne of a manor left uncultivated and unenclosed, over which the freehold and customary tenants might have rights of common. Not all manorial waste was subject to common rights.'

*Comment.* The system extended over England and Wales but not to Scotland and was an essential part of the feudal system especially from the Norman conquest onwards. Originally tenants rendered services to the lord and paid tithes—a proportion of their produce—but later money rents were substituted. With enclosure the system began to break up, some manorial lands were 'allotted' for various purposes for the benefit of the community, parts of the manorial lands were left unenclosed to provide common pasture for peasants who would otherwise have been rendered landless (see Commons). In a few places manorial courts are still held, but in most cases although 'manor-houses' still exist, the lordship of the manor is an empty title. The system was taken to parts of America during the colonial period. (L.D.S.)

### Manor-house

*O.E.D.* The mansion of the lord of a manor. Hence *manor* used loosely for a mansion, country residence, principal house of an estate.

### Manorial Courts

Report, Royal Commission on Common Land, 1955-58, H.M.S.O. 'The courts which a lord of the manor has the right to hold. Formerly there were three such courts, the court baron for the freeholders of the manor, the customary court for the copyhold or customary tenants and the court leet exercising a criminal jurisdiction. The surviving manorial courts, though usually known as "courts leet", generally seem to have assumed the responsibility for managing any remaining common land.' (p. 275)

### Mantle, Earth's

Gaskell, T. F., 1958, A Borehole to the Earth's Mantle, *Nature*, **182**, 1958, 692. 'The mantle extends about half-way to the centre of the Earth and encloses a core which earthquake seismic waves demonstrate to be fluid... the mantle lies at a depth of about 20 miles under the land but underneath the deep oceans is only about 7 miles below sea-surface.' See also Mohorovičić discontinuity.

### Mantle-map (T. Griffith Taylor, 1938)

Taylor, 1951. 'A chart depicting structure' (p. 616). Also mantle-model Taylor, T. G., *The Geographical Laboratory*, Toronto, 1938. 'Building a mantle-model of Canada. The separate "mantles" are cut out and placed over the base' (p. 34).

### Mantle Rock

von Engeln, 1942. 'Rocks subjected to weathering processes ultimately break down to fine waste, popularly referred to as soil, more accurately in geologic terms a constituent part of the cover of rock debris called *mantle rock* or *regolith*.' (p. 161)

Holmes, 1944. 'The superficial deposits which lie on the older and more coherent bedrocks form a mantle of rock-waste of very varied character.' (p. 121)

### Manufacture

*O.E.D.* (via French from med. Latin, *manufactura*; *manus*, hand+*facere*, to make). 1 b. The action or process of making articles or material (in modern use, on a large scale). c. A particular branch or form of productive industry. 2 b. An article or material produced by the application of physical labour or mechanical power. Verb. 1. trans. To work up (material) into forms suitable for use.

*Webster.* 2. The process or operation of making wares or any material products by hand, by machinery, or by other agency; often, such process or operation carried on systematically with division of labor and with the use of machinery.

Committee, List 4. 1. The making of articles or materials (now usually on a large scale) by physical labour or mechanical power. 2. The working up of material into a form suitable for use (*O.E.D.*).

Horton, B. J., 1948, *Dictionary of Modern Economics*, Washington: Public Affairs Press. 'Systematic production carried on in a shop or factory and characterized by division of labor and extensive use of machinery.'

*Comment.* The old root meaning, making by hand, is not only obsolete but the reverse of the modern meaning. In broad general terms manufacturing is the processing of raw materials or foodstuffs.

### Map

*O.E.D.* (From Latin, *mappa*, in classical Latin table cloth, napkin, but in medieval Latin used in the combination *mappa mundi*.)

1. A representation of the earth's surface or a part of it, its physical and political features, etc., or of the heavens, delineated on a flat surface of paper or other material, each point in the drawing corresponding to a geographical or celestial position according to a definite scale or projection. A hydrographical map is now more usually called a chart.

Mill, *Dict.* As *O.E.D.*

American Society of Civil Engineers, 1954, *Definitions of Surveying, Mapping and related Terms*, New York. 'Map. A representation on a plane surface, at an established scale, of the physical features, (natural, artificial or both) of a part or the whole of the earth's surface, by the use of signs and symbols, and with the method of orientation indicated. Also a similar representation of the heavenly bodies.' (p. 96)

International Cartographic Association, Commission II, 1973, *Multi-lingual Dictionary of Technical Terms in Cartography*, Wiesbaden: Franz Steiner Verlag G.M.B.H. 'A representation normally to scale and on a flat medium of a selection of material or abstract features on, or in relation to, the surface of the Earth or of a celestial body.' (p. 7)

*Comment*. Used by most writers in a loose and comprehensive sense to include diagrammatic maps or cartograms showing human as well as physical features. A map on a very large scale is generally called a plan.

Note also the verb to map, 'to make a map of; to represent or delineate on a map' and such derivatives as map-making, mapping etc.

**Map-projection**

The representation of part or whole of the spheroidal surface of the earth on a planesurface (Mill). More correctly, the arrangement of parallels and meridians which enables this to be done. No attempt has been made to deal with the complex subject of map-projections in this glossary, but see general summary under Projection.

See *inter alia* Hinks, A. R., 1921, *Map Projections*, 2nd ed., C.U.P., Steers, J. A., 1949, *An Introduction to the Study of Map Projections*, University of London Press, 7th ed.

**Maquis, Macquis** (French), **Macchia** (Italian), **Matorral** (Spanish)

Mill, *Dict*. Macchia (*pl*. Macchie). Clumps of thick-leaved and frequently prickly bushes characteristic of some of the warmer parts of the Mediterranean region (Italy); in Corsica and Mediterranean France called Maquis.

Küchler, A. W., 1947, Localising Vegetation Terms, *A.A.A.G.*, 37. A thick continuous formation of evergreen shrubs in the western Mediterranean Basin. It is anthropogenic and disappears if man, goat and fire are prevented from interfering. Maquis (French, from Italian) or Macchia (Italian, originally Corsican). (p. 204)

Miller, A. A., 1953, *Climatology*, London: Methuen, 8th ed. (of the main types of vegetation in a Mediterranean climate 'Where conditions are less favourable the forest degenerates into a low scrub, the *macquis* or *macchia* of the Mediterranean the *chaparral* of California, the *Mallee scrub* of Australia, a dense tangle of thicket made up of low-growing evergreens arbutus, laurel, myrtle, rosemary, etc with occasional taller trees.' (p. 168)

Plaisance, 1959, 'Maquis (1775). Formation dense souvent impénétrable de buissons et arbustes: généralement sur sol siliceux sous climat sec et à pluies d'hiver; formation xérophile, mais, normalement, fermée... plus serrée et plus haute que la garrigue qui, d'ailleurs, est sur sol calcaire,' p. 155.

*Comment*. 'Degenerate forest found on siliceous soils (contrast garigue) in Mediterranean lands' (E.D.L.). Similar vegetation in other lands with a Mediterranean climate is given local names—see for example chaparral, matorral and others in Carpenter 1938. The maquis formed an excellent hiding place in the days of vendetta and banditry in Corsica and again for members of underground forces in France during the German occupation of the Second World War when the term maquis became transferred to the people so hiding. (L.D.S.)

**Mar** (Swedish; *pl*. Marer)

A bay or creek, the entrances to which have been filled with silt so that its water is almost fresh. The splintered partly waterlogged topography of the skerry-guard favours the formation of such mars. (E.K

**Marble**

*O.E.D.* Limestone in a crystalline (or less strictly, also a granular) state and capable of taking a polish.

*Comment*. This word is included in the Glossary because it is commonly used in geographical writings with this broad meaning whereas a geological definition usually adds 'produced by metamorphism either thermal or dynamic'.

**March, marches, marchlands**

*O.E.D.* (Adapted from French, *marche*)
1. Boundary, frontier, border, a. The border or frontier of a country. Hence, a tract of land on the border of a country, or a tract of debatable land separating one country from another. Often collective plural especially with reference to the portions of

England bordering respectively on Scotland and on Wales, *i.e.* the Welsh Marches.

Mill, *Dict.* Marches. The border-lands between two states.

Wanklyn, H. G., 1941, *The Eastern Marchlands of Europe*, London: Philip, 1941. Deals with the countries which lie between western Europe on the one hand and the U.S.S.R. on the other.

Note also: to march with (of countries or frontiers) = to border upon or to have a common frontier with.

## Marg (Indo-Pakistan: Kashmiri)

Spate, 1954. 'High pastures above the treeline; this is the *marg* common in place-names and strictly equivalent to the Swiss *alp*' (Kashmir). (p. 377)

## Marginal fringe

Coleman, Alice, 1969, A Geographical Model for Land Use Analysis, *Geography*, 54, 48–49.

Coleman, Alice, 1976, Canadian Settlement and Environmental Planning, *Urban Prospects*, Ottawa: Department of State, Urban Affairs. 'Marginal fringe occurs where the farmer is in conflict with Nature, either pushing back her boundaries by pioneer reclamation or being pushed back himself by economic inability to hold natural forces at bay ... as long as the conflict continues, neither vegetation nor improved farmland is dominant, and instead, both are co-dominant, forming a broken pattern of interdigitating patches ... vegetation and improved farmland must be co-dominant and settlement must be subordinate.'

## Marginal land

The Committee. List 3. Land which yields an average return of no more than sufficient to cover the cost of production.

Stamp, L. D., 1948. Land which will go in or out of agricultural use according to fluctuations in economic conditions (*e.g.* pp. 30, 156, 217, 221).

Horton, B. J., 1948, *Dictionary of Modern Economics*, Washington: Public Affairs Press. Land which is just fertile enough to yield a return sufficient to cover costs of production. Strictly speaking, the return covers the costs of land and capital, but does not afford an opportunity for the payment of rent.

See also Ellison, W., 1953, *Marginal Land in Britain*.

## Marina

*Webster*. 1. A seaside promenade or esplanade. 2. *Naut.* A dock or basin providing secure moorings for power yachts and launches.

*Comment*. With the rapid growth in popularity and consequently of numbers of various types of motor boats for pleasure purposes, this word is coming to be used, especially in America, for coastal locations where facilities for such craft and their owners are provided.

## Marine

*S.O.E.D.* Of or belonging to, found in, or produced by the sea. Hence marine deposits, marine denudation, marine erosion, marine peneplanation, marine platform etc.

## Maringräns (M.G.) (Swedish)

The highest marine limit, *i.e.* the highest coastline of the post glacial sea. Norwegian: *den marine graendse* (Kjerulf 1870; G. De Geer 1888). (E.K.)

## Maritime climate

Mill, *Dict.* A climate influenced by the proximity of the sea which causes a comparatively cool summer and a comparatively mild winter on account of the different thermal capacities of land and water.

## Markaz (Sudan; Arabic)

A district of a province in the Sudan, and also its administrative headquarters. (J.H.G.L.)

## Market

*O.E.D.*

1. The meeting or congregating together of people for the purchase and sale of provisions or livestock, publicly exposed, at a fixed time and place; the occasion, or time during which such goods are exposed for sale; also, the company of people at such a meeting.

3. A public place, whether an open space or covered building, in which cattle, provisions, etc. are exposed for sale; a market-place, market-house.

5. Sale as controlled by supply and demand; hence, demand (for a commodity).

7. *The market*; the particular trade or traffic in the commodity specified in the context.

9. A place or seat of trade; a country, district, town etc. in which there is a demand for articles of trade; hence the trade of such a country etc.

Committee, List 4.

1. 'Any area over which buyers and sellers are in such close touch with one another that the prices obtained in one part affect the prices paid in another' (Benham).
2. More generally, any area in which buyers of particular products are found.

Hence *market-town*, a town in which there is a permanent or periodical market or which serves the function of 1 above.

### Market gardening, Market garden

*O.E.D.* Market-garden, a piece of land on which vegetables are grown for the market.

*O.E.D. Suppl.* Market-gardening, keeping a market garden.

Committee, List 3. The intensive production of vegetable crops, soft fruit, or flowers, for sale. Equivalent to 'truck-farming' in American usage.

Stamp, 1948. 'Market gardening is the term usually applied in Britain to a form of intensive cultivation which is concerned mainly with the production of vegetables for human consumption. When organized on a large scale and when there is a marked concentration on one or two crops it is comparable with American "truck-farming" but, speaking generally, the British market gardener concerns himself with a wider range of crops.' (p. 127)

See also Horticulture

### Marl

*O.E.D.* 1. A kind of soil consisting principally of clay mixed with carbonate of lime, forming a loose, unconsolidated mass, valuable as a fertilizer. The marl of lakes is a white chalky deposit consisting of the mouldering remains of Mollusca, Entomostraca, and partly of fresh-water algae (Geikie in *Encycl. Brit.* X, 290/2).

Mill, *Dict.* Loosely used for all friable compounds of clay and carbonate of lime, frequently formed under freshwater conditions (Lapworth). Marly Soil: a soil containing at most 75 per cent. silicate of alumina and at least 15 per cent. of lime in addition to other constituents (Supan).

Stamp, 1948, '... the very beneficial results on light land of "marling"—spreading clay on the ground to give "body" to the soil.' (p. 47)

Page, 1865. 'Any soft admixture of clay and lime is termed a marl ... but occasionally the word is used to designate soft friable clays with which not a particle of lime is intermingled just as agriculturists call any soil a "marl" that falls readily to pieces exposure to the air.'

Soils and Men, 1938. 'An earthy crumbli deposit consisting chiefly of calcium ca bonate mixed with clay or other impurit in varying proportions. It is used fr quently as an amendment for so deficient in lime.' (p. 1172)

Fay, 1920. '... Marl in America is chief applied to incoherent sands, but abro compact, impure limestones are also call marls.'

*Comment.* The word marl is currently us in several different senses but especial the two given by Page, 1865. Geologis have added to the confusion by using t word as a proper noun, as in the Keup Marls (commonly devoid of calcium carb nate) and in the Chalk Marl (an impu chalk with much clayey matter but oft hard) and the Marlstone (Middle Lias— hard ferruginous rock, worked as an iro ore, together with its associated beds).

### Marsch (German; *pl.* Marschen)

Shackleton, 1958. 'West of the River Elbe . there are three types of landscape (i) t reclaimed land called *marschen* along t coast and estuaries, (ii) the bogs call *moore*, and (iii) the heath lands call *geest* ("infertile").' (p. 257) The *marschen* carry chiefly meadow land f cattle ... rich soil supports a large farr ing population, in spite of the danger living below high tide.' (p. 258) C Polder.

### Marsh

*O.E.D.* 1. A tract of low-lying land, flood in winter and usually more or less wate throughout the year.

Mill, *Dict.* 1. An area covered partly wi water and partly with water-loving veg tation. 2. Locally applied to any meado (Tasmania).

Tansley, A. G., 1949, *Britain's Green Mantl* London: Allen and Unwin. 'The wor "marsh" is here restricted to wet areas mainly mineral soil, such as may often t found on the edges of lakes and larg ponds and on the undrained flood plair of rivers.' (p. 220) Tansley also distir guishes: Swamp:—when the summe water level is normally above the surfac of the soil; Bog:—wet acid peat so characteristically inhabited by 'bog moss (*Sphagnum*); Fen:—wet peat (*i.e.* purel organic) soil which is typically alkaline in reaction, though occasionally neutral or slightly acid.

See also Swamp.

**Massif** (French)

*O.E.D.* (From French) d. A large mountain-mass; the central mass of a mountain; a compact and more or less independent portion of a range. 1524 in *Hakluyt's Voy.* (1599), II, 1, 86... the massife of Spaine...; 1885. Geikie, A., *Text-bk. Geol.* (ed. 2), 40. A large block of mountain ground rising into one or more dominant summits, and more or less distinctly defined by longitudinal and transverse valleys, is termed in French a *massif*—a word for which there is no good English equivalent. 1899. *Nature*, 2 Nov., 20/2. The formation of a dune tract or dune massif appears to be chiefly determined by the presence of ground moisture.

Mill, *Dict.* A part of a mountainous or highland area forming a mass with relatively uniform characteristics.

Wooldridge and Morgan, 1937. '... pre-existing resistant masses, "stable blocks" or massifs, which are often the denuded relics of former mountain systems.' (p. 69)

Fay, 1920. '2. A diastrophic block, or any isolated central independent mass.' (Standard Dictionary, 1910)

von Engeln, 1942. 'Akin to horsts are the areally more extensive, and structurally less simple, elevated, ancient, geomorphic units known as *massifs*.' The Spanish *meseta*, the Central Highlands and Britanny in France and the Laurentian Highlands of North America are characteristic examples' (p. 387).

*Comment*. The word has been widely and somewhat loosely used but in the background is usually the idea of the *Massif Central* of France as a type example. (L.D.S.) The basic meaning of the French word is simply 'masse compact', which can be applied to a wide range of objects. (G.R.C.)

No reference in Joanne, 1890; Salisbury, 1907; Davis, 1909; Cotton, 1922, 1941, 1945; Holmes, 1944.

**Mass movement**

Cotton, 1941, has a chapter in *Landscape* entitled mass movement of waste. This describes the processes mentioned below under mass-wasting but mass-movement is perhaps a more logical term.

**Mass-wasting**

Thornbury, 1954. 'The gravitative transfer of material.' (p. 44) '... involves the bulk transfer of masses of rock debris down slopes under the direct influence of gravity' (p. 36). Thornbury summarizes the classification proposed by Sharpe, C. F. S., 1938 (*Landslides and Related Phenomena*, Columbia University Press) as follows:—

Four classes of mass-wasting

Slow flowage types

    Creep: the slow movement downslope of soil and rock debris which is usually not perceptible except through extended observation

    Soil creep: downslope movement of soil

    Talus creep: downslope movement of talus or scree

    Rock creep: downslope movement of individual rock-blocks

    Rock-glacier creep: downslope movement of tongues of rock waste

    Solifluction: the slow-flowing downslope of masses of rock debris which are saturated with water and not confined to definite channels.

Rapid flowage types.

    Earthflow: the movement of water-saturated clayey or silty earth material down low-angle terraces or hillsides.

    Mudflow: slow to very rapid movement of water-saturated rock debris down definite channels.

    Debris Avalanche: a flowing slide of rock debris in narrow tracks down steep slopes.

Landslides: those types of movement that are perceptible and involve relatively dry masses of earth debris.

    Slump: the downward slipping of one or several units of rock debris usually with a backward rotation with respect to the slope over which movement takes place.

    Debris slide: the rapid rolling or sliding of unconsolidated earth debris without backward rotation of the mass.

    Debris fall: the nearly free fall of earth debris from a vertical or overhanging face.

    Rockslide: the sliding or falling of individual rock masses down bedding, joint or fault surfaces.

    Rockfall: the free falling of rock blocks over any steep slope.

Subsidence: downward displacement of surficial earth material without a free surface or horizontal displacement.

**Master**
Used as a prefix in the sense of the chief, commanding, dominating as in master-joint, master-stream, master-cave etc.

**Mata, matta** (Brazil: Portuguese; *lit.* forest)
The Brazilians commonly distinguish between *mata* (forest) and *campo* (grassland) but whereas campo (especially in plural as campos) is commonly used in international geographical literature, mata is not. See James, 1959, 395.

**Mathematical Geography**
Hartshorne, 1939, 'In terms of content, the eighteenth century writers commonly distinguished three divisions: "mathematical geography", "physical geography", and what was variously called "historical" or "political geography". Mathematical geography consisted in large part of the study of the earth as an astronomical body—in practice its study was left largely to astronomers.' (pp. 42–43)

'The aspects of geography deriving from the shape, size and motions of the Earth which are capable of mathematical definition or expression' (E. G. R. Taylor in MS.).

**Matmura** (Sudan; *coll.* Arabic, *pl.* matamir; from a word meaning to bury in the soil)
Pits dug in the soil in the drier termite-free areas of the northern Sudan, in which harvested grain is stored. Rising ground on which rain water will not stagnate is always chosen as a site. (J.H.G.L.)

**Matorral** (Spanish)
Equivalent of French *maquis*, Italian *macchia*.
Fisher, W. B. and Bowen-Jones, H., 1958, *Spain*, London: Christophers. 'Much of the interior of Spain has a special and distinctive type of vegetation known as *matorral*... a complex of evergreen or thorn bushes, occasionally interspersed with dwarf oaks or conifers' (p. 46).

**Matterhorn** (Swiss-German)
Wooldridge and Morgan, 1937. 'We may select three features: the *arête*, the pyramidal peak and the *corrie* as recurrent elements in the pattern and profile of glaciated heights—inclined arêtes generally converge on one another, and at their point of meeting are situated the culminating peaks. These are generally sharply pointed and have a rough pyramidal form. The familiar Matterhorn is a magnificent example...' (p. 375).

Lobeck, 1939, refers accordingly to 'Matterhorn peaks' (p. 264), von Engeln, 1942, uses the term 'horn peak' '... the typ example is that of the Matterhorn, but other horns in the same region are of the same origin' (pp. 451–452).

**Mature—landscape**
Wooldridge and Morgan, 1937. 'A landscape has a definite life history during which shows a series of gradual changes, whereby the *initial* forms pass through a series of *sequential* forms to an *ultimate* form. We may broadly group the many successive stages into the major stages of *youth, maturity,* and *old age.* Landscape evolution is thus envisaged as a cycle... "landscape is a function of *structure, process* and *stage*".' (p. 174)

**Mature—river**
Mill, *Dict.* A river in which erosion is at minimum because the stream has acquired a normal fall in its bed.

Davis, 1909. 'When the trunk streams are graded, early maturity is reached; when the smaller headwaters and side-streams are also graded maturity is far advanced and when even the wet-weather rills are graded, old age is attained.... The so called "normal river", with torrential headwaters and well-graded middle and lower course, is therefore simply a maturely developed river.' (pp. 258–259)

Cotton, 1922. '... the establishment of grade marks the passage of a river from youth to maturity, the next stage of the cycle.' (p. 62)

See Cycle of erosion

**Mature—shoreline**
Johnson, 1919. 'The essential feature of maturity in the development of the shore profile is a condition of approximate equilibrium between erosion, weathering and transportation.... At every slope [of the profile] the slope is precisely of the steepness required to enable the amount of wave energy there developed to dispose of the volume and size of debris there in transit.' (pp. 210–211)

See also Wooldridge and Morgan, p. 351.

**Mature—soil**
*Soils and Men*, 1938. 'A soil with well-developed characteristics produced by the

natural processes of soil formation, and in equilibrium with its environment.'

*Note: cf.* mature, immature, truncated profile in Robinson, G. W., 1949, *Soils*, 3rd ed., London: Murby. A mature profile is one which has attained full development. (pp. 70–71)

*See* Soil profile

**Mature—town**—see Urban hierarchy

**Mayen** (German: Switzerland)

Mickles, T., 1932, *Europe*, London: Dent. [Of the Swiss Alps] 'In many valleys there is an intermediate shelf known as the "mayen" or "voralp" between the real "alp" and the valley floor; there the cattle make a short stay on their upward journey in May, and on their downward journey in September' (p. 20). Voralp or Mayen is mentioned but not explained in Peattie, 1936, p. 133.

**Mbuga.** (Central plateau of Tanganyika. Swahili and other Bantu languages)

A seasonal swamp developed in wide shallow valleys. The soil is a dark clay, cracking deeply in the dry season and becoming very sticky when wet, and the vegetation is normally an open grassland. (S.J.K.B.)

**Meadow**

*O.E.D.* 1. Originally a piece of land permanently covered with grass which is mown for use as hay. In later use often extended to include any piece of grassland, whether used for cropping or pasture; and in some districts applied especially to a tract of low well-watered ground, usually near a river. 2. N. America. a. A low level tract of uncultivated grass land, especially along a river or in marshy regions near the sea. 3a. 'An ice-field or floe on which seals herd.' b. 'A feeding ground of fish' (*Cent. Dict.* 1890)

*Webster.* 1. Grassland, especially a field on which grass is grown for hay; often, a tract of low or level land producing grass which is mown for hay. 2. A tract of low land, moist but not inundated, usually characterized by grasses and wild flowers, but without trees or shrubs; as, the mountain *meadows* of the Alps. 3. A feeding ground for fish; as, a cod *meadow*.

*Dict. Am.* 1. A level area of limited extent in a mountainous region, often grassy and usually dry. Frequent in place-names. *W. U.S.* 1870. 2. A feeding ground of fish. 1877.

Mill, *Dict.* A well-watered grassland in a valley or alluvial lowland.

Stamp, L. D., 1948, *The Land of Britain*, London: Longmans. 'Though the layman often uses the words interchangeably, the farmer generally restricts "meadow" to land mown for hay; "pasture" to land which is grazed.' (p. 369)

*Comment.* Confusion results from the American usage of the word as applied to marshland especially in the combination 'meadow soils' which are the soils of swamps or marshes. The normal British usage is that given by Stamp. See also Water-meadow.

**Mean information field**

Derived theoretically from the characteristics of private information fields, the mean information field expresses a negative logarithmic relationship between the probability of contact between any pair of individuals and, *ceteris paribus*, the distance separating them.

See Hägerstrand, T., 1967, *Innovation and Diffusion as a Spatial Process*, Chicago: University of Chicago Press. 'On the average, the density of the contacts included in a single person's private information field must decrease very rapidly with increasing distance.' (p. 235). See also Fig. 95, p. 236.

See also Private information field.

**Mean Sea-Level (M.S.L.)**

Moore, 1949. The average level of the sea, as calculated from a large number of observations taken at equal intervals of time. It is the standard level from which all heights are calculated. On the Ordnance Survey maps of the British Isles, the Ordnance Survey Datum (O.D.), above which the heights of all places are given, is taken as Mean Sea Level at Newlyn, Cornwall.

*Comment.* Other countries adopt other standards.

**Meander, meandering**

*O.E.D. Plural.* Sinuous windings (of a river); turning to and fro (in its course); flexuosities. Rarely in singular; the action of winding; one of such windings.

Mill, *Dict.* Meandering. The winding of a river on a plain. *Forced meanderings:* river windings due to the impact of tributaries. *Sunk meanderings:* river windings that have become comparatively stable owing to the river eroding its bed downwards where the windings occur.

Wooldridge and Morgan, 1937. 'The initial course of the stream will be slightly curved in places ... lateral corrasion will tend to enlarge the curves, under-cutting on the concave side against which the fastest current impinges. In this way a series of meanders comes into being separated by overlapping or interlocking spurs. The spurs have a gently sloping crest line or *slip-off slope* (*q.v.*), while opposite to each is a steeply cut "river cliff" (p. 156). The incision of meanders during rejuvenation ... rapid incision gives rise to "intrenched meanders" of more or less symmetrical cross-section, while if the incision is less rapid, affording time for later shift "ingrown meanders" result' (p. 225).

Thornbury, 1954. 'The first stage in the development of a valley flat involves primarily the elimination of the spurs between stream meanders ... spur trimming ... effected by lateral erosion. Continuation of this process will reduce the width of a meander spur until only a narrow *meander neck* separates the stream on the two sides of the meander. Continued impingement ... will result in a *meander cutoff*. A remnant of the spur will then be left as a *meander core* and the abandoned route round it becomes a *meander scar*' (p. 130).

Dury, G. H. 1954. Contribution to General Theory of Meandering Valleys. *Amer. Jour. Sci.*, 252, 193–224. 'The expression *meandering valley* is applied where the pattern of valley windings broadly resembles the trace of a meandering stream and where successive windings are for the most part of the same general order of size. Similarly a *valley meander* is an individual member of a series of windings' (p. 193).

Rich, J. L. 1914. Certain Types of Stream Valleys and their Meaning, *J. of Geol.*, 22, 469–497. '*Intrenched Meander Valley* is one whose stream having inherited a meandering course from a previous erosion cycle, has sunk into the rock with little modification of its original course. *Ingrown Meander Valley* is one whose stream, which may or may not have inherited a meandering course from a previous cycle, has developed such a course or expanded its inherited one as it cut down. Thus as the stream sinks its channel lower and lower into bed-rock the meanders are continuously growing or expanding. ... The term "incised meander" has apparently been used synonymously with intrenched meander in previous writing. Would it not be well to use "incised meander" as a generic term covering both the above described types and to restrict the meaning of intrenched meanders as indicated above?' (p. 470)

See also Cutoff, Oxbow, Mortlake, Scroll.

*Comment.* Note that O.E.D. statement is incorrect. A distinction should be made between stream meanders and valley meanders. G. H. Dury has commented that *forced meanders* are those cut into and confined by bedrock and doubts the explanation given by Wooldridge and Morgan.

**Meander belt**
Webster. *Phys. Geog.* That part of a valley flat, or bottom, across which a stream shifts its channel from time to time, especially in flood.

**Mear, mears**
1. Land bordering the cultivated area, used presumably for turning the plough.
2. Waste or indifferent land, for example ill-drained land between the cultivated area and a stream. Hence 'Maer Field' 'The Mears', etc. (England: Midlands). Also mere, plough land marking head lands between open fields (I.L.A.T.).

**Mearstone, meerstone**
A boundary stone used between holdings in common fields; more generally any boundary stone especially for marking limit of waste, common and wood. Trees were used for the same purpose, hence 'mear oak' (Midlands of England) (I.L.A.T.).

**Mechanical weathering**
Disintegration of rock without chemical alteration as opposed to chemical weathering.

**Median mass**
Steers, J. A., 1932, *The Unstable Earth*, London: Methuen. According to Kober, inward pressure from two forelands results in outward facing fold mountains with a less disturbed 'median mass' or 'Zwischengebirge' between (p. 146). From L. Kober, *Der Bau der Erde*, Berlin, 1921 (W.M.).

Du Toit, A. L., 1937, *Our Wandering Continents*, Edinburgh: Oliver and Boyd. 'Zwischengebirge', translated as 'Intermontane space' (Longwell), 'Median mass' (de Böckh) or 'Median area' (p. 34).

**Medical Geography**
L.D.G. The study of the distribution of human disease and causes of death,

together with the factors of the environment conducive to human health and sickness. Medical geography may deal only with deaths from disease (*mortality*) but should consider also illness not necessarily fatal (*morbidity*). It is found that some diseases seem permanently located in certain areas where they are then said to be *endemic*, there may at times be a much wider spread (*pandemic*) or sudden outbreaks (*epidemic*).

For a general introduction see L. D. Stamp, *The Geography of Life and Death*, London: Collins 1964; Ithaca: Cornell University 1965; L. D. Stamp, *Some Aspects of Medical Geography*, London: Oxford Univ. Press, 1964. For basic material see the series of *Studies in Medical Geography* issued by the American Geographical Society under the editorship of Dr. Jacques May (with series of world maps issued with the *Geographical Review*).

**Mediterranean climate**
Mill, *Dict.* A climate such as characterizes the Mediterranean region, and above all its more southerly parts; that is, one in which there is a moderate range of temperature throughout the year, and a very marked predominance of winter rains, the summers being nearly or quite rainless.
Miller, A. A., 1953, *Climatology*, 8th ed., London: Methuen. A convenient abbreviation for the Western Margin Warm Temperate Type of climate. There is great complexity within the Mediterranean. 'Yet the numerous variants in the Mediterranean basin all agree in certain essential respects, especially—1. A winter incidence of rainfall and a more or less complete summer drought. 2. Hot summers (warmest month usually above 70°) and mild winters (coldest month usually above 43°). 3. A high sunshine amount, especially in summer.' (p. 163)
*Note:* Of the major classifications, Mediterranean climate=1. Köppen: Cs (*e.g.* Trewartha, 1937), 2. Thornthwaite's First: CB's, summer dry mesothermal subhumid (*Geog. Rev.* 23, 1933, p. 439).

**Megalith**
*O.E.D.* A stone of great size used in construction, or for the purpose of a monument.
Hence megalithic, applied to such monuments as Stonehenge and to the people who built them and the period in which they were constructed (Megalithic Age).
Trent, 1959. 'Megalithic, a general description of monuments typical of the New Stone Age and early Bronze Age composed of large blocks of stone, *e.g.* the siliceous sandstone in the downland country of the South.' (p. 43)

**Megalopolis**
*Webster.* A very large city.
Taylor, 1951. 'The overgrown city' (p. 617).
Mumford, L., 1938, *The Culture of Cities*, New York. Applied to London and New York.
*Comment.* The much older term, *megapolis*, for chief city, is obsolete. See *O.E.D.* The usual pronunciation is meg-al-ŏp-ōlis; it should strictly be meg-alō-pŏlis (J. N. L. Baker in MSS.). A recent application has been to the whole urbanized area New York to Washington.

**Meifod** (Welsh)
Intermediate dwelling *i.e.* between hafod and hendref (summer and winter dwellings) (E.G.B.).

**Melt-water**
Water derived from the melting of snow or the ice of a glacier.

**Me-nam** (Thai: *lit.* big water)
The Thai word for the larger rivers. The Me-nam Chao Phraya on which Bangkok stands is 'the river' and has been wrongly described in English—tautologically—as the 'River Me-nam'.

**Mendip**
Thornbury, 1954. 'In the process of valley cutting across the cuesta some of the buried hills on the old land surface may be exposed as *inliers*. They are sometimes referred to as *mendips* from the Mendip Hills of England which were of such origin' (p. 134).
*Comment.* An American usage. See also Lobeck, 1939, 451.

**Menhir**
*O.E.D.* A tall, upright monumental stone of varying antiquity, found in various parts of Europe; also in Africa and Asia (a Breton, Cornish and Welsh word).
Trent, 1959. 'Menhir, a term used with various meanings in archaeological writing. It is now generally reserved for free-standing single-stone monuments or for individual stones in more complex monuments such as the Hele Stone of Stonehenge.' (p. 43)

**Mental maps**
Gould, P. R. and White, R. R., 1968, The Mental Maps of British School Leavers, *Regional Studies*, **2**, No. 2, Oxford: Pergamon Press. 'An individual's mental map of residential desirability, which might be thought of in cartographic terms as a "perception surface" overlying geographic space, is obviously a result of a set of unique information flows impinging upon him that may be ordered according to quite particular personality characteristics into a set of unique space preferences.' (p. 161)

*Comment.* The concept is potentially applicable across a wide range of topics of interest to human geographers. See Trowbridge, C. C., 1913, On Fundamental Methods of Orientation and Imaginary Maps, *Science*, **38**, 888–897; and Gould, P. and White, R., 1974, *Mental Maps*, Harmondsworth: Penguin Books (Pelican.) See also Cognitive map.

**Merdeka** (Malay)
Independence. Refers especially to the independence of Malaya on 31 August, 1957, when British rule came to an end.

**Mere** (rarely, meer)
Mill, *Dict.* mod. A large pond or shallow lake, especially among drumlins.

Meres occupy hollows in plains of glacial drift, especially till or boulder clays, and all stages in silting up or transition to bogs or lowland mosses may be found.

The *O.E.D.* gives many obsolete or obsolescent meanings to this old term which is ultimately linked with meer, mer, mare—the sea. At the present day it is commonly used in parts of England, notably Cheshire. Including also round ponds for cattle lined with cement.

**Meridian**
The Terrestrial meridian is the half of the great circle passing through any given place and the North and South Poles *i.e.* it is the same as the line of longitude of the place. The other half of the great circle (*i.e.* 180° from the meridian) is the antimeridian.

The Celestial meridian is the corresponding great circle in the heavens, passing through the zenith of the place and the celestial poles. When the sun crosses the meridian it is twelve o'clock or midday local time.

The *Prime* (or *Initial*) *Meridian* is that from which longitude is measured. By general international agreement the meridian o Greenwich has been recognized since 188 as the Prime Meridian (0°).

**Meridian Day, Antipodes Day**
When a vessel crosses the International Dat Line from west to east a whole day i gained. The day so gained is called meri dian day. So as to avoid having, fo example, two Wednesdays in one week th second would be called Meridian Day The term is also used for the day on whicl the International Date Line is crossed.

**Mérokarst, merokarst**
This little used term was introduced by Cjivić (see Le Mérokarst, *C.R.Ac.Sci.*, 180 1925, 757–8) in contrast to 'le Karst complet ou *Holokarst*' for regions where 'ces phenomènes sont imparfaitement developées, le Karst imparfait ou *Mérokarst.*' *Idem.* (p. 594).

**Meromixis, meromictic**
'Meromixis is characterized by an absence of complete circulation (in lakes) . . . meromictic stratification may be either thermal or chemical or both . . . the lower, stable, unmixed layer of a meromictic lake is called the monimolimnion.' Newcombe, C. L., *Abs. of Papers, Ninth Pacific Science Congress*, Bangkok, 1957, 173.

**Mesa** (Spanish; *lit.* table)
*O.E.D.* South U.S. A high table-land.

*Webster.* A natural terrace or flat-topped hill with abrupt or steeply sloping side or sides such as are common in the physiography of the southwestern U.S.

*Dict. Am.* An elevated tableland. S.W. and W. U.S. '*Mesa* is applied to two phenomena: an isolated flat-topped hill with abrupt and steep sides, and a comparatively flat plateau extending back from the abrupt ridges of a valley.' 1759.

Mill, *Dict.* 1. Plateau Mesa. A circumscribed summit the continuity of which with other areas has been destroyed by erosion. 2. Bench mesa. A level area against higher eminences and bounded partly by escarpments called cejas. 3. Bolson-mesa. A bench mesa forming the outer escarpment of a drainage valley cut through a bolson.

Knox, 1904. (Sp.) flat or level surface on the top of a hill or mountain, tableland; a landing place, *lit.* a table. (U.S.A.) a flat-topped mountain bounded on at least one side by a steep cliff.

Salisbury, 1907. 'A hard stratum of rock,

such as a lava-bed, overlying less resistant formations, such as clay or soft shale, often gives rise to buttes. If such an elevation has a considerable expanse of surface at its top, it is a mesa, though this term is also applied to wide terraces, especially if high.' (pp. 172-173)

Cotton, 1922. 'Features closely related structurally to cuestas and homoclinal ridges ... are formed by remnants of horizontal resistant strata capping weaker rocks. Large table-like forms are termed mesas, and small residuals are buttes. The level top of a mesa or butte is the upper surface of the hard stratum but little lowered by erosion. The slopes on all sides are escarpments.'

Wooldridge and Morgan, 1937, p. 202; Holmes, 1944, p. 182. Similarly to Cotton.

Comment. The meaning in physical geography of this common Spanish word has undergone an evolution. Senses 2 and 3 of Mill have been dropped and the modern usage is as in Cotton, 1922, rather than Salisbury, 1907.

**Meseta** (Spanish)
Commonly applied to the high tableland of the heart of Spain but according to Pierre Deffontaines (in MS.) it is used also in place of mesa.

**Mesoclimate**
Hogg, W. H. 1963, Classification of Agricultural Land in Britain, *Min. Agr. F.F., A.L.S. Techn. Report* No. 8. 'Mesoclimate is concerned with smaller areas... within a parish could well be considered.... usually expressed as deviates *from* macroclimatic "normals".'

**Mesophyte**
Webster. *Phytogeog.* A plant that grows under medium conditions of moisture. Contrast hydrophyte, xerophyte.

**Mesosphere**
The layer of the atmosphere between the *stratosphere* (q.v.) and the *ionosphere* (q.v.).

**Mesothermal, microthermal**
These terms as applied to climate were defined originally by C. W. Thornthwaite (*Geog. Rev.*, **21**, 1931, 646) in his concept of thermal efficiency or the T–E index. They lie between his A' (Tropical) and D' (Taiga). Later writers especially Finch and Trewartha have suggested modifications.

**Mesquite** (Spanish from Nahuatl)
Webster. A spiny, deep-rooted tree or shrub (*Prosopis juliflora*) of the southwestern U.S. and Mexico, its thickets often constitute the only woody vegetation. Its beanlike pods are rich in sugar and form an important food for stock.

*Dict. Am.* 1. A species of deep-rooted shrublike tree, *Prosopis juliflora*, often growing in dense clumps or thickets, or any species of this genus; also a tree of one of these species, or the wood of such a tree or shrub. 1759. 2. A mesquite thicket. 1834.

**Mesta** (Spanish)
The pastoral organization typical of Castile. (P.D.)

**Mestizo**
*O.E.D.* A Spanish or Portuguese half-caste; now chiefly the offspring of a Spaniard and an American Indian.

James, 1959. 'In Spanish America, the mixture of Indian and European is called a *mestizo*; the mixture of Negro and European is called a *mulatto*; and the mixture of Negro and Indian is called a *zambo*' (p. 13).

Steward, J. H., and Faron, L. C., 1959, *Native Peoples of South America*. New York: McGraw-Hill, p. 158. 'A person who wears sandals, lives in a mud-walled, thatch-covered house, believes in a "pagan" religion, and has certain other simple cultural features is classed as an "Indian". One who wears shoes, eats bread, has a tile-roofed house, practises Christianity, exhibits others features of Hispanic culture is "mestizo", or, if he has a light skin or is a near-white descendant of well-to-do criollos, he is "white".' '... the categories "Indian" and "mestizo" are cultural rather than biological ...' (p. 158).

**Metacartography**
International Cartographic Association, Commission II, 1973, *Multi-lingual Dictionary of Technical Terms in Cartography*, Wiesbaden: Franz Steiner Verlag G.M.B.H. 'The portrayal of spatial properties in maps considered in competition with other devices, such as photographs, pictures, graphs, language and mathematics.' (p. 2)

**Metal, road-metal**
Apart from the ordinary meaning of metal, applied to gold, silver, copper, iron, lead, tin etc. there is the confusing use for 'broken stone used in macadamizing roads or as ballast for a railway' (*O.E.D.*). Only certain rocks are suitable as road metals;

they must be tough, not splintery, and capable of binding naturally (see Macadam), hard enough to withstand pressure of vehicle weights.
Hence metalled-roads.

**Metallogenic, metallogenetic** (L. de Launay, 1900)
Tomkeieff, S. I., 1959, Metallogenic Maps, *Nature*, **184**, 1693. 'Due to L. de Launay, the terms "metallogenic province" and "metallogenic epoch" respectively meant a particular geographical region or a specific geological epoch characterized by a particular assemblage of mineral deposits. The representation of such a province on the map led to the construction of metallogenic maps ... such a map differs from a mineral map which only shows the distribution of one or several mineral deposits while the aim of a metallogenic map is not only this but also to relate the distribution of such deposits to the geological formations or periods and to the tectonic features of the region.'
*Webster. Geol.* Designating or pertaining to the origin of ores.

**Metamorphic**
*O.E.D.* From Greek, meta, change etc. +form. Pertaining to, characterized by or formed by metamorphism. Of a rock or rockformation; that has undergone transformation by means of heat, pressure, or natural agencies.
Mill, *Dict.* Metamorphic Rocks. Applied to rocks which have undergone pronounced alteration by water, heat, pressure, or any combination of these since the rocks were formed.
Page, 1865. '... applied to rocks and rockformations which seem changed from their original condition by some external or internal agency. In geological nomenclature the crystalline stratified rocks—Gneiss, Mica-schist, Clay-slate, etc.—are termed metamorphic and erected into a separate system.'
Holmes, 1920. Metamorphic Rocks. Rocks derived from pre-existing rocks by mineralogical, chemical, and structural alterations due to endogenetic processes; the alteration having been sufficiently complete throughout the body of the rock to have produced a well-defined new type.
*Comment.* The modern usage is as in Holmes

**Metamorphic Aureole**
Holmes, 1944. 'Rocks in contact with igneous intrusions are commonly metamorphosed by heat and migrating fluids (*contact metamorphism*). The zone of altered rock surrounding the intrusion is described as the metamorphic *aureole*.' (pp. 62–63)
See also Magma, skarn

**Metamorphism**
Holmes, 1944. 'The transformation of preexisting rocks into new types by the action of heat, pressure, stress, and chemically active migrating fluids.' (p. 31)
'Kinds of metamorphism ... by severe compressional earth movements (*dynamic metamorphism*) ... by the action of heat (*thermal metamorphism*) ... in contact with igneous intrusions (*contact metamorphism*) ... when all agencies operate together the metamorphism is described as regional.' (pp. 62–63) Also pyrometamorphism (*q.v.*).
See also Magma.

**Metasomatism, metasomatic, metasomatosis**
Replacement. Holmes, 1928. 'The processes by which one mineral is replaced by another of different chemical composition owing to reactions set up by the introduction of material from external sources.' (p. 156)

**Métayage** (French)
*O.E.D.* A system of land tenure in Western Europe and also in the United States, in which the farmer pays a certain proportion (generally half) of the produce to the owner (as rent), the owner generally furnishing the stock and seed or a part thereof. Adam Smith used 'métayer' in 1776.
*O.E.D. supp.* Metayers who, like our sharecroppers, farmed a piece of land for a stipulated portion ... of the harvest.
*Comment.* The last quotation is misleading as métayers are share-tenants, not sharecroppers. The distinction is clear, for example, in both the United States and Italy. (C.J.R.)

**Meteoric Water**—see Juvenile water

**Meteorite**
*S.O.E.D.* A fallen meteor; a mass of stone or iron that has fallen from the sky upon the earth; a meteoric stone.

**Meteorite Crater**
A crater caused by the falling of a large meteorite on the earth's surface.

**Meteorology**
*S.O.E.D.* The study of, or the science that treats of, the motions and phenomena of the atmosphere, esp. with a view to fore-

casting the weather. See also Climatology.
**Metrology**
*S.O.E.D.* The science of weights and measures.
**Metropolis**
*O.E.D.* 1. The seat or see of a metropolitan bishop.
2. The chief town or city of a country (occas. of a province or district), esp. the one in which the government of a country is carried on; a capital. The metropolis, often somewhat pompously used for 'London'. Also, in recent use, occasionally applied to London as a whole, in contradistinction to the City.
b. A chief centre or seat of some form of activity (*e.g.* metropolis of all perfection, idolatry, abuse, law, gain, religion).
c. Nat. Hist. The district in which a species, group etc., is most represented, *e.g.* Darwin (1859)
3. Greek Hist. The mother-city or parent-state of a colony. Hence occas. applied to the parent-state of a modern colony.
Mill, *Dict.* 1. The mother-city whence colonies have been formed. 2. The largest town or agglomeration of population in a district or country. 3. Often used as a synonym for capital.
Aurousseau, M., 1924, *Geog. Rev.*, **14**. 'The great unit that organises business for a wide metropolitan area, or hinterland....' (part of an undefined classification: town, city, metropolis). (pp. 444–455)
*Comment.* Census 1801 describes London as 'the Metropolis of England, at once the Seat of Government and the greatest Emporium in the known world'. (W.G.E.) See also Metropolitan.
**Metropolitan**
*O.E.D.* 1. Belonging to an ecclesiastical metropolis.
2. Of, pertaining to, or constituting a metropolis; *metropolitan city* or *town=* Metropolis (also of 'the metropolis', *i.e.* London).
3. Belonging to or constituting the mother country.
Smailes, A. E., 1953, *The Geography of Towns*, London: Hutchinson. 'The commercial and industrial cities of the Low Countries ... were also metropolitan towns, the nodes of long-distance traffic.' (p. 20) Metropolitan cities are referred to as the 'greatest cities' of their country. (p. 148)

Gras, N. S. B., 1922, *Introduction to Economic History*, New York. A city becomes metropolitan 'when most kinds of products of the district concentrate in it for trade as well as transit; when these products are paid for by wares that radiate from it; and when the necessary financial transactions involved in this exchange are provided by it'. (p. 294) Dickinson, 1947, calls this the economic metropolis and is distinguished from New York or London which are 'the central economic focus of the State, and ... an international economic and cultural depot'.
Green, F. H. W., 1950, Urban Hinterlands ... *Geog. Jour.* **116**. 'Metropolitan Centre' is the senior in a five class order of centres. There is only one—London. (p. 64)
*Comment.* Since this word is used in a variety of senses, the meaning is often only clear from the context. The special use of *O.E.D.* 3 in 'Metropolitan France' should be noted as applying to the Mother Country together with Corsica and, later Algeria. See also Metropolis.
**Metropolitan District** (United States)
Hallenbeck, W. C., 1951, *American Urban Communities*, New York: Harper. In the census of 1910 and 1920 comprised each city of 200,000 inhabitants and the civil units, all or a major portion of which fell within ten miles of the city boundary which had a density of population not less than 150 per square mile. In the census of 1930, a central city or cities together with all contiguous or surrounding civil divisions with a density of not less than 150 inhabitants per square mile which had an aggregate population of 100,000 or more. In 1940, similar to 1930 except that all cities of 50,000 or more population were included. (pp. 224–225)
U.S. Bureau of the Census, County and City Data Book, 1952. 'The following criteria ... were adopted in determining the boundaries of standard metropolitan areas:
1. Each standard metropolitan area must include at least one city of 50,000 inhabitants or more. Areas may cross State lines.
2. Where two cities of 50,000 inhabitants or more are within 20 miles of each other, they will ordinarily be included in the same area.
3. Each county included in a standard

metropolitan area must have either 10,000 non-agricultural workers, or 10 per cent. of the non-agricultural workers in the area, or at least one-half of the county's population must have been included in the 'metropolitan district' as defined by the Bureau of the Census ... In addition, non-agricultural workers must constitute at least two-thirds of the total employed labour force of the county. Each county included in a standard metropolitan area must be economically and socially integrated with the central counties of the area. A county has been regarded as integrated (a) if 15 per cent, or more of the workers living in the county work in the central county of the area, or (b) if 25 per cent. or more of those working in the county live in the central county of the area, or (c) if telephone calls from the county to the central county of the area average 4 or more calls per subscriber per month.' p. xi.

### Microclimate, microclimatology
*Met. Gloss.*, 1944. 'A modification of the general climate produced by the immediate environment ... in Ecological studies a distinction is sometimes drawn between the "local climate" (*i.e.* the climate of the habitat) and the "microclimate" (*i.e.* the climate of the immediate surroundings of the subject of study). In agriculture the term "microclimate" has been applied to the meteorological conditions within the crop; for example, among the ears and stems in a cornfield.'

Carpenter, 1938. The actual ecoclimate in which an individual lives (Uvarov). (p. 169)

No reference in Miller, A. A., 1931, but see 1953, Trewartha, G. T., 1937.

Miller, A. A., 1953, *Climatology*, 8th ed., London: Methuen. 'The slight contrasts of climate that may result from small differences of aspect, slope, and form of the ground, the colour, moisture and texture of soils and surfaces, vegetation and plant cover etc., constitute the study of microclimatology.' (p. 51)

*Comment.* A distinction should be, but is not always, drawn between local climate and micro-climate. In recent years the study of microclimatology has attracted much attention. The best treatment is in *The Climate near the Ground*, 1950, a translation from Geiger's *Handbuch der Klimatologie, Mikroklima und Pflantzenklima.* See also Local climate, Mesoclimate.

### Microcosm
*O.E.D.* The 'little world' of human nature.
*Comment.* According to F. R. Fosberg first used by S. A. Forbes in 1887 for what is now called an ecosystem (*q.v.*).

### Microgeography
The detailed study of a small area. See especially Platt, R. S., 1939, Reconnaissance in British Guiana with comments on microgeography, *Ann. A.A.G.*, **29**, 105-126; also discussion in Hartshorne 1939. Platt uses also microchoric.

### Microseism
A minor earth tremor, such as those constantly occurring and due to such natural causes as winds and waves.

### Microthermal—see mesothermal

### Migrant, migrate, migration
A migrant is a person (also applied to animals and plants) who voluntarily moves from one country to another especially for the purpose of permanent residence. He becomes an emigrant from his native country and an immigrant into the country of his choice. The emphasis is on the voluntary nature of the movement though it may be dictated by economic circumstances, hence the contrast with exile and refugee (*qq.v.*) See also expatriate.

### Migration movement
Roseman, C. C., 1971, Migration as a Spatial and Temporal Process, *Ann. Assoc. Am. Geographers*, **61**.3, 589-598. 'Since migration is a form of human movement through space, its initial definition is within the context of all human movements.' But if contrasted to reciprocal moves 'human displacements that are essentially one-way and relatively permanent may be identified as a second major movement category. These movements are defined as migration and represent removal of the center of gravity of the weekly movement cycle, the home, to a new location.' (p. 590)

See also Reciprocal movement.

### Mile
*O.E.D.* Originally the Roman lineal measure of 1,000 paces (mille passus or passuum) computed to have been about 1,618 yards (1,480 metres). Its length has varied considerably at different periods and in different localities. The legal mile in British Commonwealth and U.S.A. is

1,760 yards (1,609 metres). The old Irish mile was 2,240 yards (2,048 metres), Scottish about 1,976 (1,806 metres). The geographical or nautical mile is one minute of a great circle of the earth which varies owing to the geoid form of the earth and so has been standardized at 6,080 feet (1,853 metres) which is its value in lat. 48°.

**Mille map** (L. D. Stamp, 1948)
Stamp, L. D., 1948, *The Land of Britain*. 'The maps used may be called mille maps, in that each dot represents one-thousandth or 0·1 per cent of the total area under the crop concerned, and one thereby gains a picture of the main facts of spatial distribution ... believed to be a new basis.' (p. 99)

**Millet rains**—see **Maize rains**

**Millionaire City** (C. B. Fawcett, 1938), **Million-City**
Linton, D. L., 1958, *Geography*, 43. 'The very large city, for which the million population provides a convenient if arbitrary size definition ... Fawcett (in *Mélanges de géographie offerts par ses collègues et amis à l'étranger à M. Václaw Svambera*, ed. by B. Salamon and K. Kucher, Prague, 1936, 52–57) coined the term millionaire cities for them in 1935' (p. 253).
Comment. The term has been widely adopted but is regarded as unfortunate by some because of the normal connection of millionaire with wealth (*cf.* millionaire row = a street with residences of millionaires) hence use of 'million-cities'.

**Mine, mining**
O.E.D. Mining: the action of the verb mine in various senses. Mine (verb): 2. trans. To dig or burrow in (the earth); also, to make (a hole, passage, one's way) underground. c. To supply with subterranean passages; to make subterranean passages under. 6. trans. To obtain (metals, etc.) from a mine.
7. intr. To dig for the purpose of obtaining minerals, etc., to make a mine; to work a mine.
8. trans. To dig in or penetrate for finding ore, metals, etc. Mine (sb.): 1. An excavation made in the earth for the purpose of digging out metals or metallic ores, or certain other minerals, as coal, salt, precious stones. Also, the place from which such minerals may be obtained by excavation.
Zimmermann, E. W., 1933, *World Resources and Industries*, New York: Harper. 'Mining is the extraction of non-renewable fund resources.' (p. 160)
Bengtson, N. A. and Van Royen, W., 1950, *Fundamentals of Economic Geography*, 3rd ed., London: Constable. 'Mining may be carried on either by surface operations or by underground workings. ... The operations of oil wells, quarries, sand and gravel pits, and clay pits all are forms of mining and all are extractive industries.' (p. 50)
Comment. A distinction is frequently made between mining (by underground workings) and quarrying (by surface workings). 'Deep mining' and 'surface mining' are also used. During and since World War II much coal has been obtained in Britain from surface workings. This is invariably referred to as 'open-cast' mining (*q.v.*) and not as quarrying. The term open-cast is extended to iron-stone workings; sand and gravel and clay workings are referred to as pits, leaving quarrying to refer to non-metallic minerals and rocks, usually hard, but excluding coal. Up to the middle of the 19th Century 'quarry' was often used for a mine as well as for a surface working. For American usage see also **Strip mine**.

**Mineral**
O.E.D. 1. Any substance which is obtained by mining; a product of the bowels of the earth. In early and mod. technical use, the ore (of a metal).
5. In modern scientific use, each of the species or kinds (defined by approximate identity of chemical composition and physical properties) into which inorganic substances as presented in nature are classified.
Mill, Dict. 1. (Geog.) Generally used in the popular sense, meaning any solid inorganic matter.
2. (Geol.) An inorganic body distinguished by a more or less definite chemical composition, and usually a characteristic geometrical form.
Fay, 1920. '1. A mineral is a body produced by the processes of inorganic nature, having a definite chemical composition and, if formed under favourable conditions, a certain characteristic molecular structure, is exhibited in its crystalline form and other physical properties. A mineral must be a homogeneous substance, even when minutely examined by the microscope; further, it must have a definite chemical

composition, capable of being expressed by a chemical formula.' (From Dana, E. S.)

Himus, 1954. 'A naturally-occurring inorganic substance which possesses a definite chemical composition and definite chemical and physical properties.'

*Comment.* In geological and geographical writings a distinction is commonly drawn (and should be) between minerals as defined above by Fay and rocks which are usually mixtures of minerals. Only when rocks consist mainly or almost entirely of one mineral (*e.g.* rock-salt) does the distinction disappear. On the definition petroleum and natural gas are minerals, so also is water.

### Mineral spring
A spring containing a high proportion of mineral salts in solution.

### Mineral water
*O.E.D.* Originally water found in nature impregnated with some mineral substance, usually such as is used medicinally. Later, applied also to artificial imitations of natural mineral waters; in recent use extended to include other effervescent drinks as lemonade (familiarly termed 'minerals').

### Mineralization, mineralizers
*Webster.* Mineralization. *Geol.* The process of change or metamorphism whereby minerals are secondarily developed in a rock; especially, the formation or introduction of ore minerals into previously existing rock masses; metallization; also the resultant state.

Himus, 1954. 'Mineralizers. Magmatic gases and vapours, such as hydrogen, steam, compounds of fluorine, boron, sulphur, and carbon, which are capable of facilitating the crystallization of various minerals, causing the formation of new minerals, and which are able to extract and concentrate metallic and other compounds from the magma through which they were previously dispersed.'

### Minette (French)
*Nouveau petit Larousse*, 1952. 'Minerai de fer de Lorraine sesquioxyde phosphoreux.'

Pounds, N. J. G., 1959, *The Geography of Iron and Steel*, London: Hutchinson. 'The *minette* of Lorraine has rarely much more than 30 per cent of its weight in iron ... similar in geological age and chemical composition to the English Jurassic ores, is the most abundant ore deposit in Europe and one of the largest in the world.' (pp. 33, 38)

*Comment.* The word has also been used by geologists for an entirely different micaceous rock from the Vosges (see *O.E.D.*).

### Ming land
Lands of different proprietors lying intermixed in common fields (Midlands of England: historical) (I.L.A.T.).

### Minor intrusion
Himus, 1954. 'Igneous intrusions of comparatively small size compared with plutonic, or deep-seated intrusions. They include dykes, sills, veins and small laccoliths.'

### Miombo (Tanganyika; Swahili)
Plural of *myombo*, an important tree in this vegetation formation. An open deciduous woodland, *Brachystegia* species predominating, with a ground cover of grasses and herbs. Its main distribution is in western and in south-eastern Tanganyika, and it occupies nearly one half of the total land surface of the territory. (S.J.K.B.)

See also Keay, R. W. J., 1959, *Vegetation Map of Africa*, London, O.U.P., where it is referred to as covering large parts of south-central Africa. (p. 9)

### Mire
*O.E.D.* A piece of wet swampy ground.

*Comment.* Sometimes used in local names, *e.g.* Ha Mire and Great Close Mire, Malham Tarn, Yorkshire. See *Field Studies*, **1**, 1959, 84–85.

P. W. Richards comments 'an anglicization of the Swedish *myr*, which H. Godwin has tried to naturalize in English to cover moorland, fen and bog, *i.e.* vegetation of wet peat, whether acid or not'.

### Misfit river
Cotton, 1922. '... the volume of the river may be so reduced that it is obviously a misfit—*i.e.* it appears too small to have eroded the valley in which it flows.' (p. 120)

Wooldridge and Morgan, 1937. 'The unsuccessful consequent is thus beheaded, and, deprived of its headwaters, will dwindle in size and become a misfit or underfit river, too small for the valley in which it flows.' (pp. 196–197)

Dury, 1959. 'A river which appears once to have been far larger than it is today.' (p. 201). Dury in *Trans. I.B.G.*, 1958. argues that climatic change is the potent factor.

See also Underfit river

**Mist**
O.E.D. A cloud formed by an aggregation of minute drops of water and resting on or near the ground. In generalized sense, vapour of water precipitated in very fine droplets, smaller and more densely aggregated than those of rain. Sometimes distinguished from fog, either as being less opaque or as consisting of drops large enough to have a perceptible downward motion.

Mill, *Dict*. A cloud resting upon the ground and wetting objects exposed to it.

Met. Gloss. An obscurity in the surface layers of the atmosphere caused by particles of condensed moisture held in suspension in the air, limiting visibility to less than two but more than one kilometre.

**Mist Forest**
A hygrophilous forest found especially at higher levels in tropical regions and constantly or frequently enveloped in mist or cloud. Identical with cloud forest.

**Mistral** (French)
O.E.D. (French, mistral. The literal meaning is 'master-wind'; *cf.* Sp. *maestral* or *viento maestro* (Minshen)). A violent cold northwest wind experienced in the Mediterranean provinces of France and neighbouring districts.

Mill, *Dict*. A cold, dry and often very strong more or less northerly land-wind in Provence and the adjoining parts, blowing from the cold high plateau of the Massif Central to the relatively warm Gulf of Lions. The name Cers is applied to a similar wind in Aude (S. France).

Met. *Gloss*., 1944. A north-westerly or northerly wind which blows offshore along the north coast of the Mediterranean from the Ebro to Genoa. In the region of its chief development its characteristics are its frequency, its strength, and dry coldness. . . .

**Mitteleuropa**
*Webster*. Central Europe, especially that portion of Europe which the advocates of Pan-Germanism purposed to form into one great empire.

Sinnhuber, K. A., 1954, Central Europe Mitteleuropa, Europe centrale: an analysis of a geographical term. *Trans. Inst. Brit. Geog.*, 20, 15–40.

Fischer, E., 1956, in East and Moodie, *The Changing World*, London: Harrap. 'Mitteleuropa is used as a politico-geographical term which in the usage and mind of every author has a very definite meaning, but, unfortunately, for every author a different one.' (p. 60; in the chapter entitled The Passing of Mitteleuropa)

**Mixed cultivation**
Committee, List 3. The growing of two or more different crops intermingled on the same field or plot, especially the mixture of tree and ground crops.

**Mixed farming**
Committee, List 3. Farming in which both arable crops and livestock play considerable parts in the economy.

Stamp, 1948. 'It here means that the type of farming is such that the enterprises are both numerous and dissimilar. . . . Where there has been doubt whether an area should be classified as a mixed farming area, these considerations of integration and flexibility of enterprises have been taken into account. . . . "Dissimilar" can be understood as requiring the presence of both livestock and crop enterprises . . .' (p. 302).

Comment. It should be noted that mixed cultivation refers to a mixture of crops, mixed farming to the combination of crops and livestock.

**Model**
O.E.D. 1. Representation of structure. 1.c. a description of structure.

Harvey, D. in Chorley and Haggett, 1967. 'A model may be regarded as the formal presentation of a theory using the tools of logic, set theory, and mathematics. The use of these tools allows us to identify and eliminate inconsistencies within our theory. It also allows us to use the powerful tool of algebraic analysis to make deductive statements as regards a particular system (this method is typified in classical economics and, in some cases, to develop objective statistical tests of the relationship between the model we are using and the world. To make a model operational, therefore, we have to develop some simple system of model building).' (p. 552)

Chorley and Kennedy, 1971. 'A representation of an event, object, process or system that is used for prediction or control. By manipulating the model the effects of changing one or more aspects of the entity represented can be determined.' (p. 353)

Kendall, M. G. and Buckland, W. R., 1972, *A Dictionary of Statistical Terms*, 3rd ed., Edinburgh: Oliver and Boyd. 'A model is a formalized expression of a theory or the causal situation which is regarded as having generated observed data. In statistical analysis the model is generally expressed in symbols, that is to say in a mathematical form, but diagrammatic models are also found. The word has recently become very popular and possibly somewhat overworked.'

**Moel** (or foel; Welsh) *Lit.* bare field, bald.

Marr, J. E., 1901, The origin of moels and their subsequent dissection, *Geog. Jour.*, **17**, 63–69. 'Vegetation-clad summits, then, will possess a rounded outline, which is characteristic of certain Welsh hills, known as *moels*, the term *moel* may therefore be used as a general term for hills of this character' (p. 64).

*Comment.* Not in any of standard texts.

**Moela** (Spanish)

A plateau having the form of an elevated syncline. (P.D.)

**Mofette** (French)

Himus, 1954. An opening in the ground from which exhalations of carbon dioxide, oxygen and nitrogen arise; found in regions of former volcanic activity.

Mill, *Dict.* gives *mofetta* (Italian; *pl. mofette*): an area, usually depressed, characterized by emanations of carbonic acid gas.

See Fumarole, suffioni.

**Mofussil** (Anglo-Indian)

In India, the rural parts of a district as distinguished from the chief town or administrative centre. Obsolescent.

**Mogotes** (Spanish; Cuba)—see Hum

**Mohorovičić discontinuity, or Moho**

Gaskell, T. F., 1958, *Nature*, **182**, 1958, 692. 'The Mohorovičić discontinuity is the boundary between the mantle (*q.v.*) and the assorted surface rocks of the Earth, and it marks a very sharp change in the velocity with which earthquake waves travel—a jump from about 21,000 ft/sec. to 27,000 ft/sec. ... at depth of about 20 miles under the land but underneath the deep oceans it is only about 7 miles below sea-surface.' The abbreviated form 'Moho' is generally accepted (Gaskell in MS.).

**Molasse, mollasse** (French)

Himus, 1954. 'Deposits of soft sandstones and grey and red sandy marls in Switzerland.'

Holmes, 1944. '*The Swiss Plain*: a broad lowland, filled with soft Tertiary sediments called *molasse*, derived from the denudation of the rising Alps.' (p. 393)

Plaisance and Cailleux, 1958. '3. Les géologues alpins tendent à restreindre le mot molasse à des formations postérieures aux plissements principaux; âge tertiaire, le plus souvent miocène.' (p. 359)

**Momentum, geographical**

*O.E.D.* Momentum: in popular usage applied to the effect of inertia, in the continuance of motion after the impulse has ceased.

Committee, List 4. 'Geographical Momentum. The tendency of places with established installations and services to maintain or increase their importance after the conditions originally determining their establishment have appreciably altered.' See also Industrial inertia.

**Monadnock** (W. M. Davis, 1895)

*O.E.D. suppl.* Geol. (The name of a mountain in New Hampshire, U.S.A., having this character.) A hill or rocky mass rising above the general level of a peneplain, believed to be a remnant of erosion (quotes Davis, 1895).

Davis, W. M., 1895. The development of certain English rivers, *Geog. Jour.*, **5**. 'Most of the peneplains ... still possess residual elevations, rising somewhat above the general upland, and evidently to be regarded as unconsumed remnants of the denudation of the former cycle. ... I have fallen into the habit of calling a residual mound of this character a monadnock, taking the name from that of a fine conical mountain of south-western New Hampshire, which grandly overtops the dissected peneplain of New England.'

Mill, *Dict.* A residual mountain left standing above the general level of an old denudation plain which has been elevated and eroded.

*Comment.* For a different conception of the origin of inselberge known to Americans as monadnocks but better as '*Härtlinge*' see W. Penck, translated Czech and Boswell, pp. 170, 195.

**Monimolimnion**
The lower stable, unmixed layer of a meromictic lake (q.v.) is called the monimolimnion ... so convincing is the evidence in support of the theory of complete isolation of the bottom layer, that some investigators refer to this water as "fossil water".' Newcombe, C. L., 1957, *Abs. of Papers, Ninth Pacific Science Congress*, Bangkok, 173.

**Monoclinal block**
Thornbury, 1954. 'The Basin and Range Province of the Western United States, has become the type area for topography developed upon tilted fault blocks. Johnson (Johnson, Douglas, 1929, Geomorphic Aspects of Rift Valleys, *C.R.* 15ᵉ *Congrès Inter. Géol.*, **2**, 354–373) has referred to the type of fault blocks found in this region as *tilted* or *monoclinal blocks*.' (p. 259)
*Comment*. This is using monoclinal = homoclinal in the American sense (see Monocline).

**Monoclinal shifting** (G. K. Gilbert)
Mill, *Dict*. Lateral erosion taking place where a river flows along the strike of gently dipping rocks, and a less resisting stratum overlies one or more resistance.

**Monoclinal valley, ridge**
Mill, *Dict*. Valleys which have the strata on both sides dipping in the same direction.
Powell, 1875. '... which run in the direction of the strike between the axes of the fold—one side of the valley formed of the summits of the beds, the other composed of the cut edges of the formation' (p. 160).
*Comment*. A term of American origin, using monocline in the American sense. See also von Engeln, 1942, 327.

**Monocline**
*O.E.D.* Geol. A monoclinal fold.
1875. Geikie in *Encycl. Brit.*, X, 300/1. The strata are thus bent up and continue on the other side of the tilt at a higher level. Such bends are called monoclines or monoclinal folds, because they present only one fold, or one half of a fold, instead of the two which we see in an arch or trough.
*Webster*. Monocline *Geol*. A monoclinal fold. *Monoclinal Geol*. Having or pertaining to, a single oblique inclination; as a *monoclinal* fold or flexure. 'A *monoclinal* flexure is a single, sharp bend connecting strata which lie at different levels and are often horizontal except along the line of flexure. *W. B. Scott*.'
Wooldridge and Morgan, 1937. [In regions of tension] 'These flexures exhibit only one inclined limb, which links tracts of horizontal strata at different levels. ... The reader is reminded that the term "monocline" should not be extended to cover asymmetrical anticlines such as that of the Isle of Wight, which are not tensional features' (p. 105 and diagram).
Himus, 1954. 'A fold in which the bend is in only one direction. The rock stratum changes its dip by increasing its angle of inclination and then levels out again or resumes its original dip' (diagram).
von Engeln, 1942. 'Beds all dip in the same direction' (p. 327).
*Comment*. Very serious confusion exists because of a difference in British usage and the common American usage of beds dipping in one direction (*i.e.* a homocline). Lobeck, 1939, uses monocline in the English sense on p. 35 (photo) and p. 495 (diagram) but in the American sense on p. 588 (diagram) and p. 591.

**Monoculture, monocultural**
*O.E.D. suppl*. The cultivation or production of one kind of thing.
Committee, List 3. Cultivation in which a single crop preponderates over all others; generally used in respect of a region or large area.

**Monoglacial**
Wright, 1914. The monoglacial theory considers the Pleistocene ice-sheet to have expanded and contracted in only one general movement without any substantial 'interglacial periods' of recession followed by advance. (pp. 124–125)
No reference in Wooldridge and Morgan, 1937; Holmes, 1944; Daly, 1934. No reference to 'uniglacial' in Wright, 1914; etc.
See also works of J. K. Charlesworth.

**Monoglot**
One language, in contrast to polyglot (q.v.).

**Monolith**
*O.E.D.* A single block of stone, especially one of notable size, shaped into a pillar or monument.
See also Menhir, Megalith

**Monsoon**
*O.E.D.* 1. A seasonal wind prevailing in southern Asia and especially in the Indian

Ocean, which during the period from April to October blows approximately from the south-west, and from October to April from the north-east, the direction being dependent upon periodic changes of temperature in the surrounding land-surfaces.
2. Any wind which has periodic alternations of direction and velocity, caused by variations of temperature between the land surfaces and the surrounding ocean, or by the difference of temperature between the polar and equatorial regions. *Cf.* Trade Wind.

Maury, M. F., 1855, *The Physical Geography of the Sea*, London: Sampson Low. 'When a trade wind is turned back or diverted by over-heated districts from its regular course at stated seasons of the year, it is regarded as a monsoon.' (para. 474) And 'Monsoons, properly speaking, are winds which blow one half of the year from one direction and the other half from an opposite, or nearly an opposite, direction.' (para. 462)

Mill, *Dict*. Winds exhibiting annual monsoonal changes. The summer winds, in the case of continents affected by them, are ocean winds, and the winter winds are from the land. In the case of peninsulas and islands, in regions affected by them, this distinction is necessarily modified by the local relations of sea and land. Monsoonal Changes. The periodical reversal of wind-direction due to the alternate heating and cooling of a land surface.

*Met. Gloss.*, 1944. The name is derived from an Arabic word for 'season', and originally referred to the winds of the Arabian sea, which blow for about six months from the NE, and six months from the SW. It has been extended to include certain other winds which blow with great persistence and regularity at definite seasons of the year. The primary cause of these winds is the seasonal difference of temperature between land and sea areas.... In the countries of its occurrence the term 'monsoon' is popularly used to denote the rains, without reference to the winds.

**Monsoon Forest**

Stamp, L. D., 1959, *Asia*, 10th ed., London: Methuen. 'Where the rainfall is between about 40 and 80 inches the typical "monsoon forests" are found. They are broad-leaved forests which become leafless in the hot season, bursting into flower and then into leaf just before the rains break.. numerous species, pure stands are rare.. this is the case with the "teak forests" (p. 46)

**Montaña** (Spanish)

*O.E.D.* 1. *pl.* As the proper name of certain mountain districts in Spain. 2. In Spanish American countries: A forest of considerable extent; specifically the name of the part of Peru east of the Andes.

Holton, I. F., 1856, *New Granada*, 436 'All land covered with thicket is called monte if it be but a few miles through, and montaña if more.'

Mill, *Dict*. Mountain as distinguished from hill (Spanish America). *Cf.* Cerro and Loma.

Jones, 1930. 'The naturally rich, though economically insignificant, forest—the Montaña—on the east' [of Highland Peru] (p. 202). Jones devotes a chapter to the Montaña as a region (pp. 219–223) 'in a strict sense, the term montaña includes only the forests of the eastern plain, but here it applies to the thickly forested Andean slopes below an elevation of some 5,000 feet'.

N.R. in Küchler, 1947, *A.A.A.G.*, 37.

*Comment*. There seems to be no evidence for Mill's definition except that montaña in Spanish still denotes mountain or highland. The use of montaña in international geographical literature is essentially as the name of a specific region of the eastern plains (Amazon basin) of Peru. See also Monte.

**Montane**

*O.E.D.* Nat. Hist. Pertaining to or inhabiting mountainous country.

Carpenter, 1938, p. 175. Montane Region. Regions with vegetation more hygrophilous and less thermophilous than in the neighbouring lowlands, resembling that of the lowlands of the higher latitudes.

*Comment*. There seems to be no technical meaning to this in botany (see Schimper, Weaver, Clements etc.). Some authors prefer to refer to mountain forests, others to montane forests. It is simply an adjectival form.

**Monte** (Spanish and Italian, *lit*. mountain)

*O.E.D.* In Spanish-American countries: A more or less wooded tract; a small forest.

*Dict. Am.* 2. A chaparral region in the southern California foothills. 1851.

Mill, *Dict.* A clump of trees forming an eminence on the pampas (Argentine).

James, 1959. 'The monte is composed of deciduous broadleaf scrub trees and bushes with a marked xerophytic character. The Argentines distinguish between a *monte alto*, which grows in the wetter places and which includes trees 25 to 30 feet in height, and *monte bajo*, in which the plants are more widely spaced and seldom more than 10 or 12 feet in height ... between the trees there is a cover of short grass.' (pp. 325–326) (According to Schmieder formerly more widespread but restricted by burning. See map p. 21. W.M.)

Küchler, A. W., 1947. Localizing Vegetation Terms, *A.A.A.G.*, **37**, '... authorities like James, Kühn, Frenguelli, and others all disagree so thoroughly on location and extent of this formation that it has been omitted.' (p. 208)

See also Pampa.

**Monument**

Mill, *Dict.* More or less columnar product of denudation (N. America).

Thornbury, 1954. Somewhat related in origin to horns but detached from the main mountain range are *monuments* or *tinds* as they are called in Scandinavia.

Not in von Engeln, Cotton, 1941.

Holmes, 1944. 'To the south of the Henry Mountains on the other side of the Colorado River, is Monument Valley, so called because of its obelisks and towers and other castellated erosion remnants carved out of red Triassic rocks' (p. 422, also Plate 64 B).

*Comment.* Mill and Holmes refer to quite small erosion features: Thornbury to mountains since the type of a 'horn' is the Matterhorn. The word is best avoided. Currently used in Britain in term 'geological monument', a site of such geological interest as to deserve conservation.

**Moor, Moorland**

*O.E.D.* 1. A tract of unenclosed waste ground; now, usually, uncultivated ground covered with heather; a heath. Also, a tract of ground strictly preserved for shooting.
2. A marsh (obs.); also dialect.
1883. Grant Allen, Colin Clout's Calendar, XXVIII, 228. In Yorkshire a moor means a high stretch of undulating heath-covered rock; whereas in Somerset it means a low flat level of former marshland, reclaimed and drained by means of numerous 'rhines.' (Also elsewhere).
3. dial. The soil of which moorland consists; peat.
4. Cornwall. a. A moor or waste land where tin is found.

*Webster.* An extensive area of waste ground overlaid with peat, and usually more or less wet. In popular usage the word is restricted to the European moors, in which heather is often the prevailing plant; but similar phytogeographical areas occur elsewhere. Sphagnum moss is always characteristic of high moors.

Mill, *Dict.* Moor or Moorland. An area covered with heather, coarse grass, bracken, moss or other vegetation of low growth, in some cases including patches of pasture grasses, and in others hollows with water or bog.

Tansley, A. G., 1949, *Britain's Green Mantle*, London: Allen and Unwin. '"Moor" or "moorland", words which are applied to any open tract of "waste" land, particularly at a high elevation, which is not good pasture like the limestone or the bent-fescue grasslands.' (p. 182)

Carpenter, 1938. (Includes:) ... 2. A habitat with at least 50 cm. of peat soil of more or less decomposed plant remains ... (Clements). 3. An area underlaid by bog mass (Sphagnum) and arising on moist soil, but slightly permeable to water does not necessarily show open water' though a very damp air hangs over it (p. 175)

Tansley, 1939. 'We cannot use the English word "moor" as an ecological term in the German sense of an area covered with deep peat ... the word is now generally used for *any* tract of unenclosed land (generally elevated) with acid peaty soil and not used *primarily* as pasture ... such "moorlands" may be wet or dry' (pp. 673–674)

**Mopane** (Afrikaans)

Common broad-leaved tree (*Copalifera mopane*) of the northern bush veld regions of South Africa, adapted to comparatively low rainfall. It reaches from the Transvaal, through the northern Kalahari, into the northern part of Southwest Africa, and is found both in the form of shrubs and as full-grown trees. The latter may stand so closely together that the term *mopane forest* is sometimes used. (P.S.)

**Mor** (Soil science)
Jacks, 1954 (equated with raw humus). '$A_0$ horizon unmixed with and sharply demarcated from the underlying mineral horizon. Consists of L, F and H horizons.' (See Soil Profile and Mull.)

**Moraine** (French but now accepted as an English word)

*O.E.D.* An accumulation of debris from the mountains carried down and deposited by a glacier.

Mill, *Dict.* A pile of loose material upon the surface of a glacier or deposited along its sides (lateral moraines) or at its foot (end or terminal moraines), or underneath the glacier (ground moraines).... The term 'moraine' was once locally used in Valais, adopted by Charpentier, and thence extended.

Tyndall, J., 1860, *The Glaciers of the Alps*, London: Murray. (As Mill, but does not mention ground moraines.) (pp. 263–264)

Flint, 1947. 'The word moraine is an ancient French word long used by peasants in the French Alps for the ridges and embankments of earth and stones around the margins of the glaciers in that region. It appeared in the literature as early as 1777, and was taken up and used by Saussure and later by Venetz and Charpentier, and was given wide currency by Agassiz. The recognition, later, of a wide variety of forms of drift fashioned by large ice-sheets made it necessary to extend the original quite limited meaning of the word. Accordingly we now think of moraine as an accumulation of drift with an initial topographic expression of its own, built within a glaciated region chiefly by the direct action (deposition and thrust deformation) of glacier ice. Moraine is usually subdivided into ground moraine, end moraine, lateral moraine, medial moraine, and ablation moraine.' (p. 126) (footnote) 'In Scandinavian literature moraine is often erroneously used as a synonym for till.'

Holmes, 1944. 'Rock fragments liberated from the steep slopes above a glacier, mainly by frost shattering, tumble down on the ice and are carried away. Thus the sides of a glacier become streaked with long ribbons of debris described as *lateral moraines*, when two glaciers from adjacent valleys coalesce, the inner moraines of each unite and form a *medial moraine*.... Sooner or later part of the debris is engulfed by or washed into crevasses. Material that is enclosed within the ice is referred to as *englacial moraine*. A certain proportion reaches the sole of the glacier and there together with the material plucked or scraped from the rocky floor, it constitutes *subglacial moraine*. If the lower part of the ice becomes so heavily charged with debris that it cannot transport it all, the excess is deposited as *ground moraine* which is then overridden by the more active ice above. All the varied debris ranging from angular blocks and boulders to the finest ground-down rock flour, that finally arrives at the terminus of the glacier is dumped down when the ice meets. If the ice front remains stationary for several years an arcuate ridge is built up, called a *terminal* or *end moraine*.' (pp. 212–214)

*Comment.* There is a double usage for (a) the landform and (b) the material. (G.T.W.) See also Push moraine.

**Morass**
Mill, *Dict.* A wet swampy tract, a bog, marsh; occasionally in generalized sense, boggy land.

*Comment.* Has no exact scientific meaning.

**Morfa** (Welsh)
*Lit.* Marsh, sea-fen; frequently fronting coastal cliff scenery.

**Morphogenetic region**—see Formkreis

**Morphographic map**
Monkhouse and Wilkinson, 1952. 'American geographers, particularly E. Raisz, have devised methods of showing physiographic features on small-scale maps by the systematic application of a standardized set of conventional pictorial symbols, based on the simplified appearance of the physical features they represent as viewed obliquely from the air at an angle of about 45 degrees. Some American geographers call this a "morphographic" or "morphologic" method.'

Raisz, E., 1948, *General Cartography*, 2nd ed., New York: McGraw-Hill. 'The method had many names. Lobeck called his a "physiographic diagram", although this is the least diagrammatic of all relief methods. The author earlier called it the physiographic method. The term "morphographic", however, is preferred here as it indicates the geomorphographic origin of the method. The informal term "land form

map" seems to express the direct appeal of the method.' (p. 122)
*Comment.* First major map of the kind was A. K. Lobeck, Physiographic Diagram of the United States, 1921. Systematized by E. Raisz in *Geog. Rev.*, **21**, 1931, pp. 297–304.

**Morphological Region**
Linton, D. L., 1951, The Delimitation of Morphological Regions in *London Essays in Geography*, London: Longmans. 'Nature offers us two inescapable morphological unities and two only; at the one extreme the indivisible flat or slope, at the other the undivided continent ... we may bridge the gap and develop a related series of intermediate units. ... At the one end of the series are units ... with a high degree of homogeneity in all their morphological attributes—form, rock structure and evolutionary history. Each succeeding member of the series shows increasing diversity ... 'a hierarchy of morphological units ... site, stow, tract, section, province and continental subdivision' (pp. 215–217). See also Physiographic Region.
*Comment.* When Academician I. P. Gerassimov of Moscow was lecturing in Britain in 1958 he insisted that 'Landforms can be divided into three main groups (a) the elements of morphotexture, (b) the elements of morphostructure and (c) the elements of morphosculpture. These groups correspond approximately to what are known to English workers as First, Second and Third order landforms.'

**Morphological systems**
Chorley and Kennedy, 1971. 'These comprise the morphological or formal instantaneous physical properties integrated to form a recognisable operational part of physical reality, the strength and direction of their connectivity being commonly revealed by correlation analysis.' (p. 5) See also pp. 23–76.

**Morphology**
Committee, List 1. Morphology. The science of form, and the structures and development which influence form.
*Comment.* Sauer in his *Morphology of Landscape* attributes the origin of the word to Goethe in his scientific writings. The term is widely used in most sciences—biology, geology, geomorphology etc., as well as in such studies as linguistics. (G.T.W.)

**Morphology (Urban)**
Dickinson, 1951, *The West European City*. 'The study of the morphology of the urban habitat. ... It is concerned with the plan and build of the habitat, viewed and interpreted in terms of its origin, growth and function.' (p. 8)
Smailes, 1953, *Geography of Towns*. '... the intimately related aspects of urban morphology, function and form ...' The 'Morphology of Towns' is treated in two chapters, 1. 'Urban Regions.' 2. 'The Development of the Town Structure.' The former is largely based on function, or land use, suggesting that morphology = internal regional geography. (W.M.)

**Morphometric analysis**
Harvey, D., 1969, *Explanation in Geography*, London: Edward Arnold. 'Morphometric analysis thus provides a framework within which the geographer examines shapes and forms in space. In general the assumptions are geometric ones and this amounts to identifying a co-ordinate system suitable for discussing the particular problem in hand. In particular this allows the discussion of shape and pattern of town locations, the structure of networks, and so on.' (pp. 79–80)
See also Morphometry.

**Morphometry**
*O.E.D.* The art or process of measuring the external form of objects.
Chorley, R. J., 1958. Aspects of the Morphometry of a 'Poly-cyclic' Drainage Basin, *Geog. Jour.*, **124**, 370. The precise measurement of land forms.
Savigear, R. A. G., 1965, *Ann. Assoc. Am. Geographers*, **55**.3. 'It is derived from the Greek words morphe (form) and logos (studies). Biologists, philologists, and others use it to mean form studies, and within the subject of geomorphology its meaning may be logically restricted to surface-form studies. It should not be synonymous with geomorphology and a morphological map should represent surface form only. If some other aspect, or classification, or interpretation of the earth's surface is shown the map becomes, for example, either a geological, or a soils, or a morphographic, or a morphogenetic map.' (p. 514)

**Morphosequent** (S. W. Wooldridge, 1930)
In contrast to tectosequent applied to surface features which do not reflect the underlying geological structure.

**Morphotectonics** (E. S. Hills, 1956) see *Q. J. Geol. Soc.*, **117**, 1961, 79.

**Mortlake**
Avebury, Lord, 1902, *The Scenery of England*. 'The loop often remains as a dead river channel or "Mortlake". Such loop lakes are known in America by the special name of "Oxbows".'
von Engeln, 1942. 'An oxbow lake; called a mortlake in Great Britain' (p. 145).
*Comment*. Despite von Engeln's statement the term is practically obsolete in Britain; oxbow or cut-off has taken its place.

**Morvan** (French)
Lobeck, 1939. 'The intersection of two peneplanes,' one of which has been tilted, as in the Fall Line of U.S.A. So named by W. M. Davis after the example in Morvan, France. (p. 454)
von Engeln, 1942. 'In general their areas [of massifs] are blocked out and cut off from the surrounding lands by great fault displacements. In places, however, the declivity which affords the descent from the peneplaned uplands to the lower lands is itself a warped peneplain surface. This type of intersection of two peneplain surfaces is referred to as a *morvan* from the type locality of that name in France. It has been said that a *morvan* is not in itself a land form but is, rather, the *problem* of the intersection of two peneplains. In the United States the intersection of the Piedmont Province with the Fall Zone peneplain at the eastern border of the former is a *morvan* relationship' (p. 388).

**Mosaic** (air photo)
Hotine, M., 1931, *Surveying From Air Photographs*, London: Constable. 'A mosaic is a composite photographic representation of an area obtained by joining individual photographic prints.' (p. 182)
Raisz, 1948. 'Uncontrolled mosaics are assembled from pictures as they come from a flight without trying to adjust them to some control.... Straight-line mosaics are laid out along straight roads or railroads giving them a kind of control. In controlled mosaics a number of secondary control points are needed, at least three in each picture. The pictures are restituted in a rectifying printer for scale and tilt until they match the control. Secondary control can be obtained by radial line plotting.' (p. 193)

**Mosaic, desert**—see Desert pavement

**Moshav** (*pl.* moshavin), **moshavah** (*pl.* mushavot) (Hebrew: Israel)
'The first Jewish agricultural colonies in the land of Israel in modern times. An agricultural village where a measure of mutual cooperation prevails but where each farmer owns, works and is responsible for his own land'. *Agriculture in Israel*, April 1959, Israel Min. of Agric. pp. 8–9.
See also Kibbutz, Kvutza.

**Mosore** (Penck)
Monadnock which has survived because of remoteness from streams. See Thornbury, 1954, 181.

**Moss**
*O.E.D.* 1. A bog, swamp, or morass; a peat bog. Scottish and northern England. b. Wet spongy soil; bog.
Tansley, A. G., 1949, *Britain's Green Mantle*, London: Allen and Unwin. 'In the north of England and in some parts of Scotland the bogs are commonly called "mosses", probably because they were at one time dominated by bog moss (*Sphagnum*), as some still are' (p. 184). 'The cotton-grass "mosses" are characteristic of the summit plateaux of the southern Pennines and occupy wide stretches of dreary monotonous moorland in which the tufted cotton-grass (*Eriophorum vaginatum*) is dominant alone, sometimes with very few accompanying species.'

**Mota** (Indo-Pakistan: Hindi)
A clay pan.

**Motorway**
A specially constructed road limited by law to certain types of traffic hence distinct from highway. The first motorway in Britain M1 from London to the Midlands was opened in 1959. *Cf.* Autobahn, *q.v.* where a fuller explanation is given. See also Turnpike, Highway.

**Moulin** (French)
*O.E.D.* (French, *moulin*, lit. a mill. The term is suggested by the swirling motion of the water as it falls down the shaft.) A nearly vertical circular well or shaft in a glacier, formed by the surface water falling through a crack in the ice, and gradually scooping out a deep chasm. (Quote: Tyndall, 1860.)
Mill, *Dict*. Glacier Mill or Moulin. A vertical hole through the ice of a glacier, or ice

sheet caused by a stream of descending water.

Holmes, 1944. 'In sunny weather small pools and rills diversify the surface [of a glacier] gathering into streams which mostly fall into crevasses. By a combination of melting and pot-hole action (aided by sand and boulders) deep cauldrons called *glacier mills* or *moulins* are worn through the fissured ice' (p. 214).

## Mountain

*O.E.D.* A natural elevation of the earth's surface rising more or less abruptly from the surrounding level, and attaining an altitude which, relatively to adjacent elevations, is impressive or notable. Down to the 18th c. often applied to elevations of moderate altitude.

Mill, *Dict.* A mass of land higher than the immediately surrounding land and exceeding 1,000 feet in altitude.

Strahler, A. N., 1945, Geomorphic Terminology and Classification of Land Masses, *Jour. of Geol.* 54. Discusses the use of 'mountain' to describe disturbed structures independent of their 'topography' or relief. Examples from W. M. Davis, 1909; Douglas Johnson; A. K. Lobeck, 1926 and 1939; and R. S. Tarrand, L. Martin, 1914. This use has not been followed by Cotton, 1922; Longwell, Knopf and Flint; Preston James, Cressey, etc. except for the use of 'block mountain'. Strahler argues for separating mountain from structure and provides an alternative terminology based on geological structure rather than the initial land form. (W.M.) (pp. 32–42)

*Comment.* Under 'Hill' *O.E.D.* gives: 'In Great Britain heights under 2,000 feet are generally called hills; "mountain" being confined to the greater elevations of the Lake District, North Wales, and of the Scottish Highlands. "Mountain" may be used for lesser elevations even under 1,000 feet especially when they rise abruptly from the surrounding country (*e.g.* Conway Mountain in North Wales).'

## Mountains, classification

With the development of geomorphology the old descriptive classification of mountains into 'mountain of accumulation, mountain of circumdenudation' etc. has become outmoded.

## Mountains without roots—see Nappe outlier, Klippe

## Mountain-building—see Orogenesis

## Mountain sickness

*O.E.D.* A malady caused by breathing the rarefied air of mountain heights.

Mill, *Dict.* A feeling of nausea which attacks nearly everyone at a considerable altitude above sea level. The altitude at which this sickness is felt varies in different individuals.

*Comment.* The effect of altitude on human physiology is a complex subject where many problems still remain. Geographically it is important in assessing the value of such elevated areas as the White Highlands of Kenya for settlement of immigrants from mid-latitudes.

See also Soroche.

## Mountain-wind—see Föhn, Katabatic

## Mouza (Indo-Pakistan: Bengali)

Johnson, B. L. C., Crop-Association Regions in East Pakistan, *Geography*, 43, 1958. '... the *Union*, an administrative unit comprising a number of *mouzas*, the latter more or less equivalent to village areas ... the next larger unit to the Union in the administrative hierarchy, the *thana* (police station) ...' (p. 86).

## Mover-stayer model

Goodman, L. A., 1961, Statistical Methods for the Mover-Stayer Model, *Am. Stat. Assoc. Jour.*, 56. 'The mover-stayer model, a generalization of the Markov chain model, assumes that there are two types of individuals in the population under consideration: (*a*) the "stayer", who with probability one remains in the same category during the entire period of study; (*b*) the "mover", whose changes in category over time can be described by a Markov chain with constant transition probability matrix.' (p. 841); 'The mover-stayer model can be described as follows: industries are grouped into a finite number, I, of industrial code categories. In the $i$th code category $(i=1, 2, \ldots, I)$, there are two kinds of workers, the stayers and the movers. Let $s_i$ denote the proportion of workers in the $i$th code category in the initial quarter who are stayers $(i = 1, 2, \ldots, I)$. Then $1 - s_i$ is the proportion of workers in the $i$th category in the initial quarter who are movers. It is assumed that each stayer remains in a particular code category with probability one, and that each mover changes his code category over time in a way that can be described by the one quarter transition probability matrix

$$M = \begin{Vmatrix} m_{11} & m_{12} & . & . & m_{1I} \\ m_{21} & m_{22} & . & . & m_{2I} \\ . & . & & & . \\ . & . & & & . \\ m_{I1} & m_{I2} & . & . & m_{II} \end{Vmatrix}$$

where $m_{ij}$ is the probability that a mover who is in the $i$th code category in a particular quarter will be in the $j$th category ($i, j = 1, 2, \ldots,$ I) in the following quarter.' (pp. 842–843)

*Comment.* The mover-stayer model first introduced by Blumen, I., Kogan, M. and McCarthy, P. J., 1955, The Industrial Mobility of Labor as a Probability Process, *Cornell Studies of Industrial and Labor Relations*, VI, Ithaca, New York: Cornell University.

**Msitu** (Bantu)
A type of dense bush in Northern Rhodesia. (P.S.)

**Muck**
*S.O.E.D.* Farmyard manure. Now chiefly dial.

Howell, 1957. 1. A dark-colored soil, commonly in wet places, which has a high percentage of decomposed or finely comminuted organic matter. Compare Peat.

*Comment.* The common American usage given by Howell is almost unknown in British literature and tends to cause much confusion.

**Mud**
Himus, 1954. 'An unconsolidated rock of clay-grade often containing much water. May consist of several minerals.'

**Mud Circles**
Mackay, J. R., Fissures and Mud Circles on Cornwallis Island, N.W.T., *The Canadian Geographer*, 3, 1953, 31–37: 'Many of the gravel beaches of Cornwallis Island have a type of patterned ground which might be referred to as mud circles ... in youth, the mud circles are represented by plugs beneath conical pits. As the mud plugs rise, or are injected upwards, they increase in size and eventually break out into the pits' (p. 35). 'The cryostatic hypothesis (*q.v.*) may possibly explain the formation of these mud circles' (p. 36).

**Mudflow**
*Webster.* 3. A moving mass of soil made fluid by rain or melting snow; a mud avalanche. Cf. Earthflow.

**Mudir** (Arabic)
An administrator or a governor, according to the country, of a unit from a province to a village.

**Mudiriya** (Sudan; Arabic)
A province in the Sudan. The word is also applied to the provincial offices in the seat of government. (J.H.G.L.) In Egypt a province is called mudiriyet, a governorate is muhafazet.

**Mudstone**
Himus, 1954. 'An unlaminated, indurated sedimentary rock, consisting of clay minerals and other constituents of the clay grade.'

**Mud-volcano**
A cone of mud associated with escaping gas in a petroliferous area: rarely with escaping volcanic gases.

Kuenen, 1955. 'If the expansive force of these gases becomes excessive, they will seek some outlet ... if a layer of clay is encountered, which the mixture of gases and water whisks into a soft slurry, mud will emerge and build up a cone on the surface. Within its centre a cauldron of thin mud bubbles and splashes; every so often the bubbles lift the mud to above the edge of the crater, and it overflows onto the flanks of the cone.' (p. 203)

**Muirburn** (Scottish)
The burning of heather or moorland vegetation as a system of land management. *Adv. of Science*, 21, 1964. (p. 164)

**Mukim** (Malay)
The lowest administrative unit in Malaya, corresponding roughly with a parish.

**Mulatto**
*O.E.D.* One who is the offspring of a European and a Negro; also used loosely for any half-breed resembling a mulatto. See also Mestizo.

**Mulga scrub** (Australia)
Scrub characterized by the dominance of *Acacia aneura*, known to Australian aborigines as mulga (*antiq.* malga, mulgah, mulgam). Contrast Mallee scrub.

**Mull** (Swedish; soil science)
Mild humus; forest humus layer of mixed organic and mineral matter with a gradual transition to the underlying mineral horizon. Formerly mull was the term for a substance (humus) or a mineral soil mixed to

a great extent with humus. Müller however gave a quite different meaning to it. According to him mull is the upper layer of the mineral soil mixed with humus, and mainly thanks to the activity of earthworms is kept with a crumb or granular structure. This new meaning has however never become popular in Swedish literature. Later the term mull has been divided into the granular mull of Müller (Swedish granulärmull) and amorphous mull (amorfmull). (E.K.)

Jacks, 1954. 'Forest-humus layer of mixed organic and mineral matter with a gradual transition to the underlying mineral horizon' (contrast Mor).

**Mull** (Scottish, from Gaelic)
A promontory or headland.

**Mulola**—see Oshana

**Multi-cycle landscape, Multiple-cycle valleys**
Wooldridge and Morgan, 1937. 'Many regions give the plainest testimony of having passed through one or more former cycles of erosion, since they retain considerable relics of uplifted and dissected erosion surfaces. These marked the culmination of former cycles; by uplift they became the initial surfaces for current cycles, and while they remain recognizable elements in the landscape the two-cycle or multi-cycle character of the latter is readily apparent. The landscape as a whole has been *revived* by change of base-level, but bears traces of a former condition.' (p. 210)

*Note:* Cotton, 1922. Multi-cycle coasts; von Engeln, 1942, uses 'multiple-cycle valleys' (p. 222). Many writers now prefer to use polycyclic (*q.v.*); others refer to 'valley-in-valley' forms.

**Multiple nuclei**
Harris, C. D. and Ullman, E. L., 1945, The Nature of Cities, *Ann. Am. Acad. Polit. Soc. Sci.*, **242**, 7–17. 'In many cities the land-use pattern is built not around a single centre but around several discrete nuclei. In some cities these nuclei have existed from the very origins of the city; in others they have developed as the growth of the city stimulated migration and specialization.' (p. 13)

*Comment.* Separate nuclei and differentiated districts are due to a combination of certain activities requiring specialised facilities, certain like activities grouping together because they profit from cohesion, certain unlike activities being detrimental to each other, and other activities being unable to afford the high rents of the most desirable sites.

**Munro** (Scotland)
Darling, F. F., and Boyd, J. M., 1964. The Highlands and Islands, Collins New Naturalist Series, 'A "munro" is a hill 3,000 feet or over, separated from another by a dip of 500 feet or more; the name is from the Scottish mountaineer, H. T. Munro, who listed them. There are 276 Munros [in the Highlands], and 543 tops over 3,000 feet.' (p. 23)

**Mura-yama** (Japanese)
Matsumura, Yasukazu, *Geog. Rev. Japan*, **30**, 5, 1957, 395. 'A kind of common or waste.'

**Muri**
Evergreen scrub on podsolic soils in Guiana. Physiognomically similar to padang in Malaya. (P.W.R.)

**Murram** (East Africa, especially Uganda: original language unknown)
A lateritic ironstone formed in level topography as a result of the vertical movement of otherwise stagnant waters through a succession of seasons. Until the advent of the tarmac road, murram was the staple road surfacing material of Uganda. (S.J.K.B.)

See also Laterite, Kabouk.

**Muskeg** (Canada)
*O.E.D.* From Cree Indian; a kind of bog.
*Webster.* Of Algonquian origin. A sphagnum bog, esp. one with tussocks in it. *Northern U.S. & Canada.*

Putnam, D. F., 1952, *Canadian Regions*, Toronto: Dent. 'The Subarctic or Transition Forest Region ... this is a region of lakes and muskegs, the latter being undrained basins now filled with peat moss. Around the edges dense stands of tamarack and black spruce are found, the trees being smaller and smaller toward the centre of the bog' (p. 23).

**Myo** (Burmese)
Town: used as suffix in place names, *e.g.* Maymyo = May's Town, named in honour of General May. (L.D.S.)

# N

**Nab** (Yorkshire)
Versey, H. C., 1959. *Geology in York: A Survey*, 1959 (British Association). 'In the faulted Howardian Hills the Kellaways rock forms characteristic rounded "nabs"' (p. 6).
*Comment.* A term used in Yorkshire for a headland and also for a spur of an escarpment. See Fox-Strangways, E. and Barrow, G., 1915, *The Geology of the Country between Whitby and Scarborough*, Mem. Geol. Surv., p. 53.

**Nad** (Indo-Pakistan: ?Hindi)
Swampy land kept permanently moist by the presence of springs.

**Nadir**
O.E.D. The point of the heavens diametrically opposite to the zenith; the points directly under the observer. *Also attrib.*
*Comment.* Frequently used figuratively for the lowest point of anything.

**Nagelfluh** (German)
Mill, *Dict.* The 'conglomerates' which accompany the molasse (*q.v.*) of the Alpine region.

**Nahyad** (Iraq: Arabic)—see Liwa

**Nāi** (Pakistan: Sindhi)
A hill torrent.

**Nailbourne**
A temporary stream occupying a dry valley, Monkhouse, F. J. *Principles of phys. geogr.*, 1957, 'These are known in various chalkland areas as "bournes", "winterbournes", "nailbournes", "gypseys", and "levants", and the term "bourne" is an extremely common place-name element' (p. 89).

**Nālā** (Indo-Pakistan: Urdu-Hindi)
A dry river bed or one with an intermittent stream. Commonly anglicized as nullah (*q.v.*).

**Nappe** (French)
*Webster.* 2. *Geol.* A mass thrust over other rocks by a recumbent anticlinal fold, by thrust faulting, or by a combination of both.
Lobeck, 1939. 'The Alps and other mountains of this type are extremely complicated in detail ... their essential plan consists of a number of great recumbent, that is, strongly overturned, folds thrust from the south, one over the other. These are called *nappes*. Each *nappe* is many miles long ...' (p. 605).
Hills, E. S., 1953. *Outlines of Structural Geology*, 3rd Edn., London. Methuen. 'A *nappe* (Fr) or Decke (Ger). is a sheet of rocks, of large dimensions (of the order of miles), that has moved forward for a considerable distance (again of the order of miles) over the formations beneath and in front of it, finally covering them as a cloth covers a table. A nappe may be either the hanging wall of a great low-angle overthrust (*thrust nappe, Überschiebungsdecke*), or a recumbent fold (*fold nappe, Überfaltungsdecke*), of which the reversed limb has been completely sheared out as a result of the great horizontal translation. Classical regions of nappe structure are the Highlands of Scotland and the European Alps. ...' (pp. 54–55). 'It should be noted that the terms *nappe* and *Decke* are used in French and German respectively for any covering sheet of rock, such as a layer of gravels or a basalt flow. In English however they are used only in a tectonic sense' (p. 54).
*Comment.* Similar definitions in most standard works. In writings in English the French word nappe is invariably used and not the German Decke but Klippe is preferred to nappe outlier whilst for an associated phenomenon either Fenster or fenêtre is preferred to window. (L.D.S.) Not to be confused with alternative meaning. Hatzfeld and Darmsteter, 1932. '4. Table de plomb qu'on étend sur un terrain, un toit. *Fig.* Ce qui s'étend en couche. Une nappe d'eau souterraine,' etc.

**Nappe outlier**
Lobeck, 1939. 'A nappe outlier or Klippe is a remnant of a higher nappe spared by erosion.' (p. 605).
von Engeln, 1942. 'When a crowning nappe-fold has been thrust far forward and then almost completely destroyed by erosion, the last remnants of the structure appear

as a type of outlier called a klippe. Klippen may have the proportions of lofty mountain peaks ... the Matterhorn and Weisshorn summits in the Alps are of such origin. They are referred to as *mountains without roots* because they have been shoved so far along a nearly horizontal plane as to have lost all contact with the foundations on which they once rested.' (pp. 332-333)

## Nation

*O.E.D.* 1. An extensive aggregate of persons, so closely associated with each other by common descent, language or history, as to form a distinct race or people, usually organized as a separate political state and occupying a definite territory. In early examples the racial idea is usually stronger than the political; in recent use the notion of political unity and independence is more prominent.

5b. A tribe of North American Indians.

Mill, A complex human group with *Dict.* constituents frequently of very diverse origin, but united around a common tradition and often under a common government.

*Webster.* (1) (Ethnology) A part or division of the people of the earth, distinguished from the rest by common descent, language or institutions: a race, a stock.
(2) The body of inhabitants of a country united under an independent government of their own.

Taylor, T. Griffith, 1936, *Environment and Nation*, Univ. Toronto Press, discusses the definition at length and objects to 'nation' being taken as synonymous with race or stock [*Webster* (1)] but accepts *Webster* (2).

*Comment.* The practice now is to make nation = an independent political unit, as in United Nations. Terms such as 'English nation,' used by Hakluyt, have now dropped out of use. (G.R.C.)

## National

*O.E.D.* Of or belonging to a nation; affecting, or shared by, the nation as a whole. Peculiar to the people of a particular country, characteristic or distinctive of a nation.

*Comment.* The above apply to current usage if 'nation' is given the meaning in *Webster* (2).

## National Parks

*L.D.G.* National Parks are areas usually of great natural beauty and interest, now established in many countries, and serve especially for conservation of native flora and fauna, as well as for recreation of human beings—especially those interested in wildlife. In addition to those in the United States and Britain, especially famous are the Canadian National Parks in the Rockies (notably Banff and Jasper), the Kruger Park in the eastern Transvaal, Serengeti in Tanzania, Murchison Falls and Queen Elizabeth in Uganda, Mt. Cook in New Zealand, and Mt. Fuji in Japan.

In Britain, National Parks were set up in England and Wales (there are none in Scotland) by an Act of 1949. They are areas of natural beauty, but for the most part the land is farmed or otherwise used, and the Parks include numerous villages and other settlements and the normal life of the countryside goes on in them. They are, however, specially protected from building development. They are ten in number (1965)—Lake District, Yorkshire Dales, North York Moors, Peak District, Snowdonia, Brecon Beacons, Exmoor, Dartmoor, Cheviots, Pembrokeshire Coast. The same Act set up Areas of Outstanding Natural Beauty (A.O.N.B.) also protected. Quite distinct from the National Parks Commission is the Nature Conservancy which controls the National Nature Reserves (N.N.R.).

In the United States, they are mainly extensive wild areas, often largely uninhabited and so serve both for recreation and for conservation of wildlife. The first to be set up was Yellowstone in 1872, including parts of the States of Wyoming, Montana and Idaho, with an area of over 7,770 sq. km. (3,000 sq. miles). It is said to have more geysers than all the rest of the world. Its creation really started the 'national park movement' all over the world. The United States has 28 other national parks, some of the most famous being Crater Lake, Everglades, Glacier, Grand Canyon, Mammoth Cave, Carlsbad Caverns, Mount McKinley, Mount Rainier, Sequoia (giant trees), Yosemite and Zion. The United States also has National Historical Parks, National Recreational Areas, National Parkways and National Monuments, as well as a range of lesser areas controlled by the States themselves (State Parks etc.).

### Nationalism

*Webster*. 3. Devotion to, or advocacy of, national interests or national unity and independence; as, the *nationalism* of Ireland or China.

This word has many meanings as indicated in *O.E.D.* but is currently used in the sense of *O.E.D.* 2 'National aspiration; a policy of national independence', again using nation in the sense of *Webster* (2). See Kohn, *The Idea of Nationalism*.

### Nationality

*O.E.D.* 3. The fact of belonging to a particular nation.

There are many other meanings but this is the one most in current international use. An official distinction is commonly drawn (as in the United States) between nationality and race.

### Nationalization

*O.E.D.* 3. The action of bringing land, property, industries etc., under the control of the nation.

This meaning has almost entirely superseded the others given in *O.E.D.*

Again, nation is used in the sense of *Webster* (2).

### Natural

The Committee, List 1. 'Existing in, or formed by, nature; not artificial' (this is the *O.E.D.* definition I.6).

*Comment*. With 'natural' used in this normal sense it has not seemed necessary to give natural arch, natural bridge, natural pit, natural tunnel and other self-explanatory terms.

### Natural Landscape—see Landscape, Natural

### Natural region

Mill, *Dict*. A part of the earth's surface characterized by the comparatively high degree of uniformity of structural and climatic features within it.

Herbertson, A. J., 1905, The Major Natural Regions: An Essay in Systematic Geography, *Geog. Jour.*, **25**, 300–309. 'The question is, what are the characteristic and distinguishing elements of the areas which we may term natural regions...? These will deal, however, not merely with the mutual adjustment of drainage and land forms, but also with the well-marked zones of climate, vegetation, and even human distributions which characterize such forms when situated in similar climatic areas.... A natural region should have a certain unity of configuration, climate, and vegetation.... The mapping of human conditions has less significance in indicating the major natural geographical regions, for the factor of human development has to be taken into account as well as the possibilities of the natural environment ... political divisions ... must be eliminated from any consideration of natural regions.

Hartshorne, 1951 (discusses three principal usages) '... some students use the word in the term "natural regions" to indicate "something inherent and not arbitrarily imposed". This appears to have been Herbertson's intention...' '... to indicate that the basis of the regional division is to be found in nature as a whole, including man, in contrast to a division based on a single element, as in the case of "climatic regions", "agricultural regions" etc.' '... regions considered in terms of their non-human elements'. He concludes 'No matter how we term our system, whether "natural regions" or "regions of the natural landscape", we are fundamentally measuring the different natural criteria in terms of their importance to man. Since the relative importance of the different natural elements to man is certainly not determined by nature as distinct from man, it follows that our systems inevitably have a human basis; in this sense all might be called "artificial".' (pp. 296–305)

Geographical Association, 1937, Classification of Regions of the World, *Geography*, **23**. '"Natural regions" has been used to cover two distinct types of unit-areas of the earth's surface: (1) those which are marked out as possessing certain common physical characteristics—*e.g.* a certain kind of structure and surface relief, or a particular kind of climate—and (2) those regions which possess a unity based upon any significant geographical characteristics whether physical, biological or human, or any combination of these, as contrasted with areas marked out by boundaries imposed, frequently for political or administrative purposes, without reference to any geographical unity of the areas' (p. 253).

*Comment*. The concept of the region is a much debated problem. See, *inter alia*, Kimble, G. H. T., The Inadequacy of the

Regional Concept in *London Essays in Geography*, London: Longmans, 151–174. In the delimitation of regions, the crux of the matter seems to be 'Is the region inherent in the landscape or is it the product of criteria applied by the individual geographer?' In other words is the region an objective or subjective concept? See Bowen, E. G., *Trans. I.B.G.*, 1959, p. 1. (G.R.C.)

**Natural resources**
*Encyclopedia of the Social Sciences*. New York: Macmillan, 1933. Vol. XI. 'Resources are those aspects of man's environment which render possible or facilitate the satisfaction of human wants and the attainment of social objectives.... In the narrow sense, natural resources are original aspects of nature untransformed by man, such as air, water, sunshine and wild animals and vegetable life, which spontaneously satisfy human wants. In a wider sense natural resources comprise also the substances, forces, conditions, relationships and other aspects of Nature which are transformed by man and underlie, shape, affect or inhere in that complex mixture of natural and cultural landscapes which constitute the environment of modern man.' (pp. 290–291)

Sloan, H. S., and Zurcher, A. J., 1949, *A Dictionary of Economics*, New York: Barnes and Noble. 'Wealth supplied by Nature. Mineral deposits, soil fertility, timber, potential water power, and fish and wild life are included in the concept. The term "natural resources" is identical with the formal economic concept of "land".'

Sauer, C. O., 1952, *Agricultural Origins and Dispersals*, New York: Amer. Geog. Soc. '"Natural resources" are, in fact, cultural appraisals.' (pp. 2–3)

*Comment.* Now commonly classified as flow (those that may be depleted, sustained or increased by human activity, e.g. soils, forests); stock (non-renewable, e.g. minerals); and continuous (always available and independent of human action, e.g. solar and tidal energy; or always available but open to human modification, e.g. amenity landscape).

**Natural vegetation**
Tansley, 1939. 'By natural vegetation we mean of course that which is primarily due to "nature" rather than to man.... But a great deal of existing vegetation is intermediate between the two extremes.' (p. 195)

Unstead, J. F., 1953, *A World Survey*, London: U.L.P. '"Natural" vegetation is here regarded as that which has been but little changed by man...' (p. 115).

Hartshorne, 1939, discusses the problem at length (pp. 301–305) and concludes 'in view of the impossibility of reconstructing the natural vegetation as it would have been if man had not entered the scene, and of the extreme difficulty of reconstructing the original natural vegetation before man, it is not surprising that even those who have been most enthusiastic for a regional system based on vegetation have not developed it in a manner consistent even in major outline.' (p. 304)

*Comment.* Recent ecological research has demonstrated how little of the existing vegetation of the earth's surface is unmodified by man's activities either direct (*e.g.* burning of grass which has created savanna lands) or indirect (*e.g.* introduction of grazing animals, spread of rabbits) and that much of the so-called natural vegetation is at best only *semi-natural*. The current tendency is to include all vegetation not deliberately managed or controlled in farming activities as natural vegetation or to use the term semi-natural.

**Naturalize, naturalization**
*S.O.E.D.* 1. To admit (an alien) to the position and rights of citizenship. 3. To introduce (animals or plants) to places where they are not indigenous, but in which they may flourish under the same conditions as those which are native.

**Nautical mile**—see Mile, geographical; also Knot

**Naze**
*O.E.D.* A promontory or headland, 1826, Ewing, *Geog.* (ed. 7) 23 note Naze, ness, and mull are also used to signify remarkable portions of land stretching out into the water.

Mill, *Dict.* Outstanding peninsula, promontory or headland, often not rocky. Also called Ness, Nose, Nore.

**Nazzaz** (Soil science)
Jacks, 1954. 'A compact, impermeable pan, concretionary in character, occurring at a slight depth below the surface of red sandy soils in the Levant.' See Hard pan.

**Nearest-neighbour analysis**
Harvey, D. in Chorley and Haggett, 1967.

A technique used by plant ecologists and applied to the study of geographic problems most notably by Dacey. 'The technique requires that we select a location within our point pattern according to some rule ... and then record measurements from that point to the first, second, third ... *n*th nearest neighbour. If we repeat the procedure for a number of points according to some sampling design, we may then calculate out the mean first order distance (and its variance), the mean second order distance, and so on. These mean order distances, together with the frequency distribution of each order distance, then comprise the mathematical description of pattern.' (p. 576)
See Dacey, M. F., 1960, A Note on the Derivation of Nearest-Neighbour Distances, *Jour. Reg. Sci.*, **2**, 81-87.

## Nebular Hypothesis
*O.E.D.* The theory propounded by Kant and elaborated by Herschel and Laplace which supposes a nebula to be the first state of the solar and stellar systems.
*Webster. Cosmog.* A hypothesis to explain the process of formation of the stars and planets. As framed by Laplace, it supposed the matter of the solar system to have existed originally in the form of a vast, diffused, rotating nebula, which, gradually cooling and contracting, threw off, in obedience to physical laws, successive rings of matter, from which, by the same laws, were produced the several planets, satellites, and other bodies of the system.
Wooldridge and Morgan, 1937. 'This "nebular hypothesis" led to the conception of planets initially gaseous, assuming a liquid condition on cooling and ultimately developing a solid crust' (p. 3).

## Neck
By analogy applied to a narrow part between two larger parts, hence for example (1) a neck of land—a narrow piece of land with water on each side; an isthmus or promontory (*O.E.D.*). (2) a narrow stretch of wood (neck of the woods), ice, etc. (3) especially a volcanic neck: the aperture by which molten rock has reached the surface to a volcano, when afterwards filled with solidified lava is called a volcanic neck or plug. (4) a high level pass; especially the narrowest part (*cf*. nek).

## Nefūd (Arabia: Arabic)
The *erg* (*q.v.*) or sandy desert in Arabia; see also Ahqāf.

## Negative area
A phrase sometimes used in the sense of *anecumene*—unsuitable for cultivation and so uninhabitable at least for purposes of producing food from the soil.

## Nehrung (German; *pl.* Nehrungen)
Shackleton, 1958. 'East of the Oder, the Baltic coast is noted for its smooth outlines which are associated with the development of great dune-crowned sand-spits, here known as *nehrungen* (*cf.* French *cordons littoraux*) which were built up in front of a formerly indented coastline by the action of longshore drift and the prevailing south-west wind.' (p. 250)
Mill equates with Barrier Beach; see also Lido.

## Neighbourhood effect
Hägerstrand, T., 1967, *Innovation and Diffusion as a Spatial Process*, Chicago: University of Chicago Press. 'The rank order of acceptances, which did not follow the size of cultivated holdings in any obvious way, is quite clearly related to their location with respect to one another. We can speak of the appearance of a neighbourhood or proximity effect. It appears as a dominant feature of the innovation processes.' (p. 163)
*Comment.* Hägerstrand stressed that the neighbourhood effect was not a logical phenomenon and should not be interpreted too literally.

## Neighbourhood unit
A concept in modern town planning of a unit enclosed within a town which is relatively self-contained having the shopping, banking, postal and other services adequate for daily needs of a residential population and preferably a unit not traversed by a main road or roads. Not infrequently a neighbourhood unit is created automatically where a village is absorbed by an expanding town. (L.D.S.)

## Nek (Afrikaans; *lit.* neck)
Used figuratively as a geographical term for a low place in a mountain range, compared to the neck of an animal between the head and the shoulders. The difference between a nek and a pass is the that latter term presupposes some kind of road, while a nek is a natural feature. (P.S.)
A saddle or col. (J.H.W.)

**Nekton (Haeckel), necton**
*O.E.D. Suppl.* A collective name for all the forms of organic life found at various depths of the ocean or of lakes which possess the power to swim actively in contrast to Plankton which float or Benthos which live on the ocean floor.

**Nemoriculture (G. T. Renner)**
From Latin nemus, nemoris a glade or grove. The primitive stage of human culture when food (fruits, roots etc.) is gathered in forest glades etc.

**Neolithic (John Lubbock, 1865)**
Lubbock, John (later Lord Avebury), 1865, *Prehistoric Times*. The New Stone Age (of polished stone implements), contrast Palaeolithic, Eolithic.

**Neo-Malthusianism, Neo-Malthusianist**
The body of doctrines derived from the writings of Thomas Robert Malthus brought up-to-date and applied to the conditions of the present day. See Malthusianism.

*Webster*. Designating, or pertaining to, the doctrine that only through the limitation of births by the use of artificial contraceptives can the numbers of the population be sufficiently controlled to make possible the elimination of vice and misery and a general elevation of the standard of living.

Stamp, L. D., 1953, *Our Undeveloped World*, London: Faber (p. 28); Stamp, L. D., 1952, *Land for Tomorrow*, Indiana University Press (p. 28). 'A century of world expansion with development of new lands undreamed of by Malthus led to general disparagement of the essay and to the branding of its teaching as completely out-of-date. Now, once more, the truth of his basic assertion is very much "in date".'

**Neptunism, neptunianism**
In the latter part of the 18th century the origin of earth features was hotly disputed between the Plutonists who claimed an igneous origin and the Neptunists who claimed an aqueous origin for certain geological formations. The publication of Charles Lyell's *Principles of Geology* in 1830 finally resolved the controversy and 'Neptunianism' has become of historic interest only, though a new meaning has been found for Plutonism (*q.v.*).

**Neritic**
*O.E.D. Suppl.* Of regions or living things in the sea and in lakes; that is near to the shore or found in shallow coastal waters; opposed to oceanic.

*Neritic zone*. Variously defined but commonly used for the zone between low-water mark and 100 to 200 fathoms, or the edge of the continental shelf. See under Pelagic.

**Ness (Scotland, eastern England)**
A promontory, headland or cape. Used especially in place names in Scotland and elsewhere apparently where Scandinavian influence was strong. *Cf.* naze, nose, nore. Mill notes 'often not rocky'.

**Net, settlement**
Andrews, J., The Settlement Net and the Regional Factor, *Australian Geographer*, **2**, 1934. '... the settlement structure or "net" of a region. . . .' 'The settlement net is formed of a combination of rural and urban elements' (pp. 34–48). (Discusses population, function, pattern and form of agglomerations.)

*Comment.* This appears to be a general term in comparison to 'urban mesh' of Smailes which is purely functional. It also includes rural settlements (villages). *Cf.* Smailes, *I.B.G.*, **11**, 1946, pp. 87–101.

**Nevados (Ecuador: Spanish)**
Miller, 1953. 'In the neighbourhood of large snow fields and glaciers the force of the night wind is strengthened by the cooling of the air on contact with the ice and snow ... in places so profound as to overcome the day wind and to give rise to persistent cold downcast winds, *e.g.* the *Nevados*, an unpleasant feature of the climate of the higher valleys of Ecuador.' (p. 267)

**Névé, névé (French)**
*O.E.D.* (From Alpine dialect, from Latin, niv-, nix, snow)
1. The crystalline or granular snow on the upper part of a glacier, which has not yet been compressed into ice; = Firn. (Examples, 1853–1897.)
2. A field or bed of frozen snow.

Mill, *Dict.* Névé or Firn. The upper portion of a glacier, the top layers of which are practically in the condition of snow, and in the whole of which much air is mingled with the ice, *i.e.* it is rather frozen snow, though often hard frozen, than true ice.

*Adm. Gloss.*, 1944. Firn Snow (Névé). Old snow which has outlasted one summer at least and is transformed into a dense, heavy material as a result of frequent melting and freezing; a transitional stage to glacier ice.

Hatzfeld and Darmesteter, *Dictionnaire Général de la Langue française*, Paris: Delagrave, 1932. Couche de neige durcie près d'un glacier.

*U.S. Ice Terms*, 1952. More or less loose, granular ice in transition from snow to glacier ice. Névé, in being buried about 100 feet, becomes compacted and gradually changes to glacier ice. The upper layers of glaciers and shelf ice are usually composed of névé.

Similarly in Cotton, 1945, 1922; Holmes, 1944, p. 204; Wright, 4.

*Comment.* There is doubt as to the correct spelling of the word which would seem to be neve though many authors write névé which is given by the O.E.D. Glaciologists are tending to prefer *firn* since firnification is possible but scarcely névéification. (L.D.S., G.T.W.)

### New geography

Mackinder, H. J., 1904, The Geographical Pivot of History, *Geog. Jour.*, **23**, 25 Jan. 'Of late it has been common-place to speak of geographical exploration as nearly over, and it is recognised that geography must be diverted to the purpose of intensive survey and philosophic synthesis.' (p. 421)

Manley, G., 1966, A New Geography, *The Guardian*, 18 Mar. 'Geographers display anxiety about what a new generation ought to do; what frontiers now exist for them to cross... evoke even more anxiety among those who teach and remain entangled in conventional attitudes, in that cosy core region that so many geographers inhabit. So strong are these attitudes that some feel the need to leave conventional geography and advance into the study of process motion, and change, rather than mere description and discussion of state, however embroidered with fashionable statistical analysis and the design of analogues and models.' (p. 8)

### New World

Commonly applied to the continents of North and South America; the Western Hemisphere.

### Niai (Indo-Pakistan: Panjabi)

Highly manured land, usually near the village or settlement.

### Niaye (West Africa)

Harrison-Church, 1957. 'In the north [of Senegal] there are marshy depressions or *niayes* (meaning "clumps of oil palms") between the dunes, and parallel with the coast. Dew and intermittent streams provide water for the niayes and around their edges is luxuriant vegetation' (p. 192).

### Niche

Sears, Paul B., 1957, *The Ecology of Man*, Univ. of Oregon Press. 'Environment abounds in *niches* or opportunities for particular organisms to carry on... the course of evolution has been a process of filling these niches, meanwhile creating new ones... any species moving into a Niche has an effect or reaction upon the situation... call this its Rôle' (p. 11).

See also Range.

### Niche glacier

Groom, G. E., 1959, *Jour. Glaciology*, **3**. 'Small glaciers lying in funnel-shaped hollows on steep slopes have been variously described as wall-sided glaciers, cascade glaciers, cliff glaciers and niche glaciers' (p. 369).

### Nick point, nickpoint

von Engeln, 1942. 'The English phrase, nick points, does not exactly translate knickpunkte' (p. 265).

Baulig however gives either knick-point or nick-point as the translation of knickpunkte, so does Thornbury (p. 110 etc.).

See Knickpoint, Knickpunkte.

### Niggerhead

*S.O.E.D.* Applied to various black or dark-coloured roundish objects.

*Comment.* Examples include certain masses of vegetation or peat (especially in U.S. and in the tundra) and masses of coral.

### Night-soil

Soil in the sense of manure or excrement, especially human excreta, so called because of the old custom of removing it from cesspits under cover of darkness, a practice still extant in some eastern towns.

### Nili (Egypt)

The harvest obtained in Egypt from the higher fields, not flooded by the Nile, which have to be irrigated by some form of water lift and fertilized. The harvest is later than the chetoi.

### Nilometer

*O.E.D.* A graduated pillar, serving as a scale or gauge to indicate height to which the Nile rises during its annual floods.

*Comment.* Later more elaborate gauges were used in the Nile and the word has come

Nival Region—see Subnival Belt

**Nivation** (F. E. Matthes, 1899)

*O.E.D.* The action of frost causing disintegration of rocks in the neighbourhood of melting snow.

Matthes, F. E., Glacial Sculpture of the Bighorn Mountains, Wyoming, 21*st Ann. Report U.S. Geol. Survey*, 1899–1900. 'From this rather brief study of snowdrifts and their sites we infer:
1. That snowdrifts do not form except in the presence of favourable topographic features.
2. That the effect of their presence is to accentuate these features by frost action at their peripheries.
3. That they tend to protect their sites against aqueous erosion.
4. That they favour the formation of deposits of fine mud.
5. That they have no sliding action. . . . From our observations on existing drifts we know that stream erosion is arrested under the névé; . . . The effects of the occupation by quiescent névé are thus to convert shallow V-shaped valleys into flat U-shaped ones and to efface their drainage lines without material change of grade. These névé effects, which are wholly different from those produced by glaciation, I shall, for the sake of brevity, speak of as effects of nivation, the valleys exhibiting them having been nivated.' (pp. 167–190)

*U.S. Ice Terms*, 1952. The specific effects produced by névé in land sculpture as contrasted with those by glacier ice, called glaciation.

Martonne, E. de, 1952, *Traité de Geographie Physique*, vol. 2, Paris: Colin. 'On peut généraliser l'expression de nivation pour désigner l'ensemble des phénomènes qui transforment la surface dans ces régions où les neiges jouent un grand rôle, sans être des neiges éternelles qui donnent des glaciers.' p. 857

Similar usage of 'snow patch erosion' found in Wooldridge and Morgan, 1937, p. 402; Holmes, 1944, p. 219; Cotton, 1945, p. 310.

No reference in Wright, 1914.

**Nivation hollow**

Wooldridge and Morgan, 1937. 'Nivation is the work of snow and ice beyond the true glacial limits . . . the direct action of nivation is best seen in the formation of "snow niches", shallow amphitheatres occupied during part of the year by a small snow patch . . . such patches "dig themselves in" . . . early nivation niches no doubt began the sculpture of glaciated mountain ranges. . . . W. H. Hobbs supposes the initial hollow of a corrie is in many cases a "nivation hollow".' (pp. 377, 402)

Linton, D. L., Presidential Address to Section E, British Association, 1957, *Advancement of Science*, **14**, No. 54, 'Under periglacial conditions some pre-existing land forms were certainly modified, and the transformation of pre-glacial valley heads into "nivation hollows" has often been mentioned in southern Britain though never demonstrated.' (p. 63)

**Niveo-eolian**

Vink, A. P. A. (1949), *Bijdrage tot de kennis van loess en dekzanden.* [Contributions to the knowledge of loess and coversands] Wageningen: Veenman. 'Deposited by snowstorms during the periglacial climate' (p. 113).

**Nodality**

*O.E.D. Suppl.* Nodal quality; the degree or extent to which lines, roads or any set of things having a lineal character, approach each other or converge at a point.

1897. *Geog. Jour.*, **9**, 78. A higher degree of 'nodality', to use Mr. Mackinder's term, is found where several such furrows meet to form a well-marked though by no means deep hollow.

1902. H. J. Mackinder, *Britain and the British. Seas*, 330. A spot on which more numerous land and water roads converge . . . may be said to have a higher degree of nodality.

Mill, *Dict.* The degree to which any point on the Earth's surface is a natural meeting-place of roads.

Houston, J. M., 1953, *A Social Geography of Europe*, London: Duckworth. A site or area having an important position in relation to its location and communications.

*Comment.* The *O.E.D.* reference of 1897 is to a paper by G. G. Chisholm 'On the distribution of towns and villages in England', and the 'furrows' are shallow valleys in the chalk which are followed by routes linking-up villages.

Confusion arises since in the view of some (*cf.* Mill) nodality is independent of man-made communications.

In the printed evidence presented to the Royal Commission on the Geographical

Location of the Industrial Population a printer's error introduced the word 'nodulity' instead of nodality.

**Nodal region (Functional region)**
Grigg, D. in Chorley and Haggett, 1967. 'The early exponents of the nodal region were essentially concerned with such interconnections between a central place and the neighbouring countryside.' Upon this original and simple idea the nodal region has been used to delineate 'areas of organization and traces the functional relationships between places ... so that it is now customary to describe the nodal region as a functional region'. (p. 470)
*Comment.* Nodal regions can be attributed to a number of writers. See Galpin, J. C., 1915, The Social Anatomy of an Agricultural Community, *University of Wisconsin, Agricultural Experimental Station, Bulletin No. 35*; Park, R. E., Burgess, E. W. and McKenzie R. D., 1925, *The City*, Chicago: University of Chicago Press; McKenzie, R. D., 1933, *The Metropolitan Community*, New York: President's Research Committee on Social Trends; and Gras, N. S. B., 1922, *An Introduction to Economic History*, New York: Harper. The term 'functional region' can be found in Robinson, G. W. S., 1953, The Geographical Region: Form and Function, *Scot. Geog. Mag.*, **69**, 49–50.
See also Uniform regions.

**Node**
*O.E.D.* 6, b. A central point in any complex or system.

**Nodule, nodular**
Holmes, 1928. 'A general term for rounded concretionary bodies, which can be separated as discrete masses from the formation in which they occur.'

**Nomad, Nomadism**
*O.E.D.* Nomadism. The practice, fact, or state of living a wandering life.
1841. Emerson, *Ess.*, History. In the early history of Asia and Africa, Nomadism and Agriculture are the two antagonistic facts.
Nomad. 1. A person belonging to a race or tribe which moves from place to place to find pasture; hence, one who lives a roaming or wandering life.
Mill, *Dict.* The habit of changing one's dwelling place frequently; used especially of steppe and desert people who wander in search of pasture, etc., or for trade; also used of Alpine villagers who go up to the high pastures in summer time and come down to the more permanent villages in winter.

*Chambers's Encyclopedia.* Nomads (Sir E. H. Minns). 'Nomads are people who live by pasturing animals; since usually the pasture must be sought in different places at different times of the year nomads have to move accordingly; the essence of nomadism is movement due to the needs of cattle. True nomads must be distinguished from mere wanderers, who may be either collectors, like the aboriginal Australians, or like some depressed tribes in India and Africa and the Gipsies, practisers of some peculiar craft: nor would one willingly term nomads the Europeans who move merely from a valley to the nearest mountain pasture (transhumance).'

*Encyclopedia of the Social Sciences*, New York: Macmillan, 1933, Vol. XI. 'Nomadism involves the repeated shifting of the habitat of a people in its search for subsistence. It does not consist of unrestricted and undirected wandering but is focussed around temporary centres of operation, the stability of which is dependent upon the food supply available and the state of technical advance....' (R. Thurnwald)

**Nomen nudum** (Latin; *pl.* nomina nuda)
Lit. naked name; a name given to an animal, plant, mineral etc. without a full description. Many soil series have been named without full descriptions to enable the type to be recognized elsewhere.

**Nomothetic approach**
Abler, R., Adams, J. S. and Gould, P. R., 1971. 'Nomothetic sciences try to develop universal statements of law applicable to all cases.' (p. 52) A geographical approach concerned with explanation in terms of general theories. The polar opposite to the Idiographic approach (*q.v.*).
See Bunge, W., 1962, Theoretical, *Lund Studies in Geography*, Series C, *General and Mathematical Geography*, **1**, 7–13.
See also Idiographic approach.

**Non-parametric (distribution free) method**
Kendall, M. G. and Buckland, W. R., 1960, *A Dictionary of Statistical Terms*, 2nd ed., Edinburgh: Oliver and Boyd. 'A method e.g. of testing a hypothesis or of setting up a confidence interval, which does not depend on the form of the underlying

distribution; for example, confidence-intervals may be obtained for the median, based on binomial variation, which are valid for any continuous distribution. In the case of testing hypotheses, the expression is to be understood as meaning that the test is independent of the distribution on the null hypothesis. It is better to confine the word "non-parametric" to the description of hypotheses which do not explicitly make an assertion about a parameter.'

### Non-place urban realm
Webber, M. M. in Webber, M. M. et al., 1964, *Explorations into Urban Structure*, Philadelphia: University of Pennsylvania Press. 'An urban realm ... is neither urban settlement nor territory, but heterogeneous groups of people communicating with each other through space.' '... accessibility rather than the propinquity aspect of "place" that is the necessary condition. As accessibility becomes further freed from propinquity, cohabitation of a territorial place ... is becoming less important to the maintenance of social communities.' 'Specialized professionals, particularly, now maintain webs of intimate contact with other professionals, wherever they may be ...'; 'Spatial distribution is not the crucial determinant of membership in these professional societies, but interactivities.' (pp. 109–116)

### Non-sequence
Stamp, L. D., 1921, On Cycles of Sedimentation in the Eocene, *Geol. Mag.*, **58**. 'Non-sequence is most frequently applied to a break in sedimentation which can be detected only by a study of successive faunas.' (p. 109)
*Comment.* In a non-sequence the time interval is relatively short; in a disconformity it is longer.

### Noösphere (de Chardin, Le Roy and Vernadsky, 1920)
De Chardin, P. T., 1956, in *Man's Role in Changing the Face of the Earth*, 'at the end of Tertiary time ... our planet developed the psychically reflexive surface for which we suggested in the 1920's the name "Noösphere'... from the Greek *noos*, "mind"...' (p. 103).

### Nore, nose—see Naze, Ness

### Noria
A chain and bucket arrangement to draw water from a well, worked by an animal. In Mediterranean countries.

### Normal erosion
Davis, 1909, *Eighth International Geographical Congress*. 'Thus far it has been tacitly implied that land sculpture is effected by the familiar processes of rain and rivers of weather and water. It is certainly true that the greater part of the land surface has been carved by these agencies, which may therefore be called the prevailing or normal agencies; but it is important to consider the peculiar work of other special agencies, namely ice and wind. (pp. 150–163)
Cotton, 1922. 'The subaerial, as distinguished from marine, eroding agencies, fall into two groups, normal and special, and it is by the normal group, running water and the weathering processes, that the shaping of the land surface is mainly effected. (p. 43)
*Note: Cf.* 'normal cycle of erosion'; involves a land surface 'under ordinary climatic conditions, not so dry but that all parts of the surface have continuous drainage to the sea, nor so cold but that the snow of winter all disappears in summer.' (Davis 1909, p. 296)
Also *cf.* Mill, Normal Fall of a River-bed
*Comment.* The modern tendency is to drop the term and to regard any distinction between 'normal' and 'special' agencies as unreal.

### Normal fault—see Fault

### Normative
*O.E.D.* Establishing or setting-up a norm or standard.

### Norte (Mexico and Central America: Spanish)
Miller, 1953. 'Winds along the Mexican coast are northerly and sometimes strong (Nortes), a continuation of the American "Norther" and are occasionally felt high up on the plateau (Papagayos). These northerly winds bring cold which is intense for the latitudes and the fall of temperature is often sudden and unpleasant' (p. 131).
Mill, *Dict.* 'Similar winds at Valparaiso have the same name.'
Moore also applies to cold, northerly wind in eastern Spain during winter.

### North
*Webster.* That one of the four cardinal points of the compass, at any place (except the

poles) which lies in the plane of the true meridian, and on the left hand of a person facing due east; the direction opposite south.
See Pole, Compass.

**Norther** (= Norte)
Mill, *Dict.* Dry cold wind blowing from September to March and bringing a rapid fall of temperature in Texas and over the Gulf of Mexico.
Monkhouse, 1954. 'The *Bora* and the *Mistral*... similar are the *Southerly-burster* in New South Wales, the *Pampero* and *Friagem* in the Argentine, the *Norther* in the south-east of U.S.A., and the *Norta* and *Papagayo* in Mexico.' (p. 308)
Moore, 1949. 'The *Cold Wave* sometimes experienced in the southern United States, usually in rear of depression. The temperature may fall by 30° to 40° in 24 hours causing great damage to fruit crops. The norther often blows with great violence, is accompanied by severe thunderstorms and hail... in the Sacramento Valley of California, however, is often dry and dusty... heated and dried by descent from the mountains' (p. 122).
See also Miller, 1953, p. 173.

**Nor'wester, northwester**
Webster. A storm or gale from the northwest; a strong northwest wind.
Miller, 1953. 'Similar winds to the *foehn* occur in all mountain districts where cyclonic storms occur; the *Chinook* is exactly similar, so are the *Samun* of Persia, the *Nor'-Westers* of New Zealand off the New Zealand Alps, and many others' (p. 269). 'Storms are fairly frequent in the plains during the Hot Season... the duststorms of the north-west, the "Nor'-westers" of Bengal and the rainstorms of Assam all appear to be similar in origin, since all originate at the junction of two currents, a dry cool upper current from the north-west riding over the shallow surface flow of air drawn in by the low pressure of the plain' (p. 142).

**Nosology, nosological**
O.E.D. A classification or arrangement of diseases; systematic or scientific classification of diseases.
*Comment.* Hence nosological or nosographical map showing distribution of diseases.

**Notch (1)**
Webster. 2. A narrow passage between two elevations, as mountains; a deep, close pass; a defile; a gap. *U.S. Dict. Am.* A narrow passageway or defile between mountains, or the narrowest part of such a passage. Chiefly E. U.S. 1718.

**Notch (2)**
Dury, 1959. 'Undercutting of cliffs is usually concentrated near high-water mark where a notch is characteristically found... the cliff is undermined... old notches can be used in determining former sea-levels' (p. 99).

**Nubbins** (A. C. Lawson, 1915)
Cotton, 1942. '... originally used by Lawson for the last remnants of a mountain range surviving along its crest as the range succumbs to desert erosion' (p. 64). Cotton extends the meaning to small spur-remnants. See Lawson, A. C., 1915, The Epigene Profiles of the Desert, *U. of Calif. Pubs., Bull. Dept. Geol.*, **9**, No. 3, pp. 23–48.

**Nucleated settlement**
O.E.D. 2. Clustered together. 1897. *Eng. Hist. Rev.* April, 314. The Germanic nucleated village is distinguished from the isolated homestead. *Ibid.* Oct., 769. He draws a sharp distinction between the 'nucleated' villages of eastern and central England, and the 'hamletted' villages of the south-west.
Houston, J. M., 1953, *A Social Geography of Europe*, London: Duckworth. 'It is better to distinguish three units of (rural) settlement, the nucleated village, the hamlet, and the isolated dwelling.' (p. 81)
*Comment. Cf.* 'village aggloméré' (W.G.E.). Some authors imply a settlement with an organizational centre and not a mere cluster. (C.J.R.)

**Nuée ardente** (French)
Holmes, 1944, quoting Perret, 'an avalanche of an exceedingly dense mass of hot, highly gas-charged and constantly gas-emitting fragmental lava, much of it finely divided, extraordinarily mobile, and practically frictionless, because each particle is separated from its neighbours by a cushion of compressed gas... downward-rolling explosive blasts, one of which wiped out St. Pierre in 1902... may be either dark or incandescent, and are not always "glowing" as the French term suggests.' (p. 462)

**Nullah** (Anglo-Indian; corruption of nālā)
O.E.D. A river or stream; a watercourse, river-bed, ravine.

Mill, *Dict.* The dry bed of an intermittent stream.
*Comment.* This familiar Indian word, widely used by Europeans in India, has been taken by them to other parts of the world. It is applied in Hong Kong to wholly artificial water supply or drainage channels constructed of concrete, presumably because they may be periodically dry. See Nālā; *cf.* wadi (Arabic), chaung (Burmese)

**Nunakol**
Taylor, 1951. 'A rounded rock "island" in a glacier' (p. 617) contrasted with nunatak 'a peaked rock' island in a glacier.
Charlesworth, 1957. '"Nunakol", a compound word coined for nunataks rounded by ice, is objectionable etymologically and because such forms are difficult to distinguish in the field' (p. 576). Charlesworth quotes J. Huxley *et al.*, *Scott's Last Expedition.* 1913, p. 427.
*Comment.* Not frequently adopted.

**Nunatak** (Eskimo; introduced by Nordenskiöld)
*O.E.D.* A peak of rock appearing above the surface of the inland ice in Greenland. The Swedish plural nunatakker has sometimes been used in English works. (First English example, 1882.)

Mill, *Dict.* A hill projecting above land-ice (Greenland, etc.). Similarly *Adm. Gloss* 1953; Wooldridge and Morgan, 1937 p. 365; Wright, 1914.
Charlesworth, 1957. 'A nunatak (*nuna* lonely; *tak*, a peak)—nunataq is a variant spelling (M. Vahl. 1928)—is an Eskimo term for an island of rock or mountain peak in a sea of land-ice. A. E. Nordenskiöld introduced the word into glacial literature. "Marginal nunataks", like the Jensen and Dalager Nunataks of west Greenland have ice on three sides only and on the fourth are bounded by sea fjord or land' (p. 576).
*Comment.* The *O.E.D.* definition fails to convey the idea of size.

**Nunja** (India: Tamil; Hindi)
Wet land, as opposed to *punja*, dry land.

**Nyika** (Kenya; Swahili and other Bantu languages)
A wilderness; a barren, desolate area. The word is used to denote the semi-desert with acacia shrubs and bunch grass which occupies the hinterland of the Kenya coastal strip. (S.J.K.B.)

# O

**Oasis** (*pl.* oases)
*O.E.D.* A name of the fertile spots in the Libyan desert; hence, generally, a fertile spot in the midst of a desert.
Mill, *Dict.* A fertile area in a desert.
Moore, 1949. An area in the midst of a desert which is made fertile by the presence of water.
Lebedev, V. (translated Ivanov-Mumjiev, G. P.), 1957, *Antarctica*, Moscow: Foreign Languages Publishing House. 'The discovery in the Southern Continent of "oases", that is, patches of ice-free land, has evoked keen interest in the world of science. Several of these "oases" were found in 1946–47 by the Byrd Expedition.' (p. 78)
*Comment.* Oases may be of very varied extent —some are hundreds of square miles. The popular restriction to a small area with a few date palms is incorrect. Notice the extension by analogy exemplified by the last quotation.

**Oblast** (Russian)
Administrative division of level below republic. Each oblast surrounds a town of importance after which it is usually named.

**Oblate spheroid**—see Geoid

**Obruk** (Turkish)
Sirri Erinç. On the Karst Features in Turkey, *Review of Geographical Institute of the University of Istanbul* No 6, 1960. 'This name is used in Turkish to designate a deep natural sink in limestone. They usually have very steep walls and are often occupied by permanent lakes or ponds. They are comparable to jamas or avenes extending down below the underground water table' (p. 8).

**Obsequent, obsequent river, obsequent valley** (W. M. Davis, 1895)
*O.E.D. Supp.*, 2. *Geol.* Of streams: Flowing in the opposite direction from the 'consequent' drainage.
Davis, W. M., 1895. *Geog. Jour.*, **5**. 'These short streams have a direction opposite to that of the original consequents and may therefore be called obsequents' (p. 134; also Davis, 1909, p. 204).
Mill, *Dict.* Obsequent river. A scarp stream which is tributary to a subsequent river; a river which flows from the top of an escarpment to its base to join a subsequent river.

**Occupance**
James, P. E., 1935, *An Outline of Geography*, New York 'An obsolete word revived and adopted in geography to indicate the process of occupying or living in an area and the transformations of the original landscape which result.' (p. 31) See also Sequent occupance.

**Ocean**
*O.E.D.* 1. The vast body of water on the surface of the globe, which surrounds the land; the main or great sea. 2. One of the main areas into which this body of water is divided geographically.
*Comment.* Before 1650 commonly 'ocean sea' from the Latin *mare oceanum* the great outer sea as opposed to the Mediterranean. The original Greek ὠκεανός (okeanos) the great river supposed to encompass the earth. The oceans commonly accepted are Atlantic, Pacific, Indian, Southern and Arctic.

**Ocean basin**
Webster. *Phys. Geog.* The depression occupied by the waters of an ocean; contrasted with *continental plateau.*

**Oceanic climate**
Mill, *Dict.* A climate with a moderate range of temperature in summer and winter, and with maxima and minima occurring at a longer interval after the summer and winter solstices respectively than in the case of a continental climate; the climate characteristic of the ocean and of seaboards and islands exposed all the year round to prevailing ocean winds.

**Oceanic island**
Webster. An island in the ocean, far from any continent; contrasted with *continental island.*

**Oceanicity, oceanity; index of**
'The trend from oceanic to continental climate involves a number of factors which are important in the understanding of

plant distribution. Various attempts have been made to combine them in Indices of oceanity'... C. L. Godske [The geographical distribution in Norway of certain indices of humidity and oceanicity. *Bergens Mus. Årb.*, **8**, 1944] has discussed various indices of oceanicity which are calculable from standard meteorological measurements... Kotilainen's original index is $K = \frac{Md_t}{100\Delta}$ M. E. D. Poore and D. N. McVeen, *Jour. Ecol.*, **45**, 1957, 411, 412.

## Oceanography

*O.E.D.* That branch of physical geography which treats of the ocean, its form, physical features, and phenomena.

*Comment.* The alternatives oceanology and thalassography have never gained wide use.

## Octoroon, octaroon—see Quadroon

## O.D., Ordnance Datum, Ordnance Survey Datum

Mean Sea Level (*q.v.*) at Newlyn, from which all heights shown on British Ordnance Survey Maps are calculated. Mean Sea Level at Liverpool was formerly used.

## Œkumene, oikoumene, oecumene, ecumene

Mill, *Dict.* 1. The habitable world known to the ancient Greeks.

2. A community or district which may be considered as an entity, to some extent self-sufficing and not primarily dependent on other regions for its population or supplies. France is in some degree such an œkumene, and many semi-civilized states come under this category. *Cf.* Anœkumene.

Houston, J. M., 1953, *A Social Geography of Europe*, London: Duckworth. 'The habitable world for human settlement.' (p. 247)

Semple, E. C., 1913, *Influences of Geographic Environment*, London: Constable. 'Humanity's area of distribution and historical movement we call the Oikoumene.' (p. 171)

Vidal de la Blache, P., 1926, *Principles of Human Geography*, London: Constable. 'Ocean solitudes long divided inhabited countries (oikumenes)...' (p. 18). Also p. 49, used in the sense of the inhabited earth.

Jefferson, M., 1934, The Problem of the Ecumene, *G. Annaler.*, **16**, 1934. 'The ecumene is the utilized land of a country, the part from which the people draw life.' (From the diary of Xenophon who would record 'ecumene' when he found an inhabited area on the march. W.M.) (p. 146)

Whittlesey, D., 1944, *The Earth and the State* 2nd ed., New York: Holt. 'The ecumene i the portion of the state that supports th densest and most extended population an has the closest mesh of transportatio lines' (p. 2). Cites Jefferson as definin ecumene as 'the land within ten miles of railroad'. (W.M.)

Trewartha, G. T., 1953, A Case for Population Geography, *A.A.A.G.*, **43**. 'Distribution of people in its broadest aspect, o global scale, involves dividing the lanc portions of the earth into permanentl inhabited as compared with uninhabited or temporarily inhabited, parts. The term ecumene and non-ecumene have beer employed to represent these two major subdivisions.' (p. 92)

Vallaux, C., 1911, *Le Sol et l'Etat*, Paris: Octave Doin. 'Oecoumène. Mot employé par Humboldt sous la forme grecque, pour signifier l'étendue de la terre habitable; introduit sous sa forme actuelle dans la langue scientifique par Ratzel.' (p. 400)

Not in *O.E.D.* but œcumanic, of or belonging to the 'inhabited (earth)', the whole world.

*Comment.* Now commonly used in the senses of Whittlesey, 1944 and Trewartha, 1953. The form favoured in American writings is ecumene, in British oecumene.

## Offset

Johnson, 1919. 'In many cases the part of a bar on the up current side of an inlet is a little farther seaward than the part below the inlet, in which cases the shore is said to be offset. It is possible to have offsets where there is no inlet' (p. 38).

## Ogive

In the report of a Conference on Terminology of Glacier Bands, *Journal of Glaciology*, 1953, **2**, 229–230, 'it was agreed that the word "ogives" is descriptive of the form of a series of bands of various origins seen on the surface of a glacier' but subsequent correspondence gives various interpretations. Sigurdur Thorarinsson extends use to 'low waves on the surface of a lava stream' (lava ogives), *Ibid.*, 295.

*Comment.* This word is used for features on the surface of a glacier which stretch right across the glacier from side to side, and which, as a result of the faster flow of the centre of the glacier, take up a shape similar to that of an ogival arch with the point pointing down glacier. The term appears originally to have been used for the outcrop of the foliation structure of the ice of glaciers beneath an ice fall. This structure starts just below ice falls and persists down the rest of the length of the glacier. It is also now commonly applied to the broad stripes of dark and light ice which are also observed below ice falls, the separation between successive dark bands being approximately one year's flow. This phenomenon was originally called 'dirt bands' by Forbes and 'Forbes' Bands' by others. Another feature which occurs below ice falls is a wavy surface to the glacier also with approximately one year's flow as the wavelength. By analogy with the dirt bands, this structure has been called 'wave ogives'.

On glaciers without ice falls, there are sometimes similar features to the Forbes' bands on ogives, though their separation is less regular. These have been called Alaskan bands, but since their occurrence is not limited to Alaska, this term is not to be encouraged.

The word ogive is thus now used for any feature which originally stretches right across the glacier (or a part of the glacier between two moraines) and which is drawn out by glacier flow into ogival garlands. For a discussion of the term see *Journal of Glaciology*, **2**, 1953, 229–232. (John Glen in MS.)

**Old age**—See River

**Oldland**

Webster. Geol. An extensive area of ancient crystalline rocks reduced to low relief by long erosion.

Lobeck, 1939. An area of mature landscape, used especially with reference to the area bordered by a 'coastal plain'. (p. 449)

Waters, R. S., 1957. Differential Weathering and Erosion on Oldlands, *Geog. Jour.*, **123**, 503–509, applies the term to Dartmoor.

**Oligomict**

Read, H. H., 1958, *Proc. Geol. Assoc.*, **69**, 85. 'Two kinds of conglomerate are worth distinguishing, first the oligomict made of one stable kind of pebble of an obdurate type such as quartzite....' See Polymict.

**Oligotrophy, oligotrophic** (lakes)

Webster. Oligotrophic. Having abundant oxygen in the lower stratum in summer; said of lakes.

'Oligotrophic lakes are those whose waters are always low in plant nutrients and usually highly oxygenated, and which have relatively small amounts of slowly decaying organic matter in their bottom deposits... mostly deep with steep sides.'

W. D. R. Hunter, 1958, in *The Glasgow Region*, 106 (contrast eutrophy).

**Ombrothermic diagrams or curves** (F. Bagnouls and Henri Gaussen in French as 'diagrammes ombrothermiques', 1957; in English by UNESCO, 1959)

UNESCO, *Arid Zone*, No. 3, March 1959. 'Ecological map of the Mediterranean area.... The *bioclimatic map* will be drawn up by the ombrothermic curve method, described by Mr. Gaussen...' (p. 5).

See *Annales de Géographie*, No. 355, LXVI$^e$ année, 1957, 193–220 where the construction of the diagram is described (194–195). 'Si, sur un graphique, on porte: en abscisses les mois de l'année; en ordonnées, à droite les précipitations (en mm.), à gauche, les températures (en °C.) *à une échelle double* de celle des précipitations, quand la courbe ombrique (de ombros = pluie) passe sous la courbe thermique on a P < 2T. La surface de croisement indique alors la durée et, dans une certaine mesure, l'intensité de la période sèche. Un tel graphique est appelé *diagramme ombrothermique*.'

In this paper a completely new classification of climate with new names (in French) is introduced: Érémique (hot desert); hemiérémique (hot semi-desert); xérothérique (long dry days); xérochimènique (short dry days); bixérique (two dry periods); axérique (without dry season) divided into thermaxérique (warm), mésaxérique (cool) and cryomérique (cold), with many sub-regions with new names or adjectival descriptions not yet used in English. Although the terms, to a Greek scholar, are self-explanatory, some become very cumbersome—such as thermoxérochimènique, or xérothermoméditerranéen. Many of the terms, in English, will be found in Vol. I of *Travaux de la Section Scientifique et Technique, Institut Français de Pondichery*, 1959, 155–196. (L.D.S.)

**Omuramba** (Herero)

Dry bed of a periodical river, mostly carrying water only for a few days or weeks every year, sometimes only as a series of pools and vleis. Usage limited to central and north-eastern part of South West Africa. (P.S.)

**Onion weathering**—see spheroidal weathering.

**Onomastics**

Aurousseau, M., 1957, *The Rendering of Geographical Names*, London: Hutchinson. 'The scientific study of names as names, that is, of the human habit of naming things, is the science of onomastics.' (p. 1) See also Toponymy.

**Oölite, oolite**

*Webster.* A rock consisting of small round grains, resembling the roe of fish, cemented together. The grains are small concretions. They are usually carbonate of lime, producing a variety of limestone, but sometimes of silica or iron oxide. b. Oölite (capital), the upper part of the Jurassic in England, named from the widespread oölitic texture of its limestones.

*Comment.* The individual concretions are sometimes called ooliths. Webster gives only oölite but the modern English usage is oolite. The *O.E.D.* definition has several errors. See also pisolite.

**Ooze**

*O.E.D.* 1. Wet mud or slime; especially that in the bed of a river or estuary. b. A stretch or extent of mud; a mud-bank; a marsh or fen, a piece of soft boggy ground. 2. Ocean-sounding. White or grey calcareous matter, largely composed of remains of Foraminifera, covering vast tracts of the ocean-floor.

Mill, *Dict.* A subaqueous deposit of finely divided matter more coherent and plastic than mud, such as is formed by accumulations of minute organic remains (globigerina, pteropod, radiolarian, and diatom oozes).

*Adm. Gloss.*, 1953. (As *O.E.D.* 1, and Mill, *Dict.*)

Sverdrup, H. U., and others, 1942, *The Oceans*, New York: Prentice-Hall. A pelagic deposit containing more than 30 per cent. of material of organic origin. Terrigenous deposits are never termed oozes. (p. 972)

*Comment.* In geographical literature the *O.E.D.* 1, has become obsolete. No reference in Page, 1865.

**Open**

1. A large cavern (also self-open). 2. A natural fissure (miner's term) (G.T.W.)

**Open field**

Before the inclosure of England, mainly in the 17th and 18th centuries, each village unit cultivated its arable land in two or three common or open fields. See field system.

**Opening**

*O.E.D.* U.S. A tract of ground over which trees are wanting or thinly scattered in comparison with adjoining forest tracts. Oak-opening. U.S. An opening or thinly wooded space in an oak-forest (*Webster*, 1864). Especially 'oak-opening'.

**Opencast mining**

*O.E.D.* Open-cast, -cut, in Mining, an open working.

1851. Greenwell, *Coal Trade Terms, Northumb. and Durh.*, 37. Opencast, a cutting in stone, coal etc., at the top or bottom of an excavation already made and open to that place.

*Comment.* The equivalent U.S. term is 'stripmining' because the unwanted overburden is stripped off to expose the coal or ironstone or other mineral. In Britain the term has come into common use because of the large scale working of coal made possible by the development of large earth-moving machinery which took place especially during the Second World War. The term is generally associated with extensive open workings whereas a quarry is more limited in extent. An old term in England is 'delving' and the workings were known as 'delphs' (G.T.W.)

**Opstal** (Afrikaans)

The farm buildings as distinguished from the farm lands. On most South African farms the opstal forms a group of buildings, a small inhabited spot in the midst of a large area of agricultural or grazing land. (P.S.)

**Optimizer models**

Wolpert, J., 1964, The Decision Process in Spatial Context, *Ann. Assoc. Am. Geographers*, 54. 'To a certain extent the optimizer concept has been introduced into economic geography by the tacit assumption that men organize themselves, their production, and their consumption in space so as to maximize utility or revenue. Whenever the analyst projects his knowledge of the best or most efficient location of economic activities into an explanation of the actual distribution of phenomena and finds correspondence, he assumes, even though tacitly, that man is rational and that his objective really corresponds with the subjective reality of the actor being observed.' (p. 544)

Haggett, 1965. 'The optimizer concept has been ... introduced into human geography through the assumptions of models like those of von Thünen, Weber, Christaller and Lösch, that individuals would arrange themselves spatially so as to optimize the given set of resources and demands.' (p. 26)

See Simon, H. A., 1957, *Models of Man*, New York: Wiley.

**Optimum Population**

Smith, T. L., 1948, *Population Analysis*, New York: McGraw-Hill, p. 4. '... that number of persons which combines with the other relevant factors to produce the highest possible level of living.... It is, of course, conceivable that in some countries the goal sought after may not be the "highest possible level of living". For such cases, the optimum population would more correctly be stated in terms of the national aspirations more generally held.' (p. 4)

Houston, J. M., 1953, *A Social Geography of Europe*, London: Duckworth. 'The population estimated to be desirable for the full utilization of the natural resources at an adequate standard of living.' (p. 247)

**Ore**

Solid, naturally occurring mineral aggregate of economic interest from which one or more valuable constituents may be recovered by treatment.

*Comment.* After a long discussion this definition was agreed by the Council of the Institute of Mining and Metallurgy (*Bull.*, 1955, iii).

**Orgues** (French)

*O.E.D. Obs.* (French; orgue) 3. An organ: a series of basaltic columns like organ-pipes.

Mill, *Dict.* A mass of vertical prisms having the appearance of organ pipes, due to the exceptionally rapid cooling of a lava flow (Auvergne).

*Comment.* The term has fallen into disuse.

**Orocratic Period**

Wills, L. J., 1929, *The Physiographic Evolution of Britain*, London: Arnold. 'Periods of profound and world-wide earth-movements, accompanied by vast volcanic activity' (p. 7). Between are 'pediocratic periods'.

**Orogenesis, orogeny, orogenic, orogenetic—** Mountain building

*Comment.* The normal practice is to use orogenesis for the process of mountain building and associated phenomena and to use orogeny for the great periods of mountain building in the earth's history. See Altaid, Alpine, Armorican, Caledonian, Charnian, Hercynian, Huronian, Laramide, Malvernian, also Epeirogenesis.

**Orography, orographical, orology**

*O.E.D.* That branch of physical geography which deals with the formation and features of mountains; the description of mountains.

Mill, *Dict.* Orography or Orognosy. The knowledge and description of the relations and developments of any particular mountain system. Orology. The theory of mountain formation (Penck).

Committee. List 1. 1. The description of the relief of the Earth's surface, or part of it. 2. The representation of relief upon a map or model.

Page, 1865. Orography, Orology (Gr. oros, a mountain)—The science which describes or treats of the mountains and mountain-systems of the globe—that is, of the profiles or elevations of the earth's surface.

**Orterde** (German; Soil science)

Jacks, 1954. 'Compacted B horizon of a podsol' (see Ortstein).

**Orthogneiss**

Himus, 1954. 'A gneiss which has resulted from the metamorphism of an igneous rock' in contrast to Paragneiss, a gneiss which has resulted from the metamorphism of detrital sediments.

**Orthophotomap**

British National Committee for Geography, 1966, *Glossary of Technical Terms in Cartography*, London: The Royal Society. 'A photomap produced by the transformation of the central projection of one or more air photographs to an orthogonal projection, so eliminating the distortion of perspective.' (p. 30)

**Ortstein** (German; Soil science)

Jacks, 1954. 'Hard cemented B horizon of a podsol' (see Orterde).

*Comment.* Robinson equates with hardpan; Orterde with claypan.

**Osar**

Adapted from the Swedish åsar, plural of ås,

a ridge, and anglicized as osar because å does not exist in English.

Mill, *Dict.* An esker.

Wooldridge and Morgan, 1937 'Eskers (comparable if not identical with the Swedish "osar") ...' (p. 387).

*Comment.* Scandinavian authors use 'eskers' when writing in English, apparently as equivalent to åsar.

**Ose**—See Ås, åsar

**Oshana** (Afrikaans)

Periodic river channel in Ovamboland. Travelling from west to east, just south of the Angola border, one encounters an amazing number of these channels at distances of at most a few miles apart. Only during the highest floods is there any sign of a slow southward current in them, and soon afterwards they become chains of standing pools which quickly dry away. The reason for the absence of a well defined main channel seems to lie in the extreme flatness of the land, which has hardly any general slope, though showing slight undulations. This is the principal difference between the Oshanas of Ovamboland and the Omurambas of Hereroland, which form clearly defined river courses, though rarely having flowing water. See J. H. Wellington, 1938, *S. Afr. Geog. Jour.* (P.S.) Called in Angola *mulolas*.

**Ouklip** (Afrikaans; *lit.* old rock)

A gravel, hardened into a kind of conglomerate. (P.S.)

**Outcrop**

The coming out of a bed of rock to the surface of the ground; the part of the deposit that appears at the surface.

**Outfield, out-field**—see Infield, also Breck

**Outlier**

*O.E.D.* 2.b. *Geol.* A portion or mass of a geological formation lying in situ at a distance from the main body to which it originally belonged, the intervening part having been removed by denudation. (quot. 1833, Lyell, Princ. Geol.)

Mill, *Dict.* A mass of newer stratified rock surrounded by other rocks.

*Comment.* By analogy used for other units isolated from the main mass, *e.g.* of population and also of hills separated from a main highland even if they are not geologically outliers. A *geological* outlier is not necessarily elevated above the surrounding country.

**Outport**

*O.E.D.* 1. A port outside some defined place as a city or town; in England a term including all ports other than that of London.

2. A port of embarkation or exportation.

Mill, *Dict.* A harbour situated nearer the sea, and consequently more accessible to vessels than the chief seaport to which it is subordinate.

Morgan, F. W., 1952, *Ports and Harbours*, London: Hutchinson. 'It is a product of the increase in the size of ships which progressively find it more difficult to reach a port some distance up a river or estuary. In order to prevent a loss of trade the port undertakes the development of an "outport" nearer the sea, which can attract the larger vessels. ... A special use of the word outport occurs in Great Britain among shippers, exporters etc., who refer to all British ports outside London as "the outports", as in the eighteenth century.' (p. 76)

No reference in Sargent, 1938; Van Cleef, 1937.

*Comment.* The normal usage in geographical literature is that given in the first part of the quotation from Morgan. The definition in *O.E.D.* is archaic, though perpetuated by the special case of British export shippers.

**Outwash fan, plain**

Mill, *Dict.* A plain composed of material washed out from a glacier or ice-cap. See Valley train; also Flint, 1947, p. 133.

**Ouvala**—see Uvala

**Ova** (Turkey)

Sunken basin, occupied by marshes and mudflats. (W. B. Fisher)

**Overburden**

L.D.G. In opencast or strip mining, the overlying soil and rock which has to be removed before the seam of coal or bed of ore is exposed is called the overburden.

**Overfit river**

See Davis, W. M., 1913, Meandering Valleys and underfit rivers. *A.A.A.G.*, 3, 38.

**Overflow channel**

L.D.G. The Channel by which water escaped from a lake at a former period of higher water—commonly used of lakes held up by ice in the Great Ice Age.

**Overfold**

Webster. *Geol.* a. An overturned anticline,

or an anticline of which one limb is inverted and lies beneath the other. b. A sigmoid fold comprising an overturned anticline and a syncline.

See quotation under Holmes, 1944, Thrustplane.

**Overlap**
In stratigraphy the extension of one bed or stratum beyond that of the bed below which happens when deposits are being laid down on a surface sinking below the sea (in this sense = overstep).

**Overpopulation**
Smith, T. L., 1948, *Population Analysis*, New York: McGraw-Hill. Where population is in excess of optimum (*q.v.*); see pp. 4–5, 388–389.

Houston, J. M., 1953, *A Social Geography of Europe*, London: Duckworth. 'The condition of rural population when the redundancy of agricultural labour and therefore avoidable poverty indicates a population excessive to the existing standard of living, the economy practised and the natural resources.'

*Comment.* This is an economist's concept and does not apply necessarily where population is greatly concentrated but only where it exceeds the optimum and thus reduces average income per head (W.G.E.). On the other hand income per head seems as convenient a measure as practicable (C.J.R.). 'Overpopulated' is often used very loosely when simply 'crowded' or 'densely populated' is meant. There is no simple French word for 'crowded' which tends to get translated as '*surpopulé*' (overpopulated).

**Ox-bow, Ox-bow Lake**
*Webster.* 1. A frame, bent into the shape of the letter U, and embracing an ox's neck as a kind of collar. 2. Anything shaped like an oxbow; specifically, *Phys. Geog.* a river bend such that only a neck of land is left between two parts of the stream. The river may cut through, leaving a crescent-shaped lake. *U.S.*

*Dict. Am.* 'The term oxbow seems to have acquired in New England these two meanings: (1) the bend or reach of a river, and (2) the land enclosed, or partially enclosed, within such bend. The oxbows of the Connecticut River are well known ... having long been famous; and it is in that region that the term appears to have arisen.' (1896. A. Matthews in *Nation*, 23 July 65. Other quotations date from 1856.)

Wooldridge and Morgan, 1937. 'Flood alluvium presents no considerable resistance to the growth and movement of meanders and the short-circuiting and abandonment of meander loops in time of flood is a frequent occurrence on flood-plains. The abandoned loops form "cut-offs", "ox-bows" or "mortlakes" which, in time, become silted up....' (p. 173)

**Ox-gang** (Agricultural History)—see under Yardland.

# P

**Pa** (New Zealand)
A Maori fortified village as distinct from *Kainga*, an unfortified Maori settlement.
**Pacific Suite**—see **Atlantic Suite**
**Pacific Type Coast**—see **Longitudinal Coast**
**Pack ice**
*O.E.D.* Ice forming a pack. A large area of floating ice in pieces of considerable size, driven or 'packed' together into a nearly continuous and coherent mass (as found in polar seas).
Mill, *Dict.* The broken ice of floes driven together by winds and currents.
*Adm. Gloss.*, 1953. Any area of sea ice other than fast-ice, no matter what form it takes or how it is disposed. Pack-ice cover may be reported in tenths (see Ice Cover), or may be described as Very Close pack-ice, 10/10ths with very little if any water visible; Close pack-ice, 8/10ths to 9/10ths; Open pack-ice, 4/10ths to 7/10ths; and Very Open pack-ice, 1/10th to 3/10ths.
*U.S. Ice Terms*, 1952. Any large area of floating ice driven closely together. Note also: Packed ice: close ice—ice covering from eight to ten-tenths of a sea water area.
Bencker, H., 1931, Ice Terminology, *Hydrographic Review*, 8. Term used in a wide sense to include any area of sea ice, other than fast ice, no matter what form it takes or how disposed. Getting through is possible only at leads or other open spaces. Pack ice is a large collection of pieces of ice from broken-up floes or icebergs, which have to a certain extent closed together again. The principal difference between pack ice and floe ice is that for the formation of the pack the presence in the locality of polar ice of many years' standing is necessary, whilst floe ice and ice fields can be formed from one year ice.

**Packet, packet-boat**
*O.E.D.* Packet-boat. A boat or vessel plying at regular intervals between two ports for the conveyance of mails, also of goods and passengers: a mail-boat. Originally the boat maintained for carrying 'the packet' of State letters and dispatches. Often shortened to packet.

*Comment.* Obsolescent. It would be more usual to say ferry or ferry-boat.
**Packet port, packet station**
Morgan, F. W., 1952, *Ports and Harbours*, London: Hutchinson. 'Packet stations or ferry ports.... Train ferry ports are an even more specialized form of the packet station.' (p. 72)
No reference in Sargent, A. J., 1938.
**Padang** (Malay)
Dobby, E. H. G., 1957, *Malayan J. of Tropical Geog.*, 10. 'Treeless waste land; (loosely) treeless plain; (Kelantan) unusually wide and level stretches of sandy terrain' (p. vii).
*Comment.* P. W. Richards notes (in MS.), 'Malay equivalent of savanna, but often used for the peculiar types of scrub-like and heath-like vegetation found on podsolic (kerangas) soils in Borneo, Sumatra etc.'
**Paddock**
*S.O.E.D.* A small field or enclosure; usu. a plot of pastureland adjoining a stable.
*Comment.* Used in New Zealand for a grass field as opposed to arable.
**Pa-deng** (Thai)
A dry deciduous open forest in eastern Thailand consisting mainly of *Dipterocarpus tuberculatus* and *Pentacme suavis* equivalent to the in-daing of Burma.
**Padi** (Malay; commonly anglicized as paddy)
Dobby, E. H. G., 1957, *Malayan J. of Tropical Geog.*, 10. 'Rice (i) as a plant; (ii) in the ear; (iii) as unhusked grain.'
Hence paddy field, paddy cultivation, paddy harvest etc.
*Comment.* The term *rough rice* was formerly used internationally and by the U.S. Dept. of Agriculture for unhusked rice. FAO now uses *paddy* as against milled rice when the husk has been removed. (C.J.R.)
**Padiny** (Russian)
Depressions in dry steppe lands in which snow, rain and melt-water collect to promote soil formation with humus, permitting patches of cultivation. V. A. Kovda in MS., also in *History of Land Use in Arid Regions*, Paris: UNESCO, 1961.
**Pagoda** (Portuguese)
*O.E.D.* 'apparently a corruption of a name found by the Portuguese in India ... a

temple or sacred building (in India, China and adjacent countries); especially a sacred tower, built over the relics of Buddha or a saint, or in any place as a work of devotion.

*Comment.* In recent years especially used of the Buddhist shrines of Burma (Burmese: dagon) which are found in immense numbers all over the country and nearly all of characteristic form.

**Pahoehoe** (Hawaiian; *pron.* pa-hó-e-hó-e)

Holmes, 1944. 'The surfaces of newly consolidated lava flows are commonly of two contrasted types, described in English as *block* and *ropy* lavas, but known technically by their Hawaiian names *aa* and *pahoehoe* respectively' (p. 448).

Cotton, 1944, Volcanoes. 'When pyromagma escapes as a lava flow and its chilled surface becomes a "pahoehoe" skin (dermolithic solidification) the lava retains gas in solution as well as encircling it in round vesicles (frozen bubbles)...' (p. 27). See references under *aa* with Dana's use, 1887.

**Paint-pot**

A coloured mud-well usually associated with geysers and hot springs, as in Yellowstone National Park. Rising water throws out pink, blue, yellow, red and white mud, generally when there is too little water to form a spring and the surrounding rocks have been weathered to clay with compounds of iron.

Davidson, C. F., 1949, *A Prospector's Handbook to Radioactive Minerals*, London: H.M.S.O. 'Paint—A very thin smear of ore-mineral on a rock surface.' (p. 27)

von Engeln, 1942. 'Rise of less violent steam currents cooks the rock to pools of thin mud, called *paint pots* in recognition of the vivid colors that are induced by the chemical reactions incidental to the cooking.' (p. 611)

**Pakeha** (New Zealand: Maori)

A white New Zealander, as contrasted with a Maori. *Lit.* a stranger.

**Pakihi** (New Zealand: Maori)

A water-logged gravel flat.

**Pakka**—see pukka

**Pāko** (Pakistan: Sindhi)

Alluvial land in the lower Indus valley outside the protection bunds. (A.H.S.)

**Palaeoecology, paleoecology**

The ecology of past geological periods.

**Palæogeography, paleogeography**

Wills, L. J., 1948, *The Palaeogeography of the Midlands*, Liverpool: U.P. of Liverpool. 'Palaeogeography is an attempt to interpret the geological record as a succession of physical geographies.' (p. 1)

Rice, 1941. The geography of some former geological epoch, or of former geological time in general (*Webster*).

Moore, 1949, 'The study of the distribution of land and water etc., during earlier periods of the earth's history.'

*Comment.* Palaeogeographical maps are those which seek to show a reconstruction of past geological times. Of necessity they can only be drawn with reference to existing geographical outlines and it is difficult to indicate adequately the degree of deformation due to mountain building. See L. J. Wills, 1951, *Palaeogeographical Atlas of Europe* and the series of maps in L. D. Stamp's *Britain's Structure and Scenery*, Collins, 1946.

**Palaeolithic** (John Lubbock, 1865), **paleolithic**

Lubbock, John (later Lord Avebury), 1865, *Prehistoric Times*. The Old Stone Age (of chipped stone implements); contrast Neolithic.

**Palaeomagnetism**

Blundell, D. J., and Read, H. H., 1958. Palaeomagnetism of the Younger Gabbros of Aberdeenshire. *Proc. Geol. Assoc.*, 69, 191–204. 'At the beginning of this century it was discovered that igneous rocks possessed a remanent magnetisation of great stability. Konigsberger in 1938 was the first to point out that it was probably acquired when the rock cooled and that it was directed along the earth's field acting at that time.' (p. 191)

See also Archaeomagnetism.

**Palaeontology, paleontology**

The study of fossils, more especially of fossil animals, the study of fossil plants being designated palaeobotany.

**Pale**

*O.E.D.* A district or territory within determined bounds, or subject to a particular jurisdiction.

*Comment.* As a general term obsolete, but used especially in reference to that part of Ireland, of varying extent, over which English jurisdiction was established and known as the English Pale or, simply, The Pale. A Hebrew Pale was set up by the Russian Government in the eighteen-eighties. (W.G.E.) Note also connection with palings (for a fence; indeed 'pale' meant the surround of the early parks) and the expression 'beyond the pale'.

**Palevent** (L. J. Wills, 1956)
Abbreviation for palaeogeographical event.

**Pālēz** (Pakistan: Pashto)
Generic term for cucurbitaceous crops but used also for their bed. (A.H.S.)

**Palichnology** (Seilacher, 1964)
Seilacher in *Approaches to Paleoecology*, New York, Wiley, 1964, 'the study of fossil tracks, wiggles and burrows' (*Science*, **147**, 1965, p. 593).

**Palisade**
A high fence used, as in early European settlements in America, as a protection against wild animals and attackers, hence transferred to unscaleable cliffs so that *A.G.I.* gives 'A picturesque, extended rock cliff rising precipitately from the margin of a stream or lake and of columnar structure.'

**Palsa** (Swedish), pals
From Finnish palsa = elliptical; dome-like peat hillocks of varying form and size, up to 3 m. and more high (hillocks of 7 m. have been recorded too) with a base of 2–25 m. often surrounded by open water, found in northernmost Sweden, Finland and the subarctic elsewhere. Palsa as an equivalent of pingo (*q.v.*) is incorrect though Palsenfelder has been translated as 'Pingo fields'. J. B. Bird (in MS.) considers Thornbury is wrong in equating earth hummock and palsa.

Thornbury, 1954. 'On surfaces covered by a good growth of tundra vegetation low, rounded mounds composed of fine materials are often found. They are called *earth hummocks* or *palsen*.' (p. 89)

**Palsa bog**
Hare, F. K. 1960, *A Photo-Reconnaissance Survey of Labrador-Ungava*, Mem. 6. Geog. Branch, Ottawa, 'In these bogs, ice-heaved mounds of lichen-covered peaty soil rise a few feet above the wetter ground.' (p. 32)

**Palstage** (L. J. Wills, 1956)
Abbreviation for palaeogeographical stage.

**Palynology**
Godwin, H., 1956, *The History of the British Flora*, Cambridge: C.U.P. 'In many ways the most important advance has been the development of the technique of pollen analysis or "palynology" in application to the field of Quaternary investigations.' (p. 2)

*Comment.* Recently extended to cover the investigation of minute organic bodies highly resistant to acids found in sedimentary rocks of all ages.

**Pampa** (South America: Spanish; an adaptation of Quechua *bamba*)
*O.E.D.* The name given to the vast treeless plains of South America south of the Amazon, especially of Argentina and the adjacent countries. The similar plains north of the Amazon are known as llanos.

Mill, *Dict.* Pampas: The characteristic grasslands of Argentina.

Pampa: Applied to a stretch of level ground even if barren rock desert (Western South America).

Küchler, A. W., 1947, Localizing Vegetation Terms, *A.A.A.G.*, **37**. '... is broadly applied to the entire grassland that stretches from southern Brazil through Uruguay into the heart of Argentina.' (p. 202)

James, 1959, 'The third major division of the country (Argentina) is the *Pampas*—the great plains ... originally covered with a growth of low scrubby trees and grasses, a vegetation type known as *monte*; but toward the southeast tall prairie grasses were once probably more important than the monte. It is customary to divide the Pampas into the Humid Pampa and the Dry Pampa.' (p. 295)

*Comment.* Whatever the original meaning, which may have been the high grass-covered intermontaine plains of the Andes, in English textbooks Pampa or Pampas implies the South American temperate or mid-latitude *grasslands*. James avoids this, describing South American grassland as steppe and prairie. (p. 39)

**Pampero** (Spanish)
*O.E.D.* A piercing cold wind which blows from the Andes across the S. American pampas to the Atlantic.

Mill, *Dict.* Pamperos: Strong south-west winds in the pampas area and on the adjoining coasts.

*Met. Gloss.*, 1944. A name given in the Argentine and Uruguay to a severe storm of wind, sometimes accompanied by rain, thunder and lightning. It is a line-squall, with the typical arched cloud along its front. It heralds a cool south-westerly wind in the rear of a depression; there is a great drop of temperature as the storm passes.

**Pan** (Soil science)—see Hard Pan

**Pan** (Afrikaans; *pl.* panne)
Shallow hollow, usually of rounded form, occurring in arid and semi-arid regions, and now and then holding water during the rainy season. In some cases pans keep water all through the year. The difference between a pan and a lake is (a) less permanence (b) less depth. (P.S.)
For a classification and discussion of origin see Wellington, J. H., 1955, *Southern Africa*, I, 474–486.

**Pandy** (Welsh)
Fulling mill

**Panfan** (A. C. Lawson, 1915)
Lawson, A. C., 1915, The Epigene Profiles of the Desert, *U. of Calif. B. Dept. of Geology*, 9. The end stage in the process of geomorphic development in a desert in which relief features have been reduced to a 'sub-alluvial bench' buried beneath alluvial fans (pp. 23–45). 'The surface thus evolved is, in its ideal completion, wholly one of aggradation, a vast alluvial fan surface to which for convenience in discussion I propose to give the name panfan.... The panfan may be regarded [as] an end stage in the process of geomorphic development in the same sense that the peneplain is an end stage of the general process of degradation in a humid climate ... the panfan closes a cycle of degradation and aggradation, is evolved by both cutting and filling, and is a built surface.' (p. 33)
Wooldridge and Morgan, 1939, quote Lawson, p. 315.
No reference in Cotton, 1922, 1941, 1945; Holmes, 1944; Lake, 1952; Strahler, 1951; Moore, 1949.

**Pannage**
*O.E.D.* 1a. The feeding of swine (or other beasts) in a forest or wood; pasturage for swine; b. The right or privilege of pasturing swine in a forest; c. The payment made to the owner of a woodland for this right; the profit thus accruing. 2. Acorns, beechmast, etc., on which swine feed.

**Panneveld** (Afrikaans)
Sub-humid region covered with a great many pans. The run-off collects in these hollows, where it evaporates, thus preventing the formation of an organized drainage system. De Martonne called this condition 'humid areism'; he might also have called it 'small scale endoreism.' (P.S.)
Wellington, J. H., 1955, *Southern Africa*, I, 'An area in which a large number of pans are found is generally referred to as "panneveld".' (p. 474)

**Panplain, panplane, panplanation** (C. H. Crickmay, 1933)
Crickmay, C. H., 1933, The Later Stages of the Cycle of Erosion, *Geol. Mag.*, 70, 337–347. 'This plain, formed of floodplains joined by their own growth, may be called a panplain' (p. 345). It is the end stage of a cycle of erosion, but differs from a peneplain in being the result of lateral erosion or corrosion rather than the wasting of divides. (W.M.)
Wooldridge and Morgan, 1937. Discussion of Crickmay, pp. 185–187.
Wooldridge in Taylor, 1951. 'Panplanation— the integrated product of long-continued lateral corrasion by rivers.' (p. 170)
No reference in Cotton, 1941, 1945; Lake, 1952; Strahler, 1951; Holmes, 1944; Moore, 1949.
See also Panfan, Peneplain.

**Pantanal** (Portuguese: Brazil)
Cole, M. M., 1960, Cerrado, Caatinga and Pantanal..., *Geog. Jour.*, 126, 168–179. 'A mixture of grassland and trees... fascinating in its complexity ... a complex in which the several types of savanna and the mata are all represented.' Found on the flood plain of the Paraguay.

**Papa** (New Zealand: Maori, *lit.* earth)
Soft mudstones, siltstones and sandstones of Tertiary age covering large areas in the North Island.

**Papagayo** (Spanish)—see Norte

**Parabolic dune**
A crescentic dune facing the wind, *i.e.* the opposite to a barchan.

**Paradelta**
Not in *O.E.D.* Para, alongside of, beyond.
Strickland, C., 1940, *Deltaic Formation*, Calcutta: Longmans. 'As then it has been found impossible to define satisfactorily the term "delta", and the sequel will show that it is very essential to have epithets by which to denote the two opposing tracts, where respectively deposition or erosion has prevailed, these will hereafter be named Delta and Paradelta.' (p. 10)
Spate, 1954. 'The main bulk of the region is taken up by the true Delta and the great mass of alluvial fans—Strickland's paradelta—to the N' (p. 522). Spate's telescoped sentence does not make it entirely clear that he is referring to the northern

part of the great Ganges-Brahmaputra deltas region where the forces of corrasion are at work in contrast to the true delta in the south where deposition is active.

**Paradigm**
*O.E.D.* 1. A pattern, exemplar, example.
Kuhn, T. S., 1962, *The Structure of Scientific Revolutions*, Chicago: University of Chicago Press. 'In its established usage, a paradigm is an accepted model or pattern ... the paradigm functions by permitting the replication of examples any of which could in principle serve to replace it. In a science, on the other hand, ... it is an object for further articulation and specification under new or more stringent conditions. Paradigms gain their status because they are more successful than their competitors in solving a few problems that the group of practitioners has come to recognize as acute.' (p. 23)

**Paragneiss**—see Orthogneiss
**Parallax**
*O.E.D.* (*Astron.*). Apparent displacement, or difference in the apparent position, of an object, caused by actual change (or difference) of position of the point of observation; *spec.* the angular amount of such displacement or difference of position, being the angle contained between the two straight lines drawn to the object from the two different points of view, and constituting a measure of the distance of the object.

**Parallel, parallel of latitude**
A line of latitude is often referred to as a parallel of latitude because each is a line encircling the earth parallel to the equator, the circle getting steadily smaller from the equator towards the poles.

**Parameter**
Kendall, M. G. and Buckland, W. R., 1960, *A Dictionary of Statistical Terms*, 2nd ed., Edinburgh: Oliver and Boyd. 'This word occurs in its customary mathematical meaning of an unknown quantity which may vary over a certain set of values. In statistics it most usually occurs in expressions defining frequency distributions (population parameters) or in models describing a stochastic situation (e.g. regression parameters). The domain of permissible variation of the parameters defines the class of population or model under consideration.' (p. 211)

**Parametric tests**
Gregory, S., 1973, *Statistical Methods and the Geographer*, 3rd ed., London: Longman. '... can only be used for interval data ... By the word "parameter" we imply some quality, characteristic or value of the population data, not of the sample data. Parametric tests are therefore those that assume certain conditions in the population, and which are relevant only if such assumptions are valid.' (p. 132)
See also Non-parametric tests.

**Paramo** (Spanish), strictly páramo
*O.E.D.* Paramo (Sp. paramo, apparently from a native language of Venezuela). A high plateau in the tropical parts of South America, bare of trees, and exposed to wind and thick cold fogs.
Mill, *Dict.* 1. The high, dry plateaus of Mesozoic calcareous rock on the eastern side of Spain. 2. The interior tablelands above the limit of cultivation from about 10,000 feet to the snowline. They are mostly treeless, but in places have isolated gnarled dwarf trees (Spanish America).
James, 1959. 'Above the upper limit of forests and of agriculture but below the lower limit of permanent snow is the "Zone of Alpine Meadows", to which the people in northern South America apply the term paramos ... approximately 10,000 feet to the snow line' (p. 81). 'Mountain tall grass' (p. 39).
Shackleton, M. R., 1958, *Europe*, 6th ed., London: Longmans. 'The Northern Meseta [of Spain] ... the harder rocks, among which limestones predominate, form higher and drier tablelands, called paramos, often terminating in a cuesta or scarp overlooking the cereal lands.' (p. 93)

**Parautochthonous**—see Autochthonous
**Pargana** (Indo-Pakistan)
In most parts of northern India and east Pakistan the smallest administrative unit — a subdivision of a tahsil (*q.v.*).

**Parhelion**
*O.E.D.* A spot on a solar halo at which the light is intensified.

**Paring and burning**
A method formerly much used in Britain of preparing peatlands for cultivating whereby a layer of peat was pared off from the surface and then dried and burnt, crops being grown in the underlying peat or soil enriched by the ash. The value of the method was much disputed.

**Parish**
*S.O.E.D.* 1. In the United Kingdom, the name of a subdivision of a county. a. *orig.* A township or cluster of townships having its own church and clergyman, to whom its tithes and ecclesiastical dues are (or were) paid. b. A later division of such a parish having its own church. 2. A district, often identical with an original parish for purposes of civil government (*civil parish*). Other meanings elsewhere.

**Park, parkland (1)**
*S.O.E.D.* 1. *Law.* An enclosed tract of land held by royal grant or prescription for keeping beasts of the chase. (Distinguished) from a *forest* or *chase* by being enclosed, and from a *forest* also by having no special laws or officers. 2. Hence, a large ornamental piece of ground usually comprising woodland and pasture, attached to a country house or mansion, and used for recreation and often for keeping deer, cattle, etc. 3. An enclosed piece of ground, within or near a city or town, laid out and devoted to public recreation, a 'public park'. 4. An extensive area of land set apart as a national property to be kept in its natural state for the public benefit, as the *Yellowstone Park* in the U.S. 1871. 5. In Ireland, Scotland, etc. an enclosed piece of ground for pasture or tillage, a field, a paddock. 6. In Colorado, Wyoming etc. a high plateau-like valley among the mountains.

**Parkland (2)**
From the analogy of a typical English park of grassland with scattered trees has been used to describe tropical savanna lands especially in Africa of grassland with scattered trees.

**Parna** (Australia: aboriginal)
Butler, B. E., 1956, Parna—An Aeolian Clay, *Aust. J. Sci.*, **18**, 6, 145–151. Aboriginal word for sandy and dusty ground. Parna consists of aggregates of clay particles which have been completely windborne (T. Langford-Smith). Also parna sheet, parna aspect, duneparna etc. in Australian literature.

**Pass**
*O.E.D.* A way or opening by which one passes through a region otherwise obstructed or impassible or through any natural or artificial barrier, *esp.* a narrow and difficult or dangerous passage through a mountainous region or over a mountain range.

**Pasture**
*O.E.D.* The growing grass or herbage eaten by cattle.
A piece of land covered with grass used or suitable for the grazing of cattle or sheep; grass land; a piece of such land.
U.S. (a) that part of a deep-water weir which the fish first enter (*Cent. Dict.*, 1890). b. An inshore spawning-ground for codfish (Funk's *Standard Dict.*, 1895).
Committee, List 3. Grazing (*q.v.*); in Britain usually grassland.
Stamp, 1948. '. . . the farmer generally restricts "meadow" to land mown for hay; "pasture" to land which is grazed.' (p. 369)
Stroud, 1952. '"Pasture" is a general name for herbage, acorns, mast, and nuts, and for leaves and flowers, and for all things comprised under the name of pannage. (Britton, 1.2, Ch. 24, 1 Nichols's Ed., 371).'
*Comment.* The word is derived ultimately from the Latin, *pascere*, to feed, graze, attend to the feeding of beasts. Note the verb to pasture animals, *i.e.* to graze animals or put animals out to graze for which some writers prefer to use the somewhat pedantic 'depasture' with the same meaning.

**Pasturegate**—see **Cattlegate**

**Pat** (Pakistan: Sindhi)
Arid clay plain (A.H.S.). See also Spate, 1954, 454, 567 and 586 with meaning of little plateaus with steep sides (in Chota Nagpur).

**Paternoster lakes**
Cotton, 1942. Strings of lakes in glacial valleys dammed by morainic ridges or rock bars. (p. 256)
Thornbury, 1954. 'Descent of trough floors takes place in a series of glacial steps . . . *riser . . . riegel*, a sort of rock bar . . . *tread* . . . a reversed up-valley slope with resulting chain of lakes, which because of the similarity of their arrangement to beads on a rosary are called *paternoster lakes*.' (pp. 369–370)

**Patina, desert crust, desert varnish**
Stamp, 1953, *Africa*, New York: Wiley. 'The sun, bringing to the surface of the rocks, by capillary action, any moisture contained in the surface layer, causes the water to evaporate, and a thin hard skin of iron and manganese salts is formed to protect the rock surface. This desert *patina* or

crust is often black and suggests that fire has passed over the rocks. The process is not fully understood, and it is probable that some of the moisture comes from dew.' (p. 257)

**Pavement**
By analogy with the paved surface of a road, yard, etc. consisting of blocks of stones fitted closely together, pavement is used loosely for bare rock surfaces of varied origin, especially if more or less horizontal —desert pavement, limestone pavement, boulder pavement or lag gravel etc.

**Pawindāh, pavindā** (Pakistan: Pashto)
Afghan migratory traders who come to West Pakistan during winter. (A.H.S.)

**Pays** (French)
Mill, *Dict*. A French word meaning 'country', frequently used for a region of considerable historical unity, *e.g.* one of the old provinces of France, with special reference to its physical unity, as a natural region.
*Comment.* The concept of the 'pays' is firmly entrenched in French national thinking. Because the 'pays' is in essence a region, frequently with no administrative significance, demarcated by varied and sometimes indefinable features, many writers in English prefer to retain the French word rather than use 'natural region' which involves immediate discussion as to the meaning of 'natural'. Though very widely used the word is not in the *O.E.D*. For early use by geologists see *Geo. Jour*., **2**, 1893, 53. Hatzfeld and Darmestetter, 1932 give 1. Territoire d'une nation ... 2. Spécialt, Famil. Localité ou qqn est né.

**Paysage** (French)
*O.E.D.* A rural scene, landscape.
*Comment.* Occasionally used by writers in English who wish to avoid the controversies centring round the word 'landscape' (*q.v.*). *Paysage humanisé* is also used. See also Pays.

**Peak**
*O.E.D.* Later form of Pike, the pointed top of a mountain; a mountain having a more or less pointed summit, or of conical form. Hence *fig.* highest point, summit.

**Peak-plain**
Ängeby, O., 1955, *Medd. Lund Univ. Geog. Inst.*, Av. 30, p. 30 in an English summary gives peak-plain as equivalent to German *Gipfelflur*. See summit-plain, accordance of summit levels, Gipfelflur.

**Peasant**
*S.O.E.D.* One who lives in the country and works on the land; a countryman, a rustic. (In early use prop. only of foreign countries, antithetical to *noble*.)
*Comment.* In such phrases as peasant-farming implies small scale, family unit for subsistence as contrasted with large scale or commercialized.

**Peat**
*O.E.D.* 1. (With a and pl.) A piece of the substance described in sense 2, cut of a convenient form and size for use as fuel, usually roughly brick-shaped. (Chiefly Scottish and Northumbrian dialects.)
2. Vegetable matter decomposed by water and partially carbonized by chemical change, often forming bogs or 'mosses' of large extent, whence it is dug or cut out, and 'made' into peats (in sense 1).
Mill, *Dict*. A deposit of more or less altered vegetable matter which has accumulated through decay of plants under water.
Tansley, A. G., 1939, *The British Islands and their Vegetation*, Cambridge: C.U.P. Distinguishes 'fen peat', with an alkaline reaction, structureless and black, rich in ash and proteins, but with no cellulose left, and 'bog or moss peat', which is highly acid in reaction, rich in cellulose and humicelluloses, is often brown in colour and with plant structure more or less well preserved. The former bears fen vegetation dominated by sedges, grasses and rushes, the latter carries bog moss or sedges. (p. 99) (W.M.)
Jacks, 1954. Slightly decomposed vegetable matter accumulated in water.
Himus, 1954. An accumulation of plant debris that has remained incompletely decomposed principally due to lack of oxygen ... first stage in the series lignite, brown coal, bituminous coal. ...

**Peat-hag, Peat Hag, Peat Hagg, Hagg**
More or less vertical bank cut in upland peat by wind or water erosion. (A.C.O'D.)
Tansley, 1939. 'Peat "hagging" by the formation of deep cracks and channels is due, according to Fraser, to the cutting away of the peat along drainage channels and subsequent erosion of their sides.' (p. 708)
*Comment.* Tansley gives 'hagg' but 'hag' is more usual.

**Pebble**
*O.E.D.* A stone worn and rolled to a rounded

form by the action of water; usually applied to one of small or moderate size, less than *a boulder* or *cobble*.
*Comment.* See cobble. Sometimes wrongly applied as angular pebbles to the fragments in a breccia.

**Ped-**
From the Greek πιδον the ground, has come to be used in combination as related to soil (see Pedography, Pedology, Pedon etc., also Soil science). The word *ped* has been used by some writers for a soil particle or crumb.

**Pedalfer**—see Soil classification

**Pedestal, pedestal rock**
*Webster.* Pedestal rock. A residual or erosional rock mass balanced upon a relatively slender neck or pedestal.
Thornbury, 1954. 'Differential weathering helps to develop and modify such upstanding topographic forms as columns, pillars, pedestal rocks, earth pillars, and toadstool rocks which, because of their bizarre shapes, are sometimes collectively classed as *hoodoo rocks*' (p. 70). *Pedestal rocks*, which consist of residual masses of weak rock capped with harder rock, have been explained by wind abrasion. It is doubtful whether this process has contributed significantly to their shaping. They are more commonly the result of differential weathering aided by rainwash.' (p. 301)
Dury, 1959. 'Pedestals, reminiscent of seastacks sapped at the base by waves, are the most striking results of undercutting [by desert winds].' (pp. 191–192)

**Pediment**
*O.E.D.* Referred to Latin pes, pedem, foot, and used for: A base, foundation; a pavement.
Committee, List 2. An eroded rock platform of considerable extent at the foot of an abrupt mountain slope or face—characteristic of arid and sub-arid regions.
Bryan, K., 1925, The Papago Country, Arizona, *U.S. Geol. Survey, W.S. Paper* 499. '"Mountain pediment" has been chosen as the name for such a plain of combined erosion and transportation at the foot of a desert mountain range. The plain ordinarily surrounds and slopes up to the foot of the mountains, so that at a distance the mountains seem to be merely ragged projections above a broad triangular mass—the pediment or gable of a low-pitched roof. This metaphor was first used by McGee ... also used by Weed. ... "Mountain pediment" as a term replaces "sub-aerial platform" and in part "sub-alluvial bench" as defined by Lawson.' (pp. 93–97)
Dury, G. H., 1959. 'The broad gentle slopes, identified with the waning slope (*q.v.*) are called *pediments* after the architectural features of the same name.' (p. 64)
*Comment.* There is considerable controversy as to the origin of pediments and to their outline, *i.e.* whether they are level parallel to the mountain foot (Blackwelder) or consisting of low semi-cones (Johnson etc.).
Refs. include: Davis, W. M., 1930, Rock Floors in Arid and in Humid Climates. *Jour. of Geol.*, **38**, 1–27, 136–158; Blackwelder, E., 1931, Desert Plains. *Jour. of Geol.*, **39**, 133–140; Johnson, D., 1932, Rock Plains of Arid Regions, *Geog. Rev.*, **22**, 656–665; Wooldridge and Morgan, 1937, pp. 312–317; King, L. C., 1951, *South African Scenery*, 2nd ed.

**Pediocratic Period**
Wills, L. J., 1929, *The Physiographical Evolution of Britain*, London: Arnold. The relatively quiet period between two orocratic periods. (*q.v.*)

**Pediplain, pediplane, pediplanation** (Maxson and Anderson, 1935, and Howard, 1942)
Maxson, J. H., and Anderson, G. H., 1935, Terminology of Surface Forms of the Erosion Cycle, *Jour. of Geol.*, **43**. At the mature stage of the arid cycle, or an arid matureland: 'The recession of front and reduction of mountain mass proceeds until only narrow ridges and "inselberge" are left, surrounded on all sides by rock-cut floors which in turn are overlapped by alluvial deposits covering earlier-cut portions of the floor and the intermont basins. Widely extending rock-cut and alluviated surfaces of this type formed by the coalescence of a number of pediments and occasional desert domes may be called "pediplains." ... When the surface consists for the most part of narrow ridges and pediplains, the region may be considered to have reached old age, and its characteristic surface may be called an "arid oldland".' (p. 94)
Howard, A. D., 1942, Pediment Passes and the Pediment Problem, *Jour. of Geomorphology*, **5**. 'The writer proposes the term pediplane as a general term for all degradational piedmont surfaces produced

in arid climates which are either exposed or covered by a veneer of contemporary alluvium no thicker than that which can be moved during floods.' He quotes Maxson and Anderson, 'pediplain' and writes: 'As thus defined the term seems inappropriate inasmuch as the complex surface to which it is applied is not as a whole at the foot of a slope and is not a "plain" in the usual geomorphic sense.' 'Pediplanation may be applied as a general term to the process of formation of pediplanes. The term peneplanation would still apply to the degradation of the region as a whole.' 'The term peripediment is suggested for a pediplane which bevels an earlier basin fill. If both elements of the pediplane are present, the peripediment is always beyond and peripheral to the pediment.' (p. 11)

King, L. C., 1951, *South African Scenery: a Textbook of Geomorphology*, Edinburgh: Oliver and Boyd. 'The formation of erosion surfaces by scarp retreat and concomitant production of pediments.' (p. 240): quoted in the review by F. Dixey, *Geog. Jour.*, **118**, 1952, 75–77.

Pediplane is cited and pediplain is used in Thornbury, 1954, 292.

### Pedocal—See Soil classification.

### Pedogenesis
Jacks, 1954. 'The formation of soil from parent material.'

### Pedogenics, pedogenic
Pedogenics—the study of the origins and development of soils; found mainly as the adjective pedogenic or soil-forming. See pedology.

### Pedogeography
The geography of soils; the study of the geographical distribution of soils.

### Pedography
The description of soils.

### Pedology
Mill, *Dict.* 'The study of soils' Soil science.

*Comment*. From the Greek πέδον the ground and not to be confused with pædology, the study of the nature of children. The first reference in *O.E.D. Suppl.* to pedology is 1925 but Mill evidently knew of its use in 1900. (L.D.S.)

### Pedon (R. W. Simonson and D. R. Gardiner, 1960)
Simonson, R. W., and Gardiner, D. R., 1960, Concept and Functions of the Pedon, *Trans. 7th Intern. Cong. Soil Sci.*, Madison V.18, 127–131. Also U.S.D.A., Soil Classification; a Seventh Approximation, 1960. Harris, S. A., 1964, *Proc. Geol. Assoc.*, **74**, 1963, 'The term *weathering unit* would appear best suited for the equivalent of the pedon ... three dimensions with their lower limit at the surface of the fresh rock ... large enough laterally to permit the study of the nature of any horizon present ... one to ten square metres'. (p. 441)

### Pedosphere
Jacks, 1954. 'The part of the earth in which soil-forming processes occur.'

### Pelagic, pelagic zone
*Webster*. Pelagic. a. Of or pertaining to the ocean; oceanic; specifically *Biogeog.* of or found in the pelagic zone. b. Conducted, or conducting operations, upon the open sea; as, *pelagic* sealing; a *pelagic* sealer.

*Webster*. Pelagic zone. *Biogeog*. The realm consisting of the open ocean, especially those portions beyond the outer border of the littoral zone and above the abyssal zone, to which light penetrates. Pelagic organisms usually lead a free-floating existence at or near the surface of the sea, away from the coast.

Lake, 1958. 'The marine environment can be divided into various zones on the basis of depth. The pelagic division includes the whole mass of water and consists of a neritic province and an oceanic province the boundary between the two provinces occurs at the edge of the continental shelf, at about 200 metres depth. The oceanic province can be subdivided into epipelagic, bathypelagic, and abyssopelagic zones. The benthic division includes all the ocean floor and consists of two systems, the neritic and the deep sea, the dividing line again being at 200 metres depth. In the neritic system are the littoral and sublittoral zones. The littoral zone usually has strong wave and current action and enough light for plant growth. Below it is the sub-littoral zone starting usually at 40–60 metres depth. The tidal zone forms the upper part of the littoral zone. In the deep-sea system are the archibenthic and abyssobenthic zones, the dividing line being between 800 and 1100 metres ... the water column can be divided into three zones based on the light factor. The euphotic zone is uppermost and has sufficient light for photosynthesis. It rarely extends beyond 100 metres... below is the disphotic zone... sufficient light for animal responses... to 800 metres where aphotic zone begins... no light of biological significance.' (pp. 422–424)

**Pele's hair** (Hawaii)
Volcanic glass spun out into hairlike form not to be confused with Pelée and Peléan.

**Pele's tears** (Hawaii)
Congealed lava droplets, notably resulting when a shower of lava falls into the sea. Pele was the fire Goddess of the early Hawaiians; her home the volcano of Kilauea.

**Peléan, pelean**
*Webster. Geol.* Designating or pertaining to volcanic eruptions characterized by the violent expulsion of clouds or blasts of incandescent volcanic ash, such as occurred at M. Pelée, Martinique, in 1902.
One of the four main types of volcanic activity (Hawaiian, Strombolian, Vulcanian, and Pelean), characterized by the extrusion of a very acid and very viscous lava which tends to consolidate as a solid plug. From Mont Pelée, Martinique.

**Pelite, pelitic**
Himus, 1952. Clastic sediments composed of clay, minute particles of quartz or rock-flour. A volcanic ash of corresponding grade is known as a pelitic tuff. Hence pelitic gneiss, pelitic schist derived from metamorphism of pelitic sediments.
*Comment.* Argillaceous (*q.v.*) is sometimes regarded as synonymous but is usually used in a more restricted sense (L.D.S.).
See Psammite, Psephite.

**Pellicular**
Howell, 1957. 'A term applied to water "adhering as films to the surfaces of openings and occurring as wedge-shaped bodies at junctures of interstices on the zone of aeration above the capillary fringe" (Tolman).'

**Peneplain** (W. M. Davis, 1889), **peneplane, peneplanation**
Davis, W. M., 1889, Topographic Development of the Triassic formation of the Connecticut Valley, *Am. Jour. Sci.*, **37**, 430.
Davis, 1909. Quoting from the *Geographical Cycle*, 1899. '*Old Age.* Maturity is passed and old age is fully entered upon when the hilltops and the hillsides as well as the valley floors are graded... The landscape presents a succession of gently rolling swells alternating with shallow valleys... An almost featureless plain (a peneplain) showing little sympathy with structure, and controlled only by a close approach to base-level, must characterize the penultimate stage of the uninterrupted cycle and the ultimate stage would be a plain without relief' (p. 270).
Davis, 1922. *Bull. Geol. Soc. Amer.*, **33**, 567–598. 'As no one, I believe, proposes to call the surface of ultimate degradation a "plane", I see no reason for calling the penultimate surface a "peneplane".' (p. 587)
*O.E.D.* Peneplain, a nearly flat region, a tract of land almost a plain.
*O.E.D. Suppl.* Peneplane, a variant of peneplain
*Webster. Geol.* A land surface worn down by erosion to a condition of low relief, or nearly to a plain.
Mill, *Dict.* Peneplain. A tract of country from which denuding agents have removed all prominent irregularities, and in which there remain at most gentle undulations (W. M. Davis).
Committee, List 2. Peneplain. An almost featureless plain, showing little relation to structure, and controlled only by a close approach to base-level (W. M. Davis, The Geographical Cycle, *Geog. Jour.*, **14**, 1899, 497).
Strahler, 1951. 'The word peneplain is given to a land surface of faint relief produced in the old age stage of a cycle of denudation.' (p. 168)
Thornbury, 1954. 'The term *peneplain* was introduced by W. M. Davis in 1889 to describe the low and gently undulati plain which the processes of subaer erosion presumably develop at the penultimate stage of a geomorphic cycle... the concept has had a rather controversial history which continues down to the present, and probably more has been written pro and con regarding it than about any other geomorphic idea' (p. 177). Thornbury devotes a complete chapter to 'The Peneplain Concept'.
*Comment.* Davis used the spelling peneplain; peneplane was deliberately introduced by those who contend that 'plain' implies a region of horizontal *structure* whereas Davis was concerned with 'almost a plane,' that is almost a flat surface (Lobeck, 1939,

634). The majority of writers use peneplain but with many notable exceptions including J. W. Gregory, 1908, *Geography*; P. Lake, 1952; Lobeck, 1939. It is generally agreed however that the verb is 'to peneplane', whilst 'peneplanation' is apparently always used and never 'peneplaination'. The issue has been further complicated by the translation into English by Czech and Boswell of Walther Penck's *Morphological Analysis of Land Forms*, where a distinction is made between peneplane purely to express relief and the 'end-peneplane' (contrasted with Primärrumpf or primary peneplane) which is the Davisian peneplain. Penck's statement '... an der Endrumpfläche, den Endrumpf oder den Peneplain von Davis' is translated '... of the final surface of truncation, the end-peneplane or Davisian peneplain' (p. 144). As Wooldridge (in Taylor, 1951, p. 170) points out, future work may possibly support Crickmay's contention that 'peneplanation is in the last analysis wholly or partly panplanation.' The safe solution is to refer to the Davisian peneplain when such is meant. For discussion of multiple peneplains and peneplain remnants in the landscape see Brown E. H., 1957, The Physique of Wales, *Geog. Jour.*, **123**, 208–230.

**Peninsula**
*O.E.D.* A piece of land that is almost an island, being nearly surrounded by water; by extension, any piece of land projecting into the sea, so that the greater part of its boundary is coastline.
*Comment.* The noun is peninsula; the adjective meaning of or belonging to a peninsula is peninsular.

**Penitent** (Louis Lliboutry, 1954)
Lliboutry, L., 1954, The Origin of Penitents, *Jour. of Glaciology*, **2**. 'Penitents are spikes of old compact winter snow or of glacier ice in the Andes of Santiago ... we venture to translate the words used by Chileans ar 1 Argentinians into English: *penitentes* (noun), *campo de penitentes* (field of penitents). *Nieve penitente* is not used' (p. 331). Nieve penitente is *lit.* snow-penitent and was previously used by writers in English. Also micropenitent, pocket-penitent.

**Pentref** (Welsh)
Village, homestead.

**Peperite**
A mixture of lava and sediment; used of such deposits in the Puy de Dome district of France (M.D.).

**Pepino Hill** (Puerto Rico)—see Hum

**Perambulation**
*O.E.D.* 3. The action or ceremony of walking officially round a territory (as a forest, manor, parish, or holding) for the purpose of asserting and recording its boundaries, so as to preserve the rights of possession etc.; beating the bounds.
4. The boundary traced, or the space enclosed, by perambulating; circuit, circumference, bounds; district, precinct, extent.
*Comment.* A word much used in old documents in Britain and still in regular use in certain cases.

**Perched water table**
Rice, 1941. Ground water is said to be perched if it is separated from an underlying body of ground water by unsaturated rock. Perched water belongs to a different zone of saturation from that occupied by the underlying ground water (Meinzer).

**Percolation**
Howell, 1957. 'Movement, under hydrostatic pressure, of water through the interstices of the rock or soil except movement through large openings such as caves.'

**Perforation deposit** (J. H. Cook, 1946)
Thornbury, 1954. 'Cook (1946) thought that the term kame should be discontinued and suggested the term *perforation deposit* for individual mounds of sand and gravel and *kame complex* for areas of sag and swell topography' (p. 378). See Cook, J. H., 1946, Kame-complexes and perforation deposits, *Amer. J. Sci.*, **244**, 573–583.

**Pergelisol** (Kirk Bryan, 1946)—see Tjäle, Permafrost, Tele

**Periclinal, Pericline**
*O.E.D.* Greek, sloping all round, on all ies. In geology sloping in all directions from a central point: = Quaquaversal.
Page, 1865. Dipping on all sides from a central point or apex; applied to strata which often dip in this manner from some common centre of elevation.
*Comment.* Page's definition is clear; alternatively the term structural dome is used. Another definition is an anticline pitching sharply at both ends. The term has been applied to the Mendips and even to Castle Hill, Dudley, which are morphological rather than structural periclines. The word does not appear in many standard works.

**Perigee**
O.E.D. That point in the orbit of a planet at which it is nearest to the earth.
*Comment.* Now used especially of the moon. Opposed to apogee.

**Periglacial** (W. V. Lozinski, in German, 1909)
Thornbury, 1954. 'The term *periglacial* was introduced by Lozinski in 1909 to refer to: areas adjacent to the borders of the Pleistocene ice sheets; the climatic characteristics of these areas; and, by extension, the phenomena induced by this type of environment.' (p. 411)

Zeuner, F. E., 1946, *Dating the Past*, London: Methuen. 'This periglacial zone may be defined as the zone in which during any particular glacial phase the climate favoured permanently frozen subsoil. . . .' (p. 119)

Rice, 1941. 'Having a position marginal to but beyond the glacier. (Antevs and MacClintock.)'

*Note:* Lozinski, W. V., Die periglaziale Fazies der mechanischen Verwitterung, *Naturw. Wochenschrift*, **26**, 1911, 641–647. Also: Das Sandomierz-Opalower Lössplateau. *Globus*, **96**, 1909, 330–334.

No reference in Wright, 1914; Wooldridge and Morgan, 1937; *Adm. Gloss.*, 1914; Flint, 1937; *U.S. Ice Terms*, 1952.

International Geographical Union; Commission de Morphologie Périglaciaire. Rapports Préliminaires 1952. '*Nous proposons de comprendre, sous le terme: morphologie périglaciaire, la morphologie des régions situées à l'extérieur des glaciers présents et passés et dans les climats froids où l'eau agit, au moins partiellement, sous forme de glace.*'

**Perihelion**
O.E.D. That point in the orbit of a planet ... at which it is nearest to the sun. Opp. to *aphelion*.

**Period**—See Geological Time

**Peripediment** (A. D. Howard, 1942)
See quotation and definition under Pediplain

**Permafrost** (S. W. Müller, 1947)
Permanently frozen sub-soil. This term, advocated by S. W. Müller, *Permafrost or Permanently Frozen Ground and Related Engineering Problems*, Ann Arbor, 1947, has proved more generally acceptable than Kirk Bryan's pergelisol or such foreign words as tjäle or tele.

**Permatang** (Malay)
An old sand beach; defined on the official one-inch maps of Malaya as 'rising sandy ground, sandy ridge'.

**Permeability, permeable**
O.E.D. The quality or condition of being permeable; capability of being permeated; perviousness. Permeable: 1. Capable of being permeated or passed through; permitting the passage or diffusion of something through it; penetrable; pervious.

Rice, 1941. A permeable or pervious rock, with respect to subsurface water, is one having a texture that permits water to move through it perceptibly under the pressure ordinarily found in subsurface water. Such a rock has communicating interstices of capillary or supercapillary size (Meinzer).

Babbitt, H. E. and Doland, J. J., 1949, *Water Supply Engineering*, 4th ed., New York: McGraw-Hill. 'Permeability and transmissibility express the relative ease with which water will flow through a porous medium. The coefficient of permeability may be expressed as the gallons of water which, in one day, will flow through 1 sq. ft. cross-sectioned area of the sample of material, when the hydraulic gradient is unity and the temperature is 60°F.'

Hohlt, 1948, *Jour. Col. Sch. of Mines*. 'The permeability of a porous substance is the volume of fluid of unit viscosity which passes through a unit cross-sectional area in unit time under a unit pressure gradient.'

*Comment.* A distinction should be made between primary (pores) and secondary (joints, bedding planes etc.) permeability. (G.T.W.)

**Pervious, perviousness**
See Permeability; also Porosity.

**Petroglyph**
L.D.G. A drawing made on a rock face such as the wall of a cave by primitive man.

**Petrographic Province**
Webster. Geol. A region in which the various igneous rocks are so related as to indicate origin from a common magma.

Himus, 1954. 'A natural region in which the rocks belonging to a definite cycle of igneous activity are marked by specific peculiarities, collectively as well as individually, which differentiate them from other assemblages of rocks belonging to other regions or cycles.' The rocks in a petrographic(al) province will usually all belong to one of the three suites—Atlantic, Pacific or Spilitic.

**Petrology**
The study of rocks. It includes petrography, the systematic description of rocks, and also the study of petrogenesis or the origin of rocks. See also Lithology.

**pH**
Howell, 1957. 'The negative logarithm of the hydrogen ion activity (less correctly, concentration).' The standard measure of acidity and alkalinity (*q.v.*).

**Phacolith, phacolite**
Wooldridge and Morgan, 1937. 'Phacolites are lens-shaped masses of [igneous] rock occupying the saddles of anticlines or the keels of synclines, places where rigid rock sheets naturally gape apart in the folding process' (p. 109, with figure of the Corndon, Shropshire).
*Comment.* Phacolites, which should strictly be phacoliths, differ only from laccoliths (*q.v.*) in that the base of the latter is believed to be flat. *Cf.* lapolith.
Himus, 1954, gives phacolith or saddle reef which is incorrect. Not in *O.E.D.* See also Tyrrell, G. W., 1931, *Volcanoes*, London: Butterworth.

**Phanerophyte**
Raunkiaer, 1934. 'Plants whose stems, bearing the buds which are to form new shoots, project freely into the air . . . ; plants of those portions of the earth most favoured climatically' (p. 16).

**Phanerozoic**
*Webster*. Living unconcealed, *esp.* in daylight.
*Comment.* This word, of which the natural opposite is cryptozoic, has been adopted by geologists for the period Cambrian to present when there have been evident life forms, preserved as fossils, on the earth. Hence Phanerozoic time, Phanerozoic rocks, Phanerozoic time-scale. *The Phanerozoic Time-Scale* published by the Geological Society of London in 1964 thus deals with the development of a quantitative time-scale for this period.

**Phenetic** (H. K. Pusey, 1960)
Used by A. J. Cain and G. A. Harrison ('Phyletic weighting', *Proc. Zool. Soc. Lond.*, **135**, 1960, 1–31) to mean relationship based on the phenotype (as opposed to genotype or phylogeny).

**Phenomenal environments**
Kirk, W., 1963, Problems of Geography, *Geography*, **48**, 357–371. 'Phenomenal environments . . . it is an expansion of the normal concept of environment to include not only natural phenomena but environments altered and in some cases almost entirely created by man.' (p. 364)
See also Behavioural environment.

**Phenon** (P. H. A. Sneath and R. R. Sokal, 1962)
A numerical phenetic group, *i.e.* a group constructed by numerical taxonomy ('Numerical Taxonomy', *Nature*, **193**, 855–860). *Phenon* was used in a different sense by W. H. Camp and C. L. Gilly (*Brittonia, N.Y.*, **4**, 1943, 323–385), but has not been adopted.

**Phenology, phenological**
*O.E.D.* The study of the times of recurring natural phenomena, especially in relation to climatic conditions. Quotes Willis, 1897, *Flowering Plants* I 'The study of the periodic phenomena of vegetation.' (p. 155)
*Comment.* See also isophene. From 1891 to 1947 the Royal Meteorological Society published *The Phenological Reports* with maps showing dates of first flowering of selected common plants.

**Photic, Photic Region, Photic Zone**
*O.E.D. Suppl.* Of sea-water; that is penetrated or influenced by sunlight.
Murray, J., 1913, *The Ocean*, **7**. 'This superficial layer affected by sunlight is called the photic zone of the ocean' (p. 133).
Neaverson, E., 1955, *Stratigraphical Palaeontology*, 2nd ed., Oxford, 1955. 'Two general light regions recognised by oceanographers are based on the distribution of plants: the photic region where the intensity of the sunlight is sufficient for the process of photosynthesis in plants, and the aphotic region, where the light is insufficient for this process to be carried on' (p. 36).
See also Pelagic. Note also Disphotic.

**Photogrammetry**
*L.D.G.* The science or art of making a topographical map from a series of vertical air-photographs. The various corrections needed in plotting involve the use of highly complex and large instruments.

**Photo relief**
*L.D.G.* If a plain model is made of the relief of a tract of country and then illuminated by a light (traditionally coming from the north-west) shadows are cast which, when the whole is photographed, give a realistic representation of the distribution of hills and valleys. This old method of repre-

**Phreatic surface**
Ground-water table; water table.
Kuenen, 1955. 'Suppose we sink a number of wells and imagine all the water levels connected by a flat or undulating surface, the result will be the ground-water table or phreatic surface.' (p. 184)

**Phreatic water**
See Ground Water (nappe phréatique of French) *lit.* water which feeds wells (Greek: φρέαρ, φρέατος well).

**Phrygana**
'More or less equivalent to garrigue in the eastern Mediterranean.' (P.W.R. in MS.)

**Phylum**
*O.E.D.* A tribe or race of organisms, a series of animals or plants genetically related.
*Comment.* From the Greek φῦλον, a tribe or race. Hence also phyletic, phylogenetic, and other combinations.

**Phyochorology** (S. S. Peterson, 1961)
Paterson, S. S., 1961, *Introduction to Phyochorology of Norden*, Stockholm 'Science of the ecology and the geographic occurrence pattern of areal production.' See *G.J.*, 129, 1963, 90.

**Physical Geography**
*O.E.D.* That branch of geography which deals with the natural features of the earth's surface, as distinct from its political divisions, commercial or historical relations, etc.
*Webster.* Physical geography treats of the exterior physical features and changes of the earth, in land, water, and air.
Mill, *Dict.* 1. A synonym for physiography. 2. The natural phenomena directly related to the earth's surface.
Committee, List 1. 1. The geographical study of those features of the earth which are called natural. 2. (sensu stricto) The geographical study of the inanimate phenomena and features of the earth's surface.
Davis, 1909. Physical geography or physiography may be defined as 'the study of those features of the earth which are involved in the relation of earth and man; that is, the study of man's physical environment'. (p. 130) (1900)
Strahler, 1951. '... the descriptive study of a number of earth sciences which give us a general insight into the nature of man's environment ... a body of basic principles of earth science selected with a view to including primarily the environmental influences that vary from place to place over the earth's surface.' (p. 1)
Debenham, F., 1950, *The Use of Geography*, London: E.U.P. 'Thus Mr. Peel's book "Physical Geography" deals with the physical background; those aspects of air, land and water which, quite independently of man, affect the environment in which we live and which are almost, but not quite, beyond our control.' (p. vi)
Wooldridge, S. W., and East, W. G., 1956, *The Spirit and Purpose of Geography*, London: Hutchinson. Physical Geography deals with land, air and the ocean but excludes biogeography. (Ch. III)
Hartshorne, 1939. '(Of the 18th cent.) 'As used by nearly all the writers of the period —including Kant, Forster, and later, Humboldt—the term "physical geography" was not limited to that which later was to be called the "physical" or "natural" environment, but included races of men, and, commonly, their physical works on the earth.' (p. 219)
*Comment.* Out of ten standard textbooks of physical geography, selected at random and published between 1880 and 1955 all ten include geomorphology or the sculpturing of the earth's surface but only three deal specifically with soil; all ten include climate, nine have a section on oceanography, seven on natural vegetation but only five on zoogeography. Four include astronomical geography and two include projections. Four deal to some extent with the relation between man and nature.

**Physical geology**
*Webster. Structural geology*, treating of the form, arrangement, and internal structure of rocks and *dynamic geology*, dealing with the causes and processes of geological change are sometimes combined as *physical geology*.
Holmes, 1944. 'Physical geology is concerned with all the terrestrial agents and processes of change and with the effects brought about by them. This branch of geology is by no means restricted to geomorphology, the study of the surface relief of the present day, which it shares with physical geography.' (p. 6)
Miller, W. J., 1936, *An Introduction to Physical Geology*, London: Chapman and Hall. 'Physical geology deals with the materials of the earth; earth-crust move-

ments; the structure of the earth; and the processes and agencies by which the earth has been for many millions of years, and is being, modified, including such agencies as weather, wind, streams, glaciers, sea, organisms, volcanoes, subterranean waters and lakes' (p. 1). Geology is divided in Physical and Historical sections.

## Physiognomy, physiognomical

Cochrane, G. Ross, 1963 (April 1964), A Physiognomic Vegetation Map of Australia, *Journ. Ecology*, **51**, No. 3, 639–655, 'based on the structural characters, rather that on floristic composition. Symbols and formulae are used to indicate structural features'. (p. 650)

See A. W. Küchler, 1949, A physiognomic classification of vegetation, *A.A.A.G.*, **39**, 201–210.

Gaussen, H., 1959. The Vegetation maps, *Trav. Sect. Sci., Inst. Franc. de Pondichery*, **1**, 155–179. 'Each vegetation landscape is a physiognomical unit' (p. 155).

## Physiography, physiographic, physiographical

*O.E.D.* 1. a description of nature, or of natural phenomena or productions generally.

1828–1832. *Webster*, Physiography, a description of nature, or the science of natural objects. 1840. J. H. Green, *Vital Dynamics*. 101. The office of... Physiography is to enumerate and delineate the effects and products of nature as they appear.

2. A description of the nature of a particular class of objects.

3. Physical geography. 1873. J. Geikie, *Gt. Ice Age*, xiii, 176. To restore the physiography of the land during successive stages of the glacial epoch. 1877. *Elem. Lessons in Phys. Geog.*, 3 note. This term (Physical geography) as here used is synonymous with Physiography, which has been proposed in its stead.

*Webster*. 1. A description of nature or natural phenomena. 2. Physical geography specifically geomorphology.

Mill, *Dict*. 1. The science of natural phenomena. 2. Sometimes restricted to geomorphology (W. M. Davis).

Committee, List 1. A word variously used either as a synonym for physical geography or for the structural features of the earth's surface.

Huxley, T. H., 1877, *Physiography*, London: Macmillan. So called to distinguish his subject from Physical Geography which is accused of consisting of 'scraps of all sorts of undigested and unconnected information.' Physiography is rather the study of the causal relationships of natural phenomena or a consideration of the 'place in nature' of a particular district. (pp. vi–vii) (W.M.)

Kropotkin, P. On the teaching of physiography. *Geog. Jour.*, **2**, 1893, 350–59. 'I cannot conceive Physiography from which man has been excluded. A study of nature without man is the last tribute paid by modern scientists to their previous scholastic education.' (p. 355)

Salisbury, 1907. 'In England, physiography is often regarded as a general introduction to science, and is made to include the elements of all the physical and biological sciences. In some other quarters physiography is regarded as the physical geography of the land.... physiography has to do with the surface of the lithosphere, and with the relations of air and water to it. Its field is the zone of contact of air and water to it, and of air with water.' (pp. 3–4)

Fennenman, N. M., 1938, *Physiography of Eastern United States*, New York: McGraw-Hill. 'The term physiography as used in the United States is the approximate equivalent of geomorphology. The latter term is used more in Europe and is growing in favor in this country. Geomorphology is definitely limited to the genetic study of land forms. Physiography has often been made to include treatment of the atmosphere and ocean. In that broad sense it is essentially the scientific study of physical geography. In this book the atmosphere and the ocean are mentioned only as related to the land surface.' (p. v)

Powell, J. W., 1895. *Physiographic Processes*, National Geographic Monographs No. 1. 'Physiography is a description of the surface features of the earth, as bodies of air, water and land. In it are usually included an explanation of their origin, for such features are not properly understood without an explanation of the processes by which they are formed' (p. 1). *Note*: The review in *Geog. Jour.*, **7**, 1896, 319, comments that this definition 'is that more usually applied in this country to physical geography'. (W.M.)

Morgan, A., 1901, *Advanced Physiography*, London: Longmans. 'I have put upon Physiography the liberal interpretation given to it in the South Kensington Syllabus, in which it includes the main facts and principles of elementary physics, geology, oceanography, meteorology, and astronomy.' (p. v)

Herbertson, A. J., 1906, *Outlines of Physiography*, 3rd ed. London: Arnold. 'Physiography literally means a description of nature.... This little book is intended to give an account of the natural movements that are constantly taking place on the Earth's surface, and the forms to which these movements give rise.'

Comment. On the whole 'physiography' and especially 'physiographic' are more usual in America where British geographers use 'morphology' and especially 'morphological'. Linton, 1951. *The Delimitation of Morphological Regions* in Stamp and Wooldridge, tacitly assumes that Fenneman's Physiographic Regions are what he means by morphological regions. There is also a general assumption that whereas 'physiography' may be only descriptive, 'geomorphology' must be interpretive.

### Physique
Used by some geographers (see Wooldridge, S. W., in Darby, H. C., 1936, *An Historical Geography of England before* A.D. 1800, C.U.P., 88) to mean the relief, physical structure and organization of a region, from analogy with the physique of a human body. Not in *O.E.D.* or *Webster* in this sense.

### Phytal Zone (of a lake)
'Water which is shallow enough to permit the growth of rooted green plants.' W. D. R. Hunter, *The Glasgow Region*, 1958, 107.

### Phytoclimate
Van Steenis, C. G. G., 1957, *Abs. of Papers*, *Ninth Pacific Science Congress*, Bangkok. 'The relationship between vegetation assemblages and the climatic conditions under which these assemblages exist' (p. 254).

### Phytogeography
Plant geography; the geography of vegetation and plants.

### Piccottah (India: Tamil) also peccottah
A shaduf (*q.v.*).

### Pidjin-English, Pidgin-, Pigeon-
See Lingua franca.

### Piedmont
Not in *O.E.D.*; *O.E.D. Suppl.* gives only piedmontal, of, pertaining to, or situated at, the foot of a mountain.

*Webster. Phys. Geog.* Lying or formed at the base of mountains; a piedmont district, plain, glacier, etc.

Piedmont is now very widely used by geographers as an adjective; thus Thornbury, 1954, lists piedmont alluvial plain, piedmont depression, piedmont lake, piedmont province; Wooldridge and Morgan, 1937, have piedmont fringe ('of coarse sediments, essentially screes and alluvial fans'), and piedmont glacier. The latter term is used in a specialized sense for 'an extensive sheet of ice covering low-lying ground at the foot of a mountain range, formed by the union of several glaciers... one of the best known examples is the Malaspina glacier in Alaska'. (Moore, 1949.) The Piedmont (U.S.A.) is the plateau of ancient rocks lying east of the Appalachians and bounded on its east by the edge which gives rise to the 'Fall Line'. (L.D.S.)

### Piedmonttreppen (German: Penck)
Thornbury, 1954, 'used to describe the succession of step-like benches which are found around the flanks of the Black Forest and other mountains' (p. 203).

### Piezometric surface
An imaginary surface that coincides with the static level of water in an aquifer to which the water will rise under its full head.

### Pike—See Peak

### Pillow Lava
Lava which has consolidated so as to resemble a jumbled mass of pillows. This generally seems to have resulted from their extrusion and consolidation under water. Pillow lavas have usually, though not always, particular chemical and mineralogical characters which distinguish them from the closely allied basalts and cause them to be typical members of the Spilitic Suite, in contrast to the Atlantic and Pacific Suites. See Petrographic Province.

### Pin (Irish dialect)
A peak, but rarely used except for the Twelve Pins of Connemara. An anglicized rendering of the Gaelic *beann*, a peak; also ben (*q.v.*).

### Pin fallow
Same as Bastard fallow. See Fallow, Green. Pin fallow, Bastard fallow: land left fallow

but not an entire year—often between taking off a crop in autumn and the sowing of a different crop late the next spring. In the meantime cultivation controls weeds and restores the land to condition (I.L.A.T.).

**Pingo** (Eskimo: *lit.* conical hill; A. E. Porsild, 1938)
Porsild, A. E., Earth Mounds in Unglaciated Arctic Northwestern America, *Geog. Rev.*, 28, 1938, 46–58. 'Scattered, curious isolated more or less conical mounds, which have been commented on by nearly all travellers since the days of Franklin ... the name "gravel or earth mound" seems to be fairly well established. The Eskimo name *pingo*, which has come into universal use in the north, is here introduced as an alternative' (p. 46). 'Mounds of similar appearance but much smaller in sub-Arctic bogs of Europe and Asia in Swedish literature are known as *pals*' (p. 58).
Bird, J. B., *Zeitschrift für Geomorphologie*, 3, 1959. 'Pingos, low hills of fine unconsolidated material generally containing an ice core' (p. 165); 'some are 150 feet high' (p. 165). See also Laverdière, C., L'Origine des Pingos, *Rev. Can. Geog.*, 9, 1955, 226, Charlesworth, 1957, 576 and Müller, F., 1959, Beobachtungen über Pingos, *Meddelelser on Gronland*, 153, 127. See also Palsa, Hydrolaccolith.

**Pinnate drainage pattern**
Thornbury, 1954. 'A special dendritic pattern is the pinnate. The tributaries to the main stream are subparallel and join it at acute angles. It is believed to represent the effect of the unusually steep slopes on which the tributaries developed.' (p. 121)

**Pipkrake** (Swedish)
The layer or layers of ice needles, or single groups of needles, formed on the surface of the ground, *i.e.* the limit surface between soil and free atmosphere. The pipkrake-structure is typical of a relatively warm ground and a plentiful water-supply. Pipkrake formation is to be found on waterlogged earth, rich in mull, with alternating frost and thaw during the end of winter and spring. (E.K.) See also Charlesworth, 1957, 566, which notes the orientation of the needles and other details.

**Piracy (of streams)**—See Capture of rivers
**Pisolite, Pisolith, Pisolitic**
Pisolite is a type of limestone made up of rounded bodies, sometimes called pisoliths, larger than the ooliths of an oolite and hence the name pisolite meaning 'pea-stone'. The pisoliths are 2 mm or more in diameter. Other rocks than limestone may exhibit a pisolitic structure, notably some laterites and iron ores.

**Piste** (French, from Italian *pista*)
Literally the beaten track made by the passage of horses and other animals: applied by the French to the old-established routeways across the Sahara some of which in due course have become 'pistes automobiles' (Bernard, A., 1939, *Géographie Universelle*, 9, 352–353). As a convenient term this phrase has been used by several writers in English. Stamp, 1953, *Africa*, New York: Wiley, gives 'main desert highway (*piste automobile*)' (p. 261).

**Pitch**
*L.D.G.* When an anticline or a syncline of folded rocks slopes as a whole in one direction this is referred to as the pitch of the fold. Not to be confused with dip.

**Pitfall**
Defined in *O.E.D.* 2 as 'a concealed pit into which animals or men may fall and be captured', similarly *Webster*, but used in Northumbria and elsewhere for subsidence hollows, occupied by a pond, caused by underground mining. Henry Tegner, Birds of Northumbrian Pitfalls, *Country Life*, June 11, 1964, p. 1467.

**Piton** (French)
*Webster*. A sharp peak.
A cone; equivalent of the German Kegel in the terminology of Karst lands though the Report of the Commission on Karst Phenomena (International Geographical Union, 1956) gives three varieties of Kegelkarst *a.* karst à coupoles (Halbkugelkarst), *b.* karst à pitons (Spitzkegelkarst) and *c.* karst à tourelles (Turmkarst). Piton may be defined as a peaked limestone residual in tropical or subtropical karst regions; coupole is a cone (G.T.W.).

**Placer** (Spanish: sandbank)
*L.D.G.* This term is used by miners for a deposit of sand or gravel with alluvial gold, tin ore etc.

**Pladdy** (Northern Ireland)
A residual island drumlin awash at high tide, *e.g.* in Strangford Lough. (E.E.E.)

**Plage** (French)
*O.E.D. Suppl.* The beach, especially at a seaside resort.

*Comment.* The use of this word in English has had an interesting history. In mediaeval works it meant a region, especially a littoral region or zone but became obsolete. It has been re-introduced from the French (where it means simply beach or shore) with the special meaning of a beach developed with facilities for the enjoyment of visitors. But it is again becoming obsolescent having been replaced by the Italian word *lido* (*q.v.*).

**Plain**

*O.E.D.* 1. A tract of country of which the general surface is comparatively flat; an extent of level ground or flat meadow land; applied specifically in proper or quasi-proper names to certain extensive tracts of this character; *e.g.* Salisbury Plain, the Great Plain of England, etc.
2. Chiefly pl. In colonial and U.S. use applied to level treeless tracts of country; prairie.
3. An open space in the midst of houses. Local.

*Webster.* 1. Level land; especially, an extensive open field or broad stretch of land having few inequalities of surface. 2. *pl.* In North America and the British colonies, broad tracts of almost treeless level country; with *the*, the territory extending from North Dakota to Texas and from the Mississippi River to the Rocky Mountains.

*Dict. Am. pl.* Extensive regions of level or rolling treeless country, prairies. 1755 L. Evans *Anal. Map Colonies* 13, 'the woodless Plains over the Mississipi'.

*Mill, Dict.* 1. Any land with a flat or very slightly undulating surface. The term implies treelessness in Australia.
2. The *Plains* or the *Great Plains*, U.S.A.; specifically the high plains (3,000 to 6,000 feet) west of the prairies, characterized by a deficient rainfall. 3. Name often restricted to lands at comparatively low elevation, *e.g.* coastal plain, river-plain, plain or plane of marine denudation, peneplain; high-lying plains are strictly plateaus.

Strahler, A. M., 1946, Geomorphic Terminology and Classification of Land Masses *Jour. of Geol.*, **54**. A number of authors, led by W. M. Davis, use 'plain' or 'plateau' to imply horizontal structure, namely Davis, Lobeck, Fenneman, Howard and Spock. They are more generally used for an area of level land, irrespective of structure, *e.g.* by Powell, J. Geikie, Cotton, de Martonne, P. E. James, Finch and Trewartha, and Cressey. Strahler strongly favours the latter usage, advocating the use of 'horizontal or flat-lying strata' for the former concept. The structural implications of 'coastal plain' however should be retained. Authors of the Davis school, especially Davis himself, are not consistent in their usage. (W.M.) (pp. 32–42)

James, P. E., 1935, *An Outline of Geography*, Boston: Ginn. 'A plain is an area of low relief, generally less than five hundred feet. It is low-lying with reference to surrounding areas and is usually, but not in every case, low in altitude.' (p. 66)

Cressey, G. B., 1944, *Asia's Lands and Peoples*, New York: McGraw-Hill. 'Plains and plateaus are essentially flat, or having only gently rolling forms with slopes up to 5°. ... The difference between a plain and a plateau is that whereas the former has little or no relief, say tens or hundreds of feet at the most, plateaus are plains that are intersected by deep valleys so that the area as a whole has noticeable relief. ... Plains are near their base level, while plateaus are not; either may be at low or high elevations.' (p. 16)

*Comment.* The above quotations reveal a considerable divergence in meaning. There is a general tendency to equate plain with a 'plane' surface. See also upland plain, plateau, peneplain.

**Plan**

*O.E.D.* A drawing, sketch, or diagram of any object, made by projection upon a flat surface, usually a horizontal plane (opposite to elevation); a map of a comparatively small district or region, as a town etc. drawn on a relatively large scale and with considerable detail.

*Adm. Gloss.*, 1953. A large-scale detailed chart.

*S.O.E.D.* 3. A scheme of action, project, design; the way in which it is proposed to carry out some proceeding.

*Comment.* In recent years the word has been much used in the last mentioned sense and from it 'planning'. Note also 'Working Plan' applied to management of forests. Ordnance Survey Plans are on the scale of 1/2,500 and larger.

**Planalto** (Brazil: Portuguese)

*Lit.* high plain or plateau, applied especially to the great Brazilian plateau.

**Planation** (G. K. Gilbert, 1877)

Gilbert, 1877. 'The process of carrying away the rock so as to produce an even surface, and at the same time covering it with an alluvial deposit, is the process of planation.' (pp. 126–127)

*O.E.D. Suppl.* (not in *O.E.D.*). The production of a level surface by the action of glaciers and flowing water.

*Webster. Phys. Geog.* Process by which a stream develops its flood plain by erosion and deposition.

*Comment.* Since the actual meaning of the word is simply the making of a plane or flattened surface, the modern tendency is to ignore the second half of Gilbert's original definition and to use 'planation' as a general term distinguishing different types such as cryoplanation, pediplanation, peneplanation etc. (*qq.v.*).

**Planetary**

*O.E.D.* Belonging to, or connected with, a planet....

*Webster.* Planetary wind. Any one of the major winds, as trade wind, countertrade, prevailing westerly.

*Comment.* The *O.E.D.* does not give the normal geographical usage which is related to the earth as a planet, hence planetary wind, in contrast to local, such as mountain winds, land and sea breezes. Amiran and others refer likewise to planetary deserts in contrast to topographical deserts (*q.v.*).

**Planetesimal Hypothesis**

*Webster. Cosmog.* A hypothesis to explain ... the evolution of the solar system, propounded originally by Chamberlin and Moulton. It postulates the growth of the planets by aggregation from minute but numerous secondary bodies called planetesimals.

Wooldridge and Morgan, 1937. 'This theory was put forward by Chamberlin and Moulton in 1904 as an explanation of the genesis of planets ... not necessary to assume that the earth as a whole was ever in a molten condition ... it grew from small beginnings by the addition of planetesimal matter, rapidly at first, but with decreasing speed' (p. 4–5).

**Planèze** (French), **planeze**

A mesa; used especially of lava-capped plateaus in the Puys de Dome district of France where former cinder cones have been denuded away leaving only those parts which were protected by a lava flow —a form of inverted relief (information from Professor Max Derruau).

Cotton, C. A., 1944, *Volcanoes*, Christchurch: Whitcombe and Tombs. 'At the submature stage of dissection of a cone or of a basalt dome, dwindling sectors of the constructional surface survive on the ridges between deeply eroded major consequent valleys. Such sectors are termed "planezes".' (p. 365). Similarly in Hawaii. See also Pierre Gourou, Symposium on Man's Place in the Island Ecosystem. *Pac. Sci. Congress X*, 1961 (eroded lava field in Réunion).

**Planimetry**

*O.E.D.* The measurement of plane surfaces; the geometry of plane surfaces, plane geometry.

Mill, *Dict.* The representation on a map or plane surface of the features seen on the surface of the ground, such as rivers, coasts, roads, etc.

*Comment.* Now rarely used by geographers in the sense suggested by Mill. When used by geographers, it is applied to measurement of areas—especially by a planimeter.

**Planina** (Serbo-Croat)

Broad, featureless limestone plateau of the karst. *Pl.* planine.

**Plankton**

*O.E.D.* A collective name for all the forms of floating or drifting organic life found at various depths in the ocean, or, by extension, in bodies of fresh water.

*Comment.* See Nekton and Benthos, constituting the three main groups of marine life. It should be noted that 'plankton' applies only to minute creatures whether animal (zooplankton) or plant (phytoplankton) and is not extended to cover large floating seaweeds where such exist.

**Planosol** (Soil science: Byers, Kellogg, Anderson and Thorp, 1938)

Jacks, 1954. 'Intrazonal (*q.v.*) soil having a sharply delineated clay pan or hard pan arising from cementation, compaction or high clay content; formed under forest or grassland vegetation in mesothermal to tropical perhumid to semi-arid climates, usually but not always with fluctuating water table.'

**Plant geography**

Phytogeography: the geography of vegetation and plants.

**Plantation, plantation agriculture**
*S.O.E.D.* Plantation. 4. A settlement in a new or conquered country; a colony.
5. An estate or farm, especially in a tropical country, in which cotton, tobacco, sugarcane, coffee, or other crops are cultivated, formerly chiefly by servile labour.
Sloan, H. S. and Zurcher, A. J., 1949. *A Dictionary of Economics*, New York: Barnes and Noble. 'Plantation system. The agricultural system, firmly established in the southern States by 1840, under which a proprietor and his managerial staff, employing the labor of slaves, devoted relatively large acreages to the production of certain staples such as tobacco, cotton, or sugar.' (p. 194)
*Comment.* The present day contrast is between plantation agriculture and peasant agriculture the former large scale, devoted to the production of one or more cash crops by the employment of a large or considerable labour force. See R. O. Buchanan, 1938, *Econ. Geog.*, 23, 156.

**Plate tectonics**
Tarling, D. H. and Tarling, M. P., 1971, *Continental Drift*, London: Bell. '... vast slabs of the earth, some 80 to 100 kilometres thick, which are moving individually over the "soft" layer of the mantle. As each slab is moving at a constant rate in a particular direction there is very little geological activity within it and the world's earthquakes and volcanoes are concentrated along the margins where the slabs or plates interact. The study of the surface movements of our present-day Earth is now termed "plate tectonics" and has evolved from "continental drift".' (p. 92)
See also Continental drift.

**Plateau**
See also Plain, Upland plain.
*O.E.D.* An elevated tract of comparatively flat or level land; a table-land.
Mill, *Dict.* 1. A more or less extensive elevated area wholly or in large measure at an approximately uniform height. In rainy countries plateaus are attacked by streams which cut deeply into them, so that they become 'dissected plateaus'.
2. (Special use in the Jura) A great stretch of upland country floored by horizontal strata.
James, P. E., 1949, *A geography of Man*, Boston: Ginn. 'A plateau stands distinctly above bordering areas, at least on one side; and it has a large part of its total surface at or near the summit level. It local relief may be very great in case where it is cut by canyons; or it may have as small a local relief as a plain, from which it differs in such a case only because of its position with reference to bordering areas.' (p. 88)
Cressey, G. B., 1944, *Asia's Lands and Peoples* New York: McGraw-Hill. '... plateau are plains that are intersected by deep valleys so that the area as a whole ha noticeable relief. This may amount to hundreds or thousands of feet, but the essential feature is undissected flatland cu by steep-sided young valleys' (p. 16) They are not near their base level and may be at low or high elevations.
*Comment.* Those who regard plateau as an alien (French) word use the plural form plateaux; those who consider it naturalized use plateaus. With regard to the distinction sometimes made between plateaus and upland plains, see under Plain, also Griffith Taylor's term Elbasin.

**Plateau basalt, plateau eruption**
Basic lavas, notably basalts, are very fluid at high temperatures and when reaching the surface through fissures in the earth's crust (fissure eruptions) the lava flows often extend evenly over vast areas and build up lava plateaus, especially if there are successive eruptions. Good examples are the Deccan lavas of India (formerly called 'Deccan Traps') and the Columbia-Snake Plateau of the north-western United States, each covering over 200,000 square miles.

**Plateau Gravel**
Gravel occurring either as extensive sheets on the surface of plateaus or capping hills which represent the dissected remains of former plateaus. Such gravel deposits afford invaluable evidence, though often susceptible to different interpretations, of denudation chronology and geomorphological history.

**Platform**
*O.E.D.* 6c. A natural or artificial terrace, a flat elevated piece of ground; a table-land, a plateau.
*O.E.D. Suppl.* 6c. Also, a continental shelf.
Holmes, 1944. 'As the cliffs are worn back a wave-cut platform is left in front' (p. 289).
Shackleton, 1958. 'Built up mainly of unfolded sediments, the whole forming a slightly raised peneplain of undulating or

level relief... the lowland known geographically as the Russian platform' (p. 418).

Kirkaldy, J. F., 1954, *General Principles of Geology*, London: Hutchinson. 'A "platform" of bevelled spurs in the North and South Downs' (p. 95).

*Comment*. The word 'platform' means a flat or plain surface and has been used in geographical literature with very varied meanings, but rarely as a precise term. It is equated by Mill with continental shelf; it is commonly applied to a wavecut shelf, or to any bevelled surface. The 'Russian platform' seems to refer to the rigid peneplaned block underlying later sediments. Some writers prefer 'Russian table'.

**Plav**
Reed swamp in the Danube delta. (P.W.R.)

**Playa** (Spanish; *lit*. shore)
*O.E.D. Suppl*. In geology, more fully, playa lake, a lake which exists only in winter, being dried up in summer.
  1883. I. C. Russell in *Pop. Sci. Monthly*, **32**, 380. The Spanish word playa... has been adopted by geologists as a generic term under which the various desiccated lake-basins may be grouped.
*Webster. Geol*. The flat-floored bottom of an undrained desert basin, becoming at times a shallow lake, which, on evaporation, may leave a deposit of salt or gypsum; a salt pan.
*Dict. Am*. A broad, level place that accumulates water after a rain, but is at other times dry, a dry lake. 1854.
Mill, *Dict*. 1. A shallow temporary lake, disappearing in dry weather. 2. Various desiccated lake-basins in the western U.S.A.
Thompson, D. G., The Mohave Desert Region, *U.S. Geol. Survey*, Paper 578. Playas are regarded as expanses of clayey soil, not necessarily flooded, seasonally or less frequently. They are divided into three types by their soil characteristics, largely derived from ground water conditions. (W.M.) (pp. 64–65)
Knox, 1904. (Sp.), shore, beach, coast. (U.S.A.), an alkali-flat; the dried bottom of a temporary lake without outlet; an alluvial coastland, as distinguished from a beach.
*Comment*. 1. Usage as *O.E.D.* in Cotton, 1922; Wooldridge and Morgan ('temporary desert lakes'), 1937; Moore, 1949; Strahler, 1951.
2. 'Playa' may refer either to the lake or the dried-out lake bed.

**Pleiades**
'Correlation pleiades' is a Russian introduction still little used. It approximates to a cluster of mutually well-correlated things, and is briefly mentioned in P. H. Davis and V. H. Heywood, *Principles of Angiosperm Taxonomy*, Oliver and Boyd, Edinburgh, 1963.

**Plinth** (R. A. Bagnold, 1941)
Bagnold, R. A., 1941, *The Physics of Blown Sand and Desert Dunes*, London; Methuen (rep. 1954). 'It seems desirable to give a distinguishing name to all that lower and outer portion of the [sand] dune beyond the slip-face boundaries which has never been subjected to sand avalanches. The term "Plinth" is suggested tentatively' (p. 229). 'In the Egypt Sand Sea the plinths have grown to such a vast size that the dunes which surmount them are small in proportion. Here we find what appear to be continuous plinths 1 to 3 kilometres in width and perhaps 50 metres high, running in straight lines for distances of the order of 300 kilometres' (p. 230).

**Pleck, plek**
1. Waste or common. 2. Small individual holding or allotment of indifferent land, usually pasture or meadow (England: Midlands) (I.L.A.T.).

**Plucking**
*O.E.D.* Pluck: 1.b. Geol. To break loose and bear away in large masses; said of glaciers acting on solid rock. Contr. with Abrade.
Similarly: Holmes, 1944; Wooldridge and Morgan, 1937.
Thornbury, 1954. 'Plucking, sapping and gouging are erosional processes restricted to glaciers. *Plucking* refers to the acquisition of parts of the bedrock by a glacier when water enters cracks in the rock and subsequently freezes with resulting detachment of rock fragments as the ice moves forward. The term *sapping* implies undermining and is used by some as synonymous with plucking, but by others it is restricted to detachment which takes place at the bottoms of crevasses. The local basining of bedrock surfaces often effected by glacial erosion is sometimes called *gouging*, but this usage of the term is not widespread.' (p. 48)

See also Colloidal Plucking.
**Plug**
Himus, 1954. A roughly cylindrical mass of igneous rock occupying the vent of an extinct volcano. See also Neck. Hence plug-dome.
**Plum rains**—See Bai-u
**Plunge-pool**
The deep pool at the base of a waterfall into which the water plunges. It is often cut out to a very considerable depth by the action of the falling water in whirling round boulders and stones to produce a large pot hole. If the river should later desert its course the plunge-pool will remain as an almost circular lake, cut in bed rock and often very deep.
**Pluton**
A mass of plutonic rock, *e.g.* a granite batholith.
**Plutonic**
*O.E.D.* Pertaining to or involving the action of intense heat at great depths upon the rocks forming the earth's crust; igneous [Pluto was the Greek god of the underworld].
**Plutonic Theory, Plutonism**
The theory which attributes most geological phenomena to the action of internal heat. The battle between the 'Plutonists' and 'Neptunists' was virtually brought to an end by the publication of Charles Lyell's *Principles of Geology* in 1830. H. H. Read revived 'Plutonism' about 1940 to mean all the phenomena associated with deep-seated plutonic or igneous rocks.
**Pluvial**
*S.O.E.D.* of or pertaining to rain; rainy; characterized by much rain.
*Comment.* Has come to be much used in connection with post-glacial climatic fluctuations 'pluvial periods' having alternated with relatively dry periods. From this has come to be used as a noun— 'a pluvial' or pluvial period.
**Pluviometric coefficient**
*O.E.D. Suppl.* (quotation) 1917. McAdie, *Princ. Aerography*, 218. The term 'pluviometric' was introduced by Angot to indicate the ratio of the mean daily rainfall of a particular month to the mean daily rainfall of the whole year.
Mill, *Dict.* The relation between the average rainfall for a given month and the twelfth part of the annual rainfall for a given locality (Angot).

Miller, A. A., 1931, *Climatology*, London: Methuen. 'The amount of rainfall which would be precipitated at a given place, assuming that such rain were perfectly evenly distributed through the year, is taken as a "norm", and actual mean amount of rain is stated as a percentage of this (the pluviometric coefficient).' (p. 19)
*Comment.* Not to be confused with the humidity coefficient (Ångström, *G. Annaler*, **18**, 1936, pp. 245–254). No reference in *Met. Gloss.*, 1944; Kendrew, 1949.
**Pneumatolysis, pneumatolytic**
Pneumatolysis, the action of highly heated gases and steam, usually in the later stages of a cycle of igneous activity, resulting in the formation of new minerals, including some metalliferous ores, in surrounding rocks which are within reach especially from cracks which later become veins.

**Pocket-penitent** (W. H. Workman, 1914)
Workman, W. H., 1914, Nieve penitente and allied formations in Himalaya, *Zeit. f. Gletscherkunde*, **8**, 289–330. See Penitent, also *Jour. Glac.*, 1958, 265.
**Podu** (India: Telugu)
Shifting cultivation (*q.v.*).
**Podzol, podsol** (Russian); **podzolization**
*Webster.* White or gray ashlike soil, typically occurring in northern Russia.
*Soils and Men*, 1938. 'Podzol Soils. A zonal group of soils having an organic mat and a very thin organic-mineral layer above a gray leached layer which rests upon an illuvial dark-brown horizon, developed under the coniferous or mixed forest, or under heath vegetation in a temperate to cold moist climate. Iron oxide and alumina, and sometimes organic matter, have been removed from the A and deposited in the B horizon. From the Russian for like, or near, ash.'
'Podzolization. A general term referring to that process (or those processes) by which soils are depleted of bases, become acid, and have developed eluvial A horizons (surface layers of removal) and illuvial B horizons (lower horizons of accumulation). Specifically the term refers to the process by which a Podzol is developed, including the more rapid removal of iron and alumina than of silica, from the surface horizons, but it is also used to include similar processes operative in the formation of certain other soils of humid regions.'

acks, 1954. 'Soil with acid-humus horizon overlying B horizon of iron-oxide or iron-oxide and humus accumulation.' Jacks also lists and defines Brown Podzolic Soil, Gray Brown Podzolic Soil, Ground-water Podzol (Intrazonal), Red Podzolic Soil, and Yellow Podzolic Soil.

FAO, 1963, '*Podzolized soils* have an ABC horizon sequence, the B-horizon characteristically being one of iron and/or humus accumulation .... The humus form is raw humus or moder—the range of soils is wide ... includes well-drained soils showing weak, moderate or strong podzolization (podzol) ... in lowland sites similar soils with poorer drainage (gley-podzol); ... peat-podzol. The B-horizon may consist almost entirely of humus-coated or cemented grains (humus-podzol) or of grains cemented or only coated with iron compounds (iron-podzol) or more generally the coating is a mixture of iron and humus compounds (iron-humus podzol)....'

*Comment. Podsol* was the common early form in English but recently there has been a tendency to prefer *podzol*, the correct form of the original Russian word. Thus, Webster, 1934, gives only *podsol*, but *Webster's New Collegiate Dictionary*, 1949, gives both forms, the *American College Dictionary*, 1947, gives only *podzol*, and *Soils and Men*, 1938, gives only *podzol*. (C.D.H.)

**Point bar**—see Scroll
The term is coming into general use and tending to supersede scroll and meander scroll (comment by G. H. Dury).

**Polar front**
L.D.G. The front (*q.v.*) of the great mass of cold air (see Air-mass) which occupies the north polar region. Disturbances along the front play a major part in determining north European weather.

**Polder** (Dutch)
O.E.D. A piece of low-lying land reclaimed from the sea, a lake or a river, from which it is protected by dikes: so called in the Netherlands; rarely used of similar land in other countries.
Mill, *Dict.* Reclaimed land in the lower parts of the Netherlands and Belgium enclosed by embankments and drained by pumping.
*Comment.* The work of creating polders is often referred to as empoldering (verb—empolder) and occasionally the tract enclosed is called an empolder.

**Pole**
*S.O.E.D.* 2. Each of the extremities, north and south of the axis of the earth (North Pole, South Pole).
*Comment.* By analogy extended in use to the north and south magnetic Poles, to the 'Pole of Inaccessibility' (the most difficult point to reach in Antarctica), the Pole of extreme cold (see Cold Pole) etc.

**-polis**
*O.E.D.* Representing Greek, city, as in Metropolis, Necropolis; sometimes used (in the form -opolis) to form names or nicknames of cities or towns, *e.g.* Cottonopolis....

**Political geography**
*O.E.D.* That part of geography which deals with the boundaries, divisions, and possessions of states.
*Webster.* The geography of human governments, ... treats of the boundary of states and their subdivisions, the situations of cities, etc.
Committee, List 1. 1. The Branch of human geography which deals with the boundaries, divisions, territories and resources of States. 2. The study of the effects of political actions upon human geography.
Wooldridge, S. W., and East, W. G., 1951, *The Spirit and Purpose of Geography*, London: Hutchinson. 'Political geography focuses attention on both the external and the internal relationships of states. ... There is first the effect on the present day geography of political action; secondly, the significance of the geography behind political situations, problems and activities.' (pp. 123–124)
Van Valkenburg, S., 1939, *Elements of Political Geography*, New York: Prentice-Hall. 'Political geography is the geography of states and provides a geographical interpretation of international relations.' (p. vii)
Whittlesey. D., 1944, *The Earth and the State*, 2nd ed., New York: Holt. '... the differentiation of political phenomena from place to place over the earth is the essence of political geography.' (p. iii)
East, W. G., 1937, The Nature of Political Geography, *Politica*, **2**, 259–286.
Hartshorne, R., 1935, Recent Developments in Political Geography, *Amer. Pol. Sci. Rev.*, **29**, 785–804, 943–966.

Pounds, N. J. G., 1963, *Political Geography*, New York: McGraw-Hill. 'Political geography is concerned with politically organized areas, their resources and extent, and the reasons for the particular geographical forms, which they assume ... in particular, it is concerned with that most significant of all such areas, the State.' (p. 1)

*Comment.* As an approximate synonym for human geography, and alternatively styled 'moral geography', 'political geography' was widely used by early writers of geographical texts in and after the 18th century (*cf.* Hartshorne, 1939, pp. 42–43). (W.G.E.)

See also Geopolitik, Geopolitics, Geopacifics.

**Polje** (Serbo-Croat; *lit.* field or cultivated area)

Mill, *Dict.* (Slavonic) A large sunken area in a limestone district with a flat bottom intersected by disappearing rivers. *Cf.* Uvala.

Wooldridge and Morgan, 1937. 'The term "polje" is applied popularly to depressions of varying natures, many of which are nothing more than ordinary valleys' (p. 293). (Largest of a succession, doline, uvala, polje. See p. 289.)

von Engeln, 1942. 'Poljes are vertical-walled closed basins, commonly of elliptical outline and with flat alluvial floors ... areas up to 100 square miles ... frequently have a central lake ... may be periodically flooded ... by water brought in by *ponore*' (*q.v.*) (p. 580). 'Poljes are interpreted as grabens' (p. 580).

*Comment.* A common Slav word which has been introduced into international literature in connection with karst phenomena. See karst terminology. Notice the very special interpretation of von Engeln. In Yugoslavia it is applied to any closed or almost enclosed valley.

**Pollen analysis**—see Palynology

**Polster**

Benninghoff, W. S., 1955, *J. of Glaciology*, **2**. 'Small spheroidal, silt-packed, moss cushions, or polsters, are abundant on the terminus of Matanuska Glacier, Alaska... In outward appearance, these resemble "jökla mýs" (glacier mice), described and named from ... Iceland, by Dr. Jón Eythórsson (*J. of Glaciology*, **1**, 1951, 503). The jökla mýs described were moss-covered stones, whereas specimens from Matanuska glacier are concentric moss layers in which sandy silt and a few small pebbles have been incorporated during growth.' The longest axis ranges from 1 to 6 inches. (pp. 514–515)

**Polyclic**

Preferred by many writers to multi-cyclic. 'Polycyclic swallow holes'—swallow hole operating at different levels, *i.e.* under separate cycles, as in Manifold Valley and within a limited locality (G. T. Warwick, C. R. Ier Congrès Int. de Spéléologie 1953).

**Polygenetic**

Of many origins: applied to soils of complex origin.

**Polyglot**

Many languages; applied to persons, peoples, books etc. Contrast Monoglot.

**Polygon (soil)**

Black, R. F., 1952, Polygonal Patterns and Ground Conditions from Aerial Photographs, *Photogrammetric Engineering*, **18**. A soil polygon is a polygonal pattern to be seen on the surface of the soil in some areas, frequently where there is permafrost or very cold winters, but also in other areas where contraction has taken place, *e.g.* in playa lakes and deserts. They vary in diameter from a few millimetres to many tens of metres and may have a concave or convex surface. An extensive literature is discussed and listed and twelve hypotheses discussed by Washburn in 1950 (pp. 123–134).

See also Muller, 1947.

**Polymict**

Read, H. H., 1958, *Proc. Geol. Assoc.*, **69**. 'Two kinds of conglomerate are worth distinguishing, first the oligomict (*q.v.*) ... second, the polymict, with a variety of pebbles of unstable rocks. . . .' (p. 85)

Not in *O.E.D.*

**Polynya** (from Russian polŭinya)

*O.E.D.* A space of open water in the midst of ice, esp. in the arctic seas.

*Comment.* Large areas of open water surrounded by sea-ice and found in the same region—notably off the mouths of big rivers—every year. One side is sometimes formed by the coast.

**Pond**

*O.E.D.* A small body of still water of artificial formation, its bed being either hollowed out of the soil or formed by

embanking and damming up a natural hollow.
Often named according to its use *e.g.* millpond, fish-pond etc. and also applied in some areas to a small natural lake. A *hammer-pond* was constructed to impound water then used to work the hammer required in the early iron-smelting industry (Weald of south-east England, Sheffield district).
Also used contemptuously or humorously of the sea, especially the 'herring pond' for the Atlantic Ocean.

**Pond** (verb), **ponding, ponded**
*O.E.D.* 1. To hold back or dam up a stream into or as into a pond;
2. To form a pool or pond; to collect by being held back.

When the normal flow of a river is interrupted in some way—by an uplift of part of the stream bed, by an obstruction which may include a strong flow of water from a side valley, the water of the main stream is said to be ponded back and forms a lake or large pond.

**Ponor** (Serbo-Croat; *pl.* ponore)
Mill, *Dict.* A swallow-hole (Serbia and Croatia).
Wooldridge and Morgan, 1937. 'Vertical or highly inclined shafts, leading from swallow-holes, or direct from the surface to underground caves, are "ponor" in the Serbian karst and "avens" in France....' (p. 289)
von Engeln, 1942. 'Pipelike, lateral passages which may function alternately to pour in or draw off water [from poljes]'. (p. 580) See also Karst terminology. Compare pothole, pot.

**Pool**
*O.E.D.* 1. A small body of still or standing water, permanent or temporary; chiefly one of natural formation.
2. A deep or still place in a river or stream.
Also in special combinations: plunge-pool (*q.v.*), swimming-pool etc., see also Whirlpool. *Cf.* pond.
*London Pool:* the part of the River Thames just below London Bridge, much used by shipping which is unable to pass under the bridge.

**Poort** (Afrikaans: *lit.* gate)
Gorge in which a river breaks through a range of hills or mountains. *Cf.* the Westphalian Gate of the Weser, and the Iron Gate of the Danube. (P.S.)
A water-gap, cut by a river through a transverse ridge. (J.H.W.)

**Population geography**
G. T. Trewartha urged geographers to recognize population geography as a distinct and primary subdivision of the broader field of human geography.
Zelinsky, W., 1966, *A Prologue to Population Geography*, London: Prentice-Hall International. 'Population geography can be defined accurately as the science that deals with the ways in which geographic character of places is formed by, and in turn reacts upon, a set of population phenomena that vary within it through both space and time as they follow their own behavioural laws, interacting one with another and with numerous demographic phenomena. "Place" in this context may be a territory of any extent, from a few acres up to the entire surface of the earth. In briefer terms, the population geographer studies the spatial aspects of population in the context of the aggregate nature of places.' (p. 5)
Trewartha, G. T., 1969, *A Geography of Population*, New York: Wiley. '... this study of the spatial variations in human population ... this involves not only numbers but also population characteristics, as well as growth and mobility.' (p. 1)

**Population potential**
Concept developed by Stewart, J. Q. and Warntz, W., 1958, Macrogeography and Social Science, *Geog. Rev.*, April. 'The potential is a measure of the propinquity of people. Each individual contributes to the total potential at any place an amount equal to the reciprocal of his distance away; contours therefore are in units of "persons per mile".' (p. 171)

**Poramboke** (India: Urdu)
Unassessed waste where cultivation is prohibited; *e.g.* tank-beds, footpaths, village sites, etc. (G.K.)

**Porosity**
*O.E.D.* The quality or fact of being porous; porous consistence. Porous: full of or abounding in pores; having minute interstices through which water, air, light etc. may pass. (1879. Rutley, Stud. Rocks, 1.5. Questions of water supply hinge mainly

on the porous or impervious character of rocks. Also 1625. Carpenter.)

Wooldridge and Morgan, 1937. 'Perviousness as a rock character must be clearly distinguished from porosity. Crystalline rocks show but the slightest porosity since their crystals interlock, but they may be highly pervious in virtue of numerous joints. Conversely most argillaceous rocks, while distinctly porous, are essentially impervious. The pore spaces are too small to allow free passage of water, which is firmly held by surface tension.' (p. 268)

Babbitt, H. E., and Doland, J. J., 1949, *Water Supply Engineering*, 4th ed., New York: McGraw-Hill. 'Porosity is the volume of the pores of a substance. It is usually expressed as a percentage of its total volume. It represents the volume of water that the dry material will soak up. Perviousness is the capacity of a substance to allow water to pass through it, which is equivalent to the water available for a supply.' (p. 65)

### Port

*S.O.E.D.* 1. A place by the shore where ships may run in for shelter from storms, or to load and unload; a harbour, a haven. 2. A town possessing a harbour... Also *obsolete*: a town; perhaps specifically a walled-town or a market town.

Comment. Despite efforts at more precise definitions (see Harbour) the word port is even more loosely used than harbour. See also Outport, Treaty port.

### Portage

*O.E.D.* II. The carrying or transporting of boats and goods from one navigable water to another, as between two lakes or rivers. A place at or over which such portage is necessary, 1698.

Comment. *O.E.D.* says originally American; but probably derived from French and used by early French explorers.

### Portolan chart

A chart (*q.v.*) widely drawn and used in the 14th to 16th centuries characterized by radiating systems of loxodromes (*q.v.*).

### Possibilism (Febvre)

The philosophical concept which stresses the freedom of man to choose.

Tatham, G., in Taylor, 1951. 'Other geographers, particularly those who entered the field after training in history, instead of natural science, have tended to stress this freedom of man to choose. For them the pattern of human activity on the earth's surface is the result of the initiative and mobility of man operating within a frame of natural forces. Without denying the limits every environment sets to man's ambition, they emphasize the scope of man's action rather than these limits Febvre has named this point of view "Possibilism" and a very vigorous statement of its principles is to be found in his *Geographical Introduction to History*. The development of Possibilism is closely linked with the writings of Vidal de la Blache and Brunhes in France, and of Isaiah Bowman and Carl Sauer (among others) in the U.S.A.' (*Environmentalism and Possibilism*, p. 151)

### Possibilism, pragmatic (G. Tatham, 1951, p. 161)

Suggested by Tatham (*op. cit. sup.*) as a possible alternative name for Griffith Taylor's Stop-and-Go Determinism.

### Potamology, potamologic, potamological

*O.E.D.* The scientific study of rivers (from 1829).

Ishikawa, Y., 1959, Potamologic Geography of the Doki River, Kagawa Prefecture, *Proc. IGU Regional Conference in Japan*, 1957, Tokyo. 'The study area includes the entire drainage basin of the Doki River.' (abstract only) (p. 153).

### Potential Production Unit, P.P.U. (L. D. Stamp, 1954)

Stamp, L. D., 1954, *The Under-Developed Lands of Britain*, London: Soil Association.

Stamp, L. D., 1955, *Man and the Land*, Collins New Naturalist Series. 'I have made an attempt to measure more precisely the relative output of different types of land by introducing a Potential Production Unit (P.P.U.) equivalent to the average output of ordinary good farmland.... Best lands have a ranking of 2 P.P.U.; poor lands... only 0·1 P.P.U.' (p. 255).

See also *Geog. Rev.*, **48**, 1958, 1–15.

### Pot hole, pot-hole, pothole, pot; also potholing

*O.E.D.* 1. Geology. A deep hole of more or less cylindrical shape, especially one formed by the wearing away of rock by the rotation of a stone, or a collection of gravel, in an eddy of running water, or in the bed of a glacier. 2. See quot. 1898. *Archaeol. Jour.* Ser. II, V, 294. That the manufacture of pottery was carried on in Hayling in former times is shown by the existence of 'pot-holes', *i.e.* holes from which clay has been taken.

Mill, *Dict. Pot.* A hole produced by the action of weather, usually at the intersection of two opened joints; sometimes also applied to the more irregular of the swallow-holes, and often to the funnel-shaped hollows through the shale or drift which surmounts the shaft of these (Jura). Cauldron, Pothole, or Giant's Kettle. A natural formation suggesting a large kettle or boiler, either from shape or form, containing a fluid in a state of agitation.

Gemmell, A. and Myers, J. O., *Underground Adventurer*, Yorkshire: Dalesman, 1952. Pot. Pot-hole. Any underground system including pitches which require rope ladders for a descent. The term is also applied to individual pitches, *e.g.* Fall Pot in Lancaster Hole.

*Comment.* Notice at least four distinct and all common uses:
(a) potholes worn by running water whirling round stones;
(b) potholes which develop in a road mainly as a result of traffic;
(c) potholes or ponor in karst country;
(d) pothole as used by explorers of underground caverns in the sport of potholing by potholers.

**Pound**
S.O.E.D. An enclosure maintained by authority, for the detention of stray or trespassing cattle, and for the keeping of distrained cattle or goods until redeemed; a pinfold. An enclosure for sheltering or dealing with sheep or cattle in the aggregate.

**Power**
O.E.D. 13. Any form of energy or force available for application to work. spec. a. Mechanical energy (as that of gravitation, running water, wind, steam, electricity), as distinguished from hand-labour; often viewed as a commodity saleable in definite quantities. b. Force applied to produce motion or pressure; the acting force in a lever or other mechanical power, as opposed to the weight.

Mill, *Dict.* A nation of sufficient size, wealth, and war-like strength to maintain its influence amongst other nations.

**PPU, P.P.U.**—see Potential Production Unit

**Pradoliny** (Polish, *lit.* pra = old, doliny = valley)
Monkhouse, 1954. 'While northern and central Europe was under the continental Quaternary ice, the southern edge of the sheet lay more or less west-east. As the uplands of central Europe formed a barrier farther south, a great volume of melt-water was forced to flow westwards into the North Sea (when it was not frozen) along the front of the ice-sheet, thus carving west-east depressions, known as *Urstromtäler* ('ancient river valleys') in Germany and *Pradoliny* in Poland.' (pp. 151–2).

**Prairie** (French; *lit.* meadow)
O.E.D. (From French, prairie, a tract of meadow land; Latin, pratum, meadow). A tract of level or undulating grassland, without trees, and usually of great extent, applied chiefly to the grassy plains of North America; a savannah, a steppe. In *salt* or *soda prairie*, extended to a level barren tract covered with an efflorescence of natron or soda, as in New Mexico etc.; in *trembling* or *shaking prairie*, to quaking bogland covered with thin herbage, in Louisiana.

*Webster.* 1. A meadow or tract of grassland; specifically a. An extensive tract of level or rolling land in the Mississippi Valley, characterized in general by a deep fertile soil, and, except where cultivated, by a covering of coarse grass without trees. The lack of forests has been attributed to compactness of the soil, to Indian fires, and to low rainfall. b. Less correctly, one of the plateaus into which the prairies proper merge on the west, whose treeless state is due to dryness. c. A low, sandy, grass-grown tract in the Florida pine woods.

*Dict. Am.* A level or rolling area of land, destitute of trees and usually covered with grass. The word has been, and still is, applied variously. 1770.

Mill, *Dict.* French term for meadow; applied in N. America to the grasslands where the rainfall is adequate for cultivation.

Küchler, A. W., 1947, Localizing Vegetation Terms, *A.A.A.G.*, 37.
1. The grasslands of the area from western Indiana to eastern Colorado and from central Alberta to Texas. This is the prior usage and favoured by Küchler. 2. Any area of tall grass, as distinct from steppe,

or areas of short grass. Based on Weaver and Clements this is confusing and results in much of the Russian Steppe being termed Prairie; Küchler does not accept this usage. (W.M.) (pp. 198–199)

*Comment.* Prairie is the common word in French for grassland or meadow and confusion has resulted because the reference in French literature is often to small enclosed fields. Moreover in the international statistics of land-use published by FAO 'Prairies et paturâges permanents' is translated 'Permanent meadows and pastures'. The separation of the American prairies into Long Grass Prairie and Short Grass Prairie has been extended but Steppe is substituted for Short Grass Prairie by James in the South American Pampas. The best solution would seem to be that proposed under steppe (*q.v.*).

**Prairie soil**
Jacks, 1954. (1) Soil developed under grass in humid temperate regions and resembling chernozem, but dark brown on the surface, ordinarily with some textural profile and without a prominent horizon of accumulated calcium carbonate.
(2) A general term for all dark soils of treeless plains.

**Precinct**
*O.E.D.* 1. The space enclosed by the walls or other boundaries of a particular place or building, or by an imaginary line drawn around it; specifically the ground (sometimes consecrated) immediately surrounding a religious house or place of worship. b. esp. in pl., often applied more vaguely to the region lying immediately around a place, without distinct reference to any enclosure; the environs. 2. A girding or enclosing line or surface; a boundary or limit, a compass. 3. A district defined for purposes of government or representation; a district over which a person or body has jurisdiction; a province; also a division of a city, town or parish; specifically in U.S., a subdivision of a county or ward for election purposes.
*Precinct.* Division of the arable field of a township where the common arable was not divided into 'Fields'. Precincts were composed of furlongs and strips and were in some cases coterminous with 'shifts' (*q.v.*). Norfolk, 15th to 17th centuries (I.L.A.T.).
Forshaw, J. G., and Abercrombie, P., 1943,

*County of London Plan,* London: Macmillan. '... the area by-passed, or the precinct, round which traffic sweeps without invading it.' (p. 10)
*Comment.* The word precinct is now widely used in town planning to denote a specialized area in a town *e.g.* a shopping precinct, especially one not traversed by a main road or one accessible only on foot.

**Precipitation**
*O.E.D.* Condensation and deposition of moisture from the state of vapour, as by cooling; especially in the formation of dew, rain, snow, etc. That which is so deposited.
*Webster.* 4. *Meteorol.* A deposit on the earth of hail, mist, rain, sleet, or snow; also, the quantity of water deposited. Deposits of dew, fog, and frost are not regarded by the United States Weather Bureau as *precipitation.* Sleet and snow are melted.
*Met. Gloss.,* 1944. Used in meteorology to denote any aqueous deposit, in liquid or solid form, derived from the atmosphere.
*Comment.* Precipitation should strictly be used where most authors use rainfall since it includes snowfall, water falling as hail or deposited as dew.

**Precipitation day**
Atlas of Canada, 1959, plate 26. 'A "precipitation day" is considered to be a day on which the rainfall recorded amounts to one hundredth of an inch or more, or the snowfall measured is one tenth of an inch or more.' See Rainy day.

**Primärrumpf** (German, *pl.* Primärrümpfe)
Penck, W., *Morphological Analysis of Land Forms* (Trans. H. Czech and K. C. Boswell, 1953), London: Macmillan. A *primary peneplane*; a low featureless plain following uplift of a land mass so slow as to be matched by the rate of denudation, so there is no actual increase in relief or net rise of the surface. (p. 215)
*Comment.* Has been described as an 'old from birth peneplain'.

**Primary**
Of the first order in time or precedence—see Geological time, Industry, etc.

**Primate city**
Jefferson, M., 1939. The Law of the Primate City, *Geog. Rev.,* 29. '... once a city is larger than any other in its country, this mere fact gives it an impetus to grow that cannot affect any other city, and it draws

away from all of them in character as well as in size. It is the best market for all exceptional products. It becomes the primate city' (p. 227). 'The facts of record the world over seem to justify the law: a country's leading city is always disproportionately large and exceptionally expressive of national capacity and feeling' (p. 227). A primate city is one much larger than any other in the country and is a unique centre of trade, opportunity, culture, political affairs etc.
No reference in Van Cleef, 1937; Dickinson, 1947; Smailes, 1953.
*Comment.* Equivalent to metropolis (*q.v.*) as defined by some experts. Not a term widely adopted. (W.G.E.)
**Prime meridian**—see Meridian
**Primeur** (French)
*O.E.D.* Anything new or early, especially fruit before its ordinary season.
Stamp, L. D., 1953, *Africa*, New York: Wiley. 'Near Algiers along the fertile raised beach are carefully cultivated plots with "primeurs" or early vegetables, including tomatoes, potatoes and beans, also grapevines and fruits.' (p. 245)
**Prisere** (Ecology)
The whole evolution from a pioneer plant community to a climatic climax—see Climax.
**Private information field**
Hägerstrand, T., 1967, *Innovation Diffusion as a Spatial Process*, Chicago: University of Chicago Press. Private information is information passed from person ($A$) to person ($B$) and so on. Private information fields attempt to find 'the probable distribution of the information at different selected junctures in the process ... observe the extent to which information during a certain time ($t_0 - t_1$) flowed
from $A$ to $B$, $C$, $D$
from $B$ to $A$, $C$, $D$
from $C$ to $A$, $D$, $B$
etc. through the entire population. We designate these individual structures as $A$s, $B$s, $C$s, ..., private information fields for the period $t_0 - t_1$, or, more simply, information fields.' (p. 165)
**Probabilism**
*O.E.D.* 2. The theory that there is no absolutely certain knowledge, but that there may be grounds on belief sufficient for practical life. 1902 Baldwin *Dict. Philos.* II.344. 'The term probabilism is also used to describe the theory which mediates between a sceptical view regarding knowledge, and the needs of practical life.'
**Process-response systems (Process-response model)**
Chorley and Kennedy, 1971. 'Process-Response Systems. These represent the linkage of at least one morphological and one cascading system, so that process-response system demonstrates the manner in which form is related to process.' (p. 3)
Whitten, E. H. T. 1964, Process-response Models in Geology, *Bull. Geol. Soc. Am.*, 75. 'A geological process model is a framework for the processes that operate within a specified geographic domain, and a geological response model is concerned with the attributes of a population of samples. Commonly it is convenient to consider a process-response model as a framework within which a specified set of process factors is linked to a specific rock formation.' (p. 455)
See Chorley and Kennedy, pp. 9, 126–159.
**Producers' goods**
*Webster. Econ.* Goods that satisfy wants only indirectly as factors in the production of other goods, such as tools and raw material.
**Production**
*O.E.D.* 1.b. (quote) 1825. McCulloch, Pol. Econ. II, 1, 61. By production, in the science of Political Economy, we are not to understand the production of matter, ... but the production of utility, and consequently of exchange value, by appropriating and modifying matter already in existence. 2. That which is produced; a thing that results from any action, process or effort; a product.
Committee, List 3. Total output measured in absolute terms.
**Productivity (of land)**
Committee, List 3. 1. Actual productivity; the equivalent of yield. 2. Potential productivity; the hypothetical yield under stated conditions, especially under the best conditions possible.
*Comment.* In the latter sense 'land potential' is frequently used.
**Profile of equilibrium**
Profile of a graded river—see River profile.
*Comment.* For a discussion of the concept and its impossible realization in nature see Briault, E. W. H. and Hubbard, J. H., 1957, *An Introduction to Advanced Geography*, London: Longmans, p. 99.

**Profile, Projected**—see Projected profile
**Profile, River**—see River profile
**Profile, Soil**—see Soil profile
**Proglacial**
Wooldridge and Morgan, 1937. 'We may distinguish cases in which the lake occupies the lower end of an ice-free valley, others in which it is ponded between the ice-front and rising ground, sometimes a morainic ridge, while in a few cases the lakes rest between two morainic ridges. The term "extra-morainic" has been proposed to cover all such lakes, but they are evidently not literally extra-morainic in every case, and the term "pro-glacial" is preferable.' (p. 393)
*Comment.* 'Pro-' may refer to local position (in front of) or in the sense of 'before in time'. See *O.E.D.*
**Progradation**
Johnson, 1919. 'Just so long as the current aggrades (builds up) the sea bottom offshore, the waves will prograde (build forward) the shore. Following Davis we may call any shore which is experiencing such a long-continued advance into the sea, a prograding shore, and distinguish it from the more usual retreating or retrograding shore.'
Similarly Cotton, 1922; Wooldridge and Morgan, 1937.
**Projected profile**
Monkhouse and Wilkinson, 1952. 'Projected Profiles—It is possible to plot on a single diagram a series of profiles including only those features not obscured by higher intervening forms, this will give a panoramic effect . . . it is, in fact, an outline landscape drawing showing only summit detail. . . . The profiles should be spaced at equal intervals, but it is possible to add selected lines, running along, for example, a crest line.'
Briault, E. W. H., and Hubbard, J. H., 1957, *An Introduction to Advanced Geography*, London: Longmans. 'The projected profile is a rather special development of the cross-section, which is sometimes valuable in bringing out an accordance of summit levels or the occurrence of plateau remnants at more than one level . . . instead of a simple section along a line . . . the projected profile covers a rectangle' (p. 66). The sections are drawn along a number of parallel lines equally spaced and are superimposed.

For an early example see Miller, A. A., 1935, *Geog. Jour.*, **85**, 173.
*Comment.* The earliest use was by Barrell, J., 1920. The Piedmont Terraces of the Northern Appalachians, *Amer. Jour. Sci.* (4th Ser.), **49**, 227-258, 327-362, 407-428. The types referred to by Briault and Hubbard are called Composite Profiles by Monkhouse and Wilkinson and Zonal Profiles by Raisz. Monkhouse and Wilkinson also refer to Superimposed Profiles where all parts of straightforward profiles are placed one on top of the other—a device used by Trueman.
**Projection**
*O.E.D.* 7.b. Cartography. A representation on a plane surface, on any system, geometrical or other, of the whole or any part of the surface of the earth, or of the celestial sphere; any one of the many modes in which this is done.
Mill, *Dict.* Map-projection. The representation of lines or figures of the spheroidal surface of the Earth on plane-surfaces, the object being to find an arrangement giving minimum error for the work in hand. The following list summarizes some of the possible arrangements: . . .
*Notes on Making of maps* etc., 1937, London: H.M.S.O. 'A map projection is any systematic way of representing the meridians and parallels of the earth on a flat sheet of paper.' (p. 119)
*L.D.G.* A projection is an orderly arrangement of the parallels and meridians which allows the surface of the globe, or any portion of it, to be mapped in a convenient manner. It is impossible to map any part of a sphere on a plane, or a flat sheet, with complete accuracy; it is necessary therefore to choose the properties which are desirable for the purpose of the proposed map—correct area, shape, bearing, scale. True shape can be obtained for small areas only, and is incompatible with correct area. A map which shows correct areas cannot give true direction. No projection can give true distances over the whole surface. Certain classes, *e.g.* perspective projections, can be obtained by geometrical construction—the zenithal or azimuthal (true direction from the centre), the gnomic (shortest distance between two points is a straight line), stereographic (preserves correct shape) and the orthographic (produces effect of a globe). These

are useful features but the modified, or non-perspective, projections are in greater use since the net can be calculated to meet particular requirements. Two much-used classes are derived from the so-called developable surfaces, those of the cylinder and the cone (Mercator's is a modified cylindrical, and Bonne's a modified conic. Conventional projections include the Mollweide (giving the whole surface within an elipse) and various interrupted projections. The latter modify the central meridian or meridians to show the areas of most interest conveniently, with the result that the whole surface is not continuous (Goode). Good atlases generally discuss the projections they employ in the introductory matter, and indicate the correction to be applied to distances. Projections frequently used for maps in atlases are Mercator (bearings are straight lines and shapes correct but areas in high latitudes greatly exaggerated; also used for navigational charts), Mollweide (equal area but shapes greatly distorted on margins), Bonne (modified conic, used for great continental areas), zenithal equidistant (correct direction and distance from centre, used for Polar regions). A number of projections have transverse and oblique forms, when in place of the equator as axis, a suitable great circle is used. The principles of projections and methods of construction are discussed by A. R. Hinks, *Map Projections*, and J. A. Steers, *Study of map projections*.

### Proluvial
'The term "proluvial" is used in U.S.S.R. for dry delta sediments formed by temporarily existing rivers.' (V. A. Kovda in MS., also in *History of Land Use in Arid Regions*, Paris: UNESCO, 1961.) Kovda uses alluvial, proluvial and diluvial accumulations. (L.D.S.)

### Promontory
*O.E.D.* A point of high land which juts out into the sea or other expanse of water beyond the line of coast, a headland.
*Comment.* A cape, but not used in place names.

### Proterozoic
The older Palaeozoic systems—Cambrian, Ordovician and Silurian (*cf.* Deuterozoic).

### Province, provincial
The evolution of the many meanings of these words may be traced in the *O.E.D.* In the Roman Empire a country or territory under Roman rule but outside Italy especially Provence (France) one of the earliest provinces. Hence used of other empires and with the implication of provincial being away from the centre or capital. Later used of the major administrative divisions of a country especially if distinct historically, linguistically or in other ways (provinces of Spain, Ireland, India, etc.). Also of natural divisions of a country (physiographic provinces) especially to avoid the overworked word region (*e.g.* physiographic provinces of North America). Also extended to mean the sphere of action of a person or body of persons (*e.g.* ecclesiastical province, branch of learning). Provincial and provincialism tend to be used in a slightly derogatory sense. Attempts to give a precise definition to province are best avoided. (L.D.S.)

### Psammite, psammitic
Himus, 1954. 'Clastic sediments made up of particles of sand size.' From Greek *psammos*, sand.
*Comment.* Psammitic = arenaceous. Read, H. H., 1958, *Proc. Geol. Assoc.*, **69**, 84, gives reasons for using the three groups psephites, psammites and pelites. *O.E.D.* quotes Page, 1859, 'a term in common use among Continental geologists for fine-grained fissile, clayey sandstones in contradistinction to those which are more siliceous and gritty', which is a much more restricted use.

### Psephite, psephitic
Read, H. H., 1958, *Proc. Geol. Assoc.*, **69**. 'Rocks composed of large rock-fragments, pebbles or blocks set in a groundmass varying in kind and amount' (p. 87). '... may arise by processes other than those of sedimentation... accordingly pseudo-psephites... such rocks as fault-breccias, crush-conglomerates and tectonic melanges.' (*Ibid.*, 87–88)
*Comment.* From Greek *psephos* (ψῆφος) a pebble. Read attributes the introduction of the term to Haüy and Naumann in the mid-19th century but it has been so rarely used that it is not in most geological glossaries. Read gives reasons for its resuscitation together with psammite and pelite.

### Pseudo-cirque
Charlesworth, 1957, uses this term for arm-chair-shaped hollows resembling cirques

but of non-glacial origin found in arid lands in limestone, sandstone, and granite rocks. He attributes the term to O. W. Freeman. (G.T.W.)

**Pseudogley Soil**

FAO, 1963. '*Pseudogley* soils have ABC horizon sequence with a compact textural B-horizon. Both A and B horizons display strong mottling due to a temporary water-table above the B-horizon. The soils are very acid and destruction of clay minerals occurs in the upper part of the B-horizon into which the A-horizon may penetrate in tongues.'

**Pseudovolcanic features**

Thornbury, 1954. 'Certain topographic features resemble volcanic forms so much that, lacking a better name, we have designated them as pseudovolcanic.' (p. 515) They include bomb and meteor craters, Carolina bays, salt plugs etc.

**Psychosphere**

Sears, Paul B., 1957, *The Ecology of Man*, Univ. of Oregon Press. [After listing the lithosphere, atmosphere, hydrosphere and biosphere as parts of the environment of man to be studied] 'to these might be added *mind*—the Psychosphere, studied by psychologists, anthropologists, and other social scientists'.

**Pudding Stone**

A country name in Britain for conglomerate.

**Pueblo** (Spanish; *lit.* people, population, town, village)

*O.E.D.* A town or village in Spain or Spanish America; especially a communal village or settlements of Indians. In American Archaeology applied to a communal or tribal dwelling of the aborigines of New Mexico, etc. Pueblo Indians, partly civilized and self-governing Indians, dwelling in pueblos, in New Mexico and Arizona.

*Webster.* 1. A town or village, in Spain, Spanish America, or the Philippine Islands. 2. One of the Indian villages of Arizona, New Mexico, and adjacent parts of Mexico and Texas, built of stone or adobe in the form of compact communal houses, sometimes of several stories or terraces. . . . 3. (*cap.*) An Indian of one of the pueblos. 4. Any Indian village of the southwestern United States.

Mill, *Dict*. Collections of rock houses, often of more than one storey in height, and built of adobes or square cut stones; some were very large and were occupied by a whole clan (Spanish America).

*Comment.* An example of a simple Spanish word which has been given a special though loose meaning in America and American literature. Best avoided.

**Pukka, pakka** (Indo-Pakistan: Urdu-Hindi)

*Lit.* built of stone or masonry, hence strong, good. Pukka well, one lined with masonry.

**Pumice, Pumice stone**

Himus, 1954. 'An extremely vesicular form of lava which resembles froth, resulting from the escape of gases and vapours during consolidation.' The pumice of commerce is light-coloured and has the same composition as rhyolite; if it consolidated as a glass it would form obsidian. Pumice being so full of holes is extremely light and will float on water.

**Puna** (South America; from Quechua language of Peru)

*O.E.D.* A high bleak plateau in the Peruvian Andes; especially the tableland lying between the two great chains of the Cordilleras at an elevation of more than 10,500 feet.

Mill, *Dict*. 1. Soroche = mountain sickness. 2. A climate zone of Bolivia, etc., between 11,000 and 12,500 feet altitude, having two seasons, *viz*. cold summer and a winter; air cold and dry. Puna Brava. A climate zone of Bolivia, etc., from 12,500 up to snow limit of 15,000 feet; district bleak and inhospitable.

Küchler, A. W., 1947, Localizing Vegetation Terms, *A.A.A.G.*, 37. 'Puna (Quechua): Reaching from north-central Peru into Bolivia, this grassland is limited to plateaus of very high altitudes (above 4,000 m.) and the mountains that rise even higher. The upper limit of the Puna is between 4,600 and 5,100 m. . . . There are many herbaceous plants other than grasses in the Puna, also low-growing cacti . . . Puna is also used physiographically, and as such includes areas as far south as northwestern Argentina with its distinctly xerophytic vegetation. One must therefore beware of a confusion in the use of the term Puna.' (pp. 198–199) (Based on Weberbauer, A., 1911, *Die Pflanzenwelt der Peruanischen Anden*, Leipzig, pp. 192–227.)

James, 1959. 'Montane Short Grass' (p. 39). 'In this formation the bunch grasses are more widely spaced than in the montane grassland to the north' [*i.e.* páramos] (p. 173). 'The word Puna, designating a region [in Bolivia] should not be confused with the same word used to designate a vegetation type.' (p. 209)

*Comment.* Notice the dangerous confusion.

**Punja** (India: Tamil; Hindi)
Dry land as opposed to nunja, wet land. (G.K.)

**Purdah** (Persian; Urdu)
A curtain, especially one screening women from men or strangers, hence applied to the system of secluding Indian and other eastern women of rank.

**Purga** (Russian, from Finnish *purku*)
*Webster*. In Siberia and Alaska, a very violent snowstorm or blizzard.

**Push-moraine** (also, less frequently, push-ridge moraine)
Pre-glacial material pushed up into paralle ridges by advancing land ice. Such ridging is well seen in parts of the Netherlands and the term is commonly used by Dutch workers lecturing or writing in English (*see inter alia* A. P. A. Vink, 1949, *Contribution to the knowledge of loess and cover-sands*, Wageningen, 104).
Charlesworth, 1957. (Term introduced by Chamberlin, 1893) 'Chamberlin distinguished several morainic sub-types. (1) "dump moraine" composed mainly of superglacial and englacial material dumped at the front of the ice; (2) "lodge" or "submarginal" moraine, consisting of subglacial debris lodged under a thin ice-edge and passing into "till billows" and subject to mechanical action of oscillating and oversliding ice—they may be a few miles broad, and, according to Chamberlin, the predominant type in North America; and (3) "push moraine" (other terms are "shoved moraine", *Stau-* or *Stauchmoräne*, Dutch Stuwwal), which had two varieties, the one composed of glacial material, the other of local and non-glacial debris which the advancing ice ridged up in its path. The lodge and dump elements had long been recognized.
Push moraines, first recognized as a type in the seventies of last century are usually broad, smooth and massive. They are relatively free from small surface irregularities, transversely asymmetrical, and frequently arc-shaped' (p. 410). See Chamberlin, T. C., 1893, *Jour. Geol.*, **1**, p. 28.

**Puszta** (Hungarian; *lit.* waste)
Mill, *Dict.* The Hungarian steppe region; also called alföld.

Shackleton, 1958. 'South-east of Kecskemet the soil is impregnated with salt and there is a large area of *puszta* ( = waste), known as the Bugac steppe. . . . Hungary's largest area of *puszta* lies west of Debrecen.' (p. 363)

**Puy** (French)
Himus, 1954. 'A small volcanic cone, so named from the extinct cones in Auvergne, France.'
Thornbury, 1954. 'Several *puys* (volcanic hills) of the Auvergne region of France, are remnants of extinct plug domes.'
Mill, *Dict.* Used for any isolated hill or mountain but more particularly for cones of extinct volcanoes (Central France).
Comment. The district in Central France described in English in the classic work of G. P. Scrope, *The Geology and Extinct Volcanoes of Central France*, 2nd enlarged edition, London: Murray, 1858, is associated with the early recognition by Guetthard (1715–1786) of the *puys* as evidence of former volcanic activity but the word puy has no precise meaning. The puys include both Pelean and Strombolian types of volcano.

**Pyroclasts, pyroclastic rocks**
Holmes, 1944. 'In addition to the eruption of hot gases and molten lavas from volcanoes, vast quantities of fragmental materials are often produced by the explosion of rapidly liberated gases. These materials, collectively known as *pyroclasts*, may themselves consist of molten or consolidating lava, ranging from the finest comminuted particles to masses of scoriae and volcanic bombs of considerable size, or they may be fragments of older rocks ranging from dust to large ejected blocks. . . .' (p. 443)
*Webster*. Pyroclastic. *Geol.* Formed by fragmentation as a result of volcanic or igneous action; of rocks or their texture.'
See also Tephra.

**Pyromagma**
A fluid, highly gas-charged, very hot basaltic lava (Hawaii).

**Pyrometamorphism**
Contact metamorphism (*q.v.*).

# Q

**Qadha** (Iraq: Arabic)—see Liwa

**Qanat, kanat** (Persian; from Arabic)
An underground conduit, karez or foggara.

**Qoz** (Sudan: Arabic *coll.* for sand-dune in the northern Sudan)
The word is especially applied to the extensive area of dunes which were formed during an arid period of the Pleistocene in Kordofan and Darfur, and which are now fixed by scrub. (J.H.G.L.)

**Quadrat (analysis)**
Harvey, D., in Chorley and Haggett, 1967. 'A quadrat is used to denote a small areal unit for sampling within a particular study region. Having decided upon an appropriate size and shape, the quadrats are laid down "at random" within the study area, and the number of points which fall in any quadrat are counted and a frequency distribution of quadrats containing $0, 1, 2, \ldots, n$ points is constructed. The resultant observed frequency distribution may then be compared with a theoretical probability distribution generated by a particular set of processes which we suspect govern the distribution of points over space.' (p. 572)

Smith, D. M., 1975, *Patterns in Human Geography*, Newton Abbot: David and Charles. '... compile information for very small areas of regular size and shape, which could in their turn be amalgamated into close approximations of any larger area for which information is needed. In a number of countries numerical data are already being compiled and made available for relatively small grid squares, or quadrats.' (p. 33)

See Grieg-Smith, P., 1957, *Quantitative Plant Ecology*, London: Butterworth.

**Quadroon**
*O.E.D.* One who has a quarter of negro blood; the offspring of a white person and a mulatto.
Also used loosely and of other races. *Cf.* octoroon (one eighth negro blood).

**Quagmire**
*O.E.D.* A piece of wet and boggy ground, too soft to sustain the weight of men or of the larger animals; a quaking bog; a fen, marsh.

**Quaking bog**
A quagmire; so called because it quakes or trembles when foot is put on it.

**Quantification, quantity, quantifying, quantifier**
*O.E.D.* Quantification: the action of quantifying. Quantify: to determine the quantity of, to measure (with quotations from 1589).
*Comment.* Words resuscitated by the modern geographical school of quantifiers which insists upon exact measurement. See Haggett, 1965. See also Exceptionalism.

**Quantitative revolution**
Burton, I., 1963, The Quantitative Revolution and Theoretical Geography, *Canadian Geographer*, V.11.1. '... the quantitative revolution in geography began in the late 1940s or early 1950s; it reached its culmination in the period 1957 to 1960, and is now over. Ackerman remarks that "Although the simpler forms of statistical aids have characterized geographic distribution analysis in the past, the discipline is commencing to turn to more complex statistical methods—an entirely logical development. The use of explanatory models and regression, correlation, variance and covariance analysis may be expected to be increasingly more frequent in the field. In the need for and value of these methods geography does not differ from other social sciences."' (p. 152)
Quotation from Ackerman, E. A. taken from *Geography as a Fundamental Research Discipline*, University of Chicago, Department of Geography, *Research Paper No. 53*, 1958 (p. 11). For further views on the 'quantitative revolution' see Hartshorne, R., 1959, *Perspective on the Nature of Geography*, Chicago: Rand McNally, and Spate, O. H. K., 1960, Quantity and Quality in Geography, *Ann. Assoc. Am. Geographers*, **50**, 377–394. See also Curry, L., 1967, Quantitative Geography, *Canadian Geographer*, XI.4, 265–279.

**Quaquaversal, quâquâversal, quâ-quâ-versal**
*O.E.D.* Turned or pointing in every direction; chiefly geological in phrase quaquaversal dip.
Green, A. H., 1877, *Physical Geology*, ix. 'If the beds dip away in all directions from a centre they are said to have a quaquaversal dip.' (p. 347)
*Comment.* Obsolescent since the structure would be more simply described as a structural dome.

**Quarry, Quarrying (1)**
*O.E.D.* Quarry. An open-air excavation from which stone for building or other purposes is obtained by cutting, blasting or the like; a place where the rock has been or is being cut away in order to be utilized.
Fay, 1920. '... In its widest sense the term mines includes quarries, and has been sometimes so construed by the courts; but when the distinction is drawn, mine denotes underground workings and quarry denotes superficial workings. Open workings for iron ore, clay, coal, etc. are called banks or pits rather than quarries....'
*Comment.* Though 'pit' is commonly used, especially in the north and midlands of England, for all coal mines, 'quarry' is now restricted to open-air workings though formerly applied also to mines. A term now much used for open-air working, especially of coal and ironstone, is 'opencast'. See also Mine, mining.

**Quarrying (2)**
Lobeck, 1939. 'Young streams are able to erode their channels by (a) *corrasion*, or scraping and scratching away the bedrock; (b) *impact*, or the effect of definite blows on the bed of the stream by large boulders; (c) *quarrying*, due to the lifting effect of the water as it pushes into the cracks of rock; and (d) *solution* ... quarrying provides the larger blocks that are rolled along.' (p. 193)
Also called plucking.

**Quaternary** (geology)—see Geological time.
**Quaternary** (industry)—see Industry, quaternary.

**Quicksand**
*O.E.D.* A bed of extremely loose wet sand, easily yielding to pressure and thus readily swallowing up any heavy object resting on it. Quicksands are frequent on some coasts, and are very dangerous to travellers, stranded ships etc.
*Comment.* Also found in some river sands and near river mouths; the condition is one of a fine loose sand supersaturated with water.

**Quinta** (Spanish; Portuguese)
A country mansion; used especially in South America.

**Quograph, Quo-graph** (T. Griffith Taylor 1938)
Taylor, 1951. A graph giving same results as a slide-rule. (p. 619)
Taylor, T. G., 1938, *The Geographical Laboratory*, University of Toronto Press. 'The Quograph is a device for determining quotients (*e.g.* densities), by the intersection at the diagonals of the ordinates for Dividends (*e.g.* crops) and Divisors (*e.g.* areas). It is a simple form of Nomograph, and serves the same purpose as a Slide Rule' (p. 56–57, with example). See also *Econ. Geog.*, **14**, Jan. 1938.

**Qurer** (Sudan: Arabic)
Silt land along the river-bank just above the highest flood level attained by the Nile, and watered by *shaduf* or *saqia*. (J.H.G.L.)

# R

**Ra** (Norwegian)
Ridge-formed moraine feature, generally in or near sea, of gravel and clay with a surface layer of large stones (*esp.* South Norway). (P.J.W.)

**Rabi** (Indo-Pakistan: Urdu-Hindi)
1. The cool dry or winter season in northern India.
2. Winter sown and spring harvested crop.
See also Kharif and Hemantic.

**Race** (1)
*O.E.D.* b. A tribe, nation, or people, regarded as of common stock. c. A group of several tribes or peoples, forming a distinct ethnical stock. d. One of the great divisions of mankind, having certain peculiarities in common.
Mill, *Dict.* Best restricted to denote one of the main divisions of mankind, thus all woolly-haired people (Ulotrichi) may be said to belong to one race; but usually the negrilloes, bushmen, negroes and Papuans are spoken of as races. A heterogeneous mixed people, such as the English or the Welsh, should never be termed a race. The members of the same race have important physical characteristics in common.
Trevor, J. C., Race in *Chambers's Encyclopedia*. '... a classificatory group of mankind characterized by the possession of certain inherited physical features....'
Huxley, J. S., and Haddon, A. C., 1935, *We Europeans*, London: Cape. '... as an ethnological term, the transmission by descent of certain constant traits sufficient to characterize a distinct type is insisted upon.' (pp. 19–20)
U.N.E.S.C.O., 1952. *The Race Concept*, Paris. Statement on the nature of race and race differences by physical anthropologists and geneticists—June, 1951: 'Scientists are generally agreed that all men living today belong to a single species, *Homo sapiens*.... The concept of race is unanimously regarded by anthropologists as a classificatory device providing a zoological frame within which the various groups of mankind may be arranged and by means of which studies of evolutionary processes can be facilitated. In its anthropological sense, the word "race" should be reserved for groups of mankind possessing well-developed and primarily heritable physical differences from other groups.' (p. 11)
See also *What is Race*, in the same series.
Carpenter, 1938. The geographical variation of an Association = faciation (Braun-Blanquet and Pavillard, 1930). (p. 227)
Jackson, B. D., 1928, *A Glossary of Botanic Terms*, 4th ed., London: Duckworth.
1. A variety of such fixity as to be reproduced from seed. 2. Used also in a loose sense for related individuals without regard to rank. (p. 318)
*Comment.* A useful discussion of the term, with references, is contained in Penniman, T. K., *A Hundred Years of Anthropology*, London: Duckworth. 'The views of those who would deny the existence of racial differences in any scientific sense of the term and those who, like Gates, believe Mankind to be split into distinct species are generally rejected by geneticists and non-geneticists alike.' (p. 407)

**Race** (2)
Swiftly flowing water in a narrow channel or river; also the channel itself which may be artificial as in a mill-race. Also a swift rush of water through a narrow channel in tidal waters and caused by the tidal movement of the waters.

**Racial Geography**
Taylor, 1951, has a chapter entitled *Racial Geography* in which the scope of racial geography is discussed at length though no concise definition is given except that the opening sentence states: 'Geography is concerned primarily with distributions, and of these distributions none is more important than the distribution of man himself.'

**Radial drainage**
Thornbury, 1954. 'Streams diverging from a central elevated tract.'

**Radiolarian ooze**
A type of ooze found in the deeper parts of tropical oceans, especially the Pacific, and characterized by an abundance of the

minute siliceous shells of Radiolaria. These are less soluble than the calcareous shells of foraminifera which are apt to be dissolved by the sea water before they reach the bottom where the ocean is very deep. In the deepest parts, however, even the siliceous radiolarian shells are dissolved and the bottom deposit is Red Clay (*q.v.*).

**Radiometric** (Arthur Holmes)
Radiometric age of rocks

**Rag, ragstone**
*S.O.E.D.* A piece (mass or bed) of hard, coarse or rough stone *obs.* exc. *dial.* A name for certain kinds of stone, chiefly of a hard coarse texture....
*Comment.* Best known is 'Kentish Rag' widely used as a building and road stone of Lower Cretaceous age.

**Railway gauge**
Brees, 1841, *Glossary Civil Engineering*. The width in the clear between the top flanges or the rounded rims of the rails.
*Comment.* The Standard Gauge, in use throughout Europe except Spain and Portugal and the U.S.S.R. and certain satellites and also used throughout North America is 4 ft 8½ in. Broad gauges, such as 5 ft 3 in. and 5 ft 6 in. are used in Ireland, Spain and Portugal, U.S.S.R., parts of India, Australia and South America. Narrow gauges especially 3 ft 6 in. (South Africa, parts of Australia) and metre (3 ft 3⅜ in.) are used in many parts of the world.

**Rain**
Mill, *Dict.* The condensed water vapour of the atmosphere, in drops large enough to fall under the influence of gravity with a sensible velocity. In clouds, mist and fog the droplets are smaller and remain suspended or with a scarcely perceptible downward movement. There is no clear distinction between rain and drizzle except that in a drizzle the droplets are small.
Similarly in *O.E.D.* etc. See also Mist.

**Rain-day**—See Rainy day, Precipitation day

**Rains, The**
Mill, *Dict.* The rainy season, or monsoon season, in India and other Monsoon countries.

**Rainfall**
*O.E.D.* 2. The quantity of rain falling in a certain time within a given area, usually estimated by inches (in depth) per annum.
Mill, *Dict.* (Meteorol.) This includes, unless a restricted meaning is expressly indicated, total precipitation, *viz.* rain, snow, hail etc. (the two latter estimated as water), measured by the total amount which would accumulate on flat ground if there were neither evaporation nor percolation.
*Comment.* For the calculation of mean annual rainfall, or mean monthly rainfalls, in Britain 'the adopted standard period is the thirty-five years 1881 to 1915' (Bilham, E. G., 1938, *The Climate of the British Isles*, London: Macmillan, p. 76). This is now being revised but 35 years is commonly accepted as a good period for calculation of averages.

**Rain forest, rain-forest, rainforest**
*O.E.D. Suppl.* A forest typical of rainy tropical regions.
Mill, *Dict.* Forest mainly composed of evergreen hygrophilous trees, most abundant in the torrid zone, but extending into the temperate zone, where, however, it usually has deciduous trees as subordinate components.
Schimper, A. F. W., 1903, *Plant-Geography*, Oxford: Clarendon. 'The temperate, like the tropical, rain-forest is essentially formed of evergreen hygrophilous trees, for therein truly consists the most essential characteristic of a rain-forest. In most cases, however, periodically foliaged trees occur as subordinate components, yet they are no longer rain-green, but are summer-green trees, such as, for instance, *Fagus obliqua* in South Chile' (p. 260). Also refers to the temperate rain forest in New Zealand and on the mountains of Java and Ceylon.
Richards, P. W., 1952, *The Tropical Rain Forest*, Cambridge: C.U.P. 'The name "Rain forest" is commonly given, not only to the evergreen forest of moist tropical lowlands... but also to the somewhat less luxuriant evergreen forest found at low and moderate altitudes on tropical mountains, and to the evergreen forests of oceanic subtropical climates, in southwestern China, southern Chile, South Africa, New Zealand and eastern extratropical Australia. These other formation-types will here be called Montane Rain forest and Subtropical Rain forest respectively. In this book "rain forest" without qualification means Tropical Rain forest' (pp. 1–2). Quotes Schimper (1903, p. 260) but many would prefer a narrower sense such that 'the term Rain forest would be reserved for the almost completely non-seasonal forest of tropical climates with the most evenly distributed

rainfall; it would not be applied to the more seasonal types of evergreen forest found in areas with a marked dry season.' 'Tropical Rain forest' (*tropischer Regenwald*) was a term coined by Schimper in *Plant Geography* (1898, 1903); other terms are pluvisylva, Hylaea (Humboldt, of S. America) and sometimes Urwald.

### Rain shadow, Rain-shadow
*O.E.D. Suppl.* Rain-shadow, a region in which the rainfall is small compared with the surrounding regions. (Quote: H. J. Mackinder, Britain and the British Seas, 1902.)

Mill, *Dict.* Rain-shadow Area. The area of reduced rainfall situated on the leeside of a sheltering mass.

*Met. Gloss.*, 1944. An area with a relatively small average rainfall due to sheltering by a range of hills from the prevailing rain-bearing winds.

*Comment.* Always used as in *Met. Gloss.*; *O.E.D.* definition is inadequate.

### Rain spell
In Britain a period of at least fifteen consecutive 'rain-days' (*q.v.*) on each of which at least 0·01 inch of rain has fallen.

### Rain-wash, rainwash
Himus, 1954. The surface creep of soil and superficial rocks, lubricated by rain water, under the action of gravity. Also applied to the material which originates in this way. Hill-wash is often used as an alternative to rain-wash.

### Rainy day, rain-day, day of rain
Mill, *Dict.* (Meteorology) A day having at least one-hundredth of an inch of rain (Symons); 1 mm. is the limit for the continent of Europe, etc.

*Met. Gloss.*, 1944. Rain day. A period of 24 hours, commencing normally at 9 h., on which ·01 in. or 0·2 mm. or more of rain is recorded.

*Comment.* 'Rainy day' may be used in a general sense, but 'rain-day' has the precise meaning of the *Met. Gloss.*

### Raised beach
*O.E.D.* A former beach, now situated above sea-level.

Mill, *Dict.* Raised Beach. Strand Line, Coast Terrace. An ancient beach now elevated above sea-level.

Geikie, J., 1898. 'Former sea-margins; sometimes appear as shelves cut in solid rock; ...'

*Webster. Geol.* A beach formed by a sea or lake and subsequently elevated above high-water level; not properly applicable to a beach abandoned by a lake which is drying up or being drained.

### Rake (Scottish)
Hunter, R. F., 1964, Social behaviour in hill sheep, *Adv. of Science*, **21**. 'Each sheep has a well-defined group of patches of different types of vegetation over which it moves in a fairly regular daily fashion. This behaviour is described in the shepherds' term, a sheep's 'rake'. The size of a rake may vary greatly from sheep to sheep ... 60 to 150 acres' (p. 170). See also Home range.

### Rand (Afrikaans)
As explained under *rant* linguistic purists prefer rant but popularly the Witwatersrand is always the Rand though pronounced rant. On 14 February 1961 South Africa introduced a decimal coinage with the rand (10 shillings) divided into 100 cents.

### Randkluft (German, *pl.* Randklüfte)
See Bergschrund, ref. *Ice Gloss.*, 1958.

### Raml, ramla (North Africa: Arabic; *lit.* sand)
El Raml, the sand, *par excellence*—the desert. Professor M. Kassas (in MS.) points out that ramla or ramlah is used in Spain to mean the bed of an ephemeral river when dry.

### Rañas (Spanish)
Fanglomerate (or Saharan *reg*) consisting of a surface of rolled pebbles. (P.D.)

### Range (1)
A row or line of things especially of mountains, hence range of mountains or mountain-range often shortened to 'range'.

*Webster*. 3a. *pl.* Mountains; mountainous country. b. In Australia, sometimes, a single mountain. c. A series or chain of mountain peaks considered as forming one connected system; a ridge of mountains; as, the Appalachian *range*. d. *Western U.S.* Any stretch of open country. e. In the Lake Superior iron region, the belt of outcrop of an iron-bearing formation.

Mill, *Dict.* An elevated portion of land distinguished by a well-marked (but not always unbroken) longitudinal extension, and usually possessing an equally well-marked crest-line or 'divide' (Holdich).

*Comment.* A mountain-range is usually reserved for a single line of mountains; a chain is more complex and may include several ranges.

**Range (2)**
The difference between maxima and minima in temperature, pressure, elevation etc., the meaning made clear by the context or by qualifying words, *e.g.* mean annual range of temperature = the difference between the mean temperature of the hottest and coldest months.
*Webster.* 9. The limits of a series of actual or possible variations; also, the series or scale of variations within such limits. 22. *Statistics.* The difference between the least and the greatest values. . . .

**Range (3)**
The area (*e.g.* in plant and animal geography) within which, or the period (*e.g.* in palaeontology) within which the occurrence of something (*e.g.* a given species) is possible.
Sears, Paul B., 1957, *The Ecology of Man*, Univ. of Oregon Press. 'Every organism (except man) is confined to a limited geographical area known as its Range . . . within that Range it is usually restricted to certain favorable Habitats that afford it an appropriate Niche (*q.v.*). These patterns are set by the operation of what are called Limiting Factors' (p. 14).

**Range (4)**
*Dict. Am.* 1. An area of uncultivated ground or wild country over which domestic or wild animals range for food, now especially a cattle range. 1626. 'plentiful range for cattle.'
An area, space, or stretch of ground over which ranging (*i.e.* wandering or roaming of men or animals) takes place or is possible. Hence, in the United States an extensive stretch of grazing or hunting ground; natural or semi-natural grazing ground.

**Range (5)**
*Webster.* 6a. In the public land system of the United States, a row or line of townships lying between two successive meridian lines six miles apart.
*Dict. Am.* In the public land survey, a row of townships lying between two meridians six miles apart, the rows being numbered in order from east to west from the principal meridian. For a discussion of the origin of this system of surveying land see F. C. Hicks, ed. *A Topographical Description* . . . by T. Hutchins, Cleveland, 1904, pp. 37 ff. 1785 *Jrnls. Cont. Congress* XXVIII. 376 'The geographer shall designate the townships . . . by numbers progressively from south to north, always beginning each range with number one.'

**Range (6)**
*Webster.* 19. *Mining.* A mineral belt; as, the Mesabi *range.*

**Range Type**
'A Range Type may be defined as a type of country which, because of a certain uniformity of climate, soil and vegetation, may be submitted to uniform land-use.' *Report on the Range Classification Survey of the Hashemite Kingdom of Jordan.* London: Hunting Technical Surveys Ltd., 4 Albemarle St., 1956 (p. 16). (Also uses 'Range specialist', 'Range maps', 'Range Managements' etc.)
*Comment.* This is a recent specialized extension derived from the use of range in the western United States in sense 4 and now taken by American workers to other parts of the world.

**Ranger**
One living on range land in sense 4; now used especially of an official such as a game warden or forest officer.

**Rank-size rule**
Haggett, 1972. '. . . if we arrange settlements in order of size, for some regions the population sizes are related. In its simplest form the relationship is such that the population of the $n$th city is $1/n$ the size of the largest city's population. This inverse relationship between the population of a city and its rank within a set of cities is termed the rank-size rule.' (p. 282) More generally, a plot of log (population) against log (rank) is a straight line.
*Comment.* The first positive recognitions of regularity came from a German geographer, Auerbach, F., 1913, Das Gesetz der Bevölkerungskonzentration, *Petermann's Mitteilungen*, **59**, 74–76. See also Jefferson, M., 1939, The Law of the Primate City, *Geog. Rev.*, **39**.2, April, 226–232.

**Ranker** (Soil Science)
FAO, 1963. '*Rankers* are soils with AC horizons developed on silicate rocks or sediments. The content of organic matter in the A-horizon is considerably higher than in the (A)-horizon of non-calcareous lithosols or regosols.'

**Rant** (Afrikaans; *pl.* rante)
*Antiq.*: *Rand*, preserved in proper names like Witwatersrand, *i.e.* the Ridge of White

Water. Range of hills, especially when forming an escarpment.

Rante can hardly be considered as indicating a special type of pasture. Any form of relief, however, accompanied by steep slopes, diminishes the value of the veld for farming, and for this reason the names of relief forms may sometimes be used to denote degrees of usefulness of the pasture. (P.S.) See also Rand.

The diminutive form *rantjie*, *pl. rantjies*, is used for a narrow ridge of rocky hills.

**Rape**

O.E.D. One of the six administrative districts into which Sussex is divided, each comprising several hundreds.

*Comment.* Recorded in Domesday Book. For a discussion of the word and of the division see A. Mawer and F. M. Stenton, *The Place Names of Sussex*, 1, pp. 8-10, Cambridge, 1929.

**Rapids** (rarely **rapid**)

Mill, *Dict.* Portions of a stream with accelerated current where it descends rapidly but without a break in the slope of the bed sufficient to form a waterfall.

**Rasputitsa** (Russian)

Swayne, 1956. 'The period of spring in Siberia and northern Russia, characterized by floods and mud resulting from the thaw.'

Ushakov (see Appendix I, Russian words). 'Time when the roads become difficult or impossible to use because of mud, state of the road at that time.'

**Rath** (Irish; now pron. rä)

O.E.D. Antiq. An enclosure made by a strong earthen wall and serving as a fort and place of residence for the chief of a tribe; a hill-fort.

*Comment.* Literally a mayor's settlement; used of an isolated farm settlement as contrasted with a clachan.

**Rauk** (Swedish; *pl.* raukar)

A sea stack; a lofty, isolated, columnar rock in the sea, isolated by wave erosion. Røk, *pl.* røkir or drangur on the Faroe Islands. (E.K.)

**Ravine**

*Webster.* 2. A small, narrow, steep-sided valley, larger than a gully and smaller than a canyon, and usually worn down by running water; a gorge or gulch.

Mill, *Dict.* (1) in general, any steep-sided valley. (2) A stream with a slight fall between rapids (North America).

**Ravinement** (French; *lit.* ravining, gullying)

*Webster.* Act or result of ravining.

Evidence of a geological unconformity on a small scale is sometimes seen when overlying beds have 'scooped down into' (ravined) the underlying beds. French geologists use raviner and ravinement in this special sense and the words were introduced into English by Stamp. See Stamp, L. D., 1921, *Proc. Geol. Assoc.*, **32**, 61; *Geol. Mag.*, **58**, 1921, 109.

Stamp, 1923. 'Although the bedding of the upper series is parallel to that of the lower, there are signs that the sea, in depositing the upper series, has slightly eroded the top of the lower series. Such an erosion line is termed a "ravinement".' (p. 17)

**Raw material**

O.E.D. Raw. 2. In a natural or unwrought state; not yet subjected to any process of dressing or manufacture. (Examples include 'raw materials' in quotations, 1796 and 1868.)

Committee, List 4. Raw materials. The tangible commodities transformed in a given industry or industrial process into a further product; sometimes loosely used to include the source of energy employed.

**Reach**

A straight section of a river, especially a navigable river, between two bends.

**Rean, rein**

Mill, *Dict.* An old terrace or ridge produced by cultivation (Northumbria). See also shine.

**Recent**

In the Geological time scale (*q.v.*) equivalent to Holocene. Broadly, the time since the last Great Ice Age.

**Recessional moraine**

Mill, *Dict.* The successive terminal moraines marking temporary limits of a glacier which on the whole is receding. Similarly Salisbury, 1905; Cotton, 1945.

No reference in Wright, 1914; Wooldridge and Morgan, 1937. The latter use stadial moraine (*q.v.*).

**Reciprocal movement**

Roseman, C. C., 1971, Migration as a Spatial and Temporal Process, *Ann. Assoc. Am. Geographers*, **61**.3, 589-598.

'... reciprocal movements of individuals, which begin at the home or dwelling, proceed to one or more specific locations, and return to the home. A reciprocal movement cycle is defined by aggregating all reciprocal movements of a person over a period of time.' (p. 590)

*Comment.* These reciprocal movement cycles can be daily, weekly, monthly, etc. See also Migration movement.

**Recumbent Fold**—see Fold, folding

Also quotation under Holmes, 1944, Thrust-plane.

**Red Clay**

Mill, *Dict.* A mineral deposit of volcanic and other fragments which accumulates in the deepest parts of the oceans where the water is so deep that the calcareous shells of foraminifera and even the siliceous shells of radiolaria are dissolved before reaching the bottom.

Also used for any red clay.

**Reddish-brown soil**

FAO, 1963. '*Reddish-brown soils and brown soils* have an AC or an ABC profile. The A1-horizon is reddish brown to red in color, the B-horizon is red or reddish brown, and a Ca-horizon that is often hardened is normally present. The brown soils have the same horizon sequence and similar properties but have a brown color.'

**Red Earth**

Jacks, 1954. Tropical soil usually leached, red, deep, clayey, and moderately low in combined silica (included with latosol in U.S. usage).

Jacks separates Red Loam: Tropical soil usually leached, red, deep, friable, and low in silica.

*Comment.* Red Earth should not be used as a translation of terra rossa.

**Redir** (Arabic; *pl.* redair)

Mill, *Dict.* Patches of temporary water frequently left in wadi (*q.v.*) after the torrent has ceased to flow.

Knox, 1904. A natural reservoir of rain water; a sheet of water; a temporary sea.

**Red-line areas (of building societies)**

Boddy, M. J., 1976, The Structure of Mortgage Finance: Building Societies and the British Social Formation, *Trans. Inst. Brit. Geog., New Series*, **1.1**. Areas viewed by mortgage controllers as 'being unstable in social and economic terms and thus any property they contain is considered unsuitable as mortgage security.... All mortgage allocators interviewed named specific areas, considered "blacked" or "red-lined", in which no property would be accepted as loan security.' (p. 67)

See also Weir, S., 1976, Red-line Districts, *Roof*, July, London: Shelter.

**Red Mediterranean Soil**

FAO, 1963. '*Red Mediterranean Soils* have an ABC horizon sequence. The profile has a textural B-horizon of high chroma (*i.e.* red or yellow) .... thick continuous clay coatings on ped surfaces in the B-horizon ... formed on calcareous and non-calcareous parent materials.' See also Brown Mediterranean soils.

**Red-yellow podzolic soil**

FAO, 1963. '*Red-yellow podzolic soils* have an ABC horizon sequence. The profile has an A2 horizon unless it has been removed by erosion. The textural B-horizon has a blocky structure, clay coating on ped surfaces and the base saturation of the clays is less than 40% or decreases with depth.'

**Reef**

*O.E.D.* 1. A narrow ridge or chain of rocks, shingle, or sand, lying at or near the surface of the water. 2. Gold-mining (orig. Austral.) a. A lode or vein of auriferous quartz. b. The bed-rock.

Mill, *Dict.* 1, A vein of auriferous quartz (N. America and Australia). 2. A submarine land form less than 35 feet below tide-level.

*Adm. Gloss.*, 1953. An area or rocks or coral, detached or not, of considerable extent, which may dry, or nearly dry, in places. Also sometimes used for a low rocky or coral area some of which is above water.

Fay, 1920. 1. (Aust.) A lode or vein. A word introduced into mining by sailors who left their ships to participate in the rush to Ballarat and Bendigo in 1851. To them a rock projecting above the water was a reef, and the term was therefore applied to quartz outcrops on land. 2. (So. Afr.) In the diamond mines, the barren shales, etc., limiting, like an oval funnel, the soft diamantiferous breccia.

*Comment.* See also Coral reef. When the context allows, coral reef can be understood though the reference may be only to 'reef' —as in fringing reef (a coral reef fringing the shore), barrier reef (separated from the shore by a lagoon and acting as a barrier between the open ocean and the sheltered waters of the lagoon). On a much larger scale the Great Barrier Reef for the thousand miles of its length acts similarly for the whole coast of Queensland. See Lake, 1958, 206.

**Reef knoll**
In certain limestones, especially the Carboniferous Limestone of the Craven district of Yorkshire, fossil coral reefs occur, building up masses which have proved more resistant to weathering than the surrounding beds and so form prominent rounded hills or knolls. These reef knolls vary in size from a few feet in diameter to masses hundreds of yards across. There is more than one type of reef; many consist of massive unbedded limestones, or show rudimentary beds dipping away from their centres. The term originated with Tiddeman, R. H., 1890, On the Formation of Knoll reefs, *Rept. Br. Ass. for* 1889, p. 600 and a general discussion of terminology is to be found in Bond, G., 1950, The Nomenclature of Lower Carboniferous 'Reef' Limestones in the North of England *Geol. Mag.*, **87**, 267–278. (G.T.W.)
See also Bioherm.

**Reference sheet**, see Characteristic sheet

**Reforestation, reafforestation**
The planting of trees over an area previously forested but which has been cut over or, sometimes, burnt over. The right word is obviously reforestation and it is now replacing reafforestation which was widely used in Britain.

**Refugee**
*S.O.E.D.* One who, owing to religious persecution or political troubles, seeks refuge in a foreign country; originally applied to the French Huguenots who came tc England after the revocation of the Edict of Nantes in 1685.
For the modern problem see Proudfoot, M. J., 1957, *European Refugees*, London; Faber.
See also exile, expatriate.

**Reg** (North Africa: Arabic)
Stony desert with a surface of gravel, gravel desert.
Mill, *Dict.* Stretches of desert surface on which small stones and gravel have been exposed by the blowing away of fine sand.
Knox, 1904. (Arabic, N. Africa), firm level ground generally without vegetation, a barren, naked plain. Another form is Rek.
Fitzgerald, W., 1950, *Africa*, London: Methuen, 7th Ed. 'An intermediate stage between rocky and sandy desert is represented by the gravel and pebble wastes that are the detritus of the eroded "hamada". This type is the "areg" or "reg"' (p. 60).
*Comment.* Fitzgerald confuses 'areg' (the plural of 'erg' or sandy desert) with reg, the gravel desert, and his error has unfortunately been copied in some textbooks. As Augustin Bernard notes in the *Geographie Universelle*, XI, 311, *Reg* is a desert of the plains, *hamada* of the plateaus and regs may be formed either *in situ* or of transported material.

**Regelation**
Mill, *Dict.* The freezing again of ice which has melted under pressure, or the amalgamation of portions of ice by subsequent release of exerted pressure.
Regelation plays an important part in the behaviour of glaciers.

**Régime** (French; anglicized as **regime**)
*O.E.D.* A manner, method, or system of rule or government.
Hence transferred in geography especially to rainfall or climate generally and especially to rivers.
*Webster. Hydraul.* The condition of a river with respect to the rate of its flow, as measured by the volume of water passing different cross sections in a given time.
Mill, *Dict.* The general cycle of activities with respect to inflow of rain-water, evaporation, outflow of river-water, movement of solid contents and mixture with the sea.
Kendrew, 1953. 'Precipitation. The seasonal distribution or "régime" is independent of the total amount.' (p. 17)
Lake, 1958. 'River Régimes. Régime signifies the totality of phenomena relating to the alimentation of rivers and streams, and their variations in outflow. The volume of water in a river may vary much in the course of a year.' (p. 326)
Ahlmann, H. W., 1948, *Glaciological Research on the North Atlantic Coasts*, London: *Roy. Geog. Soc.*, 'By the regime, or material balance' of a glacier is meant the total accumulation volume during one accumulation season and its total gross ablation volume during the following ablation season. Their sum is the glacier's balance-sheet total; it is best expressed in cubic metres per square kilometre. The regime comprises one "budget year". If in such a year the accumulation volume is greater than the ablation volume, the

regime is positive; otherwise it is negative. ... The regime is illustrated by a regime diagram, showing by curves the total volumes of accumulation and net ablation in relation to the areas of the altitude intervals.'

¹C. Troll (*Geol. Rundschau*, **34**, Stuttgart 1943) translates the word 'regime' by 'Massenbilanz' or 'Materialshaushalt' ...' (p. 49).

*Comment.* Mill is a little too wide.

### Regimen

*O.E.D. Suppl.* quotes *Geog. Jour.* **11**, 1898. 'The angle of the slope of the ridge of a shingle beach depends on the materials of which it is chiefly composed. *Regimen* is attained when the assistance which gravity gives to transport with the back-wash makes the seaward equal to the shore-ward transport.' (p. 634)

*Comment.* Rarely used; grade is used in this context or Johnson's 'profile of equilibrium'.

### Region

*O.E.D.* 1b. A large tract of land; a country; a more or less defined portion of the earth's surface, now esp. as distinguished by certain natural features, climatic conditions, a special fauna or flora, or the like. d. An area, space, or place, of more or less definite extent or character. 4a. One of the successive portions into which the air or atmosphere is theoretically divided according to height. Also similarly of the sea according to depth.

Mill, *Dict.* Any portion of space considered as possessing certain characteristics.

Committee, List 1. An area of the Earth's surface differentiated (from adjoining areas) by one or more features or characteristics which give it a measure of unity. According to the criteria employed in differentiation, regions are termed physiographic, political, economic, etc.

Wooldridge and East, 1951. A distinction is made between (a) special regions—each one being unique, and (b) generic regions containing a number of the same type. A second distinction is made between (1) homogeneous region—alike in all its parts and (2) synthetic region—made up of a number of contrasting though related parts. (pp. 144–147)

Schimper, A. F. W., 1903, Plant-Geography (Trans. W. R. Fisher). Oxford: Clarendon.

On ascending a mountain, three regions are possible, basal, montane and alpine regions. These are similar to but distinct from zones of vegetation which are related to latitude. (p. 702).

Braun-Blanquet, J., 1932, Plant Sociology. (Trans. Fuller and Conard). New York: McGraw-Hill. 'Following the floristic divisions of Engler, Flahault, Diels, etc., we recognize six regional units of different ranks: region, province, sector, sub-sector, district and subdistrict.... The region is the most comprehensive unit, characterized by numerous well-defined climax communities and many transition communities. It shows long-standing endemism in groups of high taxonomic rank (families, subfamilies, sections, etc.). The unity of the region is shown by the occurrence throughout, or nearly throughout, of identical or closely allied species of high importance (examples: Euro-Sibero-North American Region, Mediterranean region, Capetown region, oceanic (thalamic) region).' (p. 355)

Herbertson, A. J., 1904. The Major Natural Regions: an essay in Systematic Geography, *Geog. Jour.*, **25**. 'A natural region should have a certain unity of configuration, climate and vegetation. The ideal boundaries are the dissociating ocean, the severing mass of mountains or the inhospitable deserts.' (p. 309)

James, 1935. 'An area homogeneous with respect to announced criteria.'

*Comment.* See also Natural region, The study of regions other than those delimited by state boundaries was advanced in the 18th century in France (Buache), Holland and Italy but especially in Germany by Gatterer. (See Hartshorne, 1939, p. 37.) Its revival in Britain owes much to Mackinder and Herbertson. For a modern criticism see G. H. T. Kimble, The Inadequacy of the Regional Concept, in *London Essays in Geography*, London: Longmans, 1951, 151–174.

### Regional geography

Committee, List 1. The geographical study of particular areas or regions of the Earth's surface.

Mackinder, H. J., in discussion on Herbertson's paper (see Region, *supra*) '(Mr. Herbertson) seeks, as I believe, the best compromise of criteria for determining the natural regions of the world—in other words the "method" appropriate to regional geography.' (p. 312)

*Comment.* 1. It appears implied in Wooldridge and East, 1951, Chapter VIII that there are two types of regional geography: a. The study of special regions, which equates with 'special geography' of Varenius. b. The study of generic regions. This could be regarded not as regional geography but a form of systematic geography.

2. 'Regional' implies large area, in contradistinction to 'local' in geology, *e.g.* 'regional metamorphism'. See Fay, 1920; Holmes, 1920; Rice, 1941.

For a general discussion of the character of regional geography see Hartshorne, 1939, 436–459.

### Regionalism

*O.E.D.* Tendency to, or practice of, regional systems or methods; localism on a regional basis.

*Comment.* In geographical literature 'regionalism' is used with a wide variety of meanings:—

1. Regionalism in planning and the concept of regional planning or selecting a region as a basis for future development: often a heterogeneous unit comprising both urban and rural interests.

2. Regionalism in the sense of the French movement in the later 19th century to revive regional identities and feeling, sometimes with political overtones.

3. Regionalism as a local feeling of group consciousness associated with a particular area, *e.g.* the South, the West Country, etc. (C.T.S.).

See especially Dickinson, R. E., 1947, *City, Region and Regionalism*, London: Kegan Paul; Gilbert, E. W., 1939, Practical Regionalism in England and Wales, *Geog. Jour.*, **94**, 29–44; Gilbert, E. W., 1948, The Boundaries of Local Government Areas, *Geog. Jour.*, **111**, 172–206; Taylor, E. G. R., 1942, The Geographical Aspect of Regional Planning, *Geog. Jour.*, **99**, 61–80.

### Regional Metamorphism—see Metamorphism

### Regionalization, regionalize

Closely connected with the varied uses of regionalism are those of regionalization. Gilbert, 1939, listed above, quotes its use by D. J. Ledbury, 1937 (Post Office Regionalization), and Mary Marshall (in MS.) suggests it is used with three different meanings:

1. Division into regions (a) administrative, (b) academic.
2. Regional organization, *i.e.* regional systems.
3. Regionalism, *i.e.* development of regional character.

Grigg, D. in Chorley and Haggett, 1967. '... regionalization is analogous to classification and if it is remembered that the fundamentals of classification are based upon the principles of formal logic then it may be profitable to examine regional systems in the light of these principles.' (p. 485)

Hilhorst, J. G. M., 1971, *Regional Planning*, Rotterdam: Rotterdam University Press. 'As major purposes of regionalization we have (1) that of analysis and (2) that of planning. As criteria for regionalization, we have again two, namely that of interdependency and that of similarity.' (p. 54)

See Grigg, D. B., 1965, The Logic of Regional Systems, *Ann. Assoc. Am. Geographers*, **55**, 465–491, and Hettner, A., 1908, Die Geographische Einteilung der Erdoberfläche, *Geogr. Ztschr.*, **14**.

### Regional multiplier concept

Keeble, D. E., in Chorley and Haggett, 1967. 'Broadly speaking, the regional multiplier concept concerns the way in which a rise in income, production or employment in one group of economic activities in a region stimulates the expansion of other groups, through an increased demand from the former group and its workers for the goods and services produced by the latter. The rise is typically induced by changes external to the region.' (p. 275)

*Comment.* An early example of the importance of the multiplier process in industry can be found in Barford, B., 1938, *Local Economic Effects of a Large-Scale Industrial Undertaking*, Copenhagen: The Economic Research Department of Aarhus Oliefabrik.

### Regional planning

Hall, P., 1974, *Urban and Regional Planning*, Harmondsworth: Penguin (Pelican). '... looking at the writings of Howard Geddes and Abercrombie we saw that, increasingly from 1900 to 1940, the more perceptive thinkers came to recognize that effective urban planning necessitated planning on a larger than urban scale—the scale of the city and its surrounding rural hinterland, or even several cities forming a conurbation and their common overlapping hinterlands. Here, the development of the idea of regional planning in one commonly used sense of the word begins.' '... there is another common meaning to the term "regional planning" in modern usage ... assumed prominence during the 1930s, as the result of the great economic depression ... Refers specifically to economic planning with a view to the development of regions which, for one reason or another, are suffering serious economic problems, as demonstrated by indices such as high unemployment or low incomes in relation to the rest of the nation.' (p. 81). Hall notes that the two definitions are related although they present problems that require different expertise. The economic region is differently designed, and different in size, from the 'region' of the city-region planners.

### Regional science
Isard, W., 1960, *Methods of Regional Analysis*, New York: Wiley. '... the vast field of regional science ... must probe the area of theory—theory which has regional and interregional structure and function at its core; theory which cuts into and generalizes the system and subsystem interdependencies...' (pp. 757–758)

Grigg, D., in Chorley and Haggett, 1967. 'Some regional scientists consider that regional studies should consider the relationships between society and environment whilst there are still many geographers ... where geographical determinism is anathema, who regard regional geography as the study of interrelationships between economic life and environment.' (p. 477)

### Regional systems
Abler, Adams and Gould, 1971. 'Regional systems are areal classifications lying somewhere between maximum uniqueness and maximum generality. There are two types of regional systems: general and specific. In general (or generic) regional systems, types of places resemble one another according to a certain set of attributes like climate, language ... and so forth. Attributes ... selected according to purpose of the classification. The important feature of general regional systems is that places of each type in the system may be located at widely separate absolute locations.... Specific regional systems are defined not only by combinations of intrinsic attributes, but by location as well ... in specific regional systems all the parts of a homogeneous region must be spatially contiguous.' (p. 183)

### Regolith
*Webster. Geol.* The mantle of loose material consisting of soils, sediments, broken rock, etc., overlaying the solid rock of the earth.

Robinson, G. W., 1936, *Soils*, 2nd ed., London: Murby. 'Regolith (or better, rhegolith), the fragmental unconsolidated débris, mantling the rocks of the earth's crust.'

Rice, 1943. 'General term for the superficial blanket of denudation products which is widely distributed over the more mature "solid" rocks ... includes weathering residues, alluvium and aeolian and glacial deposits.' (p. 341)

*Comment*. Rhegolith is rarely if ever used. See Mantle rock.

### Regosol (Soil science)
Jacks, 1954. 'Soil without definite genetic horizons developing from deep unconsolidated rock or soft mineral deposits.'

FAO, 1963. '*Regosols* have (A)C profiles but differ from lithosols in that the calcareous or non-calcareous rocks from which they are formed are coarse-textured or soft sediments or unconsolidated deposits. The (A)-horizon is dominantly mineral but is not stony and may be moderately thick.'

### Regur (Indo-Pakistan)
Black cotton soil of India.

Jacks, 1954. Dark coloured, usually calcareous tropical soil that swells when wet and cracks deeply on drying.

### Reh (Indo-Pakistan: Panjabi; Hindi)
Saline soil, but differing in composition from Kallar (A.H.S.). Saline lands (G.K.)

### Reilly's law of retail gravitation
Reilly, W. J., 1931, *The Law of Retail Gravitation*, New York: William J. Reilly Company. 'Two cities attract retail trade from any intermediate city or town in the vicinity of the breaking point, approximately in direct proportion to the populations of the two cities and in inverse

proportion to the square of the distances (distance via most direct improved automobile highway) from these two cities to the intermediate town.' (p. 9)
See also Gravity model.

### Rejuvenation
*Webster.* Rejuvenate. 3. *Phys. Geog.* a. To stimulate, as by uplift, to renewed erosive activity; said of streams. b. To develop youthful features of topography in (an area previously worn down nearly to base level).

Mill, *Dict.* Rejuvenated Rivers. Rivers which by changed circumstances start a fresh cycle of erosion or are marked by the reappearance of an earlier phase of that cycle in a long-eroded region.

Committee, List 2. The development of younger surface forms, appropriate to the earlier stages of the cycle of erosion, which occurs when a comparatively well advanced cycle is interrupted by an increase in the rate of erosion.

Thornbury, 1954. 'Rejuvenation may result from causes which are dynamic, eustatic, or static in nature. *Dynamic rejuvenation* may be caused by epeirogenic uplift of a land mass, with accompanying tilting and warping.... *Eustatic rejuvenation* results from causes that produce world-wide lowering of sea-level ... two types ... *Diastrophic eustatism* resulting from variation in capacity of the ocean basins, whereas *glacio-eustatism* refers to changes in sea-level produced by ... accumulation or melting of ice sheets.... At least three changes may produce *static rejuvenation* ... decrease in load, increase in runoff because of increased rainfall, and increase in stream volume through acquisition of new drainage.' (pp. 142-144)

*Comment.* The older authors appear to use 'revived' in the same meaning, *e.g.* Geikie, 1898; Davis, 1909.

### Rejuvenation-head—see Knickpoint

### Relict boundary
Pounds, N. J. G., 1963, *Political Geography*, New York: McGraw-Hill. 'Relict boundaries are boundary lines which have been abandoned for political purposes, but which, nevertheless, remain discernible in the cultural landscape.' (p. 64)

### Relict mountain, Relic mountain
A mountain of circumdenudation, a mountain representing the remains or relic of a pre-existing plateau or range. Monadnocks and inselberge are relict mountains.

### Relief
*O.E.D.* 3b. Phys. Geog. The contour of some part of the surface of the earth considered with reference to variations in its elevation.

*Webster.* 15. *Phys. Geog.* The elevations or inequalities of a land surface, considered collectively; also, the difference in elevation between the hill tops or summits and the lowlands of a region.

Mill, *Dict.* The relative vertical inequality of the land surface.

Committee (as *O.E.D.*).

James, P. E., 1935, *An Outline of Geography*, Boston: Ginn. 'The relief of an area measures the difference in elevation between highest and lowest level. Relief is different from altitude....'

*Comment.* Relief refers to the physical shape of the surface of the earth and is thus more than just altitude; it is the physical landscape in the narrower sense of physical. Many writers use 'topography' when they mean 'relief', despite the root meaning and older usage of the word. The tendency is to use 'relief' as a loose general term: Whilst the relief is made up of a number of 'land-forms' that word suggests a classification or inference of origin. Some authors use 'Relative Relief' in the sense of relative altitude; other 'Available Relief'. Following Lake, 1958, it is desirable to restrict the terms 'high relief' and 'low relief' to areas which respectively show much or little variation in altitude.

### Relief map
A map depicting the surface relief of an area. A contour map is one type of relief map but there are other types, including photo-relief maps which are either photographs of a model or diagrammatic maps simulating a photograph.

### Relief model
A model depicting the surface relief of an area, but not necessarily to true scale. It is usual practice to exaggerate the vertical scale compared with the horizontal so as to accentuate mountains and plateaus.

### Relief rainfall
*L.D.G.* Relief rainfall is that resulting from the relief of the land; hills and mountains cause air to rise resulting in cooling, condensation of moisture and rain.

**Rendzina, rendsina** (Polish; soil science) Soils and Men, 1938. Rendzina soils. An intrazonal group of soils, usually with brown or black friable surface horizons underlain by light gray or yellowish calcareous material; developed under grass vegetation or mixed grasses and forest, in humid and semi-arid regions from relatively soft, highly calcareous parent material. From a Polish peasant term for productive calcareous soils.

Jacks, 1954. 'Rendzina: humus-carbonate soil. Dark, calcareous, usually shallow soil formed on soft limestone.'

FAO, 1963. 'Soils with AC horizons occurring on calcareous materials. Calcium carbonate is usually distributed throughout the profile and the humus form is generally mull.'

*Comment.* Well developed in England on the Chalk, but also found on harder limestones. (G.T.W.) Though a Polish word it came into English through Russian. Barkov (see Appendix II, Russian words) defines as a soil high in humus and lime, developed under grass on limestones in the podzolized zone. See also Xerorendsina.

**Replat** (French)
A bench; shoulder or tilted terrace above the steep side of a U-shaped valley (E.D.L.). Baulig translates as bench or shelf, wider than a ledge. Plaisance and Cailleux give it the much wider meaning as a stretch of land, more or less horizontal, interrupting a slope.

**Representative Fraction**—see R. F.

**Resam** (Malay)
A type of vegetation shown on the official one-inch maps of Malaysia. It is given without translation or mention in the glossary of Malay terms but applies to areas of fern or bracken.

**Resequent drainage, Re-consequent drainage**
Cotton, 1922. 'Synclinal valleys and anticlinal ridges simulating consequent features but developed from a subsequent drainage pattern are termed resequent.' (p. 88) Similarly Wooldridge and Morgan, 1937, p. 212.

No reference in Salisbury, 1907; Davis, 1909 but see Davis, The Mountains of Southernmost Africa, *Bull. Geol. Soc. Amer.,* **38**, 1906, 593–629.

Lake, 1958. 'After an intervening period of peneplanation or a change in the river pattern by some agency such as capture the rivers might return to the synclines This drainage pattern might be termed re-consequent or resequent drainage (p. 320)

**Resequent fault-line scarp**
Cotton, 1945. '... a resequent fault-line scarp faces, or descends towards, the structurally depressed (downthrown) side of the fault.' (p. 179) 'A fault-line, a distinct from a fault-scarp, is the result of differential erosion on either side of a fault It is resequent if it faces in the same direction as the initial fault. If it slopes in the opposite direction is is obsequent.'

*Comment.* The distinction between a fault line scarp and a fault scarp was not made in Cotton, 1922, where there are references to resequent and obsequent fault scarps No reference in Davis, 1909; Wooldridge and Morgan, 1937. See also Thornbury, 1954, 248.

**Residual deposits**
Himus, 1954. Accumulations of rock-waste which result from disintegration of rocks *in situ*, covering the complete range of grain-size from boulders to clays.

**Residual mapping**
Thomas, E. N., 1968, Maps of Residuals from Regression, in Berry, B. J. L. and Marble, D. F. *Spatial Analysis: A Reader in Statistical Geography,* Englewood Cliffs, New Jersey: Prentice Hall (originally a monograph first printed in 1960). A statistical map portraying 'the spatial distribution of deviations from regression (henceforth referred to as residuals from regression, or merely as residuals).'

*Comment.* Developed by Thomas, E. N. using McCarty, H. H., 1956, Use of Certain Statistical Procedures in Geographic Analysis,, *Ann. Assoc. Am. Geographers,* **46**. An early formulation of the technique is attributed to Mill, J. S., 1874, *A System of Logic,* New York: Harper and Row (pp. 284–285). Thomas, 1960, states that the mapping and use of residuals from regression is merely Mills's 'method of residues' reformulated within a probabilistic framework.

**Residual soil**
Jacks, 1954. 'Soil resting on the material from which it was formed.' Also called sedentary soil: contrast with Secondary or Transported soil.

### Resistance, resistant rock
*S.O.E.D.*, 3b. The opposition offered by one body to the presence or movement of another. Hence applied to rocks which resist the forces of mountain building movements.

### Resource
*O.E.D.* 1. A means of supplying some want or deficiency; a stock or reserve upon which one can draw when necessary. Now usually pl. 2. pl. The collective means possessed by any country for its own support or defence. (quots. inc.: 1870. Yeats, Nat. Hist. Comm. 2. In speaking of the natural resources of any country we refer to the ore in the mine, the stone unquarried, the timber unfelled, etc.).
*Encyclopedia of the Social Sciences*, New York: Macmillan, 1933, Vol. XI. 'Resources are those aspects of man's environment which render possible or facilitate the satisfaction of human wants and the attainment of social objectives....' (pp. 290–291)
Zimmermann, E. W., 1933, *World Resources and Industries*, New York: Harper. Features of the environment which are, or are considered to be, capable of serving man's needs; they are given utility by the capabilities and wants of men. (W.M.) (p. 1)
*Comment*. Usually expressed as natural resources (*q.v.*) but note resource conservation.

### Response surfaces
Haggett, 1965. 'The resulting map completely describes the area in two-dimensional form. Like contour values for terrain, it could be converted to a three-dimensional plaster model but in any case can be regarded conceptually as a three-dimensional trend surface.... This surface may be thought of statistically as a response surface. That is the height (i.e. degree of forest cover) at any one point may be regarded as a response to the operation of that complex of "... geology, topography, climatic peculiarities, national composition, economic disparities, and local and regional history" (Köstler, J., 1956, *Silviculture*, Edinburgh: Oliver and Boyd) which together determine forest distribution.' (p. 269)
See Box, G. E. P., 1954, The Exploration and Exploitation of Response Surfaces: Some General Consideration and Examples, *Biometrics*, **10**, 16–60.

### Retrogradation
Johnson, 1919. '... the more usual retreating or retrograding shore.' (p. 223)
Wooldridge and Morgan, 1937. '... the steepening of the profile [the beach-profile] by cutting away at the breaker line is retrogradation.' (p. 332)
*Comment*. This term appears principally as the logical antithesis of progradation.

### Reversed fault—see Fault

### Revived river
Geikie, J., 1898. 'When the rivers of a region have succeeded in cutting their channels down to the base level, they have a slight fall and flow sluggishly. Should the whole region then be elevated while the direction of its slopes remains unchanged the erosive energy of the rivers is renewed, and they are said therefore to be revived.' (p. 306)
Davis, 1909. '... at any time ... the drainage area of a river system may be bodily elevated. The river is then turned back to a new youth and enters a new cycle of development.... Such rivers may be called revived.' (p. 441)
*Comment*. Mill has: Revived Rivers. This is now obsolete; the current term is Rejuvenated (*q.v.*). Davis does not use 'rejuvenated rivers' in 1909 essays.

### R.F., Representative Fraction
Swayne, 1956. The fraction expressing the ratio between the linear measurement on the map and the corresponding measurements on the ground.
*Comment*. For convenience an R.F. of 1/1,000,000 is written 1:1,000,000 or for example 1:1m. meaning that one inch on the map represents one million inches or about 16 miles on the ground. A map on the scale of one inch to one mile has an R.F. of 1:63,360 because there are 63,360 inches in one mile. The term 'natural scale' applies to round numbers.

### Rhine, rean, rine
A man-cut water course usually in association with a water mill.

### Rhizosphere
'The rhizosphere (root zone of plants) harboured a bacterial flora which is physiologically more active than that of a soil a short distance (5–10 inches) away from the root.'

Katznelson, H., *Abs. of Papers, Ninth Pacific Science Congress*, Bangkok, 1957, 236.

Jacks, 1954. 'The immediate neighbourhood of plant roots.'

**Rhos** (Welsh; *pl.* rhosydd)
Rough moorland; contrasted with ffridd (*q.v.*)

**Rhumb Line**
*Admiralty Navigation Manual*, 1938, vol. II. 'A line on the earth's surface which cuts all the meridians at the same angle' (p. 17). A line of constant bearing. Also referred to as loxodrome, loxodromic curve.

**Ria** (Spanish; properly *ría*)
*O.E.D.* Geol. (Spanish) A river mouth.
*Webster. Geog.* A long narrow inlet, with depth gradually diminishing inward; a creek.
Mill, *Gloss.* A relatively narrow bay, in some cases wedge-shaped (narrowing from mouth to head) in others winding, bordered by a rocky coast, and for the most part gradually increasing in depth from the head to the mouth. A ria differs from a fiord in being shorter and without the irregularities of depth which characterize the latter.
Holmes, 1944. 'Along coasts of "Atlantic" type, that is, where the trend lines of an orogenic belt are transverse to the coast, drowning gives an alternation of long promontories and estuaries. The latter are called rias, the name given them in Spain, where they occur south of Cape Finisterre.' (p. 301)
*Comment. O.E.D.* definition is inadequate; but the simple Spanish word has been given a specialized meaning in geographical literature. It is restricted by most writers to *drowned* river valleys (*cf.* Wooldridge and Morgan, 1937, 334). C. A. Cotton distinguishes between rias *sensu lato* and *sensu stricto*—see *Geog. Jour.*, **122**, 1906, 360–364.

**Ribbon-development**
*O.E.D. Suppl.* The building of houses along a main road, extending outwards from a town.

**Ridge (1)**
*O.E.D.* A long and narrow stretch of elevated ground; a range or chain of hills or mountains.
*Comment.* Ridge is often applied to a small feature in a mountain range, but scarcely to the range itself and never to a chain.

**Ridge (2) and furrow**
*O.E.D.* Ridge—A raised or rounded strip of arable land, usually one of a series (with intermediate open furrows) into which a field is divided by ploughing in a special manner.
Stamp, L. D., 1955, *Man and the Land* London: Collins. 'The Anglo-Saxons.. faced the difficulty ... of the management of the clay soils ... gradually the fields were ploughed into broad ridges with hollows between which served to drain off surplus water.... In later times, especially in the latter half of the nineteenth century many of these old ploughlands went down to permanent grass but the fields bear the marks of their Anglo-Saxon managers. (p. 42)
But see Mead, W. R., 1954. Ridge and Furrow in Buckinghamshire, *Geog. Jour.*, **120** 34–42. Much ridge and furrow is clearly not pre-enclosure strip but belongs to the post-enclosure period and is the result of purposeful ploughing for drainage. Also Aitken, R., 1954, *Geog. Jour.*, **120**, 260. 'Whenever an asymmetric plough, turning a sod to one side only, is used, a "ridge and furrow" pattern results usually—on practical grounds not connected with drainage.'

**Riegel** (German)
Cotton, 1922. Of the floor of glaciated valleys: 'Associated with such steps— immediately above them—low transverse ridges of rock, which were overridden by the glaciers, occur in the valleys of the Alps, and convert the treads of the steps into basins. These barriers are termed verrous or Riegel.' (p. 296)
Fischer, E. and Elliott, F. E., 1950, *A German and English Glossary of Geographical Terms*, New York: American Geog. Soc.
1. Natural rock dam formed by resistant rock across a valley.
2. Rock bar, rock threshold (glacial).
No reference in Fay, 1920; Wooldridge and Morgan, 1937; Holmes, 1944; Rice, 1941; Lake, 1952.

**Rift**
*O.E.D.* A cleft, fissure or chasm in the earth.
Applied especially to a narrow cave passage of considerable height eroded along a joint, steeply inclined bedding plane or fault plane. (G.T.W.)

**Rift valley**
*O.E.D.* A valley with steep parallel walls, formed by the subsidence of a part of the earth-crust.
Mill, *Dict.* A long trough-shaped valley formed by insinking and bordered by steeply rising ground on both sides (Richthofen).
Himus, 1954. A valley formed by the sinking of land between two roughly parallel faults.
*Comment.* See also Graben and Senken. Whereas a rift valley is definitely a relief feature, a graben is a structural feature which may or may not coincide with a valley.
See also Willis, Bailey, 1936, *Rift Valleys and Plateaus of East Africa*.
**Right ascension**—see Ascension
**Rights of Common**—See Common, Rights of
**Rill**
Mill, *Dict.* A small brook or rivulet
Also applied to a small erosion channel in soil erosion.
**Rill marks**
Swayne, 1956. 'Small furrows, often branched, made on surfaces of sand or mud by little streams of water following a retreating tide or oozing from springs; fossil marks of this kind.'
**Rillenstein (German)**
Howell, 1957. 'Solution-grooved rocks; grooving generally imposed on Karren.'
**Rillstone**—See Ventifact
**Rimaye (French)**—see Bergschrund
**Rime**
*Webster. Meteorol.* An accumulation of granular ice tufts on the windward sides of exposed objects, slightly resembling hoar-frost (which see), but formed only from undercooled fog or cloud and always built out directly against the wind.
**Rimstone**
Howell, 1957. 'A term suggested by W. M. Davis to designate calcareous deposits formed around the rims of overflowing basins.'
**Rio (Portuguese and Spanish)**
River; usually applied to a permanent stream. In Spanish contrast arroyo.
**Rip current, rip tide**
*O.E.D.* Rip. 1. A disturbed state of the sea, resembling breakers; an overfall. 2. A stretch of broken water in a river.
*Webster.* Rip. a. A body of water made rough by the meeting of opposing tides or currents. b. A portion of a current roughened by passing over a shoal; applied especially to tidal currents, more rarely to land streams.
Cotton, 1945. In compensation for landward drift of surface water where waves are at right angles to the shoreline, 'the water streams back seaward also as rapidly-flowing narrow tongues reaching to the surface, which are called rip currents'. (pp. 402–403)
U.S. Coast and Geodetic Survey, Spec. Pub. No. 228, *Tide and Current Glossary*, 1941. A strong surface current of short duration flowing outward from the shore. It usually appears as a visible band of agitated water and is the return movement of water piled up on the shore by incoming waves and wind. With the outward movement concentrated in a limited band its velocity is somewhat accentuated. Frequently called Rip tide.
*Comment. cf. Adm. Gloss.*, 1953. Overfalls or Tide-Rips. Turbulence associated with the flow of strong tidal streams over abrupt changes in depth, or with the meeting of tidal streams flowing from different directions. Rip current discussed by F. P. Shepard in *Science*, **84**, 1936, 181–182.
**Riparian**
*O.E.D.* Of, pertaining to, or situated on the banks of a river.
Although *O.E.D.* gives also 'riverine', riparian refers strictly to the river *banks*.
**Ripple, sand ripple, ripple-mark**
*Webster.* Ripple marks. A system of small, more or less parallel ridges produced, especially on sand, by wind, by the current of a stream, or by the agitation of waves.
Mill, *Dict.* Ripple-marks are formed by waves on a beach and on the sandy bottom of a stream by the current, and by wind on the land.
Bagnold, 1941, devotes a chapter to 'Small-scale sand forms. Transverse Ripples and Ridges'. 'A sand ripple is merely a crumpling or heaping up of the surface, brought about by wind action, and cannot be regarded as a true wave in a strict dynamical sense. The similarity lies only in the regular repetition of surface form.' (p. 144)
**Ripple-till (sometimes ribble)**

Patterned till-sheets, in which sinuous ripple-like ridges, 200 yards to 2 miles long, run *across* the direction of ice movement, very roughly at right angles. The 'ripples' are typically low (20–50 feet above intervening depressions) and are smooth-topped. The ripple-till belts are normally elongated parallel to the ice-movement, being 2 to 10 miles wide, and sometimes 50 miles long. In general they are flanked by drumlin fields, and in a few areas drumlins can be seen fused together transversely, indicating that the mechanics of drumlinization and of ripple-till formation are closely akin. Ripple-till belts are extensive in Labrador-Ungava, Keewatin and parts of northern Ontario (definition supplied by Professor F. Kenneth Hare, June, 1956).

### Rise or Swell
Thornbury, 1954. 'Extensive, long, broad elevations which rise gently from the deep-sea floor.' (p. 477)

### River
O.E.D. A copious stream of water flowing in a channel towards the sea, a lake, or another stream.

Webster. A natural stream of water larger than a brook or a creek. A *river* has its stages of development, youth, maturity, and old age. In its earliest stages a river system drains its basin imperfectly; as valleys are deepened, the drainage becomes more perfect, so that in maturity the total drainage area is large and the rate of erosion high. The final stage is reached when wide flats have developed and the bordering lands have been brought low.

Swayne, 1956. 'A large body of fresh water which flows with a perceptible current in a certain definite channel or course, usually uninterruptedly throughout the year.'

Comment. River is a very general term and the O.E.D. definition is too narrow; the water is often far from 'copious' and many rivers are reduced to a string of pools in the dry season. Many great rivers lose themselves in desert sands before reaching the sea or a lake. For river régime, see Régime.

### River capture, river piracy
Himus, 1954. The action of a river in acquiring the headwaters of another river by enlarging its drainage area at the expense of the other.

Swayne, 1956. The diversion of the upper waters of a stream to another river system due to the one with greater erosional power cutting off the headwaters of another.

Comment. River capture plays a large part in the development of river systems and is considered at length in most text-books. See, *inter alia*, Lake, 1958, 312–315; Wooldridge and Morgan, 1937, 192–196; Thornbury, 1954, 152–156; von Engeln, 1942, 126–127. The stream whose headwaters have been captured is said to be *beheaded*; the point at which capture takes place is the *elbow of capture*; capture may result in *misfit streams* and *wind-gaps* (*qq.v.*). When a river shortens its own course, the phenomenon is referred to as *autopiracy* (*q.v.*).

### River-cliff—see Meander

### River profile
Swayne, 1956. A section showing the slope of a river from source to mouth.

Comment. This is the longitudinal profile or long-profile of a valley (see Wooldridge and Morgan, 1937, 159–161) or talweg (*q.v.*). There are several related types of profile. The true river profile takes the actual length of the centre of the stream and the height of the surface is taken at the mean level, and measurements are adjusted for minor variations of level. Some authors prefer to take the height of the flood plain, especially when comparing the present profile with reconstructed profiles. In such cases it is usual to take a mean path, ignoring minor windings since such features cannot be reconstructed for past phases of the river at higher levels. (G.T.W.)

### River terrace
Himus, 1954. Flat, step-like strips, occurring on the flanks of river valleys at various levels above the present channels. Each terrace consists of deposits of fluviatile gravels and sands which were laid down by the river when it was running at the level of the terrace. The higher the level of the terrace above the present river, therefore, the older it is.

Comment. Strictly speaking rock benches and gravel-covered terraces are both classified as terraces, but usually the term may be taken to refer to gravel-covered features. The surface may not be perfectly flat, especially if the terrace is a constructional one, still preserving the remains of swales, bars etc. For further discussion of the different types of river terrace see Cotton, C. A., 1941, Chapter XIII and Cotton, C. A., 1940, Classification of River Terraces, *Jour. of Geom.*, 3, 27–37 (G.T.W.).

**Riverain, riverine**
*Webster.* Riverine. The region immediately adjoining a river.
*Mill, Dict.* Pertaining to a river or its vicinity, situated on the banks of a river.
Note that riverain has a wider meaning than riparian (*q.v.*).

**Riviera** (Italian)
*Mill, Dict.* A shore or strand region; especially applied to the shore regions which are popular places of resort.
Not in *O.E.D.* or *O.E.D. Suppl.*
*Comment. The* Riviera is the coastal strip bordering the Mediterranean from Marseille in southern France to Genova in Italy with its numerous resorts. By analogy the word has been applied to other resort coasts in different parts of the world, e.g. Cornish Riviera, the southern coast of Cornwall.

**Rivulet**
*Mill, Dict.* A small river.

**Road-metal**—see Metal

**Roadstead, roads, road**
*O.E.D.* A place where ships may conveniently or safely lie at anchor near the shore.
*Mill, Dict.* Open anchorage.
*Adm. Gloss.*, 1953. Roads: An open anchorage which may, or may not, be protected by shoals, reefs, etc.; sometimes found outside harbours. A roadstead. Open roadstead: An anchorage unprotected from the weather.

**Roaring Forties**
Miller, 1953. 'In the southern hemisphere the disturbance of the planetary winds is much less; "Roaring Forties" and the "Brave West Winds" blow all the year round with considerable force and with a high degree of dependability.' (p. 199)
Mill, *Dict.* The latitudes between 40°S. and 50°S. where the 'brave west winds' blow.

**Robber-economy**
Zimmermann, E. W., 1933, *World Resources and Industries*, New York: Harper. 'As here interpreted, "robbing" means the needless destruction, for the sake of immediate profit or out of sheer folly, of the future possibilities of recovery' (p. 162).
*Comment.* The phrase is often used in elementary books to cover the working of capital or non-renewable resources, especially minerals, in contrast to the development of renewable resources such as waterpower and forests. For example Stamp, 1940–1958, *Introduction to Commercial Geography.* 'Most mineral products are obtained by *robber economy*. Once the coal or the oil ... has been removed from the ground it does not grow again' (p. 3). Probably derived from the German *Raubwirtschaft*—see F. Ernst, *Pet. Mitt.*, 50, 1904, 68–69, 92–95.

**Rock-drumlin, rocdrumlin**
Charlesworth, 1957. 'The core may swell until the whole drumlin is wrought in solid rock. These "rock-drumlins", "rockdrums", "drumlinoids" or "false drumlins" (Ger. *Rundhöckerdrumlins*) are sometimes found at the proximal ends of drumlin fields. Their shape is almost indistinguishable from true drumlins if they coincide with the strike of the rock but is generally less symmetrical or regularly sloped, is larger and narrower and has an impact face which is liable to be more abrupt and uneven. Transitions proving a common origin link rock-drumlins with true drift-drumlins on the one hand and on the other with roches moutonnées and crags and tails.' (p. 391)
*Comment.* There seems no justification for rocdrumlin used by some writers since a 'roc' is a large mythical bird.

**Roche moutonnée** (French; *lit.* sheep-like rock)
Charlesworth, 1957. 'De Saussure (1779–1796) gave the name roche moutonnée to the distinctive rounded forms which abound in glaciated terrain (he himself failed to associate them with ice) and give the effect of a thick fleece or the wavy wigs styled *moutonnées* in his day (they were slicked down with mutton tallow). It was again used by B. Studer (1840) and became general after Agassiz (*Études sur les Glaciers*, Neuchatel, 1840) adopted it in 1840. *Roches bosselées* and *roches nivellées* have been suggested for the large planed surfaces that resulted, it was thought, from more rapid flow ...' (p. 252).

Lake, 1958. 'Hillocks of rock, rounded on the upstream side and rough on the other, are called roches moutonnées and are to be seen in almost any glaciated valley... looking down the valley we see the smoothed surfaces... looking up the valley it is the rough sides that we see and the general effect is one of ruggedness.' (p. 347)

Moore, 1949. '... smoothed and marked with *striae* by the glacier....'

### Rock

*O.E.D.* 3f. Geol. One of the stratified or igneous mineral constituents of which the earth's crust is composed, including sands clays, etc.

*Webster*. 3. *Geol*. Solid mineral matter of any kind occurring naturally in large quantities or forming a considerable part of the earth's mass; also, a particular mass or kind of such material. Rock may be consolidated or unconsolidated, and composed of one mineral or, more commonly, of two or more; or it may be to a greater or less extent of organic origin, as coal. In geology, granite, sand, gravel, clay, and glacial ice are rocks. Rocks may be broadly classified as *igneous, sedimentary,* and *metamorphic*.

Mill, *Dict*. 1 (Geog.). Generally used in the ordinary sense to denote the harder portions of the Earth's crust when present in large mass. 2 (Geol.). The meaning is less restricted, and may be applied to any mass of mineral matter whether hard, relatively soft, or incompact, and whether composed of a single mineral or, as is more usually the case, of several.

Holmes, 1920. 'As a geological concept, rock may be defined as (a) any formation of natural origin that constitutes an integral part of the lithosphere, and that cannot be referred to a single fossil, or to a single individual of a mineral species; or (b) a representative specimen of such a formation.'

*Comment. Cf.* mineral. Should be used as in Holmes; *O.E.D.* definition is inadequate. The use of rock as an adjective should be noted, *e.g.* rock basin. In ordinary speech rock denotes something hard, usually massive and large, especially if isolated. Note also rock-ribs, outcrops of bare rock.

### Rock-flour

*Webster*. Finely powdered rock material produced by grinding action, as of a glacier on its bed; called also *glacial meal* The finely ground rock debris produced when a glacier with rocks frozen into it mass abrades its bed. The action is a mechanical one and little or no chemical action is involved, hence the rock-flour does not differ in mineralogical composition from the rocks of which it is composed

### Rocking stone

A large boulder, the result of circumdenudation or an erratic so poised that it can be rocked. See Logan stone.

### Rognon

Llibouty, L., 1958, *Jour. Glaciology,* 3 'around a *rognon* (rounded nunatak)' (p. 264) *Cf.* nunakol.

### Rondawel, rondavel (Afrikaans)

Cylindrical hut with conical roof used by many Bantu tribes. This form is often imitated by white people for buildings carried out in brick, wood or galvanized iron, and serving as private pavilions for visitors in mountain hotels, seaside resorts, etc. (P.S.)

### Rooikalk (Afrikaans; *lit.* red lime)

Wellington, J. H., 1955, *Southern Africa,* I. 'Where the rainfall is from 12 to 15 inches the surface layer consists of a thin light brown sandy loam... under this is... a layer of extremely hard sandy loam cemented by a siliceous material, generally referred to as "*rooikalk*"' (p. 322).

### Ropy lava, corded lava

A lava which has solidified so that the surface of the flow is glassy and smooth with surface shapes which resemble ropes or cords. Also called pahoehoe (*q.v.*).

### Rotation (agriculture)

*O.E.D.* 1c. Agric. A change or succession of crops in a certain order on a given piece of ground, in order to avoid the exhaustion of the soil.

Mill, *Dict*. 2 (Agric.). Sequence; especially in connection with types of cultivated plants which replace one another annually.

Committee, List 3. A systematic succession of different crops on the same land.

*Comment.* The above definitions refer to crop rotation. The term land rotation is also used where successive pieces of land are regularly cropped in turn and then allowed to lie fallow.

### Rotation (temporary) grassland

Committee, List 3. Grass grown for one or more years in rotation with other crops.
See Ley.

**Rough grazing, Rough pasture**
Committee, List 3. Unimproved grazing, including *e.g.* moorland, scrubland, saltmarsh, mountain pasture.
*Comment.* In Britain the distinction between rough grazing, often or indeed usually unenclosed, and improved grazing, generally in enclosed fields, is quite clear and is always separated in official statistics. In many countries, however, where improved and enclosed grassland for grazing is almost if not entirely unknown, statistics separate simply 'arable' and 'pasture'. Figures for grassland, pasture or grazing, even those published by FAO, are not therefore internationally comparable (L.D.S.).

**Roxen lake**
Davis, W. M., 1925, A roxen Lake in Canada, *S.G.M.*, **41**. '... occupies a basin that has been excavated in a second cycle of erosion after post-faulting peneplanation had been reached in a first cycle and ... the basin is limited along one side by a fault-line scarp ...' (p. 74). (From the type example, Roxen, a lake in Sweden.)
Cotton, 1942. Rock rimmed lake formed by selective glacial action in fault angles. (p. 151)
*Comment.* Not in common use.

**Rubrozem** (Soil Science)
FAO, 1963. '*Rubrozem soils* are ABC soils with a dark-colored A1-horizon of low base saturation overlying a prismatic reddish brown or red-colored textural B-horizon.'

**Rudaceous**
The texture of a rock in which the grain is coarser than that of sand. Hence 'rudytes' of Grabau; more usual term is psephite (*q.v.*).

**Rug** (*pl.* rûens) (Afrikaans)
The back of an animal. As a geographical term used in the plural, rûens means a landscape consisting of a low plateau, strongly cut up by river erosion into a large number of rounded ridges of equal height, comparable to the backs of a herd of large animals. (P.S.)

**Rumpf** (German; *pl.* Rümpfe: *lit.* rump)
Usually used in combinations: Primärrumpf, Endrumpf, Rumpffläche, Rumpftreppe (K.A.S.).

**Rundale** (Ireland)
*O.E.D.* A form of joint occupation of land, characterized by dividing it into small strips or patches, a number of which, not contiguous to each other, are occupied and cultivated by each of the joint holders. Frequently in phrase *in rundale*. Also land occupied in this manner. The current term in Scotland is runrig (*q.v.*).
*Comment.* Rundale mountain—mountain or hill pasture grazed in common.

**Runddorf, Rundling** (German; *pl.* Runddörfe, Rundlinge: *lit.* round, circular, village)
A village type largely found in Central Germany, the first areas to be colonized during the German eastern colonization. Possibly of Slav origin but also used by German settlers. The farmhouses are arranged in a circle leaving a central green with originally only one means of access (K.A.S.).

**Rundhäll** (Swedish)
A Swedish term for roche moutonnée. (E.K.)

**Run-off**
*O.E.D.* 1. U.S. (see quot.). 1892–3, 14th Rep. U.S. Geol. Surv., 149. The run-off, that is, the quantity of water flowing from the land.
*Webster.* Runoff. That which runs off. Specifically: a. *Hydrog.*, the water which is removed from soil by either surface or under drainage; flowoff.
Mill, *Dict.* 1. The volume of water discharged by a river at its mouth or any other point. 2. The flow of surface water from the land.
*Met. Gloss.*, 1944. The portion of the rainfall over a drainage area which is discharged from that area in the form of a stream or streams.
Cotton, 1922. 'When rain falls, generally some of the water runs off the surface immediately. The proportion of this run-off....' (p. 36)
*Comment.* Moore, 1949, distinguishes between immediate run-off, along the surface, and delayed run-off, which returns to the surface by seepage and from springs. This complicates the otherwise simple relationship that total rainfall or precipitation equals the sum of percolation, evaporation (or evapotranspiration) and run-off. In contrast to *Webster*, run-off is normally restricted to surface drainage. See Meyer, A. F., 1942 in Meinzer, O. E., *Hydrology*, New York: McGraw-Hill,

## Runrig

p. 478, with distinction of surface run-off and ground-water or effluent run off.

## Runrig, run-rig (Scotland etc.)

*O.E.D.* 1. A ridge of land lying among others held by joint tenure, rare. 2. A form of land tenure = Rundale. Rundale. 1. A form of joint occupation of land, characterized by dividing it into small strips or patches, a number of which, not contiguous to each other, are occupied and cultivated by each of the joint holders.... Used esp. to designate this mode of occupation as practised in Ireland; in Scotland... the current term is Runrig. 2. Land occupied in this manner, or a share in such land.

Grant, I. F., 1930, *The Social and Economic Development of Scotland before* 1603, Edinburgh: Oliver and Boyd, (As *O.E.D.* 2.) p. 97.

Oppé, A. S., *Wharton's Law Lexicon*, 14th ed. London: Stevens, 1938. Runrig Lands. Lands in Scotland where the ridges of a field belong alternately to different proprietors.

## Rural

*O.E.D.* Of persons: Living in the country; having the standing, qualities, or manners of peasants or country-folk; engaged in country occupations; agricultural or pastoral. Of, pertaining to, or characteristic of the country or country life as opposed to the town.

Stevens, A., 1946, The Distribution of Rural Population in Great Britain, *Trans. I. B. G.*, 11. 'It is generally implied in geographical analysis that rural population is that which is directly, or at one remove only, maintained by the exploitation of the intrinsic resources of the land' (p. 27).

Aurousseau, M., 1921, The Distribution of Population: A Constructive Problem, *Geog. Rev.*, 11. 'those sections of the people who are spread over the countryside and are engaged in the production of the primary necessities from the soil' (p. 568).

**Comment.** In British official statistics the 'rural population' is defined as that living in the administrative units known as 'Rural Districts' which may in fact comprise towns or suburbs of considerable size. See comment and references under Urban, Urban-rural continuum. Apart from this British official or administrative definition 'Rural population' may be defined as: 1. Maintained by the exploitation of the intrinsic resources of the land (a) agrarian (b) agrarian and mining (functional). 2. Living on a non-'built-up' area (landscape-sociological). 3. Size agglomeration or density of population (statistical). 4. Many 'primary' social contacts (socio-pyschological).

## Rural Population

Literally population living in the country as opposed to the towns but actually given various technical meanings especially as in England those living in the administrative units known as 'rural districts' which may in fact include considerable towns. For discussions of classification of rural population see Stevens, A., 1946, The Distribution of the Rural Population of Great Britain, *Trans. Inst. Brit. Geog.*, 11, 23–5. Vince, S. W. E., 1952, *Ibid.*, 18, 53–76.

**Rural-urban continuum**—see Urban-rural continuum

## Rurban fringe

Coleman, Alice, 1969, The Planning Challenge of the Ottawa Area, *Geographical Paper No. 42*, Ottawa: Department of Energy, Mines and Resources. 'The rurban fringe is the frontier of discontinuity between city and country.... The term "rurban" was introduced by Galpin (1915) to describe rural land in process of conversion to urban, and subsequent synonyms include "fringe", "urban fringe", "rural urban fringe", "urban sprawl" and "urban shadow". It is here proposed to compound the term "rurban fringe" to denote the area involved and to restrict "sprawl" to denote one of its chief characteristics.'

Coleman, Alice, 1976, Canadian Settlement and Environmental Planning, *Urban Prospects*, Ottawa: Department of State, Urban Affairs. 'The rurban fringe develops where settlement advances upon the countryside in less than dominant, but more than subordinate, amounts. Town and country occur intermixed in co-dominant patches, which are both capable of inflicting serious mutual damage. Settlement must be one of the co-dominant elements. The other may be either farmland, which is exceedingly vulnerable when invaded by urban sprawl; or vegetation, as for example when a city such as Sudbury has advanced into the wildscape and fragmented it; or alternatively, there may be a codominance of all three of these, settlement, farmland *and* vegetation, where sprawl has occupied marginal fringe, or where some of the farmland has reverted to weeds or scrub after settlement pressures have rendered it economically unviable.'

See also Functional (rurban) fringe.

**Rurbanisation (H. H. Balk, 1945)**

Balk, H. H., 1945, Rurbanisation of Worcester, *Econ. Geog.*, 21. A word created to cover the Rural-Urban Continuum; the example described envisaged eight satellites with links to the parent city of Worcester, Massachusetts.

**Ryot, rayat** (Indo-Pakistan: Urdu-Hindi) (various spellings)

Tenant cultivator, hence ryotwar, ryotwari, the system of tenant farming in India characterized by direct settlement between the government and cultivators without the intervention of a zemindar or landlord.

# S

**Saaidam, zaaidam** (Afrikaans: *lit.* sowing dam)
Wellington, J. H., 1955, *Southern Africa*, I. 'Saaidam irrigation is practised along the Sak river ... the sowing "dam" is a basin bordered by low earthen walls into which the flood waters are diverted ...' (p. 385).

**Sabzbār** (Pakistan: Pashto)
Autumn crop.

**Saddle**
A col or pass or any land-form recalling in shape a saddle, hence
Mill, *Dict.* (following Richthofen): A mountain pass sloping rather gently on both sides from the summit.
*Comment.* Also used by miners for anticline, anticlinal or saddleback and in Australia for saddle-reef.

**Saddle-reef**
Geikie, J., 1912, *Structural and Field Geology*, Edinburgh: Oliver and Boyd. 'The abrupt plication of the rocks has caused lenticular spaces to occur between adjoining beds in the cores of the anticlinal and synclinal folds. These spaces subsequently filled with quartz forming the so-called "saddle-reefs". Each reef is thickest along the middle line or axis of a fold ... good example the Bendigo Goldfield.' (pp. 262–263)
*Comment.* Wrongly equated by some writers with phacolith, which is an igneous intrusion of the same form.

**Saeter**, also **sæter, seater, seter, setr, setter**
*O.E.D. Suppl.* 1. Shetland and Orkney: a meadow; a pasturage attached to a dwelling.
2. In Scandinavia, a mountain pasture where cattle remain during the summer months.
*Comment.* In Norway (current spelling seter, old form sæter) 'simple farm in high mountain districts used only in the summer in connection with grazing on mountain pasture'. (P.J.W.)
See also Seter.

**Sag and Swell Topography**
von Engeln, 1942, 'Till sheets have an undulating surface to which the phrase *sag and swell* is applied. In the Middle West of the United States this type of topography extends unbroken for thousands of square miles.' (p. 489)
Thornbury, 1954, used but not defined (p. 378).
See also Swell and Swale topography

**Sagebrush, sagebrush desert;** also **sage-bush; sage grass**
A semi-desert type of vegetation in western North America dominated by sagebrush (*Artemisia tridentata*).
Korstian, C. F., 1926, in *Naturalists' Guide to the Americas*, Baltimore. [Of Nevada] 'practically all the valleys and the plateau country between the mountain ranges were originally a sagebrush desert, the dominant species being sagebrush (*Artemisia tridentata*) associated with other small-leaved shrubs' (p. 560).

**Sahara** (Arabic)
Knox, 1904. A desert, a plain; a corrupt European form of Sahra with the same meaning.
Bernard, A., 1939, *Le Sahara*, Géographie Universelle, Paris: Colin. 'Le mot *Sahara*, féminin de *Ashar*, signifie primitivement "fauve" "rougeâtre"; puis ce terme a pris le sens de "plaine no cultivée" et enfin celui de "désert".' (p. 285)

**Sahel, sahelian**
A term adopted from French writers for the savanna country south of the Sahara in West Africa.
Harrison-Church, 1957. 'Where Thorn Woodland was the original climax, where thorn trees finely divided in their leaf structure are now the dominant vegetation, acacias being the most common, and where there are also thorn shrubs and discontinuous wiry and tussocky grasses.' (pp. 77–8)

**Sailābā** (Indo-Pakistan: Panjabi)
The same as kachchi; flood-plain, the area actually flooded by the river; used also of the area irrigated by flood-waters. (A.H.S.)
See also Abi-sailābā. For more exact use see bārāni.

**Salina** (Spanish)
*O.E.D.* (from Spanish). A salt lake, pond, well, spring, or marsh; a salt-pan, salt-works.
*Webster.* A salt marsh, pond, or lake, enclosed from the sea.
*Dict. Am.* A salt spring, lick, pond, etc.; a salt mine or saltworks. 1844.
Mill, *Dict.* Orifices from which saline water and mud issue.
Thornbury, 1954. 'The waters of some playa lakes are brackish or salty. Those having a high concentration of salts are called *salinas*.' (p. 284)

**Saline**
1. As adjective, *lit.* salt.
2. An anglicized form of salina, *q.v.*
3. A salt spring.

**Salpausselkä** (Finnish)
Shackleton, 1958. 'steep though not very high ... one of the best defined end-moraines in the world ... in places a double wall runs in a great arc parallel with the coast at a distance of some forty miles inland. Within occurs the type of landscape which gave Finland its native name of *Suomi*, meaning "Lakeland" or "Swampland".'
*Comment.* More correctly described as a recessional moraine.

**Salt dome, salt plug**
*Webster. Geol.* A domical anticline, in sedimentary rocks, of which the core is a mass of rock salt, which has probably been forced up in plastic form by earth stresses from an underlying bed of salt.
Swayne, 1956. 'A great mass of rock salt (or other salt) which has moved upward from a great depth breaking through overlying strata.'
Holmes, 1944. 'Curious structures occurring in great numbers along the Gulf Coast of the United States and in other regions where salt deposits have been deeply buried. Being plastic under high pressure, the salt is squeezed towards places of weakness in the sedimentary cover' (pp. 349-350).
See also Thornbury, 754, p. 212.

**Salt Glacier, salt corrie**
Thornbury, 1954. 'The so-called salt-glaciers extend down slopes [of salt plugs] as tongues of salt similar in many respects to coulees produced by lava flows.' (p. 521)
'Salt corries are cirque-like hollows resulting from solution; they somewhat resemble craters or calderas.' (p. 521)

**Salt lick**
*Webster.* A place where salt is found on the surface of the earth, to which wild animals resort to lick it up; also, a salt spring or a salt brook. *Local U.S.*
*Dict. Am.* A place to which animals resort to lick the ground for impregnated saline particles. 1751. 1796. Morse, *Amer. Geog.* I., 663. 'The terms Salt Lick and Salt Spring are used synonymously, but improperly, as the former differs from the latter in that it is dry.'

**Saltation** (W. J. McGee, 1908; G. K. Gilbert, 1914)
*O.E.D.* (Latin, saltare, to dance, to leap)
1. Leaping, bounding or jumping; a leap
Gilbert, G. K., 1914. *The Transportation of Débris by Running Water*, U.S. Geol. Svy Prof. Paper. 86. Washington.
In the summary he says 'Some particles of the bed load slide; the multitude make short skips or leaps, the process being called saltation. Saltation grades into suspension.' (p. 11)
Holmes, 1928. A mode of transportation of debris by running water, in which the particles make intermittent leaps from the bed of the stream; a form of movement intermediate between rolling or sliding, and suspension. G. K. Gilbert: U.S.G.S. Prof. Pap., 86, 1914, p. 15.
Bagnold, 1941. 'It seems physically possible, therefore, that the phenomenon of sand-driving can be explained by this bounding motion of the grains. In his classic paper on the movement of sand under water, Gilbert described the same type of motion, and called it *saltation*. I shall use the name "saltation" for the motion of sand in air, but without prejudice to the question of whether or not the mechanism which causes the grain to jump from the surface is the same in the two fluids.' (pp. 19-20)

**Saltings**
*O.E.D.* Salting. 3. Chiefly pl. Salt Lands; in some parts spec., lands regularly covered by the tide, as distinguished from salt-marshes. Local.
Jackson, B. D., *A Glossary of Botanic Terms*, 4th ed., London: Duckworth, 1928. Salt marshes, the grass being overflowed at high-water, leaving numerous muddy channels.

Steers, J. A., 1946, *The Coastline of England and Wales*, Cambridge Univ. Press. 'The natural marshes outside the sea-walls usually show two distinct levels... the higher flats, the saltings, extend from the high water mark of spring tides to about 3 feet below that level... the outer edge is usually a sharp cliff 3 or 4 ft. high... beyond lie the mud-flats with seaweeds....' (p. 393) [Essex coast].

Comment. Distinction from salt marshes not always made.

Tansley's 'salt-marsh formation' is that of the saltings described by Steers; he does not mention saltings.

**Salt marsh, salt-marsh**
O.E.D. Marsh overflowed or flooded by the sea; specially one in which the sea-water is collected for the manufacture of salt.

Comment. See Salting; when a distinction is drawn a sea-marsh is generally protected by some form of sea wall and used for grazing. But sea-marsh is the general term; salting is local.

**Salt pan**
Webster. Geol. Any undrained natural depression, as an extinct crater, tectonic basin, or the like, in which water gathers and leaves a deposit of salt on evaporation.

Pan in which the deeper rock-layers are impregnated with salt which is brought to the surface by capillary action during the dry season. As soon as the pan is dry, the salt forms a snow-white crust on the soil. Where the salt-content is high, exploitation may be profitable; most table salt in South Africa is obtained in this way. The largest salt pans in the country are those of the Makarikari Depression in the northern Kalahari, and the Etosha Pan in South West Africa. The interesting Pretoria Salt Pan is a small caldera, the youngest volcanic phenomenon in South Africa, possibly even of Pleistocene age. (P.S. referring to South Africa)

**Samsam (Malay)**
A person of mixed Siamese-Malay origin, especially characteristic of the State of Kedah under Siamese suzerainty from 1821 to 1909.

**Samun**
Miller, 1953. 'Similar winds to the foehn occur in all mountain districts, where cyclonic storms occur; the *Chinook* is exactly similar, so are the *Samun* of Persia, descending from the mountains of Kurdistan...' (p. 269).

**Samun, samoon**—see Simoom

**Sand**
O.E.D. 1. A material consisting of comminuted fragments and water-worn particles of rocks (mainly siliceous) finer than those of which gravel is composed; often spec. as the material of a beach, desert, or the bed of a river or sea.

Mill, *Dict*. Mineral detritus finer than gravel; siliceous in composition.

*Soils and Men*, 1938. Small rock or mineral fragments having diameters ranging from 1 to 0·05 mm.; coarse sand 1 to 0·05; sand, 0·5 to 0·25; fine sand, 0·25 to 0·1; very fine sand, 0·1 to 0·05. The term 'sand' is also applied to soils containing 90 per cent. or more of all grades of sand combined. Although usually made up chiefly of quartz, sands may be composed of any materials or mixtures of mineral or rock fragments.

Glentworth, 1954. Mineral particles between 2 and 0·2 mm. diameter. Coarse sand particles 2–0·2 mm. Fine sand 0·2–0·02. The term sand is applied to soils containing 90 per cent. or more of sand.

Bagnold, 1941. 'In the finest wind-blown sands the predominant diameter is never less than 0·08 mm. Usual values lie between 0·3 and 0·15 mm.... any substance consisting of solid non-cohesive particles which lie between these limits of size may be classed as "sand".' (p. 6)

Comment. Used currently in several senses. 1. Comminuted rock fragments—not necessarily siliceous. Thus coral sand consists of fragments of coral. 2. The mineral particles in a soil between 0·2 mm. and 2 mm. diameter. This is now the agreed international standard. 3. A soil as defined above. See also Soil texture.

**Sand bar**—see Bar, Spit

**Sand drift**
Bagnold, R. A., 1941, *The Physics of Blown Sand and Desert Dunes*, London: Methuen (rep. 1954). 'a *drift* is formed in the lee of a gap between two obstacles, and is due to "funnelling", or the concentration of the sand stream on the windward side from a broad front to a narrower one' (p. 191).

**Sand dune**
A ridge of sand piled up by the action of wind on sea coasts or in deserts. A general

**Sand key, sandkey, sand cay**
Shepard, 1952. '*Sandkey or sand cay*—A small sand island which is notably elongate parallel with the shore.... The word "key" is used for various types of islands along the Florida coast, including elevated coral reefs and barrier islands. The British use the word "cay" for the same thing and "sand cay" in British publications refers to the feature for which "sandkey" is used here' (p. 1909). See also Key, cay.

**Sand levee**
A whaleback (1) *q.v.*

**Sandplain** (Australia)
Especially in Western Australia there are large sand-covered plains of uncertain origin interrupted by inselbergs which are referred to as 'sandplains'. (E.S.H.)

**Sand shadow**
Bagnold, R. A., 1941, *The Physics of Blown Sand and Desert Dunes*, London: Methuen (rep. 1954). 'deposits caused directly by fixed obstructions in the path of sand-driving wind' (p. 188). '... as its name implies, an accumulation formed in the shelter of, and immediately behind, an obstacle' (p. 191).

**Sand sheet**
Thornbury, 1954. 'Bagnold applied the name sand sheet (more commonly called sand drift) to a sand area, marked by an extremely flat surface.' (p. 310)

**Sands, singing, whistling, booming, roaring**
Bagnold, R. A., 1941, *The Physics of Blown Sand and Desert Dunes*, London: Methuen (rep. 1954). 'sound-making sands are found in two types of locality: on the sea shore and on the slip-faces of desert dunes and drifts' (p. 247). For discussion see pp. 247–256.

**Sandur** (Icelandic), **sandr**
G.S.G.S., 1944. Short Glossary of Icelandic. Sandur (*pl.* Sandar), sandy ground, sand flat, sand-bank.
Alluvial outwash plain. The Icelandic word sandur signifies sand, but in scientific literature it has been adopted for the large sandplains which are formed by glacier streams on their way from the edge of the glacier to the sea. The form sandur and not the old fashioned sandr is used in Iceland and ought to be generally adopted. E.K.)

**Sandveld** (Afrikaans)
Sandy soil covered with a vegetation of some use for grazing. (P.S.)

**Sanindo, sanyōdō** (Japanese)
The 'shady', *i.e.* wet, cloudy western side of Japan (sanindo) contrasted with the 'sunny', *i.e.* drier eastern side of central Japan (sanyōdō). Really region names.

**Santa Ana, Santa Anna** (California)
Miller, 1953. 'The *Santa Annas* of southern California and the *Northers* of the Sacramento Valley are hot, dry winds charged with dust, and, like the sirocco at its worst, owe their high temperatures largely to adiabatic heating during their descent of the mountain slopes' (p. 173).

**Sapping**—see Plucking
*Comment.* Sapping also refers to spring-head sapping—the undermining of the hillside at the back of a spring causing small slips and thus resulting in the retreat of the valley head.

**Saprolite**
*Webster. Petrog.* Disintegrated rock, usually more or less decomposed, which lies in its original place.
Thornbury, 1954. 'A deeply weathered rock cover of *saprolite* extending down many tens of feet' (p. 193) *i.e.* lying *in situ.*

**Sapropel**
*L.D.G.* Sludge or mud which collects in swamps or shallow marine basins, rich in organic matter.

**Saprophyte**
A plant living on decayed or decaying organic matter, *e.g.* a fungus living on dead wood. In contrast to a parasite subsisting on living organic matter.

**Saqia, saqiya, sakya, sakiya** (Arabic; Sudan)
Water wheel or Persian wheel used to raise water from a river for purposes of irrigation. The term used in the Sudan hence saqia irrigation and saqia cultivation. (J.H.G.L.)

**Sardar**—see Sirdar

**Sarn** (Welsh: *pl.* sarnau)
*Lit.* causeway. This word has come into geographical literature because of the disputed origin of the features concerned for which both natural and artificial origins have been argued. See Steers, J. A., 1946, *The Coastline of England and Wales*, C.U.P., 118, 148. (E.G.B.)

**Sarsen, sarsen stone**
Himus, 1954. Irregular masses of hard sandstone occurring in the Reading and Bag-

shot Beds of the Eocene in southern England found as residual masses when the softer sands have been eroded away. They are popularly known as *greywethers* from their fancied resemblance to grazing sheep.

*Comment*. The word is the same as *saracen*, brought back to England from the Crusades and applied generally to the Moslems, strangers, infidels. It may have been applied to the stones because they lie scattered over the chalk downs, far from any similar beds, being the only remnants of the former Tertiary cover and are thus 'strangers'.

### Sastrugi, Zastrugi (Russian)

Mill, *Dict*. Undulations or furrows on a snow surface caused by wind action.

*Webster*. Sastrugi, zastrugi *pl*. Wavelike ridges of hard snow formed on a level surface by the action of the wind, and with axes at right angles to the wind; a formation often observed by arctic and antarctic explorers. The ridges frequently terminate in a perpendicular wall with overhanging crest, and have a long even slope to windward.

*Comment*. Zastrugi is the correct transliteration of the Russian form of the word (C.D.H.).

### Satellite town, satellite city

Barlow Report. Cmd. 6153, 1940. 'The term Satellite Town arose somewhat later (than Garden City) and was taken to indicate a development on garden city lines, but in the neighbourhood of, and to some extent dependent on, an existing large town. (para. 271)

According to the Marly Committee, 1935, these two terms had been confused and it was 'questionable whether at this date there is any great value in the maintenance of the expression in any definitive sense'.

McKenzie, R. D., 1935, *The Metropolitan Community*, New York: McGraw-Hill. 'Some satellites are primarily agglomerations of commuters' dwellings, while others are almost independent cities.' (p. 175)

Osborn, F. J., 1945, *Green Belt Cities*. London: Faber and Faber. (First used in G.B. in 1919 as an alternative description of Welwyn Garden City and has been used by some to describe an industrial Garden Suburb) 'It is better reserved for a Garden City or country town, at a moderate distance from a large city, but physically separated from that city by a Country Belt.' (p. 182) In early publicity, as applied to Welwyn, it meant 'a detached town, dependent on local industry and girdled by a country belt, but having economic linkages with London'. (p. 40)

Van Cleef, E., 1937, *Trade Centers and Trade Routes*. New York: Appleton-Century. 'In the continuous hinterland are trade centres whose dominant interests are associated with the major center. Such centers are referred to as satellites of the major center.' (p. 34)

*Comment. O.E.D. Suppl.* quotes *Times*, 1929, 'satellite towns.' Discussed in E. G. R. Taylor, *Satellite Cities*.

### Satisficing behaviour

Wolpert, J., 1964, The Decision Process in Spatial Context, *Ann. Assoc. Am. Geographers*, 54.4. '. . . the satisficer concept or the principle of bounded rationality . . . requires nothing which is beyond the capacity of the human organism. The decision maker merely classifies the various alternatives in his subjective environment as to their expected outcomes whether satisfactory or unsatisfactory. If the elements of the set of satisfactory outcomes can be ranked, then the least satisfactory outcome of this set may be referred to as the level-of-aspiration adopted by the decision maker for that problem. His search is complete and the action is taken. . . . aspiration levels tend to adjust to the attainable, to past achievement levels, and to levels achieved by other individuals with whom he compares himself.' (p. 545)

*Comment*. Some consider this level as optimal in terms of the person's information and willingness to make efforts and take risks.

### Saturated

L.D.G. Saturated is applied to air containing as much water vapour as it will hold; if it is then cooled condensation results giving mist, cloud or rain.

### Savannah, Savanna, Savana

*O.E.D.* (From Carib zavana, via Spanish).
1. A treeless plain, properly one of those found in various parts of tropical America (from 1535).
2. 'Magnificent pine-woods—the Savan-

nahs of the South' (U.S.). [This use is obsolete—C.D.H.]

*Webster.* 2. A treeless plain; an open, level region; used chiefly with reference to the southern United States, especially Florida. 3. *Biogeog.* A tropical or subtropical grassland containing scattered trees, the undergrowth being chiefly of the xerophilous type, such as the campos of Brazil or similar regions in Africa where the baobab is the dominant tree. They pass on the one hand into steppes, on the other into savanna woodland.

Mill, *Dict.* Xerophilous grassland containing isolated trees (Schimper).

Küchler, A. W., 1947. Localizing Vegetation Terms. *A.A.A.G.*, **37**. Savanna is no longer a localizing vegetation term (p. 208).

Carpenter, 1938. 2. A tract of damp level land with a growth of grass or reeds (Southern U.S.). 3. = parkland.

Miller, A. A., 1953, *Climatology*, 5th ed., London: Methuen. 'the selvas gradually pass into woodland and scrub with scattered trees in a dominant setting of tall grass (the savanna).' (p. 126)

Comments. 1. A discussion is contained in Waibel, L., Place Names as an Aid in the Reconstruction of the Original Vegetation of Cuba. *Geog. Rev.*, **33**, 1943, 379–382. The derivation is not from Spanish, sabana (a sheet) which would place emphasis on the level terrain but from the Carib which is primarily in antithesis to woodland.

2. Spelling: 'Savanna' in: Küchler, *A.A.A.G.*, 1947; Moore, 1949; Platt, 1943; Stamp, Africa, 1943; Strahler, 1951. 'Savana' in: Gourou, 1953. P. W. Richards uses Savanna.

3. Whatever the original meaning, the term has come to indicate essentially grassland with scattered trees or bushes—the characteristic vegetation of much of tropical Africa. Whether tropical savannas are a natural climax vegetation or the result of human interference is arguable. Locally in the West Indies and South America savannah is used for an open space (*e.g.* the Savannah or Queen's Park 'a fine open space of about 199 acres' at Port-of-Spain, Trinidad).

4. In contrast to Küchler who would drop the term others would extend it to cover a type of climate; still others use 'savanna cycle' for sub-tropical cycle. See Pelzer, K., 1949, *A.A.A.G.*, **39**; see also Richards,

P. W., 1952, *The Tropical Rain Forest*, C.U.P.

**Savanna woodland**

*Webster. Biogeog.* A parklike woodland in which the undergrowth is of the xerophytic type.

**Sawah** (Indonesia)

Robequain, C., *Le Monde Malais* (translated by Laborde, E. D., 1954). 'The *Sawah* is a flat field enclosed by little embankments that prevent water running off. It is essentially a swamp-rice plantation' (p. 95).

**Saylo** (Indo-Pakistan: Kashmiri)

Ubac. See Tailo.

**Saza** (Uganda)

An administrative division corresponding to a county in Buganda. This Luganda word has been applied throughout Uganda, though the alternative county is often used. (S.J.K.B.)

**Scabland**

*Webster.* An area with a plateau surface characterized by numerous low, flat-topped hills of bare rock. *Northwestern U.S.*

Wooldridge and Morgan, 1937. 'In semi-arid regions... unchecked by the distributing action of a turf cover, myriads of water channels arise and the country is carved into the condition of typical "bad lands" or "scab-lands".' (p. 304)

Bretz, J. H., 1928, The Channelled Scabland of Eastern Washington. *Geog. Rev.*, **18**. '... elongated tracts of bare, or nearly bare, black rock carved into mazes of buttes and canyons.' (p. 446)

Rice, 1941. Scab-land topography. Rough surface of large parts of the Columbia Plateau, showing dry falls, pot-holes, abandoned channels, cut in basalt.

No reference in Salisbury, 1907; Von Engeln, 1942; Cotton, 1945; Moore, 1949.

*Comment.* Whereas scabland is developed typically on basalt, bad lands are characteristically associated with soft sediments.

**Scalded flats** (Australia)

'Scalding' is used as a pedological term indicating impregnation by salts, hence scalded plains are plains rendered of little use because of a high salt content in the soil (E.S.H.).

'Red-brown earth that has lost part or all of its "A" horizon. The term is traditional but is employed by N. C. W. Beadle (1948) *The Vegetation and Pastures of Western New South Wales*, Sydney: Govt. Printer.' (H.C.B. in MS.)

*Comment.* Dr. Brookfield's definition sent in reply to a direct query suggests that a widely used term in Australia is being given a specialized meaning by pedologists.

## Scale (of a map)

*Webster.* A ladder, a series of steps, hence, anything graduated, esp. when employed as a measure, or rule, or marked by lines at regular intervals. Specif: a series of spaces marked by lines, representing proportionately larger distances as, *a scale* of miles, feet etc. for a map or plan ... A divided line on a map, chart etc., used to measure distances on it.

*O.E.D.* III. 9. A set or series of graduations (marked along a straight line or a curve) used for measuring distances ... *spec.* the equally divided line on a map, chart, or plan which indicates its scale (sense 11) and is used for finding the distance between two points.
11. The proportion which the representation of an object bears to the object itself.

*Hinks,* 1942. 'The scale of a map in a given direction at any point is the ratio which a short distance measured on the map bears to the corresponding distance upon the surface of the earth.'

*Comment.* A rough distinction is commonly made between small-scale maps (e.g. 1:50,000) in which a large area is represented and large-scale maps (e.g. 1:10,000) or plans. See also R.F.

## Scallop (T. Griffith Taylor)

Taylor, 1951. 'Scallop shores are found along Lake Erie' (p. 619).

## Scallop (2)

Warwick, G. T., 1953, *British Caving*, London, Routledge, Kegan Paul. 'Oval-shaped hollows, ... forming patterns on the walls of caves and stream beds, were used by Bretz to determine the direction of flow of turbulent water. ... These scallops (or "flutes" of American authors) have an asymmetric cross-section along their main axis, being steeper on the *upstream* side of the scallop' (p. 55). See also J. H. Bretz, 1942, Vadose and Phreatic Features of Limestone Caverns. *Jour. Geol.,* **50**, 675–811; Coleman, J. C., 1949, An Indicator of Water-flow in Caves, *Proc. Univ. Bristol. Spel. Soc.,* **6**, Pt. 1, 57–67 and Maxson, J. H., 1940, Fluting and Facetting of Rock Fragments. *Jour. Geol.,* **48**, 717–751.

## -scape

From Dutch -schap; also in Old English and Old High German and in other forms.

*O.E.D.* Scape. A view of scenery of any kind, whether consisting of land, water, cloud or anything else. Also as the second element in combinations formed in imitation of *landscape,* as sea-scape, cloud-scape and various nonce-words.

*Comment.* Geographers have introduced Townscape, Wildscape (*q.v.*); proliferation of electric pylons and poles with wires for various purposes is *wirescape*.

## Scar

*O.E.D.* 2. A lofty, steep face of rock upon a mountain-side; a precipice, cliff. 3. A low or sunken rock in the sea; a rocky tract at the bottom of the sea.

Mill, *Dict.* A crag, cliff, hilltop, or other place where the bare rock is well exposed to view. The absence of soil or weathered debris is essential, hence the term is sometimes applied to the bare rock in a river bed or to the rocky portions of a foreshore (N. England).

Knox, 1904 (from Nor. Skar). Glen, gap, notch in a mountain, *e.g.* Scarborough.

Cotton, 1945. The concave fronts of terraces cut by a meandering river. A meander-scar. (p. 246)

*Comment.* Still much used in the north of England (notably the former West Riding of Yorkshire) for bare limestone faces—hence the Great Scar Limestone as the name for the most massive bed of the Carboniferous Limestone. The older spelling scaur is sometimes used.

## Scarp

*O.E.D.* (From Italian, scarpa).
2. The steep face of a hill; =Escarp. 1802. Playfair, *Illustr. Huttonian Theory,* 410. 'The scarps of the hills face indiscriminately all points of the compass.' Note also Escarp. 1. 'A steep bank or wall immediately in front of and below the rampart ... generally the inner side of the ditch.' (Adm. Smyth.) 2. trans. 'A natural formation of a similar kind.'

Mill, *Dict.* Scarp—Escarpment (*q.v.*). Scarp-ridge: see Cuesta. Scarp Slope: see Côtes.

Committee, List 2. 'An abrupt cliff-like face or slope terminating an elevated surface of low relief. In particular the steep contrary slope to the dip-slope of a gently tilted bed of resistant rock; the whole feature constitutes a cuesta or scarped ridge.'

Fay, 1941. 1. An escarpment, cliff, or steep slope along the margin of a plateau, mesa, terrace, or bench. The term implies a certain amount of linearity and should not be used for a cliff or slope of highly irregular outline. (La Forge.)

**Scarp land, scarpland, scarplands**

Committee, List 2. A region characterized by a number of parallel or sub-parallel scarped ridges, separated by vales.

**Schattenseite** (German)—see Ubac.

**Schist**

Himus, 1954. A foliated metamorphic rock which can be split into thin flakes or flat lenticles. The schistose structure is controlled by the prevalence of lamellar minerals such as mica, chlorite, talc, etc. Schists and gneisses are two of the commonest of metamorphic rocks: gneiss being much coarser. Schists are usually named from the dominant mineral, e.g. mica schist.

**Schiste** (French)

Plaisance, G. and Cailleux, A., 1958. 'Roche sédimentaire silico-alumineuse, susceptible de se débiter en feuillets lisses au toucher.'

This French word is included here because it is so often, like the German word *schiefer*, wrongly translated into English as *schist*. Schiste includes shales and slates as well as schists and does not necessarily indicate a metamorphic rock whereas the English schist definitely does. 'Schiste crystallin' is the equivalent of the English schist and for mica-schists the term 'schiste lustré' is employed. (L.D.S.)

**Schneebrett** (German)

Kuenen, 1955. In describing avalanches says, 'Wherever the snow cover is well hardened and consolidated the sliding and subsequently tumbling movement of the large aggregations, continues to be the main feature. This accounts for Schneebretter, which might be translated as "floe avalanches".' (pp. 133-134)

**Schratten** (German)

The same as Karren (*q.v.*). See also Karst terminology.

**Schrund line** (G. K. Gilbert, 1904)

Gilbert, G. K., 1904, Systematic Asymmetry of Crest Lines in the High Sierra of California. *Jour. Geol.*, **12**. 'Usually in viewing a cirque it is possible to trace about its wall a somewhat definite line separating a cliff or steeper slope above from a gentler, usually scalable, slope below. This line I conceive to mark the base of the berschrund at a late stage in the excavation of the cirque basin. I have called it in my notes "the schrund line". It can usually be traced for some little distance beyond the cirque, and sometimes for several miles on one wall or other of the glacial trough.' (p. 582)

Similarly Cotton, 1942, p. 179, but Gilbert's theory is not fully acceptable and the term is rarely used.

**Schuppenstruktur** (German)

Imbricate structure, the *schuppen* being the individual thrust masses in between the thrust planes in the imbricate structure, which is caused by a large number of small thrusts.

**Schwingmoor** (German)

A term adopted by many British ecologists for the 'mosses' or bogs largely of *Sphagnum* replacing former glacial lakes or meres as in the Cheshire plain. A quaking bog.

**Scirocco**—see Sirocco

**Sclerophyll**

A plant having leaves with a hard leathery surface with few stomata and hence resisting loss of water by excessive transpiration: a sclerophyllous woodland is typical of a Mediterranean climate where the shrubs and trees are evergreen but designed to withstand the hot dry summer.

**Scoria** (*pl.* scoriae)

O.E.D. 1. The slag remaining after the smelting out of a metal from its ore. 2. Rough clinker-like masses formed by the cooling of the surface of molten lava upon exposure to the air, and distended by the expansion of imprisoned gases.

Himus, 1954. Vesicular masses of volcanic rock which resemble clinker from a furnace.

See also Cotton, 1944, *Volcanoes*, pp. 138-139.

**Scour**

O.E.D. The action of a current or flow of water in clearing away mud or other deposit; in *civil engineering* an artificial current or flow produced for this purpose. Note especially *tidal scour*.

**Scree**

*O.E.D.* A mass of detritus, forming a precipitous, stony slope upon a mountain-side. Also the material composing such a slope.

Mill, *Dict.* Accumulations of loose stones lying on the slope or at the foot of the cliffs, hills, or precipices from which they have been detached. Many local terms are in common use, as Clatter, Clitter, Eboulis, Glitter, Glyders, Glydrs, Screef, etc.

See also Talus, Mass-wasting.

**Scroll, flood-plain**

Davis, 1909. Narrow areas of flood plain added to the outer end and the down-valley side of spurs between enclosed meanders. (p. 536) (1902)

Lobeck, 1939. Illustration: Marks left by abandoned meanders on the flood plain. (p. 227)

Cotton, 1941. Similar to Davis.

Thornbury, 1954. 'Meander belt deposits. The term *point bar* was applied to the bar that develops on the inside of a meander bend and grows by the slow addition of individual accretions accompanying migration of the meander. It is roughly equivalent to what has been called a meander bar, meander scroll (Davis, 1913) or scroll meander (Melton, 1936).' (p. 168)

**Scrub**

*O.E.D.* (var. of Shrub). 1. A low stunted tree. cf. Shrub. 2. Collectively stunted trees or shrubs, brushwood; also a tract of country overgrown with 'scrub.'

*Webster.* Vegetation consisting chiefly of dwarf or stunted trees and shrubs, often thick and impenetrable, growing in poor soil or in sand; also, a tract of country covered with such vegetation, especially a palmetto barren of the southern United States, or the 'bush' of Australia and South Africa.

*Dict. Am.* 5c. Designating trees that are small or dwarfed, or a hill or land overgrown by such trees. 1779.

Mill, *Dict.* Xerophilous vegetation composed mainly of dense masses of a great variety of evergreen shrubs, 4 to 6 feet high, with thick leathery entire leaves mostly of a glaucous green colour, intermingled with numerous bulbous and tuberous plants, but deficient in grasses; these, where they occur, grow only in isolated tufts.

Moore, 1949. A dense mass of low-growing evergreen plants, about four to six feet high, with occasional taller trees.

Raunkiaer, C., 1934. *The Life Forms of Plants.* Oxford: Clarendon Press (trans.). Describes deciduous, hygrophilous scrub. (pp. 320–322)

Tansley, A. G., 1939. *The British Islands and their Vegetation.* Cambridge: C.U.P. 'Communities dominated by shrubs or bushes, generally known collectively as "scrub." ... In this book heath is not counted as scrub ...' (p. 472).

Vestal, A. G., 1944. Use of Terms Relating to Vegetation. *Science*, **100**. 'Some have considered that a vegetation of shrubs is a scrub. This is true only if the shrubs are scrubly (in the usual connotation of sparse, dwarfed or malformed). Scrubly trees and scrubly bushes likewise are prevalent in certain areas. ... Scrub forest is composed of reduced and perhaps gnarled trees. Scrub woodland is an open cover of such trees.' (p. 100)

*Comment.* 'Scrubby' is more usual than 'scrubly'. P. W. Richards (in MS.) notes use for rain forest in Queensland.

**Sea**

Mill, *Dict.* 1. Oceans: (the sea—the great body of salt water in general as opposed to land). 2. One of the smaller divisions of the oceans, e.g. North Sea, China Sea. 3. Sometimes applied to some large body of inland salt water without any communication with the ocean, e.g. Caspian Sea, Sea of Aral, Dead Sea, Salton Sea.

**Sea breeze**—see Breezes, land and sea.

**Sea-fret**

A dialect word, especially in south-western England, for a salt sea-mist often very destructive of vegetation. (L.D.S.)

**Sea-level**

*O.E.D.* 1. The mean level of the surface of the sea, the mean level between high and low tide. 2. A level or flat surface of the sea.

Mill, *Dict.* 1. The assumed mean level of the surface of the ocean, whether that corresponding to the surface of a geoid or that corresponding to the surface of a rotation ellipsoid. 2. The actual mean level between high and low tide at any station. 3. The assumed mean level between high and low water with reference to which the official survey maps of any country are constructed.

*Adm. Gloss.*, 1944. Mean Sea Level. The

mean level of the sea throughout a large number of complete tidal oscillations, or the level which would have existed in the absence of tide-raising forces ($Z_0$ in Harmonic Notation).

Comment. Ordnance Datum (O.D.) from which heights are measured and recorded is mean sea-level at Newlyn, Cornwall; previously at Liverpool.

### Sea mile
Nautical or geographical mile—see Mile.

### Sea-mill
Kuenen, 1955. 'On the island of Cephalonia off the west coast of Greece, for instance, are to be found swallow-holes close to the sea, towards which sea-water flows, disappearing into the limestone.... The inhabitants use this force for the operation of water-mills; hence the name.' (p. 206)

Comment. A unique case; recently affected by earthquakes. Also known as tide-mill.

### Seamount
Thornbury, 1954. 'Lesser topographic features on the deep-sea floor are volcanic islands, seamounts and guyots... in the main they seem to be truncated volcanic islands that rise 9000 to 12,000 feet above the sea floor with summits 3000 to 6000 feet below sea level.' (p. 479)

### Search theory
Gould, P. R., 1966, *Space Searching Procedures in Geography and the Social Sciences*, Hawaii: *Social Science Research Institute Working Paper No. 1*, University of Hawaii. Concerned with how men search geographic space, 'they search in the very process of geographic exploration; they search for better economic and employment opportunities, and frequently choose to move or stay according to the results of their search; they search for alternatives, some of which may be highly restricted by their perception of space; they search for information and they search for recreation and pleasure.' This search behaviour can explain 'the patterns, relationships and processes that we see on the face of the Earth...' (pp. 3–4)

### Season
O.E.D. A period of the year: any one of the periods into which the year is naturally divided by the earth's changing position in regard to the sun; specifically spring, summer, autumn, winter. Under tropical conditions other names may be given to such well marked divisions of the year as the Rainy Season, Hot Season, etc.

### Seaway
O.E.D. A way over the sea; a channel made for the sea.

Comment. Has recently come to be used in the meaning of ship-canal; a canal large enough for sea-going vessels, *e.g.* St. St. Lawrence Seaway, opened 1959. It is used of a way for ships across the open ocean in contrast to estuaries or sheltered water. (E.O.G.)

### Sebka, Sebhka (North Africa: Arabic)
Mill, *Dict.* (1) country traversed by a network of dry valleys, (2) a temporary brackish lake or saline marsh (*cf.* shott).

Harrison-Church, 1957. 'salt encrusted mud-flats, marshy only after rare rains' (p. 231). (West Africa.)

Comment. *Cf.* playa (*q.v.*).

### Second
A sixtieth part of a minute in angular measurement (as of time) hence one/three-thousand-six-hundredth part of a degree. A second of latitude is thus about 31 metres (101 feet 4 inches) (see Mile, nautical).

### Secondary
See Geological Time, Industry etc.

### Secondary soil, transported soil
Jacks, 1954. 'Soil formed on transported material.' Contrast residual or sedentary.

### Section—see Morphological Region, Agricultural Region

### Sector model (of land use)
Garner, B.J., in Chorley and Haggett, 1967. 'Models of this type are developed on the assumption that the internal structure of the city is conditioned by the disposition of routes radiating outwards from the city centre. Differences in accessibility between radials cause marked sectoral variation in the land value surface and correspondingly an arrangement of land uses in sectors.' (p. 341)

Haggett, 1965. '... approach to urban growth patterns was put forward by Hoyt as the sector model. Studies of rent levels in American cities led him to argue that the different types of residential areas tend to grow outwards along distinct radii, with new growth on the outer arc of a sector tending to reproduce the character of earlier growth in that sector. Hoyt's model is clearly an improvement on the earlier Burgess model in that both distance and direction from the city-centre are taken into consideration...' (pp. 178–179)

Morrill, R. L., 1970, *The Spatial Organization of Society*, Belmont, California: Wadsworth Publishing Co. 'The tendency for sectors or wedges projecting outward from a city's center to be devoted to different uses and social classes.' (p. 244)

See Hoyt, H., 1939, *The Structure and Growth of Residential Neighbourhoods in American Cities*, Washington, D.C.: U.S. Printing Office.

For Burgess's theory see Concentric-zone growth theory.

**Sedentarization** (H. Capot-Rey in French, 1961, D. H. K. Amiran, 1963)

Amiran, D. H. K., and Ben-Arieh, Y., 1963, Sedentarization of Beduin in Israel, *Israel Exploration Journal*, 13, No. 3, 161–181. 'Transition from nomadism to semi-nomadism to settled life... de-nomadization of nomadic peoples.' (p. 161)

**Sedentary soil**
Soil formed directly from the decay and decomposition of the solid rocks on which it lies. Contrast secondary or transported soil.

**Sediment, sedimentary** (rocks),

*O.E.D.* Sediment. 1. Matter composed of particles which fall by gravitation to the bottom of a liquid. 2. Specifically, in Geology, etc. Earthy or detrital matter deposited by aqueous agency. Sedimentary. 2. Geol. Formed by the deposition of sediment.

*Webster.* Sedimentary rock. *Geol.* Rock formed of sediment, mechanical, chemical, or organic; especially: (1) clastic rocks, as conglomerate, sandstone, and shale, formed of fragments of other rock transported from their sources and deposited in water. (2) rocks formed by precipitation from solution, as rock salt and gypsum, or from secretions of organisms, as most limestone.

Mill, *Dict.* Sedimentary. Rocks composed of sediment; contrasted with volcanic, plutonic, or igneous rocks that have consolidated from a fused state.

Holmes, 1928. A general term for loose and cemented sediments of detrital origin, generally extended to include all exogenetic rocks (residual, detrital, organic, and solution deposits).

Rice, 1943. 'In the singular... material in suspension. In the plural... all kinds of deposits from the waters of streams, lakes or seas, or in a more general sense to deposits of wind and ice. Consolidated sediments... are called sedimentary rocks. Sedimentary—formed by deposition or accretion of grains or fragments of rock-forming material...' (p. 366).

*Comment.* Many geologists would exclude (2) of *Webster* which are described as chemically formed and organically formed rocks.

**Seepage**

*O.E.D.* Percolation or oozing of water or fluid; leakage; also that which oozes.

*Comment.* Applied especially to a small escape of oil which affords indication of oil-bearing rocks below.

**Seiche**

*O.E.D.* An occasional undulation of the water of lakes, like a tide wave, sometimes to the height of five feet, supposed to be caused by the unequal pressure of the atmosphere.

Mill, *Dict.* A somewhat sudden oscillation in the surface-level of a lake or a partly enclosed sea-area. It may be due to changes in barometric pressure, earthquakes, or other causes.

*Met. Gloss.*, 1944. 'The name given to the quasi-tides which were first observed to occur in the Lake of Geneva.... Among the important causes of seiches are winds, small earthquakes which tilt the bed of the lake, and probably the atmospheric oscillations which are recorded as waves by the microbarograph.... Another type of seiche, called a temperature seiche, was discovered by Watson and Wedderburn in some of the Scottish lakes... (see G. H. Darwin, The Tides, *Trans. R. Soc. Edinburgh*, from 1905, papers by Chrystal, White, Watson and Wedderburn).'

*Adm. Gloss.*, 1953. Seiche (Morrobia). Oscillations in the level of bays, estuaries, and lakes, of periods of a few minutes brought about by abrupt changes in barometric pressure and wind.

**Seif-dune**

Bagnold, R. A., 1941, *The Physics of Blown Sand and Desert Dunes*, London: Methuen (rep. 1954). 'the longitudinal dune, whose Arabic name *seif* or "sword" will be used in default of any universally recognised name' (p. 189). 'The seif dune occurs when the wind régime is such that the strong winds blow from a quarter other than that of the general drift of sand' (p. 195). Bagnold also uses seif-dune chain (or seif dune chain) and 'tear-drop' form of seif dune chain.

**Seismology**
*O.E.D.* The science and study of earthquakes, and their causes and effects and related phenomena.
*Seismography*—the descriptive science of earthquakes. Not common.

**Seistan**
Swayne, 1956. 'A strong northerly wind blowing in eastern Iraq during the four summer months, which occasionally reaches hurricane force.'

**Selenomorphology** (Alice Coleman, 1952)
Coleman, A., 1952, Selenomorphology, *Jour. of Geol.*, **60**. 'It is hardly permissible to use the term "geomorphology" for extra-terrestrial studies; nor does the well established "selonomorphology" quite suit the case, as it deals with the distribution of land forms rather than with their dynamic evolution. However, "selenomorphology" (Greek σεληνη, "moon", μορφη, "form", λογος, "a discourse") is a perfectly clear and apt term and has been adopted here.' [As applied to the study of the land forms of the moon and their origin.] (p. 451)

**Selion**
*S.O.E.D.* A portion of land of indeterminate area comprising a ridge or narrow strip lying between two furrows formed in dividing an open field, a 'narrow-land'.
Butlin, 1961. 'The basic unit of ploughing in the common arable fields ... synonymous with "land" ... "stitch", "stetche," "ridge", and "rig" ... the term "strip" has occasionally been used synonymously'.

**Seluka** (Arabic; Sudan)
Ahmed El-Sayed Osman, *Report of a Symposium held at Makerere College*, I.G.U. 1956. 'The word seluka means a digging-stick with foot-rest and it also applies to land cultivated by the seluka. Seluka cultivation is limited to the areas left annually by the Nile flood after it subsides' (p. 67). Used especially on *gerf* land (*q.v.*).

**Selva** (Portuguese and Spanish) also *erron.* silva
*O.E.D.* A tract of densely wooded country lying in the basin of the river Amazon. Usually *pl.*
Mill, *Dict.* Selva or Silva. 'The equatorial forest of the Amazon valley.'
Küchler, A. W., A Geographic System of Vegetation, *Geog. Rev.*, **37**, 1947. (As an example of a proposed classification, selva is described as 'Btej', which indicates: broadleaf evergreen woody vegetation, minimum height 25 metres, epiphytes occur in abundance, lianas are conspicuous.) (pp. 233–240)
Küchler, A. W., Localizing Vegetation Terms, *A.A.A.G.*, **37**, 1947. (Selva is no longer a localizing vegetation term.) (p. 208)
Moore, 1949. 'Selvas. The dense Equatorial Forest region of the basin of the River Amazon in South America.'
James, 1959. 'Selva. Tropical evergreen broadleaf forest' (p. 38).
*Comment.* It is noticeable that selva is used (a) for the region, (b) for the type of vegetation in the Amazon basin, (c) for equatorial forests generally. For full descriptions see P. W. Richards, *The Tropical Rain Forest*, Cambridge Univ. Press, 1952. For sylva cycle see Karl Peltzer, *A.A.A.G.*, **39**, 1949. See also Hylea, Mata.

**Senescent deserts**
Cotton, 1942. 'Those that still retain the strongly undulatory profiles of recently coalesced sloping pediments.' (Apparently coined by Cotton.) (p. 67)

**Senile river**
*O.E.D.* Senile: 3. Phys. Geog. Approaching the end of a cycle of erosion.
Mill, *Dict.* A river-system in which the contributory streams have approached the profile of equilibrium; all slopes have become worn down until any regular flow is impossible, and products of decomposition have accumulated until they cloak the irregularities of surface so that the great flood-plains may become marshes with no regular outflow, *i.e.* it is a river on a completely formed peneplain.
*Comment.* The Davisian cycle of erosion recognizes three stages—youth, maturity and old age.

**Senile town**—see Urban hierarchy
**Senke, Senkung** (German)
Some authors have introduced into English

these common German words, Senke meaning a depression or hollow, Senkung subsidence or sinking together with such combinations as Bruchsenke (faulted basin), Vorsenke (fore-deep), sometimes giving them specialized meanings which are not justified. See Baulig, 1956, for English and French equivalents.

## Sensible temperature
*Webster.* The temperature as felt, being essentially that of the wet-bulb thermometer.

Mill, *Dict.* The degree of warmth or cold felt by the human subject. It depends on relative humidity as well as on absolute temperature.

Klimm, Starkey *et al.*, 1956. *Introducing Economic Geography.* New York: Harcourt, Brace, 3rd ed. 'The sensible temperature—how hot or cold a person feels—cannot be stated with exactitude. Sensible temperatures differ from dry-bulb temperatures largely because of air movements and unusually high or low relative humidities.' (p. 91)

*Comment.* For a discussion of the problems involved see Lee, D. H. K., *Climate and Economic Development in the Tropics*, New York: Harper.

## Sequanian Type (of rivers)
Mill, *Dict.* Rivers which do not rise in mountains and which diminish in volume in summer on account of evaporation and the demands of vegetation; so named, by French geographers, from the river Seine.

*Comment.* Little used.

## Sequential form
*O.E.D.* 1. That follows as a sequel to. Of two or more things: Forming a sequence. c. resultant, consequent.

Davis, 1909. '... we must always expect to find some greater or less advance in the sequence of developmental changes, even in the youngest known land forms. "Initial" is therefore a term adapted to ideal rather than to actual cases, in treating which the term "sequential" and its derivatives will be found more appropriate.' (p. 257) (1899)

Wooldridge and Morgan, 1937. '... the initial forms pass through a series of sequential forms to an ultimate form.' (p. 174)

## Sequent occupance (D. S. Whittlesey, 1929)
Whittlesey, D. S., 1929, Sequent occupance, *A.A.A.G.*, **19**, 162–165. See Historical geography, occupance.

*Comment.* The term introduced by Derwent Whittlesey to describe a chronological series of cross-sections of the geography of an area. He compared it to plant-succession in botany. (H.C.D.)

## Serac, sérac (French)
*O.E.D.* From Swiss-Fr. sérac, orig. the name of a kind of white cheese; the transferred application was doubtless suggested by similitude of form. A tower of ice on a glacier formed by the intersection of crevasses.

*Webster.* A pinnacle of ice among the crevasses of a glacier; also, one of the blocks into which a glacier breaks on a steep grade.

Mill, *Dict.* One of the needle-like masses into which a glacier is broken where it passes over a bed with an exceptionally steep slope.

Tyndall, 1860. Of a glacier crossing the summit of a steep gorge: '... over this summit the glacier is pushed, and has its back periodically broken, thus forming vast transverse ridges which follow each other in succession down the slope. At the summit these ridges are often cleft by fissures transverse to them, thus forming detached towers of ice of the most picturesque and imposing character ... to such towers the name séracs is applied.' (p. 51)

*Comment.* Both the French form sérac and the anglicized form serac are in regular use.

## Serai, Sarai (originally Persian; in many eastern languages).
A building for the accommodation of travellers, an inn. *Cf.* caravanserai.

## Serendipity (Horace Walpole, 1754)
*O.E.D.* The faculty of making happy and unexpected discoveries by accident ... looking for one thing and finding another. See also *Science*, **143**, 1964, 196–197; *Science*, **140**, 1963, 1177.

## Sericulture
*O.E.D.* The production of raw silk and the rearing of silkworms for the purpose.

## Series (geology)—see Geological Time

## Serir (Egypt and Sahara: Arabic)
Gravel desert (*cf. reg*, *q.v.*)—a plain with millions of pebbles of every sort and kind from the size of a fist to that of a pea (A. Bernard).

**Serozëm** (Russian, *pl.* Serozëmy)
Barkov (see Appendix II, Russian words) gives 'desert soils, the upper layers of which are colored light grey'. See also sierozem.

**Serra** (Portuguese)
The Portuguese equivalent of the Spanish *sierra*, range of mountains (*q.v.*) but also used in a somewhat specialized sense in Brazil (see quotation).
Jones, 1930. 'The physical landscape of Northeast Brazil consists of three major divisions: marginal lowlands; *sertão*, parched uplands; *serra*, elevated mountain zones, islands of comparatively luxuriant tropical vegetation and supporting advanced agricultural districts.' (p. 471)

**Sertão** (Brazil: Portuguese; *pl.* sertões)
Mill, *Dict.* Sertaos or Chapadaos. High woodlands. Brazil.
Schimper, A. F. S., 1903, *Plant-Geography.* Oxford: Clarendon, 'In contrast with the southern portion, the middle part of Central Brazil, the so-called Sertão district, possesses a xerophilous woodland climate' (*i.e.* it is distinguished from the campos areas. W.M.) (p. 275).
Knox, 1904. Sertões (Brazil) 'backwoods', suggestive of waste land, wilderness, rather than woodlands, and applied both to Taboleras and Chapadas.
Carpenter, 1938. Half deserts covered with caatinga (N.E. Brazil) (Hardy, *Geog. of Plants*, 1925, p. 141).
Jones, C. F., *South America.* New York: Holt, 1930. 'The physical landscape of Northeast Brazil consists of three major divisions: marginal lowlands, forested in part, with relative abundance of rainfall; sertão, parched uplands of brushwood and grasses interspersed with patches of hard-leaved *caatinga*; *serra*, elevated mountain zones.... Owing to the high density of population, the sertão constitutes one of the important famine zones of the world.' (p. 471)
James, P. E., *Brazil.* New York: Odyssey Press, 1946. '... the back country ... the thinly peopled wilderness beyond the frontiers of concentrated settlement;' '... its area can be roughly, but not exactly, delimited as the zone with a population density between two and ten per square mile.'
James, 1959. 'Some Brazilian writers now insist that the real Brazil is only to be found in the back country—in the thinly peopled wilderness beyond the frontiers of concentrated settlement: in the land which the Brazilians call the sertão' (pp. 406–407)
*Comment.* The word has obviously been misunderstood by earlier writers, including Mill, and there is confusion between a physical and a sociological interpretation as well as other inconsistencies in description. (L.D.S.)

**Set**
*S.O.E.D.* The direction in which a current flows or a wind blows.

**Seter** (Norwegian)
*O.E.D. Suppl.* A wave-cut terrace in rock.
Mill, *Dict.* A rock terrace, the base of a raised beach that has been worn away (Norway).
Geikie, A., 1903, *Text Book of Geology,* 4th ed., I. 'The same strandline in one part of its course, along an exposed promontory, may be a rock-terrace ("seter" of Norway)' (p. 383).
*Comment.* Not to be confused with sæter (*q.v.*); now also spelled seter and best avoided because of the confusion. Seter as used by Geikie and Mill is the plural of sete, a bench.

**Settlement**
*S.O.E.D.* 1. The act of peopling or colonizing a new country or of planting a colony. 2. An assemblage of persons settled in a locality: hence a small village or collection of huts or houses.
*Comment.* Used by geographers to cover all groups of human habitations even single dwellings, *e.g.* Settlement Geography.

**Settlement hierarchy**
Morrill, R. L., 1970, *The Spatial Organization of Society*, Belmont, California: Wadsworth Publishing Co. 'The concept that urban places, together with their trade areas, may be grouped into distinctive levels of functional importance, and that the individual consumer will travel to smaller, closer places for everyday purchases and to larger more distant places for less-demanded goods.' (p. 242)
*Comment.* First developed by Christaller, W., *Central Places in Southern Germany,* Englewood Cliffs, New Jersey: Prentice-Hall (translated by Baskin, C. W., from Christaller, W., 1933, *Die Zentralen Orte in Süddeutschland,* Jena: Fischer) and

Lösch, A., 1954, *The Economics of Location*, New Haven: New Haven University Press (translated by Woglom, W. H., and Stopler, W. F. from Lösch, A., 1940, *Die Räumliche Ordnung der Wirtschaft*, Jena: Fischer).
See also Central place theory.

**Seven seas**
The Arctic, Antarctic, North and South Atlantic, North and South Pacific and Indian Oceans; in classical literature referred to seven supposed salt-water lagoons on the east coast of Italy including the lagoon of Venice (cut off from the Adriatic by the Lido, *q.v.*).

**Shachiang** (Chinese)
Poorly drained tight clay soils of north China resembling some of the high-lime soils of the Indus and Ganges lowlands: the B-horizon usually has lime concretions derived from calcium in the ground water. See Cressey, G. B., 1955, *Land of the 500 Million*, p. 112.

**Shādūf, shadouf, shadoof** (Arabic)
A counter-balanced dipper widely used in many eastern countries for raising water from a river or shallow well for purposes of irrigation. It consists of a long pole moving in a vertical plane about a point near one end. The short arm is heavily weighted with stones, pieces of iron, etc., from the long arm a bucket is suspended by a long rod or chain or rope. The bucket is pulled down by hand till it is below the water surface. The counterpoise weights then do most or all of the work of lifting the filled bucket.

Similar dippers or sweeps are used in many countries—notably in south-eastern Europe and as far north as Norway (svängel in Swedish).

**Shakehole, Shackhole**
*O.E.D.* Shake. 9. A natural cleft or fissure produced during growth or formation. b. in rock, mineral strata, etc. Water shake. one in which a stream empties itself. 1802. J. Mawe, *Mineral. Derbysh.* iii, 38. In this limestone stratum are frequently found openings or caverns, which are commonly called shakes, or swallows. 1823. Buckland, *Reliq. Diluv.* 6 note, open fissures, locally called shake-holes, or swallow holes, from their swallowing up the streams that cross the limestone districts.

Mill, *Dict.* Shake or shakehole. A large and deep form of swallow-hole formed close under the shales which overlie the limestone of the Pennine District.

Peel. R. F., 1952, *Physical Geography*, London: E.U.P. 'Scattered about the hillsides and bare grassy slopes are often to be found round or funnel-shaped hollows, in the Pennines known as "shakeholes", which reflect more concentrated solution down vertical pipes, or subsidence of the surface over underground cavities.'

*Comment.* A Derbyshire term, transferred to Yorkshire and now spreading to other areas (see *British Caving*, 1953). Swallowhole is best reserved for a water swallow. Also in Yorkshire dialect called *shacks* or *shackholes.* (G.T.W.)

**Shale**
Himus, 1954. A laminated sediment, the constituent particles of which are of the clay grade.

*Comment.* The distinctions should be noted between clay (non-laminated, plastic when wet), mudstone (non-laminated but non-plastic), shale (laminated), slate (metamorphosed and hardened and with a slatey cleavage not coincident with the bedding planes).

**Shamal**
Swayne, 1956. A north-westerly wind blowing in Iraq principally during the summer. See Kendrew, 1953, 252–6.

**Shamba** (East Africa: Swahili)
A cultivated plot, sometimes a temporary cultivation clearing in contrast to lusuku (*q.v.*).
From the French *champ*, and probably introduced into Zanzibar from Mauritius with clove cultivation at the end of the eighteenth century. A plantation, estate, farm, garden, or any plot of cultivated ground. (S.J.K.B.)

**Shamilat** (Indo-Pakistan: Urdu)
Village common land.

**Sharav, sharaf** (Israel: Hebrew)
The hot desert wind occurring especially in spring: Khamsin.

**Shatter belt**
Taylor, 1951. Rocks shattered along a zone of movement. Swayne, 1956. A belt of rocks which have been shattered along a zone of movement.

*Comment.* The rock found in a shatter belt is known as a fault-breccia.

## Shaw
Mill, *Dict.* A small wood or grove on a hillside; a thicket (Yorkshire and Lancashire).

*S.O.E.D.* Specially a strip of wood or underwood forming the border of a field.

## Shearing
Mill, *Dict.* The bending, twisting, or drawing out of a layer of rock near a fault or thrust plane (*q.v.*).

See also *O.E.D.* and refs., Hills, 1953, p. 30.

## Sheikh, sheik, and many spellings (Arabic: *lit.* old man)
*O.E.D.* The chief of an Arab family or tribe; the headman of an Arabian village; an Arab chief; an Eastern governor, prince, king. Now also used among Arabs as a general title of respect.

## Sheikhdom, sheikdom, shaykhdom
*O.E.D.* The status or office of a sheikh, the territory ruled by a sheikh.

## Shelf
*O.E.D.* A ledge, platform or terrace of land, rock etc. (may be submerged).

*Comment.* Applied especially to the *Continental Shelf*—the broad area surrounding many parts of the continental masses over which the water is usually less than 100 fathoms (600 feet) deep. The outer edge, the *Continental Slope*, sinks rapidly to the deep ocean-floor.

## Shield
Webster. *Geol.* The pre-Cambrian nuclear mass of a continent, around which and to some extent upon which the younger sedimentary rocks have been deposited. The term was originally applied to the shield-shaped pre-Cambrian area of Canada, but is now also used for the primitive areas of other continents, regardless of shape.

Rastall, R. H., 1941. *Lake and Rastall's Textbook of Geology*, London: Arnold. '... immense areas of very ancient rocks, which in the earlier stages of earth-history have been compacted into solid masses that apparently reacted as units towards later crust disturbances, whatever may have been the cause of these. Such resistant blocks are now usually called shields.' (p. 197)

*Comment.* A 'shield volcano' is of shield shape, not a volcano associated with a shield—Hawaiian type volcano. The best known example of a shield is the Canadian Shield. See also Craton; and Hills, 1953, who refers to the larger cratons with cores of pre-Cambrian rock being called shields.

## Shieling (Scottish; many spellings)
*O.E.D.* 1. A piece of pasture to which cattle may be driven for grazing. 2. A hut of rough construction erected on or near such a piece of pasture (cf. Shiel, with similar meanings, from which the town of Shields).

Darling, F. F., 1947, *Natural History in the Highlands and Islands*, London: Collins New Naturalist Series. 'The shielings of Lewis are the summer dwellings of a pastoral people taking advantage, for their cattle and sheep, of the short spell when the peat grows its thin crop of sedge and drawmoss. The people lived on the little knolls as on islands, bringing their cattle up to them twice a day for the milking... the shieling life is mostly gone but the green knollies in the sea of rock and peat remain' (p. 48).

*Comment.* The *O.E.D.* definition fails to explain that the shieling was summer pasture, on the hills, with accompanying settlement inhabited only in summer and where butter-making and other activities were carried on. It was in fact an essential part of the transhumance of the Scottish Highlands and Islands. (C.J.R.)

## Shift
A division of the common arable where it was not divided into fields; the basis of crop rotation and fallowing. Norfolk 15th to 17th centuries (I.L.A.T.).

## Shifting cultivation, shifting agriculture (see also Swidden, Land Rotation, Taungya)
Conklin, H. C., *Hanunóo Agriculture*, FAO Series on Shifting Agriculture, Vol. II, 1957, 1. 'Minimally, any agricultural system in which fields are cleared by firing and are cropped discontinuously (implying periods of fallowing which always average longer than periods of cropping).'

Worthington, E. B., 1938, *Science in Africa*, London: O.U.P. 'Shifting cultivation may be defined as any form of agriculture in which a patch of ground is cultivated for a short period of years until the soil shows signs of exhaustion or the land is overrun by weeds, after which the land is left to the natural vegetation while cultivation is carried on elsewhere. In due course the original site is usually planted again after the natural growth has restored fertility and checked the weeds of cultivation.' (p. 376)

*Comment.* It is said that at least a hundred and fifty names are used in different languages to indicate some form of shifting agriculture. The FAO volume quoted above notes the terms may be applied sometimes to the plot, sometimes to the system, and lists: *milpa* in Central America, *coamile* in Mexico, *conuco* in Venezuela, *roca* in Brazil, *masole* in the Belgian Congo, *chitemene* in other parts of Central Africa, *tavy* in Madagascar, *djum, bewar, dippa, erka, jara, kumari, podu, prenda, dahi* or *parka* in different parts of India, *chena* in Ceylon, *taungya* in Burma, *tamrai* in Thailand, *ray* in Indochina, *karen* in Japan, *ladang* in Indonesia, *humah* in Java, *djuma* in Sumatra and *kaingin* in the Philippines.

Because of these numerous vernacular words commonly used, it has been proposed to revise the old English dialect word for 'burned clearing' *swidden* (*q.v.*). Other descriptive terms in English are slash-and-burn, field-forest rotation, shifting-field agriculture, brand tillage (*cf.* German *Brandwirtschaft*) and fire economy.

All these terms, including the modern *swidden* obscure the distinction which is usually possible between (*a*) true shifting cultivation of nomadic tribes (*b*) a regular system of land rotation or bush fallowing practised by peoples who usually have a fixed central village and (*c*) a shifting cultivation associated with certain cash crops whereby land is abandoned when yields begin to drop below an economic level. (L.D.S.)

### Shingle, shingle bank
*Webster.* Coarse, rounded detritus or alluvial material, as on the seashore, differing from ordinary gravel only in the size of the stones, which may be as large as a man's head. *Chiefly Brit.*

Mill, *Dict.* A ridge or terrace of pebbles (shingle) heaped up on the shore at the wave-limit.

### Ship Canal
Mill, *Dict.* A canal of sufficient dimensions to admit ocean-going steamers, *e.g.* Manchester Ship Canal. See also Seaway.

### Shippon, shippen
*O.E.D.* A cattle-shed, a cowhouse.

### Shire
*O.E.D.* In Old English times, an administrative district, consisting of a number of smaller districts ('hundreds' or 'wapentakes'), united for purposes of local government and ruled jointly by an ealdorman and a sheriff, who presided in the shire moot... current as a literary synonym for county (chiefly restricted to those counties that have names ending in shire).

As the terminal element in names of counties (as Berkshire, Derbyshire) and of certain other districts (as Hallamshire, Bedlingtonshire, Islandshire, Norhamshire, Hexhamshire) which have from early times been regarded as separate entities.

*The Shires.* A name often applied to those counties which have names ending in shire, especially by those living in other parts of England.

*The grassy shires.* Those counties of the Midlands of England which have a high proportion of permanent grass and fields divided by hedgerows. These are also the well known 'hunting shires' famous for their fox-hunts.

*Comment.* Those counties of England which do not carry the suffix -shire were in the main separate kingdoms (Kent, Sussex, Essex, East Anglia or Norfolk and Suffolk) whereas the larger of the old Anglo-Saxon kingdoms, notably Mercia and Wessex, are now represented by a number of counties or shires. Some writers have suggested that the shires were shares of the Kingdom, but this connection seems to have no etymological justification. The old meaning shire was an official charge, hence a district or province under a governor.

### Shitwi (Sudan: Arabic)
Wintry, from Ar. *shita*, winter. Applied in the northern Sudan to the cooler season (December–January). (J.H.G.L.)

### Shore
*O.E.D.* 1. The land bordering on the sea or a large lake or river. Often in a restricted sense more or less coinciding with the legal definition. b. In Law usually defined as the tract lying between ordinary high and low water marks. 3. Scotch. A part of the seashore built up as a place for lading and landing; a landing place.

*Webster.* Shore, coast, beach, strand, bank

*Shore* is the general word for the land immediately bordering on the sea, a lake, or a large stream; *coast* denotes the land along the sea only, regarded especially as a boundary; *beach* applies to the pebbly or sandy shore washed by the sea or a lake; *strand* is ... poetical for *shore* or *beach*; *bank* denotes the steep or sloping margin of a stream.

Mill, *Dict*. 1. The part of a coast-line or river-bank within the limits of the rise and fall of the sea or river, or within the reach of sea-waves. 2. = Serir of Sahara; a gently undulating rocky surface in a desert uncovered by deflation.

*Adm. Gloss.*, 1953. The meeting of sea and land considered as a boundary of the sea. Interchangeable with coast when used in a wide sense to denote land bordering the sea as seen from a vessel.

Johnson, 1919. The zone from low water mark to the base of the cliff. This comprises the foreshore, between the tide levels, and the backshore, covered by water during exceptional storms only. This is the definition in Roman law. (pp. 160–161)

Wooldridge and Morgan, 1937. As Johnson, 1919 (p. 321).

Burke, J., 1953, *Stroud's Judicial Dictionary*. London: Sweet and Maxwell, 3rd ed., Vol. 4. '... that ground that is between the ordinary high-water and low-water mark' (Hale, *De Jure Maris*, ch. 4).

Note: *Adm. Gloss.*, 1953. Foreshore. The part of the shore lying between the high and low water lines of mean spring tides.

### Shoreface
Wooldridge and Morgan, 1937. 'An additional zone between the shore and the off-shore region .... risk of confusion with cliff-face.' (p. 322).

### Shoreline, Shore-line
*O.E.D.* The line where shore and water meet.

Mill, *Dict*. 2. The line along which the coast dips under the sea. W. M. Davis recognizes two types: (a) a smooth and simple line bordered by shallow water, generally due to relative emergence of the land; (b) an irregular and complicated line bordered by water of variable and locally considerable depth, due to relative submergence of the land.

*Adm. Gloss.*, 1953. Shoreline. Another name for coastline in a more general sense. See Coastline.

Johnson, 1918. 'When the term "shoreline" alone is used in the text, low tide shoreline is to be understood.' (p. 161)

*Comment.* The terminology of coasts and shores is very confused. Whilst shoreline and coastline are often used as synonymous, there is some tendency to regard coastline as the landward limit, fixed in position at least for considerable periods of time, whereas the shoreline moves as suggested by Johnson's reference to 'low tide shoreline'.

### Shott (North Africa: Arabic)
*O.E.D.* A shallow brackish lake or marsh in Northern Africa, usually dry in the summer and covered with saline deposits.

Mill, *Dict*. Temporary brackish lakes or areas of saline marsh (*cf*. Sebkha) of Algerian plateau and in valleys south of the Atlas; used also for the depressions existing after the very temporary lake has disappeared; characterized by efflorescences of salt, and frequently by absence of vegetation.

*Comment. Cf*. Salina.

### Shott (2)
Alternative name in some places for 'furlong', a unit in the open-field system. See Orwin, C. S. and C. S., 1954, *The Open Fields*, Oxford Univ. Press., pp. 5, 96 and 100.

Butlin, 1961. 'A block of arable land, consisting of several or many selions, all running in the same direction, and having at either end a headland on which the plough-team could turn ... various words have been used ... "furlong", "flatt", "wong", etc.

### Sial
From the initial letters of *Si*licon and *Al*umina, major constituents in the rocks relatively light in colour and density (*e.g.* granite) which are believed to underlie the great continental masses in contrast to sima which underlies the oceans.

*E.B.* 14th Ed., 1929, 6. 'The visible part of the earth's crust consists chiefly of lighter or more acid rocks, and beneath this it has been supposed that there lies a layer of the denser and more basic rocks. The interior core must be denser still. In formulating these ideas Suess proposed the names *sal, sima* and *nife* for the three concentric regions respectively. The term sal is now usually replaced by *sial*, and with this modification Suess's nomenclature is widely adopted. It has generally been supposed that the sial covers the whole globe. It may be thinner under the oceans and thicker on the continents, but it is present everywhere. Wegener believes that it is discontinuous. The floor of the ocean is formed of sima, and the continents are sheets of sial floating in the sima... Wegener's view is that the sheets of sial are not only separate and floating in the sima but also moving laterally, and that their positions relatively to one another have altered in the past and are still altering.' (p. 333)

**Sierozem, sierosem** (Russian: Soil science)
Jacks, 1954. 'Gray Desert Soil; brownish gray soil overlying a calcareous horizon or lime pan.'
FAO, 1963. '*Sierozem soils* are AC soils with a weakly developed A1-horizon of light grayish brown or gray color which grades into a calcareous layer that is generally hardened.'
*Comment.* The preferred Russian spelling is serozëm (*q.v.*).

**Sierra** (Spanish)
*O.E.D.* In Spain and parts of America formerly Spanish: a range of hills or mountains, rising in peaks which suggest the teeth of a saw. In general use: a mountain-range of this description.
*Comment.* The Latin word is *serra*, a saw, but the word sierra has come to be applied in Spanish to almost any range of mountains and in English has also been extended to 'the mountains' or 'mountain region'. Thus C. F. Jones, *South America*, 1930, divides Peru into three main regions—the coastal desert, the sierra, and the montana and in the chapter on 'The Sierra' uses such terms as 'sierra agriculture'. *Cf.* Portuguese *serra*.

**Sieve map, sieve method**
An approach to land planning developed especially by E. G. R. Taylor whereby areas suitable for a particular type of development are revealed by 'sieving out' all areas unsuitable for one reason or another. The results are obtained by using a series of maps printed as transparencies.

**Sike**
*O.E.D.* 1. A small stream of water, *esp.* one flowing through flat or marshy ground, and often dry in summer. 2. A gully, dip, or hollow. 3. A stretch of meadow; a field. *Obs.*

**Sikussak**—See Ice terminology
**Siliceous sinter**—see Geyserite
**Sill**
*O.E.D.* 4. b. A bed, layer, or stratum of rock, esp. of an intrusive igneous rock.

Mill, *Dict.* 1. Sheets or beds of lava intruded parallel to, *i.e.* between, other strata. 2. A northern miner's term for a bed of rock.
Holmes, 1928. A tabular sheet of igneous rock injected along the bedding planes of sedimentary or volcanic formations.
Rice, 1943. 'Intrusive sheet of igneous rock of approximately uniform thickness, which is slight compared with the lateral extent, forced between level or gently inclined beds.' (p. 34). When the molten mass swells out it becomes a laccolite (*q.v.*) or phacolite (*q.v.*). Because a sill, in contrast to a dyke, does not cross the bedding planes of the rocks into which it is intruded it is said to be concordant.
*Comment.* In scientific writing, sill should be given the restricted meaning of Holmes. In modern geological usage the *O.E.D.* definition is definitely wrong, since bed and stratum are not used of igneous rocks. Mill is also incorrect, since lava is restricted to the outpouring of a magma at the surface.

**Silt**
*Webster.* 1a. Unconsolidated or loose sedimentary material whose constituent rock particles are less than $\frac{1}{16}$ millimeter in diameter. b. *Esp.* Such sedimentary material suspended in running or standing water. c. A deposit of sediment.
*Comment.* The *O.E.D.* definition is unsatisfactory. The accepted size for silt particles is 0·002–0·02 mm. See grade

**Silviculture, sylviculture**
The culture of forest trees, from the Latin *silva* a wood.

**Sima**
From the initial letters of *Si*licon and *Ma*gnesium, major constituent elements in the rocks relatively dark in colour and relatively heavy (*e.g.* basalt) which are believed to underlay the great oceanic areas. The continental masses are pictured as gigantic rafts of sial floating on the sima. See Sial.

**Simoom, simoon** (Arabic)
Miller, 1953. 'The *simoom*, fairly frequent in the northern Sahara during the hottest months, is the most fearsome type (of desert cyclonic storms)—a swirling rush of scorching air (120°–135°), laden with dense clouds of blistering sand through which it is impossible to see more than a few yards' (p. 257).

## Simulation (models)

Chorley and Kennedy, 1971. 'Simulation ... a way of representing real systems in an abstract form for purposes of experimentation, so that there is a close relationship between the experimental situation and the real-life situation which remains unaffected. Simulation is commonly performed by means of mathematical or physically-constructed models.' (p. 355)

*Comment.* Simulation models can be deterministic (based on a direct cause and effect, consisting of a set of mathematical assertions from which consequences can be derived by logical mathematical argument) or stochastic (based on probability rather than on mathematical certainty).

## Sink, sink hole

*O.E.D.* Sink. 7. A flat, low-lying area, basin etc., where waters collect and form a bog, marsh or pool, or disappear by sinking or evaporation. Now *U.S.* = Sink-Hole. Chiefly *U.S.* Sink-hole. 2. A hole, cavern, or funnel-shaped cavity made in the earth by the action of water on the soil, rock or underlying strata, and frequently forming the course of an underground stream; a swallow-hole. Chiefly U.S. 3. U.S.c. A soft place in a marsh, remaining unfrozen in winter.

*Webster.* 8. *Geol.* A depression in the land surface, especially, one having a central playa or saline lake with no outlet; as, *Carson Sink,* in Nevada; specifically, one of the hollows in limestone regions (*limestone sink*), often communicating with a cavern or subterranean passage, so that waters running into it are lost; called also *sinkhole, swallow hole,* etc.

Mill, *Dict.* Sink or Sink-hole. A swallow hole (America).

Thornbury, 1954. 'By far the most common and widespread topographic form in a karst terrain is the sinkhole ... a sinkhole is a depression that varies in depth from a mere indentation to a 100 feet or even more ... a few square yards to an acre or more ... two major classes: solution beneath a soil mantle and collapse of rock above an underground void ... the term *doline* will be used to designate the first type and *collapse sink* the other ... surface waters enter some sinkholes through openings called *swallow holes.*' (pp. 321–322)

See swallow hole, karst terminology.

## Sinter

Mill, *Dict.* A chemical deposit of silica or carbonate of lime formed by geysers or other hot springs.

*Comment.* Note distinction between calcareous sinter (travertine) and siliceous sinter (geyserite, *q.v.*). To sinter means to cause to become a coherent solid mass (*i.e.* like a cinder) by heating without complete melting, hence 'sinter' is used of such substances as powdery iron ore treated in this way.

## Sirdar, Sardar

A word of Persian origin, formerly used in India for a military chief or leader and taken by the British army elsewhere (*e.g.* for the commander in chief of the Egyptian army). Currently used in India for a landlord of old type.

## Sirocco, Scirocco (Italian)

*O.E.D.* An oppressively hot and blighting wind, blowing from the north coast of Africa over the Mediterranean and affecting parts of southern Europe (where it is also moist and depressing).

*Webster.* 1a. A hot, oppressive, dust-laden wind from the Libyan deserts, experienced on the northern Mediterranean coast, chiefly in Italy, Malta, and Sicily. b. A warm, moist, oppressive southeast wind in the same region. 2. In general, any hot or warm wind of cyclonic origin, blowing from arid or heated regions, including the harmattan of the west coast of Africa, the hot winds of Kansas and Texas, the khamsin of Egypt.

## Site

*O.E.D.* 2. The situation or position of a place, town, building, etc., esp. with reference to the surrounding district or locality. 3. The ground or area upon which a building, town, etc. has been built, or which is set apart for some purpose. Also, in mod. use, a plot, or number of plots, of land intended or suitable for building purposes.

Smailes, A. E., 1953, *The Geography of Towns,* London: Hutchinson. '... the ground upon which a town stands, the area of earth it actually occupies' (p. 41).

Van Cleef, E., 1937, *Trade Centers and Trade Routes,* New York: Appleton-Century. 'The term location is used in this chapter to convey the idea of regional position whereas the term site may be employed with reference to topographic position' (p. 15).

Burke, J., 1953, *Stroud's Judicial Dictionary*. London: Sweet and Maxwell, 3rd ed., Vol. 4. 'The term "site" in relation to a house, building, or other erection, shall mean the whole space to be occupied by such house, building, or other erection, between the level of the bottom of the foundations and the level of the base of the walls' (Metropolis Management Act, 1878 (41 and 42 Vict., c. 32), s. 14).

Carpenter, 1938. An area considered as to its physical factors with reference to forest-producing power, the combination of climatic and soil conditions of an area; use essentially similar to habitat, with reference to wood-producing capacity (*Ecological Soc. of America*, 1934, List R-1) (p. 245).

*Comment.* Some soil scientists currently refer to 'site factors' embracing elevation, slope and aspect in discussing soil formation, and 'site' can be used as the smallest unit in a system of morphological regions (*q.v.*), the concept being that a number of similar or related sites would form a stow (*q.v.*).

**Situation**

*O.E.D.* (From F., situation; L. situs, site). 1. The place, position, or location of a city, country etc. in relation to its surroundings. 3. a. Place or position of things in relation to surroundings or to each other. b. A place or locality in which a person resides, or happens to be for the time.

Smailes, A. E., 1953, *The Geography of Towns*, London: Hutchinson. '... the situation of a town, its position in relation to its surroundings.'

Van Cleef, E., 1937, *Trade Centers and Trade Routes*. New York: Appleton-Century. Situation is used as a synonym for location which refers to regional, rather than topographical, position. (p. 15)

*Comment.* This word has been included because of the emphasis some writers have placed on the need to distinguish between site and situation.

**Skare** (Swedish)
Crust on the snow. (E.K.)

**Skarn** (Swedish)
Grout, F. F. (1932), *Petrography*..., New York: McGraw-Hill. 'A Scandinavian term for gangue or country rock of contact magnetite ores' (p. 360). 'A silicate complex associated with iron ores of contact origin'. (p. 370)

**Skauk** (T. Griffith Taylor)
Taylor, 1951. An extensive field of crevasses.

**Skavle** (Norwegian; *pl.* skavler)
The same as sastrugi (*q.v.*). Although skavler was the preferred term in the *Polar Record* list (see Ice terminology) the word sastrugi has been generally adopted.

**Skär** (Swedish)
Skerry, rocky islet; common term for smaller islands. Other identical terms are har and for smallest islets båda, kobb and klabb. In the inner skerry-guard and in lakes the usual term for small islands is holme. (E.K.)

**Skärgård** (Swedish)
From Swedish skär, skerry and gård, yard, enclosure in the meaning fenced in, or barred, thus really an enclosure by skerry-formation. It is however commonly translated as skerry guard, a belt or fringe of skerries off the coast. It is most typically developed along flat glaciated coasts where no or very little sediment exists to hide the relief of the bedrock surface. One speaks of inre (inner) and yttre (outward) skärgård. In the outward skerry guard can be found fjärds, well enclosed by splintered clusters of islands, usually termed utskärgårdar. On the border of the open sea the outermost skerries form the havsband. (E.K.)

**Skärtråg** (Swedish)
Sickle trough; a flat crescent-shaped rock basin belonging to glacially sculptured landscapes, characteristic of some regions. Their position in regard to the place of origin of the ice-current is the same as that of the moon-sickle to the sun. Found and described by E. Ljunger and proved to exist in Bohuslän (1921), Alps (1924), Andes (1928) and the Scandes (the Scandinavian part of the Caledonides). First tvåflikig gryta, but later skärtråg, (German: Sichelwanne).

**Skeletal soil**
Jacks, 1954. 'Soil consisting of nearly unweathered rock fragments.' (Jacks gives as U.K. term equivalent to U.S. term lithosol, *q.v.*).

**Skerry** (Scottish; Orkney dial., from O.N. sker; also Northern Ireland)
*O.E.D.* A rugged insulated sea-rock or stretch

of rocks, covered by the sea at high water or in stormy weather; a reef. Used with reference to Scotland, especially those parts of it formerly under Scandinavian influence, but also generally.
Mill, *Dict.* A rocky islet; often one over which the sea breaks at high tide or in stormy weather; anglicized from Scandinavian.
Knox, 1904. Skär (Sw.), Skjer (Da.), Skær (Nor.), a skerry, a rock at times covered by water.

**Skerry-guard**
The common, though incorrect, translation of the Norwegian *skjergaard* or Swedish *skärgård* (*q.v.*). The root word *gaard* or *gård* is the same as that of garden, an enclosure, and refers properly to the area of water enclosed by a line of skerries and *not* to a line of skerries acting as a guard (information from Professor William William-Olsson). The term skerry-guard is thus best avoided as it raises difficulties.

**Skid row** (United States)
*Dict. Am. U.S.* A place in a city where derelicts and petty criminals congregate and where cheap saloons, rooming houses, etc. abound. 1944 *N. & Q.* Nov. 120/2. 'A skidrow... is not a red light district. It is the district (mostly in western cities) where unskilled workers, in ordinary times gather to look for jobs—a district of employment agencies, cheap flop-houses, etc.'
This term is applied to the street or district in an American town which has become the resort or refuge of the 'down and out' —those who have become either temporarily or permanently destitute. The origin of the term is not very clear. It is said by some to derive from the logging term, to skid, by others the ease of descent with the skids on.

**Skjer, skjaer** (Norwegian)
Same as skär (Swedish) or skerry, *q.v.*

**Skjergaard, skjaergaard** (Norwegian)
Same as skärgård (Swedish), *q.v.*

**Slack (1)**
*O.E.D.* 1. A small shallow dell or valley; a hollow or dip in the ground; a depression in a hill-side or between two stretches of rising ground. 2. A hollow in the sand- or mud-banks on a shore. 3. A soft or boggy hollow; a morass.
Nature Conservancy, *Report* 1959–1960. 'Dune and slack vegetation of Newborough Warren... between the dunes lie damp hollows known as dune slacks.'

**Slack (2)**
*O.E.D.* Small or refuse coal
*Comment.* In order to encourage the sale of such coal, the British National Coal Board in the nineteen fifties introduced the term 'nutty slack' suggesting that the small coal had larger 'nuts' included within it.

**Slade**
Mill, *Dict.* 1. A hollow in the side of a valley corresponding to the cwm (*q.v.*) at its head; often marks the early stages of the rejuvenation of a tributary of the main valley. The slades of S. Wales are often well wooded. 2. Also used for a plain.

**Slash (1)** (United States)
*O.E.D.* A piece of wet or swampy ground overgrown with bushes. Hence slash-pine (Cuban pine).
*Webster.* Swampy bottom land; a wet swale; marsh; usually *pl.*, Local *U.S.*
*Dict. Am.* 1. A low, wet, swampy or marshy area, often overgrown with bushes, canes, etc. Usually *pl.* 1652. 2. Slash pine. Any one of various pines that grow in slashes, or low coastal regions, as *Pinus caribaea*, and the loblolly and shortleaf pines. 1882.

**Slash (2)**
*S.O.E.D.* The debris of felled trees.
Hence slash-and-burn: the clearing of ground by felling trees and burning the slash. See Shifting agriculture.
*Webster.* 4. In a forest, an open tract strewn with debris, as from logging, wind, fire, or the like; also, such debris.

**Slate**
A metamorphic rock derived from shales or mudstones which has acquired the property of being fissile into thin slabs (slates) along planes quite independent of the original bedding and which is due to the formation (especially under dynamometamorphism) of such minerals as micas forming thin flakes at right angles to the direction of pressure.

**Slickenside(s)**
*S.O.E.D.* A polished (and occasionally striated) surface on the wall of a mineral lode or on a line of fracture in a rockmass; a smooth glistening surface produced by pressure and friction. From 'slick' in the sense of smooth, glossy, sleek.

**Slip face**
Applied to the windward face of a sand dune.

## Slip-off slope

Cotton, 1922. The relatively gentle slope of a spur in a winding valley, less steep than the outside or undercut slope in the curve of the river.

Wooldridge and Morgan, 1937. Of meander spurs: 'The spurs have a gently sloping crest line or *slip-off slope*, while opposite to each is a steeply cut "river-cliff".' (p. 156)

*Comment.* Refers to meanders in a young valley and not to stream meanders in an alluvial plain. (G.T.W.)

## Slobland

Swayne, 1956. Muddy ground, especially land which has been reclaimed. The term is chiefly used in Ireland e.g. in Belfast.

## Sloot (Afrikaans)

Narrow water-channel, either artificial (for irrigation or drainage) or natural (a shallow erosion gully). Not used for a really deep ravine. Sometimes spelt sluit in English. (P.S.)

## Slope (of a stream)

Gilbert, G. K., 1914. The Transportation of Debris by Running Water *U.S.G.S. Prof. Paper*, 86. 'The inclination of the water surface in the direction of flow is known as the slope of the stream. It is the ratio which fall, or loss of head, bears to distance in the direction of flow. Percent slope ... is numerically 100 times as great, being the fall in a distance of 100 units.' (p. 35b)

*Comment.* The term gradient is perhaps more frequently used in England—see A. O. Woodford, 1951—Stream Gradients and Monterey Sea Valley, *Bull. Geol. Soc. Am.*, 62, 799–852. 'Gradient is used for slope expressed in per cent, feet per mile (p. 802b) or other convenient units.' (G.T.W.)

## Slope terminology

See waning slope, waxing slope, slip-off slope, gravity slope, haldenhang and references there.

## Slough

*O.E.D.* 1. A piece of soft, miry, or muddy ground; especially a place or hole in a road or way filled with wet mud or mire and impassable by heavy vehicles, horses, etc. U.S. = Slew. A marshy or reedy pool, pond, small lake, backwater, or inlet.

*Webster.* 1. A place of deep mud or mire; a mudhole; a road so full of mudholes as to be impassable. 3. Also slew, sloo, slue. A wet or marshy place; a swamp; a marsh land creek; also a. A side channel or inlet as from a river; a bayou. b. *Chiefly Local U.S.* In the Mississippi Valley and in California, a tide flat or bottom-land creek.

*Dict. Am.* 1. A comparatively narrow stretch of backwater; a sluggish channel or inlet, a pond. 1665. (The spellings slow, slew, sloo, slue are obsolete.)

Mill, *Dict.* A small marsh or bay (E. Eng. etc.).

*Comment.* Rarely used except in literary or poetic sense in Britain, though common in America where it is used for a sluggish channel or backwater.

## Sluggy, Slugga (Ireland)

Sweeting, M. M., The Landforms of Northwest County Clare, Ireland, *Trans. Inst. Brit. Geog.*, 21, 1955, 44. 'These swallow-holes being at the water-table, are water-logged, and in the east of the district are known as *sluggys* or *sluggas*. The upper Fergus disappears into one.'

Not in *O.E.D.*; see also Turlough.

## Slump, slumping

Thornbury, 1954. 'The downward slipping of one or several units of rock debris usually with a backward rotation with respect to the slope over which movement takes place.'

Swayne, 1956. 'Land-sliding on the surface of a curved slip-plane.'

See Dury, 1959, p. 10; Holmes, 1944, p. 148.

*Comment.* Submarine slumping down the steep slopes of the continental slope is believed to be very important—see Holmes, 1944, 362. See also Mass-wasting.

## Slump-fold

A fold in strata produced, not by earth-building movements but by soft sediments, especially mudstones, sliding or slumping down a slope such as the edge of the continental shelf.

## Small Circle

Any hypothetical circle on the earth's surface whose plane does *not* pass through the earth's centre (contrast Great Circle). The equator is a Great Circle, but the circles of latitude to north and south are all Small Circles, decreasing in size towards the Poles.

## Smallholding

When spelt as two words a small holding is any holding (*q.v.*) small in size; but spelt as one word smallholding has been given a legal definition in certain British Acts, e.g. Agriculture Act, 1947—under 50 acres in extent or under a certain rental value.

**Smog**
Contraction of smoke-fog.

**Smonitza** (Soil science)
Jacks, 1954. 'Hydromorphic black or dark gray soil of Yugoslavia, usually derived from calcareous clay overlying sand. The surface is leached of lime.'

**Snag**
*Webster.* 3. A tree or branch embedded in a river or lake bottom and not visible on the surface, forming thereby a hazard to boats.
*Dict. Am.* A tree trunk or large branch embedded in the bottom of a river, bayou, etc. in such a way as to be dangerous to boats. 1804.
Mill, *Dict.* Timber obstructing river navigation, either by accumulating in masses and arresting other floating matter so as to block up the river, or by being partially embedded in a river bottom and projecting upward to or nearly to the surface of the water (U.S.A.).

**Snout** (of glacier)
Mill, *Dict.* The lower extremity.

**Snow-limit**
*O.E.D.* The limit (towards the equator) for the fall of snow at sea level.
*Webster.* Snow line or limit. 2. The extreme limit from the equator within which no snow falls unmelted, varying according to elevation, nearness to the sea, and other physical conditions. Canton, China, 23°N., and Brownsville, Texas, 26°N. represent extreme points at sea level in the Northern Hemisphere.
*Comment.* Not in general use as a technical term.

**Snow line, snow-line**
*O.E.D.* 1. The general level on mountains, etc., above which the snow never completely disappears; the lower limit of perpetual snow, or (more rarely) of snow at a particular season. 2. 1898. Morris, *Austral. Eng.,* 425. In pastoralists' language of New Zealand 'above the snow line' is land covered by snow in winter, but free in summer.
Mill, *Dict.* The height at which the snow which falls in winter is in excess of that which is removed in summer by melting and evaporation. Snow does not accumulate everywhere above the snow-line, for the slope in many places is too steep to allow the snow mass to lie, *i.e.* the climatic snow-line differs from the orographical snow line. The actual or orographical snow lines vary from sea level in high latitudes to about 16,000 feet on the north slope of Ruwenzori (lat. 3°). The height varies in the same latitude and is dependent on aspect, precipitation, and humidity. The average height in about the middle latitude of the Alps (46½°N.) is 9,000 feet, *i.e.* commencement of High Alps.
Miller, 1953. 'The term "snow-line" refers to the line above which the snow does not melt in summer ... it is a line of some biological significance, no independent life being possible above this line.' (p. 273)

**Snow niche**—see Nivation hollow

**SNU, S.N.U.**—See Standard Nutrition Unit

**Soak, Soakage**
Mill, *Dict.* 1. A depression holding moisture after rain. 2. Damp or swampy spots round the base of granite rocks (W. and Central Australia).

**Socage**
*O.E.D.* The tenure of land by certain services other than Knight-service.
*Webster.* The status, tenure or holding of a sokeman. Originally the tenure was by service fixed in amount and kind, generally agricultural; but, with later commutation, *socage* came to include also any such tenure paying a money unit only and not burdened with any military service.

**Social geography**
Gilbert, E. W., and Steel, R. W., 1945, Social Geography and its Place in Colonial Studies, *Geog. Jour.,* 106. Social geography has four main branches: 1. The distribution of population over the earth's surface. 2. The distribution and form of rural settlements. 3. The geographical study of towns and cities, or 'urban geography'. 4. The distribution of social groups and their 'way of life' in different environments, including a geographical analysis of the housing, health, and conditions of labour of different communities. (p. 118)
Taylor, 1951. '... the identification of different regions of the earth's surface according to associations of social phenomena related to the total environment.' (p. 482)
Houston, J. M., 1953. *A Social Geography of Europe,* London: Duckworth. 'Social geography is defined as the study of rural and urban settlements, together with population studies.' (p. 13)
Fitzgerald, W., 1945, The Geographer as Humanist, *Nature,* 156, Sept. 22. 'Like

every aspect of the subject ... social geography is concerned with the spatial arrangement or pattern over the world of certain phenomena—in this case the phenomena which are of social, as distinct from political and economic, significance to man. They are treated, not in isolation, but in their interacting relationships with the total environment of man—an environment subject to, and undergoing constant, transformation.' (p. 356)

Huntington, C. C., and Carlson, F. A., 1931, *Environmental Basis of Social Geography*, New York: Prentice-Hall. Geography may be defined as the science which studies the reciprocal relations between man and his environment ... social geography examines it (the relationship) from the point of view of man and his activities, that is, the social aspect. (pp. 6–7) '... social geography, having to do with the environmental aspects of man and his affairs ...' (p. 10). 'Social' is preferred to 'human' because humans are considered not as individuals but rather in groups or numbers in relation to the earth of which they form an integral part.

*Comment.* According to J. M. Houston, 1953, p. 27, first used by Camille Vallaux in *Géographie Sociale: La Mer* (1908) as equalling Human Geography. See also R. H. Kinvig, Presidential Address to Section E of the British Association, *The Advancement of Science*, 10, No. 38, 1953, 157–168. When the Chair of Social Geography was created by the University of London at the London School of Economics in 1948 and the Editor of this Glossary became first Professor, the title implied no more than a sphere of work appropriate to a School concerned with the Social Sciences. Compare Cultural Geography, the more usual American term.

**Soffione, suffione** (Italian pl. suffioni)

Mill, *Dict.* Orifices in the old volcanic districts of Italy through which steam and sulphurous vapours are emitted.

*Webster.* Soffione, *pl.* suffioni. A jet of steam, usually accompanied by other vapors, that issues from the ground in a volcanic region.

**Soil**

*O.E.D.* 1. The earth or ground; the face or surface of the earth. 6. The ground with respect to its composition, quality, etc.; or as the source of vegetation. 7. Without article: mould; earth. 8. With a and pl. A particular kind of mould or earth. Also refuse matter, sewage, excrement, manure.

*Webster.* 1. Firm land; earth; ground. 2. The upper layer or layers of earth which may be dug, plowed, excavated, etc.; specifically the loose material of the earth in which plants grow, in most cases consisting of disintegrated rock with an admixture of organic matter and soluble salts. 3. The surface earth of a particular place with reference especially to its composition or its adaptability to the ends of the farmer, builder, engineer, etc.

Mill, *Dict.* Surface layers of disintegrated rock usually mixed with humus. The result of excess of weathering over transport.

Soils and Men, 1938. 'The natural medium for the growth of land plants on the surface of the earth. A natural body on the surface of the earth in which plants grow, compound of organic and mineral materials.' (p. 1177)

*Comment.* The word is a very old one, connected ultimately with Latin *solum*, the ground and, as the *O.E.D.* shows, has and has had many meanings. Its most common present day use is in the sense of the medium in which plants grow. Because of the processes of pedogenesis it is distinct from the regolith, the fragmental debris, mantling the rocks of the earth. Further 'a soil' has both chemical and physical constitution and a structure (*e.g.* Robinson, 3rd ed., 1949, ch. 1). Civil engineers, however, use 'soil' (especially in the term 'soil mechanics') as covering the surface layers of the earth down to any depth to which they may be concerned in the preparation of foundations for building. In the special sense of sewage, manure, or excrement, the use of the word 'soil' is becoming obsolete except in the phrase 'night-soil' (*q.v.*).

**Soil association**

Jacks, 1954. 'A group of soils, differing mainly in their degree of natural drainage, associated geographically on one relatively uniform parent material' (U.K.).

'A group of defined and named taxonomic units regularly associated geographically in a defined proportional pattern' (U.S.).

*Comment.* Some soil scientists use soil associations, others do not. An anomalous position was reached when soil associations were used on the published maps of the Soil Survey of Scotland but not on those of the Soil Survey of England and Wales. The Soil Survey Research Board, which supervises both, thereupon insisted that the position should be clarified and accepted at its meeting in October, 1959, the following Statement:—

*American Definition*
A soil association is an area in which different soils occur in a characteristic pattern or a landscape which has characteristic kinds, proportions and distribution of component soils.

*Scottish Definition*
A soil association is a group of soil series developed on parent material derived from similar rocks or combinations of rocks.

**Soil classification**
Most soils which develop in the *solum* (*q.v.*) fall into one or other of two great groups, the lime-rich *pedocals* (Jacks, 1954, 'soil containing an accumulation of calcium carbonate') or lime-poor *pedalfers* (soil containing accumulations of aluminium, Al and iron, Fe, compounds).

From another point of view soils fall into three world groups: *zonal* ('soil having a profile which shows a dominant influence of climate and vegetation on its development'); *azonal* ('soil lacking a profile dominantly influenced by climate, vegetation, etc.') and *intrazonal* ('well developed soil whose morphology reflects the influence of some local factor of relief, parent material or age rather than of climate and vegetation'). Most soil scientists recognize the existence of *great soil groups* at the one end of the scale and *soil series* (*q.v.*) or *soil types* as the units for description and mapping but the intermediate *soil families* and *soil associations* (*q.v.*) are differently interpreted.

In 1960 the Soil Conservation Service of the U.S. Department of Agriculture put forward a new classification and a new nomenclature. The terms used have not yet been included in this Glossary. See *Soil Classification: A Comprehensive System,* 7th Approximation U.S.D.A.

**Soil creep or Solifluxion, solifluction**
Jacks, 1954. 'The slow movement of soil material under the force of gravity'.
*Comment.* Some authors distinguish between soil creep and solifluxion—see discussion under mass wasting and references.

**Soil erosion**
The removal of soil by erosion (*q.v.*). Several main types of soil erosion are commonly distinguished:—
*gully erosion* in which deep channels (gullies) are cut into the land;
*sheet erosion,* 'the gradual uniform removal of surface soil by water'. (Jacks 1954);
*rill erosion,* 'formation of small channels by the uneven removal of surface soil by running water'. (Jacks, 1954);
*wind erosion.*

**Soil mechanics**
An engineer's term covering the behaviour of both soils and surface rocks especially in relation to foundations of buildings.

**Soil nomenclature**
The following grouping of soils is given by Jacks, 1954. Under each group certain outstanding soils or those with distinctive names not self-explanatory have been selected for inclusion in this Glossary.

Arid and semi-arid soils—chernozem, prairie soil, black earth, blackturf soil, smonitza, chestnut soil (kastanosem), sierozem (serosem).
Gley and meadow soils—Wiesenböden, meadow soils, gley soils.
Intrazonal and Azonal Soils—rendzina, mountain soils, skeletal soils (lithosol), secondary (transported) soil, residual soil, regosol, planosol.
Organic and peat soils.
Podzolic soils—brown earth (brown forest soil, braunerde), podzol (and allied types).
Saline and alkaline—solonchak, solonetz, solod, szik soil.
Tropical and sub-tropical soils—laterite (latosol), terra rossa, red earth, regur, tirs.

For a large number of other terms see Kubienà, W. L., 1953, *The Soils of Europe,* London: Murby. In addition to most of the above he uses protopedon (subaqueous raw soil), sapropel (fetid slime), rambla (raw warp soil), rutmark, anmoor, paternia (grey warp soil), borovina (rendsina-like warp soil), vega (warp soils), rawmark (raw soil of the cold desert), yerma (dry desert rawsoil), syrosem (raw soil of temperate zone), ranker, burosem (brown desert steppe soil), terra fusca (allied to terra rossa).

Most of the soil names ending in -sol are Americanizations, most of those ending in -sem have been coined in Europe under Russian-Polish influence.

FAO, 1963 describes the 21 soil units used to form the 33 soil associations shown on the 1:2½ m. Soil Map of Europe. 'The units approximate to great soil groups and represent the trend of development of the soils rather than a specific soil ... the descriptions include only the minimum essential properties of the soils in the units.' The units include: lithosols, regosols, alluvial soils, organic soils, rankers, rendzinas, brown forest soils, acid brown forest soils, gray brown podzolic soils, podzolized soils, red-yellow podzolic soils, red Mediterranean soils, brown Mediterranean soils, chestnut soils, reddish-brown and brown soils, sierozem soils, rubrozem soils, and pseudogley soils. Each of these types will be found listed in the *Glossary*.

### Soil phase
Jacks, 1954. 'A subdivision of any class of any category in the U.S. system of soil classification, based on features important to soil use and management, but not significant in the natural landscape for native plants.'

### Soil profile
Jacks, 1954. 'Vertical section of soil showing sequence of horizons from surface to parent material.' The study of the profile is the essential basis of modern soil science; it may be carried out in the field or on *monoliths* (vertical sections taken from soil). An *horizon* is a 'soil layer with features produced by soil-forming processes'. The A horizon is 'the uppermost eluvial (*q.v.*) layers of a soil profile'; B. horizon, the 'illuvial (*q.v.*) layers'; C horizon 'the weathered rock material little affected by biological soil-forming processes'; D horizon 'unweathered rock below the C horizon'. The distinction of A, B and C horizons is standard for most soils; also in some cases to be distinguished are: F layer or F horizon, 'layer of forest soil consisting of partly decomposed plant residues; G horizon, 'the horizon at which gley occurs'; H layer or H horizon, 'organic layer of forest soils with dark coloured structureless humus'.

Soils may also be classed as *mature*, having a fully developed profile; *immature*, lacking a well-developed profile; *truncated*, having lost all or part of the upper horizons.

FAO, 1963, follows the practice of describing the main soils groups of Europe by indicating whether or not all three horizons are present: thus chernozem is typically AC, a lithosol is (A)C—neither shows a B-horizon.

### Soil science
The development of soil science or pedology has given rise to an extensive specialist terminology. In the literature some common words (texture, structure) have been given restricted meanings; many foreign words have been introduced into international literature (podzol, chernozem, regur) many special terms have been created (pedalfer, pedocal, lithosol) whilst some authors have developed a whole terminology of their own which may or may not have been generally adopted. Inevitably many of the words concerned are now commonly used in geographical literature and an attempt has been made to include the most important in this Glossary. An invaluable guide is afforded by the *Multilingual Vocabulary of Soil Science* edited by G. V. Jacks and published by Food and Agriculture Organization of the United Nations, Rome, 1954 (referred to as Jacks, 1954). This has been made the basis of words included and entered under the following separate headings:—

    Soil association
    Soil classification
    Soil erosion
    Soil nomenclature
    Soil phase
    Soil profile and horizon
    Soil series
    Soil structure
    Soil texture
    Soil type
    Soil water

The terminology of pedology continues to grow. Roy Brewer in his *Fabric and Mineral Analysis of Soils* (New York: Wiley, 1964) uses and defines cutans (*q.v.*), peds, pedality, pedotubules (*q.v.*), glaebules, and crystallaria.

### Soil series
Jacks, 1954. 'A group of soils having horizons similar in distinguishing characteristics and arrangement in the soil profile, except for the texture of the surface soil,

and formed from the same parent material.' (The lowest unit in the British system of classification though textural phases may be distinguished.)

Robinson, G. W., 1938, *Soils*, London: Murby. 'Soils with similar profiles derived from similar material under similar conditions of development are conveniently grouped together as series.' (pp. 341–342) In the early soil survey of Wales, division was into 'suites', *i.e.* soils derived from the same or similar parent material and divided into 'series' according to the mode of development, as reflected in the profile. In U.S.A. textural variation in series give 'types'.

Soils and Men, 1938. A group of soils having genetic horizons similar as to differentiating characteristics and arrangement in the soil profile, except for the texture of the surface soil, and developed from a particular type of parent material. A series may include two or more soil types differing from one another in the texture of the surface soils.

## Soil structure
Jacks, 1954. 'The arrangement of soil particles into aggregates', an *aggregate* being a compound particle of soil. A 'crumb' is a rounded porous aggregate up to 10 mm in diameter. *Crumb structure* is 'consisting of small, soft, porous aggregates of irregular shape'. The word *granule* is also used for a 'friable rounded aggregate of irregular shape up to 10 mm in diameter' and *granular structure* when the soil consists of such granules.

## Soil texture
Jacks, 1954. 'Composition of soil in respect to particle size (U.K.); classification of soil based on the relative amounts of the various size groups of individual soil grains (U.S.).' The internationally recognized particle sizes are:—

| | |
|---|---|
| Gravel | 2–20 mm in diameter |
| Coarse sand | 0·2–2 mm in diameter |
| Fine sand | 0·02–0·2 mm in diameter |
| Silt | 0·002–0·02 mm in diameter |
| Clay | less than 0·002 mm in diameter |

American usage for silt is 0·002–0·05 and sand 0·05–0·2

The 'clay fraction' is clay distinguished from coarser particles. Mechanical analysis is the term applied to particle-size analysis.

## Soil type
Jacks, 1954. 'A group of soils having horizons similar in distinguishing characteristics and arrangement, and developed from a particular kind of parent material. The lowest unit in the U.S. system of soil classification' (though phases may be distinguished) (*cf.* Soil series).

## Soil water
Many special terms are used in connection with water in the soil (see Jacks, 1954). *Moisture* is 'water that can be removed from soil by heating to 105°C'. *Field capacity* is 'water held in a well-drained soil after excess has been drained away and rate of downward movement has materially decreased'. *Gravitational water* is 'water that moves in the soil under force of gravity', whereas *film water* is 'water retained in layers thicker than one or two molecules on the surface of particles in unsaturated soil'. *Waterlogging* is the 'state of being saturated with water'.

## Soke
*O.E.D.* A district under a particular jurisdiction; a local division of a minor character.

*Webster.* c. A jurisdiction or franchise... d. The district over which such jurisdiction or franchise extended. Such districts are retained, in some cases, as modern administrative divisions; as, the Soke of Peterborough.

## Sokeland
See reference under foldage.

## Sokeman
See Socage.

## Solano (Spanish)
*O.E.D.* In Spain a hot south-easterly wind.

*Webster.* A hot, oppressive east wind of the Mediterranean, especially on the eastern coast of Spain; also a cloudy, rain-bringing wind of the same locality and direction.

## Solfatara (Italian)
*O.E.D.* The name of a sulphurous volcano near Naples. A volcanic vent from which only sulphurous exhalations and aqueous vapours are emitted, encrusting the edge with sulphur and other minerals.

Mill, *Dict.* A fumarole (*q.v.*) emitting acid, especially sulphurous, vapours.

## Solfataric stage
Himus, 1954. A dormant or decadent phase of volcanic activity, characterized by the

emission of gases and vapours of volatile substances.

**Solifluxion, solifluction** (J. G. Anderson, 1906)

Anderson, J. G., 1906, Solifluction, a component of subaerial denudation, *J. Geol.*, 14, 91-112.

*O.E.D. Suppl.* A gradual downward movement or slide of particles of the earth's surface; soil-flow.

Wooldridge and Morgan, 1937. Soil creep, whether aided by nivation or not (p. 403). Spelling: solifluction.

Von Engeln, 1942. 'Where rainwash is exaggerated to the degree that it becomes a major translocation process, it is referred to as solifluction, that is, soil flow.' (p. 246)

*Comment.* Solifluction is the spelling used by Holmes, 1944; Cotton, 1945; Robinson, 1949. Some authors distinguish between solifluction and soil creep—see discussion under mass wasting.

**Solod, soloth** (Russian: Soil science)

Jacks, 1954. 'Leached saline soil (degraded solonetz) having a pale $A_2$ horizon and a degraded fine-textured B horizon.'

*Comment.* Solodi (*pl.*) is the Russian for degraded solonetz soils (C.D.H.).

**Solonchak** (Russian: Soil science)

Jacks, 1954. 'Saline soil without structure.'

*Comment.* Barkov (see Appendix II, Russian words) gives 'soils of arid and semiarid regions'.

**Solonetz** (Russian: Soil science)

Jacks, 1954 'Formerly saline soil from which the salts have been leached, with cloddy prismatic or columnar B horizon.'

*Comment.* Russian is solonets, *pl.* solontsy: soils with surface rock salt.

**Solstice**

*O.E.D.* One of the two times in the year, midway between the two equinoxes, when the sun is farthest from the equator and appears to stand still *i.e.* about 21st June (the summer solstice) and 22nd December (the winter solstice).

**Solum**

Jacks, 1954. 'The part of the earth's crust influenced by climate and vegetation' (Soil science).

**Somatology**

*Webster. Anthrop.* The comparative study of the structure, functions and development of the human body—practically restricted to statistical treatments of bodily measurements and to comparative descriptions of those traits which chiefly distinguish races and populations.

**Somma** (Italian)

Mill, *Dict.* Originally, the rampart remaining from the old crater of Vesuvius and forming an arc around one side of the new one. The name is sometimes extended to similar formations in other volcanoes.

*Webster.* The rim of a volcanic crater.

*Comment.* The derivation from Monte Somma has been misunderstood by some writers and the term has been applied, quite wrongly, to small subsidiary craters inside a caldera—almost the exact reverse of the proper meaning. See Inter. Geog. Union, 1958, Regional Conference in Japan, *Report*, 1959.

**Sonnenseite** (German: *lit.* sunny-side)—see Adret

**Soroche**

*O.E.D.* Native name in the Peruvian Andes for mountain sickness.

**Sotch** (French)

Mill, *Dict.* Doline (*q.v.*) (Causses, France). The term 'cloup' is used in Aquitaine.

Plaisance, G. and Cailleux, A., 1958. 'Bas-fond en terrain calcaire, rempli de terre rouge de décalcification.'

**Soum** (Irish and Scottish)

Unit of stock in common grazing. (E.E.E.)

*O.E.D.* 1. The amount of pasturage which will support one cow or a proportional number of sheep or other stock.

2. The number of sheep or cattle that can be maintained on a certain amount of pasture. A *soum of sheep*, a number varying from four to ten.

*Comment.* Compare stint.

**Sound**

Mill, *Dict.* English form of Scandinavian sund = strait or inlet.

*Webster.* 1b. A long passage of water connecting two larger bodies but too wide and extensive to be termed a strait, as a passage connecting a sea or lake with the ocean or with another sea, or a channel passing between a mainland and an island; as, the sound between the Baltic and the North Sea.

**Source**

Mill, *Dict.* The point where a river rises. The term, which is also used for headwater, or one of the headwaters of a stream, should be restricted to the first meaning.

**Southerly Burster** (Australia: New South Wales; New Zealand)
Kendrew, W. G., *Climates of the Continents*, 4th ed., 1953. 'In the east of New South Wales strong cold winds from the south known as Southerly Bursters.... The average number is 30 a year, most of them in spring and summer' (pp. 552–553).

Miller, 1953. 'In the absence of a great continental mass in high latitudes there can be no great reservoir of cold air in the southern hemisphere such as exists in Siberia and Canada to supply the "Cold Waves" and "Northers" of the United States and the "Buran" of Siberia. Compared with these the "Southerly Bursters" of Australia and New Zealand and the "Pampero" of the Argentine are relatively mild' (p. 43).

**South-Wester, sou-wester**
Mill, *Dict*. A strong south-west wind.

**Sovkhoz** (Russian)
A State farm. See Kolkhoz.

**Sowbacks**
Geikie, J., 1898. '... in some of the broader dales of Scotland the configuration of the boulder-clay becomes strongly defined, the accumulation being arranged in a well-marked series of long parallel banks known as "drums" or "sowbacks".' (p. 196)
(These are referred to separately from drumlins, eskers or lateral and terminal moraines. W.M.)

ay, 1920. Same as hogback or horseback; a kame or drumlin.
*Comment*. Rarely used.

**Sowneck**
Mill, *Dict*. A gentle rise of land forming a fairly narrow boundary between two expanses of water or of lowland.
*Comment*. Rare.

**Spa**
The name of a watering place near Liège, Belgium, celebrated for the curative properties of its mineral springs. Hence used in a general sense and applied to similar mineral springs and watering places elsewhere.

**Sparselands** (T. Griffith Taylor)
Taylor, 1951. Semi-desert pastoral regions (p. 620).

**Spatial analogue**
Chorley, R. J. in Chorley and Haggett, 1967. 'Spatial analogues associate one set of phenomena with others on the assumption that observations at another place are easier to make or simpler in character than those of the original, or that comparison with other areas believed to be in some way similar will enable one to make more confident and meaningful generalizations about the original area... The most common form of spatial analogue model is that in which adjacent contiguous areas are grouped together on the assumption that each unit can be better understood in terms of generalizations about some larger region of which it forms a part.' (p. 62)
An early example of this method is in Johnson, D. W., 1919, *Shore Processes and Shoreline Development*, New York: Wiley.

**Spatial analysis**
Berry, B. J. L. and Marble, D. F., 1968, *Spatial Analysis*, Englewood Cliffs, New Jersey: Prentice-Hall. Spatial analysis a deviation from the philosophy of area differentiation being the main concern of geography. Spatial analysis has 'an awareness of the interdependence of geography and geometry—of the science of spatial analysis and the mathematics of space.' '... it was workers in other fields, not geographers, who provided many of the initial examples of applications of modern statistical methods to geographic problems.... Spatial analysis remains basic to quantitative plant ecology and to much of epidemiology, and from these fields have come many of the ideas used in the study of pattern in point distributions and of spatial diffusion processes. (pp. 2–3)

**Spatial autocorrelation**
Cliff, A. P., Haggett, P., Ord, J. K., Bassett K. A. and Davies, R. B., 1975, *Elements of Spatial Structure: a Quantitative Approach*, Cambridge: Cambridge University Press. '... the relative locations of the points or areas to which the data refer can provide some information about the spatial pattern of variation in these data. That is, the data exhibit spatial autocorrelation. In general, if high values of a variable in one area are associated with high values of that variable in neighbouring areas, we say that the set of areas exhibits positive spatial autocorrelation with regard to that variable. Conversely, when high and low values alternate, the spatial autocorrelation is negative.' (p. 145)

**Spatial equilibrium**
Morrill, R. L., 1970, *The Spatial Organization*

*of Society*, Belmont, California: Wadsworth Publishing Co. 'A theoretical state of stability. Any deviation from this state would decrease efficiency or profitability equilibrium prices are values (for goods, labour, capital, or land) corresponding to these conditions.' (p. 242)

**Spatial interaction**
Abler, Adams and Gould, 1971. 'Flows of various kinds of commodities and information from one place to another.' (p. 74)
Morrill, R. L., 1970, *The Spatial Organization of Society*, Belmont, California: Wadsworth Publishing Co. 'The interrelation of locations usually in terms of people or communications . . .' (p. 244)

**Spatial margin**
Rawstron, E. M., 1958, Three Principles of Industrial Location, *Trans. Inst. Brit. Geog.*, 25. The spatial margins are those which delimit the area in which industry can be located if economic viability is to be achieved. '. . . the economically determined margin bounds the area or areas within which viability can be attained and that the margin may be deemed to apply to locational costs as a whole or, with certain provisos, to one component or part of a component in the cost structure.' (p. 138)

**Spatial organization**
Morrill, R. L., 1970, *The Spatial Organization of Society*, Belmont, California: Wadsworth Publishing Co. 'The aggregate pattern of use of space . . .' (p. 244)
For a discussion of alternatives for the spatial organization of future urban society see Wurster, C. B. in Wingo Jr., L., 1963, *Cities and Space*, Baltimore: The Johns Hopkins Press.
See also Spatial structure.

**Spatial process**
Abler, Adams and Gould, 1971. 'The emphasis in contemporary geography on spatial structure is somewhat misleading, because it overemphasizes distributions to the neglect of the spatial processes which interact causally with them. What we call spatial processes are mechanisms which produce the spatial structures of distributions.' (p. 60)
See also Spatial structure.

**Spatial structure**
Abler, Adams and Gould, 1971. 'In recent years internal relative location has often been called "spatial structure". . . . The spatial structure of a distribution is both the location of each element relative to each of the others, and the location of each element relative to all the others taken together.' (p. 60)
See Wurster, C. B. in Wingo Jr., L., 1963, *Cities and Space*, Baltimore: The Johns Hopkins Press, pp. 72–101.
See also Spatial processes and Spatial organization.

**Spatter cone**
Lobeck, 1939. Volcanoes. 'Where gases sputter out through the side of the dome, a *spatter* or *driblet cone* may be built up 10 to 15 feet above the ground.' (p. 675)

**Specific region**—see Regional system.

**Spelaeology, speleology**
The scientific study of caves.

**Speleothem** (G. W. Moore, 1952)
Howell, 1957. A secondary mineral deposit formed in caves.

**Sphere of influence**
*O.E.D.* 7. d. Sphere of action, influence, or interest, in recent use, a region or territory (esp. in Africa or Asia) within which a particular nation claims, or is admitted, to have a special interest for political or economic purposes.
Mill, *Dict.* A portion of territory reserved without annexation for or by one power from the political interference of other Powers.
Smailes, A. E., 1953. An urban sphere of influence in the area whose inhabitants depend on the town for various services. It is equated with the 'urban field' and the 'commuting area'. (p. 143)
Glass, R., 1948, *The Social Background of a Plan*. London: Routledge. Reference is made to 'spheres of influence' or organisation within the town, e.g. clubs, hospitals, schools, shops, etc.
Van Cleef, E., 1937. *Trade Centers and Trade Routes*. New York: Appleton-Century. '. . . a trade center and its hinterland, that is, the territory of its sphere of influence. . . .' (p. 100)
*Comment.* With the world-wide spread of nationalism the old concept of 'sphere of influence' as defined by Mill and the *O.E.D.* has almost disappeared and the phrase is avoided for political reasons. In the field of urban geography most writers prefer umland or urban field or even urban hinterland (*q.v.*).

**Spheroid, oblate**—see Geoid

**Spheroidal weathering**
Holmes, 1944. 'The separation of shells of decayed rock is distinguished as *spheroidal weathering* (plate) best developed in well-jointed rocks like many basalts and dolerites. Water penetrates the intersecting joints and this attacks each separate block from all sides at once... as each shell breaks loose, a new surface is presented to the weathering solutions... a round core... still left in the middle' (pp. 118–119).
Also called onion weathering.

**Spilitic suite**—see Atlantic suite

**Spill-bank, spill-slope, spill-hollow**
Spill-bank is the term used by British and Indian engineers engaged in river training for the bank of coarse alluvium spilled over by a river in flood (C. C. Inglis). Arthur Geddes has added the terms 'spill-slope' and 'spill-hollow' to describe the loamy slope terminating in a clayey, stagnant hollow, characteristic of alluvial plains due to deposition. To describe a spill-bank the term levee is ambiguous and requires the qualification 'natural'. (C.J.R.) See Levee.

**Spill-way**
The area below a dam or natural obstruction over which excess water from the reservoir or lake above is allowed to drain away.

**Spinifex desert** (Australia)
O.E.D. Spinifex 'one or other of a number of coarse grasses (now classed in the genus *Tricuspis*) which grow in dense masses on the sand-hills of the Australian deserts, and are characterized by their sharp-pointed, spiny leaves'.
*Comment.* The Spinifex deserts of Australia are characterized by large tufts of *Spinifex* separated by bare ground not necessarily sandy. (L.D.S.)

**Spinney, spinny**
Mill, *Dict.* 1. A thorny plot; a place full of briers. 2. A small wood with undergrowth; a clump of trees; a small grove or shrubbery.

**Spit**
O.E.D. 6. A small, low point or tongue of land, projecting into the water; a long narrow reef, shoal, or sandbank extending from the shore.
Mill, *Dict.* A protruding narrow tongue of sand or silt—the prolongation of a beach or bar, often running across a small bay from headland to headland if there is a current along the shore. Such spits are also called Barrier Beaches.
*Adm. Gloss.*, 1953. A long narrow shoal (if submerged) or a tongue of land (if above water) extending from the shore and formed of any material.
Johnson, 1919. 'So long as an embankment has its distal end terminating in open water, it is called spit.' (p. 287)
Thornbury, 1954. 'The term *bar* may be used in a generic sense to include the various types of submerged or emergent embankments of sand and gravel built on the sea floor by waves and currents. One of the commonest types of bar is a *spit*, defined by Evans (Evans, O. F., 1942, The origin of spits, bars, and related structures, *Jour. Geol.*, 50, 846–863) as "a ridge or embankment of sediment attached to the land at one end and terminating in open water at the other"... growth may be deflected landward with the creation of a *recurved spit* or *hook*. Several stages of hook development may produce a *compound recurved spit* or *compound hook*.' (p. 445)
*Comment.* The term is now used as defined by Evans so that both *O.E.D.* and Mill are misleading.

**Spitskop** (Afrikaans)
Hill with sharply pointed top, as distinguished from Tafelkop, hill with flat top, caused by a layer of hard rock such as dolerite. The spelling Spitz is influenced by German and occurs in South-West Africa. (P.S.)

**Splays, floodplain**
Thornbury, 1954. '... a term applied to materials spread over the floodplain through restricted low sections, through breaks in natural levees, or along distributary channels....' (p. 172)

**Spoil bank**
A bank or mound consisting of waste material from mining or quarrying operations; a tip-heap.

**Spot height**
On British Ordnance maps marked by a dot against which is placed a figure representing the height above O.D. (Ordnance Datum or Mean Sea Level) of the ground level (such as that of the crest surfaces of a road). In contrast to the accurately marked Bench Mark (*q.v.*) it is not marked on the ground.

**Spring** (1)
*S.O.E.D.* The season between winter and summer, reckoned astronomically from the spring or vernal equinox (March 22) to the summer solstice (21 June) in the northern hemisphere; in popular usage commonly used for March, April and May, or by some mid-February to mid-May or even February, March and April.

**Spring** (2)
*O.E.D.* A flow of water rising or issuing naturally out of the earth.
Especially such a flow of water having special mineral properties (mineral spring) and notably of curative quality.

**Spring alcove**—see Alcove

**Spring-line**
Where the underground water-table reaches the surface, as it may do under certain circumstances along a steep scarp slope, a line of springs may occur roughly at the same level. Where such springs are copious and constant they afforded a reliable water supply and led to the siting of villages, hence the 'spring-line village'.

**Spruit** (Afrikaans)
Mill, *Dict.* A small river or rivulet (S. Africa).
*Comment.* Used especially of streams which may be quite dry for periods and then subject to sudden floods. The use of the word has spread to East Africa.

**Squattocracy** (Australia)
'Describes the class of wealthy landowners who obtained their land in the period before the Free Selection Acts. In its nature the group is not unlike the English "squirearchy".' (H. C. Brookfield in MS. 1959.) Used by T. Langford-Smith in unpublished thesis, A.N.U., 1958, and in wide use in speech.

**Stack**
*O.E.D.* 7. dial. (cf. Faeroese, stakkur, 'high solitary rock in the sea'). A columnar mass of rock, detached by the agency of water and weather from the main part of a cliff, and rising precipitously out of the sea.
Mill, *Dict.* An isolated rock mass or pinnacle which has been separated from the main mass by ordinary erosive processes usually off a sea coast and possessing steep sides; also applied to outstanding mountain masses.

**Stadial moraine**
Wooldridge and Morgan, 1937. 'In strict usage, "terminal moraines" are those marking maximum extension of the ice, while "stadial moraines" mark stages of retreat, though the former is often used in a general sense covering both cases.' (p. 378)

**Stage** (1)
The word stage has numerous meanings. A basic idea is a division of space, time, journey, process.
Mill, *Dict.* Stage or Series (Geol.). The narrowest subdivision of strata; a succession of strata in which the fossil remains have the closest affinity to one another = French Étage and German Stufe. The corresponding term referring to time is 'age'.
Fay, 1920. In the nomenclature adopted by the International Geological Congress, the stratigraphic subdivision of the fourth rank; a division of a series. The chronologic term of equivalent rank is age (La Forge).

**Stage** (2)
Committee, List 2. The cycle of erosion can be divided into early, middle and late stages; the surface features appropriate to each of these stages may be described as young, mature and old (or senile).
von Engeln, 1942. 'When a geomorphic feature, most commonly a valley cross section, is said to be in a certain stage, it is meant to indicate that its development by process has proceeded to a given one of the characteristic points in the series of changes that must ensue between a beginning and an end condition. The reference by such designation is to the form arrived at, and has nothing to do with the measure of time required to bring about that stage' (p. 76).
*Comment.* 'Stage' does not stem from Davis, 1899, *Geog. Jour.* in which he refers to structure, process and *time*. Later writers have changed this to structure, process and *stage*.

**Stage** (3)
Carpenter, 1938. The result of every transformation of vegetation is called a stage, if there is an appreciable change in the floristic composition or in an evident extension of some species; three stages are recognized: pioneer, transition, final (Braun-Blanquet and Pavillard, 1930) (p. 253).

**Stage** (4)
This concept has application to political

geography: cf. references to geopolitically mature and immature states in Whittlesey, D., 1939, *The Earth and the State*, p. 18, New York: Holt & Co.

**Stage-coach**
In the days when travel was by road and horse-drawn public coaches a 'stage' was as much of a journey as was performed without stopping for a rest or change of horses. A public coach which proceeded by such stages was a stage-coach, often shortened to stage.

**Staith, staithe**
Mill, *Dict*. An elevated wharf with a chute for shipping coal (N. England).

**Standard Nutrition Unit, S.N.U.** (L. D. Stamp, 1958)
Stamp, L. D., 1958, The Measurement of Land Resources, *Geog. Rev.*, **48**, 1958, 1-15. 'We may say that 1,000,000 Calories a year must be produced in order to provide adequately for each human being. I propose to call this a *Standard Nutrition Unit* which we can use to measure actual diets and which we can relate to the production of different crops and different types of land... 1,000,000 Calories of food *produced* or 900,000 *consumed*.'

**Standard time**
Because of the difficulties which result if each place keeps its own local time (*q.v.*) most civilized countries adopt one or more standard times. Standard time is usually the local time of a meridian or line of longitude near the centre of the country or zone and is commonly chosen so as to be a number of complete hours or half hours behind or ahead of Greenwich Mean Time —the local time of 0°. In large countries, such as the United States, falling into a number of time zones each is given a name. Thus North America has Atlantic, Eastern Standard, Central, Mountain and Pacific Times, respectively 4, 5, 6, 7, and 8 hours behind Greenwich Mean Time.

Many countries in mid-latitudes adopt in the summer months what is called Summer Time or Daylight Saving Time, normally one hour in advance of the local Standard Time. This has the effect of giving office and factory workers, leaving their work at nominally the same hour throughout the year, an extra hour of daylight in the evening.

**Staublawine** (German)
Kuenen, 1955. In describing avalanches says 'If a great deal of powdered snow on a steep slope is carried along, it will be tossed up into a thick cloud hiding all other movement from view.... It was long before the nature of these snow-slips, known as Staublawinen (=dust avalanches), was properly understood... anyone overtaken by such a cataclysm must inevitably inhale large quantities of snow-dust which instantly melts and the unfortunate victim is asphyxiated. It is very much the same thing as drowning.' (pp. 132-133)

*Comment*. Since there is no dust but only finely powdered snow, 'dust-avalanche' seems an unfortunate translation. Fischer and Elliott, 1950, *German and English Glossary*, give 'Staublawine = powdery avalanche, dry avalanche.'

**Steady state**
Chorley and Kennedy, 1971. 'The balance between inputs and outputs of energy and/or mass of a system maintained in such a manner so as to maintain its level of integration.' (p. 356)

**Steilwand** (German; Penck)
Equivalent to Bösche or gravity slope.

**Stenohaline**
Lake, 1958. 'Marine plants and animals can be divided into two groups according to their tolerance to changes in salinity. *Stenohaline* organisms are sensitive to relatively small changes,... animals in the open ocean... *Euryhaline* organisms have a tolerance to a wide range of salinity... characteristic of coastal areas and estuaries.' (p. 425)

**Stenothermic, Stenothermy**
Lake, 1958. Marine 'organisms can be divided into two groups depending on their degree of tolerance to temperature changes. *Stenothermic* forms are tolerant only to small temperature changes whilst *eurythermic* forms can tolerate large changes.' (p. 426)

**Stentorg** (Swedish)
From Swedish folk name sten, stone, boulder and torg, square, market place. Well-defined stone or boulder field mostly on the crests of oses but may be extended to their flanks. They often are striped by former shore lines and beach-ridges. (E.K.)

**Step-faults**
Mill, *Dict*. Faults resulting in a series of sudden step-like changes of level of strata.

**Steppe**
*O.E.D.* (From Russian) 1. One of the vast comparatively level and treeless plains of south-eastern Europe and Siberia. 2. An extensive plain, usually treeless.

Mill, *Dict.* An area adapted to Xerophytic grasses and shrubs, but not to trees (Schimper).

Knox, 1904. (from Russ.) A vast treeless plain, prairie; the Russian form is Step (pronounced 'stepp'), and is applied generally to grassy, saline, and sandy tracts.

Küchler, A. W., 1947, Localizing Vegetation Terms, *A.A.A.G.*, 37. 'The vast grasslands that reach from the lower Danube to central Asia bear the name Steppe... the term is all inclusive and implies treeless grassland... the abuse of the term has been so great and the meaning of the term so variable as actually to lose its significance. Allan lists 54 uses of this overworked term and claims he can add more. These ones carry entirely different and even contradictory implications. This author agrees with Allan that the Steppe best remains in its area of origin.' (pp. 206–207) (See Allan, H. H., 1946, Tussock Grassland or Steppe? *The N.Z. Geog.*, 2, 223–234.)

*Comment.* Although the *O.E.D.* and numerous authorities equate steppe with prairie, there is a general tendency to regard steppe as implying grassland drier than prairie. Probably the most satisfactory solution is that commonly used in elementary textbooks many of which state quite correctly 'the mid-latitude grasslands have received different names in different continents—Steppes in Asia and Europe, Prairies in North America, Pampas in South America, High Veld in South Africa, and Downland in Australia—but they are very similar throughout.' (Stamp, L. D., 1959, *The World*, 16th ed., London: Longmans, pp. 127–128.)

Russian is *step'*, *pl. stepi*. Ushakov (see Appendix II, Russian words) gives 'treeless and usually waterless stretches with a level surface, covered with grass vegetation'.

**Steppe-heath** (steppenheide of R. Gradmann) At the end of last century, the German geographer R. Gradmann suggested that the settlement of early man in Europe was encouraged by relatively open areas of *steppenheide* where the vegetation was easier to clear than in the thicker forests. What came to be known as the 'steppenheide theory' was based on botanical studies of the Swabian Jura and was restated in modified form on later occasions by Gradmann. The criticisms of Grahame Clark (*Econ. Hist. Rev.*, 17, 1947) and Alice Garnett were directed to early statements of the theory. (L.D.S. *fide* K.A.S.)

**Steppenheide** (German; *pl.* steppenheiden: *lit.* steppe-heath)
Coined by R. Gradmann to describe a plant association consisting of a number of herbaceous plants usually found in the steppes of Southern Russia and some gravel wastelands in Bavaria (there called 'Heiden'), mixed with scrub and even badly developed trees. Heather is however completely absent. It is found on sunny sites, under relatively dry conditions, on calcareous soils neither worked nor manured. (K.A.S.)

**Steptoe**
Knox, 1904. '(U.S.A.) island-like areas in a sea of lava'.

von Engeln, 1942. [Of the Columbia Lava Plateau] 'The lava, when emitted, was very fluid... keeping its appropriate level the lava wraps around the spurs of mountains and extends between them as bays. Isolated summits were converted to islands, called steptoes, in the seas of molten rock.' (p. 592)
See also Kipuka.

**Still-stand, stillstand**
*Webster. Geol.* To remain stationary with respect to sea level or with reference to the center of the earth; said of continents, islands, and other land areas. The fact or condition of stillstanding.

Wooldridge and Morgan, 1937. 'The cycle of erosion can only move uninterruptedly to its close if it coincides with a period of *still-stand*, i.e. of unvarying base-level.' (p. 218)

**Stint**
Butlin, 1961. 'The number of animals which a holder of common right is entitled to put on to common land... the stint is always for a specific number of beasts. The equivalent term in the north of England is gate or gait, hence cattle gait etc.'

**Stitchmeal, stitch-meal**
*O.E.D. obs.* In separated pieces; in 'stitches' of land.
Stamp, 1948. 'Devon and Cornwall . . . there land was held in "stitch-meal", wide balks of earth and stones separating the cultivated strips' (p. 46).
See also Selion.

**Stochastic**
Haggett, 1965, 'In human geography . . . it was not till 1957 that Neyman introduced a growth theory in which chance or *stochastic* (Greek στόχος = aim, guess) processes play a major part'. (p. 27)
Comment. The reference is to Neyman J., and Scott, E. L., On a mathematical theory of population conceived as a conglomeration of clusters. *Cold Spring Harbor Symposia on Quantitative Biology*, **22**, 1957, 109–120.

**Stock**
Stock or boss: a small batholith (*q.v.*). See Wooldridge and Morgan, 1937, p. 108.

**Stone stripes, stone polygons**
Miller, R. *et alia*, 1954, *Geog. Jour.*, **120**. 'The rock fragments are more or less compacted and arranged in the patterns known as "stone stripes"' (p. 217).
See also Hollingworth, S. E., 1934, Some Solifluction phenomena in the northern part of the Lake District, *Proc. Geol. Assoc.*, **45**, 167–188; Hay, T., 1936, Stone stripes, *Geog. Jour.*, **87**, 47–50.

**Stony rises** (Australia: Victoria)
Extensive stretches of lava are often characterized by ridges or stony rises separated by valleys resulting from the different flow of the lava when liquid. Isolated hillocks or lava blisters may also occur. (E.S.H.)

**Stop-and-Go Determinism** (T. Griffith Taylor)
—see Determinism

**Stoping, magmatic stoping**
*Webster. Geol.* The process whereby intrusive igneous magmas are thought to make space for their advance, by detaching and engulfing fragments of the invaded rocks; magmatic stoping.
Himus, 1954. The mode of emplacement of some igneous rock bodies in which blocks of country rock are wedged off and sink into the advancing magma.

**Store cattle**
Cattle bought and then kept for fattening as fat cattle destined for the butcher.

**Stoss end, Stossend** (German and English)

*Webster. Stoss. Geol.* Facing toward the direction from which an overriding glacier impinges or impinged; said of the side of a hill or knob of rock. Opposed to *lee*.
Lobeck, 1939. 'Forms of drumlins. Their *stoss end*, facing the glacier, is usually blunter and steeper than the tail or *lee-side*' (p. 306).

**Stow** (J. F. Unstead, 1933)
*O.E.D.*, obsolete: 1. = Place in various senses; a place on the surface of the earth or in space; occasionally a place in a book or writing.
Knox, 1904. (England), a place, a stockaded place, from A.S. stow = a place.
Unstead, J. F., 1935, *The British Isles*, London: U. of L.P. 'A stow may be shortly defined as the smallest unit-area of geographical study: a region of the first or lowest order.' (Stows can be combined into 'tracts') (pp. 12–13). (See also *Geog.*, **18**, 1933.)
Comment. An old word resuscitated by Unstead and given by him a specialized geographical meaning. D. L. Linton (The Delimitation of Morphological Regions in *London Essays in Geography*, 1951) has suggested it might be made one of a hierarchy of morphological regions (*q.v.*) —site, stow, tract, province.

**Straate** (Afrikaans; *lit.* streets)
Wellington, J. H., 1955, *Southern Africa*, I. 'In the south-west [of the Kalahari] between the dunes the troughs or *straate* are often floored with a clayey sand' (p. 325). 'Known to the Hottentots as *Kivas*' (p. 481).

**Strait, straits**
A narrow passage of water connecting two larger bodies of water.

**Strand** (German)
Shore or beach.
The word 'strand' is only used in parts of England and the common German word is much more comprehensive so that some writers have tended to use the word in the German sense, with such numerous combinations as Strand-line (raised beach, *fide* Mill), Strand-wall (shingle beach, bar). See Baulig, 1956, for some of the numerous combinations with their French and English equivalents.

**Strandflat** (Norwegian)
*Webster. Geol.* An elevated wave-cut terrace.
Holmes, 1944. 'The wave-cut platform off the rocky coast of west and north-east

Norway—there known as the strandflat—has reached an exceptional width—locally up to as much as 37 miles. It now stands slightly above the present sea level (because of recent isostatic uplift) and innumerable stacks and skerries rise above its surface' (pp. 229–230).

Shackleton, M. R., *Europe*, 4th ed., London: Longmans, 1950. 'Accompanying the fjord coast of Norway is a somewhat discontinuous shelf of land reaching to about 100 feet high and of rounded and dissected form ... the so-called "strandflat"...' (p. 229). (An origin as a plain of marine denudation is doubted by some because they are also found inside the fjords.)

**Strassendorf** (German; *lit.* street-village)
Houston, J. M., 1953, *A Social Geography of Europe*, London: Duckworth. 'The street village or *strassendorf*' (p. 86). 'This type of nucleated settlement is commonly associated with colonisation in forested lands' (p. 105). A village plan in which the houses are ranged on either side of a main street.

**Strath** (Scottish, from Gaelic; also northern Ireland)
*O.E.D.* A wide valley; a tract of level or low-lying land traversed by a river and bounded by hills or high ground.
Mill, *Dict.* A wide valley with a river meandering through an alluvial flat; an area where the valley is becoming aggraded (Scotland).
*Comment.* Often contrasted in Scotland with a glen which is narrower or smaller and usually without the flat floor associated with a strath. *Cf.* also carse. Strath in the Scottish sense has been applied to broad open valleys in various parts of the world. (L.D.S.)

**Strath, strath terrace, strath valley** (O. D. von Engeln, 1942)
von Engeln, 1942. 'When an uplift induces rapid downcutting in the floors of middle or late mature valleys ... some portion of the open valley floor is left undissected .. such a remnantal flat may be called a *strath* or sometimes, together with the valley shoulder, a *berm*. The valley shoulder and the strath are the two distinctive form elements of the *two-storey valley* or the *two-cycle valley*' (p. 221). 'If the strath level has greater extension than that of a narrow ribbon along one valley, the term *strath terrace* may be applied' (p. 222). 'If a stream course is actually dislocated, suffers an intervention rather than an interruption, a *strath valley* is left behind' (p. 224) (also illustration).
*Comment.* This use of strath seems to have no relation to the Scottish word and so is treated as a new term.

**Stratified, stratification**
Mill, *Dict.* Rocks formed in layers (strata), usually as accumulations of sediments.

**Stratigraphy**
Stamp, 1923. 'Stratigraphy is another name for Historical Geology. It is so called because it is chiefly concerned with the stratified rocks which by careful study, can be made to tell the story of the earth ... the study of stratigraphy has for one of its principal objects the tracing of the changes in the geography of the globe through the different periods of its history' (p. 1).
*E.B.* 14th ed., 1929. 'The study of the relative position and order of succession of deposits containing or separating archaeological material ... as fundamental to archaeology as the equivalent study of the superposition of rocks and their fossil contents is to geology.' (G. Caton-Thompson)
*Stratigraphical Geology.* The section of geology dealing with the chronological succession of rock formations.

**Stratosphere**
*Webster. Meteorol.* The upper portion of the atmosphere, above 11 kilometers, more or less (depending on latitude, season, and weather), in which temperature changes but little with altitude and clouds of water never form, and in which there is practically no convection; originally, and still often, called the *isothermal region.*
Hare, 1953. 'This upper atmosphere of constant temperature is called the *stratosphere.*' (p. 15)

**Stratum** (*pl.* strata)
Mill, *Dict.* A bed or layer of a sedimentary rock.

**Stray**
*O.E.D.* The right of allowing cattle to stray and feed on common land (hence transferred to the common land itself).
*Comment.* In the first sense the word is obsolete but in parts of the north of

**Stream**

England it is retained in the names of sections of common land, *e.g.* The Stray at Harrogate, the Strays of York.

**Stream**

Mill, *Dict.* 1. Any river, brook, rivulet, or course of running water. 2. A steady current in the sea or in a river; especially the middle or most rapid part of a tide or current.

**Stream ordering**

Morisawa, M., 1968, *Streams: their Dynamics and Morphology*, New York: McGraw-Hill. 'For purposes of comparison within and among drainage areas, a hierarchy of streams has been set up wherein streams are ranked according to order. Although other methods of stream ordering have been suggested, that proposed by Strahler is more objective and straightforward. According to his system, fingertip tributaries at the head of a stream system are designated as first order streams. Two first-order streams join to form a second-order stream segment, two second-order streams join, forming a third order, and so on. It takes at least two streams of any given order to form a stream of the next higher order.' (pp. 152–153)

See Strahler, A. N., 1957, Quantitative Analysis of Watershed Geomorphology, *Trans. Am. Geophys. Union*, **38**, 913–920.

**Street**

In deserts *cf.* Straate.

Bagnold, R. A., 1941, *The Physics of Blown Sand and Desert Dunes*, London: Methuen (rep. 1954). 'Individual chains [of sand dunes] are separated from one another by gaps . . . so that the point of one chain is separated from the stem of the next by a "street" of bare desert floor' (pp. 229–230).

**Street-village**

The English translation of *Strassendorf* (German) which most geographers prefer to retain.

**Striæ**

Mill, *Dict.* Scratches made on ice-smoothed rock surfaces by pieces of grit or other rocks frozen into the ice-mass.

*Comment.* Rarely used in the singular. Hence striated, striation. Sometimes clarified by reference to 'glacial striae,' as similar striae may result from soil-creep and in other ways.

**Strike**

O.E.D. 8. Mining and Geol. The horizontal course of a stratum; direction with regard to the points of the compass.

Mill, *Dict.* An imaginary horizontal line at right angles to the direction of dip of a stratum.

Geikie, J., 1898. '. . . the general direction or run of the outcrops of strata.' (p. 307)

Himus, 1954. The horizontal direction at right angles to the direction of dip of a rock.

*Comment.* The miner's semi-slang use of the word 'strike' for the discovery of a deposit of gold, oil or other valuable mineral will often be found in geographical literature.

Note also the use of strike as an adjective, *e.g.*, strike fault, strike joint, strike valley etc., when the features concerned are roughly parallel to the strike.

**Strike slip, strike-slip faulting**

A slip or fault in rocks characterized by horizontal movement parallel to the strike.

**Strip field**

A field cultivated in long narrow strips, usually in separate ownership or occupation. See common field, field systems.

A short strip or ridge at right angles to other ridges, probably where ploughs were turned, was known as a butt, or butte—used of land on the edge of open fields. Also small enclosure attached to farmhouses of open field villages (England: Midlands) (I.L.A.T.).

See also Selion.

**Striped farm**—see Farm, ladder

**Strip map**

Webster. A map showing only a narrow band of territory in which the user is interested, as one made for an airman. It is usually made up in the form of a roll.

**Strip mine**

Webster. A mine that is worked by stripping, *i.e.* removing the overburden; especially, a coal mine situated along the outcrop of a flat dipping bed; called also *strip pit*.

*Comment.* The ordinary British term is opencast.

**Strombolian eruption, Strombolian projections**

Mill, *Dict.* Masses of fused and glowing lava ejected from volcanoes, especially characteristic of Stromboli. The glowing lava solidifies in the air and falls as a volcanic bomb.

Webster. *Geol.* Designating or pertaining to volcanic eruptions like those of Stromboli, which explode violently and eject incandescent dust, scoria, and bombs, with little water vapor.

*Comment.* Strombolian is one of the four types of volcanism or volcanic activity commonly distinguished, the others being Hawaiian (with the most basic and highly fluid lava extending over large areas as a 'shield volcano'), next Strombolian (with somewhat less basic lava), then Vulcanian or Vesuvian, finally Pelean with very acid and sticky lava which solidifies as a cone.

## Structure

*Webster.* 7. *Geol.* The attitude and relative positions of rock masses consequent upon deformative processes such as folding, faulting, and igneous intrusion; as, an anticlinal *structure*; a basin-and-range *structure*; an alpine *structure*. 8. *Petrog.* The arrangement of a rock mass as regards the larger features, such as jointing, columnar and platy parting, bedding, etc.; distinguished from *texture*.

Mill, *Dict.* Sometimes applied to rock features exhibited on a large scale as opposed to texture which refers to features exhibited on a small scale, *e.g.* in a hand specimen; but the term has frequently a more general application.

Davis, 1909. The arrangement of strata as a result of earth movements, or their absence. (p. 249)

Holmes, 1928. A term applied (a) to the morphological features of rocks due to fracture, *e.g.* columnar structure, perlitic structure; and (b) to the appearance of a heterogeneous rock in which the textures or composition of neighbouring parts differ from one another, *e.g.* spherulitic structure, orbicular structure, bedded structure, gneissose structure, banded structure.

Cotton, 1922. 'As generally used in geomorphology the term structure is more comprehensive than in geological writings. As used in Davis's all-inclusive formula for explanatory descriptions of landforms—"structure, process, and stage"—it "indicates the product of all constructional agencies. It includes the nature of the material, its mode of aggradation, and even the form before the work of erosive agencies begins. In other words, it stands for that upon which erosive agents are, and have been, at work." (Fenneman).'

*Comment.* As most commonly used in geographical writings structure refers to the disposition of the rocks of the earth's crust, structural being the same as tectonic, *e.g.* when applied to 'structural maps'.

## Structural landscape

A landscape adjusted to tectonic structure.

Cotton, 1941, *Landscape*, C.U.P., uses structural bench, structural escarpment, structural terrace.

The same as tectonic landscape (see tectonic as used by Mill), but Cotton draws a distinction between the two—see *Geog. Jour.*, 119, 213–222.

*Cf.* also Tectosequent.

## Struga (Serbo-Croat)

A passage formed along a bedding plane in karst country.

## Sty

Mill, *Dict.* A steep side of a hill (Cumbrian Lakes).

## Sub- (*lit.* under)

*Webster.* 1a. In many words, mostly of Latin origin, *sub-* has the sense of *under, beneath, below, down,* as in *sub*terranean, underground. 6. *Geog. & Geol.* Denoting *near the base of*; *bordering upon*; as in *sub*alpine, *sub*arctic, *sub*arid.

*Subaerial.* Under the atmosphere. Applied to atmospheric weathering, deposits transported by wind and laid down on land.

*Subaqueous.* Under water.

*Submarine.* Under the sea.

*Sub-* (in sense of partly)
Sub-Arctic.
Sub-infantile.
Sub-tropical.

## Subaerialism

Stoddart, D. R., 1960. Colonel George Greenwood, the Father of Modern Subaerialism, *S.G.M.*, 70, 108–110. 'Greenwood's main thesis was the supremacy of rainwash as an agent of land sculpture.'

*Comment.* Not in *O.E.D.* (which gives however subaerialist) or *Webster.*

## Submarine Canyon

Steep-sided canyons in the continental shelf have been demonstrated to exist in many parts of the world and their origin has given rise to much discussion. See von Engeln, 1942, 49–54.

## Submerged forest

Applied specifically to certain forest remains now found below sea level around the coasts of Britain and elsewhere. Modern studies by pollen analysis or palynology have shown that the submerged forests were formed over a considerable range of time. See Clement Reid, *Submerged Forests*, C.U.P., 1913; H. Godwin, *The History of the British Flora*, C.U.P., 1956.

**Subnival Belt** (Carl Troll, 1957) also **Nival Region**

Troll, Carl, 1957, *Abs. of Papers, Ninth Pacific Science Congress*, Bangkok. 'Instead of "Alpine belt", not very suitable for tropical conditions, the term "Subnival belt" is suggested' (p. 255). Troll suggests six altitudinal vegetation zones for mountains in the tropics including Subnival Dense Tussock Grassland, Subnival Frigid Desert and Nival Fern Region.

**Subsequent Falls**

Mill, *Dict*. Falls which have originated through the uncovering of hard rock behind soft rock by a stream while excavating its valley.

Cotton, 1945. '... falls ... developed when down-cutting streams encounter rocks of varying hardness.'

**Subsequent international boundary**

Boundaries which have been marked off since the development of the regions. From Hartshorne, see Boggs, S. W., 1940, *International Boundaries*, New York: Columbia U.P., p. 29.

**Subsequent tributaries, rivers** (J. B. Jukes, 1862)

*O.E.D.* 2. d. Phys. Geog. 1895. W. M. Davis in Geog. Jour., V, 131. The peculiarity of subsequent streams is ... that they run along the strike of weak strata; while consequent streams run down the dip, crossing harder and softer strata alike. 1898. I. C. Russell, River Developm. vii, 185. Streams originate, the directions of which are regulated by the hardness and solubility of the rocks. Such streams appear subsequently to the main topographic features in their environment, and are termed subsequent streams.

Mill, *Dict*. Subsequent rivers. Rivers developed along the strike of (usually soft) strata, or along the crush zones of faultlines, subsequent to the formation of consequent rivers of which they are usually the tributaries (Jukes).

Committee, List 2. Subsequent tributaries. Tributaries developed under the guidance of weak outcrops or other lines of weakness (e.g. faults).

*Comment*. The concept originated from Jukes, J. B., 1862, Formation of River Valleys in the South of Ireland, *Quart. Jour. Geol. Soc.*, **18**.

**Subsistence agriculture**

Committee, List 3. Agriculture the final products of which are not primarily for sale, but are mainly consumed in the farmer's household; the opposite of commercial agriculture.

**Sub-soil, subsoil**

*O.E.D.* 1. The stratum of soil lying immediately under the surface soil.

Mill, *Dict*. The material covered by soil, which is formed by the decomposition of rock or the alteration of deposits.

Glentworth, 1954. 'That part of the soil profile commonly below plough depth and above the parent material.' (p. 165)

Jacks, 1954. 'Part of a soil between the layer normally used in tillage (see Topsoil) and the depth to which most plant roots grow.'

*Comment*. In the older concept of soil studies the division was into soil or topsoil, subsoil and bedrock in the sense indicated by Mill. The *O.E.D.* definition is obviously ambiguous. Modern pedologists tend to drop the term subsoil altogether or equate it with the 'C' horizon. (L.D.S.)

**Subtopia** (Ian Nairn, 1955)

Nairn, I., 1955. 'A compound word formed from Suburb and Utopia.' In an article published in *The Architectural Review*, 1955, afterwards issued as a book entitled *Outrage* (Architectural Press, 1955) Ian Nairn described a journey by two men in a car from Southampton to Carlisle in which he stressed the chaos, ugliness and devastation caused by urban expansion and development despite the existence of a comprehensive system of planning. For the 'death by slow decay, a creeping mildew that already circumscribes all our towns' he created the word subtopia. It conveys the idea of a utopia promised by planning reduced to nought by suburban sprawl.

*Comment*. Although introduced as a satirical comment, the word soon gained wide use. See *Jour. Town Planning Institute*, **41**, 1955, 287–288; also R. W. Holland, *Jour. Roy. Soc. Arts*, **104** (4965), 1955, 6. 'The best definition of this hateful newcomer is "the idealization and universalization of suburbia".'

## Sub-tropical

Miller, 1953. Used as synonymous with Warm Temperate and including the Mediterranean as well as Eastern Margin Warm Temperate climates.

*Comment.* As a term sub-tropical has no precise meaning: it is sometimes used to imply that conditions are tropical for part of the year, at other times equivalent to 'near-tropical'. The word 'extra-tropical' is preferred when the reference is to something just outside the tropics, *e.g.* the extra-tropical high pressure belt. In South Africa this term is as a rule used in a more restricted sense than in Europe. It denotes the warmest part of the temperate climate only: *e.g.* the coastal districts of Natal and the Low Veld of the eastern Transvaal. It is never applied to the High Veld, the Karoo, or the South-western Cape Province (Mediterranean climate). (P.S.)

## Suburb, suburban, suburbia

*O.E.D.* 1. The country lying immediately outside a town or city; more particularly, those residential parts belonging to a town or city that lie immediately outside and adjacent to its walls or boundaries. 2. Any of such residential parts, having a definite designation, boundary, or organization.

Mill, *Dict.* A portion of a city lying outside the central area, but intimately related to the municipality.

Osborn, F. J., 1946. *Green-Belt Cities*, London: Faber and Faber. 'The word suburb is conveniently reserved for an outer part of a continuously built-up city, town or urban area, implying that it is not separated therefrom by intervening country land. Thus a district so placed, and containing, besides dwellings, businesses serving only the local population should be termed a Dormitory Suburb, or Residential Suburb; and one so placed having industry as well, an Industrial Suburb.' (p. 181)

*Comment.* The word, like the concept, is a very old one and is used by Chaucer and even earlier. Suburbia, equivalent to the suburbs, especially of London, tends to be used contemptuously. See also Urban-rural continuum.

## Succession (invasion and succession)

Hawley, A. H., 1950, *Human Ecology*, New York: Ronald Press. 'Areas of homogeneous use are formed and reformed through a more or less routine process called succession. This term refers to the sequence of changes by which units of one land use or population type replace those of another in an area.' (p. 400) 'The succession approaches its culmination with the achievement of numerical preponderance by the invading population or land use type. It reaches completion with the importation of the customary institutions and services of the new occupant. Control of the area has thus passed to the invader and a condition of relative equilibrium is established.' (pp. 401–402)

*Comment.* This term used to explain the outward growth in the concentric-zone model of the city. See also Concentric-zone growth theory, and Urban fringe.

## Sudan, Soudan (Arabic *Sudan*, *pl.* of *suda*, black)

Applied originally by Arab peoples to the country and the people south of their own territories inhabited by 'blacks', hence the country mainly grassland and savana south of the Sahara but north of the equatorial forests; now the specific name of certain countries in that belt—former French Sudan and the Republic of the Sudan.

## Sudd (Sudan: Arabic)

*O.E.D.* An impenetrable mass of floating vegetable matter which obstructs navigation on the White Nile; also a temporary barrier or dam.

Applied also in the Sudan to the marshes of the White Nile between Bor and Malakal where navigation is obstructed by masses of floating vegetation. (J.H.G.L.)

## Sugar loaf

Mill, *Dict.* A rounded conical hill.

*Comment.* The form in which sugar was once retailed was in a conical mass, known as a 'sugar-loaf'. As these sugar-loaves have long since disappeared from the shops, the aptness of the term is no longer apparent though it is still used as a topographical name.

## Suk, souk—see Suq

## Sukhovey (Russian, *pl.* sukhovei)

Barkov (see Appendix II, Russian words). Hot dry wind, blowing during the summer, mainly from the southeast, in the south-eastern part of European USSR and Kazakhstan.

Lydolph, Paul E., 1964, 'The Russian Sukhovey. *A.A.A.G.*, **54**, 291–309. '"Sukhovey" refers to individual spells of hot, dry flows of air which have immediate and profound effects on vegetation.' (p. 291)

*Comment.* A hot, dry, usually southeast, wind blowing in the summer in the Southeastern part of the Soviet Union in Europe and also in Kazakhstan. The temperature of the air may rise to 95–105°F. and relative humidity drop to 15 per cent or less. The sukhovei by causing excessive evaporation seriously injures vegetation and crops. (C.D.H.)

**Sulung** (Agricultural History)—see under Yardland.

**Sumatra**
Miller, 1953. 'Local disturbances of the Malacca Straits, sudden squalls with violent thunder and lightning and heavy rain which occur, always at night, during the south-west monsoon' (p. 112).

**Summer**
*O.E.D.* Reckoned astronomically from the summer solstice (21 June) to the autumnal equinox (22 or 23 Sept.); in popular use in the northern hemisphere the period from mid-May to mid-August, or the months of June, July, August. Paradoxically the first day of summer (21 June) is also called Midsummer Day.

*Comment.* Note also the use of summer (in contradistinction to winter) as meaning the warmer half of the year.

**Summer Time**—see under Standard Time

**Summit level**
*Webster.* The summit, or highest point, of a road, railroad, canal or the like.

Swayne, 1956. The highest of a series of elevations over which a canal or railway is carried.

**Summit plane**
The plane passing through a series of accordant summits and so inferred to be the level of a former peneplained surface. See also Gipfelflur.

**Sump**
Mill, *Dict.* A hole, generally subterranean, into which the drainage of an area can collect.

**Sundri, sundari** (Indo-Pakistan: Bengali)
The mangroves, *Heritiera fomes* and *H. littoralis*, which make up the bulk of the swamp forests (hence sundarbans) of the Ganges delta. The swamp forests of the Irrawaddy delta are also mainly of *Heritiera fomes*.

**Sungei** (Malay: usually abbreviated to S. in place names)
River.

**Supergene**
Howell, 1957. 'Applied to ores or ore minerals that have been formed by generally descending water.'

**Superglacial till**
Ablation moraine.

**Superimposed boundary**
Pounds, N. J. G., 1963, *Political Geography*, New York: McGraw-Hill. 'Superimposed boundaries were established, like subsequent boundaries, after the territory to be divided had been settled and developed but, unlike the latter, they ignore completely the cultural and ethnic characteristics of the area divided.' (p. 63)

For examples of superimposed boundaries see de Blij, H. J., 1973, *Systematic Political Geography*, 2nd ed. New York: Wiley, pp. 182–185.

**Superimposed** (J. W. Powell, 1875), **Superposed** (McGee) **drainage**.
*O.E.D.* 1. b. Phys. Geog. Applied to 'a natural system of drainage that has been established on underlying rocks independently of their structure' (Funk's Stand. Dict., 1895) 1898. I. C. Russell, *River Development*.

*Webster.* 2. *Phys. Geog.* Pertaining to or noting a river or a drainage system let down by erosion through the formations on which it was developed, into underlying formations of different structure, unconformable beneath.

Mill, *Dict.* Applied first by Maw to the valleys termed 'epigenetic' by Richthofen. Epigenetic Valleys: Erosion valleys conceived as having originated when the existing mountains in which the valley is cut were covered by sedimentary deposits the slope of which determined the direction of the valley, a direction maintained owing to the fact that the valley was already worn down sufficiently into the existing mountains before the removal by denudation of the overlying deposits (v. Richthofen).

Gilbert, 1877. Systems of drainage in relation to structure are (A) consequent, (B) antecedent and (C) superimposed. Superimposed drainage may be (a) by sedimentation, or subaqueous deposition (b) by alluviation, or subaerial deposition and (c) by planation. (p. 144)

Powell, 1875. '... the present courses of the streams were determined by conditions not found in the rocks through which the channels are now carved, but that the beds in which the streams had their origin when the district last appeared above the level of the sea, have been swept away.' (p. 165)

Committee, List. 2. Superimposed (or epigenetic) drainage. A drainage system established on an upper rock-mantle or simpler structure than the underlying formations now exposed by erosion.

Comment. 1. The alternative, 'superimposed', is used by Davis and Cotton and now generally in America. 2. Gilbert's 'superimposed by planation' might be now changed to 'superimposed from a peneplain'. See Campbell, *A.A.A.G.*, **18**, 1928, 35. See also Drainage.

**Superposition, Law of**

Stamp, 1923. 'The Law of Superposition: where one bed of rock rests on another it is presumed that the upper bed has been laid down after the lower and hence is younger. This is true (a) for all sedimentary rocks, (b) for effusive igneous rocks—lavas which have been poured out at the surface and beds of ashes, always provided that the original order has not been reversed as a result of folding or faulting or other disturbance' (p. 4).

**Suq** (Arabic: *lit.* a market), also **suk, sook, souk, sôk,** etc. (*pl.* aswaq)

A market-place or market; applied to markets both in towns and villages as well as to periodic markets. A word widely current in North Africa and Arab countries generally; also as a place-name element.

**Surazo** (Brazil)

Miller, 1953. 'In the campos of Brazil as in the middle Amazon, a feature of the climate is a winter anticyclone bringing cold waves known as "friagems" or "surazos", when the temperature may fall below 50° and cause considerable discomfort.' (p. 130)

**Survey**

*S.O.E.D.* 1. The act of viewing, examining or inspecting in detail, *esp.* for some specific purpose. b. A written statement or description embodying the result of such examination. 5. The process of surveying a tract of land, coast-line etc. ... a plan or description thus obtained; a body of persons or a department engaged in such work.

**Suspension**

Gilbert, G. K., 1914. The Transportation of Running Water, *U.S.G.S. Prof. Paper*, **86**. 'With swifter current leaps are extended, and if a particle thus freed from its bed be caught by an ascending portion of a swirling current its excursion may be indefinitely prolonged. Thus borne it is said to be **suspended**, and the process by which it is transported is called *suspension*. There is no sharp line between saltation and suspension....' (p. 15a) See also Saltation, traction.

**Swadeshi** (India: Bengali; Hindi; *lit.* home-country things)

The pre-Partition national movement in India in favour of home goods, especially cottons, and against foreign goods.

**Swag** (English Midlands)

A shallow, water-filled hollow on the surface due to subsidence resulting from underground mining. Cf. flash (1), pitfall.

**Swale**

*Webster.* 2. A piece of meadow; often, a slight depression or valley, as in a plain or moor, marshy and rank with vegetation.

*Dict. Am.* A marshy or moist depression in a level or rolling area. 1667.

*O.E.D.* (Origin unknown, probably conveyed to the United States from the eastern counties of England where it is still in use). A hollow, low place; especially United States, a moist or marshy depression in a tract of land especially in the midst of rolling prairie.

*Comment.* Used especially in 'Swell and Swale topography' (*q.v.*).

**Swallow, Swallow-hole, swallet**

*S.O.E.D.* A deep hole or opening in the earth; a pit, gulf, abyss. Obsolete except as meaning an opening or cavity, such as are common in limestone formations, through which a stream disappears underground.

*Webster.* = sink, *Eng.*

*Comment.* In karst terminology there is a general tendency to make the word 'sink' the comprehensive term and to limit 'swallow-hole' to those into which water disappears. There may be swallow holes

## Swamp

*O.E.D.* First recorded as a term peculiar to the N. American colony of Virginia, but prob. in local use before in England.
1. A tract of low-lying ground in which water collects; a piece of wet spongy ground; a marsh or bog. Orig. and in early use in the N. American colonies, where it denoted a tract of rich soil having a growth of trees and other vegetation, but too moist for cultivation.

*Webster.* Wet, spongy land; soft, low ground saturated, but not usually covered with water; specifically, *Ecol.*, such a tract, sometimes inundated, and characteristically dominated by trees or shrubs. It differs from a bog in not having an acid substratum. *Cf.* marsh.

Mill, *Dict.* A marsh so saturated with water as to be unfit even for pastoral purposes.

*Adm. Gloss.*, 1953. Wet, spongy, ground usually with stagnant pools of water and coarse undergrowth, which it is possible to traverse with care.

Tansley, A. G., 1939, *The British Islands and Their Vegetation*, Cambridge: C.U.P. 'Swamp is used for the soil-vegetation type in which the normal summer water level is above the soil surface....' The last term of aquatic vegetation, giving way to land vegetation of marsh or fen. Usual vegetation is reed. (p. 634)

*Comment.* American usage is different from British in that it includes forested areas, e.g. the Great Dismal Swamp of Virginia and the extensive swamp forests of the south-eastern states dominated by such trees as the swamp-cypress. The essential point is that the water level is above surface of ground. In modern ecological freshwater studies (*e.g.*, Twigg, H. M., 1959, Freshwater Studies in the Shropshire Union Canal, *Field Studies*, 1, 116–142) the habitats shown from the centre of a body of freshwater to the margins are (1) aquatic, (2) swamp, (3) marsh, agreeing with Tansley 1939. The surface of a marsh is only periodically below the surface water level, of a swamp permanently so. See Marsh.

## Swash, send

The rush of water up a beach from a breaking wave.

## Swash channel, swash-way

*O.E.D.* 'A channel across a bank, or among shoals, as the noted instance between the Goodwin Sands.' 1867.

## Swell—see Rise

## Swell and Swale topography

Thornbury, 1954. 'Clay tills give rise to flat till plains or undulating surfaces which have been described as *swell and swale topography*' (p. 388). See also Sag and Swell, also Swale.

## Swidden farming

In view of the very large number of terms used for shifting agriculture (*q.v.*), certain American writers have advocated the use of swidden agriculture, or swidden farming. Swidden is stated to be an old English dialect word, but is not to be found in the *O.E.D.* See *Report of a Symposium, Ninth Pacific Science Congress*, Bangkok, 1957. Not in *Webster*.

## Swing of the winds

The term applied to the northward and southward migration of the planetary winds (*q.v.*) with the seasons.

## Sylviculture—see Silviculture

## Symbiosis, symbiotic

Association of two different organisms, usually two plants or a plant and an animal, physically attached and interdependent.

## Symbolic model

Haggett, 1972, termed these models as the third stage of abstraction from the real world. 'Abstraction is pushed still further in the third type of model, the symbolic model, in which the original properties of the real world are represented by symbols.' (p. 15)
See also Iconic and Analogue models.

## Symbols, cartographic

Conventional signs used on maps to represent specified objects; they are usually explained on the face of or below the map; if very numerous reference may be made to a 'characteristic sheet' (*q.v.*).

## Syncline, synclinal

*O.E.D.* Synclinal. As adjective Geology: applied to a line or axis towards which strata dip or slope down in opposite directions; also said of the fold or bend in such strata, or of a valley, trough or basin so formed. As substantive: a synclinal line, fold, or depression.

Mill, *Dict.* Synclinal (adj.) 1. Dipping towards an inner axis, as synclinal strata. 2. Formed of or in strata dipping to an inner axis, as synclinal valleys, synclinal ranges. (Noun) A fold, on the two sides of which strata dip from opposite directions towards a common axis.

Page, 1865. Syncline, synclinal (Gr. *syn*, together, and *clino*, I bend). Applied to strata that dip from opposite directions inwards, like the leaves of a half-opened book; or which incline to a common centre, forming a trough or basin-shaped hollow.... (p. 425)

Holmes, 1944. 'When the beds are downfolded into a trough-like form the structure is called a *syncline*, because in this case the beds on either side "incline together" towards the keel.'

*Comment.* 'Synclinal' is also still used by some authors as a noun; which is the older use as given by *O.E.D.* and Mill, but the majority now prefer syncline.

### Synclinal valleys
Powell, 1875. '... which follow synclinal axes.' (p. 160)

### Synclinorium
Holmes, 1944. 'Puckers and smaller folds are superimposed on broad anticlinal and synclinal folds of a much larger order. An anticlinal complex of folds of different orders is called an *anticlinorium*, and the complementary complex a *synclinorium*.' (p. 74)

Himus, 1954. 'A major structure which in the main is synclinal, but on which numerous minor folds are superimposed.'

*Comment.* Some writers have confused synclinorium with geosyncline. An old attempt to anglicize synclinorium as synclinore has not been followed (see *O.E.D.*).

### Synecology
The ecology of a group, in contrast to *autecology*.

### Syngenesis, syngenetic
*Webster.* Reproduction in which two parents take part.

### Synopsis, synoptic
A brief or condensed statement, a conspectus, Synoptic is applied especially to weather charts giving a general view of meteorological conditions.

System (geology)—see Geological Time

### Systematic geography
Hartshorne, 1939. The study of the areal differentiation of categories of individual features, *e.g.* political boundaries, as contrasted with regional geography. It is widely used by American authors and finds ample precedent in the writings of many German geographers. It is also commonly called 'general geography', from Varenius's 'geographia generalis'. The most usual equivalent in German is 'allgemeine Geographie' and in French, géographie générale. (pp. 406–408)

*Comment.* Herbertson, A. J., 1905, The Major Natural Regions: An Essay in Systematic Geography, *Geog. Jour.*, **25**, 300–312. Despite his sub-title, Herbertson wrote: 'Geography is not concerned with the distribution of one element on the Earth's surface, but with all.' In the discussion, Mackinder opposed the use of 'systematic geography' as an inappropriate borrowing from botany in which it is applied to the classification of large numbers of species. There are only a few species of regions and Herbertson's paper was concerned not with 'systematics' but the method appropriate to regional geography.

The simple meaning of 'systematic' is arranged according to a system, plan or organized method and in this sense systematic geography would seem to be a better term than general geography.

### Systems analysis
Harvey, D., 1969, *Explanation in Geography*, London: Edward Arnold. 'System analysis provides a framework for describing the whole complex and structure of activity. It is peculiarly suited to geographic analysis since geography characteristically deals with complex multivariate situations.' (p. 81)

See Berry, B. J. L., 1965, Cities as Systems within Systems of Cities, *Papers Reg. Sci. Assoc.*, **13**, 147–163; and Chorley, R. J., 1962, Geomorphology and General Systems Theory, *Prof. Papers U.S. Geol. Surv.*, **500-B**.

### Szik soil
Jacks, 1954. 'Saline or alkaline soil of Hungary.'

# T

**Tabetisol**—see Talik (K. Bryan, 1945)

**Tabki** (Nigeria: Hausa)
A small, semi-permanent pond in a saucer-shaped depression with an impervious fine clay bed which holds up rain-water, found in the savanna areas of northern Nigeria. (J.C.P.)

**Table cloth** (South Africa)
Humorous name given at Cape Town to the white cloud covering the flat top of Table Mountain during a Southeaster. The cloud is formed on the windward side, rolls over the top and comes down the northern precipices like a waterfall, evaporating by adiabatic heating before reaching the lower slopes. From the streets of Cape Town the violent motion in the cloud is clearly visible, but seen from a distance to the north it seems stationary, and the likeness to a table cloth, hanging down along the sides of the table, is striking. See Hann, III, p. 447, for a good description of this 'Tafeltuch'. It is not a lenticular cloud. (P.S.)

**Tableland**
*O.E.D.* An elevated region of land with a generally level surface of large or considerable extent; a lofty plain; a plateau.
b. Without a or pl.: elevated level ground.
Mill, *Dict.* A highland area where the rocks on the whole lie horizontally with a fairly uniform surface-level, into which valleys are cut as canyons or gorges.
Page, 1866. Table-Land. In Geography, any flat or comparatively level tract of land considerably elevated above the general surface of a country.
Tarr, R. S., 1914, *College Physiography*, New York: Macmillan. 'Both buttes and mesas so abound in many plateau lands as to have suggested the term tableland.' (p. 503)
No reference in Lyell, 1830; Salisbury, 1907; Davis, 1909; Cotton, 1922; Wooldridge and Morgan, 1937; Holmes, 1944; von Engeln, 1942; Strahler, 1951; Lake, 1952. Largely superseded by plateau, though tableland is sometimes restricted to a plateau with abrupt cliff-like edges rising very sharply from surrounding lowlands.

**Taboleiro** (Brazil: Portuguese)
James, 1959. 'The coastal *taboleiros* are flat-topped mesa-like forms held up by a cover of relatively recent sedimentary strata.' (p. 412)

**Tafelberg** (Afrikaans)
Tafelkop of big dimensions. The best known is Tafelberg at Cape Town (Table Mountain) whose top consists of a very hard sandstone; elsewhere in South Africa the cap-rock is commonly dolerite. (P.S.)
A mesa (J.H.W.).

**Tafelkop** (Afrikaans)
A kop (*q.v.*) or isolated hill with a flat top, a butte. (J.H.W.)

**Tafoni** (Corsica: Italian)
Cotton, 1942. Small and large recesses in rock faces conspicuous in dry regions and in seaside cliffs, the result of differential sapping. Term adopted by Penck from the Corsican.

**Tafrogenesis** (E. Krenkel in German, 1922; B. G. Escher in English, 1934)
Tafrogenese was used for the first time by E. Krenkel in his book, *Die Bruchzonen Ostafrikas*, Berlin, 1922, defined as follows: 'Die Schollenzerlegung durch Zerrung, das zeitliche Gegenstück der Orogenese, sei deshalb Tafrogenese genannt' (p. 181). The same author uses the word repeatedly in his *Geologie Afrikas*, vol. I, Berlin, 1925, but in the second volume (1928, p. 636) writes 'taphrogenese'. For B. G. Escher, 1934, *Algemeene Geologie*, Amsterdam, 437, tafrogenesis means formation of rift phenomena in general, not restricted to those which can be attributed to tension only. (Information from Professor Leo Peeters in MS.)

**Tahsil, tehsil** (Indo-Pakistan: Urdu-Hindi)
A political division comprising several villages. (A.H.S.)
Spate, 1954. 'The States of India are divided (if large enough) into Divisions and these into Districts... roughly equivalent to English counties.... Some Districts go back to Moghul times, but the great majority are merely *ad hoc* units for administrative convenience. Districts are

subdivided into *taluks* (*taluqs*) or *tahsils* (*tehsils*) normally from 3 to 8 to a district. In Bengal, however, the next unit to the District is the *thana* or police-station area, which is much smaller than the average *tahsil* or *taluk*' (p. xxiii).

In the explanation of the Administrative map of the *National Atlas of India* in Hind (edited by Professor S. P. Chatterjee of Calcutta), 1958, 'each district has been divided into tahsils in North India, taluks in South India and subdivisions in East India (Assam, West Bengal, Bihar and Orissa) . . . each subdivision of these four states has been subdivided into police-stations.' On the map 2,768 'tahsils, taluks and police-stations' are shown and listed by name in the Republic of India's 14 states. In East Pakistan (East Bengal) there are three divisions, seventeen districts, 54 subdivisions and 409 *thanas* (Ahmad, 1958, p. 2). Since 1955 West Pakistan has been organized in 10 Commissioner's Divisions and the Federal Capital, the divisions into Deputy-Commissioner's Districts and the Districts into tahsils.

**Taiga** (Russian from the Yakut word for forest)

*O.E.D. Suppl.* (Russ.) A (Siberian) pine-forest.

*Webster*. The vast, swampy, coniferous forest region of Siberia, beginning where the tundra ends; hence, a similar region in Europe or North America.

Mill, *Dict*. Undisturbed natural forest of the cold, fairly dry zone immediately south of the tundras (*q.v.*) (N. Russia and America).

Küchler, A. W., 1947, Localizing Vegetation Terms, *A.A.A.G.*, **37**. This term refers to the vast coniferous forests that span the globe in high latitudes south of the Tundra. Like Tundra, Taiga is a Russian word, applied at first only in Eurasia but later extended to very similar plant associations in North America.

*Comment*. Russian sources (see Appendix II, Russian words) give: Ushakov (from Yakutsk word for forest). Thick, little penetrated predominantly coniferous forest stretching in a broad zone in the north of Europe and Asia as far as the Okhotsk Sea. Barkov. Siberian name for coniferous forest. Murzaev. coniferous forest. Siberian, Turkic-Mongolian word adopted by the world geographical and botanical literature for the designation of the zone of coniferous forests of the Temperate Zone of Eurasia. There seems to be no authority for the spelling taïga. Etymologically tayga is more correct.

**Tailo** (Indo-Pakistan)

Spate, 1954. 'The importance of sunshine is also emphasised by the distinction between cultivated *tailo* slopes—the sunny or *adret* side of the valleys—and the forested *saylo* (=Alpine *ubac*)' (p. 403) (in Kumaon, Himalaya).

**Tail-race**

An artificial channel through which waste water is led back to a natural channel after having operated a water mill, or, latterly, having passed through a generating plant.

**Takyr** (Russian)

Area of barren alkaline soil with heavy unstructural clay soil. 'In summer their surface forms a solid crust underneath which the soil retains an increased quantity of salts soluble in water. In winter they turn into a marshy bog.' (Kovda in MS.)

*Comment*. Dr. S. Erinç informed the Editor that this is a Turkish word meaning dry and hard and is applied to the surface crusts of salt and clay in arid regions.

**Takyrisation**

Formation of clay-salt crust in arid regions. See *Information U.S.S.R.*, ed. A. Maxwell, 1962, 63.

**Tala** (as term, Berkey and Morris, 1924)—see Gobi

**Talik** (Russian)

Howell, 1957. '1. A layer of unfrozen ground between the seasonally frozen ground (active layer) and the permafrost. 2. An unfrozen layer within the permafrost. 3. The unfrozen ground between the permafrost.' See Muller, 1947.

**Tālluqā** (Indo-Pakistan: Urdu; anglicized as taluka)

A political division of a 'district' larger than a tahsil (A.H.S.), but see under Tahsil.

**Talus**

*O.E.D.* A sloping mass of detritus lying at the base of a cliff or the like, and consisting of material which has fallen from its face; also, the slope or inclination of the surface of such a mass.

Mill, *Dict*. A scree. Submarine talus (Meerhalde of Germany); a talus formed on the seaward slope of a beach.

Rice, 1943. 'The sloping heap of loose rock fragments lying at the foot of a cliff or steep slope. A heap of coarse rock-waste at the foot of a cliff or a sheet of waste covering a slope below a cliff; same as scree, which is more commonly used in Great Britain, whereas talus is more commonly used in the United States, but is often incorrectly used for the material composing the talus.' (p. 406)

*Comment.* Originally used in a military sense for the outside of a rampart. Also used by archaeologists for the natural and human debris outside a cave or shelter.

**Talweg** (German), **Thalweg** (older spelling)
Mill, *Dict.* A valley line. The term has been adopted from the German and is sometimes used in English, French and other writings. The lowest line of a valley; generally the deepest line in the river traversing a valley.

Cassell's *German Dictionary*, 1952. Road through a valley; channel of a river.

Cissarz, A., and Jones, W. R., 1933, *German-English Geological Terminology*, London: Murby. 'The whole course of a river, from its upper to its lower course, is known as its Talweg (Thalweg).'

Wooldridge and Morgan, 1937. Thalweg—longitudinal profile or long profile of a valley. (p. 154)

Rice, 1941. The natural line of drainage over the underground surface of a bedrock along which the underflow takes place may be called the underground thalweg (German, valley way). (Pirsson and Schuchert, *Textbook of Geology*, Part 1, p. 156. John Wiley and Sons, edition of 1924.)

Adami, V., 1927, *National Frontiers in Relation to International Law* (trans. T. T. Behrens), London: O.U.P., pp. 16–19. (Four differing interpretations are given: 1. The stream line of the fastest current. 2. The line of greatest depth of the river. 3. The line most favourable to downstream navigation during the period when the waters are at their lowest. 4. The median line of a river. The latter however is also contrasted with the 'Thalweg'.)

*Comment.* 1. Alternatives for the use of Wooldridge and Morgan would be: Längsprofil and Gefällskurve; see Cissarz and Jones, p. 19. In Germany the tendency is to drop the term (C.T.).
2. As a term in international law, see also: Calvo, C., *Dictionnaire de Droit International*, Berlin: Puttkamer and Mühlbreth, 1885, vol. 2, pp. 256–7, and Boggs, S. W., 1940, *International Boundaries*, New York: Columbia U.P., p. 184.

**Talwind** (German; *lit.* valley-wind)
A wind blowing up a valley in contrast to Bergwind. (C.T.)

**Tangi** (Pakistan: Baluchi)
Spate, 1954. 'Transverse clefts, often only a few yards wide, by which the streams penetrate the longitudinal ridges' [in Baluchistan] (p. 425).

**Tank**
*L.D.G.* The word tank is used in India in the special sense of a reservoir for irrigation made by damming a valley to retain the monsoon rain which is released later.

**Taphrogeosyncline**
Graben or rift-valley; used thus of the Gulf of Suez by Rushdi Said, *The Geology of Egypt*, New York, 1962; see *Science*, 5 Apr. 1963, 41.

**Taphrogenesis**—see Tafrogenesis

**Tarn**
Mill, *Dict.* A small lake among the mountains, usually of glacial origin. (W. Britain)

*Comment.* Applied especially to the almost circular lakes which occupy corries or cirques and are fed by rainwater from the surrounding steep slopes rather than by any distinct feeder streams. Originally a local term in the north of England, now used widely by geographical writers in Britain but Americans prefer *cirque lake* (Thornbury, 1954, p. 368).

**Taung** (Burmese; also—**daung**)
Mountain.

**Taung-ya, taungya, toungya** (Burmese; *lit.* mountain or hill field or plot)
1. Temporary clearings on hill-sides made by the hill-tribes of Burma and constituting a form of shifting cultivation to be compared with the chena of Ceylon. When deserted the taung-ya are invaded by bamboo and other forest 'weeds' and do not easily revert to high forest. (L.D.S.)
2. A system of tropical forest management developed in Burma from 1856 and now widely practised in south-east Asia, central and south America and elsewhere whereby shifting cultivators are encouraged to plant seeds of useful forest trees with their last food-crops before deserting a clearing. The derivation is obvious. See *A World Geography of Forest Resources*, New York, 1956, pp. 198–9, 486.

**Tautochrone**
*O.E.D.* That curve upon which a particle moving under the action of gravity (or any given force) will reach the lowest (or some fixed) point in the same time, from whatever point it starts.

*Comment.* The term is used in rather a different sense by Geiger, R., 1950, *Climate near the ground*, University Press, Harvard 29, 34 (Fig. 15). J. A. Taylor (University College of Wales, Aberystwyth) writes that in using the term he defines a tautochrone 'as a line joining together points of varying condition or value referring to a particular moment or period of time, *e.g.* soil temperatures at varying depths at 4.00 p.m. on April 1st.'

**Tectogene**—see geotectocline

**Tectonic**

*O.E.D.* Geology. Belonging to the actual structure of the earth's crust, or to general changes affecting it.

*Webster.* c. *Geol. & Phys. Geog.* Of, pertaining to, or designating, the rock structures and external forms resulting from the deformation of the earth's crust.

Mill, *Dict.* Pertaining to or arising from movements of the earth's crust, as tectonic theories, tectonic earthquakes, basins, valleys, and terraces.

Rice, 1941. 'Pertaining to rock structures and external forms resulting from the deformation of the earth's crust.' (p. 409)

**Tectosequent** (S. W. Wooldridge, 1930)

Literally following the structure. Applied to surface features, for example valleys, which reflect the underlying structure in contrast to morphosesequent.

**Tegal** (Java)

Robequain, C., *Le Monde Malais* (translated by Laborde, E. D., 1954). 'Land worked by natives falls into three main classes: *sawahs* or irrigated fields, *tegals* or fields without irrigation, and gardens' (p. 198).

**Tektite**

Small glassy bodies found usually in groups in scattered areas of the earth's surface, believed by many to be of extra-terrestrial origin, possibly from lunar craters.

O'Keefe, J. A., *et alii*, 1958, Origin of Tektites, *Nature*, **181,** 172–4. Wrey, H. C., 1958, Origin of Tektites, *Nature*, **182,** 1078. Called australites in Australia.

**Tele** (Norwegian)

Frozen ground. (Older form: tæle; Swedish: tjäle.) Often wrongly used for permanently frozen ground (permafrost, *evig* tele) (P.J.W.). See Tjäle.

**Teleconnection**

From Gr. *tele*, far and Lat. *connectere*, to bind together. In Swedish telekonnektion or fjärrkonnektering, identifying of varves, especially over great distances along synchronous lines of the ice-front, also at different ice-centres (G. De Geer, 1916). (E.K.)

**Tell** (Israel: Hebrew)

Literally a hill but used especially for a hillsite many times occupied or over successive periods. Also in place names and as tel.

**Temperate**

*O.E.D.* Tempered, not excessive in degree, moderate. Spec. of the weather, season, climate etc., moderate in respect of warmth; of mild and equable temperature.

*Webster.* Having a moderate climate or temperature; mild.

*Comment.* In geographical descriptions 'temperate' may be used in this sense but more often as equivalent to mid-latitude. As some of the climates in mid-latitudes are of great extremes, the word temperate is best dropped in favour of mid-latitude.

**Temperate rain forest**

*Webster. Ecology.* A woodland found in cool, but mostly frost-free regions of heavy annual rainfall. It is characterized by having many tree species, but, unlike tropical rain forests, there is usually a dominant one. . . . Temperate rain forests occur, in any considerable quantity, only in the region of Tierra del Fuego, southeastern Australia, New Zealand, and parts of Japan. See *rain forest.*

**Temperate Zone**

Middle latitudes, between the Tropic of Cancer ($23\frac{1}{2}°$N.) and the Arctic Circle ($66\frac{1}{2}°$N.) in the Northern Hemisphere and between the Tropic of Capricorn and the Antarctic Circle in the Southern Hemisphere. See under Zone.

Mill, *Dict.* They are really zones of illumination rather than of temperature as this element of climate varies greatly throughout the zones; but the term corresponds fairly well to actual conditions as observed in Western Europe where it originated.

**Temperature**

L.D.G. Temperature is measured by thermometers which may be graduated according to different systems of which the two chief are Centigrade and Fahrenheit. On the Centigrade scale (also called Celsius from Swedish astronomer Anders Celsius who invented it in 1742, though his numbering was reversed) the freezing point of pure water is zero degrees (0°C.) and the boiling point of pure water at sea-level with a standard pressure of the atmosphere of 760 mm. is 100 degrees (100°C.). This is the internationally used scientific scale, but the Fahrenheit scale is widely used, especially in Britain and America. The

Prussian physicist, Fahrenheit (1686–1736), mistakenly thought he had discovered the lowest possible temperatures which he accordingly called zero (0°F.). Actually absolute zero calculated by Kelvin is −273.16°C. The freezing point of pure water is 32 degrees (32°F.) and the boiling point of pure water again at sea-level and normal atmospheric pressure is 212°F. Thus between freezing and boiling points there are 100 degrees in the Centigrade scale and 180 in the Fahrenheit, or 1 degree Centigrade equals 1·8 Fahrenheit. The number 1·8 is the key to conversion.

To convert Fahrenheit to Centigrade subtract 32 and divide by 1·8; example 68°F.: 68−32=36, divided by 1·8=20°C. To convert Centigrade to Fahrenheit, multiply by 1·8 and add 32; example 30°C.: 30 × 1·8 = 54, add 32 = 86°F. Some useful equivalents are:

0°C. = 32°F.;   10°C. = 50°F.;   20°C. = 68°F.;   30°C. = 86°F.

Ordinary temperature readings are obtained by a thermometer with a bulb freely exposed to the air (dry bulb). But if the bulb is covered with a wet cloth (wet bulb thermometer) the moisture evaporates quickly in a dry atmosphere and lowers the temperature considerably, whereas it evaporates very slowly in a wet atmosphere and the temperature is only slightly lower. Thus the relative humidity of the atmosphere can be measured by the difference between the two—the wet and dry bulb thermometers. See also Sensible temperature.

**Temperature anomaly**
Moore, 1949. The difference between the mean temperature of a place and the mean temperature along its parallel of latitude, both temperatures being reduced to sea-level; the anomaly is positive if the place is warmer, negative if it is cooler. The greatest temperature anomaly exists over the north-east Atlantic ocean, including the British Isles, in January, when a considerable area has a positive temperature anomaly of over 20°F. See also Isanomalous line.

**Temporal modes**
Harvey, D., 1969, *Explanation in Geography*, London: Edward Arnold. 'It is but a short step from causal explanations to causal-chain explanations which stretch back over a long period of time. The general mode of explanation which follows this tack will be termed temporal. The assumption is that a particular set of circumstances may be explained by examining the origin and subsequent development of phenomena by the operation of process laws.... They provide us with one dimension through which we may comprehend geographical distributions—a dimension which, by its insistence on the study of temporal change, inculcates a deep awareness of the nature of temporal processes.' (pp. 80–81)

**Temporales** (Spanish)
Strong south-west winds of monsoon type blowing in summer on the Pacific coasts of Central America.

**Temporary grass**—see Rotation (temporary) grassland

**Tent-hill** (Australia)
An Australian term for a butte or kop, so called because of a resemblance to a canvas tent though the top of the tent-hill is often a flat remnant of a former plateau surface. (E.S.H.)

**Tephigram**
*Met. Gloss.* 'A diagram on which is represented the condition of the atmosphere at different levels in terms of its temperature (t) and entropy (phi), hence the name tephigram' M.O. 225, ii (A.P. 897).
See also *Meteorology for Aviators*, H.M.S.O., and Watt, I. E. M., *Equatorial Weather*, 25–34.

**Tephra** (Sigurdur Thorarinsson, 1944)
'The present writer has proposed that the term tephra be used as a collective term for all material ejected through the air by a volcano in the same way as lava is a collective term for all molten material flowing from a crater.' *Resumenes de los Trabajos Presentados*, XX Congreso Geologico International, Mexico, 1956, p. 20 *fn.*
*Note.* Thorarinsson derived this term from Aristotle and introduced it in Tefrokronologiska Studier på Island, *Geografiska Annaler*, 1944 (with English summary). He uses tephrochronological and other compounds.

**Terai** (Indo-Pakistan: Urdu-Hindi)
Swampy ground at the outskirts of bhabbar. (A.H.S.)
Spate, 1954. 'In these all but featureless alluvial expanses [of the Indo-Gangetic plain] ... there are only three really important variations from the norm: *bhabar, terai, bhur.* ... The *bhabar* (='porous') is simply the great detrital

piedmont skirting the Siwaliks, where the stream profiles suddenly flatten out and the coarser detritus—boulders and gravels—is deposited. In this tract the smaller streams, except when in spate during the rains, are lost in the loose talus, to seep out again where the slope is still flatter and finer material is deposited in the marshy and jungly *terai* below. Originally the terai covered a zone perhaps 50–60 miles wide.... Much has been so altered by settlement that the true terai is now confined to a relatively narrow strip parallel to the bhabar... *Bhur* is a generic term for patches of sandy soil' (pp. 497–498). In Bengal the equivalent of terai is duārs.

**Teras** (Sudan; *coll*. Arabic, *pl*. turus)
A ridge constructed with hand tools for restraining run-off after rain, in the drier parts of the northern Sudan, on an area selected for cultivation. (J.H.G.L.)

**Terminal Moraine**
Alternatively, end moraine. The moraine (*q.v.*) which marks the farthest extent of a glacier; a moraine at the end of a glacier.

**Terms of trade**
Committee, List 4. 'The export-price index as a percentage of the import-price index, that is, the amount of other countries' products that a nation gets in exchange for a unit of its own products.' (Hicks, *The Social Framework*.)
Comment. Also used in similar calculations as between primary commodities and manufactures. For a comprehensive and critical discussion see Kindleberger, C. P., *The Terms of Trade*, 1956. (C.J.R.)

**Terra** (Latin)
Literally earth or land as opposed to water hence terra firma, firm earth or land, terra incognita, an unknown land etc.

**Terrace**
*O.E.D.* Adopted from French, terrace, rubble, a platform, a terrace... from Latin, terracea, earthen, of the nature of earth, earthy... hence the primary sense, useless earth, heap of earth or rubbish, whence earthen mound made for a purpose.
1. A raised level place for walking, with a vertical or sloping front or sides faced with masonry, turf or the like, and sometimes having a balustrade; especially a raised walk in a garden, or a level surface formed in front of a house on naturally sloping ground, or on the bank of a river, as 'The Terrace' at the Palace of Westminster.
2. A natural formation of this character.
a. A table-land. b. Especially in Geol., a horizontal shelf or bench on the side of a hill, or sloping ground.
*Webster*. 5. *Geol*. A level and ordinarily rather narrow plain, usually with a steep front, bordering a river, a lake, or the sea; a topographic bench. Many rivers are bordered by a series of terraces at different levels, indicating former flood plains formed at different stages of erosion or deposition.
Mill, *Dict*. A nearly level strip of land, extending along the edge of a sea, river, or lake, or on the sides of a hill or valley. It is bounded above and below by rather abrupt slopes.
Comment. A vocabulary, French-English-German, as used in 1951 by the Commissions of Terraces and of Erosion Surfaces of the IGU states: 'Une terrasse peut être définie: une surface sensiblement plane, sensiblement horizontale, qui, d'un côté, domine le terrain avoisinant par un talus et, de l'autre, est souvent, mais non toujours, dominée par une pente plus fort. Il s'agit essentiellement d'une forme et non des matériaux qui la constituent.'
On the question of distinguishing between terraces as morphological units and the deposits which compose them see *inter alia* Hare, F. K., The Geomorphology of a Part of the Middle Thames, *Proc. Geol. Assoc.*, **58**, 1947, 294–339.
In geographical literature, unless the context shows otherwise, terrace usually means river-terrace except when referring to cultivation.

**Terrace cultivation**
Swayne, 1956. 'A system of cultivation in terraces formed by man on the sides of hills and mountains, the soil being retained by artificial walls [or earth banks]. Irrigation is sometimes used to supplement precipitation.'
Hence paddy-terraces in rice cultivation.

**Terrace epoch**
*Webster. Geol*. The time just after the last ice advance of the early Quaternary glaciation, when streams developed terraces from valley plains aggraded by river deposits during glaciation.

**Terracing**
The work of constructing terraces for cultivation purposes; hence paddy-terracing etc.

## Terracette
Dury, 1959. 'On very many steep slopes a kind of ribbed pattern appears on the surface of the creeping waste, with little steps a foot or two in height running horizontally. These steps are called terracettes. An alternative name is sheep-tracks, but this title is grossly misleading. Terracettes can be found where no sheep have ever been.' (p. 13)
See also Lynchet.

**Terra-firme** (Brazil: Portuguese)—see Várzea

## Terrain, Terrane
*O.E.D.* 1.b. Standing ground, position. 2. A tract of country considered with regard to its natural features, configuration, etc.; in military use esp. as affecting its tactical advantages, fitness for manœuvring, etc.; also an extent of ground, region, district, territory. 3. Geol. (usually spelt terrane). A name for a connected series, group or system of rocks or formations; a stratigraphical subdivision.

*Webster*. 2. A tract or region of ground immediately under observation; environment; milieu. 3. *Geol.* = terrane. A formation, or a group of formations; the area or surface over which a particular rock or group of rocks is prevalent. 4. *Mil.* An area or ground considered as to its extent and topography in relation to its use for a specific purpose, as for a battle or the erection of fortifications.

Mill, *Dict.* 1. A district or tract of land considered with reference to its fitness or use for a special purpose. 2. (Geol.) Formerly used in French as the equivalent of system, and sometimes applied to smaller subdivisions. 3. Topographical features as represented on a map (Germany).

Fay, 1920. A variation of terrane. 1. A group of strata, a zone, or a series of rocks. This word is used in the description of rocks in a general, provisional or non-committal sense (Winchell).

Plaisance, G. and Cailleux, A., 1958. '1. Amas de terre, alluvion. 2. Étendue de terre considérée dans sa quantité. 3. La terre considérée comme ayant des qualités particulières, la prédisposant à une certaine utilisation, en partiuclier à la culture.'

*Comment*. The geographical usage is in the sense of *O.E.D.* 2. The geological usage has now become unusual.

This word, long familiar in military use, has become popular with geographers, especially in America where the spelling terrane is often used. 'Terrain studies' are regional studies usually with an emphasis on relief and essential physical features—soil, vegetation, and drainage.

## Terrain analogues
Chorley, R. J., in Chorley and Haggett, 1967. 'Another type of geomorphic spatial analogue is the so-called "natural model", in which what are believed to be characteristic assemblages of land-form units are identified and presented as type assemblages. ... Van Lopik and Kolb (1959) have employed the idea that one varied and accessible region can be divided into "component landscapes" which are each defined in terms of a characteristic association of four measureable terrain factors (characteristic slope, relief, plan profile and occurrence of steep slopes), to form "terrain types" with which other regions can be compared and classified. This scheme of desert "terrain analogues" is based on the landscape around the Yuma Test Station.' (pp. 62–63)

See Van Lopik, J. R. and Kolb, C. R., 1959, *A Technique for Preparing Desert Terrain Analogs, U.S. Army Engineer Waterways Experiment Station, Technical Report*, Vicksburg, Mississippi.

## Terral
The land breeze along the coasts of western Peru. See Virazan.

## Terra rossa (Italian)
*Webster*. The red-colored residue from the weathering and partial solution of certain rocks, especially limestone.

Mill, *Dict.* Terra Rossa. Red earth (Italy). The name has been extended to other regions.

*Soils and Men*, 1938. 'The term has been widely applied to red soils developed under the warm-temperate Mediterranean type of climate, marked by wet and dry seasons. Many writers have preferred to limit Terra Rossa to soils developed on Limestones, while some would have it include any red soil in a Mediterranean climate. ... At present its only distinction lies in its color.' (p. 991)

Robinson, G. W., 1932, *Soils*, London: Murby. 'Terra rossa is the name given to

a red soil which occurs commonly in the countries bordering the Mediterranean Sea. Typically it is associated with limestone... we follow A. Reifenberg in restricting the term to such soils.' (p. 287)
Jacks, 1954. Red base-saturated clayey soil formed from hard limestone in the Mediterranean climate.

**Terra roxa** (Brażil: Portuguese) *pr.* ro-shah
James, 1959. 'Among the soils of the Paraná Plateau, the *terra roxa* (*lit.* purple soil) formed on the outcrops of the diabase... is a deep, porous soil containing considerable humus, which can be easily recognized by its dark reddish purple color—when wet it becomes so slippery and sticky that travel is very difficult, and in dry weather it gives off a powdery red dust which stains everything' (p. 474).

**Terrestrial**
Of or pertaining to the earth, hence
*Terrestrial deposits:* those laid down on land as opposed to marine deposits; most geologists include aeolian, riverine and lacustrine.
*Terrestrial magnetism:* the magnetic properties possessed by the earth as a whole.
*Terrestrial radiation:* heat given out by the earth.

**Terrier,** also **Terrar** and other spellings
A register of landed property especially a book complete with maps, plans, details of tenancies, land use etc. Such a record was usually kept for any important estate, including for example the lands of a manor and such terriers in MSS. when preserved from the Middle Ages have material of immense interest to the historical geographer.

**Terrigenous**
Mill, *Dict.* Derived from the land.
Applied to marine deposits originating from the erosion of the land as opposed to pelagic or deep sea deposits. Deposited in the shallower parts of the sea—on the continental shelf.

**Territorial waters, Territorial**
Webster. *Internat. Law.* The waters under the territorial jurisdiction of a state, including (1) its *marginal sea* (called also *marine belt* or *territorial sea*), that part within three miles of its shore as measured from mean low water mark or from the seaward limit of a bay or mouth of a river, and (2) its *inland waters,* those inside its marginal sea and those within its land territory.
Swayne, 1956. (a) The strip of water, sea bed and subsoil between the coast and the high sea, over which a maritime State retains absolute jurisdiction. (b) In the United Kingdom, limited by a line drawn three miles from low water mark, or in the case of bays and estuaries, from a closing line drawn at the first point where they narrow to 10 miles in width.
*Comment.* The three-mile limit has been widely recognized internationally but in the years after World War II a number of countries sought to extend the limit to 10, 12, or 15 miles, with serious results especially in connection with fisheries. Iceland sought in this way to restrict fishing grounds long used by British fishermen.

**Territory**
*Webster.* 2. An extent of land and waters belonging to, or under the jurisdiction or sovereignty of, a prince, state, or government of any form, or any given portion of it. 3. Any definite or particular portion of the area of a state considered by itself, as an area or tract of a state not invested with full rights of sovereignty, but governed or ruled as a dependency or subject area, or having a legal system more or less peculiar to itself; as Tanganyika Territory. 4. A large extent or tract of land; a region; district. 6 (*cap.*) a. In the United States, a portion of the country not included within any State, and not yet admitted as a State into the Union, but organized with a separate legislature, under a Territorial governor and other officers appointed by the President and Senate. b. In Canada and Australia, a similarly organized portion of the country not yet formed into a Province or State.
Mill, *Dict.* 1. Any portion of land, of considerable extent. 2. An unorganized or only partly autonomous province or state.
Until granted 'statehood' in 1959 Alaska and Hawaii were territories of the United States.

**Tethys**
*Webster.* 3. *Geol.* A paleogeographic sea (ancestral to the present Mediterranean), which coming into existence in early Permian time, connected with the Pacific and Arctic oceans and separated the Permian continents Angara on the north from Gondwana on the south.
Himus, 1954. The name given to the geosyn-

cline, initiated during Triassic times, in which the sediments now forming the Alpine-Himalayan systems of mountains were deposited.

**Tetrahedral Theory** (Lothian Green, 1875)
The theory suggested by the apparent symmetry of the great land masses that the crust of the earth in cooling was warping towards a tetrahedral shape. Now discarded.

**Texture**
*O.E.D.* Of inorganic substances, as stones, soil etc.: Physical (not chemical) constitution; the structure or minute moulding (of a surface).
*Webster.* 1. Act or art of weaving; hence, intricate composition. *Obs.* 7. *Petrog.* Generally, the smaller features of a rock, which depend upon the size, shape, arrangement, and distribution of the component minerals. The larger features, such as folding, cracking, faulting, jointing, etc., are known as *structures*. Among the common textures are granular, ophitic, porphyritic, hypidiomorphic-granular, xenomorphic-granular, trachytic, and fluidal. 4. The disposition or manner of union of the particles or smaller constituent parts of a body or substance; fine structure; as, the *texture* of earthy substances or minerals; the *texture* of a plant or bone, [the texture of soil or topography].
Holmes, 1928. (Rock) The appearance, megascopic or microscopic, seen on a smooth surface of a homogenous rock or mineral aggregate, due to the degree of crystallisation (crystallinity), the size of the crystals (granularity), and the shapes and interrelations of the crystals or other constituents (fabric).
Glentworth, 1954. (Soil) 'The proportion of the various size groups of individual soil grains. The presence of organic matter or calcium carbonate both affect the texture' (p. 165). See also Soil texture.

**Texture (topography)**
Lobeck, 1939. 'Coarse- and Fine-textured Topography. In regions of massive and resistant rocks the elements of a dissected plateau are large and bold, the rivers being far apart. This constitutes coarse topography.' The opposite is fine textured topography which, if extremely fine, becomes true badland topography. (W.M.)

Cotton, 1922. (Topography) 'Mature topography is of coarse or fine texture according as the stream-lines are widely or closely spaced.' (p. 67)
*Comment.* Not widely used; no reference in Davis, 1909; Salisbury, 1907; Cotton, 1941, 1942; Wooldridge and Morgan, 1937; Mill, *Dict.*; Lake, 1952; Worcester, 1939; Holmes, 1944. Compare A. A. Miller, *Skin of the Earth*, pp. 74–75 where 'texture of dissection' and 'texture of drainage' are discussed.

**Thal** (Pakistan: Panjabi)
Sandy waste; desert (*cf.* Sindhi—thar) (A.H.S.). Applied especially to the central section of the Sind-Sagar doab between the Jhelum-Chenab and the Indus.

**Thalassic, thalassography**
Oceanic, oceanography
*Webster.* Pertaining to the sea or ocean; sometimes distinguished from *oceanic*, as applying to seas, gulfs, etc. rather than to oceans. Thalassography. Oceanography.
*Thalassostatic*—related to a period of static sea-level.

**Thānā** (Indo-Pakistan: Urdu)
A political division of a district which is under the jurisdiction of a single police-station so that a thana is really a police-station area but see under *tahsil*.

**Thar** (Pakistan: Sindhi)
Desert, sandy waste. (A.H.S.)
Especially the thar *par excellence*, the Thar or Great Indian Desert.

**Thaw depression, Thaw lake, Thaw sink**
Hopkins, D. M., 1949, Thaw Lakes and Thaw Sinks in the Imuruk Lake Area, Seward Peninsula, Alaska, *Jour. Geol.*, 57, 119–131.
*Thaw Depressions*—depressions which result from subsidence following the thawing of perennially frozen ground.
*Thaw Lakes*—lakes which occupy thaw depressions (synonymous with 'cave-in lakes').
*Thaw Sinks*—closed depressions with subterranean drainage, believed to have originated as thaw lakes. (p. 119)
See also Thermokarst.

**Thematic**
*O.E.D.* Of or pertaining to a subject or topic of discourse or writing.
*Comment. O.E.D.* notes 'rare' but this word has been revived and is often applied, *e.g.* to maps which illustrate a particular topic or theme.

## Thermal equator
An imaginary line on a world map drawn through the centre of the belt with the highest temperature. Its position varies with the season, but it lies mainly north of the true equator.

*Webster. Meteorol.* The region of the earth enclosed within the annual isotherm of 80°, including the northern part of South America and the greater part of Africa and India; also, the middle line of this belt.

## Thermal metamorphism—see Metamorphism

## Thermal strain diagram
A climagram or climograph (*q.v.*) elaborated to show effects of climate, especially heat and moisture, on human beings. See D. H. K. Lee, *op. cit.*

## Thermograph, thermogram
A thermograph is a self-recording type of thermometer, a continuous trace of the air temperature being made on a thermogram fixed to a rotating drum actuated by clockwork.

## Thermiosopleth diagram (F. Erk, 1885 in German; Carl Troll 1957 in English)
A diagram in the Cartesian coordinates months and hours of the day, in which isopleths of temperature at a station are drawn.

Erk, Fritz, 1885, 'Ueber die Darstellung der stündlichen und jährlichen Vertheilung der Temperatur durch ein einziges (Thermo-Isoplethen-) Diagramm und dessen Verwendung in der Meteorologie,' *Meteorol. Ztsch.*, **2**, 281–286. Erk cites for the introduction of the term 'isopleth' Ch. A. Vogeler, *Anleitung zum Entwerfen graphischer Tafeln*, Berlin, 1877.

Troll, C., 1957, *Abs. of Papers, Ninth Pacific Science Congress*, Bangkok, 1957, 254; *Petermann's Mitt.*, 1943; see also Troll, C., 1958, *Oriental Geographer*, Dacca, **2**, 1958, 143. 'The most complete picture of the temperature conditions of a station is given by the so-called thermoisopleth diagram which shows the yearly as well as the daily temperature curve.'

## Thermokarst
Muller, 1947, 'Karst-like topographic features produced by the melting of ground-ice and the subsequent settling or caving of ground.' (p. 223) 'Some of the commonest land forms which are produced by the thermokarstic processes are: 1. Surface cracks. ... 2. Cave-ins and funnels. ... 3. Sinks and saucers and shallow depressions. ... 4. "Valleys", gulleys, ravines and sag basins. ... 5. Cave-in lakes, windows and sag ponds. ....' (p. 84)

*Comment.* This term is not widely used in English geomorphological literature but is favoured by Russian writers. See also Thaw depression.

## Thermosphere
The earth's outermost atmosphere.

## Therophyte
Raunkiaer, 1934. 'Plants which complete their life cycle within a single favourable season and remain dormant in the form of seed throughout the unfavourable periods' (p. 19). An annual.

## Thicket
A wood, usually small, with dense undergrowth and closely set trees, but a term of no precise ecological meaning.

## Third World
The term *tiers monde*, coined by Alfred Sauvy in the 'cold war' period of the 1950s, referred to areas committed to neither the Western 'capitalist' (the First World) nor the Eastern 'communist' (the Second World) power blocs; but now generally used as a synonym for 'less-developed' or 'developing' countries, defined primarily in economic terms though still perhaps seen as areas peripheral to both capitalist and communist 'worlds'. Conventionally includes most of Asia, Africa, and Latin America, but opinions differ as to whether China, Vietnam, Cuba, South Africa, Israel and the oil-rich nation states of the Middle East should be included.

## Thixotropy
King, C. A. M., 1959, *Beaches and Coasts*, London: Arnold. 'A state which shows a decrease in viscosity upon agitation or a decreased resistance to shear when the rate of shear increases.' (p. 4)

## Tholoid
A dome-shaped volcanic plug.

## Thorn forest
*Webster.* A tropical, xerophytic woodland of the savanna type, commonly dominated by thorny trees.

A general term covering a tropical or subtropical forest or wood of small xerophilous thorny trees.

## Three-field system
The system of cultivation, widespread in Europe and probably introduced into

Britain by the Anglo-Saxons, whereby the arable land was divided into three parts (open fields). Each of the three fields in turn was rested for a year, *i.e.* allowed to lie fallow while the other two were cropped with grain, wheat or rye, and barley or oats. The system disappeared with enclosure and with the discovery that clovers enrich the soil and fallow is largely unnecessary.

**Thrust-plane, Thrust fault, Overthrust fold**
*Webster.* Thrust plane. *Geol.* The surface, never strictly a plane, along which dislocation has taken place in the case of a reverse, or thrust, fault.
Mill, *Dict.* Thrust-plane. The plane along which blocks of strata have been thrust (often more or less horizontally) across others.
Geikie, J., 1898. 'Thrust-plane: a reversed fault, the hade or inclination of which approaches horizontality; a common structure in regions of highly-flexed rocks.' (p. 307)
Holmes, 1954. 'An overturned fold, or *overfold*, has one of its limbs inverted, and if the latter approaches a horizontal attitude the overfold is described as recumbent. Further development results in the rocks of the upper limb being pushed bodily forward along... a *thrust plane* and the structure becomes an *overthrust fold*.' (p. 75) See also Fault.

**Tibba** (Pakistan: Panjabi)
Sand or sand hills, hence desert. (A.H.S.)

**Tidal**
*O.E.D.* 1. Of, pertaining to, or affected by tides; ebbing and flowing periodically. Hence tidal creek, tidal river, tidal basin, tidal stream (*naut.*).

**Tidal range**
Tidal amplitude is the elevation of high water above mean sea-level; tidal range (*cf.* range (2) statistics) the difference between highest and lowest tides which varies from day to day. The fortnightly neap tides have a small range, the corresponding spring tides a larger range (E.G.R.T.). See also Tide, High water, O.D., Mean Sea-Level.

**Tidal wave**
*O.E.D.* The high water wave caused by the movement of the tide = *tide-wave*; *Erron.* an exceptionally large ocean wave caused by an earthquake or other local commotion.
Comment. Although *O.E.D.* notes it as an erroneous use, 'tidal wave' is very commonly (Webster says popularly) used for the giant destructive wave caused by an earthquake. See Tsunami.

**Tide**
*Webster.* The alternate rising and falling of the surface of the ocean, and of gulfs, bays, rivers etc., connected with the ocean. The tide ebbs and flows twice in each lunar day. It is occasioned by the attraction of sun and moon... when in conjunction... a spring tide... when in apposition a smaller high tide than usual ... a neap tide.

**Tide-mill**—see sea-mill

**Tierra caliente** (Spanish = hot land)
Mill, *Dict.* The littoral and comparatively low lying regions of Spanish tropical America up to an altitude of 3,000 feet. The zone has a moist climate and great heat (mean annual temperature 75° to 83°F.), characterized by tropical products.
James, 1959. 'In high mountain regions of the low latitudes the characteristics of the various elevations are so distinct that the local inhabitants recognize general "vertical zones". The lowest zone in Spanish America is known as the *tierra caliente*, or hot country, which may also be called the "Zone of Tropical Products". The average annual temperatures are mostly between 75° and 80°, with a difference between the average of the coldest... and warmest months not more than three or four degrees... upper limit in Venezuela is about 3,000 feet.' (p. 80)

**Tierra fria** (Spanish = cold land; strictly *fría*)
Mill, *Dict.* The zone of vertical distribution of plants which lies immediately higher than the Tierra Templada and below the Paramos in Spanish tropical America. It varies in height from 6,500 feet to 10,000 feet, and has a mean annual temperature of from 65°F. to 54°F. It is characterized by products resembling those found at sea-level in the higher latitudes of the temperate zone, *e.g.* wheat, vegetables, and northern fruits.
James, 1959 (after tierra caliente and templada *q.v.*). 'Between approximately 6,000 and 10,000 feet in elevation is a zone known as the *tierra fria* or cold country, which we may call the "Zone of the Grains". Average annual temperatures are between 55° and 65° and there is practically no difference in temperature from one month to another.' (p. 80)

**Tierra templada** (Spanish = temperate land)

Mill, *Dict*. The zone of vertical distribution of plants (3,000 feet to 6,500 feet) immediately higher than the Tierra Caliente, and lower than the Tierra Fria in Spanish tropical America (mean annual temperature 65° to 75°F.).

James, 1959 (after tierra caliente *q.v.*). 'Between 3,000 and 6,000 feet in elevation is a cooler region the *tierra templada* or temperate country which may also be designated the "Zone of Coffee". Average annual temperatures vary between 65° and 75°, but the ranges between the coldest and the warmest months are a little less than those in the tierra caliente.' (p. 80)

**Tilā** (India-Pakistan: Bengali)
Chatterjee, S. P. 'A low isolated hill lying at the foot of an escarpment' (MS. communication).

**Till**
*O.E.D.* 1. A term applied to a stiff clay, more or less impervious to water, usually occurring in unstratified deposits, and forming an ungenial subsoil. Originally a term of agriculture in Scotland.
b. In the majority of cases this clay belongs to the Glacial or Drift period, and in geological use 'till' has the specific sense 'boulder clay'.

*Webster*. 2. *Geol*. Unstratified glacial drift, consisting of clay, sand, gravel, and boulders intermingled in any proportions; called also *boulder clay*.

Mill, *Dict.* See Boulder Clay. Boulder clay. Stiff clay containing irregular boulders of all sizes, which are frequently marked by glacial striae. Called Till (Scotland).

Fay, 1920. That part of a glacial drift consisting of material deposited by and underneath the ice, with little or no transportation and sorting by water; it is a generally unstratified, unconsolidated, heterogeneous mixture of clay, sand, gravel, and boulders. Also called Boulder-clay (La Forge).

*Comment*. American writers generally use till or tillite, British writers prefer boulder clay.

**Tillage**
*O.E.D.* 1. The act, operation, or art of tilling or cultivating land so as to fit it for raising crops; cultivation, agriculture, husbandry.
b. The state or condition of being tilled or cultivated. 2. concr. Tilled or ploughed land; land under crops as distinct from pasturage; the crops growing on tilled land.

Committee, List 3. 1. Land ploughed or hoed during the current year.
2. The process of cultivating the land.

Stamp, 1948. '... in the wartime statistics the word "tillage" is used to indicate arable land excluding rotation grass and clovers.' (p. 83)

Burke, J., 1953, *Stroud's Judicial Dictionary*, 3rd ed., London: Sweet and Maxwell, Vol. 4. 'Tillage' and 'agricultural' land are synonymous, and they exclude use as a garden or orchard. With reference to tithes.

**Tillite**
*Webster*. *Geol*. Rock formed of consolidated or lithified till and generally a record of a glacial epoch older than that of the Quaternary.

Consolidated till or boulder clay, especially used of such deposits of earlier geological ages (*e.g.* late Carboniferous age in South Africa, India, Australia). See Holmes, 1944, 499.

**Tilth**
*Webster*. 1. Act or occupation of tilling; cultivation of the soil; specifically, a plowing or harrowing; as, land in good *tilth*. 3. Cultivated, or tilled, land, as distinguished from pasture, etc. 4. The state of being tilled; condition when tilled; as, land in good *tilth*. 5. Surface soil as prepared for sowing or planting; the depth of friable earth.

Jacks, 1954. State of aggregation of soil in relation to its response to cultivation implements. (Soil science.)

*Comment*. This is a very old word and *O.E.D.* gives many meanings; the most appropriate to present day usage are 'the condition of being under cultivation or tillage hence good or bad condition of land under tillage' and 'the prepared surface soil'.

**Timber line**
Mill, *Dict*. The limit of growth, in numbers sufficiently close together, of such trees as furnish timber as a commodity. See Tree limit.

*Webster*. On mountains and in the frigid regions, the line above which there are no trees.

*Dict, Am*. Timberline. In cold or mountainous regions, the line above which timber does not grow. 1867, *cf*. cold, dry timberline. 1903 *Amer. Geol. Soc. Bull.* XIV, 556. 'On the mountains of central Idaho, the cold timberline is sharply drawn at an elevation of about 10,000 feet, while the dry timberline, equally well defined, has an elevation of 7,000 feet.'

*Comment.* The dry timberline is the lower limit of tree growth on mountains in arid regions, in which precipitation decreases with descent down the mountain slope.

**Time**—see Apparent time, Local time, Standard time.

**Time-space (cost space)**—see Cost space.

**Time zones**—see under Standard time.

**Tinaja** (Spanish)
McGee, W. J., 1896, Expedition to Seriland, *Science*, N.S. 3. 'A natural bowl or bowl-shaped cavity, specifically the cavity below a waterfall, especially when partly filled with water; in a more general way it is extended to temporary pools....' (p. 494)

**Tinajita** (Spanish)
Howell, 1957. 'Small, shallow flutes developed on flat surfaces of limestone (S. W. Texas).'

**Tind** (Norwegian)
Thornbury, 1954. A 'horn' detached from the main mountain range, formed where lateral cirque recession cuts through an upland spur between two glacial troughs. (p. 373)
*Cf.* Matterhorn and horn which have displaced the term. One of the first to use tind in English was W. H. Hobbs *Characteristics of Existing Glaciers.* (G.T.W.)

**Tir comin** (Welsh: *lit.* true common)

**Tir cyd** (Welsh: land grazed in common)
*Report Royal Commission on Common Land*, 1955–58, H.M.S.O., 1958, Cmd. 462. 'In certain of the northern counties of England ... there is a distinction between areas 'believed to be true common' and 'areas of moorland grazed in common'. This interpretation applied largely to Wales: 'true common' is 'Tir Comin' in Welsh and moorland grazed in common is 'Tir Cyd' (p. 252). See Common.

**Tirr**
Jacks, 1954. 'The loose, slightly decomposed, uppermost layer of a raised moss.'

**Tirs**
Jacks, 1954. 'Black clay soil of North Africa, resembling regur.'
*Comment.* 'Tersified' and 'tersoid' are used of soils turning into or resembling this. See also Grumusol.

**Tithe**
Literally tenth or a tenth part from Old English but obsolete in this general sense. *O.E.D.* The tenth part of the annual produce of agriculture etc., being a due or payment (orig. in kind) for the support of the priesthood, religious establishments etc. Hence tithable or subject to tithe; tithebarn, where the produce was delivered and stored. See also parish.

**Tjäle** (Swedish; Norwegian; tæle or tele), anglicized as tjaele and taele
This word, taken from the common speech of Sweden, designates frozen soil or ground frost showing usually a peculiar structure with layers of lenses of pure ice within the soil, a phenomenon described first in the year 1765 by E. O. Runeberg. In fact tjäle designates not only frozen soil but also the condition of the ground. There has been a movement, mostly among Scandinavian, German, English and French authors to introduce tjäle (tjaele) as a short single word for perennially frozen ground. S. W. Muller (1945) has sought another single word for permanently frozen ground by coining permafrost, made by contracting permanent and combining it with the English word frost, none of whose meanings refer to the ground. Kirk Bryan (1946) again ventures to suggest Pergelisol (fr. Lat per, through, gel from gelare to freeze and sol, from solum, ground) equivalent to the Swedish ständig (permanent) tjäle, or Norwegian *evig tele.*
Zeuner, F. E., 1946, *Dating the Past*, London: Methuen. 'Permanently frozen subsoil.' (p. 119)
Bryan, K., The Erroneous Use of Tjaele as the Equivalent of Perennially Frozen Ground, *Jour. of Geol.*, **59**, 1951. Introduced by B. Högbom in 1914 as a substitute for German, Eisboden.

**Toft**
*S.O.E.D. Orig.* a homestead, the site of a house and its outbuildings; a house site. Often in *toft and croft*, the whole holding of homestead and attached land. Also an eminence, knoll or hillock in a flat region. Now *local.*
See also Bulin, 1961.

**Toich** (Sudan: Dinka)
Annually-flooded marshlands close to watercourses, which offer valuable pasturage for cattle during the dry season. (J.H.G.L.)

**Toll, toll gate, toll bridge**
Toll. A payment exacted by an individual or authority for protection, especially for permission to pass along a road (through a toll gate) or over a bridge (toll bridge).

**Tombolo (Italian)**
Mill, *Dict.* A storm beach or coastal dyke connecting the mainland with an island or outstanding rock (Italy).
Gulliver, F. P., 1899, Shoreline Topography, *Proc. Am. Acad. Arts and Sci.*, **34**. Bars connecting islands with the mainland, the English plural to be tombolos. (p. 89)
Thornbury, 1954. 'Islands may become tide together or to land by growth of one or more spits. Such bars are called tombolos.' (p. 446)

**Topography, topographic, topographical**
O.E.D. (From Greek, topo-, place) 1. The science or practice of describing a particular place, city, town, manor, parish, or tract of land; the accurate and detailed delineation and description of any locality. b. A detailed description or delineation of the features of a locality. c. Localization, local distribution; the study of this (quots. 1658, 1835). 2. The features of a region or locality collectively.
*Webster.* 2. The art or practice of graphic and exact delineation in minute detail, usually on maps or charts, of the physical features of any place or region, especially in a way to show their relative positions and elevations. *Cf.* chorography.
4. The configuration of a surface, including its relief, the position of its streams, lakes, roads, cities, etc.; as a map showing the topography of Ohio; hence, loosely, natural or physical features collectively; lay of the land.
*American College Dictionary* 3. The relief features or surface configuration of an area.
Mill, *Dict.* 1. The detailed description or representation of a small area of land. 2. The description of the antiquities and family histories of a neighbourhood.
Committee, List 1. The detailed description, especially on a map, of a locality; including its relief and any relatively permanent objects, whether natural or of human origin, thereon.
James, P. E., 1935, *An Outline of Geography*, Boston: Ginn. 'In many writings the word is now used to refer only to the landforms or the character of the surface features. In this book, however, its original meaning will be retained. It will refer to the details which comprise the landscape of a small area, the landforms and also all the other features occurring together on the earth's surface, as on a topographic map.... A topographic study refers to the study of a small area.... Chorography refers to studies of larger regions; geography, to a study of the world or of its larger parts." (p. 145)
Rice, 1941. The general configuration of the land surface; the sum total of the results of erosion and deposition on the physiographic features of a region.
Glentworth, R., 1954, *The Soils of the Country Round Banff, Huntly and Turriff*, Edinburgh: H.M.S.O. 'In the soil survey connotation it is used for the features revealed by the contour lines on a map.' (p. 27) 'The undulations or inequalities of the land surface, the slope and pattern of these.' (p. 165) 'The expression "high frequency topography" refers to a land form which has more ridges per mile with more numerous and shorter slopes than low frequency topography.' Six main topographical classes are recognized by the Soil Survey of Canada, largely based on slope. (p. 27)
*Comment.* In the majority of geographical writings the word has come to be used in the sense of Rice or Glentworth above, which is obviously a wide departure from the original and the derivative meaning. In consequence many writers are endeavouring to avoid the word. Others regard the shift of meaning to be fully established and the general use is in the sense of Rice.
*American Comment.* A topographic map is a map of a local area. Such maps, produced by great national surveys, depict relief. Apparently by transfer the word topography came to refer to the relief shown by the map. Topography in the sense of relief or surface configuration is thoroughly established both in general speech and in technical geomorphological literature. (C.D.H.)

**Topographic adolescence or youth**
*Webster. Phys. Geog.* The condition of a district soon after the beginning of erosion by streams, when main branches have well-developed narrow valleys, but the areas between the streams are little modified.

**Topographic desert**
A term used by D. H. K. Amiran and others

to describe a local desert due to relief and descending air currents in contrast to planetary deserts found in the arid zones of the world.

**Topographic infancy**
*Webster. Phys. Geog.* The condition of a district freshly exposed to the action of surface waters, when the original hollows are still occupied by lakes and ponds and the plains are imperfectly dissected by narrow stream gorges.

**Topographic map**
*Webster.* A map intermediate between a general map and a plan, on a scale large enough to show roads, plans of towns, contour lines, etc.
*Comment.* Common scales are 1:25,000, 1:50,000, or 1:100,000.

**Topographic maturity**
*Webster. Phys. Geog.* The condition of a district in which the land is reduced to slopes, the original upland has been completely dissected, and a new plain of erosion has scarcely begun to appear. Many of the individual river valleys are mature but some of the headwaters of the tributaries may still be in the youthful stage.

**Topographic old age**
*Webster. Phys. Geog.* The condition of a district reduced by erosion nearly to base level.

**Topographic youth**
*Webster. Phys. Geog.* = *Topographic adolescence, q.v.*

**Toponymy**
Toponym: a place-name; toponymy: the study of place-names.
See Aurousseau, M., 1957, *The Rendering of Geographical Names*, London: Hutchinson, p. 3.

**Top-set beds**—see Delta structure

**Top soil**
Jacks, 1954. 'The layer of soil moved in cultivation; the A horizon.'
*Comment.* Used by agriculturalists rather than soil scientists. Roughly the cultivated soil, whatever horizons were originally involved. Surface soil as distinguished from subsoil. (*Webster*).

**Tor**
*O.E.D.* 1. A high rock; a pile of rocks generally on the top of a hill; a rocky peak; a hill. In proper names of eminences or rocks in Cornwall, Devon, Peak of Derbyshire; also sporadically in some other counties, *e.g.* Glastonbury Tor in Somerset.

b. Locally in Scotland, applied to an artificial mound; a burial mound.
Mill, *Dict.* An isolated mass of weathered rock, usually granitic, left prominent upon the sides or summit of a large rounded hill (S.W. and W. England and Pennines).
Linton, D. L., 1955, The Problem of Tors. *Geog. Jour.*, **121**, 470–487.
Palmer and Neilson, 1962, The origin of granite tors on Dartmoor, *Proc. Yorkshire Geol. Soc.*, 'A residual upward projecting fragment of naked bedrock, the result of differential weathering and mass movement.'

**Tornado**
*O.E.D.* Originally ternado, probably a bad adaptation of Sp. tronada, thunderstorm, with spelling later modified by treating it as a derivative of Sp., tornar, to turn, return. 1. A term applied by 16th c. navigators to violent thunderstorms of the tropical Atlantic, with torrential rain, and often with sudden and violent gusts of wind. Now rare or passing into 2. 2. A very violent storm (now without implication of thunder), affecting a limited area, in which the wind is constantly changing its direction or rotating; a whirling wind, whirlwind; loosely, any very violent storm of wind, a hurricane. spec. a. On the west coast of Africa, a rotatory storm in which the wind revolves violently under a moving arch of clouds; b. In the Mississippi region of U.S., a destructive rotatory storm under a funnel-shaped cloud like a water-spout, which advances in a narrow path over the land for many miles.
*Met. Gloss.*, 1944. 1. In West Africa, the squall which accompanies a thunderstorm; it blows outwards from the front of the storm at about the time the rain commences.... 2. A very violent counterclockwise whirl of small area averaging a few hundred feet in diameter, which gives wind velocities estimated to exceed 200 mi/hr. in some examples....
Mill, *Dict.* A destructive local whirlwind occurring in the hot moist equatorial section of a low-pressure system during a thunderstorm, accompanied by a cloud of funnel form; characteristic of Mississippi and Ohio basins. The term was originally restricted to a severe squall on the coast of Senegambia and Guinea.
For a modern explanation of the American tornadoes see Hare, 1953, 134.
See also Tropical Revolving Storms.

**Torrid Zone**—see Zone

**Toun, Township**
Scottish spelling preferred by some writers for Town in sense of *O.E.D.* 6a. For crofting tounships, see Croft.

**Tourelle** (French)
*Lit.* a little tower; applied to a type of Karst surface characterized by little hills. See Karst terminology.

**Tourism.**
*O.E.D.* The theory and practice of touring; travelling for pleasure. Usually deprecatory.
*Webster.* The guidance and management of tourists as a business or government function.
*Comment.* Now commonly applied to what has claims to be the world's largest industry. The chief industry of many towns is frequently given as 'tourism'. It thus covers the whole business of providing hotels and other accommodation and amenities for those travelling or visiting, or living primarily for pleasure and, for want of a better term, is extended to the catering for residents, especially retired people, in places like Bournemouth.

**Towan**
Coastal sand dunes (especially Cornwall). *Cf.* Welsh tywyn.

**Town**
*O.E.D.* Old English, tuwn, tun. The sense in O.H.G. was 'fence', 'hedge'. In O.E. the sense 'fence, hedge' does not occur, only that of 'enclosed place', as in sense 1, and its developments in senses 2 and 3, in which it was frequently used to render L., villa. The modern sense 4 is later than the Norman conquest, and corresponds to F., ville, 'town, city' . . .)
1.b. spec. The enclosed land surrounding or belonging to a single dwelling; a farm with its farmhouse (still in Scottish dialect). . . .
2. The house or group of houses or buildings upon this enclosed land; the farmstead or homestead on a farm or holding. Now esp. Scottish.
4. Now, in general English use, commonly designating an inhabited place larger and more regularly built than a village, and having more complete and independent local government; applied not only to a 'borough', *i.e.* a corporate town, and a 'city', which is a town of higher rank, but also to an 'urban district', *i.e.* a non-corporate town having an 'urban district council' with powers of rating, paving and sanitation more extensive than those possessed by a parish council or the administrative body (where such exists) of a village. Sometimes also applied to small inhabited places below the rank of an 'urban district' which are not distinguishable from villages otherwise, perhaps, than by having a periodical market or fair (market town), or by being historically 'towns'.
5. As a collective sing. a. The community of a town in its corporate capacity; the corporation; b. the inhabitants of a town, the townspeople.
6. U.S. A geographical division for local or state government.
a. A division of a county, which may contain one or more villages (in sense 4); a township; also, the inhabitants of such a division as a corporate body (esp. in the New England states).
b. A Municipal corporation, having its own geographical boundaries (as distinct from a.), considered either in reference to its area or as a body politic.
*Webster.* 5. In general, a place which is a population and business center and is recognized as such geographically and politically. Hence: a. A center of population, larger and more fully organized and developed than a village, but not incorporated as a city. b. *Eng.* A village without urban characteristics or not an episcopal see, but with a periodical fair or market; more fully *market town*. c. Any large closely populated place, as a city, a borough, or an urban district. 6. Specifically in the United States: a. In the New England States, a municipal corporation of less complex character than a city. b. In other States, a unit of rural administration more or less like the New England town: a township.

Mill, *Dict.* An agglomeration of population under the control of an administrative authority of its own.

Dickinson, R. E., 1947, *City, Region and Regionalism*, London: Kegan Paul. 'The town in western Europe and North America may be defined as a compact settlement engaged primarily in non-agricultural occupations.' (p. 25)

Smailes, A. E., 1953, *The Geography of Towns*, London: Hutchinson. 'A town may be regarded first and foremost as a community of people pursuing a distinctive way of life... or it may be considered as part of the earth's surface differentiated from rural surroundings by a particular type of human transformation with buildings and other distinctive structures.' (p. 32)

**Townland**
O.E.D. In Ireland, a division of land of varying extent; also a territorial division, a township. In Scotland, the enclosed or infield land of a farm.

**Townscape**
Smailes, 1953. 'The physical forms and arrangements of the spaces and buildings that compose the urban landscape, or townscape, as it may be called' (p. 84).
Smailes, A. E., 1955, Some Reflections on the Geographical Description and Analysis of Townscapes, *Trans. I.B.G.*, **21**, 99–115.

**Township**
O.E.D., 2c. Each of the local divisions of, or districts comprised in, a large original parish, each containing a village or small town, usually having its own church (formerly a chapel of the mother church of the original parish, whence such divisions were also known ecclesiastically as chapelries).
3. trans. Often rendering L. pagus, Gr. δημος (Deme), and thus applied to independent or self-governing towns or villages of ancient Greece, Italy and other lands, and sometimes to foreign towns or villages of medieval or modern times. 4. Sc. A farm held in joint tenancy. 5. U.S. and Canada. A division of a county having certain corporate powers of local administration; the same that in New England is called a town. In the newer states, in which the divisions were laid off by government survey, a township is a division six miles square, and is so called even when still unsettled. The name is similarly used in the western provinces of Canada, from Ontario to British Columbia, and in Eastern Quebec and Prince Edward Island. 6. In Australia, a site laid out prospectively for a town. 8. By some 19th century historical writers, adopted to designate what they consider to have been the simplest form of local or social organization in primitive Old English times.

Mill, *Dict.* 1. An area containing a number of small towns or villages subject to one common local authority (U.S.A.). 2. A village, a possible future town (Australia).

*Webster.* 4. In the United States, a primary unit of local government of varying character in different parts of the country. See *Webster*, p. 2680. 5. A geographical rather than a political division; specifically: a. In surveys of the public lands of the United States, a division of territory that is, with certain exceptions, six miles long on its south and east and west boundaries, which follow meridians, and so slightly less than six miles on the north. It contains 36 sections, and has often formed the basis of a later political township. b. In Canada, a subdivision of certain of the provinces. c. *Australia*. A townsite, *i.e.* a tract of land laid out with streets and subdivided into lots for the development of a town; also, the temporary settlement on such a site.

*Comment.* The current use in Scotland is in the crofting counties. There 'the characteristic settlement is the crofting township. Each township comprises two main elements, the individually held croft land and the common grazings... the croft land is separated from the common pasture by the township dyke, a stone or turf wall.' The number of crofts ranges widely (from half a dozen to fifty or more). See Caird, J. B., 1959, *Park: A Geographical Study of a Lewis Crofting District*, Nottingham: Geographical Field Group.

**Tract** (J. F. Unstead, 1933)
Unstead, J. F., 1933, A System of Regional Geography, *Geography*, **18**, 175–87. 'Stows would be combined into tracts... relief, structure and soils, either singly or in combination, are generally the distinguishing criteria of a tract.' (pp. 178, ) 181

*Comment.* One of a proposed hierarchy of regions; see stow, morphological region.

**Tract, census**
Canada, Ninth Census, 1951, *Population and housing characteristics by census tracts*, Halifax: Dominion Bureau of Statistics, Bulletin: CT–1. 'These statistical units are designed with a view to approximate uniformity in size and population, and to the inclusion of an area which is fairly homogeneous with respect to economic status and living conditions.' (p. 3)

*Comment.* Also in U.S. Census.

**Traction**
Gilbert, G. K., 1914, The Transportation of Debris by Running Water. *U.S.G.S. Prof. Paper*, **86**. 'In other transportation including saltation, rolling and sliding, the efficient factor is the motion parallel with

the bed and close to it. This second division of current transportation is called by certain French engineers entraînement but has received no name in English. Being in need of a succinct title, I translate the French designation, which indicates a sweeping or dragging along, by the word *traction*, thus classifying hydraulic transportation as (1) hydraulic suspension and (2) hydraulic traction.' (p. 15a)

Twenhofel, 1939. 'Transportation ... is accomplished in each of three ways of traction, suspension, and solution.' (p. 186)

**Trade, Balance of**—See Balance of Trade

**Trade wind**

*O.E.D.* From German, trade, track, App. originating in the phrase: to blow trade. The name had in its origin nothing to do with trade in the sense 'commerce'.
1. Any wind that 'blows trade', *i.e.* in a constant course or way; a wind that blows steadily in the same direction. Obs. 2. Applied to the monsoon winds of the Indian Ocean. Obs.
3. Now specifically the wind that blows constantly towards the equator from about the thirtieth parallel, north and south; its main direction in the northern hemisphere being from the north-east, and in the southern hemisphere from the south-east.

Mill, *Dict.* Constant N.E. and S.E. winds which blow from the belts of tropical calms towards the doldrums (Halley).

*Met. Gloss.*, 1944. 'This is the name given to the winds which blow from the tropical high pressure belts towards the equatorial region of low pressure, from the N.E. in the northern hemisphere and S.E. in the southern hemisphere. The name originated in the nautical phrase "to blow trade", meaning to blow in a regular course. . . .'

**Trail**

Head (*q.v.*).

**Tramontana** (Spanish and Italian)

Wind descending towards the sea from cold dry plateaus; a form of föhn. (P.D.) Also in Italy, particularly in Rome and Florence, a north wind. (C.J.R.)

**Transfer costs**

Ohlin, Bertil, 1933, *Interregional and International Trade*, Cambridge: Harvard U.P. 'Costs of transfer' include 'transportation costs as well as the costs of overcoming other obstacles (to commodity movements) such as tariff walls'. (p. 142)

**Transfluence** (A. Penck, 1909, in German)

Linton, D. L., 1949, Watershed Breaching by Ice in Scotland, *Trans. Inst. Brit. Geog.*, 15. 'Penck recognised that watersheds breached by the growth of neighbouring corries and the consequent intersection of their walls might come to be overrun by ice and to such cases he applied the term *transfluence*.' (p. 2)

**Transgression**

1. The invasion of a land surface by the sea and the strata associated with such a movement.
2. Of igneous rocks: an intrusion cutting across bedding planes of sedimentary rocks from one horizon to another.

**Transhumance**

*O.E.D.* via Fr. and Sp. from L. trans, across +humus, ground, soil. Migrating between regions of differing climates.

*Webster.* The seasonal moving of livestock from or to the mountains.

Newbigin, M. I., 1911, *Modern Geography*. 'Transhumance, still well developed in Spain, is the periodic and alternating displacement of flocks and herds between two regions of different climate.' (p. 179)

Mill, *Dict.* The periodic movement of large flocks of sheep between their summer and winter pastures.

James, P. E., 1935, *An Outline of Geography*, Boston: Ginn. '... the periodic, or seasonal, movement of flocks or herds of domestic animals between two areas of different climatic conditions' (pp. 330–331). Similar movement of human groups is 'seasonal semi-nomadism'.

King, H. W., 1949, *The Pattern of Human Activities*, Sydney: Australasian Pub. 'Transhumance, or migration of population groups. . . .' (p. 105) Used as equivalent to nomadism, *e.g.* 'perennial transhumance'.

*Comment.* Should not be limited to sheep (Mill). See also Shieling.

**Transition zone**

Harris, C. D. and Ullman, E. L., 1954, The Nature of Cities, *Ann. Am. Acad. Pol. Soc. Sci.*, **242**, 7–17. 'Encircling the downtown area is a zone of residential deterioration. Business and light manufacturing encroach on residential areas characterized particularly by rooming houses. In this zone are the principal slums, with their submerged regions of poverty, degrada-

tion, and disease, and their underworlds of vice. In many American cities it has been inhabited largely by colonies of recent immigrants.' (p. 12)
See also Concentric zone theory.

**Transport**
*O.E.D.* The action of carrying or conveying a thing or person from one place to another; conveyance. A means of transportation or convenience; originally a vessel employed in transporting soldiers, etc.
*Comment.* Transport and transportation are largely interchangeable, British usage favouring transport where possible but American favouring transportation. In physical geography the sequence erosion, transportation, deposition is used, not transport. British writers use means of transport, *e.g.* land, sea or air transport, not transportation.

**Transportation**
Mill, *Dict.* The process of transferring (either in solution or in suspension) weathered rock to another place.

**Transported soil**—see Secondary soil

**Transverse coast**
Wooldridge and Morgan, 1937. 'Cutting across the structural grain'; a category of von Richthofen (p. 352).
*Comment.* Cotton, 1945, has 'coasts of transverse deformation' 'where deformation (whether warping or block faulting) takes place along lines transverse to the coast' (p. 441). Also called a Discordant coast or Atlantic type coast in contrast to a Longitudinal or Pacific type coast.

**Transverse valley**
Powell, 1875. '... having a direction at right angles to the strike.' May be diaclinal, cataclinal or anaclinal (p. 160).

**Trap**
*Webster.* Also traprock. *Geol. & Petrog.* Any of various dark-colored, fine-grained, igneous rocks, including especially basalt, amygdaloid, etc., used especially in road making; a convenient field term.
Himus, 1954. 'An old Swedish name for igneous rocks which were neither coarsely crystalline, like granite, nor cellular and obviously volcanic... included basalts, dolerites, andesites... name derived from *trappa*, a stair, since the hills formed by the denudation of such rocks were often terraced or had a stair-like profile.'
*Comment.* As the knowledge of geology has progressed and the origin of the various trap rocks is understood the word is now usually dropped in favour of a more precise term. Thus the Deccan Traps which cover 200,000 square miles in India are properly called the Deccan Lavas or Deccan Basalts.

**Travertine**
A deposit of calcium carbonate from hot springs.
See also Geyserite.

**Treaty port**
A sea or river port, later also an inland city opened by treaty to foreign trade, especially used of certain ports in China, Japan and Korea. In 1842 the first in China were opened with extra-territorial rights to foreigners; these rights were surrendered by Great Britain and the United States in January, 1943. See Extra-territoriality.

**Tree limit, Tree line**
Mill, *Dict.* The line marking the limit of the growth of trees.
*Comment.* It can be used both of altitude in mountainous regions and of latitude in poleward extension. See Timber line. Some authorities do not distinguish between tree line and timber line. *Webster* simply gives tree line = timber line.

**Tref** (Welsh)
Homestead, hamlet, town. See also Hendref or Pentref, Maerdref.

**Trek** (Afrikaans)
*Noun.* 1. A journey made by trekking (The Great Trek). 2. The distance so covered (a trek of 10 miles). 3. Company of people trekking together (the trek decided to settle in Natal).
*Verb.* To travel, especially by means of oxwagon; also to depart, to start travelling (we trekked at 8 o'clock). (P.S.)

**Trekker** (Dutch)
Robequain, C., *Le Monde Malais* (translated by Laborde, E. D., 1954). 'Among the foreign elements, European and Chinese, a distinction was often made (in Indonesia) between sojourners, or *trekkers* on the one hand and, on the other, the *blijvers,* or those who no longer expected to retire one day to their native land' (p. 85).

**Trellis drainage, Trellised drainage**
Wooldridge and Morgan, 1937. Trellised drainage is distinguished by a regularity in its pattern with structural elements—rock outcrops, fracture systems, master

jointing—picked out and emphasized by subsequent streams. 'Whether the network of major valleys is rectangular or rhomboidal, the drainage plan may be described as "trellised".' (p. 192)

**Trench**
1. A long narrow valley between two mountain ranges, especially a rift valley or a U-shaped valley.
2. A long narrow steep-sided submarine valley.

**Trend-lines**
Mill, *Dict.* The main structural lines, *e.g.* lines of folding and faulting, in a region.

**Trend surface**
Chorley, R. J. and Haggett, P., 1965, Trend-surface Mapping in Geographical Research, *Trans. Inst. Brit. Geog.*, **37**. 'Areal data always present such ambiguities and the most obvious manner in which to treat them is to attempt to disentangle the smooth, broader regional patterns of variation from the non-systematic, local and chance variations; and then to ascribe mechanisms or causes to the different components. Thus the regional effect is commonly viewed as a smooth, regular distribution of effects termed a trend surface or trend component (Whitten, E. H. T., 1950, Composition Trends in a Granite: modal variation and ghost stratigraphy in part of the Donegal Granite, Eire, *Jour. Geophys. Res.*, **64**, 835–846) which are too deep, too broad, and too great in "relief" to admit of the purely local explanation which is reserved for the residuals or deviations. . . . Trend surfaces can thus be considered as response surfaces from which aspects of origin, dynamics or process can be inferred, wherein variations in form may be thought of as responses to corresponding areal variations in the strength and balance of controlling factors.' (pp. 47–48)
Chorley and Kennedy, 1971. 'A mathematical expression fitted to data distributed in space, by means of a least sum-of-squares regression, to produce a three dimensional surface.' (p. 358)

**Triangulation**
*O.E.D.* The tracing and measurement of a series or network of triangles in order to survey and map out a territory or region.
*Webster.* The series or network of triangles into which any portion of the earth's surface is divided in a trigonometrical survey; the operation of measuring the elements (mainly angles, with a theodolite) necessary to determine these triangles, and thus to fix the positions and distances apart of their vertices. The measurement of the *base-line*, to which all other measurements and calculations are referred, is no part of the triangulation proper.
See also Base-line, Trigonometrical point.

**Tributary**
*O.E.D.* Furnishing subsidiary supplies or aid; subsidiary, auxiliary, contributory; also said of a stream or river which flows into another.
*Comment.* Generally used as a noun, applied to a river which joins a larger one.

**Trigonometric(al) point, trigonometrical survey**
Trigonometrical Survey: *O.E.D.* a survey of a country or region performed by triangulation and trigonometrical calculation.
Trigonometrical point: a fixed point determined (astronomically) with great accuracy in the triangulation, the vertex of a triangle.

**Trip generation**
Abler, Adams and Gould, 1971. 'A trip generation analysis attempts to identify and quantify trips beginning and ending . . . within a . . . study area. Trip generation analysis, therefore, establishes the functional relationship between trip volumes at the ends of the trips—the origins and destinations—and the land use and socio-economic character of the trip ends.' (p. 214)
See also Desire lines.

**Troglodyte**
An inhabitant of a cave or rock shelter, man or animal.

**Tropic, Tropics**
1. Each of the two parallels of latitude, respectively the Tropic of Cancer and Tropic of Capricorn, roughly $23\frac{1}{2}°$N. and $23\frac{1}{2}°$S. (actually 23° 28') north and south of the equator.
2. The Tropics, the region between these two parallels, the Torrid Zone (see Zone)—the belt within which the sun shines vertically at noon for at least two days in the year.
3. The Tropics—roughly as the last but by some writers including adjacent areas since major climatic and other changes take place more nearly at 30°N. or 30°S., than at $23\frac{1}{2}°$. Some writers exclude as distinct the Belt of Calms or Doldrums or the Equatorial Belt.

## Tropical
1. Of or pertaining to the Tropics in sense (1) above.
2. More commonly in reference to the zone as (2) or (3) above.

## Tropical Air Masses
Bergeron, T., 1928, distinguished as two main air masses:—
Maritime Tropical (mT), warm and moist, originating in the Trade-wind belt and subtropical waters of great oceans.
Continental Tropical (cT), hot, very dry and unstable, originating in low latitude deserts, especially the Sahara and Australian deserts. mT air masses affect lands far outside the tropics.
See Hare, 1953, 59–64.

## Tropical Climate
As a climatic type it is usual to exclude the Equatorial.
Miller, 1953. 'Beyond the swing of the equatorial rainfall belt, in the latitudes where the trade winds blow throughout the year, are the greatest deserts of the world. Between these and the equatorial climates lies a belt which, for part of the year, is under the influence of the trades, but for the rest of the year experiences an invasion by the belt of convectional rains; here are found the tropical climates with alternate trade-wind and doldrum influences... two fundamental types of tropical climate, continental and marine, the one with, the other without, a pronounced dry season' (p. 118).

## Tropical Forest
Mill, *Dict.* The natural vegetation covering those wooded parts of the tropics which have a dry season; towards the end of the dry season the leaves fall, and it is during this season that flowering occurs. The lianes and epiphytes are generally xerophilous. Stages intermediate between the tropical forest and the equatorial forest occur in the damper spots.
James, P., *An Outline of Geography*, Boston: Ginn, divides Tropical Forest into Tropical Rain Forest, Tropical Semi-deciduous and Tropical Scrub Forest (pp. 51–52).
*Comment.* Mill, like Miller (see above under Climate) distinguished thus Tropical from Equatorial.

## Tropical Grasslands—see Savanna, Llanos, Campos

## Tropical Revolving Storms
Hare, 1953. 'Though violent storms occur quite frequently in the doldrum belt, they are usually of a localized character... greatest by far in significance are the tropical revolving storms, which are probably the most intense storms affecting the surface of the earth, though they are quite small. They occur in tropical rather than equatorial latitudes, chiefly over the western sides of the great oceans. In each area affected by the storms there is a special local name for them. In the Atlantic they are "hurricanes"; in the western North Pacific. "typhoons"; in the South Pacific "hurricanes", "cyclones" or "willy-willies" (the latter north and west of Australia); in the Indian Ocean and the Bay of Bengal, "cyclones". The typical storm is small but deep, with very low central pressure. An intense cyclonic circulation is set up about the centre, winds attaining 70–80 m.p.h. quite frequently and over 100 m.p.h. on occasion. Near the centre is an area of calm, the "eye" of the storm.... Thunder may occur, but it is not very common. Areas directly on the track of the storm may very easily get 5 inches of rain in 24 hours.... Tropical revolving storms usually move relatively slowly... may cause untold damage... what mechanism leads to the formation is doubtful... seasonal... commonest during the late summer or early autumn' (pp. 111–112).
Hare notes that the 'haboobs' of the Sudan and the line-squalls or 'tornadoes' of West Africa are different.

## Tropicality—see Humid Tropicality

## Tropopause
The upper limit of the troposphere (*q.v.*) where temperature ceases to fall with increasing height; see also Stratosphere.

## Tropophyte
*Webster. Ecology.* A plant thriving under alternating periods of dryness and moisture or of heat and cold, as the deciduous trees of temperate regions, which drop their leaves in winter, and those of the tropics which are without foliage in the dry season.

## Troposphere
*Webster. Meteorol.* All that portion of the atmosphere below the stratosphere. It is that portion in which temperature generally rapidly decreases with altitude, clouds form, and convection is active.
Hare, H. K., 1953, *The Restless Atmosphere*, London: Hutchinson. 'This lower part of the atmosphere characterized by positive lapse rates is called the *troposphere*.' (p. 15)

## Trough
1. An elongated region of low barometric pressure between two areas of higher (meteorology).
2. The hollow between two waves.
3. A syncline (geology).
4. An elongated valley or trench.

## Trough fault
A structure resulting from two parallel normal faults, between which a block of country has been let down ... a Graben (q.v.).

## Truck farming
*O.E.D.* Truck (From French, troquer, to truck, shop, barter, exchange)
4c. U.S., Market-garden produce; hence as a general term for culinary vegetables. 1891. N.Y., Weekly Witness, 22 Apr.2/2. A distinction is made between truck farming and what is known as market-gardening ... Truck-farming is defined as the production of green vegetables on tracts remote from market.
*Webster.* Truck farm. A farm on which vegetables are raised for the market.
*Dict. Am.* Truck. Vegetables, garden produce. 1805 Parkinson *Tour* 161 'what in that country is called truck, which is garden produce, fruit, &c.' Truck farm. 1866. 'A truck garden, a truck farm, is a market-garden or farm.'
Whitbeck, R. H. and Finch V. C., 1935. *Economic Geography*, New York: McGraw-Hill. Truck farming is the production, usually specialized, of vegetables for sale in areas particularly suitable, at a distance from markets but connected by good transportation. Market gardening however is less specialized and nearer to market. (pp. 49–51)
Taylor. 'Horticulture some distance from markets.' (p. 621)
*Comment.* Whatever the origin of the word 'truck' in truck-farming and how it came to be used for vegetables in the United States, there is no doubt that truck-farming is now associated with the use of motor-trucks for the transportation of the produce and that truck-farming implies a greater specialization on a few commodities, grown at a greater distance from markets, than is associated with market-gardening (q.v.).

## Truncated soil (Soil science)
Jacks, 1954. 'Having lost all or part of the upper horizons' (*i.e.* by erosion).

## Trusteeship Territory, Trust Territory
S.Y.B., 1959. 'The Charter of the United Nations provides for an international trusteeship system to safeguard the interests of the inhabitants of territories which are not yet fully self-governing ... these are called trust territories.' They include the former mandated territories (q.v.) not yet independent and certain additions resulting from the Second World War. They are: New Guinea (Australia), Nauru (Australia, New Zealand and U.K.), Western Samoa (New Zealand), Cameroons (France and U.K.), Togoland (France and U.K.), Tanganyika (U.K.), Ruanda Urundi (Belgium), Somalia (Italy), Pacific Islands formerly under Japanese mandate (U.S.A.).

## Tsunami (Japanese)
Lobeck, 1939. A large wave at sea caused by earthquakes or a volcanic explosion (p. 685).
*Comment.* 'A seismic sea wave or a sea wave generated by an earthquake.' (G.T. in MS.) Popularly but erroneously called a tidal wave.

## Tufa
A porous, concretionary or compact formation of calcium carbonate, deposited around a mineral spring (compare Sinter).

## Tuff
A rock formed of fragments of volcanic origin, dust, ashes, etc. thrown out of a volcano during an eruption, frequently hardened to a rock. Hence tuff-cone, a volcanic cone built up of such material.

## Tumulus
Thornbury, 1954. 'Tumuli—bulging mounts on the surface of a lava flow with cracks in their crests through which lava may have been extruded. Tumuli do not seem to be gas blisters but rather are domings of the lava crust produced by the resistance which the lava surface has offered to the spreading of more fluid lava below and are thus much like laccoliths in origin.' (pp. 492–493) See R. A. Daly, *Igneous Rocks and their origin*, New York, McGraw-Hill, 1914, 133–134).
Ollier, C. D., 1964, Tumuli and Lava Blisters of Victoria, Australia. *Nature*, 202 (4939), 1284–1286. 'In volcanic terminology, tumuli are small mounds or hummocks on a lava plain. They are not gas blisters, but are full of solid lava ... up to 30 ft. high and 60 ft. across the base. They often rise very steeply from the plain but gently sloping domes are also found.' (p. 1284)

## Tumulus (2)
*S.O.E.D.* An ancient sepulchral mound, a barrow.

**Tundra** (Russian from Lap)
*O.E.D.* (From Lap, tundra) One of the vast, nearly level, treeless regions which make up the greater part of the north of Russia, resembling the steppes farther south, but with arctic climate and vegetation. Also applied to similar regions in Siberia and Alaska.
*Webster.* One of the level or undulating treeless plains characteristic of northern arctic regions in both hemispheres. The tundras mark the limit of arborescent vegetation; they consist of black mucky soil with a permanently frozen subsoil, but support a dense growth of mosses and lichens, as the reindeer moss, and dwarf cespitose herbs and shrubs, often showy-flowered.
Mill, *Dict.* A treeless plain with small lakes scattered over it, and having a special, generally scanty, vegetation (N. Russia and America).
Küchler, A. W., 1947, Localizing Vegetation Terms, *A.A.A.G.*, 37. '... all vegetation on the polar side of timber line ... the boundary between the Taiga and the Tundra is frequently a broad transition zone which the Russians call Taibola. It is entirely acceptable to use the term Tundra in circumpolar fashion and this practice has been generally adopted.' [But not to be used of vegetation in high altitudes in low latitudes.] (pp. 205-206)
Weaver, J. E. and Clements, F. E., 1938, *Plant Ecology*, New York: McGraw-Hill. An arctic and alpine vegetation climax; the *Carex-Poa* formation. (pp. 481-487)
*Comment.* Given as a 'treeless plain' etc. in *Met. Gloss.*, 1944; Moore, 1949; *Adm. Gloss.*, 1953. In North America those who, led by Stefansson, see an important economic future for the tundra there have sought to popularize the name 'Arctic Prairies' in place of the old 'Barren Lands'.
Russian sources (see Appendix II, Russian words) give:
Ushakov. Treeless expanses of the subpolar regions, in the zone of permafrost, usually marshy, mossy, stony, or covered with low vegetation.
Murzaev. From the Finnish word *tunguri*. Literally a treeless level upland or high mountain. The term entered scientific literature to designate the geographical zone of the Far North of Eurasia.
Barkov. Type of landscape including treeless stretches north of the boundary of the zone of forests.
Vasmer. From Finnish *tunturi* or Lappish *tundar*.

**Tundra Soil**
Jacks, 1954. Dark coloured soil with highly organic surface horizon and a frozen subsoil.
*Comment.* Subsoil is typically but not necessarily frozen.

**T'ung** (Chinese)
The wood-oil tree, *Aleurites cordata*; tung-oil is a distinctive Chinese product used for fuel and a variety of other purposes.

**Tung oil**
*Webster.* a. A poisonous, pungent fixed drying oil obtained from the seeds in the nuts of the tung tree.... It is used as a substitute for linseed oil in paints, varnishes, and linoleum, as a waterproofing agent, etc.

**Turbary**
Mill, *Dict.* A place where peat has been dug.
*Turbary rights*, or rights of turbary: rights to cut peat for fuel. See Common rights.

**Turbidity currents**
Kuenen, 1955. 'They arise when sediment is locally churned up, raising the density of the water to higher values than that of the surrounding clear water. Under the influence of gravity, the heavy water will flow down any available slope and spread out on a horizontal floor. Turbulence due to the flow will tend to keep the sediment in suspension until the flow comes to a standstill.' (p. 52) 'A huge turbidity current charging down the continental slope, faster than the swiftest rivers on land....' (p. 52) 'Allowing for the far greater dimensions of a torrent of this kind on the sea floor, the hypothesis that deep valleys were cut in the continental slopes by them becomes comprehensible.' (p. 312)
See also R. A. Daly, 1926, *Amer. Jour. Sci.*, **231**, 401-420.

**Turlough** (Ireland; from Gaelic turloch)
*O.E.D.* Ground covered with water in winter and dry in summer. See also Sluggy.
Sweeting, M. M., The Landforms of Northwest County Clare, Ireland, *Trans. Inst. Brit. Geog.*, **21**, 1955, 47-48. 'Periodically-flooded hollows, depending upon the fluctuations of the water-table, are known locally as turloughs.' Also used on plate facing p. 37.

**Turnpike**
*O.E.D.* A spiked barrier fixed in or across a road or passage as a defence against sudden attack, especially of men on horseback (Historical). A barrier, later a gate

or gates, placed across a road to stop passage till the toll is paid; a toll gate. Hence a turnpike-road; a road on which turnpikes were erected and tolls collected.

Comment. Many were constructed and maintained in Britain in the 18th century. Later the road itself became known as a turnpike and the word has been resuscitated in America. The modern turnpikes of the United States are typically toll highways without any crossroads at grade and with traffic in the opposite directions separated by a median grassy strip. It is possible to drive from New York to Chicago on such turnpikes without stop lights or cross traffic.

**Tussock-grass**—see Bunch-grass

**Twilight fringe**
Coleman, Alice, 1969, The Planning Challenge of the Ottawa Area, *Geographical Paper No. 42*, Ottawa: Department of Energy, Mines and Resources. 'The wilderness margin of the farmscape... Isaiah Bowman spoke of the "pioneer fringe" which refers to its active expansionary phase while the succeeding static phase was designated "chronic pioneer development". When the static condition gave rise to widespread agricultural retreat there was much use of the term "marginal agricultural area" but this is too wide in scope as it covers both the rurban and the wildscape margins. A more satisfactory term, "agricultural twilight" was introduced by Maxwell (1966) and this has been used as the basis of "twilight fringe". "Twilight" emphasises the duality of competing farm and forest uses, and also conveys a sense of the "evening of life" which is true both for the worn-out marginal land and for the predominance of elderly people left behind in this zone of out migration.'

**Two-field system**
A primitive system of agriculture in which half the land was cultivated, half left in fallow each season.

**Twyn** (Welsh)
*Lit.* hillock or knoll; usually earthy. From the same root as tywyn or towyn but the meanings have diverged.

**Tywyn, towyn** (Welsh)
Hillocks along the coast, usually sandy. *Cf.* old Cornish towan.

**Tyddyn, ty'n** (Welsh)
Small farm, holding; the dwelling of a gwely (*q.v.*).

**Type of farming**
Stamp, 1948. The Economics Branch of the Ministry of Agriculture and Fisheries prepared a Types of Farming Map based on farm organisation and practice. The type of farming was determined first by the proportion of land which was under the plough or under permanent grass and secondly the dominant enterprise. (pp. 298–300)

Comment. For the United States see *Generalized Types of Farming in the United States*. Prepared by the Bureau of Agricultural Economics. Agriculture Information Bulletin No. 3 (Washington, 1950), and F. F. Elliott, *Types of Farming in the United States* (Washington, Bureau of the Census, 1933).

**Typhoon**
*O.E.D.* Two different oriental words are included here. i. Urdu, tūfān, a violent storm of wind and rain, a tempest, hurricane, tornado, commonly referred to Arab., tāfa, to turn round, but possibly an adoption of Greek, Typhon; ii. Chinese, tai fung, from ta, big, and feng, wind.
a. A violent storm or tempest occurring in India (occas. with reference to other localities); b. A violent cyclonic storm or hurricane occurring in the China seas and adjacent regions, chiefly during the period from July to October.

Mill, *Dict.* A prolonged cyclonic storm of great intensity, occurring in the China seas and their environs in July, August, September and October.

For a modern explanation see Hare, 1953, p. 110, quoted under Tropical Revolving Storm.

**Typology**
*Lit.* The study of types; has been applied in various fields but especially to printer's type, in archaeology and as equivalent to symbolism. A recent geographical usage is relative to towns; see Urban typology.

# U

**Ubac** (French; dialect)
Mill, *Dict.* Ubac or Envers. The mountain slope facing more or less northward and to a large extent left by man in the forest condition (Alps).
*Comment.* Ubac, like adret (*q.v.*), is firmly established in international geographical literature. For a detailed study see Garnett, A., 1937, Insolation and Relief, *Inst. Brit. Geog.*, Pub. 5.

**Ubehebe**
A crater formed by the expulsion of volcanic ash, lapilli etc. round a volcanic vent. From the Ubehebe craters of the northern end of Death Valley, California. See Cotton, 1944 and von Engeln, 1935.

**Uinta Structure**
Mill, *Dict.* A broad flattened flexure from which the strata descend sharply on either flank and then run horizontally again (Uinta Mountains, U.S.A.).
Holmes, 1944. 'A classic example of subsidence and uplift on a stupendous scale.' (p. 422, with section)

**Uitlander** (Afrikaans; *lit.* outlander, foreigner)
A word which came into prominence in South Africa during the Anglo-Boer war when all foreigners or aliens were given this designation in the Boer Republics.

**Ultrabasic rock**—see Basic rock

**Umland** (German; *pl.* not usual; *lit.* around land)
(Urban Hinterland; sphere of influence.)
Van Cleef, E., 1941, *Geog. Rev.*, **31**. 'The area contiguous to a trade centre (extending to and including its suburbs or "urblets") whose total economic and cultural activities are essentially one with those of the primary centre.' (p. 308) (Umland is traced in German dictionaries to 1883, but without a precise, technical, meaning. It was used by A. Allix for the 'Economic domain' round an inland city, equivalent to a hinterland, which he would reserve as appertaining to ports. A. Allix, *Ann. de 'Univ. Grenoble*, **26**, 1914, 359–394, and *Geog. Rev.*, **12**, 1922, 532–569.)
Singh, R. L., 1955, Banaras: *A Study in Urban Geography*, Banaras. 'The area in which the region and the city are culturally, economically and politically interrelated forms the "umland" of the particular town or city.' (p. 116) Singh discusses the origin and use of the word at length.
Not in *O.E.D.*

**Unaka**
Lobeck, 1939. 'Groups of monadnocks are called unakas after the Unaka Mountains in the Southern Appalachians.' (p. 633)
Thornbury, 1954. 'Occasionally the unreduced remnants of the original surface rather than existing as single isolated hills form sprawling masses or groups. Such forms have been called unakas from the Unaka Mountains of North Carolina. . . .' (p. 181)

**Unconformable, unconformity**
*Webster.* Unconformity. 3. *Geol.* a. Lack of continuity in deposition between strata in contact, corresponding to a period of diastrophism and erosion and consequently to a gap in the stratigraphic record. It is often, but not always, marked by a want of parallelism between the strata of the two series.
In geology unconformable implies literally that the disposition of one set of bedded rocks does not conform in dip and strike with another underlying set. The plane or the division between the two sets is an unconformity and implies a break in the geological record. The lower set of beds was laid down, then uplifted, tilted, warped or folded and denuded to a greater or less degree before the deposition of the upper series.

**Undation theory** (R. W. van Bemmelen; in Dutch 1932, in English 1933)
Van Bemmelen, R. W., 1933. The undation theory of the development of the earth's crust. *Proc. 16th Int. Geol. Congr., Washington*, **2**, 965–982 (pub. 1935). The theory is discussed in P. H. Kuenen's *Marine Geology*, Chapter 2 and in L. U. de Sitter's *Structural Geology*, 489–500. The basic idea is the creation of crustal waves as a result of magma differentiation in the deeper parts of geosynclines.

### Undercliff
*O.E.D.* A terrace or lower cliff formed from landslips caused by the action of rain and sea.

*Comment.* The famous example is along the south coast of the Isle of Wight where the Cretaceous rocks are dipping gently seawards. The surface of the wet gault clay is permanently slippery with the result that great masses of the well-jointed chalk above it continue to slip and give rise to a large unstable area of rock debris below a cliff face. It is neither a 'terrace' nor a 'lower cliff'.

### Undercut slope, undercutting
Cotton, 1922. The steeper slope on the outside of the curve of a winding stream; opposite to the slip-off slope. (p. 111)

*Comment.* Undercutting of slopes, cliffs etc. can be the work of a meandering stream; it is also associated with desert erosion where winds laden with sharp sand particles undercut exposed rocks and is especially associated with wave action along cliffed sea coasts. See Holmes, 1944, 162, 258, 287.

### Underdeveloped, undeveloped, lands
In his Inaugural Address of 20 January 1949 President Truman enunciated his famous Point IV in these words 'Fourth, we must embark on a bold new program for making the benefits of our scientific advances and industrial progress available for the improvement and growth of underdeveloped areas'. What had previously been words with a very general meaning 'under-developed' and 'undeveloped' took on a special significance. The words were substituted for 'backward', held to cast something of a slur and in due course themselves gave way in part to 'less-developed'. Geographically 'the obvious interpretation of under-developed is that natural resources have not been developed to the full extent possible' (Stamp, L. D., 1953, *Our Undeveloped World*, London: Faber, p. 19) but an official definition used to determine countries qualifying for American Point IV aid was national income per capita. The position in 1959 was summarized by C. Langdon White (The World's Underdeveloped Lands, *Focus*, New York, 10): 'The three criteria are per capita income (less than £150 per annum), daily consumption of calories (less than 2,500 per day) and annual utilization of inanimate energy (only about 1/40 as much as the more advanced countries).' See also Stamp, L. D., 1960, *Our Developing World*, London: Faber.

From an economist's angle J. Viner (*International Trade and Economic Development*, Oxford: Clarendon Press, 1953) discusses the problem of definition at length and concludes '... a country which has good potential prospects for using more capital or more labour or more available natural resources, or all of these, to support its present population on a higher level of living, or, if its per capita income is already fairly high, to support a larger population on a not lower level of living.' (p. 98) He rejects equation of 'under-developed' with (a) youth, (b) low density, (c) low ratio of industrial to total output and (d) scarcity of capital shown by high interest rates.

### Underfit (River)
Cotton, 1922. A river which appears too small to have eroded the valley in which it flows because of loss of volume due to seepage, or 'underflow', through thick alluvium of the flood plain. (p. 120)

Wooldridge and Morgan, 1937. A beheaded consequent will 'become a misfit or underfit river, too small for the valley in which it flows.' (p. 185)

Cotton, 1945. Equivalent to misfit, a term not used, without the distinction made in Cotton, 1922. (W.M.) (pp. 106–109)

Davis, W. M., 1913. Meandering Valleys and Underfit Rivers, *A.A.A.G.*, 3, '... the peculiar relation that is frequently observed between the small-curved meanders of a river and the larger-curved meanders of its valley. ...' (p. 3)

No reference in Davis, 1909; Salisbury, 1907; Geikie, 1898; von Engeln, 1942; Strahler, 1951; Rice, 1941; Fay, 1920. See also misfit river.

### Uniclinal
Applied to strata dipping steadily in one direction. The same as monoclinal as used by some authors but more commonly monocline is applied to a *fold* with only one limb.

### Uniclinal shifting
Wooldridge and Morgan, 1937. The process of asymmetrical development of a valley due to a stream which follows the line of the geological strike tending to cut sideways in the direction of the dip. (p. 159)

No reference in Davis, 1909; Salisbury,

1907; Geikie, J., 1898; Cotton, 1922; von Engeln, 1942; Holmes, 144; Strahler, 1951.

*Comment.* In a footnote Wooldridge and Morgan explain their preference for 'the word "uniclinal", implying a uniform dip, in one direction' to monoclinal which is liable to cause confusion. A diagram is given and the Lea Valley north of London quoted as an example.

### Uniform regions

Hartshorne, R., 1959, *Perspective on the Nature of Geography*, published for the Association of American Geographers, Chicago: Rand McNally. 'Until quite recently major attention was focused on the concept of regions as areas of homogeneity—that is, of approximate uniformity of character ... homogeneous regions differ in concept in at least three logical respects: (1) whether determined on the basis of one or more independent elements or in terms of demonstrated integrations of two or more elements; (2) whether considered in unique terms for each specific region or by generic criteria which may be applicable to any number of similar regions and (3) whether part of a system dividing the entire area under study, or the world, into discrete parts, or one which takes account only of certain parts of the area, omitting other parts.' (pp. 131–132)

Abler, Adams and Gould, 1971. 'Areal classifications of static phenomena usually produced regions which were composed of homogeneous phenomena and which were relatively uniform throughout.' (p. 85)

### Uniformitarianism

Mill, *Dict.* The theory that all changes in and on the earth's crust are due to the fairly uniform action of the forces now at work.

*Comment.* The principle was finally established by Charles Lyell through the publication of his *Principles of Geology* and *Elements of Geology* in the 1830's. The opposite view was Catastrophism or Convulsionism; compare also Neptunism and Plutonism.

### Uninverted Relief

A term used of country where the surface relief reflects the underlying geological structure, where hill ridges mark the position of anticlines and valleys of synclines.

### Upland

Mill, *Dict.* The higher ground of a region contrasted with the valleys and plains. Used in contrast to lowland or lowlands (*q.v.*)

### Upland Plain

Wooldridge, S. W., 1950, The Upland Plains of Britain: their origin and geographical significance, *Pres. Address, Section E, British Ass. Adv. Science*, also in *The Geographer as Scientist*, London: Nelson, 1956, 124–149. 'It is usual to refer to the features in question as "erosion surfaces" or "erosion platforms" and in ordinary working parlance these terms are unobjectionable and too well entrenched to be easily replaced ... in using this term one begs no question at all and merely states an observed fact.' (pp. 129–130)

### Uplands

Mill, *Dict.* A denudation highland, the heights above the valleys being hills, not mountains.

*Comment.* Especially Scotland, as in Southern Uplands, but both upland and uplands are used very loosely. There is usually an implication of subdued relief and frequently also of an inland situation, away from the coast.

### Urban

*O.E.D.* (From Latin, urbanus, urbs, city. Rare before the 19th century)
1. Pertaining to or characteristic of, occurring or taking place in, a city or town; Constituting, forming, or including a city, town, or burgh, or part of such.
2. Exercising authority, control, supervision, etc., in or over a city or town; Residing, dwelling, or having property in a city or town.

Smailes, A. E., 1953, *The Geography of Towns*, London: Hutchinson. 'For his particular purpose the geographer must regard as urban a particular man-made type of landscape.' (p. 33)

Dickinson, R. E., 1932, Distribution and Functions of the Smaller Urban Settlements of East Anglia, *Geog.*, **17**. [An urban settlement] '... is usually defined as a nucleated settlement in which the majority of the occupied inhabitants are engaged in non-agricultural occupations, *i.e.* retail and wholesale trades, handicrafts, industries and commerce, and some arbitrary population figure is taken to distinguish an urban from a rural settlement. But the

definition of an urban settlement is fundamentally a question of function; not of population.'
*Comment.* Statistical definitions, etc. are given at length in: United Nations, Dept. of Social Affairs, Population Studies No. 8; Data on Urban and Rural Population in Recent Censuses. New York, 1950. Principal classifications are based upon number of inhabitants, type of local government, administrative functions. The International Statistical Institute, 1938, classified as 'urban' communes with less than 40% agricultural population. In England and Wales the 'urban' population is that which resides in cities, county boroughs, other boroughs and urban districts, leaving as 'rural' the population which resides in 'rural districts' even if they include towns of considerable size. For Scotland see 'landward'. (L.D.S.)

**Urban economic base (basic and non-basic concept)**
Mayer, H. M. in James, P. E. and Jones, C. F., 1954, *American Geography—Inventory and Prospect*, published for the Association of American Geographers, Syracuse: Syracuse University Press. 'The functions which are performed in cities, and which lead people of certain culture areas to group themselves close together, include production, processing, exchange, and distribution of goods, and the provision of a variety of services for the people of a wide area outside of each city. The measurement of a city's economic base is an important phase of urban geography . . .' (p. 150) . . . 'Homer Hoyt . . . developed a two-fold classification of urban functions which he distinguished as basic and non-basic. The basic functions serve the areas outside of the respective cities or metropolitan areas; the non-basic functions serve the inhabitants of the respective cities or metropolitan areas. The extent to which a given activity is basic may be measured by finding the ratio of persons engaged in that activity to the total population of the city, and then comparing this with a similar ratio for the country as a whole.' (p. 151)
Alexander, J. W. in Mayer, H. M. and Kohn, C. F., 1959, *Readings in Urban Geography*, Chicago: University of Chicago Press. 'A city's basic activity links the settlement with other portions of the earth's surface; non-basic endeavours link the settlement with itself . . . it classifies economic functions fundamentally on the basis of space relationships, it reveals one group of economic ties which bind a city to other areas, it permits a classification of and comparative analysis of settlements, and it provides additional method for classifying individual economic activities within a city.' (p. 100)
See Hoyt, H., 1944, *The Economic Status of the New York Metropolitan Region in 1944*, New York.

**Urban fence**
Stamp, 1948. 'A useful device which was developed by the Planning Branch of the Ministry of Agriculture was the concept of the urban fence or urbanised land. It is possible to draw around any town or other large settlement a line including within it all land where urban influences are dominant—where the land is covered with industrial works, public buildings, offices, ships and housing, including the houses with gardens, and where such open land as exists is devoted to primarily urban uses—as playing fields, parks and cemeteries. The future development of land within the urban fence is thus a matter essentially for the town planner in that agricultural and rural interests are little if at all affected.' (p. 195)

**Urban field, urban region**
Smailes, A. E., 1947, *Geog.*, **32**. 'the territory functionally linked with a town. . . .' Used in preference to umland (unnecessary importation), sphere of influence (circumlocution), catchment area (incomplete description). (pp. 151–161)
Mayer, H., 1954, *Econ. Geog.*, **30**. (Review of Smailes, *Geography of Towns*) 'The "Urban Field" is a term used by Smailes to mean the hinterland or trade area, which in this country is commonly called the urban region.' (p. 278)
Gilbert, E. W., 1950, *Geog. Jour.*, **116**. Prefers 'urban field' to the alternatives. (p. 88)
See also Umland.

**Urban fringe**
Pahl, R. E., 1964, *Urbs in Rure: the Metropolitan Fringe in Hertfordshire*, London School of Economics and Political Science Geographical Papers No. 2. 'The outer part of a metropolitan region is seen, then, as a frontier of social change, moving over

communities, re-evaluating their spatial relationships with other communities and creating, as it were, new places, which in turn form the bases for different types of communities. These new communities are not easily defined . . .' (p. 73). 'The fringe, then, is defined in relation to the city and spatially exists in the agricultural hinterland where land use is changing. The population density is increasing rapidly and land values are rising. Ecologically it could be said to be an area of invasion . . . In addition to the changing land use, the occupations of the people living there are predominantly non-agricultural and the incidence of commuting is high.' (p. 74)

Conzen, M. P., 1971, *Frontier Farming in an Urban Shadow*, Wisconsin: Worzalla Publishing Co. for the Dept. of History, University of Wisconsin. 'Proximity to the town site . . . a potent factor in the transfer of public domain to private ownership. Once in private hands, land near the city became subject to the free play of different market forces. A growing city stimulates land-use conflict on its fringe between agricultural and urban utilization. The higher bidding power of urban uses for these peripheral sites results inevitably in the territorial recession of farm land.' (p. 17)

## Urban fringe belt
Conzen, M. R. G., 1962, The Plan Analysis of an English City Centre, in Norborg, K. (ed.), *Proc. I.G.U. Symposium in Urban Geography, Lund 1960*, 383–414. 'Thus a characteristic accumulation of latecomers such as Nonconformist chapels, new community services and growing industries reorganized on larger sites, together with existing open spaces, formed now a more or less continuous, rapidly expanding urban fringe belt . . .' (pp. 391–392)

*Comment.* For the original concept of the *Stadtrandzone* or urban fringe belt see Louis, H., 1936, Die geographische Gliederung von Gross-Berlin, in Louis, H. and Panzer, W., 1936, *Länderkundliche Forschung*, Stuttgart: S. Engelhorns Nachf., pp. 146–171.

## Urban Geography
A study of site, evolution, pattern, and classification in Villages, Towns and Cities. This is the sub-title of Griffith Taylor, *Urban Geography*, London: Methuen, 1949.

See also Griffith Taylor's chapter on Urban Geography in Taylor, 1951 and M. Aurousseau, *Geog. Rev.*, **14**, 1924.

## Urban Hierarchy
Smailes (1946—see Urban Mesh) distinguishes on a map of England and Wales between Major Cities, Cities, Minor Cities or Major Towns, Towns and Sub-towns.

Taylor, T. Griffith, 1949, *Urban Geography*, London: Methuen, discusses various classifications of towns in particular his 'seven ages'—infantile, juvenile, early mature, mature, late mature, and senile (apparently only six—pp. 76–77). See also megalopolis. In the chapter on Urban Geography in Taylor 1951 (pp. 524–525) he describes briefly the Seven Ages of a Town (introduced in an address to the American Association of Geographers in 1941) as sub-infantile, infantile, juvenile, adolescent, mature, late mature, and senile.

## Urban mesh
Smailes, A. E., 1946, The Urban Mesh of England and Wales, *Trans. Inst. Brit. Geog.*, No. 11. The geometrical pattern of the relations of urban centres. The term is derived from Christaller's theoretical pattern for '*zentralen Orte*' which, according to the '*Versorgungsprinzip*' would be arranged in symmetrical fashion in the form of a hexagonal mesh. Other factors, *e.g.* '*Verkehrsprinzip*' and '*Absonderungsprinzip*' would modify the hexagonal shape. (W.M.) (p. 87)

## Urban region
Van Cleef, E., 1937, *Trade Centers*. 'A trade center plus its continuous hinterland may be referred to as an urban region.' (p. 31)

Smailes, A. E., 1953, *Geography of Towns*, 1953. [On the internal geography of towns] 'Differences in either or both these intimately related aspects of urban morphology, function and form, give a basis for the recognition of urban regions.' (p. 84)

Mayer, H., 1954, *Econ. Geog.* (In review of Smailes, 1953) '"Urban Regions" as used by Smailes refers to relatively homogeneous land-use areas within cities, and not to the region for which the city serves as the focus, as the term is customarily applied in America.' (The American 'urban region' is equivalent to Smailes's 'urban field'.) (p. 278)

*Comment.* Not to be confused with the City-region of Geddes which = conurbation.

As used by Smailes the *internal* region of a city would seem to be the same as 'quarter' —see Wise, M. J., 1949, On the Evolution of the Jewellery and Gun Quarters in Birmingham, *Trans. Inst. Brit. Geog.*, 15, 59–72.

**Urban tract** (R. E. Dickinson, 1947)

Dickinson, R. E., 1947, City, Region and Regionalism. 'The urban tract is used to define the compact and continuous urban built-up area in preference to the term "conurbation".' [To remove consideration of administrative units and to include some of the outlying places, on a functional rather than a morphological criterion.] pp. 168–169.

*Comment.* In view of the use of 'tract' by J. F. Unstead, it is difficult to avoid confusion and 'urban tract' has not been widely adopted.

**Urban Typology**

Some writers prefer to use 'urban typology' rather than 'urban hierarchy' on the grounds that 'hierarchy' implies some degrees of dependence or interdependence whereas 'typology' is non-committal—the study of types.

**Urbanism, urbanity**

*O.E.D.* urban character.

*O.E.D. Suppl.* Quotation of 1929 'it denotes town-planning etc.'

*Comment.* 'The term *urbanism* is not greatly used. It would refer to the typical condition or town character. Care should be taken not to confuse the French term *l'urbanisme*, the science or art of town planning, with the English *urbanism*' (A. E. Smailes in MS.).

**Urbanistics** (Eugene Van Cleef, 1957)

Van Cleef, E., 1957, *The Professional Geographer*. 'Urbanistica (Italian) means "the technique of city planning". By substituting "s" for the ending letter "a" in Urbanistica the word "urbanistics" is derived, a word which conveys a sense of vigorous action and fulfils the desirability of an expression which truly reflects the spirit of the field of urbanism which ... has attracted the attention of ... geographers. ...' (p. 2)

**Urbanization**

*O.E.D.* The process of investing with an urban character; the condition of being urbanized. 1888 *Advance* (Chicago) 8 Mar. 152 One of the most remarkable characteristics of the time is 'the urbanization of the country'.

Reissman, L., 1964, *The Urban Process— Cities in Industrial Societies*, New York and London: The Free Press, Collier-Macmillan. 'Urbanization, Wirth argued, meant much more than the attraction of people to live in the city. It had to mean, as well, the accentuation of the urban way of life and its influence upon people.' (p. 114). 'Radfield argued that urbanization meant individualism. The individual would become more self-centered and self-concerned at the expense of his community ties and obligations' (p. 135). 'Urbanization means not only the transformation of rural, agricultural, or folk society, but also the continuous change within the industrial city itself. Urbanization does not stop but continues to change the city into ever different forms.' (pp. 155–156). See also pp. 208–209 for Reissman's four stages of urbanization.

See Wirth, L., 1956, Human Ecology, in *Community Life and Social Policy*, Chicago: University of Chicago Press.

Carpenter, D. B. and Queen, S. A., 1953, *The American City*, New York, McGraw-Hill. Urbanization is the process of transition or the position on a continuum from a rural to an urban social psychological condition, *e.g.* from few, permanent, 'primary' social contacts to many, impermanent, 'secondary' social contacts. From this definition, the size of a population aggregate or its density, etc., are only measures of urbanization in so far as they reflect the character of social contacts. However an index of urbanism, based on size of population groups, is provided on pp. 29–31. (W.M.) (pp. 19–20)

Smailes, A. E., 1953, *The Geography of Towns*, London: Hutchinson, 1953. Becoming town dwellers, or the extent to which a population live in towns. But a town may be considered as a community following a distinctive way of life, or a distinctive part of the earth's surface. The latter must be the geographer's definition. (W.M.) (pp. 32–33)

*Encyc. of Soc. Sciences*, New York: Macmillan, 1935. 'Urbanization is characterized by movement of people from small communities concerned chiefly or solely with agriculture to other communities, generally larger, where activities are primarily centred in government, trade, manufacture or allied interests.' (p. 189) Taylor, 1951. 'Shift of people from country to city.' (p. 621)

*Comment.* 'The term *urbanization* denotes the concentration of an increasing proportion of the community in towns, with their resultant physical extension... used to refer both to the process *and* the state reached. (A. E. Smailes in MS.)

**Urban-rural continuum**
Smailes, 1953. 'There is no longer either socially or physically a simple clear-cut dichotomy of town and country; rather it is an urban-rural continuum.., there is no definite point where rural ends and urban begins.' (p. 33)
United Nations Population Studies, No. 8, 1950. 'The distribution is not really a twofold one, in which one part of the population is wholly rural and the other wholly urban, but a graduated distribution along a continuum... consequently, the line that is drawn between urban and rural for statistical or census purposes is necessarily arbitrary.' (p. 2) See Rurban.

**Urblet**
Van Cleef, E., 1937, *Trade Centers and Trade Routes*, New York: Appleton-Century. A suburb or trade center within the umland of a major centre. 'The urblet is more closely associated with the major center than is the satellite.' (pp. 35, 89–90).

**Urstromtal** (German; *pl.* Urstromtäler: ancient river valley)
The wide but shallow valleys excavated by the melt-water in the North German Lowland corresponding to static periods in the retreat of the edge of the Scandinavian ice sheet; four principal Urstromtäler are distinguished (K.A.S.). See Monkhouse, 1954, p. 152; called *Pradoliny* in Poland (*q.v.*).

**U-shaped Valley**
A valley which, in cross section has the shape of a U in contrast to the V-shaped section of a young river valley. The U shape is usually due to the work of a valley glacier; 'where valleys have not been entirely filled by ice, the upper slopes meet the ice-steepened walls in a prominent shoulder... the cross profile is like a U sunk in a V'. (Holmes, 1944, 220)

**Usar** (Indo-Pakistan: Hindi)
Saline lands.

**Uvala, ouvala, vala** (Serbo-Croat)
Mill, *Dict.* A sunken area larger than a dolina, having a broad bottom with irregular surface traversed by streams, and occurring in a limestone region of the Karst type (Slavonic).
Sanders, E. M., 1921, The Cycle of Erosion in a Karst Region (after Cvijić), *Geog. Rev.*, 11. 'As time goes on, the divisions between neighboring dolines are broken down; and larger depressions, called "uvalas" or "ouvalas" are created. A uvala is usually more than one kilometre in diameter....' (p. 600)
See Karst terminology. A uvala does not necessarily contain a stream. (G.T.W.)

# V

**Vadose water**
*Webster.* Vadose. *Geol.* Pertaining to or due to water or solutions in that part of the earth's crust which is above the level of permanent ground water; as, *vadose circulation*; *vadose* deposits.
Wooldridge and Morgan, 1937, 'The "wandering" or "vadose" water above the water-table varies greatly in amount and position.' (p. 268)

**Val** (French; *lit.* valley)
Used in the specialized sense of a synclinal valley in a range of fold mountains, originally in the Jura (E.D.L.).

**Vala**
The Montenegrin name for the uvala (*q.v.*) of Bosnia.

**Vale**
*O.E.D.* A more or less extensive tract of land lying between two ranges of hills, or stretches of high ground, and usually traversed by a river or stream; a dale or valley, especially one that is comparatively wide and flat.
*Comment.* The word is used in the above sense in place names *e.g.* the Vale of St. Albans, and in some cases applies simply to a gently undulating lowland *e.g.* the Vale of Glamorgan. It is also used poetically and is best avoided as a geographical term. (L.D.S.)

**Valley**
*O.E.D.* A long depression or hollow lying between hills or stretches of high ground and usually having a river or stream flowing along its bottom.
The extensive stretch of flattish country drained or watered by one or other of the larger river-systems of the world (*e.g.* Amazon).
*Webster.* An elongate depression, usually with an outlet, between bluffs, or between ranges of hills or mountains. A *river valley* is the depression made by the stream, and by the various processes which precede and accompany the development of a stream. A *structural valley* is a relatively long and narrow depression, produced by movements of the surface. Thus a downfold produces a *synclinal valley*, while a *rift valley* is caused by down faulting. The *valley flat* is the low flat land bordering a stream's channel.

**Valley axis**
Woodford, A. O., 1951, Stream Gradients and Monterey Sea Valley, *Bull. Geol. Soc. Am.*, **62**, 799-852. 'The term commonly used *talweg* . . . is here replaced by valley axis, because talweg has at least two other meanings. . . . The valley axis is the surface profile along the center line of the valley' (p. 803a).

**Valley-floor basement**
Davis, W. M., 1930, Rock Floors in Arid and in Humid Climates, *Jour. Geol.*, **38**. (Faintly sloping bedrock basements of valley floors developed in humid climates, invisible because covered with slow-creeping soil and flood-plain deposits. They underlie the lateral valley-floor strip which is formed between the flood plain and the valley-side slopes. They are developed by lateral extension of valley bottoms at the expense of the inclosing slopes.) pp. 1-27, 136-158
Wooldridge and Morgan, 1937, in criticism of Davis's concept of valley-floor basements, refer to it as 'the gently sloping ground above humid river flood-plains . . .' (p. 314).
See also Pediment.

**Valley glacier**
A glacier which occupies a valley as distinct from other types of glacier (*q.v.*).

**Valley line, valley-line**
The German word Talweg, as a geographical term, is sometimes rendered into English as valley line.

**Valley train**
Wooldridge and Morgan, 1937. 'The meltwater from the ice escapes as a stream or a series of streams which breach the terminal moraine, and redistribute its material downward along the valley in more or less distinctly stratified deposits known as valley trains, outwash plains or frontal aprons.' (p. 364)

Cotton, 1922. Similar to Wooldridge and Morgan, 1937 (p. 329).

Rice, 1941. Stratified glacial debris occurring in a narrow belt along the axis of a valley, outside the region occupied by a glacier. Fluvial plain in a valley built of outwash material by meltwater; used in Denmark for esker chains of a melt water valley system (Antevs and MacClintock).

**Valley wind, valley breeze**
An anabatic movement of air up a valley usually during the day and often alternating with a katabatic mountain wind or bergwind by night. It is the result of the differential heating of the mountains above and the plains below. See also Talwind.

**Valloni**—see Canali

**Vardarac**
Kendrew, W. G., 1953, *Climates of the Continents*, Oxford: Clarendon Press, 4th Ed., 'a wind of the mistral type, which sweeps down the valley of the Vardar to the Aegean in winter'. (p. 347)

**Varigradation**
Woodford, A. O., 1951, Stream Gradients and Monterey Sea Valley, *Bull. Geol. Soc. Am.*, **62**, 799–852. 'W. J. McGee (1891, pp. 261–267) stated that in Iowa and elsewhere alluviated meadows form along stream valleys, primarily due to damming by landslides or the fans of tributaries. In time each barrier and then the meadow above is channeled by the stream, leaving terraces. But as long as the new channel is cutting in the barrier it will have a relatively steep gradient, and so a new deposit is formed where the velocity is checked at the lower end of the little gorge. This deposit will in turn be trenched from below, making a series of nicks in the long profile. McGee called this process varigradation, and stated that it is realized in the pools and rapids of every mountain brook and in great rivers as well.' (p. 815b)

**Variscan orogeny**
A phase of the Armorican orogeny (*q.v.*) which has been distinguished like Malvernian (Trueman, 1947) as the complexities of the Armorican earth-movements have been realized. From Variscia (Latin) the Vogtland of Germany.

**Varv** (Swedish)
Older form hvarf, årshvarf, from Swedish varv, hvarf = turn round, revolution, row, tier, course, layer, varve, annual varve, distinctly marked annual deposits of sediment regardless of its origin. (G. De Geer) (E.K.)

**Varve** (from Swedish *varv*, layer)
*Webster. Geol.* An annual layer of silt as deposited in a lake or other body of still water. As individual layers differ in thickness and character, a succession of such layers forms a characteristic group which can be identified as of contemporaneous deposition in whatever deposit it may be found. It is thus possible, by combining different sections, to measure the time involved in the deposition of the entire group of sediments and to construct a time scale in a manner similar to that employed in the study of annual rings in trees.

A thin lamina found in sediments, especially those deposited in lakes during the retreat of the glaciers of the Great Ice Age, which consists of a thin very finely-grained layer (the result of the slow melting of the ice in winter) and a thicker coarser layer (produced by the more rapid melting of the ice in summer). Each varve thus represents a year and by counting the varves an exact chronology in years may be established as was done by De Geer in Sweden. (L.D.S.) Not in *O.E.D.*
Hence varve clays, varved clays, varved sediments.

Varves with bedding of graded particles are diatectic varves; with bedding of floculated, not graded, materials are syminct varves. See Boswell, P. G. H., 1961, *Muddy Sediments*, Cambridge: Heffer.

**Várzea** (Brazil: Portuguese)
Int. Geog. Congress, Brazil, 1956, Excursion Guidebook No. 8, Amazonia. 'Only a relatively reduced parcel of its immense total area is formed by the alluvial floodplains—the várzeas—discontinuously arranged along its rivers, the rest of the vast Amazon plain is free from flood waters, and consists of well drained uplands called *terra-firme* in the regional nomenclature' (p. 9). Also many other regional terms such as teso, barranco, igarapé, igapó, etc.

**Vasques, Plate-forme à** (French)
Guilcher, A., 1958, Coastal Corrosion forms in Limestone around the Bay of Biscay. *S.G.M.*, **74**, 137–149. 'In the spray and storm-wave zone, a pitted topography with jagged *lapiés* and intervening shallow ponds with overhanging sides ... whole zone, termed *plate-forme à vasques* in French, is gently sloping towards the sea.' (p. 138) No translation is given.

**Vauclusian spring**
Wooldridge and Morgan, 1937. 'An actively eroding underground stream [in Karst

country]... emerging at the foot of the steep valley walls as a "Vauclusian spring" from the famous Fontaine de Vaucluse in South France' (p. 291).

**Vega** (Spanish)—see Huerta

**Vegetation Terms**

In a paper entitled Localizing Vegetation Terms, *Ann. Ass. Amer. Geog.*, 37, 1947, 197–208, A. W. Küchler gave the following list of terms, all of which have been included in the Glossary:—caatinga, chaparral, garique (garrigue), heath, llano, loma, mallee, moor, muskeg, pampa, páramo, prairie, puna, puszta, sage brush, steppe, taiga, terai, tundra, veld. See also Küchler, A. W., A Geographic System of Vegetation, *Geog. Rev.*, 37, 1947, 233–240.

The following additional vegetation terms have been included in the Glossary: campo, carr, ceja, chaco, hylea, indaing, lalang, landes, maquis, montana, monte, mopane, mulga, pa-deng, savana, selva, spinifex desert, yunga.

See also under Ombrothermic.

**Vein**

The same as lode: the general term for a crack or fissure in the rocks of the earth's crust in which highly heated waters from below have deposited from solution crystalline minerals (especially vein quartz) and, under certain circumstances, metallic minerals of economic importance. The term is also used loosely by miners in various extended senses.

**Veld** (Afrikaans; *antiq.* veldt)

*O.E.D.* In South Africa, the unenclosed country or open pastureland.

*Comment.* Etymological equivalent of Nederlands veld, German Feld, English field. In Afrikaans the meaning has undergone a definite change. It indicates any form of natural vegetation more or less fit for grazing purposes, thus excluding cultivated lands, forests, or absolute desert. The term is used with many qualifications indicating:

(a) Topographical situation or configuration—High Veld, Middle Veld, Low Veld, mountain veld, bankeveld (*i.e.* veld with long parallel ridges of scarped hills). (b) Soil conditions—sand veld, hardeveld (hard veld), sour veld (deficient in lime), sweet veld. (c) Special forms of vegetation —grass veld, bush veld, Karoo veld. Rooigras-veld means red grass veld.

Many of these terms may be written with or without capital letters, according to their use as proper names (for regions where a special type of veld predominates), or generic names; *e.g.* bush veld is a type of vegetation; Bush Veld is a region in the northern Transvaal.

It should further be noted that there is a good deal of overlapping, different terms being applied to the same stretch of country according to an author's point of view. Sand veld may carry grass, bush, or semi-desert shrubs; but each of these forms of vegetation may occur on other soils as well. Grass veld may be sweet or sour, but sour veld may also carry a vegetation of shrubs, etc. In Afrikaans a composite term of this kind is always written as one word (bosveld); in English mostly as two words (bush veld). (P.S.)

**Vendavales** (Spanish)

Kendrew, W. G., 1953, *Climates of the Continents*, Oxford: Clarendon Press, 4th Ed., 'The Spanish coast, where the prevailing winds are NW., often has strong SW. "Vendavales", at times of gale force, which may give heavy rain and high seas.' [in winter] (p. 353)

*Webster.* An autumnal thunder squall on the coast of Mexico.

**Vent**

The orifice through which molten lava reaches the crater of a volcano. Eventually it may become choked as the lava solidifies to form a plug or neck.

**Ventifact**

*Webster. Geol.* A stone worn, polished, or faceted, by wind-blown sand; called also *glyptolith*, *rillstone*.

Himus, 1954. 'Pebbles shaped by the wind, generally under desert conditions. They are characterized by being bounded by several more or less flat facets, which meet at fairly sharp angles.'

Holmes, 1944, 'Where... the desert... is exposed to blown sand... isolated pebbles or rock fragments are bevelled on the windward side until a smooth face is cut. If the direction of the wind changes seasonally, or if the pebble is undermined and turned over, two or more facets may be cut, each pair meeting in a sharp edge. Such wind-faceted pebbles... their surfaces polished are known as *dreikanter* or *ventifacts*.' (p. 258)

*Comment.* Strictly *dreikanter* should only be

applied to those with three facets, hence ventifact is preferable as a general term.

**Vents d'Autan or Autan**
Mill, *Dict*. Strong, hot, dry winds blowing out from southern France towards the lows in the Bay of Biscay. See Kendrew, 1953, 261.

**Verano, veranillo** (Tropical South America; Spanish)
Respectively the long dry season and the short dry season (which breaks the rainy season). Verano is literally the summer, Veranillo the little summer. The latter is a period of drier and brighter weather which breaks the wet season as around Christmas in Ecuador.

**Verrou** (French)
See Riegel (German). Baulig gives as English equivalent simply rock-bar but see Baulig 291.

**Versant** (French)
The slope, side or descent of a mountain or mountain chain; often applied to a whole area with this general character *e.g.* the Pacific versant of the United States.

**Vertical Zones** (Latin America)—see Tierra caliente, fria, templada

**Vesuvian**—see under Strombolian

**Vidda** (Norwegian)
High, treeless undulating area. Used of wide areas of this general type of country (*e.g.* Hardangervidda), and more precisely of the old land surface predominant in East Norway. (P.J.W.)

**Vill**
*O.E.D.* A territorial unit or division under the feudal system, consisting of a number of houses or buildings with their adjacent lands, more or less contiguous and having a common organization: corresponding to the Anglo-Saxon tithing and to the modern township or civil parish.
Houston, J. M., 1953, *A Social Geography of Europe*, London: Duckworth. 'A territorial unit of lands and buildings having a common social organization.' (p. 249)
Fawcett, C. B., 1944, *A Residential Unit for Town and Country Planning*, London: U.L.P. A 'vill' is suggested as a residential unit, either urban or rural, for town and country planning. It would have a population of 1,200–2,400, sufficient to maintain a nursery school, primary school, community centre, café and 'pub'. It appears to be equivalent to 'neighbourhood unit', although smaller than many suggested. (W.M.)

No reference in Smailes, 1953; Taylor, 1949; Dickinson, 1947.
*Comment*. F. W. Maitland (*Domesday Book and Beyond*, Cambridge, 1897), speaks of 'the mere vill or rural township' and frequently uses the word in the sense of village. (H.C.D.)

**Villa**
*O.E.D.* Originally a country mansion or residence, together with a farm, farm buildings or other houses attached, built or occupied by a person of some position and wealth; a country seat or estate.
Latterly, any residence in the suburbs of a town or in a residential district, such as is occupied by a person of the middle class; also, any small better-class dwelling house.
*Comment*. When reference is to the country mansions established by the Romans in rural Britain the word is commonly in italics and the latin plural *villae* is used. Applied to more modern dwellings the word has fallen into disfavour and is best avoided. (L.D.S.)

**Village**
*O.E.D.* A collection of dwelling-houses and other buildings, forming a centre of habitation in a country district; an inhabited place larger than a hamlet and smaller than a town, or having a simpler organization and administration than the latter.
Mill, *Dict*. An agglomeration of population (usually small) with no municipal government.
*Webster*. 2. Specifically: a. In England, a tract of land with some houses, forming a unit for purposes of national police and taxation, and corresponding to the present civil parish. b. In the United States, such a collection incorporated as a municipality. c. Any of various territorial divisions incorporated as 'villages' under statutory authority, as under various civil codes in the United States, in some Provinces of Canada, etc..
Osborn, F. J., 1946, *Green Belt Cities*, London: Faber and Faber. 'The word village implies small scale, detachment, and (I suggest) a basis which is primarily agricultural. Garden village has been used as a name for a small settlement containing a factory and an associated openly-planned housing estate; it should not however be used generically for such settlements if in a suburban situation.' (p. 181)
Burke, J., 1953, *Stroud's Judicial Dictionary*

London: Sweet and Maxwell, 3rd ed., 'In the United States, "village" has been defined as any small assemblage of houses for dwellings or business, or both, in the country, whether they are situated upon regularly laid out streets and alleys or not. (Illinois Central Railway v. Williams, 27, Ill., 49).' Vol. 4.

Comment. Dickinson, R. E., *Geography*, 1932, p. 21 distinguishes between 'rural villages', mainly engaged in agriculture, and 'urban villages' with a larger proportion of non-agricultural workers.

In order to be a viable unit a village must include representatives, it may be individuals, of a variety of trades and professions and usually of several levels of society but working together for the general good of the small and often isolated group. Such a small close-knit community may also exists as an 'island' in an urban environment and comes close to the modern concept of a 'neighbourhood unit' (*cf.* Fawcett's vill) and is vaguely recognized by such district names as Greenwich Village in New York city and Dulwich Village in London. (L.D.S.)

**Ville** (French)

*Lit.* town. Used in English as a suffix; ville in place names derived from a proper name or product (*e.g.* Sharpville, Waterlooville, Coalville) and in popular language to denote a place devoted to a main industry (*e.g.* a 'leadville') *Cf.* opolis.

**Virazon**

Kendrew, W. G., 1953, *Climates of the Continents*, Oxford: Clarendon Press, 4th ed., 'The sea-breeze (virazon) and land-breeze (terral) are regular and prominent; the sea-breeze is often so strong on summer afternoons at Valparaiso and other places that boat-work is stopped.' (p. 481) (coast of Chile)

**Virgate**—see Yardland.

**Virgation, virgating**

*O.E.D.* From latin, virga, a twig. A system of faults branching out like twigs from a bow. The Western Balkans form in their southern part six ranges, the orographical expression of a geological 'virgation'. (*Geog. Jour.*, **9**, 1897, 87)

*O.E.D.* Suppl. A divergent arrangement of fault-lines has been termed a 'virgating system' by American geologists. (*Q.J.G.S.*, **55**, 1899, 576)

Wills, L. J., 1929, *The Physiographical Evolution of Britain*, London: Arnold. 'bunching together of earth folds near an obstacle' (p. 178).

**Virgin**

As an adjective applied to soil, earth, forest etc., untouched by man, not hitherto cultivated, exploited or used. The belief that there remains on the surface of the earth large areas of 'virgin forest' has been severely shaken of recent years by repeated discoveries of former human occupation.

**Viticulture**

The cultivation of the grape vine; vine-growing.

**Vlei** (Afrikaans; *antiq.* vley)

A shallow lake or swamp. A vlei differs from a pan by (a) a more permanent water supply (b) a more irregular shape (c) by being often connected with a river system. (P.S.)

A depression not well drained, becoming marshy after rain (J.H.W.).

**Vloer** (Afrikaans; *lit.* floor)

Clay flats which develop where the arid plateau surface has so little slope that no well defined river course can exist. Any rain which falls spreads out in a thin sheet, moving very slowly and disappearing rather by evaporation than by real run-off; this causes the clay to have a high salt content. A vloer shows more horizontal development and less depth than a pan, but the distinction is not easy to make, and the names are sometimes interchanged. American equivalent—playa. (P.S.)

Wellington, J. H., 1955, *Southern Africa* I, 'Great shallow hollows in the desert pavement zone, filled with water on the rather infrequent occasions when the rivers come down in high flood.' (p. 323)

**Voe** (Scottish: Orkney and Shetland dialect)

*O.E.D.* A bay, creek or inlet.

Mill, *Dict.* A narrow gully cut in a cliff, at the end of which there is often a cave or tunnel with a blow hole (Orkneys and Shetlands).

**Volcanic**

Literally of or pertaining to a volcano or volcanoes, hence used in a number of specific senses.

*Volcanic ash or ashes*

    Himus, 1954. 'A pyroclastic rock, made up of finely comminuted fragments of rock and lava which have been ejected explosively from a volcano.' The word 'ash' is a misnomer; it dates from the time when a volcano was believed to be a 'burning mountain'.

*Volcanic breccia*
As the last, also a pyroclasic rock, but consisting of large fragments usually angular. The breccia is more or less consolidated.

*Volcanic cinders*
Also, live ashes, a misnomer and better called lapilli (*q.v.*); fragments of intermediate size.

*Volcanic bomb*
Moore, 1949, a lump of lava, usually rounded in shape, which has been thrown out of a volcano in the liquid state, solidifying as it fell.

*Volcanic dust*
The finest particles thrown out of a volcano in eruption. The dust may be shot high into the air by the force of the explosion and may then be carried immense distances by wind. Dust from the eruption (1883) of Krakatoa in Indonesia is said to have been carried round the earth three times before it finally settled.

*Volcanic mud and sand*
Himus, 1954. Deposits occurring round volcanic islands and coastlines. Also ash washed down by the heavy rains which often accompany an eruption.

*Volcanic Neck* or *Plug*
Neck is strictly the orifice through which lava reached the surface and in which the lava eventually solidified as a plug, also itself called a neck.

*Volcanic Rocks*
Igneous rocks fall into three broad classes (a) those which solidify far underground as plutonic rocks; (b) those intruded as sills, dykes etc. or hypabyssal rocks and (c) those poured out as lavas at the surface as volcanic or extrusive rocks.

**Volcano**
*O.E.D.* A more or less conical hill or mountain, composed wholly or chiefly of discharged matter, communicating with the interior of the globe by a funnel or crater, from which in periods of activity steam, gases, ashes, rocks and frequently streams of molten materials are ejected.

*Webster.* A vent in the earth's crust from which molten or hot rock, steam, etc., issue; also, a hill or mountain composed wholly or in part of the ejected material. Such a mountain is more or less conical in form, and often has a depression or crater in its top. A volcano is called *active* while it is in eruption, *dormant* during a long cessation of activity, and *extinct* after eruptions have altogether ceased. Most volcanoes are in the sea or near it, and many are in groups or linear series. Volcanoes include many of the most conspicuous and lofty mountain peaks of the earth.... The character of volcanic eruptions varies from the quiet outpouring of fluid lava, as in Hawaii, to violent explosions, accompanied by showers of volcanic ash, like that of Krakatao in 1883.

*Comment.* The *O.E.D.* definition is inadequate, inaccurate and out of date. It refers only to one type of volcano, confuses the vent or neck with the crater and places the emphasis on the cone which is often absent. For an adequate definition and discussion reference should be made to a standard work such as C. A. Cotton's *Volcanoes* (1944). The brief statement by G. W. Himus (1954) may serve to correct the *O.E.D.*

Himus, 1954, 'Rifts or vents through which magma, consisting of molten material highly charged with gases and vapours, from the depths of the earth is erupted at the surface as flows of lava, or as explosive clouds of gases and volcanic ashes. Volcanoes may be of the *central* or the *fissure* type. In the former, eruption takes place through a more or less cylindrical feeder pipe; in the latter, lava emerges through a fissure which may be many miles in length. The cone of a central volcano may consist, as in some of the puys of Auvergne, merely of a pile of fragmentary materials which have been thrown high into the air and have fallen more or less uniformly round the vent. Sometimes, as at Vesuvius, the cone may be *composite*, and consist of fragmentary material, or *cinder* penetrated in all directions by dykes, some of which (as at Etna) may serve as feeders for small subsidiary volcanoes or *parasitic* cones. Lavas rich in silica are often highly viscous from the vent; in this case, a cone may be built up over the vent by the slow upheaval of stiff lava occupying the conduit, as with the *mamelons* of Reunion. Highly fluid basaltic lavas spread out in thin sheets give rise to *shield volcanoes*, such as that of Mauna Loa in Hawaii ... in the Eifel ...

the vents merely consist of holes in the ground (often occupied by lakes)....'
Thornbury, 1954. 'The most commonly used classification is that originally proposed by Lacroix in 1908... there are four principal types of eruptions—the Hawaiian, Strombolian, Vulcanian and Peléan....'

**Volcano, mud**—see Mud-volcano

**Von Thünen model**
Found, W. C., 1971, *A Theoretical Approach to Rural Land-Use Patterns*, London: Edward Arnold. 'Von Thünen pointed out the tendency for economic rent to decline for a given land use as the distance to the market increased. The decline was due to rising costs of transportation since longer distances to the market had to be covered.' (p. 57)
See Von Thünen, J. H., 1826, *Der Isolierte Staat in Beziehung auf Landwirtschaft und Nationalokonomie*, translated by Wartenburg, C. M. for English edition, Hall, P., 1968, *J. H. von Thünen's Isolated State*, Oxford: Pergamon; and Ricardo, D., 1817, *Principles of Political Economy and Taxation*, Everyman ed., London: Dent. Ricardo and Von Thünen put forward the theory of economic rent independently of each other.

**Voortrekker** (Afrikaans)
One who travels in front. Name given to the pioneers of the Great Trek of 1835–37. (P.S.)

**Voralp** (German; especially Switzerland)
The lower pastures of an alpine valley, *i.e.*, those that are found before the alpine pasture proper.

**Voyageur** (French, *lit.* a traveller)
In North American literature used with special reference to the French explorers of the continent but applied originally to any traveller, including those employed by trading companies.

**V-shaped valley**
In contrast to a U-shaped or mature valley, one that results from down-cutting usually by a stream so that in cross-section it has the shape of a V—commonly an indication of youthfulness.

**Vug, vugh**
A miners' term, originating from Cornwall, for a cavity in a metalliferous vein the walls of which are often lined with well-formed crystals. Although sometimes of considerable size, the second meaning of 'cave or hollow' given by *O.E.D.* seems difficult to substantiate.

**Vulcanian**
The Lipari Islands, in the Mediterranean between Sicily and Italy were formerly known as the Vulcanian Islands and the word 'vulcanian' is sometimes used of a type of volcanic eruption observed in the volcanoes there in which the viscous, pasty lava quickly crusts over and is later disrupted with great violence by accumulated gases. (L.D.S.). See Strombolian.

**Voyvod** (Polish; also most Slav languages with various spellings voivode, vojevoda etc.)
A governor, hence the anglicized voyvodship for governorship or governor's province, the chief administrative division of Poland and other east European countries.

**Vulcanism, vulcanicity**
*O.E.D.* Volcanic action or condition.
Mill, *Dict.* The action of the internal heat of the Earth upon the crust of the Earth.
Wooldridge and Morgan, 1937. 'The term "vulcanicity" covers all those processes in which molten rock material or *magma* rises into the crust, or is poured out on its surface, there to solidify as a crystalline or semi-crystalline rock.' (p. 106)
Thornbury, 1954. '.... includes the movement of molten rock or magma on to or toward the earth's surface.' (p. 52)
Monkhouse, F. J., 1954, *The Principles of Physical Geography*, London: U. of London Press. 'The term vulcanicity includes in its widest sense all the processes by which solid, liquid or gaseous materials are forced into the earth's crust or escape on to the surface.' (p. 43)
*Comment.* Vulcanism covers much more than the phenomena associated with volcanoes.

**Vulcanology**
The study of volcanoes.

# W

**Wächte,** (German, *pl.* Wächten)
Kuenen, 1955. 'Snow-drifts of a kind which the Swiss call Wächten are formed along ridges. They resemble the crests of breakers curling towards the steeper slope and result from the anchorage of snowflakes swept over the ridge by a strong wind. The canopy thus formed by degrees curls downwards under its own weight. Successive accumulations in the course of the winter months complicate its structure.' (p. 131) Kuenen gives a section, Fig. 66.

**Wadden** (Dutch, *pl.* Waddens)
A tidal flat. Used by J. T. Møller, Vadehavskysten Emmerlev-Ballum, *Folia Geographica Danica*, **8**, 1961, who translates De Danske Vade og Marskunder Søgelser as The Danish Wadden and Salt-Marsh Investigation and uses wadden throughout his English summary. Not in *O.E.D.* or Webster. Cf. watte (German).

**Wadi** (Arabic)
*O.E.D.* In certain Arabic-speaking countries, a ravine or valley which in the rainy season becomes a watercourse; the stream or torrent running through such a ravine.
Mill, *Dict.* Valleys in desert regions which are only occasionally traversed by streams (N. Africa).
Knox, 1904. Wad, Wadi, *pl.* Widan (Arab.), a watercourse, dry in summer; a valley. (Morocco), a river, not a dry river-bed.
*Comment.* In those parts of Spain which came under Moorish domination the word wadi has become guadi in river names. Guadiana is thus River Ana, already tautologous as *ana* signifies river. Often on British maps the river is marked as River Guadiana and has even been recorded as the river Rio Guadiana which means river-river-river-river using four languages, English, Spanish, Arabic and Latin.

**Waldhufendorf** (German; *pl.* Waldhufendörfer: *lit.* forest village, though used as synonym)
This term is used in the special sense of a village which originated during the clearing period (9th–14th century), where the farmhouses follow in double (or single) row a stream at a valley bottom and the farm land is consolidated and stretches behind each farmhouse to the parish boundary. This kind of village lacks a common (K.A.S.).

**Wallace's Line**
Mill, *Dict.* (modified). A line which follows the deep water channel separating the islands of Bali and Lombok (Indonesia) and which, as first demonstrated by the naturalist A. R. Wallace, separates two great zoogeographical regions, the Oriental and the Australian.

**Waning slope, waxing slope** (Alan Wood, 1942)
Wood, A., 1942, The Development of Hillside Slopes, *Proc. Geol. Assoc.*, **53**, 128–140. Introduces the terms and discusses the concepts at length.
Dury, 1959. 'Four possible elements of hillside slope are recognized: the waxing slope at the top—the curve-over which steepens as the hillside is worn back; the free face—a vertical facet cut in bare rock; the constant slope—the straight slope of the lower hillside; and the waning slope, which becomes less steep as it develops and is really the slope of the widening valley floor' (p. 64).

**Wapentake**
A subdivision of certain English counties—in Yorkshire and the Midlands where Danish influence was strong. Elsewhere the corresponding division is the hundred and, in a few cases, the soke.

**Ward**
*O.E.D.* 19. An administrative division of a borough or city; originally a district under the jurisdiction of an alderman; now usually a district which elects its own councillors to represent it on the City or Town Council. Also, the people of such a district collectively.
20. In Cumberland, Northumberland, and some Scottish counties: one of the administrative districts into which these counties are divided.
21b. Sc. 'A small piece of pasture ground,

inclosed on all sides, generally appropriated to young quadrupeds.' (Jamaica) Mill, *Dict*. 1. A certain division, section, or quarter of a town or city under the charge of an alderman, or constituted for the convenient transaction of local public business through committees appointed by the inhabitants. 2. A territorial subdivision of some English counties, as Durham, Westmorland, Cumberland; equivalent to the hundred of the midland counties. 3. The division of a forest.

**Warm front**
The boundary at the surface between an advancing mass of warm air and the colder air over which it rises. Formation of cloud usually results, with precipitation.

**Warp, warp-clay**
Mill, *Dict*. Alluvium laid down on a tidal stream.
Wooldridge and Morgan, 1937. '... the deposits of the vanished Yorkshire lakes, as in the Vale of York (warp clays) and the upper part of the Esk Valley' (p. 395) (which are varved).
*O.E.D.* The silt or alluvial matter deposited by the sea or a tidal river.
*Comment.* It would seem that the term should be limited to material deposited by the process of warping (1) whether the warping is natural or man-controlled. In this case it is doubtful whether the warp-clays instanced by Wooldridge and Morgan should be included.

**Warping (1)**
*O.E.D.* 5a. The process of flooding low-lying land near a tidal river so that the muddy alluvium may be deposited when the water is withdrawn. Also warping up, the process of filling up hollows by deposit of alluvium.

**Warpland, warplands**
Land that has been built up by the process of warping (1).
Not in *O.E.D.*

**Warping (2), warped**
Mill, *Dict*. The surface undulations produced by slow movements which cause one part of the land surface to be higher yet continuous with another; they are usually reciprocal.
Lobeck, 1939, does not define but uses the term 'The Great Lakes region exhibits several domes and basins, formed by warping during uplift' (p. 453, also diagrams).
Rice, 1941. A word which has come into use to replace folding, where the effect is simply to depress or to raise the surface. The flexures that result may be either downwards or upwards (Willis).
*Comment.* Whereas 'folding' suggests the formation of anticlines and synclines 'warping' can be applied as well to more gentle deformation of the rocks, hence its growing use. It is frequently used but not defined (*e.g.* Wooldridge and Morgan; not in Thornbury, Holmes).

**Warren**
Originally a piece of land enclosed and preserved for breeding game; later a piece of land appropriated for the breeding of rabbits (rabbit-warren) and at times hares. In Medieval Britain there were certain legal 'rights of warren', there were legal definitions of 'beasts and fowls of warren', and keepers known as warreners. With changing social conditions and the spread of wild rabbits warrens and warreners disappeared and the word came to mean simply a piece of uncultivated land in which rabbits, until their practical extermination by myxomatosis in 1953–6, bred wild in burrows. See Stamp, L. D., 1955, *Man and the Land*, Collins New Naturalist Series, pp. 53, 212.

**Wash (1)**
The surging movement of the sea or other body of water.

**Wash (2)**
Mill, *Dict*. (modified). A miner's term for either (a) surface washes or valleys in the north-eastern coalfield or (b) small valleys or wash-outs (*q.v.*) in coal seams.

**Washlands**
Low lying lands bordering a river, usually part of the natural flood plain, over which flood waters are permitted to flow so that damage may be prevented elsewhere and the flood waters later controlled by pumping or in other ways. Where such land lies between the normal river bank and a main levée or dyke the term river foreland has been used but in view of other numerous uses of foreland is not to be recommended (L.D.S.).

**Wash-out**
*Webster*. 5. *Geol*. A channel cut by erosion in one sedimentary deposit and filled with the material of a younger deposit.
Holmes, 1944. 'Coal... seams are locally interrupted by "wash-outs", that is, by the sandstone-filled channels of streams that

flowed through the forest-swamps, like the distributaries in modern deltas' (p. 341).

**Wash-over**

Lobeck, 1939. 'During storms waves break over a low portion of a bar and carry material back into the lagoon, depositing wash-overs. Almost every bar bears on its lagoon side a row of deltas thus formed.' (p. 349) Also called wave-delta.

**Wash slope**—See Haldenhang (German)

*Comment.* This translation of Haldenhang is used by Meyerhoff, H. A., 1940, Migration of Erosional Surfaces, *A.A.A.G.*, 30, 247–254. See also Thornbury, 1954, p. 200.

**Waste, waste land**

*O.E.D.* 1. Uninhabited (or sparsely inhabited) and uncultivated country; a wild and desolate region, a desert, wilderness. Somewhat rhetorical. b. trans. Applied, *e.g.*, to the ocean or other vast expanse of water (often waste of waters, water waste), to land covered with snow, and to empty space or untenanted regions of the air. 2. A piece of land not cultivated or used for any purpose, and producing little or no herbage or wood. In legal use spec. a piece of land not in any man's occupation, but lying common.

Mill, *Dict.* A wild, uninhabited, uncultivated, desolate place.

Stamp, 1948. 'The term "waste" was used by older writers to indicate the little-used common land, usually on light soil, which failed to yield a return to the medieval and later cultivator. The term "waste" for such land has now disappeared because in so many cases these are the commons so valued as open spaces.' 'The present problem of waste land is usually one of land which has been previously used but which has been abandoned and for which no further use has yet been found.' (p. 433) '. . . . the inland sandy "wastes" or heathlands.' (p. 365)

Burrows, R., *Words and Phrases Judicially Defined*, London: Butterworth, 1945. 'The word "waste" means desolate or uncultivated ground, land unoccupied, or that lies in commons. This is the plain and common acceptation of the word.'

**Waste of the Manor, Manorial Waste**

Report of the Royal Commission on Common Land, 1955–1958. London, H.M.S.O., Command 462. 'Part of the demesne of the manor left uncultivated and unenclosed, over which the freehold and customary tenant might have rights of common. Not all manorial waste was subject to common rights.' (275)

**Waste-mantle**

Dury, 1959. 'Weathered material—broken and rotted rock—is known collectively as the waste-mantle' (pp. 5–6).

**Wat** (Thai)

A Buddhist shrine, wrongly translated 'temple', in Thailand.

**Water-bearing strata**—see Aquifer

**Watercourse**

*O.E.D.* A stream of water, a river or a brook; also an artificial channel for the conveyance of water.

**Waterfall**

*O.E.D.* A more or less perpendicular descent of water from a height over a ledge of rock or precipice; a cascade or cataract.

Mill, *Dict.* Strictly a perpendicular descent of a stream, but frequently confused with rapids.

Himus, 1954. A sudden descent of water over a step in the bed of a river.

*Comment.* In place names commonly shortened to Fall or Falls, *e.g.* Niagara Falls. The classical example of a waterfall, with an almost horizontal bed of hard rock overlying softer beds is afforded by Niagara but the completely contrasted example of Victoria Falls in the Zambezi, descending from a high level of the African plateau to a zigzag gorge excavated along fault lines, indicates that waterfalls may originate in various ways.

**Water-gap**

Mill, *Dict.* A cutting or opening made by a river flowing athwart a ridge.

Moore, 1949. A narrow gorge cut by a stream through a ridge of hard rock.

*Comment.* Used in contrast to wind-gap (*q.v.*). Water-gaps tend to be closely associated with antecedent drainage (*q.v.*) and may be cut by either consequent or obsequent streams. Whereas water-gaps are occupied by streams which have continued to form part of the drainage system, wind-gaps represent channels deserted as a result of river capture.

**Water Hemisphere**

The opposite half of the earth to the Land Hemisphere (*q.v.*). Its centre lies near New Zealand and water occupies six-sevenths of its surface.

**Water hole** (especially Australia, Africa etc.

*Webster.* 1. A natural hole or hollow

containing water, specifically, one in the dry bed of an intermittent river; a spring in a desert; also, any pool, pond, or small lake.
2. A hole in a surface of ice.

Mill, *Dict.* 1. Any pond, natural or artificial.
2. Depression in the bed of an intermittent river.

## Watering place

*O.E.D.* 1. A place on a river or a lake where animals are brought to obtain water.
2. A place where a ship's company goes to fill the ship's casks with fresh water.
3. A resort of fashionable or holiday visitants, either for drinking or bathing in the waters of a mineral spring, or for sea bathing.

Comment. In all three senses the word is obsolescent and the *O.E.D.* definitions, especially 2 and 3, are out-of-date in wording.

## Water meadow, water-meadow

*O.E.D.* A meadow periodically overflowed by a stream.

Committee, List 3. Meadow naturally or artificially inundated for part of the year.

Stamp, 1948. 'Meadows by the side of streams are sometimes loosely but wrongly called water meadows; this term should be restricted to those meadows which are specially irrigated—usually by an elaborate series of miniature canals and drains—to promote an early and rich growth of grass.' (p. 28)

Comment. Formerly widespread in southern Britain but many have fallen into disuse through high maintenance and labour costs. See the full description by Moon, H. P. and Green, F. H. W., 1940, *The Land of Britain: Hampshire* (Part 89), 373–390.

## Water-parting

Webster. A summit or boundary line separating the drainage districts of two streams or coasts; a divide or watershed. See *watershed*.

Mill gives as synonymous with watershed or divide.

## Water power, water-power

*O.E.D.* The power of moving or falling water employed to drive machinery; a fall or flow of water which can be thus utilized.

Comment. Whereas water power was formerly used directly to drive water-mills etc., it is now used almost always to generate electricity. The measurement of such power resources was formerly in installed or potential horse-power (HP); now it is more usual and more accurate to express resources in actual and potential output of energy—in kilowatt-hours or 'units' of electricity.

## Waterscape

Coleman, Alice, 1976, Canadian Settlement and Environmental Planning, *Urban Prospects*, Ottawa: Department of State, Urban Affairs. 'Waterscape... is formed of those water-bodies that are too large to be incorporated into any of the basic five (land use territories).... For example, farm ponds, ornamental ponds in city parks, and small, natural lakes do not rank as waterscape but form subordinate land uses in farmscape, townscape and wildscape respectively. Waterscape occurs only when a body of water exceeds a certain threshold size and so becomes dominant at this level of classification. It may itself contain small areas that are not water islands and peninsulas may rank as permissible subordinates in waterscape if they do not reach certain dimensions.'

## Watershed

Mill, *Dict.* (gives as same as water-parting or divide). The line from which surface-streams flow in two different directions; the line separating two contiguous drainage areas.

*Anomalous watershed* (in a mountain region). One that does not run along the crest of the highest range of a mountain chain.

*Normal watershed.* One that does run along such a crest.

*O.E.D.* 1. The line separating the waters flowing into different rivers or river basins; a narrow elevated tract of ground between two drainage areas = water-parting.
2a. loosely the slope down which the water flows from a water-parting.
2b. the whole gathering ground of a river system.

Under the last heading all references in *O.E.D.* are American, the earliest 1874.

Webster. 1. Water parting; a summit or boundary line separating the drainage districts of two streams or coasts; a divide.
2. The whole region or area contributing to the supply of a river or lake; drainage area; catchment area or basin.

*Dict. Am.* The entire gathering ground of a river system. 1874.

Comment. This word is causing great confusion because of the complete difference between British and American usage. The British usage is equivalent to water-parting, the American equivalent to river basin. Through such international agencies as UNESCO and FAO (Food and Agriculture Organization) the American meaning has spread widely in recent years.

In the United States there are now Watershed Conservancy Districts, Watershed Managers and Watershed Associations. Syracuse University has a Watersheds Institute. In all these cases Watershed is used in the sense of River Basin. (L.D.S.)

### Water-spout
*O.E.D.* A gyrating column of mist, spray and water, produced by the action of a whirlwind on a portion of the sea and the clouds immediately above it.

*Webster.* 3. A slender funnel-shaped or tubular column of rapidly rotating, cloud-filled wind usually extending from the under side of an ordinary cumulus or cumulo-nimbus cloud down to a cloud of spray torn up by the whirling winds from the surface of an ocean or lake. It is sometimes straight and vertical and sometimes inclined and tortuous, as it moves along. The funnel cloud is of fresh water.

### Waterstones
Himus, 1954. Term applied to the upper part of the Keuper sandstone of the English Midlands.

*Comment.* Though some water is available from springs emanating from the Keuper Waterstones they are not a markedly good aquifer. The term should *not* be used in place of aquifer (*q.v.*).

### Water table, water surface
Mill, *Dict.* The subterranean surface or level below which the rocks of any given area are saturated with water.

*Webster.* 2. The upper limit of the portion of the ground wholly saturated with water; used chiefly in hydraulic engineering. This may be at or very near the surface or many feet below it.

*Comment.* In areas with a pervious soil and pervious subsoil-rocks the water table tends to follow in general, though not in detail, the form of the land surface. Where the water table is below the surface its height corresponds to the level of water in wells and, like that level, fluctuates seasonally. Where the water table reaches the surface a spring results; fluctuations in the water table explain the intermittent flow of bournes (*q.v.*). A permanent marsh or lake results when the theoretical water table is above the land surface. In certain circumstances a regular water table may be absent. This is the case where underlying rocks are irregularly fissured and so applies to great areas of the ancient metamorphic plateau of Africa. In other cases there is the phenomenon of a perched or suspended water table: if the water-bearing stratum is pierced non-saturated rocks may be found below. See also Ground water, Phreatic water, Vadose water and Wells.

### Waterway
Mill, *Dict.* A line of water (river etc.) which can be utilized for communication or transport.

### Watte (German; *pl.* Watten)
Shackleton, 1958. 'West of the River Elbe ... between the coast and the dune-covered islands are the tidal flats or *watten* which are dry at low water except for channels....' (p. 258)

### Wave-base
Holmes, 1944. 'The greatest depth at which sediment on the sea floor can just be stirred by the oscillating water is called the "*wave-base*"' (p. 284).

Wills, L. J., 1929, *The Physiographic Evolution of Britain*, London: Arnold. 'The depth at which material of any particular grade ceases to be agitated and kept in suspension by wave- or current-motions' (p. 45).

### Wave-cut platform—see abrasion platform

### Wave-delta—see Wash-over

### Waxing slope—see Waning slope

### Weald
The old English word "weald" meant forest (*cf.* German, Wald) and was applied to the tract of country covered in Saxon times by the Forest of Andred which lies between the North and South Downs in the counties of Kent, Surrey and Sussex. Formerly applied, especially poetically, to similar areas elsewhere, it is now only a regional name—a *nom du pays*, not a geographical term (*cf.* Wold). (L.D.S.)

### Weather
*O.E.D.* The condition of the atmosphere (at a given place and time) with respect to heat or cold, quantity of sunshine, presence or absence of rain, hail, snow, thunder, fog etc., violence or gentleness of the winds.

Also, the condition of the atmosphere regarded as subject to vicissitude.

Mill, *Dict*. A general term for the atmospheric condition; the state of the atmosphere with respect to its temperature, pressure, humidity, electrification, motions, or any other meteorological phenomena at any given instant.

*Met. Gloss.*, 1944. The term weather may be taken to include all the changing atmospheric conditions which affect mankind but by meteorologists it is more commonly used in a limited sense to denote the state of the sky and whether rain, snow or other precipitation is falling. Atmospheric obscurity in the form of fog or mist is also included.

*Comment*. Weather refers essentially to short periods of time, climate to longer.

## Weathering

*O.E.D.* a. The action of the atmospheric agencies or elements on substances exposed to its influence; the discoloration, disintegration, etc. resulting from this action. b. The action of the elements (on land, clay, etc.) as a beneficial agency; the state of being pulverized and rendered workable by this action.

Mill, *Dict*. The process which tends to change the exposed surfaces of rocks, and renders them more or less disintegrated and adapted for transportation.

Geikie, J., 1898. 'applied to the decomposition, disintegration, and breaking up of the superficial parts of rocks under the general action of changes of temperature, and of wind, rain, frost, etc.' (p. 308).

Strahler, A. N., *Physical Geography*, New York: Wiley, 1951. 'The term weathering refers to all processes whereby rocks are decomposed or disintegrated because of exposure at or near the earth's surface.' (p. 128)

*Comment*. Some writers refer always to 'Atmospheric Weathering' but in fact weathering, especially under tropical conditions of high temperature and high humidity and where well-jointed rocks permit the easy penetration of circulating waters from the surface, may take place to considerable depths, even hundreds of feet and so far out of the reach of the weather as such. The general term 'weathering' is thus preferable.

**Weathering Front** (J. A. Mabbutt, 1961)—see Basal surface.

## Weberian analysis

Morrill, R. L., 1970, *The Spatial Organization of Society*, Belmont, California: Wadsworth Publishing Co. 'A classical theory of firm location was formulated by A. Weber in 1909. Optimum location was seen primarily as the point where the transport costs of bringing the necessary raw materials and of supplying goods to the necessary market were at a minimum. Because of transport costs, orientation to resource or markets was considered the normal case. However, if variations in other costs, particularly labour, were sufficiently great, a location determined solely by transport costs might not be the optimal one.' (p. 87)

See translation of Weber, A., 1909, in Friedrich, C. J., 1929, *Alfred Weber's Theory of the Location of Industries*, Chicago: University of Chicago Press.

See also Location (least cost), Location (optimum) and Location theory.

## Weir

Mill, *Dict*. 1. A dam, erected across a river to impound and raise the water for the purpose of taking fish, of conveying a head to a mill, or of maintaining the water at the level required for navigation or for purposes of irrigation.
2. A fence of stakes placed in a river or harbour used for catching fish.

*Comment*. The word is now restricted to smaller works, the larger are called barrages and dams.

## Welfare approach

Smith, D. M., 1977, *Human Geography: A Welfare Approach*, London: Edward Arnold. '... embraces all things contributing to the quality of human existence. ... Our spatial concept of welfare incorporates everything differentiating one state of society from another. It includes all things from which human satisfaction (positive or negative) is derived, and also the way in which they are distributed within society ... welfare concerns not only the what but also who gets it where.' (p. 8)

## Welfare maximization

Smith, D. M., 1977, *Human Geography: A Welfare Approach*, London: Edward Arnold. '... welfare maximization depends on optimality with respect to three distinct conditions, all of which are interrelated in the real economic world: production technique, combination of goods and services produced, and distribution among the population—individuals, groups and areas ... welfare maximization depends on both purely technical relationships of how to get the most output from given resources and the social relationships behind the preferences that are actually implemented.' (p. 65)

## Well
Although originally applied to a natural spring or pool fed from a spring, well has come to be restricted to a pit dug in the ground to obtain water, also oil. Usually the excavation is lined with brick or masonry but in hard rock or other special cases may be unlined. Normally the well fills with water up to the level of the water table and the surface of the water fluctuates seasonally with the height of the water table. The water is raised to the surface by various means. In artesian basins the water rises to the surface and may be under considerable pressure (artesian well). Deeper wells may be drilled by modern methods to reach deep seated supplies and are then lined with pipes or tubes (hence tube well). In the case of oil wells they may reach many thousand feet in depth. (L.D.S.)

## WE-ocratic (T. Griffith Taylor, 1936)
Taylor, 1951. 'Human control as opposed to environmental control.' (p. 621)

Comment of the author (in MS.) 'Geocratic and WE-ocratic result from my studies in geographical philosophy after 1936 in Toronto. The last is the brashest verbal hybrid that I have met—but none the less expresses the brash idea of the Possibilist that he can do what *he* likes and Nature can go hang.'

## Westerlies
The winds which, in mid-latitudes, blow predominantly from the south-west in the Northern Hemisphere and north-west in the Southern Hemisphere. They were once called the Anti-Trades, but this misleading term has been dropped.

## Wet and Dry Bulb thermometers, temperatures
The temperature recorded by a thermometer of which the bulb is covered by wet muslin is lower than that recorded by a thermometer with a dry bulb, much lower if the air is dry. See Sensible temperature.

## Wet-day
In Britain is applied officially to a day of 24 hours beginning at 0900 hours on which at least 0·04 inch or 1 mm of rain falls.

## Wet-point settlement
A settlement owing the selection of its site to the availability of a water supply, especially a constant spring (contrast Dry point settlement in wet lands liable to flood).

## Wet spell
Moore, 1949. In the British Isles, a period of at least fifteen consecutive days each of which has had 0·04 inch of rain or more; the definition has not been internationally accepted.

## Whaleback (R. A. Bagnold, 1941) (1)
Bagnold, R. A., 1941, *The Physics of Blown Sand and Desert Dunes*, London: Methuen (rep. 1954). 'coarse-grained residues or platforms built up and left behind by the passage of a long-continued succession of seif dunes along the same path. I have called them *whalebacks*' (p. 189). 'Whalebacks or sand levees' (p. 230).

## Whaleback (2)
Linton, D. L., 1955, The Problem of Tors, *Geog. Jour.*, **121**. '*Whaleback* is a term often used to describe granite masses, especially in the tropics; some British whalebacks are undoubtedly *roches moutonnées*, some others, and probably many tropical examples, are genetically related to tors, others may be shown to be of yet other origins.' (p. 476)

## Whare (New Zealand)
A Maori sleeping house. Often now used of any small and usually old house. Also name given to shearer's or general hand's accommodation on a farm.

## Whirlpool
Mill, *Dict*. A circular eddy or current in a river or the sea produced by the configuration of the channel, by winds meeting tides, by the meeting of currents or similar means.

## Whirlwind
*O.E.D.* A whirling or rotating wind; an atmospheric eddy or vortex, a body of air moving rapidly in a circular or upward spiral course around a vertical or slightly

inclined axis which has also a progressive motion over the surface of land or water.

*Webster.* A rotating windstorm of limited extent, marked by an inward and upward spiral motion of the lower air, followed by an outward and upward spiral motion, and usually a progressive motion at all levels; a vortex of air; applied by some meteorologists to the larger rotary storms also. *Cf. cyclone.*

### White man's grave
An outmoded expression for the hot, humid coastlands of West Africa, especially Sierra Leone. Since the conquest of such former killing diseases as yellow fever and since reforms in the habits of white men, the former reputation of these lands for unhealthiness has had to be revised.

### Wiesenboden (German: soil science) or wet meadow soil
Jacks, 1954. 'Poorly drained soil with humus—rich $A_1$ horizon grading into gray gleyed mineral soil.'

### Wilderness
In modern nature conservation 'wilderness areas' are those left in a wild state as natural habitats in contrast to those nature reserves which may require careful management to maintain small communities.

### Wildscape
Coleman, Alice, 1969, A Geographical Model for Land Use Analysis, *Geography*, **54**. 'Wildscape is the area that is dominated more by nature than by man.' (p. 46)
Coleman, Alice, 1976, Canadian Settlement and Environmental Planning, *Urban Prospects*, Ottawa: Department of State, Urban Affairs. 'Wildscape is dominated by natural or semi-natural vegetation together with other cover types such as rock outcrops, swamps, water or glaciers.'

### Wildschnee (German)
Literally 'wild snow' but used as a technical term especially in Swiss literature and for which there is no exact English equivalent.
Kuenen, 1955. 'Among high mountains, a fall of powdery snow will often happen in calm weather—the Swiss Wildschnee. It accumulates in a very light mass, its specific gravity being no more than 1/50 to 1/100; hence a layer, say, half a metre thick shrinks on melting to at most one centimetre of water. Ordinary flaky snow on the ground has one-tenth the density of water.' (p. 130)

### Williwaw
*Webster.* A sudden violent gust of cold land air, common along mountainous coasts of high latitudes.
Mill, *Dict.* (modified). A sailor's term for a sudden violent squall, originally in the Strait of Magellan.

### Willy-willy (Australian)
*O.E.D.* Also willi-willi (Native name). In North-West Australia, a cyclonic storm or tornado.
Miller, A. A., *Climatology*, 8th Ed., 1953. 'The west-coast hurricanes' (p. 161).

### Wilting point (Soil science)
The moisture content of soils at which permanent wilting of plants takes place.

### Wind
*Webster.* Any movement of air, usually restricted to natural, horizontal movements; air in motion with any degree of velocity. The vertical or inclined movement of the air often is spoken of as a current or as a wind with a vertical component.
Mill, *Dict.* Air naturally in motion and travelling with an appreciable velocity; a current of air.
*O.E.D.* Air in motion.

### Wind-break
*O.E.D.* Something, especially a line of trees used to break the force of the wind and serving as a protection against it.
Wind breaks are specially important where a strong cold wind, such as the mistral of southern France, would otherwise cause damage to crops.

### Windchill (Paul A. Siple, in thesis 1939; published 1945)
Howe, G. M., 1962, Windchill, Absolute Humidity and the Cold Spell of Christmas 1961, *Weather*, **17**, 349–358. 'The term "windchill" first introduced in 1939 (Siple) represents the cooling power of wind and temperature combinations on shaded dry human skin. As originally developed, windchill was simply the product of wind speed in metres per second and temperature in degrees Centigrade below zero ... new formula developed by Siple and Passel (*Proc. Amer. Philos. Soc.*, 1945, **89**).' Howe gives the formula which 'measures the cooling power of wind and temperature in complete shade without regard to evaporation.' (p. 349).

**Wind gap, wind-gap**
*Webster. Phys. Geog.* A notch in the crest of a mountain ridge; a pass not occupied by a stream; an air gap.
Mill, *Dict.* A water gap (*q.v.*) which has been deserted by the stream which cut it.
Himus, 1954. A notch in an escarpment, originally cut by a stream, from which the water has now disappeared.
Wooldridge and Morgan, 1937, explain how a dry col or wind gap arises as a result of river capture and why the floor of a wind gap will be at a higher level than that of a neighbouring water gap (p. 195).

**Window**—see Fenêtre

**Wind rose**
*Webster.* 2a. A diagram showing, for a given place, the relative frequency, or frequency and strength, of winds from different directions. b. A diagram showing, for a different place, the average occurrence of other meteorological phenomena, as rain, sunshine, etc., with winds from different directions.
Mill, *Dict.* A diagram constructed to show the relative number of wind observations from the eight chief points of the compass; the respective force of each is occasionally added.

**Windward**
*O.E.D.* Situated towards the direction from which the wind blows; facing the wind.

**Winged headland**
With spits on both sides.

**Winter**
1. When used in contrast to summer, the colder half of the year.
2. When used of mid- and high-latitudes as one of the seasons: in popular usage, the three months of December, January and February (June, July and August in the Southern Hemisphere).
3. When reckoned astronomically, from 22nd December (after the winter solstice, 21st December, also paradoxically called midwinter day) to 20th March (or 22nd June to 21st September in the Southern Hemisphere).
4. Loosely the cold season, *e.g.* in such expressions as 'climates with a very long winter'. A certain degree of cold is implied and in the tropics it is usual to drop the word winter and refer if needed to a cool season.

**Winterbourne** (southern England)
A bourne (*q.v.*) flowing only in winter—mainly used as a place name.

**Wirescape**—see -scape

**Witness Rock**—see Zeuge

**Woina-dega, Voina-dega, Voina dega** (Ethiopia)—see Dega

**Wold**
Originally derived from the old English word 'weald' meaning forest (*cf.* German wald) and applied to forest or wooded upland it came later to mean open land, gradually being restricted mainly to elevated tracts of open country. Since about 1600 it has been used specifically for certain tracts of open rolling upland—the Yorkshire Wolds and Lincolnshire Wolds on chalk and the Cotteswolds (Cotswolds) of Gloucestershire and neighbouring counties. Though sometimes applied to similar country elsewhere it is not strictly a geographical term (*cf.* Weald). (L.D.S.)

**Wong**—see Shott (2).

**Wood**
*O.E.D.* A collection of trees growing more or less thickly together (*esp.* naturally, as distinguished from a plantation), of considerable extent, usually larger than a grove or copse (but including these), and smaller than a forest; a piece of ground covered with trees, with or without undergrowth.
Burke, J., *Stroud's Judicial Dictionary.* London: Sweet and Maxwell, 3rd ed., 1953. '"Woods and forests", in Forestry (Transfer of Woods) Act, 1923 (13 and 14 Geo. 5, c. 21), s. 7, includes any land used or capable of being used for afforestation or for purposes in connection therewith.' (Vol. 4)
*Comment.* See the notes under Woodland. When the meaning is that given in the *O.E.D.* definition it is usual to refer to 'a wood', with the indefinite article.

**Woodland, woodlands**
*O.E.D.* Land covered with wood, *i.e.* with trees; a wooded region or piece of ground.
*Comment.* In official British usage 'woodland' is the comprehensive term. In the *Report on Census of Woodlands*, 1924 (Forestry Commission), and similar Reports published in later years the 'total woodland area' was divided into High

forest, Coppice and Coppice-with-standards, Scrub, Felled or Devastated and Uneconomic. The British tendency is to reserve the word 'forest' for the more closely wooded areas or with larger trees. International usage, exemplified by the tables of land use published in the *Yearbook of Food and Agricultural Statistics* (FAO) is to avoid the word 'woodland' and to class the whole as 'Forested Land' (in French: *terrains boisés*). The Commission on World Land Use Survey of the International Geographical Union, after exhaustive enquiries, gave up the attempt to distinguish between forest and woodland. In view of the fact, however, that forest is used by some writers in a restricted sense it seems better to use either 'woodland and forest', or 'woodlands' when all types of 'wooded lands' (*terrains boisés*) are meant. (L.D.S.)

**World geography**

Committee, List 1. The regional geography of the world.

*Comment.* This early and very brief definition published by the Committee was criticized as inadequate. It was pointed out by correspondents that world geography refers to two types of study:—
1. A consideration of the world as a whole, *i.e.* as a unit.
2. A book in which the whole world is considered but in which the arrangement of material may vary frequently having a general section and then taking up the continents and countries in turn.

**World-island** (Mackinder, 1919)—see also Heartland

H. J. Mackinder's term for the world's largest land-mass—the combined continents of Europe, Asia and Africa. Being surrounded by water this vast land-mass is, by the usual definition, an island. In the same way the two Americas constitute an island as do the continents of Australia and Antarctica. See *Democratic Ideals and Reality*, Chap. IV.

**Wrench fault, wrench faulting**

*A.G.I.* A nearly vertical strike-slip fault.

*Comment.* But see *Proc. Geol. Assoc.*, 1963, 265–288.

# X

**Xenolith, also xenoblast, xenocryst, xenomorphic, xenophobic, etc.**
The Greek word *xenos* means a stranger or a foreigner and this is the idea in the many terms, used especially by geologists, which begin with xeno-. In geographical literature xenolith may be found occasionally; it applies to a fragment of rock picked up by a molten magma, partly dissolved, but remaining as a conspicuous 'stranger' when the rock, such as granite, has solidified. (L.D.S.)

**Xerophyte, xerophytic, xerophile, xerophilous, xerosere**
The Greek word *xeros* means dry and the various scientific terms which have been introduced with the prefix xero- all indicate an association with dryness or drought. Xerophilous is literally 'dry-loving' and is applied by botanists to plants which are adapted to a dry climate or habitat, or to conditions of scanty available moisture. A xerophilous plant is termed a xerophyte (from the Greek phyton, a plant) and this is the word or its adjectival form xerophytic which enters most commonly into geographical literature. When vegetation is established on a dry sandy soil the earlier stages in the succession are markedly xerophilous so that reference is made to a *xerosere*. Later in the succession as humus accumulates and plant roots penetrate to lower wetter layers in the soil the vegetation becomes less xerophytic. For a general discussion see Dansereau, Pierre, 1957, *Biogeography*, New York: Ronald Press. (L.D.S.)

*Webster.* Xerophyte. *Ecology.* A drought-resistant plant; a plant structurally adapted for growth with a limited water supply. The term is generally applied not only to actual desert plants, but to those inhabiting salt marshes or alkaline soils or bogs, where water absorption is slow or difficult because of the excess of salts or acids in solution (*cf. halophyte, bog*). Xerophytes exhibit many modifications of structure which limit transpiration, as a thickened epidermis, waxy or resinous coatings, dense pubescence, copious aqueous tissue, etc. *Cf. hydrophyte, mesophyte.*

**Xerorendsina** (Soil science: W. L. Kubiëna, 1953)
Kubiëna, 1953. 'Synonyms: ashy rendsina, dry rendsina... very similar to the *serosem*... one calls a soil on solid parent rock with strongly varying structure and composition occurring in primarily mountainous places, a *xerorendsina*... a soil of the dry steppe with distinct mull formation and distributed over wide areas, on predominantly loose parent material, ought to be called a *serosem*.' (p. 189)

**Xerothermic index**
See reference under Physiognomy. 'The xerothermic index allows the appreciation of the intensity of biologic drought' (*Ibid.*, p. 184). See also Ombrothermic.

# Y

**Ya** (Burmese)
Field or plot.

**Yardang** (from ablative of Turki *yar*, steep bank, cliff)
Mill, *Dict.* Yardangs. The narrow crests sculptured by the wind in the clay-soiled desert of the depression of Lob Nor (Tibet).
Wooldridge and Morgan, 1937. '... described by Sven Hedin from the Central Asiatic desert. These are steep-sided rock ribs, up to 20 feet high and from 30 to 120 feet in width, separated from one another by grooves or corridors cut in the desert floor. Though irregular in form and with undercut sides, they maintain a rough parallelism over considerable areas. It cannot be doubted that they owe their origin to veritable æolian corrasion, under the influence of steady winds.' (p. 298, with diagram)

**Yardland** (Agricultural History)
Butlin, 1961. 'The yardland or virgate is not a unit of exact land measurement, but represents a tenement of varying size measured in customary acres, and including arable with appurtenant meadow and pasture. The yardland or virgate might be subdivided into OX-GANGS (half of a virgate) or FERLINGS (quarter-virgates), both terms being primarily fiscal and areally inexact. In areas of Danish influence, different nomenclature was used, within the wapentake (hundred) the CARUCATE replaced the hide, and the BOVATE was the equivalent of the OX-GANG. In south-eastern England the SULUNG was the equivalent of the hide and was subdivided into four YOKES'.

**Yazoo** (River type tributary)
Lobeck, 1939, 'Tributaries of the Mississippi ... because the Yazoo River is a good case, it is taken as the type example, and tributaries which run for some distance parallel to the main stream are called Yazoo rivers.' (p. 225 and diagrams) (A deferred tributary junction.)

**Yellow Earth**
The loess (*q.v.*) of northern China.

**Yellow ground** (South Africa)
Oxidized upper zone of soft Kimberlite in the South African diamond mines, lying above the blue ground (*q.v.*). (P.S.)

**Yeoman**
Among the many meanings of yeoman is one which is extensively used, especially in works dealing with the historical geography of Britain:—a man holding a small freehold estate but not a member of the nobility (or 'landed gentry' in the old sense) hence a commoner or countryman of standing, especially one who cultivates his own land.

**Yield**
*O.E.D.* 4. The action of yielding crops or other produce, production; that which is produced, produce; especially amount of produce.
Committee, List. 3. Output or production expressed in relation to units of land or livestock, or to units of capital or labour applied.
*Comment.* The Committee's earlier definition is somewhat too narrow in that yield is applied in a similar sense to yield of water from springs or wells, yield of minerals from mines, etc. Note should also be taken of contrasts between actual and potential yields or actual and theoretical yields, etc.

**Yoke-pass**
Mill, *Dict.* A mountain pass with a long approximately level summit between two parallel slopes (the *Joch* or *Wallpass* of Richthofen).
*Comment.* No recent use has been traced. It would seem to be an anglicized spelling of *Joch*.

**Yoma** (Burmese; older form Yomah)
Mountain range.

**Young mountains**
Geikie, J., 1898. 'Such then are the several stages through which a region of mountain-uplift must pass. First comes the stage of youth, when the surface-configuration corresponds more or less closely with the underground structure.' (p. 101)

Taylor, 1951. 'Young Mountains: Have risen during Tertiary times.' (p. 622)

Staats, J. R., and Harding, G. E., *Elements of World Geography*. New York: Nostrand, 1951. 'Young mountains... have so recently formed that agents of erosion have not had time to subdue their summits.' (p. 248)

No reference in Davis, 1909; Salisbury, 1907; Cotton, 1922; Wooldridge and Morgan, 1937; Holmes, 1944; von Engeln, 1942; Strahler, 1951.

*Comment.* The phrase 'young fold mountains' is frequently found in geographical texts referring to the Alpine orogeny in contrast to Armorican, Caledonian and other earlier orogenies.

**Youth, youthful, young** (Stage of erosion cycle)

Mill, *Dict.* Young River. One that has done little of the work required to wear away its bed.

Davis, 1909. 'There will be a brief youth of rapidly increasing relief...' (p. 256) (1899).

Salisbury, 1907. 'The topography of a drainage basin is youthful when its river system is youthful....' 'In an area of youthful topography, much of the surface has not yet been much affected by erosion ...' (p. 152).

Cotton, 1922. 'While considerable areas remain undissected the surface is still in the stage of youth; but when dissection is complete, the sloping sides of newly-cut valleys intersect one another to form well-defined divides, and no trace of the initial form remains, the surface is mature... a district may be maturely dissected by streams which are still young.' (pp. 64–65) (Note; in edition of 1945, p. 61, this is footnoted: '... the transition between youth and maturity is sometimes otherwise defined.')

Fay, 1920. Young; Youthful. Being in the stage of increasing vigor and efficiency of action: said of some streams; also, being in the stage of accentuation of and a tendency toward complexity of form: said of some topography resulting from land sculpture. Contrasted with Mature and Old. (La Forge.)

*Comment.* Subsequent authors have usually used the word in a general way, or as including one or more of the above, *i.e.* 1. Stage when rivers are youthful; 2. Stage when divides are youthful, *i.e.* with upland surfaces not yet attacked by erosion or corrasion; 3. Period of increasing relief.

In addition to the application of the concept to both land and rivers it has been applied to marine erosion (D. W. Johnson) and to the Periglacial Cycle (see Peltzer, *A.A.A.G.*, 39, 1949).

**Yudokuchi** (Japanese)

Small mounds occupied by mites which carry the disease scrub-typhus: mite-islands.

**Yungas**

Mill, *Dict.* A climate zone in Bolivia, etc., comprising the lowlands, etc., up to 5,000 feet; temperature tropical; atmosphere humid. The zones in order are Yungas, Valle, Cabezeva de valle, Puna, and Puna brava.

Jones, C. F., 1930, *South America*. New York: Holt, 'The term "Yungas" is being used in the sense of the deep valleys and ridges of the eastern slopes of the Cordillera system, by some authors called, from a geographical standpoint, the Yungas, and not in the narrower use of the term to apply locally to the Yungas of Inquisivi and La Paz.' (p. 252)

James, P. E., *Latin America*, London: Cassell, 1941. 'This rainy and heavily forested north-eastern slope of the Eastern Cordillera, which is the equivalent of the Eastern Border Valleys Region of Peru, is known in Bolivia as the Yungas.' (p. 192) (Also p. 206. It is not clear whether the region is delimited by vegetation, rainfall or land form. (W.M.))

Fay, 1920. (Bol.) A region of low plains; an alluvial basin, often containing rich placers (Halse). (Halse, E., *A dict. o, Spanish, Spanish-American Portuguese and Portuguese-American mining, metallurgical and allied terms*, 2nd ed., 1914.)

No reference in Knox, 1904; not in *O.E.D.*

**Ywa** (Burmese)

Village.

# Z

**Zaaidam**—see Saaidam

**Zāid-rabi** (Indo-Pakistan: Urdu)
An additional *rabi* crop, sown about April–May and harvested in June–July—usually melons or cucumbers.

**Zambo**
James, P. E., 1959, *Latin America*, 3rd ed., New York: Odyssey. 'In Spanish America, the mixture of Indian and European is called a *mestizo*; the mixture of Negro and European is called a *mulatto*; and the mixture of Negro and Indian is called a *zambo*.' (p. 13)

**Zariba** (Sudan: Arabic, *pl.* zaraīb)
*Webster*. Zariba, Zareba, Zareeba. An improvised stockade, especially of thorn bushes, etc.; a fortified camp; originally an African use.
An enclosure for domestic animals in the northern Sudan, usually in or near villages, and fenced by branches cut from thorny shrubs, which are either inserted into or laid upon the ground. (J.H.G.L.)

**Zastrugi**—see Sastrugi

**Zawn** (Cornwall—Land's End)
A little inlet of the sea.
Not in *O.E.D.*

**Zechstein** (German)
The upper of the two divisions of the Permian of Germany and continental Europe. The Permian of north-eastern England is probably all of this age.

**Zemindār, zamindār** (Indo-Pakistan: Urdu-Hindi) (various spellings)
A word, which formerly meant a collector of revenue, that has long been used to mean landlord, usually a large landowner from whom the peasants hold their lands and who is responsible for payment of taxes on the whole estate. Hence zemindary—the system of land tenure involved in contrast to ryotwary (*q.v.*).
A zemindar may however own only a small piece of land, and the word then means peasant proprietor or owner-occupier.

**Zenith**
*O.E.D.* The part of the sky directly overhead; the highest point of the celestial sphere as viewed from any particular place; the upper pole of the horizon (opp. to *nadir*).
Admiralty *Navigation Manual*. 'The point where the line joining the Earth's centre to the observer's position cuts the celestial sphere'. See Nadir.

**Zenith distance**
*O.E.D.* The angular distance of a heavenly body from the zenith (the complement of its *altitude* or angular distance from the horizon).

**Zeuge** (German, *pl.* Zeugen)
Mill, *Dict.* Witness Rock or Zeuge. Isolated tabular masses or columns or heights in the desert.
Wooldridge and Morgan, 1937. Tabular masses of some harder stratum on a pedestal of shale, mudstone, etc., which have been undercut by corrasion of wind-blown sand. They vary from 5 to 150 feet in height. (*Cf.* p. 298 and diagram.)

**Zeugenberg** (German)
A hill or mountain of similar form and origin to a Zeuge but on a larger scale. *Cf.* butte, kop. See Butte témoin.

**Zeyat** (Burmese)
A shelter or rest house for travellers erected by the wayside by pious Buddhists as a work of merit.

**Zodiac**
*O.E.D. Astr.* A belt of the celestial sphere extending about 8 or 9 degrees on each side of the ecliptic, within which the apparent motions of the sun, moon, and principal planets take place; it is divided into twelve equal parts called *signs*.
*Signs of the Zodiac:* the twelve equal parts into which the Zodiac is divided, and through one of which the sun passes in each month or twelfth part of the year; they are named after the twelve constellations (Aries, Taurus, Gemini, Cancer, Leo, Virgo, Libra, Scorpio, Sagittarius, Capricornus, Aquarius, Pisces) with which at a former epoch they severally coincided approximately.
*Comment.* The signs in English are thus the Ram, the Bull, the Heavenly Twins, the Crab, the Lion, the Virgin, the Scales,

the Scorpion, the Archer, the Goat, the Waterer, and the Fish. Owing to the precession (*q.v.*) of the equinoxes, the point at which the Sun crosses the Celestial Equator (the Spring Equinox), although still termed the First Point of Aries, is no longer in that sign (E.G.R.T.).

## Zonda

Kendrew, 1953. In Argentina, in strong contrast to the pampero, and 'often preceding it, is the Zonda of the Western Region, a strong W. wind, hot, dry, and dusty, which acquires its föhn character in its descent from the Andes; it is most frequent in spring'. (p. 512)

Miller, 1953. 'The *Zonda* of the Argentine is a hot humid wind which brings a feeling of complete prostration ...' (p. 182).

## Zone (geographical), zonal

*O.E.D.* 1. Each of the five 'belts' or encircling regions, distinguished by differences of climate, into which the surface of the earth (and, in ancient cosmography, the celestial sphere) is divided by the tropics (of Cancer and Capricorn) and the polar (arctic and antarctic) circles; viz. the torrid (burning, burnt, hot) zone between the tropics, the (north and south) temperate zones extending from the tropics to the polar circles, and the frigid (frozen, cold) zones (arctic and antarctic) within the polar circles.

The arctic and antarctic zones are strictly not 'belts' but circular 'caps' with the poles in the centre.

2. Any region extending around the earth and comprised between definite limits, *e.g.* between two parallels of latitude. Also applied to a similar region in the heavens or on the surface of a planet or the sun.

3. More or less vaguely: a region or tract of the world, especially in relation to its climate; also figuratively.

4. A definite region or area of the earth, or of any place or space, distinguished from adjacent regions by some special quality or condition (indicated by a defining word or phrase); also figuratively.

5. In geology and Physical Geography: a region, or each of a series of regions, comprised between definite limits of any kind, *e.g.* of depth or height, and distinguished by special characters, especially by characteristic fossils or forms of animal and plant life.

*Webster.* 9. *Biogeog.* An area or part of a region characterized by uniform or similar animal and plant life; a life zone; as, littoral *zone*; Austral *zone*, Boreal *zone*, etc. The *zones*, or *life zones*, commonly recognized for North America are the Arctic, Hudsonian, Canadian, Transition, Upper Austral, Lower Austral, and Tropical. See also *abyssal zone, littoral zone, pelagic zone.*

Mill, *Dict.* 1. One of the five great divisions of the Earth, bounded by circles parallel to the equator and named according to the temperatures prevalent in each: Arctic, North Temperate, Torrid, South Temperate, Antarctic. 2. An area of the Earth's surface; generally used in connection with some characteristic of such an area.

*Comment.* From the first and second uses (O.E.D) come zonal as applied for example to zonal soils which follow more or less climatic zones; from (4) above come urban zone or zones of urban influence; from (5) see photic zone, aphotic zone, etc. The concept of the five belts or zones dates from classical times and has tended to retard constructive thinking. For example the 'Temperate Zone' has some of the least 'temperate' climates to be found in the world. 'Torrid' as applied to a zone is obsolescent.

## Zone, zoning (in planning)

*Encyclopedia of the Social Sciences*, 1935, New York: Macmillan, 15. 'Zoning as commonly understood refers to the legal regulation by districts or zones of the use of private property ... it has to do not only with the use to which the land is put but also with the height of buildings erected thereon and the percentage of ground space which they may occupy' (p. 538).

*Comment.* As commonly used in present day practice, 'to zone' is to allocate land for major purposes in a plan for future development. Thus land can be 'zoned for housing', 'zoned for industry', etc.

## Zone, zoning, zonation (geological)

Himus, 1954. A group of strata of limited but variable thickness, characterized by a definite assemblage of fossils, which distinguishes it from all other deposits. Usually one or more of the species is confined to the zone or may be abundant therein and rare outside its limits. The zone is named after one of the characteristic species, for example, the zone of *Micraster coranguinum* in the Upper Chalk.

*Comment.* The strata comprised in a zone were deposited during a period of *time* which has been called a hemera. From the above use of zone come zone-fossil; zonal assemblage (of fossils) and zonal indices. Geologists have gone further and distinguish 'sub-zones'. See discussion in L. D. Stamp, *An Introduction to Stratigraphy*, 1st ed., 1923, and later editions. Geologists also have many other uses of 'zone', *e.g.* zone of contact around an igneous mass such as granite in which an alteration of the country rock has taken place (= metamorphic aureole); also different layers or parts of the earth's crust such as zone of weathering; deeper in the crust, zone of fracture and then zone of flow.

**Zones, vertical**—see Terra caliente, etc.

**Zoogeography, zoology, zooplankton, etc.**

The prefix zoo- (from the Greek zoon, animal) is used in the construction of a large number of scientific terms especially in contrast to *phyto-* (plants) or *bio-* (animals and plants). Zoology is the comprehensive word for the study of animal life; zoogeography the study of the distribution of animals. The planktonic life of the ocean can be separated into phytoplankton and zooplankton. (L.D.S.)

**Zoonosis** (*pl.* Zoonoses)

*Webster*, a. A disease communicable from animals to each other or to man. b. A disease due to animal parasites.

*World Health Organization*, 'those diseases which are naturally transmitted between vertebrate animals and man'. W.H.O. As quoted in *The Practitioner*, **191**, Nov. 1963, No. 1145. Quoted also in *Nature*, **201**, Jan. 18 1964, No. 4916, p. 246.

See also *Science*, **143**, 1964, 1464–1466.

**Zoophyte**

A plant-like animal such as a coral or sponge.

# APPENDIX I

## GREEK AND LATIN ROOTS COMMONLY USED IN CONSTRUCTION OF TERMS

This Appendix gives the more common prefixes, suffixes and syllables derived from classical Greek and Latin which have been used in the construction of geographical terms. It should enable a number of terms not included in the Glossary to be interpreted and understood. Some root words are common to Greek and Latin; it is generally agreed that a mixture of Greek and Latin derivatives in a single word is undesirable though such a mixture is by no means uncommon. For help in the revision of this list the Committee is indebted to my good friend the Rev. Canon Walter Prest, M.A., Dip. Th., sometime Vicar of Bude Haven.

L.D.S.

In most cases one or two typical examples are given; many others will be found in the glossary.

### GREEK

a- (used before a consonant), an (before a vowel), from ἀ, ἀν (a, an) = without, not, -less. Equivalent to in- (Latin), un- or non-
    abyssal, aclimatic, azonal, anaerobic, axeric
aer- from ἀηρ (aer), the air (also Latin)
    aerology
agri-, agro- from ἀγρός (agros), a field; also Latin, *ager*
    agronomy
agrost- from ἀγρωστις (agrostis), a certain wild grass
    agrostology
allo- from ἄλλος (allos), another, strange
    allogenic
ana- from ἀνά- (ana-), up, in place or time, back, again, anew
    anabranch
anemo- from ἄνεμος (anemos), wind
    anemometer
anthropo- from ἄνθρωπος (anthropos), man
    anthropogeography, anthropoid
anti, ant-, anth- from ἀντί, ἀντ-, ἀνθ (anti-, ant-, anth-), opposite, against
    anticline
apo-, ap- from ἀπό- (apo-), off, from, away, detached
    apogee, aphelion
archaeo-, archeo- from ἀρχαῖος (archaios), ancient, primitive
    archaeology, archaean
arch- (1) from ἀρχί-, ἀρχός (archi; archos), chief
    archipelago, lit. the chief sea
-arch (2) from ἄρχω (archo), to command, to rule
    autarchy, monarch

# APPENDIX I

argill- from ἄργιλλος (argillos), clay; also Latin
    argillaceous
aster, astro- from ἀστήρ (aster), a star, pertaining to stars; also Latin
    astronomy
astheno- from ἀσθένεια (astheneia), from α and σθενος (asthenos), lack of strength
    asthenosphere
aut-, auto-, auta- from αὐτός (autos), self; by oneself, independently
    autarchy, autochthon
    Note: automobile, a vehicle mobile by itself, i.e. not drawn by animals; shortened to auto and then used in such combinations as autobahn.

bar, baro- from βάρος (baros), weight
    barometer, isobar, millibar
batho-, bathy- from βάθος (bathos), depth
    batholith, bathymetric
benthos, benthic from βένθος (benthos), poetical for βάθος, the depth of the sea
    benthos
bio- from βίος (bios), life, course, way of living
    biosphere, biogeography
boreal from βορέας (boreas), the north wind, hence the north and Latin
    borealis, pertaining to the north
    boreal forests
brachy- from βραχύς (brachus), short
    brachycephalic, brakeph
brady- from βραδύς (bradus), slow
    bradyseism
bysma-, -bysm, -byssal from βυσσός (bussos) or βυθός (buthos), the depth, the sea, the bottom
    bysmalith, abysm, hypabyssal

caino-, caeno-, ceno-, kaino- from καινός (kainos), recent
    cainozoic
cata-, kata-, cat-, cath- from κατα-, κατ-, καθ- (kata-, kat-, kath-), down, away, entirely
    catabatic (katabatic), katothermal
ceph-, cephal, keph- from κεφαλή (kephale), the head
    cephalic index, brachycephalic, brakeph
chalyb- from χαλυβηίς (chalubeis)—the Chalybes, an ancient nation in Asia Minor famed for their work in iron and steel hence chalybeate, impregnated or flavoured with iron, applied to mineral springs.
choro- from χώρα (chōra), a place, a district
    chorography, choropleth
chrom-, chromo- from χρῶμα (chrōma), colour
chrono- from χρόνος (chronos), time
    chronology, isochrone, isophytochrone, tautochrone
clima from κλίμα (clima), slope
    climate

-cline, clino-, -clinal from root κλιν- (klin-), sloping
    anticline, syncline, isoclinal, clinometer
coeno- from κοινος (koinos), common
    coenosis, biocoenosis
-cole, -colous from κόλον (kolon), fruit, juice; τὸ κόλον, fodder; but see also Latin colere, to inhabit
    calcicole (growth on limestone), (inhabiting limestone)
copro- from κόπρος (kopros), dung
    coprolite
cosmo-, cosmic from κόσμος (kosmos), the world considered as an organized entity
    cosmography, cosmopolitan
crat-, crato-, -crat from κράτος (kratos), authority, rule, sovereignty, power
    craton, orocratic, autocrat
cryo-, from κρύος (kruos), frost
    cryology, etc.
crypto- from κρυπτός (kruptos), hidden, secret
    cryptocrystalline
cryst-, crystal- from κρύσταλλος (krustallos), clear ice, rock crystal
    crystalline
cyclo- from κύκλος (kuklos), circle
    cycle, cyclone

dasy- from δασύς (dasus), hairy, hence thick, dense
    dasymetric
demo- from δῆμος (dēmos), the people
    demography, demopleth
dendro- from δένδρον (dendron), a tree
    dendritic (shaped like a tree with branches)
deutero- from δεύτερος (deuteros), second, hence secondary
    deuterozoic
di- from δι- (di-) for δίς (dis), twice, two (see also Latin dis-)
    dimorphous, dicotyledon, diarchy
dia- from δι- (di-) for διά- (dia-), through, during, across
    diachronism, diaclinal
dolicho- from δολιχός (dolichos), long
    dolichocephalic, dokeph
dys- from δυς- (dus-), ill, bad; used as a negative with a sense of pain or hardness in fulfilment
    dysgeogenous, dysgenic, dystrophy
dynamo- from δύναμις (dunamis), power
    dynamo-metamorphism

eco-, oeco-, ek from οἶκος (oikos), a house; οικονομία (oikonomia), house management (also Latin).
    ecology, economic, ekistics
ecto- from ἐκτο- (ekto), outside; but exo- is more often used
    ectogenic
edaph- from ἔδαφος (edaphos), basis, floor
    edaphic

endo- from ἔνδον (endon), within
    endogenic
eo- from ἠώς (ēos), the dawn
    eozoic, eocene
ep-, epi- from ἐπί (epi-), on, upon, over, in addition, near; to, towards
    epigenic, epicycle, epicontinental
epeiro- from ἐπειρύω (epeiruō), to pull to, to collect
    epeirogenetic
erem- from ἐρῆμος (erēmos), desolate, lonely; ἐρημία (erēmia), a solitude, a desert, hence ἐρημίτες (erēmites), belonging to the desert. Usually applied to hermits as dwellers in solitude but applied by Gaussen to deserts
ethni-, ethno- from ἔθνος (ethnos), a nation
    ethnology
eu- from εὐ (eu), well, good, easy
    eustatic, entrophy
ex- from ἐξ (ex-), out (also Latin)
    exogenous

-gam from γάμος (gamos), marriage
ge-, geo- from γῆ (ge), the earth
    geography, geology
-gen, -genic, -genous, -geny from γεννάω (gennao), to produce; γενεά (genea), race, stock, family; γένεσις (genesis), origin, source. -gen: that which produces, or is produced. Also γένος (genos), kind
    endogenous, cratogen
-glot from γλῶττα (glotta), tongue
    monoglot, polyglot
-glyph, glyphic from ἡ γλυφή (gluphē), a carving, incision and γλύφω (gluphō), to carve, write
    petroglyph, hieroglyph
-gon from γωνία (gōnia), an angle
    polygon, agonic
-gram from γράμμα (gramma), something written; also τὸ γράφος (to graphos)
    cartogram, diagram
-graph, -graphy from ἡ γραφή (graphe), written; γράφω (graphō), to write, to draw
    geography, topography, barograph

hal-, halo- from ἅλς (hals), salt; αλο- (halo-)
    halophyte, isohaline
helio-, -hel from ἥλιος (helios), the sun
    isohel, aphelion
hemi- from ἡμι (hēmi), half (semi- in Latin)
    hemisphere
hetero- from ἕτερος (heteros), another, the other of two. Prefix denoting difference as opposed to resemblance
hiero- from ἱερος (hieros), sacred
    hieroglyph

hol- holo- from ὅλος (holos), whole, entire
    holism, holokarst
homo- from ὁμός (homos), same: opposite of hetero-
    homocline, homologue
hydra-, hydro- from ὕδωρ (hudōr), water
    hydraulics, hydrology, hydrography
hyet-, -hyet from ὑετός (huetos), rain
    isohyet
hygro from ὑγρός (hugros), wet
    hygrophyte, hygrophilous
hypa-, hypo- from ὑπό (hupo), under, within
    hypogene
hyper, hypa- from ὑπέρ (huper), above, beyond (Latin super)
    hypabyssal (better hyperbyssal)
hypso- from ὕψι (hupsi), high
    hypsometry

-id from -ις (is), -ιδα (-ida); also -ίδης (idēs), son of; a member of a group
    altaid, altaides, caledonids, caledonides
iso-, is- from ἴσος (isos), equal
    isopleth, isohyet, etc.

kata-, kaino-, kephalo-, kosmo-, etc.; see cata-, caino-, cephalo-, cosmo-, etc.

lacco- from λάκκος (lakkos), a hole, a pit, a reservoir
    laccolith
limn- from λίμνη (limnē), a marsh
    limnology, monimolimnion
litho-, -lith, -lite from λίθος (lithos), a stone
    lithology, megalith, coprolite, laccolite
-logy, -ology from λόγος (logos), that which is spoken or said, story of, a treatise about
    geology, zoology, phytology
-lysis, from λύσις (lusis), a setting free, loosing

macro- from μακρός (makros), long, large as opposed to μικρός (micros), small
    macrogeography
mega- from μέγας (megas), great, large; the combining form is mega-, megal-, megalo-
    megalith, megalopolis
-mene from μήν, μηνός a month
    isohyetomene
mero- from μέρος (meros) part, fraction
    merokarst
meso-, mes- from μέσος (mesos), middle
    mesozoic, mesoclimate
meta- from μετά- (meta-), a prefix with various meanings but especially change (of place, order, condition or nature); cf. Latin trans.
    metamorphism

meteor from μετέωρος (meteōros), raised above the earth, soaring in the air, and μετεωρὸν (meteōron), a meteor
    meteorology
metro-, -meter from μέτρον (metron), a yardstick, that by which something is measured
    dasymeter
metro- from μήτρος (mētros), genitive singular of μήτηρ (mēter) a mother
    metropolis
micro- from μικρός (mikros), small
    microclimate
mio- from μεῖον (meion), less
    miocene
mono-, mon- from μόνος (monos), alone, single
    monocline, monolith, monoculture
morphe-, morph-, -morph from μορφή (morphē), form, shape
    morphology, pseudomorph, geomorphology

nem- from νῆμα (nema), a thread
    nematoid
neo- from νέος (neos), new
    neogene, neomalthusianism
-nomy (1) from νόμος (nomos), established custom, usage or law or (2) νέμω (genitive) (nemō), administration
    agronomy
noso- from νόσος (nosos), disease
    nosopleth
-nym from ὄνομα (onoma), a name
    exonym, toponymy

oec-, see eco-
    oecology now ecology
-oid from εἶδος (eidos), form, appearance, shape. A suffix used to denote resemblance
    caledonoid
oligo- from ὀλίγος (oligos), few, little
    oligocene, oligomict, oligotrophy
-ology, see -logy
-ombro from ὄμβρος (ombros), rain
    ombrothermic, isothermombrose
oro-, oreo- from ὄρος (oros), a mountain
    orography
ortho- from ὀρθός (orthos), straight, upright, regular
    orthogneiss

pachy-, -pach from παχύς (pachus), thick
    isopach, isopachyte
palaeo-, paleo- from παλαιός (palaios), ancient
    palaeogeography, palaeomagnetism
pan- from πᾶν (pan), neuter of πᾶς (pas), all. As prefix denoting all, everything, altogether, everyway
    panplanation, panfan

# APPENDIX I 537

para from παρά (para), from the side of, alongside; *cf.* parallel
  paragneiss
ped- from πίδον, the ground hence related to soil. See also Latin pes, pedis, a foot (see pod-). Not to be confused with Greek παῖς, παιδός, a boy (pais, paidos) used in pedagogue
  ped, pedology
pelag- from πέλαγος (pelagos), the sea, the ocean (also pelagus in Latin)
  pelagic
peri- from περί- (peri-), around, about, near (*cf.* Latin circum)
  periglacial
petro-, petra-, petri- from πέτρα (petra), a stone, rock (also Latin)
  petrology
phaco- from φακός (phakos), a lentil bean, anything shaped like a bean, lenticular
  phacolith
phanero- from φανερός (phaneros), visible, evident (opposite of crypto-, hidden, secret)
  phanerophyte, phanerozoic
pheno- from φαίνειν (phainein) to show
  phenocryst
-phil, philo- from φίλος (philos), loving, fond of, cultivating
  hygrophilous, anglophil
-phobe-, -phobus from φόβος (phobos), fear
  calciphobe
photo-, phot- from φώς (phōs), light (genitive singular φωτός (phōtos))
  aphotic, photic
-phyll from φύλλον (phullon), a leaf
  schlerophyll
phylo- from φύλον (phulon), a tribe or race
  phylogenetic, phylum
physic-, physio- from φυσικός (phusikos), pertaining to nature, natural; φύσις (phusis), nature
  physiography, physical geography
phyto-, -phyte from φυτόν (phuton), that which has grown, a plant, also a creature
  halophyte, mesophyte, xerophyte, phytogeography
plat- from πλατύς (platus), broad, flat
  platform
-pleth from πλέθρον (plethron), a measure
  isopleth
plio- from πλειών (pleiōn), more
  pliocene
pleisto- from πλεῖστος (pleistos), most
  pleistocene
pluto- from πλούτων (Ploutōn), Plouton or Pluto, God of the Underworld
  plutonic rocks, plutonism
pneumat- from πνευματικός (pneumatikos), belonging to the air or wind or gases
  pneumatolysis
pod-, podo-, from πούς, πόδος (pous, podos), a foot

# APPENDIX I

-polis, -opolis, -politan from πόλις (polis), a city, city-state; πολίτης (politēs), a citizen
    metropolis, cosmopolitan
potamo- from ποταμός (potamos), a river, marsh
    potamology
poly- from πολύς (polus), many, much
    polycyclic, polygon
pro- from πρό- (pro-), before (also Latin)
    proglacial, progradation
protero- from πρότερος (proteros), before
    proterozoic
proto-, prot- from πρῶτος (prōtos), first
    prototype, proto-Thames, etc.
psamm- from ψαμμός (psammos), sand
    psammitic rocks
pseudo-, pseud- from ψευδής (pseudēs), false, having a deceptive resemblance
    pseudomorph
psycho from ψυχή (psuchē), breath, the soul
    psychosphere
pyro-, pur- from πῦρ (pur), fire
    pyrometamorphism, pyroclastic

rheo-, -reic from ῥέω (rheo) to flow
    rheology, areic, endoreic
rhiza-, rhizo-, rhiz- from ρίζα (riza), a root
    rhizosphere

sapro- from σαπρός (sapros), putrid, rotten
    saprophyte, saprolite
seismo-, -seism from σεισμός (seismos), an earthquake
    seismology, isoseismic
spher-, sphaer-, -sphere from σφαῖρα (sphaira), a ball or sphere
    spheroidal weathering, lithosphere
stadia from στάδιον (stadion), a measure of length
    stadial moraine
stat- from στατικός (statikos), at a standstill; from στα- (sta), the root of to stand
    isostasy, isostatic
steno-, sten- from στενός (stenos), narrow, within narrow limits
    stenohaline
-strophe from στροφή (strophē), a turning; στρόφος (strophos), a twisted band
    catastrophism
syn- from σύν (sun), with, together, similarly, alike; also in form σύμ (sum)
    syncline, synecology, symbiosis

tauto- from ταυτό (tauto) (for τὸ αὐτό- to auto), the same
    tautochrone
taxi-, -taxis from τάξις (taxis), order, arrangement, line of battle
    taxonomy

# APPENDIX I

tecton- from τέκτων (tektōn), a carpenter, hence builder; and τεχτονικός (techtonikos), relating to construction
    tectonic
thalass- from θάλασσα (thalassa), the sea, ocean
    thalassography
-them from θέμα (thema), that which is laid down, a theme
    cyclothem
therm-, -therm from θέρμη (therme), heat; θερμός (thermos), hot
    isotherm
-tone from τόνος (tonos), strain
    ecotone
topo-, -tope from τόπος (topos), a place
    topography, ecotope
trach- from τραχύς (trachus), rough, hairy
    trachyte
-trope from τρέπω (trepo) to turn; τροπή (tropē) a change, variation, turn; see also τρεφειν, to nourish
    entropic
tropo- from τρόπος (tropos), turning
    tropophyte, troposphere
-trophe from τροφή (trophē), nourishment, rearing; τρεφειν, to nourish
    oligotrophy, eutrophy
xeno- from ξένος (xenos), a stranger
    xenolith
xero- from ξηρός (xeros), dry
    xerophyte
zoo- from ζωον (zōon), an animal, life
    zoology, azoic, eozoic

## LATIN

ab-: off, away, from
    ablation, abrade, abrasion
-acy, -cy from -acia, atia. Used as suffix to change an adjective of quality state, or condition into a noun
    pirate, piracy; potent, potency
ad-: (becomes ac- before c, k and qu; af- before f; ag- before g; al- before l; ar- before r; as- before s; at- before t; to, with sense of motion to, change into, addition, or intensification
    acclimatize, accumulate, advection, afforestation, agglomerate, association, attrition
aer-, air from aër, the air (also Greek)
    aeration
agri-, agro- from ager, a field, agris of or pertaining to a field (also Greek)
    agriculture
-al from -alis, of the kind of, pertaining to. Used as suffix to change a noun to an adjective
    fluvial, spherical
alti-, alto- from altus, high
    altitude
ambi- from ambo, both

# APPENDIX I

-an, -ian from -anus, -ana, -anum (also -ianus), of or belonging to. Suffix to change a noun (especially a country or place) to an adjective. Occasionally -ane; also -ian
    America, American; Paris, Parisian

annum: year, hence per annum, p.a., also annual; perennial = through the years

annular from annulus or anulus, a ring
    annular drainage

ante-: before (in time or place)
    antecedent

aqua: water
    aqueous, aquifer or aquafer

ara: a plough; arare to plough; arabilis, able to be ploughed
    arable

arena: sand
    arenaceous

argilla: clay (also Greek)
    argillaceous

arti-, arte- from ars, artis art, human skill as opposed to nature
    artefact

aster, astro-: a star, pertaining to stars (also Greek)
    astronomy

auri-, auro- from aurum, gold
    auriferous

auster, austral-: the south wind, hence the south; australis, pertaining to the south (wind)
    austral, australia(n)

balnea-, balneo- from balneum, a bath
    balneology

bi-: twice, doubly, having two, two- (see also di-, and Greek)
    bipolar

boreal: borealis, pertaining to the north (wind) (also Greek)
    boreal forests

calc- from calx, calcis, lime; calcarius, of or pertaining to lime
    calcareous, calcicole, calcifuge, pedocal, calcrete

capilla: a hair, capillarius, hair-like
    capillary fringe, capillarity

carbo-, carboni- from carbo, carbonis, coal
    carboniferous

carta-, carto-, chart from charta, carta, a map (also Greek)
    cartography, chart

catena: a chain, a connected series
    catena, catenary

centrum: centre; hence centri-, centro- (also Greek)
    centrosphere

centum: a hundred
    per centum, usually shortened to per

circum-: around, round about
    circumdenudatio

# APPENDIX 1    541

-cide, -cision from caedere, to cut; decidere, to cut down
    incised, incision, deciduous
cis-: on this side of; as opposed to trans- or ultra- on the other side of
    cis-Alpine
co-: col- (before 1), com- (before b, p, m, etc.), con-, cor- (before r) from Latin cum-, together, together with, in combination or union
    confluent, conformable, congelifraction, conglomerate
contra-: against, in opposition to, opposite
    contraposed shoreline
creta: chalk
    cretaceous
cult from cultus, worship
    cultural geography
-culture from cultura, culture
    agriculture, silviculture
cultivate from cultivare, to till, plough
    cultivated land
cumulus: a heap, a pile
    accumulation mountains

de-: down, down from; as a prefix to undo the action
    degrade; deglaciation
demi-: half from Latin dimidius, through French demi
dexter, dextra from dexter, dextr- right; dextra the right hand
di-, dis-: used with a variety of meanings but in general to express negation; away from, denoting the opposite or lack of the characteristic in question
    discordant, divagation
digit- from digitus, finger
    digitate
diluvi- from diluvium, a flood
    diluvium, diluvial
dis-: prefix expressing the opposite of the thing or characteristic in question
    disconformity
dom- from domus, a house, home
    domestic
duro-, dura- from durus, dura, durum hard; duro to last
    duricrust

en-, -in- from French en-, Latin in-, Greek ἐν (en), in
    enclosure, inclosure
equi- from aequus, equal
    equinox
erode, erosion from erodo, erodere, to gnaw away
erro, errare; to wander; erratum, wandered
    erratic block
-escens, -escent: a Latin suffix which conveys the sense of 'becoming';
    crescent
        obsolescent = becoming obsolete; senescent = becoming senile, etc.
ex-: ex- out (also Greek)

## APPENDIX I

extra-: beyond, farther than, except
    extraterritorial

-etum: a suffix of Latin form used by ecologists to designate a plant association dominated by a single genus, *e.g.* calluna, callunetum; sphagnum, sphagnetum

-fact, -faction, -fication from factum a thing done (neuter singular of factus from verb facio, facere (French—faire), to do). Also factio, a doing, and -ficare
    Conveys the idea of something done or made
        artifact, ventifact
    Conveys the idea of doing or making
        petrifaction, petrification

-fer, -ferous from fero, ferre, to bear; bearing, carrying, producing
    aquifer, conifer, coniferous (cone-bearing)

ferrum, ferrous, from ferrum, iron

fluvi-, fluo-, fluction from fluvius, a river; fluo, fluere to flow. Compare French fleuve.
    fluvial, fluviatile, interfluve, solifluction, affluent

for-, fore- from foris, outside
    forest
    More often the prefix fore- is from the Anglo-Saxon fore meaning before, as in fore-deep, fore-set beds

fossa from fodio, fodere, to dig; fossus, dug, hence fossa terra, land dug up and fossa, a ditch; fossilis, dug up
    fossa, fossil

-fract, -fraction from frango, frangere to break, fractus broken, something broken, the act or result of breaking
    congelifraction

-fuge from fugio, fugere, to flee; fleeing from
    calcifuge

gel- from gelo, gelare, to congeal; gelu, frost
    congeliturbation, regelation

glac- from glacies, ice
    glacial, glaciation, deglaciation

glob- from globo, globere, to make into a ball; globus, a globe
    global

glomer- from glomero, glomerare, to collect into a ball, glomus, a ball
    agglomerate, conglomerate

grad- from gradus, a step
    aggrade, degrade, grade

gran- from granum, a grain
    granite, granular

grav- from gravis, heavy
    graviplanation

haema-, haemat-, haemato-, hema-: Latin from Greek αἷμα blood
    haematite

horti- from hortus, a garden
    horticulture

# APPENDIX I

humi- from humus, the ground; humi, on the ground; also
    humidus, umidus, moist

igne-, igni- from ignis, fire
    igneous rocks

in-, il-, im-: the Latin negative in-, which becomes il- before l; im- before m, etc. With many words the Anglo-Saxon negative un- is commonly preferred
    immature

infra-: within, below

insula: island
    peninsula, insularity

inter-: between
    interfluve

intra-: on the inside, within
    intratelluric

inver-: inversus, inverted, from in- and verto, to turn

lac-: lacus, a lake
    lacustrine

lam(m)ina, lamella: a thin plate
    lamination

later (1) lateral from latus, a side, lateralis
    unilateral

later (2) a brick
    laterite

lav-, luv-, -luv from lavo, lavare, to wash; also luere
    eluvial, illuvial

litor-, littor- from lit(t)us, the shore; littoralis, pertaining to the shore
    littoral deposits

loc-, loco- from locus, a place
    location, localization

mal- from malus, bad, ill

man-, manu- from manus, the hand
    manufacture

mar-, marine from mare, the sea; marinus of the sea
    marine, maritime

medi- from medius, middle
    Mediterranean

mil-, mill-, milli-, mille- from mille, a thousand
    mille map, mile

minut- from minutus, small
    minute

mort- from mors, mortis, death
    mortlake

mult-, multi- from multus, many
    multi-cycle landscape

navi- from navis, ship; navigo, navigare, to navigate
    navigation

niv-, nif from nix, nivis, snow
  nivation, isonif, niveo-eolian
non-: negative prefix
  non-ferrous
nud-, nudo- from nudus, nude, naked
  denudation
optimum: optimus, best
  optimum population
ob-: a common prefix with many meanings, sometimes intensive, sometimes to denote inversion or on the back of. Becomes oc-, of-, op-, etc., before c, f, p
oper- from opus, operis (*pl.* opera), work
ordin- from ordo, ordinis, order arrangement
oro-: usually from the Greek oros, a mountain, but also Latin os, oris the mouth
-origine from origo, originis, origin, a beginning
  aborigine

ped- from pes, pedis, a foot
  pedology, pediment
pelag- from pelagus, the sea, ocean (also Greek)
  pelagic
pen-, pene- from pene, almost
  peninsula, peneplane
per: by, through
  per annum, per cent(um), perennial
petra: a rock, stone (also Greek)
  petrifaction
pinna: a feather, a fin
  pinnate drainage
plan-, plano-, planus: level, flat
  plan, planation, peneplane
plen: plenus, full; plenarius, entire
  plenary
pluvi: pluvia, rain
  pluvial period
post-: after, behind, since
  post-glacial
pre-, prae: before (in time, place, etc.)
  pre-glacial
prima-, primo-: primus, first
  primate city
pro-: before (also Greek)
  proglacial

re- (i) ablative of Latin res, thing. Referring to, in the matter of (also 'in re')
  (ii) prefix denoting repetition of an action
    resequent, rejuvenation
ripa: a bank of a river
  riparian

# APPENDIX I   545

rur- from rus, ruris, country (as opposed to urbs, urbis, town)
    rural
retro-: back, backward
    retrogradation

sal: salt
    saline
salto, saltare, saltatio: to leap, leaping, jumping
    saltation
sect-, secto-: seco, secere to cut; sectus, cut
    transect diagram
semi-, sem-: half
senile: senex, senilis, old; senescens, becoming old
    senile topography
sequent, sequence from sequens, following; sequentia, a following
    resequent, consequent, obsequent
silva-, silvi- from silva, a wood
    silviculture
socio-, social from socius, a companion; socio, to accompany
    sociology
sol: the sun
    insolation
sol from solea, the sole of the foot: hence the ground, or soil it touches
    solifluxion, latosol
spelaeum: a cave (also Greek σπήλαιον, spēlaion)
    spelaeology, usually speleology
stratum, strata (*pl.*): layer, layers (*lit.* that which is laid flat)
    stratigraphy, stratosphere
sub-: under; also, to some degree
    submarine, subsoil, subnival, suburb
super-: above
    superposition
syn-: a Latinized form of Greek σύν (sun), with, together
    synecology

tellus: the earth
    telluric
tempor-: tempus, temporis, time
terra: land, the earth
    Mediterranean, terrigenous
trans-: across, beyond, through, with idea of change
    transgression, trans-Alpine, transhumance

ultra-: beyond
    ultrabasic rocks
un-: an English prefix denoting negation frequently used with words of Latin origin, though the Latin is in-
    unconformity
unda: a wave; undula, a small wave
    undation theory

uni-: unus, una, unum, one
  unilateral shifting
urb-: urbs, urbis, a town
  urbanism

vado-: vado, vadare, to wander
  vadose water
vect- from veho, vehere, to carry; vectus, carried; hence advect-, convect-
  advection, convection
vent- (1) from venio, venire, to come; ventus, come; ad-venio, to come to
  adventitious
vent-, venti- (2) ventus, wind
  ventifact
vitri-, from vitrum, glass
  vitrifaction
vulcan: Vulcanus, the god who presided over the smelting of metals
  vulcanism

# APPENDIX II

## LISTS OF WORDS IN FOREIGN LANGUAGES WHICH HAVE BEEN ABSORBED INTO ENGLISH LITERATURE

### AFRICAN LANGUAGES

International geographical literature now includes a number of words derived at least ultimately from the languages of Africa south of the Sahara. Some of the better known, such as *donga* and *dambo*, are also included in the Afrikaans list as they have been introduced via South Africa and are almost as much Afrikaans words as, in this case, Bantu. Certain words and terms long used by Europeans in East Africa have received a wider currency with the growth of geographical studies. The Committee is greatly indebted to Professor S. J. K. Baker of Makerere College, University College of East Africa, and his staff for a carefully considered list, with definitions, covering East Africa. In West Africa the writings of J. C. Pugh, K. M. Buchanan, R. J. Harrison-Church, E. A. Boateng and others have brought into common use a number of words and the Committee is indebted especially to Professor Pugh and Professor Harrison-Church for their personal advice and help. It is interesting to note that many words have been borrowed by one language from another: both original language and correct spelling are often uncertain.

Aftout (Mauritania)
Balleh (Somali)
Banto faro (Gambia)
Boli (Sierra Leone)
Boma (Swahili)
Dambo (Bantu)
Dega (Ethiopia)
Donga (Bantu)
Fadama (Hausa)
Fako (Hausa)
Firki, firiki (Nigeria)
Gombolola (Luganda)
Kivas (Hottentot)
Kolla (Ethiopia)
Lusaka (Bantu)

Lusuku (Luganda)
Mbuga (Swahili)
Miombo (Swahili)
Msitu (Bantu)
Murram (East Africa)
Niaye (West Africa)
Nyika (Swahili)
Omuramba (Hottentot)
Saza (Luganda)
Sebkha (West Africa)
Shamba (Swahili)
Tabki (Hausa)
Toich (Dinka)
Voina (Woina) Dega (Ethiopia)

### AFRIKAANS

The Committee is deeply indebted to Professor P. Serton of the University of Stellenbosch for providing a list of geographical terms in common use in South Africa, some of Afrikaans, others of Bantu, origin, together with definitions of each His definitions are acknowledged in the text by the

# APPENDIX II

letters P.S. A few additional terms have been derived from Professor J. H. Wellington's *Southern Africa* (2 vols., C.U.P., 1955).

| | | |
|---|---|---|
| Africander | Kaffir corn | Panneveld |
| Afrikaner | Kaffir farming | Poort |
| Apartheid | Karoo, Karroo | Rant (rand) |
| Banke | Karroid vegetation | Rondawel (rondavel) |
| Banket | Kimberlite | Rooikalk |
| Berg | Kivas (Hottentot) | Rug, ruens |
| Bergwind | Kloof | Saaidam |
| Biltong | Kop | Salt pan |
| Black turf | Koppie | Sandveld |
| Blue ground | Koup | Sloot |
| Bosveld | Kraal | Spitskop |
| Brak soils | Kraaling | Spruit |
| Bush | Krans | Straate |
| Bushveld | Laagte | Table cloth |
| Cape Doctor | Landdrost | Tafelberg |
| Dambo (Bantu) | Lusaka (Bantu) | Tafelkop |
| Dans | Mopane | Trek |
| Donga (Bantu) | Msitu (Bantu) | Uitlander |
| Dorbank | Mulola | Veld |
| Dorp | Nek | Vlei |
| Drif | Omuramba (Herero) | Vloer |
| Fontein | Opstal | Voortrekker |
| Hardebank | Oshana | Yellow ground |
| Hardeveld | Ouklip | |
| Kaffir | Pan | |

## ARABIC

For several reasons a large number of Arabic words have been used in international geographical literature. A few, including words of doubtful origin such as *wadi* used, somewhat differently, as far apart as Morocco and Iraq may date from the days of the great Arab empires and Arab geographers. Since Arabic is used over vast areas from the Atlantic borders of North Africa to the Indian ocean it is natural that travellers finding words such as *shaduf* and *suq* in wide use should adopt them. Further, in the great deserts are features unknown in more humid lands and it is natural that words like *wadi*, *nefūd* and *harrah* should be used in describing them. Finally, much excellent geographical work has been published in English by workers whose mother tongue is Arabic and they have almost unconsciously incorporated local words.

There is naturally much variation in the Arabic language from one region to another and in the terms used. The Committee is deeply indebted to the late Professor J. H. G. Lebon (helped by Abdel Rahman al-Nasri) of the University of Khartoum for a full and carefully annotated list of words current in the northern Sudan. He was also Professor in Baghdad and helped us with Arabic words from Iraq and elsewhere. His definitions are acknowledged by the initials J.H.G.L.

# APPENDIX II

In reaching a final decision it has been necessary to eliminate words of purely local usage, as well as cutting out, as in other languages, names of crops and plants.

We are greatly indebted to Professor M. Kassas of the University of Cairo for finally checking the list and definitions. He pointed out that many are useful terms and widely used but that bādōb, balagh, bugr, ferik, fula, gemma, 'idd, mudir, mudiriya and qurer are mainly restricted to north Sudan.

| | | |
|---|---|---|
| Ahqāf | Hafir | Raml, ramla |
| Areg | Hamada | Redir |
| Azbeh | Hariq | Reg |
| Bādōb | Harmattan | Sahara |
| Balagh | Harrah | Samun, samoon |
| Bat furan | Harratin | Saqiya, saqia, sakiya |
| Bugr | 'Idd | Sebkha |
| Chetoi | Jebel | Seif |
| Chili | Kasba | Seistan |
| Dahabiya | Khamsim | Seluka |
| Dahanah | Kharif | Serir |
| Daia | Khirba | Shādūf, shadoof |
| Dhow | Khor | Shamal |
| Erg | Liwa | Sheik |
| Fellah | Markaz | Shitwi |
| Felucca | Matmura | Shott |
| Ferik | Monsoon | Simoom, simoon |
| Foggara | Mudir | Sudan |
| Fula | Mudiriya | Sudd |
| Fulji | Nahyad | Suq |
| Gemma | Nefūd | Teras |
| Gerf | Nili | Tirs |
| Gezira | Qadha | Toich (Dinka) |
| Gibli | Qanat | Wadi |
| Goz | Qoz | Zariba |
| Habūb, haboob | Qurer | |

The following Arabic words appear widely in place names in Egypt and elsewhere:

*bab* (defile, gate or door), *barq* (spur, bluff), *beit* (house), *bir* (well, rock, cistern), *birba* (temple), *dar* (palace or large villa), *deir* (convent or monastery), *gebel* (hill or mountain), *geziret* (island, peninsula), *hallet* (village), *kasr* (castle or fortress), *khan* (caravanserai), *kom* (hill or mount), *medina* (town), *midan* (square), *nag* (village or encampment), *qal'a* (fortress or citadel), *qaret* (low hill), *ras* (cape or head), *tell* (hill).

The following words, though not geographical terms, may appear in geographical writings:

*amir* (nobleman), *diwan* (office or ministry), *hatiye* (area of vegetation), *ibn* (son), *mar* (governor), *melek* (king), *wali* (governor).

## AUSTRALIAN

Certain terms, some derived from aboriginal words, have become widely current in Australian geographical literature and to some extent have been used elsewhere. I am indebted in particular to Professor E. S. Hills, University of Melbourne (E.S.H.) and to Dr. T. Langford-Smith for help in compiling the following list and for suggesting definitions.

| | | |
|---|---|---|
| Abo | Gilgai | Southerly Burster |
| Australoid | Gums | *Spinifex* Desert |
| Billabong | Lunette | Squattocracy |
| Brickfielder, Brick Fielder | Mallee scrub | Stony rises |
| | Mulga scrub | Tent-hill |
| Brumby | Parna | Willy-willy |
| Crab hole | Sandplain | |
| Gibber plain | Scalded flats | |

## BURMESE

Burmese is a monosyllabic, tonal language (*cf.* Chinese) and place names are usually descriptive. The only terms widely used without translation in English literature are *taung-ya, chaung* and *indaing*.

| | | |
|---|---|---|
| Chaung (yaung, young) | Kwin | Ya |
| | Myo | Yoma (yomah) |
| Hpoongyi, hpongyi | Taung (daung) | Ywa |
| Indaing | Taung-ya | Zeyat |
| Kyaung | | |

## CHINESE LANGUAGES

With regard to terms of Chinese origin which have been introduced into geographical writings in English the late Professor George B. Cressey of the University of Syracuse was consulted. He considered *argol, gobi, junk, mai yu, shachiang, soy, tung* and *kaoliang* should be included. Advice was sought in the University of Hong Kong from Professor S. G. Davis and Dr. T. R. Tregear. Though there are few Chinese words as such which are used as geographical terms, most Chinese names are descriptive—*e.g.* Hong Kong is fragrant port (Cantonese), Kowloon is Nine Dragons (Cantonese) and so on. A brief bibliography of National and Cantonese words of this type with English equivalents is included in Dr. Tregear's *Land Use in Hong Kong and the New Territories*, 1958, pp. 72–73. Chinese towns are classified according to their status in an administrative hierarchy and it has always been a matter of some discussion whether the status *-fu, -hsien*, etc., is part of the place name or not. In Hong Kong it was also pointed out that certain Anglo-Indian words have been obviously imported by English-speaking workers with Indian experience but have acquired local meanings, *e.g. nullah, bund, godown* (warehouse); similarly with *paddy* (from Malaya) and *praya*. The final list of words from Chinese languages included in the Glossary is as follows:

| | | | |
|---|---|---|---|
| Argol | Gobi | Kaoliang | Shachiang |
| -chow | -hsien | Kaolin | Tala |
| Feng-shui | -king | Mai yu | T'ung |
| -fu | | | |

In the interpretation of place names the following may be noted: *tung* (east), *hsi* or *si* (west), *nan* (south), *pei* (north), *chiang* (*kiang*) and *ho*

# APPENDIX II

(river), *hu* (lake), *shan* (mountain), *ling* (mountain range), *pu* (port or mart) and *chou* (island). Tala and gobi are of Mongolian or Turkish origin.

## DANISH-NORWEGIAN

Many words of Scandinavian origin appear in British place-names and a number of terms of Scandinavian origin have been anglicized and incorporated in English geographical literature. As explained under Swedish, some authors prefer to use the original form and for this reason many of the words below have been included. The Committee is indebted to Dr. Peter J. Williams (P.J.W.) for his help in compiling the list.

| | | |
|---|---|---|
| Fiord, fjord | Ra | Tele |
| Fjeld | Seter | Tind |
| Fly | Skjer, skjaer, | Vidda |
| Föhrdes (fordes) | skjergaard | |
| Isblink | Strandflat | |

## DUTCH

Although a large number of words of Dutch origin have come into geographical literature through Afrikaans, surprisingly few have been incorporated direct from Dutch into English literature. An exhaustive treatment will be found in J. F. Bente, *A Dictionary of the Low Dutch Element in the English Vocabulary*, O.U.P., 1938. I am indebted to Professor C. H. Edelman for calling my attention to this as well as for his comment that a number of Dutch terms still appear on maps of the eastern United States, e.g. *Kil* (a kind of creek), but these are place names rather than terms. The following have been included in the glossary:

| | | | |
|---|---|---|---|
| Blijver | Maelstrom | Polder | Trekker |
| Wadden | | | |

## ESKIMO

At least two Eskimo words have become firmly established in geographical literature: *Nunatak* (introduced long ago by Nordenskiöld) and *pingo*.

## FRENCH

It is only to be expected that a large number of French words should appear in English geographical literature. Some, like glacier, hachure and plateau are so firmly entrenched in the language that we may forget their French origin; others are common French words which have been given specialized meanings in English, such as *bocage*, *pays* and *massif*; others are derived ultimately from other languages such as *doline* or from other lands such as *levée*; others are terms created in French such as *paléontologie* and given an anglicized form. Few people have had a wider experience of the use of French geographical terms than Dr. E. D. Laborde in the course of his translation of so many standard French works into English. In the first instance the Committee is indebted to him for a list with definitions of forty-one words he considered should be included in the Glossary. There are others, already included in the glossaries we have studied, so that the final list is an extensive one. For help with certain words I am indebted to Professor Pierre Deffontaines, Professor Hans Boesch, Professor Pierre Gourou, Professor Pierre Pruvost and to my colleague Professor R. J. Harrison-Church.

| | | |
|---|---|---|
| Abime | Creu | Morvan |
| Abri | Crevasse | Moulin |
| Adret | Cuvette | Nappe |
| Aiguille | Débâcle | Nevé, névé |
| Alio | Débris | Nuée ardente |
| Alp | Demoiselle | Orgues |
| Arête | Doline | Pays |
| Armorica | Éboulis | Paysage |
| Autan, vents d' (Auge) | Emposieu | Penitent (also |
| Avalanche | Enclave | Spanish) |
| Aven | Ennoyage | Piste |
| Ballon | Entrepôt | Piton |
| Bayou | Étang | Planèze |
| Bise, bize | Exclave | Plage |
| Bocage | Falaise | Plateau |
| Boulevard | Fenêtre | Prairie |
| Butte | Garrigue | Primeur |
| Butte-témoine | Gendarme | Puy |
| Campagne | Glacier | Ravinement |
| Carapace latéritique | Glacière | Régime |
| Carénage (Careenage) | Grau | Replat |
| Carrefour, carfour, | Habitant | Rimaye |
| carfax | Hachure | Roche moutonnée |
| Cartouche | Landes | Schiste |
| Causse | Lapié | Sérac |
| Champaign | Levée | Sotch |
| Cirque | Liane | (Steppe) |
| Cluse | Limon | Terrain |
| Col | Maquis | Transhumance |
| Colmatage | Massif | Ubac |
| Corniche | Métayage | Val |
| Côte | Minette | Vasque |
| Coulée | Mistral | Verrou |
| Coulisse | Mofette | Versant |
| Couloir | Molasse | Ville |
| Creole | Moraine | Voyageur |

## GERMAN

The following list of terms of German origin which have been frequently used without translation in geographical works in English is based first on one compiled with the help of Dr. K. A. Sinnhuber of the Wirtschaftsuniversität, Wien. My good friend Professor Hans Boesch of the University of Zürich then kindly arranged a special meeting of colleagues to discuss the problem. Of the twenty-four words submitted as important seven were additions to the previous list. Later the whole was discussed with Professor Carl Troll of the University of Bonn, who made several

# APPENDIX II

additions and finally with my colleague Professor T. H. Elkins of the University of Sussex.

| | | |
|---|---|---|
| Angerdorf | Graben | Rumpf |
| Aufeis | Haff | Runddörfer |
| Ausland | Haldenhang | (Rundling) |
| Autobahn | Haufendorf | Schattenseite |
| Berg | Heide | Schneebrett |
| Bergwind | Hinterland | Schratten |
| Bergschrund | Horst | Schuppenstruk- |
| Bodden | Inselberg | tur |
| Boerde, Börde | Kar | Schwingmoor |
| Bornhardt | Karre | Senke |
| Bösche | Karrenfeld | Sonnenseite |
| Bund | Kegelkarst | Staublawine |
| Decke | Klippe | Steilwand |
| Deckenschotter | Knickpunkt | Steppenheide |
| Dreikanter | Landschaft | Stossend |
| Drubbel | Lebensraum | Strand |
| Einzelhof | Loess, Löss | Strassendorf |
| Eis | Maar | Talweg |
| Endrumpf | Marsch | (Thalweg) |
| Felsenmeer | Matterhorn | Talwind |
| Fenster | Mayen | Umland |
| Firn | Mitteleuropa | Urstromtal |
| Flysch | Nagelfluh | Voralp |
| Föhn | Nehrung | Wächte |
| Föhrde, Forde | Orterde | Waldhufendorf |
| Formenkreis | Ortstein | Watte |
| Geest | Piedmonttreppen | Wiesenboden |
| Geopolitik | Primärrumpf | Wildschnee |
| Gewann | Randkluft | Zechstein |
| Gewanndorf | Riegel | Zeugen |
| Gipfelflur | Rillenstein | Zeugenberg |

## HAWAIIAN

| | | |
|---|---|---|
| Aa | Kipuka | Pahoehoe |
| | Kona | Pele's hair |

## HEBREW

With the resuscitation of the Hebrew language in Israel and the development of active geographical research in that country, Hebrew words are beginning to appear in the literature—notably *makhtesh*. Many Arabic words, current in the area, such as *Khirba* and *'Azbeh*, are however being used. The Committee is indebted to Professor D. H. K. Amiran (Hebrew University of Jerusalem) for information on several points.

| | | |
|---|---|---|
| Kibbutz, kibutz | Makhtesh | Sharav, sharaf |
| Kvutzah, kvutza | Moshav, moshavah | Tell, tel |

# APPENDIX II

## HUNGARIAN (MAGYAR)

A few words, descriptive of features particularly developed in Hungary, have become current in literature in English.

Puszta      Szik

## ICELANDIC

At least one word of Icelandic origin—*geyser*—has become an integral part of the English language. A few others are used occasionally in international geographical literature and certain terms, notably *tephra*, have been introduced by Icelandic geographers.

Dyngja      Geyser      Jökla-mýs      Jökull
Sandur

## LANGUAGES OF THE INDIAN SUBCONTINENT

For several reasons many words from various languages of the Indian subcontinent have been absorbed into geographical writings in English. Early British workers in the area naturally became familiar with features aptly described by local names: often these were features unknown in their homeland, for which no English words existed. With the establishment of schools of geography in many universities of the subcontinent there arose a generation of scholars, writing in English and publishing their geographical researches in that language, but finding themselves naturally using many familiar local terms. English writers using their material or undertaking local field work themselves inevitably absorbed their phraseology. Thus O. K. Spate in his *India and Pakistan: A General and Regional Geography* (1954) uses over a hundred vernacular terms without, unfortunately, indicating the language from which each is derived or giving a glossary. For help in compiling the list which follows I am especially indebted to many colleagues and former students. In the first place Professor George Kuriyan of the University of Madras and Vice-President of the International Geographical Union during my Presidency provided me with a fully annotated basic list. Dr. A. H. Siddiqi (acknowledged A.H.S.) furnished a list of seventy-two words with a careful definition of each and comments which have proved exceptionally helpful. Others who have helped with lists or individual words include Professor S. P. Chatterjee (University of Calcutta and Editor of the National Atlas), Professor Nafis Ahmad (many of the definitions are from his standard work, *An Economic Geography of East Pakistan*, 1958); the late Dr. P. Sengupta (author of *The Indian Jute Belt*, 1959), Calcutta; Dr. U. Singh (Varanasi), Professor M. Shafi (Aligarh), Professor R. L. Singh (author of *Banaras: A Study in Urban Geography*, 1955), University of Banaras; Professor Kazi Ahmad (University of Lahore), Dr. S. D. Misra and Dr. R. P. Singh (Ranchi). Dr. Misra kindly checked the entries in proof.

Many of the words suggested for inclusion have been eliminated. Except in special cases Indian names of trees and crops have been excluded, so have terms of local use or equivalents in the less used languages of terms in Urdu or Hindi. Even so, the total of words included is over 120.

# APPENDIX II

Abi
Abi-sailābā
Amān
Amin
Anicut, annicut, anaicut
Aūs
Aÿacut
Baid, Baidlands
Bāngar, bhāngar
Banjar
Bār, bārlands
Bārāni
Batāi
Bel, bhel
Bét, bet lands
Bhabbar, bhābar
Bhadoi
Bhīl, bīl, bheel
Bhit
Bhūr
Bhūrā
Bōrō
Bund, bundis
Busti, bustee
Chachār, Kachār
Chāhi, chāi
Châk
Char
Chaur
Cheri, chāri
Chena (Ceylon: Sinhalese)
Chō
Dāk bungalow
Dāmān
Daryākhurdi
Dasht
Demb
Dhānd
Dhāyā
Dhōrō, dhōrū
Diara
Do fasli harsala
Doab

Duār
Dūn
Ek fasli harsala
Firka
Gauthānā
Ghāt
Gher
Gorich
Hamūn
Hāor, haūr, hrād
Hemantic
Jhil, jheel
Jilla, zila, jila
Jhum
Kabouk (Ceylon: Sinhalese)
Kachchi
Kallar
Kankar
Karewa
Kanat
Kās
Kārez
Khad
Khādar, Khaddar, Khuddar
Khaderā, Khuddera
Khāl
Khari
Kharif
Khās mahal
Khoai
Khud Kasht
Khushkābā
Kizdhi
Kucha, Kachha
Kumri
Kursai
Lak
Lōō
Lōra, look
Maidan
Marg
Mota
Mouza

Nad
Nāi
Nālā (nullah)
Niai
Nunja
Pāko
Pālēz
Pargana
Pat
Pawindah
Piccottah
Podu
Poramboke
Pukka, pakka
Punja
Purdah
Rabi
Regur
Reh
Ryot
Ryotwari
Sabzbār
Sailābā
Saylo
Serai, sarai
Shamilat
Sundri
Sundarbans
Swadeshi
Tahsil, tehsil
Tailo
Talluqa, taluka
Tangi
Terai
Thal
Thānā
Thar
Tibba
Tilā
Usar
Zāid-rabi
Zemindār, Zamindār
Zemindary

## INDONESIAN LANGUAGES

Many words of Indonesian origin have long been familiar in Dutch and French geographical literature. With the publication of more works in English on the Indonesian regions and notably the translation into

English of Charles Robequain's *Le Monde Malais* (1954) many of these words have become familiar to British and American readers. The following much used terms have been included in the Glossary:

| | | |
|---|---|---|
| Banjir | Kampong | Ladang |
| Bohorok | Kumbang | Sawah |
| Caingin (Philippines) | Lahar | Tegal |

## IRANIAN (FARSI)

| | | | |
|---|---|---|---|
| Dasht | Kavir | Serai | Sirdar |

## IRISH

For help with the list of Irish words the Committee is much indebted to Professor E. Estyn Evans, Queen's University, Belfast, who was also able to add comparative notes relative to Scottish and Welsh terms. A small number of words originating from Ireland has become an integral part of the international geographical terminology—notably *drumlin*, *esker* and *lough*. Others have a more local use, such as *rundale* and *turlough*. Many Scottish terms are equally common in the north of Ireland—such as *brae*, *burn*, *clachan*, *glen*, *inch*, *skerry* and *strath*. The final list included in the Glossary is:

| | | | |
|---|---|---|---|
| Booly, booley | Curragh | Lough | Rundale |
| Cist | Drumlin | Pin | Soum |
| Clachan | Esker | Pladdy | Slugga |
| Conacre | Eiscir | Rath | Turlough |
| Cottier | Lis, liss | | |

Whilst they can scarcely be described as geographical terms, a number of words appear very commonly in place names and, as noted under Welsh, are descriptive of geographical features:

| | |
|---|---|
| Ard (hill, high place) | Moss (a bog) |
| Bally (tun or town) | Pad (a path) |
| Dun (circular earthwork, stone enclosure) | Pot (corrie) |
| | Rig (cultivation ridge) |
| Kesh (wattle bridge) | Slieve (mountain summit) |
| Kil (church or wood) | Strand (sandy shore) |
| Knock (or crock) (a hill) | |

## ITALIAN

A comparatively small number of Italian words have entered into international geographical literature. In a number of cases they duplicate words in French or other languages which have been more generally adopted (*adritto*, Fr. *adret*; *macchia*, Fr. *maquis*).

| | | |
|---|---|---|
| Adritto | Lapilli | Solfatara |
| Bora, borino | Lava | Soffioni, suffioni |
| Breccia | Lido | Somma |
| Breva | Lingua franca | Tafoni (Corsican) |
| Canale | Macchia | Terra rossa |
| Carso | Maestrale | Tombolo |
| Fumarole | Riviera | Valloni |
| Galeria | | |

# APPENDIX II

## JAPANESE

The Committee consulted Professor Glenn T. Trewartha of the University of Wisconsin on the problem and he noted that although a number of Japanese words might be used in geographical descriptions the number which could be called geographical terms is very small. Eliminating proper names and such words as *sake, tofu, hibachi*, the list became only *tsunami* and *bai-u*. Subsequently in the course of a number of journeys in Japan I discussed the problem with my colleague in the International Geographical Union, Professor Fumio Tada, and with other Japanese geographers. The final list becomes:

| Bai-u | Maki-Hata | Sanyōdō |
| Genya | Mura-yama | Tsunami |
| Ha-ta | Sanindo | Yudo-Kuchi |

The following words, appearing on American and other maps, have not been included: *shi* (city); *ku* (ward); *machi* (town); *mura* (village); *gun* (country).

Commenting on the list in the First Edition Professor S. Kiuchi (University of Tokyo) has suggested the following additions: *heiya* (plain); *sanchi* (mountains); *yama* (mountains); *kawa* (river); *kata* (lagoon); *susono* (volcanic slope); *iriaichi* (communal land); *cho* and *machi* (town with combinations *joka-machi* castle town, *ichiba-* market, *shukuba-* station, *monzen-* temple or shrine); *wan* (bay); *hise* (fringing coral reef) and *jori* (a grid land division used since the 8th century). *Ginza* the main shopping street of Tokyo is applied to similar shopping streets elsewhere and *Fuji-san* both to Mt Fuji and similar cone-shaped volcanoes.

## MALAY

A large number of Malay terms are used in E. H. G. Dobby and others: 'Padi Landscapes of Malaya', *Malayan Journal of Tropical Agriculture*, **10**, 1957, and are explained in a three-page glossary (pp. v–vii). The following Malay words have been more widely used and have been defined above:

| Bahru | Kuala | Mukim | Resam |
| Belukar | Lalang | Padang | Samsam |
| Glam | Lembah | Padi (paddy) | Sungei |
| Kampong | Merdeka | Permatang | |

## MAORI AND OTHER NEW ZEALAND WORDS

In New Zealand a number of Maori words are in common use and consequently appear in geographical literature. In addition some common English words (bush, gumland, paddock) are used with special local meanings:

| Bach | Gumland | Pa | Pakihi |
| Bush | Kainga | Paddock | Papa |
| Crib | Kiwi | Pakeha | Whare |
| Downland | | | |

The Department of Geography of the University of Auckland set up a Committee to advise on terms to be included.

## APPENDIX II

### POLISH

A few words, commonly regarded as of Russian origin and introduced by the soil scientists, are in fact of Polish origin. The best known is *Rendzina*.

Pradoliny  Rendzina  Voyvod

### PORTUGUESE (AND BRAZILIAN)

Portuguese words which have come into English geographical literatures have done so in the main via Brazil where they are descriptive of landscape types and vegetation regions:

| Caatinga | Leste | Quinta | Surazo |
| Caldeira | Mata | Rio | Taboleiro |
| Campo | Pagoda | Selva | Terra-firme |
| Cerrado | Pantanal | Serra | Terra roxa |
| Chapada | Planalto | Sertão | Varzéa |
| Friagem | | | |

### RUSSIAN

Five groups of Russian words occur in English geographical literature. The first are the well-known main features of the country—*tundra, taiga, steppe*—the second are the names of the great soil types and their use is a reminder of the pioneer work of Russians in soil science—*podzol, chernozem, rendzina*—the third group comprises words from Russian territories such as *barchan* or *barkhan* from Turkestan; the fourth are winds cold or dry—*purga* and *sukhovei*; the fifth are economic types, *kolkhoz* and *sovkhoz*.

The definitions of Russian words have been checked by Professor Chauncy Harris using the following sources:

Ushakov. *Tolkovyi Slovar' Russkogo Iazyka.* Sostavili G. O. Vinokur, Prof. B. A. Larin, S. I. Ozhegov, B. V. Tomashevskii, Prof. D. N. Ushakov. pod redaktsiei Prof. D. N. Ushakova. Moskva: Gosudarstvennyi Institut 'Sovetskaia Entsiklopediia,' 4 vols. 1935-1940. (Complete Dictionary of the Russian Language, edited by Prof. D. N. Ushakov.)

Barkov  Barkov, A. S. *Slovar' Spravochnik po Fizicheskoi Geografii.* Posobie dlia uchitelei geografii. Moskva: Uchpedgiz, 4th ed. 1958. (Dictionary and Reference Book in Physical Geography, by A. S. Barkov.)

Murzaev  Murzaevy, E. and V. *Slovar' Mestnykh Geograficheskikh Terminov.* Moskva: Uchpedgiz, 1959. (Dictionary of Local Geographical Terms by E. and V. Murzaev.)

Vasmer  Vasmer, Max. *Russisches Etymologisches Wörterbuch.* Heidelberg: Carl Winter, Universitätsverlag, 3 vols. 1953-1958. (Russian Etymological Dictionary by Max Vasmer.)

| Barchan, Barkhan | Chernozem | Kustar, Koustar |
| Bass | Kair | Kum (Turkestan) |
| Boolyanyakh, | Kolkhoz | Liman |
| Bulgunyakl | Krotovina | Oblast |

## APPENDIX II

| | | |
|---|---|---|
| Padiny | Serozëm | Sukhovey |
| Podzol, podsol | Solod, soloth | Taiga |
| Polynia | Solonchak | Takyr |
| Purga | Solonets | Talik |
| Rasputitsa | Sovkhoz | Tundra |
| Rendzina (Polish) | Steppe, Step | Zastrugi, sastrugi |

### SCOTTISH (INCLUDING GAELIC)

Geographical writings in English have been greatly enriched by numerous terms which have their origin in Scotland. Some are familiar Scottish words which have acquired specialized meanings such as *corrie*, *kame* and *strath*; others describe conditions peculiar to or characteristic of Scotland such as *croft*, *fermtoun*, *infield-outfield*, *kirktoun*, *run-rig* and *shieling*; others show the influence of ancient Scandinavian invasions, such as *saeter* and *skerry*. Some Scottish words, especially those derived from Gaelic, are almost universally used in Scotland, often in northern Ireland, but rarely elsewhere —such as *ben*, *brae*, *burn*, *carse*, *clachan*, *firth*, *howe*, *law*, *loch* and *shieling*. The list of words finally included in the Glossary is the result of many discussions, such as during the meetings of the British Association at Glasgow in 1958 but among those who made special contributions are Dr. C. J. Robertson (University of Edinburgh), Professor Ronald Miller (University of Glasgow), the late Professor A. C. O'Dell (University of Aberdeen) and Professor E. Estyn Evans (Queen's University, Belfast).

The list finally included is:

| | | |
|---|---|---|
| Ben | Glen | Loch |
| Brae | Geo | Lochan |
| Burgh | Gloup, gloap | Machair |
| Burn | Haughland, haughs | Mull |
| Carse | Head-dyke | Ness |
| Clachan | Howe | Run-rig |
| Cleugh, clough | Inch | Saeter |
| Corrie | Infield-outfield | Shieling |
| Cottar, cotter | Kame | Skerry |
| Croft, crofter, crofting | Kirktoun | Skerry guard |
| Den | Kyle | Strath |
| Fermtoun | Law | Toun, township |
| Firth | Links | Voe |
| Gja, gia | Linn | |

Use has been made of the *Scottish National Dictionary* as far as published.

### SOUTHERN SLAV LANGUAGES (SERBO-CROAT)

The pioneer work of Cvijić in describing the *karst* phenomena of what is now Yugoslavia is largely responsible for the wide use of certain Slav (Serbo-Croat) words, now having a far more restricted meaning than in the country of origin.

| | | | |
|---|---|---|---|
| Bogaz | Jama | Ponor | Voivode |
| Dolina | Planina | Struga | |
| Hum | Polje | Uvala | |

## APPENDIX II

### SPANISH

A number of words of Spanish origin have become so firmly entrenched in the English language that their Spanish ancestry is almost forgotten and the same is true of such geographical terms as *mesa, cuesta, cordillera* and *ria*. Many Spanish words have come into our vocabulary via Latin America and even from those parts of the United States formerly Spanish-speaking. A few words, notably *vega*, may have come from Spain but have an older Moorish ancestry. For a basic list of Spanish words which have entered into the international geographical vocabulary I am greatly indebted to my good friend Professor Pierre Deffontaines, Director of the Institut Français de Barcelone (acknowledged as P.D.). The Spaniards have published their own Geographical Glossary by Novo Chicarro issued in 1950 by the National Geographical Society of Madrid and in this will be found authentic definitions of all terms used in the peninsula. Of the following list it will be noted many of the words are of South or Central American origin.

| | | |
|---|---|---|
| Adobe | Estancia | Papagayo |
| Altiplano | Galeria | Paramo |
| Arroyo | Garúa | Penitente |
| Bajada, bahada | Grao | Playa |
| Baguio | Guano | Pueblo |
| Balma | Hacienda | Puna |
| Bañada, bañado | Huerta | Quinta |
| Barranca, barranco | Levante, levanter | Rañas Ría |
| Boca | Leveche | Rio |
| Bochorno | Llano | Salina |
| Caliche | Loma | Savaña |
| Calina, calima | Matorral | Sierra |
| Cañada | Mesa | Solano |
| Caldera | Meseta | Soroche |
| Cañon (canyon) | Mesquite | Temporales |
| Ceja | Mesta | Tierra |
| Cenote | Mestizo | Tinaja |
| Chaco | Moela | Tinajita |
| Chaparral | Mogotes | Tramontana |
| Chaperdos | Montaña | Vega |
| Cinglos, cingles | Monte | Verano |
| Congost | Mulatto | Veranillo |
| Cordillera | Nevados | Yunga |
| Corral | Norte | Zambo |
| Criolle | Pampa | |
| Cuesta | Pampero | |

### SWEDISH

The Committee is greatly indebted to Professor Edgar Kant, Director of the Geographical Institute of the University of Lund, who has in prepara-

## APPENDIX II

tion an international geographical dictionary and who kindly loaned the cards relative to the many Swedish words which have been used in works written in English. Some of these were local terms appearing only in regional descriptions and have not been included in the Glossary. A number of Swedish terms have been anglicized or given English spellings (partly to avoid the diacritical marks usual in Swedish) but some writers prefer to use the correct Swedish form. Thus *skärgård* is commonly but not rightly translated as *skerryguard*, and *fjell* has a more precise meaning than *fell*. In the case of several words I also derived much help from discussions with Professor William William-Olsson and the late Professor Hans W:son Ahlmann. The final list selected for inclusion is as follows:

| | | |
|---|---|---|
| Alvar | Hällanalys | Rauk |
| Dråg | Havsband | Rundhäll |
| Dy | Ishinna | Skare |
| Eohypses | Isobase, isoanabase | Skarn |
| Fjäll, fjell | Klint | Skär |
| Fjärd | Lagg | Skärgård |
| Fjordtopografi | Mar | Skärtråg |
| Flark | Marin gräns (M.G.) | Stentorg |
| Förna | Mull | Teleconnection |
| Fors | Ose (Ås), Åsar | (telekonnektion) |
| Glint | Palsa (Finnish) | Tjäle |
| Gyttja | Pipkrake | Varv |

### THAI

Only a few words from the Thai language have been adopted in international geographical literature. Four words have been included in the Glossary—*klong, menam, pa-deng* and *wat*.

### TURKISH

Obruk. See also Chinese

### WELSH

The Committee is greatly indebted to Professor E. G. Bowen (acknowledged as E.G.B.), University College of Wales, Aberystwyth, for help in preparing the list of Welsh words used in geographical literature and especially for calling attention to *A Gazetteer of Welsh Place-Names*, edited by Elwyn Davies and published by the University of Wales Press, Cardiff, 1957. This Gazetteer, with text in both Welsh and English, was prepared by the Language and Literature Committee of the Board of Celtic Studies of the University of Wales and in addition to including an authoritative list of place-names which in future will be used on Ordnance Survey maps, has a valuable glossary of the chief elements in Welsh place-names which include many geographical terms. Unless otherwise stated the English equivalents given below are taken from this glossary.

In geographical literature in English certain common Welsh words have been given a specialized meaning, notably *cwm* and *moel* with, more recently, *cors, llan, llyn* and *morfa*. Other Welsh words have exact meanings not easily translated into English and so have been adopted as terms—notably *gwely, hafod, hendref, maerdref* and *tyddyn*.

After consultation with Dr. Elwyn Davies, the final list of words included in the glossary is:

| | | |
|---|---|---|
| Blaen (*pl.* blaenau) | Hafod (hafoty) | Rhos |
| Bro | Hendref (hendre) | Sarn (*pl.* sarnau) |
| Cors (gors) | Llan | Tir comin |
| Crib | Maerdref | Tir cyd |
| Cwm | Meifod | Tref |
| Esgair | Moel (foel) | Twyn |
| Ffridd | Morfa | Tywyn, towyn |
| Gwaun | Pandy | Tyddyn (ty'n) |
| Gwely | Pentref | |

Whilst they cannot be described as geographical terms and in writings in English would normally be given in their English equivalents, the following Welsh words, very common in place-names, are descriptive of geographical features:

Aber (estuary, confluence)
Afon (river)
Ban, bannau (peak, crest, bare hill)
Banc (bank, hill, slope)
Bryn (hill)
Bwlch (pass, gap)
Caer or gaer (fort, stronghold)
Carn (cairn, rock, mountain)
Carnedd, *pl.* carneddau (cairn, barrow, tumulus, mountain)
Cefn (ridge)
Coed (wood, woodland)
Craig (deep valley)
Craig (rock)
Crib (narrow serrated ridge)
Din, dinas (hill fortress)
Dol (water meadow)
Dyffryn (valley)
Eglwys (church)

Garth (hill, enclosure)
Glas (stream, brook)
Glyn (deep valley, glen)
Llyn (lake)
Maes (field, plain)
Mynydd (mountain)
Nant (brook)
Pant (hollow, valley)
Penrhyn (promontory)
Plas (hall, mansion)
Pont (bridge)
Porth (gateway, harbour)
Rhaiadr (waterfall)
Traeth, draeth (strand, beach, shore)
Trwyn (point, cape, *lit.* nose)
Ynys (island, holm, water meadow)
Ystrad (valley floor, strath)

# APPENDIX III

## SOME STRATIGRAPHICAL TERMS

Geographical works very frequently include references to geological horizons. It is obviously impossible to list the innumerable names, often very local in application, of rock groups and deposits, but where these are referred to internationally used horizon names it has seemed useful to list these. The names given below are those in use particularly in Britain and North-West Europe with some particularly important ones from North America.

## THE GEOLOGICAL TIME SCALE

| Era | Period or System | Approximate duration in million years | Cycles of Earth movement |
|---|---|---|---|
| QUATERNARY | Holocene (Recent) Pleistocene | | |
| TERTIARY or CENOZOIC | Pliocene Miocene Oligocene Eocene Paleocene | 65 | Alpine |
| SECONDARY or MESOZOIC | Cretaceous Jurassic Rhaetic Triassic | 70 45 10 35 | |
| PRIMARY or PALAEOZOIC | Permian | 50 | |
| | | | Armorican (Hercynian or Variscan) |
| | Carboniferous Devonian | 65 60 | |
| | | | Caledonian |
| | Silurian Ordovician Cambrian | 40 60 100+ | |
| PRE-CAMBRIAN or EOZOIC | Torridonian Archaean | | |

Note: The end of the Cambrian approximately 500 million years ago.
See *The Phanerozoic Time-scale* published by the Geological Society of London, 1964.

# APPENDIX III

Aalenian (Aalen, Württemberg, Germany)—see Jurassic
Acadian (Acadia, Canada). Middle Cambrian
Acheulian, Acheulean (St. Acheul, France). A cultural stage in the evolution of man characterized by a certain type of chipped stone implements. The more usually accepted stages are:

    Neolithic or New Stone Age

    Palaeolithic
- Azilian–Tardenoisian
- Magdelanian
- Solutrean
- Aurignacian
- Mousterian
- Acheulian
- Chellian, Chellean or Abbevillian

    Eolithic. Strepyan, pre-Chellean
    In East Anglia Clactonian is contemporaneous with Acheulian. Levalloisian with Mousterian

Albian. Gault Clay—see Cretaceous
Algonkian (Algonquin, Canada). Torridonian
Altaid—see Glossary
Alpine—see Glossary
Ammanian (Ammanford, South Wales)—term introduced by Dix and Trueman in 1935 for division of Westphalian. See Morganian.
Amstelian—see Pliocene
Animikie. A division of the pre-Cambrian of the Canadian Shield
Anthropogene. Quaternary or Ice Age.
Antillean. Miocene orogenesis of West Indies
Appalachian. Permo-Triassic orogenesis of America
Aptian (Aptia, France) Lower Greensand = Vectian—see Cretaceous
Aquilonian (Aquilonia, France) uppermost Jurassic; Purbeckian—see Jurassic
Aquitanian (Aquitaine, France)—see Oligocene
Archaean. Pre-Cambrian
Arenigian (Arenig, North Wales)—see Ordovician
Argovian (Argovie, Switzerland)—see Jurassic
Arkansan. Middle and Upper Pennsylvanian orogenesis of Mississippi Valley
Armorican (Armorica or Brittany)—see Glossary
Aryan. Group of India, Carboniferous to Recent
Ashgillian (Ashgill, English Lake District)—see Ordovician
Aurignacian—see under Acheulian
Autunian (Autun, France)—see Permian
Auversian (Auvers, France) = Ledian—see Eocene
Avonian (River Avon, England)—see Carboniferous
Azilian—see under Acheulian
Azoic (*lit.* without life)—applied to pre-Cambrian

Bajocian—see Jurassic
Barremian (Barrême, France)—see Cretaceous
Bartonian (Barton, Hampshire, England)—see Eocene
Bathian, Bathonian (Bath, Somerset, England)—see Jurassic

# APPENDIX III 565

Bernician (Bernicia, old regional name in north-east England—Scotland)
—see Carboniferous
Bolderian. Bolderberg, Belgium—Upper Miocene of Belgium
Bononian (Bononia = Boulogne, France)—see Jurassic
Bradfordian (Bradford, Wiltshire, England)—see Jurassic
Brunswickian. Middle and Upper Devonian orogenesis in eastern North America
Bunter. Lower division of the Trias
Burdigalian (Burdigalia = Bordeaux, France). Lower Miocene
Butleyan (Butley, East Anglia)—see Pliocene

Caledonian (Caledonia = Scotland)—see Glossary
Callovian (Callovia = Kelloway, England)—see Jurassic
Cambrian (Cambria = Wales)—see Glossary
Caradocian (Caradoc, Wales)—see Ordovician
Carboniferous (*lit.* coal-bearing)—see also Glossary

| | | |
|---|---|---|
| | | Stephanian (not in Britain) |
| Upper or Coal Measures | Westphalian | Radstockian / Staffordian / Yorkian / Lanarkian |
| Middle or Millstone Grit | Lancastrian, Namurian | |
| Lower or Carboniferous Limestone | Avonian (Dinantian) | Viséan / Tournaisian |

In Northumberland and the Scottish Border the Tuedian is the sandy facies of Tournaisian and the Bernician of the Viséan.

Carinthian (Carinthia, Austria). Upper Keuper of Alps
Casterlian (Casterlé, Belgium)—see Pliocene
Cenomanian, Lower Chalk—see Cretaceous
Charmouthian (Charmouth, Dorset, England)—see Jurassic
Charnian (Charnwood Forest, Leicestershire, England)—see Glossary
Chattian (Chatti, tribe in Hesse, Germany)—see Oligocene
Chellian, Chellean—see under Acheulian
Chillesfordian (Chillesford, East Anglia)—see Pliocene
Clactonian (Clacton, East Anglia)—see under Acheulian
Coblenzian (Coblenz, Germany)—see Devonian
Corallian—see Jurassic
Couvinian (Couvin, Belgium)—see Devonian
Cretaceous (*lit.* chalky)—see also Glossary

| | | | |
|---|---|---|---|
| Upper | Danian / Maestrichtian | not represented in Britain | |
| | Senonian | Upper Chalk | |
| | Turonian | Middle Chalk | |
| | Cenomanian | Lower Chalk | |
| Lower | Albian | Gault, Upper Greensand | |
| | Aptian (Vectian) | Lower Greensand | |
| | Neocomian | Barremian / Hauterivian / Valanginian | 'Wealden' in south of England |

Cromerian (Cromer, Norfolk, England)—see Pliocene, but international agreement has been reached to include it in the Pleistocene
Cuisian (Cuise-la-Motte, France)—see Eocene

Dalradian. A division of the pre-Cambrian crystalline rocks of Scotland
Danian (Dania = Denmark) Highest Cutaceous, not represented in Britain
Deuterozoic, Deutozoic—see Glossary
Devonian (Devon County, England)

| | | |
|---|---|---|
| Upper Devonian and Upper Old Red Sandstone | Famennian Frasnian | |
| Middle Devonian and Middle O.R.S. | Givetian Eifelian (Couvinian) | |
| Lower Devonian and Lower O.R.S. | Emsian Siegenian | Coblenzian |
| | Dittonian Downtonian | Gedinnian |

Diestian (Diest, Belgium)—see Pliocene
Dinantian (Dinant, Belgium)—see Carboniferous
Dittonian (Ditton Priors, Herefordshire, England)—see Devonian
Divesian (Dives, France)—see Jurassic
Domerian (Mt. Domero, Italy)—see Jurassic
Downtonian (Downton, Herefordshire, England)—see Devonian
Dravidian. Of India: Cambrian to Carboniferous
Dyas. Old name of Permian

Eifelian (Eifel, Germany)—see Devonian
Emsian (Ems, Germany)—see Devonian
Eocene (*lit.* dawn of recent)

| | | |
|---|---|---|
| Bartonian (Marinesian) (Ludian in upper part) | Barton Beds | |
| Ledian (Auversian) Lutetian | Bagshot and Bracklesham Beds | |
| Ypresian (Cuisian) | London Clay Blackheath Beds | Londinian |
| Lamdenian | Woolwich and Reading Beds Thanet Sands | Sparnacian Thanetian |
| Montian | Wanting | |

Eolithic. The age of the earliest stone implements of man—see under Acheulian

Famennian (Famenne, Belgium)—see Devonian
Franconian (Franconia, district, Germany) = Muschelkalk or Middle Trias
Frasnian (Frasne, Belgium)—see Devonian

Gedgravian (Gedgrave, East Anglia)—see Pliocene
Gedinnian (Gédinne, Belgium)—see Devonian

# APPENDIX III

Givetian (Givet, France)—see Devonian
Gotlandian, Gothlandian (Island of Gotland, Sweden). Alternative name for Silurian (modern sense, formerly upper Silurian, hence reason for alternative name)
Gondwana. System of India stretching from Carboniferous to Jurassic
Grampian. (Grampian Mountains, Scotland) = Dalradian
Günz. The first or oldest of the four glaciations distinguished in the Great Ice Age in the Alps

Hauterivian (Hauterive, Switzerland)—see Cretaceous
Hebridean (Hebrides, Scotland) = Lewisian
Hercynian (Harz Mountains, Germany)—see Glossary
Hettangian (Hettange, France)—see Jurassic
Holocene or Recent—see Geological Time Scale
Huronian (Lake Huron, North America)—see Glossary

Icenian (Iceni, ancient British people of East Anglia)—see Pliocene
Iowan. Fourth American Glaciation
Irrawaddian (River Irrawaddy, Burma) Pliocene of Burma.

Jurassic (Jura Mountains, France and Switzerland). A widely accepted classification of Jurassic rocks in Britain is as follows:

| | | | |
|---|---|---|---|
| Upper Jurassic | Purbeckian—Aquilonian | Purbeck Beds | |
| | Portlandian | Portland Beds | |
| | Bononian | | |
| | Kimmeridgian | Kimmeridge Clays | |
| | Sequanian | | |
| | Rauracian | Corallian | |
| | Argovian | | |
| | Divesian (Oxfordian) | Oxford Clay | |
| | Callovian | | |
| | Bathonian (Bradfordian or Bathian) | Great Oolite | |
| | Vesulian | | |
| | Bajocian | Inferior Oolite | |
| | Aalenian | | |
| Lower Jurassic | Yeovilian | Upper Lias | |
| | Whitbian | | |
| | Domerian | Middle Lias | |
| | Charmouthian | | |
| | Sinemurian | Lower Lias | |
| | Hettangian | | |

There are of course innumerable local developments

Kansan. Second American Glaciation
Keewatin. Lowest division of Pre-Cambrian of the Canadian Shield
Keuper. Upper division of the Trias
Keweenawan. Upper division of Pre-Cambrian of the Canadian Shield; also Algonkian. *Cf.* Torridonian

Kimmeridgian, Kimeridgian (Kimmeridge, Dorset, England)—see Jurassic

Lanarkian (Lanark County, Scotland)—see Carboniferous
Lancastrian (Lancaster County, England)—see Carboniferous
Landenian (Landen, Belgium)—see Eocene
Laramian. Cretaceo-Tertiary orogenesis of Rocky Mountains and Europe
Lattorfian, Latdorfian (Latdorf, Germany). Lower Oligocene of Germany—see Oligocene
Laurentian. Applied to the early orogenesis which produced great granite masses in the Canadian Shield
Ledburian (Ledbury, Herefordshire, England) = Downtonian (obsolete)
Ledian (Lede, Belgium)—see Eocene
Lenhamian (Lenham, Kent, England), Lenham Beds: Mio-Pliocene—see Pliocene
Levalloisian—see under Acheulian
Lewisian (Island of Lewis, Scotland). A division of the Pre-Cambrian of Scotland
Lias—see Jurassic
Llandeilian (Llandeilo, Central Wales)—see Ordovician
Llandoverian (Llandovery, Central Wales)—see Silurian
Llanvirnian (Llanvirn, Central Wales)—see Ordovician
Londinian (London, England)—London clay—see Eocene
Longmyndian (The Longmynd hills, Shropshire, England). Pre-Cambrian sedimentary rocks of Shropshire
Ludian (Ludes, France)—see Eocene
Ludlovian (Ludlow, Shropshire, England)—see Silurian
Lutetian (Lutetia = Paris, France)—see Eocene

Magdelanian—see under Acheulian
Malvernian (Malvern Hills, England)—see Glossary
Menevian (Menevia = St. Davids, Wales). Middle Cambrian, Wales
Mesabian. The Huronian orogenesis of the Lake Superior region
Mindel. The second of the four glaciations distinguished in the Great Ice Age in the Alps
Miocene—see Tertiary
Mississippian. United States term for Lower Carboniferous
Moinian. A division of the crystalline pre-Cambrian rocks of Scotland
Montian (Mons, Belgium)—see Eocene. Lowest Eocene not represented in Britain
Mousterian—see under Acheulian
Morganian (Glamorgan, South Wales)—term introduced by Dix and Trueman in 1935 (*C. R. 2 Cong. Strat. Carb., Heerlen*, 185–201) for division of Westphalian.

Namurian (Namur, Belgium)—see Carboniferous
Neocene, Neogene. Late Tertiary (Miocene and Pliocene)
Neocomian (Neocomium = Neuchâtel, Switzerland)—see Cretaceous
Neolithic. The age of polished stone implements: the newer Stone Age

## APPENDIX III

Nevadian. Middle and Upper Jurassic orogenesis of western North America
Newbournian (Newbourne, East Anglia)—see Pliocene
New Red Sandstone. Permian and Trias
Niagaran (Niagara Falls, North America). Middle Silurian
Norian (Noria, Alps). Lower Keuper of Austria and Mediterranean
Nubian (Nubia, North Africa). Important series of sandstones, etc., in North Africa. Upper Cretaceous—Tertiary
Nummulitic. Equivalent of Palaeogene, from prevalence of nummulites.
Old Red Sandstone. Continental phase of Devonian
Oligocene—see also Tertiary

| France | Belgium |
|---|---|
| Chattian | — |
| Stampian | Rupelian |
| Sannoisian | Tongrian |

    Lattorfian is the lowest (transgressive) Oligocene in Germany. Aquitanian is Oligocene–Miocene of Southern France
Oolites—see Jurassic
Ordovician (Ordovices, ancient British tribe of the Welsh border):
    Ashgillian
    Caradocian
    Llandeilian
    Llanvirnian
    Skiddavian (Arenigian)
Oregonian. Cretaceous orogenesis of western North America
Oxfordian (Oxford, England)—see Jurassic
Paleocene. The earliest of the periods comprised in the Paleogene (not included in the table on p. 531)
Paleogene. Early Tertiary (Eocene and Oligocene with Paleocene if recognized)
Pampalozoic. Alternative general term ('Primeval life') for Pre-Cambrian
Pebidian. Pre-Cambrian of Pembrokeshire, Wales—term obsolete
Peguan. Oligo-Miocene oil-bearing beds of Burma
Pennsylvanian (Pennsylvania, State, U.S.A.). U.S. term for Upper Carboniferous
Penokean. An orogenesis of the Lake Superior region between the Keweenawan (pre-Cambrian) and Cambrian
Peorian. Fourth American Interglacial
Permian
    Thuringian. Zechstein or Magnesian Limestone
    Saxonian. Upper Rothliegende
    Autunian. Lower Rothliegende
Pleistocene—see Geological Time Scale
Pliocene

| | East Anglia | Belgium and Holland |
|---|---|---|
| Icenian { Cromerian | Cromer Beds | |
| Weybournian | | |
| Chillesfordian | | |
| Icenian proper | Norwich Crag | |

## APPENDIX III

Pliocene—*continued*

| | | |
|---|---|---|
| Butleyan | ⎫ | ⎫ Amstelian |
| Newbournian | ⎬ Red Crag | |
| Waltonian | ⎭ | ⎰ Poederlian |
| | | ⎱ Scaldisian |
| Gedgravian | Coralline Crag | Casterlian |
| Lenhamian | | Diestian |

Note: The Icenian is now internationally agreed as being Lower Pleistocene.

Poederlian (Poederlé, Belgium)—see Pliocene
Pontesfordian (Pontesford, Shropshire, England). A group of Pre-Cambrian volcanics. Local
Portlandian (Portland, Dorset, England)—see Jurassic
Proterozoic. Older Palaeozoic: Cambrian, Ordovician and Silurian
Purbeckian (Purbeck, district, Dorset, England)—see Jurassic

Radstockian (Radstock, Somerset, England)—see Carboniferous
Rauracian (Rauracia, Jura)—see Jurassic
Rhaetic, Rhaetian (Rhaetic Alps)—see Geological Time Scale
Riss. The third of the four glaciations distinguished in the Great Ice Age in the Alps
Rupelian (Rupel, Belgium)—see Oligocene

Salopian (Salop or Shropshire, county, England)—see Silurian
Sannoisian (Sannoise, France)—see Oligocene
Saxonian (Saxony, Germany)—see Permian
Scaldisian—see Pliocene
Senonian (Senones, an ancient tribe in France). Upper Chalk—see Cretaceous
Silurian (Silures, ancient British tribe of Welsh border)—see also Geological Time Scale

| | |
|---|---|
| Ludlovian | ⎫ Salopian |
| Wenlockian | ⎭ |
| Llandoverian or Valentian | |

Sinemurian (Sémur, France)—see Jurassic
Siwalik. Mio-Pliocene of India
Skiddavian (Skiddaw, mountain, Lake District, England)—see Ordovician
Solutrean—see under Acheulian
Sparnacian (Épernay, France)—see Eocene
Staffordian (Stafford, county, England)—see Carboniferous
Stampian (Étampes, France)—see Oligocene
Stephanian—see Carboniferous
Strepyan—see under Acheulian

Taconic, Taconian (Taconic Mountains, North America). An early phase of the Caledonian orogenesis (between Ordovician and Silurian)
Tardenoisian—see under Acheulian

## APPENDIX III 571

Tertiary—see also Geological Time Scale

Neogene { Pliocene / Miocene

Alpine orogenesis

Palaeogene { Oligocene / Eocene / Montian

Thanetian (Isle of Thanet, Kent, England). Thanet Sands—see Eocene
Thuringian (Thuringia, district, Germany)—see Permian
Tithonian. Deposits of the Alpine-Mediterranean Sea of Jurassic-Cretaceous times (Tethys)
Toarcian (Thouars, France)—see Jurassic
Tongrian (Tongres, Belgium)—see Oligocene
Torridonian (Torridon Lake, Scotland). The higher sedimentary rocks of the Pre-Cambrian of Scotland; Keweenawan and Algonkian of America
Tournaisian (Tournai, Belgium)—see Carboniferous
Tremadocian (Tremadoc, North Wales). Highest Cambrian of North Wales
Trias, Triassic—see Geological Time Scale
Tuedian (River Tweed Scottish-English border)—see Carboniferous
Turonian (Touraine, France). Middle Chalk—see Cretaceous

Uriconian (Uriconium, Roman City, Shropshire, England). Pre-Cambrian volcanic rocks of Shropshire

Valanginian (Valangin, Switzerland)—see Cretaceous
Valentian (Valentia, Roman name for Southern Scotland)—see Silurian
Variscan. Alternative name for Armorican or Hercynian orogenesis
Vectian (Vectis, Roman name for Isle of Wight, England). Alternative name for Aptian
Vesulian (Vesoul, France)—see Jurassic
Viséan, Visean (Visé, Belgium)—see Carboniferous

Wallachian (Wallachia district, Roumania). Late Tertiary orogenesis in Europe and western North America (there called Santa Barbaran)
Waltonian (Walton, East Anglia)—see Pliocene
Waulsortian. A phase of Lower Carboniferous in Belgium
Wealden (Weald, district in south-eastern England)—see Cretaceous
Wenlockian (Wenlock, Shropshire, England)—see Silurian
Westphalian (Westphalia, district, Germany)—see Carboniferous
Weybournian (Weybourne, East Anglia)—see Pliocene
Whitbian (Whitby, Yorkshire, England)—see Jurassic
Wisconsin. Fourth American Glaciation
Würm. The fourth or youngest of the four glaciations distinguished in the Great Ice Age in the Alps.

Yarmouthian (Yarmouth, Iowa). Second American Interglacial
Yeovilian (Yeovil, Somerset, England)—see Jurassic
Yorkian (Yorkshire, county, England)—see Carboniferous
Ypresian (Ypres, Belgium)—see Eocene